Interaction of
Charged Particles with
Solids and Surfaces

NATO ASI Series

Advanced Science Institutes Series

A series presenting the results of activities sponsored by the NATO Science Committee, which aims at the dissemination of advanced scientific and technological knowledge, with a view to strengthening links between scientific communities.

The series is published by an international board of publishers in conjunction with the NATO Scientific Affairs Division

A	Life Sciences	Plenum Publishing Corporation
B	Physics	New York and London
C	Mathematical and Physical Sciences	Kluwer Academic Publishers
D	Behavioral and Social Sciences	Dordrecht, Boston, and London
E	Applied Sciences	
F	Computer and Systems Sciences	Springer-Verlag
G	Ecological Sciences	Berlin, Heidelberg, New York, London,
H	Cell Biology	Paris, Tokyo, Hong Kong, and Barcelona
I	Global Environmental Change	

Recent Volumes in this Series

Series B: Physics

Interaction of Charged Particles with Solids and Surfaces

Edited by

Alberto Gras-Marti

Universitat d'Alacant
Alacant, Spain

Herbert M. Urbassek

Technische Universität
Braunschweig, Germany

Néstor R. Arista

Centro Atómico Bariloche e Instituto Balseiro
Bariloche, Argentina

and

Fernando Flores

Universidad Autónoma de Madrid
Madrid, Spain

Plenum Press
New York and London
Published in cooperation with NATO Scientific Affairs Division

Proceedings of a NATO Advanced Study Institute
on the Interaction of Charged Particles with Solids and Surfaces,
held May 6–18, 1990,
in Alacant, Spain

Library of Congress Cataloging in Publication Data

NATO Advanced Study Institute on the Interaction of Charged Particles with
Solids and Surfaces (1990: Alacant, Spain)
   Interaction of charged particles with solids and surfaces / edited by Alberto
Gras-Martí. . . [et al.].
      p.   cm.—(NATO ASI series. Series B. Physics; v. 271)
   "Proceedings of a NATO Advanced Study Institute on the Interaction of Charged
Particles with Solids and Surfaces, held May 6–18, 1990, in Alacant, Spain"—
T.p. verso.
   "Published in cooperation with NATO Scientific Affairs Division."
   Includes bibliographical references and indexes.
   ISBN 978-1-4684-8028-3   ISBN 978-1-4684-8026-9 (eBook)
   DOI 10.1007/978-1-4684-8026-9
   1. Solids—Effect of radiation on—Congresses. 2. Stopping power (Nuclear
physics)—Congresses. 3. Surfaces (Physics)—Effect of radiation on—Congresses.
4. Ion bombardment—Congresses. I. Gras-Martí, Alberto. II. Title. III. Series.
QC176.8.R3N38   1990   91-25574
530.4'16—dc20   CIP

ISBN 978-1-4684-8028-3

© 1991 Plenum Press, New York
Softcover reprint of the hardcover 1st edition 1991

A Division of Plenum Publishing Corporation
233 Spring Street, New York, N.Y. 10013

FOREWORD

Early in 1989, while most of us were gathered in the Mediterranean five-centuries-old city of Alacant, the idea of a school on stopping and particle penetration phenomena came to our minds. Later that year when discussing this plan with some of the participants in the *13th International Conference on Atomic Collisions in Solids* in Aarhus, we were pleased to note that the proposal was warmly welcomed indeed by the community. An Advanced Study Institute on this or a related subject had not been organized in the last decade. Because of the progress made particularly in the interaction of high energy beams with matter, and the many applications which the general subject of the stopping of charged particles (ions and electrons) in matter enjoys, a Study Institute appeared a worthy enterprise. Even though several international conference series cover developments in these areas, they miss tutorial introductions to the field.

The title chosen was *Interaction of Charged Particles with Solids and Surfaces,* and the objectives were stated as follows: "to cover theory and experiments, including selected applications and hot topics, of the stopping of charged particles (ions and electrons) in matter. The emphasis will be on outlining the areas where further effort is needed, and on specifying the basic needs in applications. Fundamental concepts will prevail over applications, and the character of the Institute as a school will be stressed."

The school was directed by Fernando Flores (Spain), Herbert M. Urbassek (Germany), Néstor R. Arista (Argentina) and Alberto Gras-Martí (Spain). The organizing committee consisted of R.H. Ritchie (USA), P. Sigmund (Denmark), G. Falcone (Italy), W. Lennard (Canada), W. Heiland (Germany), F. Saris (Netherlands) and H.H. Andersen (Denmark).

We are happy that it was possible to bring together at the Institute a good number of experts in this field, lecturing both on fundamental aspects and on applications. A special acknowledgement goes to the participation of Profs. Jens Lindhard and Rufus H. Ritchie. In addition to the dozen series of lectures, each extending over two to four hours, the participants enjoyed also listening to around the same number of hot topic speakers who led to the frontiers of research in their specialized areas of expertise.

We had more than 100 participants from a large number of countries: Argentina, Austria, Belgium, Canada, Czechoslovakia, Denmark, Finland, France, Germany, Greece, Hungary, Italy, Japan, Mexico, Netherlands, Portugal, Spain, Sweden, Turkey, UK, USA and USSR.

A special flavor of the Institute consisted in the presentation of the interaction of both ions and electrons with solids. Furthermore, the physics of the interaction of swift ions with the electrons of the penetrated medium showed up like a red thread in virtually all lectures and presentations of the Institute. Participants could learn to appreciate the subtleties of this subject which we are certain will fertilize many research projects in the future. The discussions were very lively thanks to the active participation of the Institute students, in particular of Dr. G. Schiwietz.

Introductory books on this subject are not available with the comprehensive approach presented here. The concepts, theoretical tools, and experimental techniques and results discussed in the Study Institute, and presented in these Proceedings, are of interest to a vast multidisciplinary community of scientists and engineers. Among others, researchers involved in surface physics and characterization, materials science or ion beam modification of materials will find in this volume a sound introduction to the field of charged particle surface interaction. A personal view by one of the participants, on currently open problems in the field, concludes the contributions to these proceedings.

A few words on the table of contents follow. There are three types of manuscripts in these Proceedings, according to the presentations that took place in the Institute: Mainstream, Invited, and Contributed papers. We have classified them roughly in the following chapters: Stopping of Ions, Stopping of Electrons, Low Energy Phenomena, and Applications. In every category we have tried to put theory before experiment.

We are grateful to the Scientific Affairs Division of NATO, without whose support our ideas certainly could not have materialized. (It is good to see the end of the cold war in Europe and the rest of the World: hopefully, in the future even more support can be given to multi-national Institutes like this, and old fashioned restrictions and rules making it difficult for Central- and East-European colleagues to be funded and attend in unrestricted numbers, will be dropped).

Further funding by the Spanish Ministerio de Educación y Ciencia made possible the participation of a good number of students from many non-NATO countries. The reception sponsored by the Ajuntament d'Alacant, in the Fortress of Santa Barbara, was essential for breaking the initial ice among participants. The staff in Hotel Maya and Viajes Ecuador made life easier to all of us.

During the conference, we could faithfully rely on the help of the local committee: Mario M. Jakas, Juan Carlos Moreno, Isabel Abril, Mikel Forcada, Jorge Calera-Rubio, and Javier Castellà.

With only one persistent exception we had all the manuscripts retyped to help homogenize the presentation and render the reading easier. The editors are grateful to all the authors, who patiently went through the various versions of their manuscripts. In this connection, we hope that no significant errors in the text or in the figures were introduced during this process. We thank Isabel Abril, Vicente Esteve Guilabert and Adelardo Victoria Montero for their untiring efforts and, specially Inmaculada Velázquez Añón and Agueda, Adrián and Azalea for their patience, encouragement and collaboration in technical matters, usually at the expense of family hours. The staff in Plenum is recognized for their collaboration and understanding of the delays in completing this volume.

Alberto Gras-Marti, Herbert M. Urbassek,
Néstor R. Arista, and Fernado Flores

Alacant, April 3, 1991

CONTENTS

MAINSTREAM LECTURE SERIES

INVITED LECTURES

CLOSING REMARKS

MAINSTREAM LECTURE SERIES

DYNAMICAL INTERACTION
OF CHARGES WITH CONDENSED MATTER

F. Flores

Departamento de Física de la Materia Condensada
Universidad Autónoma de Madrid
Cantoblanco. E-28049 Madrid, SPAIN

INTRODUCTION

The use of swift ions as probes of the static and dynamic properties of matter dates from the earliest days of modern physics. In a pioneering paper [1], Bohr calculated the slowing down of swift alpha particles in matter. The interaction of ions with matter depends crucially on the ion velocity.

In the high-speed limit, $v \gg v_0 \, Z_1^{2/3}$, (v_0 is the Bohr velocity and Z_1 the ion atomic number), the ion loses all its electrons and the ion-matter interaction can be described using the dynamic response of the electronic system. Fermi [2] was the first to indicate this way to the analysis of the ion energy losses at high speed, and his idea was later developed by Williams [3] and Weizsäcker [4]. Following this, the random-phase approximation (RPA) electronic function of an isotropic electron gas was derived by Lindhard [5]. This enabled a unified description of single particle and collective aspects of a model system that has had wide utility in solid state physics. The first part of these lectures is related to this approach: in particular, we shall analyze the classical interaction of charges with condensed matter, with specific emphasis on the dynamic dielectric response of the electronic system.

At projectile speeds such that $v_0 \leq v \leq v_0 \, Z_1^{2/3}$ an ion may lose electrons to, and capture electrons from, the medium. Then, capture and loss of electrons by the projectile must be considered. A useful criterion for estimating the number of electrons bound to the projectile was proposed by Bohr [6]. Brandt [7] has used this criterion together with the Thomas-Fermi statistical model for the electrons associated with the projectile to write the expression

Interaction of Charged Particles with Solids and Surfaces
Edited by A. Gras-Martí *et al.*, Plenum Press, New York, 1991

3

$$Z_1^* = Z_1 \left\{ 1 - \exp\left[- \frac{v}{v_0} Z_1^{2/3} \right] \right\},$$ (1)

for the effective charge Z_1^* of the ion (see also ref.[8]).

A complete a priori theory of the effective charge and the dynamic screening that an ion experiences in condensed matter must describe the charge-changing processes associated with its capture and loss of electrons. Part 2 of these lectures is concerned with this problem, and will show how to relate the ion charge states with the dynamic response of the electron gas [9].

Finally, in part 3 we address the problem of calculating the stopping power of ions at intermediate, $v_0 \leq v \leq Z_1^{2/3} v_0$ and low, $v \leq v_0$, velocities. In our discussion, we show how the charge-changing processes can be incorporated in a very general approach of the stopping power problem; our discussion will concentrate on the case of two light ions, He and H, that have been recently analyzed in ref.[10].

1. CLASSICAL INTERACTION OF CHARGES WITH CONDENSED MATTER

1.1 Inelastic Collision Processes. Semiclassical Analysis

Semiclassical arguments provide a simple and useful way to calculate the inelastic collision processes of a charge interacting with a polarizable medium. We begin by writing Poisson's equation for the scalar potential ϕ $(\vec{r}, t)$ generated at $\vec{r}$ and at time t by an applied charge density $\rho^0(\vec{r}, t)$ in a medium characterized by a dielectric constant ε,

$$\vec{\nabla} (\varepsilon \vec{\nabla} \phi) = - 4\pi \rho^0(\vec{r}, t),$$ (2)

(atomic units will be used in these lectures). Expressing all quantities as Fourier integrals of the form

$$f(\vec{r}, t) = \int \frac{d\vec{q}}{(2\pi)^3} \int_{-\infty}^{\infty} \frac{d\omega}{2\pi} e^{i(\vec{q} \cdot \vec{r} - \omega t)} f_{\vec{q}, \omega},$$ (3)

and considering $\varepsilon = \varepsilon(q, \omega)$, eq.(2) leads to

$$\phi_{\vec{q}\omega} = \frac{4\pi \rho^0_{\vec{q}, \omega}}{q^2 \varepsilon(\vec{q}, \omega)}.$$ (4)

A bare particle with charge Z_1 may be considered to give rise to a density $\rho^0(\vec{r}, t) = Z_1 \delta(\vec{r} - \vec{v}t)$ if it moves with constant velocity $\vec{v}$. Then, $\rho^0_{\vec{q}, \omega} = 2\pi Z_1 \delta(\omega - \vec{q}\vec{v})$ and the Fourier component of the induced scalar electric potential, $\phi^{ind}_{\vec{q},\omega}$ may be written

$$\phi^{ind}_{\vec{q},\omega} = \frac{8\pi^2 Z_1}{q^2} \delta(\omega - \vec{q}\cdot\vec{v}) \left(\frac{1}{\varepsilon(q,\omega)} - 1\right). \tag{5}$$

The rate of energy loss per unit time, $\dot{W}$, i.e. the power loss, is now obtained from the induced electric field $\vec{E}^{ind} = -\vec{\nabla}\phi^{ind}$ at the projectile position, as

$$\dot{W} = -Z_1 \vec{v}\cdot\vec{E}^{ind}(\vec{r} = \vec{v}t,t). \tag{6}$$

Using the even and odd character of the real and imaginary parts of the retarded response functions as they depend on ω, $\dot{W}$ can be written as

$$\dot{W} = \int \frac{d^3q}{(2\pi)^3} \int_0^\infty \frac{d\omega}{2\pi} 2\omega Z_1 \, \mathrm{Im}\left\{-\phi^{ind}_{\vec{q},\omega}\right\}. \tag{7}$$

From now on we denote $\int d^3q/(2\pi)^3 \int_0^\infty d\omega/2\pi$ as $\int d\vec{x}$. Eq.(7) allows us to define $P(\vec{q},\omega)$ as the probability per unit time of losing energy ω and momentum $\vec{q}$, thus creating a real excitation in the electron gas,

$$\dot{W} = \int d\vec{x} \, \omega \, P(\vec{q},\omega), \tag{8}$$

where

$$P(\vec{q},\omega) = 2 Z_1 \, \mathrm{Im}\left\{-\phi^{ind}\right\} =$$
$$\frac{16\pi^2 Z_1^2}{q^2} \, \mathrm{Im}\left\{-\frac{1}{\varepsilon(\vec{q},\omega)}\right\} \delta(\omega - \vec{q}\cdot\vec{v}), \tag{9}$$

which can be used to define a mean inelastic collision time, τ, as

$$\frac{1}{\tau} = \int d\vec{x} \, P(\vec{q},\omega). \tag{10}$$

It is also convenient to define $P(\omega)$, the probability per unit time of losing energy ω as

$$P(\omega) = \int \frac{d^3q}{(2\pi)^3} P(\vec{q},\omega), \tag{11}$$

Then

$$\dot{W} = \int \frac{d\omega}{2\pi} \omega P(\omega), \tag{12}$$

and

$$\frac{1}{\tau} = \int \frac{d\omega}{2\pi} P (\omega) \tag{13}$$

1.2 Energy Loss Spectra for Charges Moving in an Uniform Electron Gas

1.2.1 Dielectric Function; Different Approximations. Quasiparticles.

The loss function, $P(\vec{q},\omega)$, has two main ingredients: one in the δ-factor associated with the energy conservation law; the other one, Im $\{1/\varepsilon(q,\omega)\}$, is related to the probability of creating an excitation of momentum $\vec{q}$ and energy ω in the electron gas.

The energy conservation law relates the energy of the excited mode, ω, to the energy loss of the particle

$$\omega = -\frac{1}{2M} (\vec{P}_i - \vec{q})^2 + \frac{1}{2M} P_i^2 \simeq \vec{q}\cdot\vec{P}_i/M \tag{14}$$

where $\vec{P}_i$ is the momentum of the incoming particle, M its mass, and $\vec{P}_i$, in the classical approximation, has been assumed to be much bigger that $\vec{q}$, the transfer of momentum from the incoming particle to the electron gas.

A good description of the electron gas response, $\varepsilon(q,\omega)$ can be obtained using the Lindhard (or RPA) dielectric function [5]. Without going into many details let us only comment that this dielectric function incorporates in its expression the appropriate quasiparticles that are

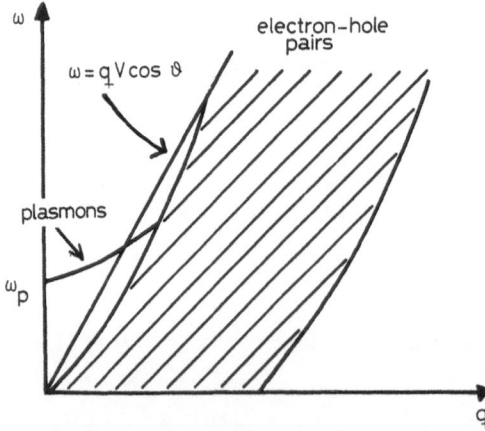

Figure 1. Dispersion relation for the different quasiparticles of a uniform electron gas: plasmons and electron-hole pairs. The straight line $\omega = q\,v\,\cos\theta$ defines the energy conservation law: $\omega = qv$.

excited in the electron gas by the incoming particle. Fig.1 shows the dispersion relation of these different quasiparticles. The electron-hole pairs are related to the excitation of an electron of momentum $\vec{k}$ ($k < k_F$) to a state of momentum $(\vec{k} + \vec{q})$, (with $|\vec{k} + \vec{q}| > k_F$). The electron-hole pair excitation is given by

$$\frac{1}{2}(\vec{k} + \vec{q})^2 - \frac{1}{2}k^2 = \vec{k}.\vec{q} + \frac{1}{2}q^2. \tag{15}$$

By varying k, with $k < k_F$, one gets the continuum electron-hole pairs spectrum shown in fig.1.

Plasmons are collective excitations having the energy

$$\omega_p^2 = 4\pi n, \tag{16}$$

(where n is the electron density) for $q \rightarrow 0$. Typically, for $q \neq 0$, the plasmon dispersion relation behaves as shown in fig.1; for a given momentum, q_C, the plasmon mode coalesces into the electron-hole pair spectrum: this indicates that plasmons are only a well-defined quasiparticle for low momentum, $q < q_C$. In this region only, collective excitations have a well-defined physical meaning, since for higher momentum the plasmons quickly decay into single pair excitations [11].

Keeping in mind this quasiparticle spectrum, one can propose different approximations to the Lindhard dielectric function, ε_{RPA}. The simplest approximation is the classical one, ε_{CL}, appropriate in the high-velocity limit; in this case, one writes

$$\varepsilon_{CL}(q,\omega) = 1 - \frac{\omega_p^2}{\omega^2}\,\Theta(q_c-q), \tag{17}$$

where $\Theta(q)$ is the unit step function. This dielectric function only yields plasmon quasiparticles of constant frequency, ω_p, for $q < q_c$, as the following result shows

$$\text{Im}\left\{\frac{-1}{\varepsilon_{CL}(q,\omega)}\right\} = \frac{\pi}{2}\,\omega_p\left\{\delta(\omega - \omega_p) - \delta(\omega + \omega_p)\right\}, \tag{18}$$

with a momentum cut-off at q_c. A better approximation to ε_{RPA} is obtained using the following plasmon-pole dielectric function [12]

$$\varepsilon_{pp}(\vec{q},\omega) = 1 - \frac{\omega_p^2}{\omega^2 - \beta^2 q^2 - \frac{1}{4}q^4}. \tag{19}$$

This equation yields the following plasmon dispersion relation

$$\omega^2(q) = \omega_p^2 + \beta^2 q^2 + \frac{1}{4} q^4, \tag{20}$$

where β^2 can be fitted to the plasmon mode of low-momentum (see fig.1), and $\frac{1}{4} q^4$ yields the appropriate electron-hole pair behavior at high momentum. Notice that the plasmon mode becomes an electron-hole pair mode for high momentum, when the plasmon branch of fig.1 coalesces into the pair excitations.

Using eqs.(17,18) and eqs.(9-12) one gets the following probability functions (for simplicity, we take $\beta = 0$ in eq.(20)),

$$P_{CL}(\omega) = \frac{2\pi \, Z_1^2}{v} \, \omega_p \, \ln\!\left(\frac{q_c v}{\omega_p}\right) \, \delta(\omega - \omega_p), \quad \text{for } v \gg \frac{\omega_p}{q_c}, \tag{21}$$

$$P_{pp}(\omega) = \frac{Z_1^2 \, \omega_p^2}{v} \frac{1}{\omega^2 - \omega_p^2}, \quad \text{for } \omega_p^2 + \frac{\omega_p^4}{4v^4} < \omega < 2v^2; \tag{22}$$

in both cases, the probability loss function, $P(\omega)$ is only due to the plasmon-mode excitation; while in the classical approximation the plasmon modes are assumed to have a constant frequency, ω_p, in the plasmon pole approximation the plasmon frequencies extend from ω_p to infinity.

Replacing eqs.(21) and (22) into eq.(12) the stopping power per unit length $\frac{dW}{dx} = \frac{1}{v}\frac{dW}{dt}$, can be straightforwardly obtained

$$\left(\frac{dW}{dx}\right)_{CL} = Z_1^2 \, \frac{\omega_p^2}{v^2} \, \ln\!\left(\frac{v q_c}{\omega_p}\right), \quad \text{for } v > \frac{\omega_p}{q_c}, \tag{23}$$

$$\left(\frac{dW}{dx}\right)_{pp} = Z_1^2 \, \frac{\omega_p^2}{v^2} \, \ln\!\left\{\frac{2v^2}{\omega_p}\right\}, \quad \text{for } v^2 \gg \omega_p; \tag{24}$$

the only difference between both results appears in the logarithmic function, since we recover the plasmon-pole behavior by replacing q_c in (23) by $2v$.

Eq.(24) yields the appropriate behavior for the stopping power of an homogeneous electron gas in the limit of high velocities. A similar equation was obtained by Bethe [13]

$$\frac{dW}{dx} = Z_1^2 \, \frac{\omega_p^2}{v^2} \, \ln\!\left(\frac{2v^2}{I}\right), \tag{25}$$

where I is a mean excitation potential of the possible transition eigenenergies associated with the homogeneous electron gas.

In order to analyze the probability loss function, $P(\omega)$, in the low velocity limit, approximations (17) and (18) for the Lindhard dielectric functions are not appropriate since then, the hole-electron pairs determine the excitation spectrum in the electron gas. We should mention that in this low-velocity limit, the linear response theory discussed here does not yield the appropriate answer to the problem of calculating the excitation spectrum: the point to be noticed is that, in this limit, the ion introduces important non-linear perturbations in the electron gas [14]. It is convenient, however, to discuss the results of the linear theory discussed here.

In the low-velocity limit, only low frequencies contribute to the excitation spectrum. Then, it is convenient to write

$$\text{Im}\left\{-\frac{1}{\varepsilon(q,\omega)}\right\} \simeq \frac{\varepsilon''}{\varepsilon'^2}, \tag{26}$$

where $\varepsilon(q,\omega) = \varepsilon'(q,\omega) + i\varepsilon''(q,\omega)$. The RPA dielectric function yields

$$\varepsilon' = 1 + C(q)f(q), \tag{27a}$$

where

$$C(q) = \frac{4\pi}{q^2}\frac{q_F}{\pi^2}, \tag{27b}$$

$$f(q) = \frac{1}{2}\left\{1 + \frac{4q_F^2 - q^2}{4qq_F}\ln\left|\frac{q+2q_F}{q-2q_F}\right|\right\} \tag{27c}$$

and

$$\varepsilon'' = C(q)\frac{\pi\omega}{2qq_F}, \qquad \text{for } q < 2q_F. \tag{27d}$$

Different approximations have been proposed to calculate $P(\omega)$ and $\frac{dW}{dx}$ in the low-velocity limit. Ritchie [15] has proposed to take $f(q) = 1$ in eq.(27a), instead of using (27c). Using this approximation one gets

$$P(\omega) = \frac{4\omega}{q_{FT}^3}z_1^2\left\{tg^{-1}\left(\frac{2q_F}{q_{FT}}\right) + \frac{2q_F/q_{FT}}{1+(2q_F/q_{FT})^2}\right.$$

$$\left. - tg^{-1}\left(\frac{\omega}{vk_{FT}}\right) - \frac{\omega/vq_{FT}}{1+(\omega/vq_{FT})^2}\right\}, \tag{28}$$

and

$$\frac{dW}{dx} = Z_1^2 \frac{2v}{3\pi} \left\{ \ln\left(1 + \pi q_F\right) - \frac{1}{1 + \frac{1}{\pi q_F}} \right\}, \qquad (29)$$

where $q_{FT}^2 = 4q_F/\pi$.

Lindhard and Winther [16] have proposed to approximate $f(q)$ by

$$f(q) \simeq 1 - \frac{1}{3} \left(\frac{q}{2q_F}\right)^2, \qquad (30)$$

obtained by expanding eq.(27c) up to second order in q. This yields the following stopping power

$$\frac{dW}{dx} = Z_1^2 \frac{2v}{3\pi} \left\{ \ln\left(\frac{1 + \frac{2}{3} a}{a}\right) - \frac{3-a}{3+2a} \right\} \Big/ \left(1 - \frac{a}{3}\right)^2, \qquad (31)$$

for $v \ll v_F$, where $a = \frac{1}{\pi q_F}$.

Let us finally comment that the linear theory for the stopping power of a charge Z_1, as given by eqs.(9-12), interpolates between the limits of high, eq.(24), and low, eqs.(19) or (31) velocities. In fig.2, we show

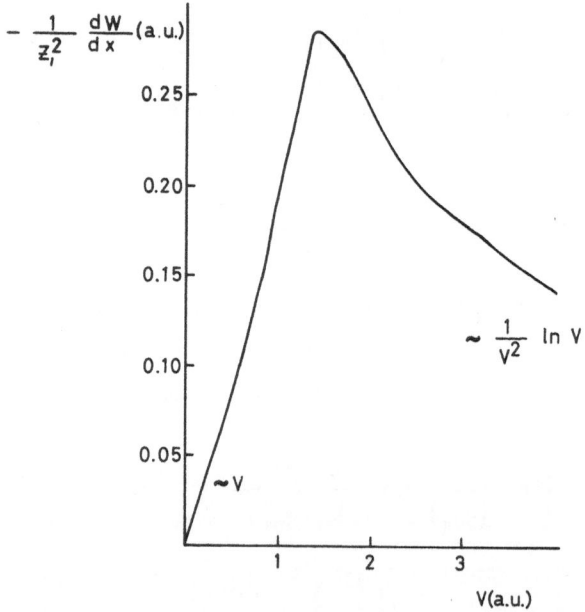

Figure 2. Stopping power for a charge Z_1 as calculated using a linear model and the full RPA dielectric function ($r_s = 2$).

$\dfrac{1}{Z_1^2}\dfrac{dW}{dx}$ as a function of v as calculated numerically using the full Lindhard dielectric function in eq.(9).

1.2.2. Energy Loss Spectrum. In previous sections we have seen, for a classical particle moving in a solid, how to obtain the stopping power, $\dfrac{dW}{dx}$, and the probability per unit time of losing an energy ω, $P(\omega)$, by means of a linear-theory approach. In experiments, the energy-loss spectrum for a beam of particles entering the solid and emerging after moving along a length, L, is currently obtained. We wish to discuss now how to calculate this spectrum using the convolution method [17] and previous results.

Consider, in a first step, the classical dielectric function of eq.(19) and the corresponding probability function, $P_{CL}(\omega)$, of eq.(21). In this approximation the incoming projectile only excites quasiparticles (plasmons) of a given energy, ω_p. This is a very convenient starting point to introduce the convolution method. From eqs.(21) and (13) we can calculate straightforwardly the mean inelastic collision time, τ, and the mean free path $\lambda = v\tau$ of the incoming particles (see fig.3).

Now, we assume that the different inelastic collision processes of the projectile and the solid are independent from each other: this means

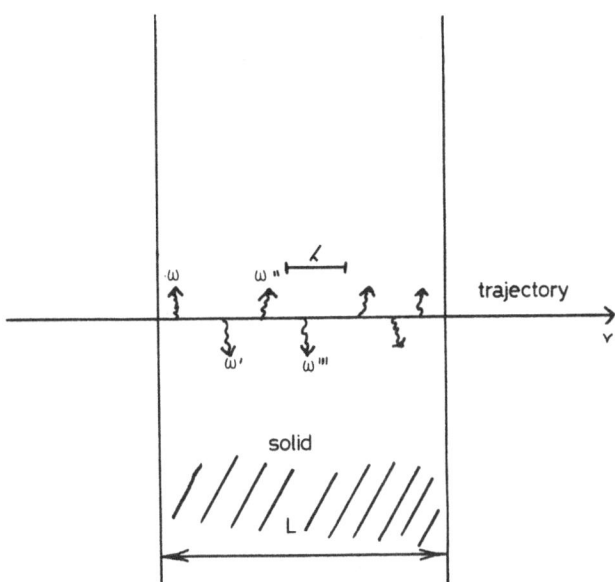

Figure 3. A particle crosses a solid along a trajectory of length L and excites quasiparticles of energy ω; λ is the mean free path between consecutive collisions.

that the projectile keeps no memory of the previous collision events. Then, the probability for having n-collisions (and losing an energy $n\omega_p$) is given by

$$P_n = \frac{P_0^n}{n!} e^{-P_0}, \tag{32}$$

where $P_0 = L/\lambda$. Eq.(32) yields a Poisson distribution for the energy loss processes. In particular, the energy-loss spectrum can be expressed as follows

$$S(\omega) = \sum_{n=0}^{\infty} e^{-P_0} \frac{P_0^n}{n!} \delta(\omega - n\omega_p), \tag{33}$$

where $\delta(\omega - n\omega_p)$ takes into account the energy loss, $n\omega_p$, associated with the creation of n plasmons. It is well known that in the limit $P_0 \to \infty$, eq.(33) goes over a Gaussian distribution. Thus,

$$S(\omega) \to K \exp\left\{ - \frac{1}{2} \frac{(\omega - P_0\omega_p)^2}{P_0 \omega_p^2} \right\}, \tag{34}$$

which shows that, in this limit, the mean energy loss is given by $P_0\omega_p = \frac{L}{\lambda} \omega_p$, and the gaussian half width by

$$(\Delta\omega^2)^{1/2} = P_0^{1/2} \omega_p. \tag{35}$$

Notice that P_0 is defined, using the probability function, $P(\omega)$, as follows

$$P_0 = \frac{L}{v} \int_0^{\infty} P(\omega) \frac{d\omega}{2\pi}. \tag{36}$$

In a more general case, instead of inelastic processes whereby a plasmon of frequency ω_p is created, a continuum spectrum appears with electron-hole pairs and plasmons contributing to the probability function, $P(\omega)$. For this case, we can generalize the results of eq.(33) by introducing the function, $\tilde{P}(\omega)$

$$\tilde{P} = \frac{L}{2\pi v} P(\omega). \tag{37}$$

Then, eq.(33) is modified in the following way: (i) $\exp(-P_0)$ is replaced by $\exp(-\int_0^{\infty} \tilde{P}(\omega)d\omega)$; (ii) $P_0^n \delta(\omega-n\omega_p)$ is replaced by the following convolution function

12

$$\tilde{P}_n(\omega) = \int_0^\infty d\omega' \int_0^\infty d\omega'' \ldots \int_0^\infty d\omega^{(n)} \; \tilde{P}(\omega-\omega') \; \tilde{P}(\omega'-\omega'')$$

$$\ldots \; \tilde{P}(\omega^{(n-1)} - \omega^{(n)}). \tag{38}$$

With these substitutions, the energy loss spectrum reads

$$S(\omega) = \sum_{n=0}^\infty \exp\left\{-\int_0^\infty \tilde{P}(\omega)d\omega\right\} \frac{\tilde{P}_n(\omega)}{n!} . \tag{39}$$

We mention that the convolution defined by eq.(38) takes into account that in each inelastic collision there is a given distribution function for the energy losses measured by the probability function $\tilde{P}(\omega)$.

It is instructive to consider again the limit of $P_0 = \int \tilde{P}(\omega)d\omega$ going to infinity. Let us also assume that, in a first approximation,

$$\tilde{P}(\omega) = \frac{P_0^n}{2\sqrt{\pi}\,\omega_0} \exp\left\{-\frac{1}{2} \frac{(\omega - \bar{\omega})^2}{\omega_0^2}\right\}, \tag{40}$$

with a mean energy, $\bar{\omega}$, and a halfwidth ω_0. Then, it is easy to find that $\tilde{P}_n(\omega)$ also behaves as a gaussian function going like

$$\tilde{P}_n(\omega) = \frac{P_0^n}{2\omega_0\sqrt{\pi n}} \exp\left\{-\frac{1}{2} \frac{(\omega - n\bar{\omega})}{n\omega_0^2}\right\} . \tag{41}$$

Then, the energy loss spectrum given by eq.(39) takes the form

$$S(\omega) = \sum_{n=0}^\infty \exp(-P_0) \frac{P_0^n}{n!} \frac{1}{2\omega_0\sqrt{\pi n}} \exp\left\{-\frac{1}{2} \frac{(\omega - n\bar{\omega})^2}{n\omega_0^2}\right\} . \tag{42}$$

This is equivalent to eq.(33) with $\delta(\omega-n\omega_p)$ replaced by

$$\frac{1}{2\sqrt{\pi n}\,\omega_0} \exp\left\{-\frac{1}{2} \frac{(\omega - n\bar{\omega})^2}{n\omega_0^2}\right\}.$$

For $P_0 \to \infty$, eq.(42) behaves as the convolution of the gaussian function (34), where ω_p is replaced by $\bar{\omega}$ and $\frac{1}{2\omega_0\sqrt{\pi P_0}} \exp\left\{-\frac{1}{2} \frac{(\omega - P_0\bar{\omega})^2}{P_0\omega_0^2}\right\}$. This shows that the mean energy loss is given by $P_0\,\bar{\omega}$, and the corresponding halfwidth by

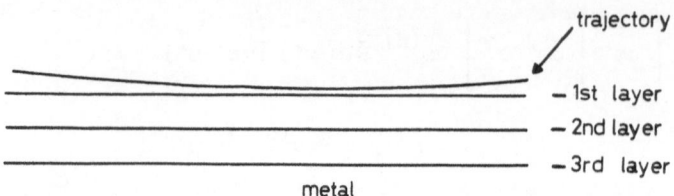

Figure 4. A charge moving near a metal surface.

$$(\Delta\omega)^2 = P_0(\bar{\omega}^2 + \omega_0^2).$$ (43)

This equation relates the straggling per unit length [18] to the second moment of the function $P(\omega)$ since using eqs. (13) and (40) one finds

$$\frac{(\Delta\omega)^2}{L} = \frac{1}{v} \int_0^{\infty} \omega^2 \, P(\omega) \, \frac{d\omega}{2\pi} \,.$$ (44)

This shows that the total straggling is the result of two contributions: one comes from the Poisson distribution, the other one from the intrinsic width of $P(\omega)$.

1.3 Energy Loss Spectra for Charges Moving Near a Metal Surface

Fig. 4 shows the physical case we discuss now. A given charge, Z_1, is scattered off a metal surface following a trajectory almost parallel to the surface. In this case, the important physical quantity is $\tilde{\pi}(\omega)$, the analogous of the probability function $\tilde{P}(\omega)$ introduced in eq. (37); $\tilde{\pi}(\omega)$ represents the total probability per unit time of exciting a quasiparticle mode of energy ω. In particular, the total inelastic energy loss, W_s, is given by

$$W_s = \int_0^{\infty} \omega \, \tilde{\pi}(\omega) d\omega.$$ (45)

Different models have been proposed to calculate $\tilde{\pi}(\omega)$ and the reader is referred to refs. [19,20]. In this paper, we assume that $\tilde{\pi}(\omega)$ can be calculated using a linear-response theory, equivalent to the one leading to eq. (11) in the metal bulk. This linear theory can be used within the convolution method, as discussed in sect. 1.2, to obtain the energy loss spectrum for the particles scattered off a metal surface.

The argument follows closely eqs. (33-39). In particular, once we define the probability function $\tilde{\pi}(\omega)$ we can introduce the following energy-loss spectrum [20], $S_\sigma(\omega)$ such that

$$S_\sigma(\omega) = \sum_{n=0}^{\infty} \exp\left\{ -\int_0^\infty \tilde{\pi}(\omega) \; d\omega \right\} \frac{\tilde{\pi}_n(\omega)}{n!} \; , \tag{46}$$

where

$$\tilde{\pi}_n(\omega) = \int_0^\infty d\omega' \int_0^\infty d\omega'' \ldots \int_0^\infty dw^{(n)}$$

$$\tilde{\pi}(\omega - \omega') \; \tilde{\pi}(\omega' - \omega'') \; \ldots \; \tilde{\pi}\left(\omega^{(n-1)} - \omega^{(n)} \right) \; . \tag{47}$$

Eqs.(46) and (47) yield the surface counterpart of eqs.(38) and (39) calculated in sect.1.2 for the metal bulk.

We complete this section by mentioning that these equations can be rewritten in a more compact way as follows

$$S_\sigma(\omega) = \int_{-\infty}^{+\infty} dt \; e^{i\omega t} \exp\left\{ \int_0^\infty (1 - e^{i\omega' t}) \; \tilde{\pi}(\omega') d\omega' \right\} \; . \tag{48}$$

Eqs.(46) and (48) can be shown to be identical by writing:

$$\exp\left\{ \int_0^\infty (1 - e^{i\omega' t}) \; \tilde{\pi}(\omega') d\omega' \right\} =$$

$$\exp\left\{ \int_0^\infty \tilde{\pi}(\omega') d\omega' \right\} \exp\left\{ -\int_0^\infty e^{i\omega' t} \; \tilde{\pi}(\omega') d\omega' \right\} \; , \tag{49}$$

and expanding the last exponential in a power series.

We advise the reader to compare eq.(48) with some of the results presented by A. Lucas in this volume for the interaction of charges and surfaces.

2. QUANTAL INTERACTION OF ELECTRONS WITH CONDENSED MATTER. CHARGE TRANSFER BETWEEN IONS AND CONDENSED MATTER. CHARGE STATES

In this second lecture we shall discuss how to introduce the quantal character of the projectile in its interaction with condensed matter. The case of ions and electrons will be analyzed independently and, finally, the charge transfer between ions and a metal will be considered. This analysis is preliminary to the discussion presented in the third chapter about how to calculate, from a fundamental approach, the stopping power of different projectiles for intermediate velocities, i.e., for $v \sim z^{2/3} v_0$. This section is based on ref.[21].

15

2.1 Quantal Interaction Between a Moving Charge and an Electron Gas

The Hamiltonian describing an external charge of mass M and charge Z_1 interacting with an electron gas in written as [22]

$$H = H_0 + T_0 + H_I. \tag{50}$$

Here H_0 is the Hamiltonian of the homogeneous electron gas, $T_0 = -\nabla_I^2/2M$ is the kinetic energy operator for the external charge and H_I represents the interaction Hamiltonian between the external charge and the medium electrons, i.e.,

$$H_I = \sum_j \frac{Z_1}{|\vec{r}_j - \vec{R}_I|} = \int \frac{Z_1\, \rho(r)}{|\vec{r} - \vec{R}_I|}\, d^3r. \tag{51}$$

The coordinate of the jth electron in the electron gas is described by $\vec{r}_j$ and $\vec{R}_I$ denotes the coordinate of the incident charge. $\rho(r)$ is the particle operator of the electron gas $\rho(\vec{r}) = \sum_j \delta(\vec{r} - \vec{r}_j)$. The state of the charged particle is described by a plane wave $|i\rangle = \exp\{i\vec{k}_i\vec{R}_I\}$ normalized to unit volume, where $\vec{k}_i$ is the initial momentum of the particle. We denote by $|n\rangle$ the many-body eigenvectors of the homogeneous electron gas such that $H_0|n\rangle = E_n|n\rangle$. From first-order perturbation theory the probability per unit time for a transition from the initial state $|i\rangle|0\rangle$ to a final state $|f\rangle|n\rangle = \exp\{i\vec{k}_f\vec{R}_I\}|n\rangle$ is

$$P_{i \to f} = 2\pi \sum_n \left| \langle n|\langle f|H_I|i\rangle|0\rangle \right|^2 \delta(E_n - E_0 + \frac{(k_f^2 - k_i^2)}{2M}) =$$

$$= 2\pi \sum_n \int d^3r \int d^3r'\, \langle 0|\rho^+(\vec{r})|n\rangle \langle n|\rho(\vec{r}')|0\rangle$$

$$\times \langle e^{i\vec{k}_f\vec{R}_I} |\frac{Z_1}{|\vec{r} - \vec{R}_I|}| e^{i\vec{k}_i\vec{R}_I} \rangle \langle e^{i\vec{k}_i\vec{R}_I} |\frac{Z_1}{|\vec{r}' - \vec{R}_I|}| e^{i\vec{k}_f\vec{R}_I} \rangle$$

$$\times \delta(E_n - E_0 + \frac{(k_f^2 - k_i^2)}{2M}). \tag{52}$$

If we denote $q = \vec{k}_i - \vec{k}_f$ and $\omega_{no} = E_n - E_0$ and realize that the matrix elements of $Z_1/|\vec{r} - \vec{R}_I|$ are given by $v_q \exp\{-i\vec{q}\,\vec{r}\}$, where $v_q = 4\pi\, Z_1^2/q$, eq. (52) becomes

$$P_{i \to f} = 2\pi \sum_n \langle 0|\rho^+(\vec{q})|n\rangle \langle n|\rho(\vec{q})|0\rangle \left(\frac{4\pi\, Z_1}{q^2}\right)^2 \delta(\omega_{no} - (k_f^2 - k_i^2)/2M). \tag{53}$$

where $\rho(q) = \sum_j e^{-i\vec{q}\,\vec{r}_j}$ is the Fourier transform of $\rho(\vec{r})$, i.e., it represents fluctuations about the average particle density. The symbol + in ρ denotes complex conjugation. Thus $P_{i\rightarrow f}$ is proportional to the Fourier transform of the density-density correlation function $|<n|\rho(q)|0>|^2$ a result first derived by Van Hove [23]. By noting that the imaginary part of the inverse of the electron-gas response function can be written, for positive ω, as [22]

$$\text{Im}\left(\frac{-1}{\varepsilon(q,\omega)}\right) = \frac{4\pi^2}{q^2} \sum_n \left|<n|\rho(q)|0>\right|^2 \delta(\omega - \omega_{n0}), \tag{54}$$

with $\omega > 0$, $P_{i\rightarrow f}$ can be written as

$$P_{i\rightarrow f} = \int_0^\infty \frac{d\omega}{2\pi} P'(q,\omega), \tag{55}$$

where $P'(q,\omega)$ is given by

$$P'(q,\omega) = \frac{16\pi^2 z_1^2}{q^2} \text{Im}\left(\frac{-1}{\varepsilon(q,\omega)}\right) \delta\{\omega + (k_f^2 - k_i^2)/2M\}, \tag{56}$$

which is very similar to eq.(9). Notice that instead of $\delta(\omega - \vec{q}\,\vec{v})$ we now find $\delta(\omega + \Delta)$, with:

$$\Delta = (k_f^2 - k_i^2)/2M = \left\{(\vec{k}_i - \vec{q}) - \vec{k}_i^2\right\}/2M = q^2/2 - \vec{k}_i\,\vec{q}/M. \tag{57}$$

For high incident momenta, Δ can be replaced by $-\vec{q}\,\vec{v}$, and then $P'(q,\omega)$, given by eq.(56) coincides with the classical expression of eq.(9), the classical limit. The quantum effect of the moving projectile only amounts to introducing in the energy conservation law, $\delta(\omega - \vec{q}\,\vec{v})$, the recoil term $q^2/2$, thus yielding $\delta(\omega + \frac{q^2}{2} - \vec{q}\,\vec{v})$.

2.2 Quantal Interaction for an External Electron in an Electron Gas

When the external particle is an electron an important difference arises with respect to the above treatment [24] because the incoming particle is indistinguishable from the electrons of the medium.

Consider an electron in the excited state $\vec{k}$ (fig.5). Due to its interaction with the electron gas, it can jump to a state $\vec{k}-\vec{q}$, creating an excitation pair of energy ω and momentum $\vec{q}$. As a first approximation we could proceed as in sect.2.1 by writing a Hamiltonian equivalent to the one leading to eq.(56) for $P'(q,\omega)$, with $\vec{k}_i = \vec{k}$ and $\vec{k}_f = \vec{k}-\vec{q}$, but now

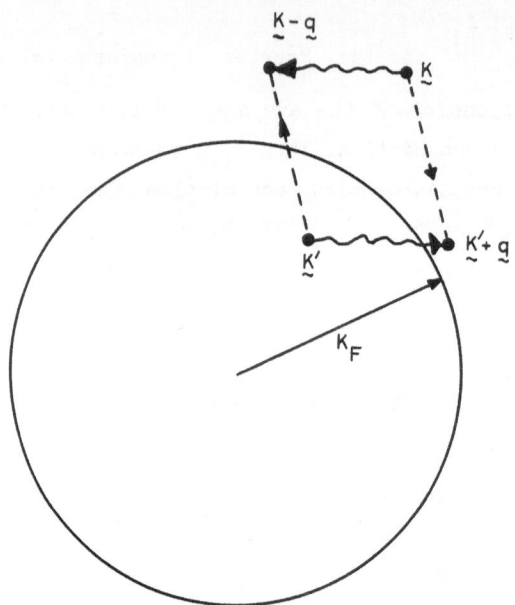

Figure 5. External electron of momentum $\vec{k}$ interacting with an electron of momentum $\vec{k}'$ inside the Fermi sea. The final states are denoted by $\vec{k}-\vec{q}$ and $\vec{k}'+\vec{q}$. Such a final state can be reached in two different ways. In the direct process (wavy line) the momentum transfer is $\vec{q}$. The dashed line (---) represents an exchange process associated with momentum transfer $\vec{k}'+\vec{q}-\vec{k}$. This second process, and therefore its interference with the direct one, has been neglected in the derivation of eq.(58).

subject to the important restriction that $|\vec{k}_f|$ has to be greater than k_F, the Fermi wave number of the electron gas. Thus

$$P'_{\vec{k}\to\vec{k}-\vec{q}} = \frac{16\pi^2}{q^2}\; Im\left(\frac{-1}{\varepsilon(q,\omega)}\right)\; \delta\left\{\omega + \frac{1}{2}\,(\vec{k}-\vec{q})^2 - k^2\right\}\; \Theta(|\vec{k}-\vec{q}|-k_F). \qquad (58)$$

One word of caution should be added here. In obtaining eq.(58) we have neglected exchange processes [24,25] interfering with the direct process shown in fig.5. In an exchange process, the transition from the initial state $|\vec{k}\rangle|0\rangle$ can be performed in a different way from the direct one described previously: the electron in state $\vec{k}$ can make a transition to the state $\vec{k}'+\vec{q}$ while the electron in $\vec{k}'$ goes to the state $\vec{k}-\vec{q}$ (see fig.5). These two processes interfere, modifying our previous result of eq.(58). We shall neglect this interference, however. This approximation is equivalent to the GW approximation in the language of Feynman diagrams [22,26].

18

Eq. (58) defines the following lifetime, τ, for an electron having initial momentum k,

$$\frac{1}{\tau} = \int_{-\infty}^{\infty} \int_{-\infty}^{\infty} \frac{d^3q}{(2\pi)^3} \frac{d\omega}{2\pi} P'(q,\omega). \tag{59}$$

The quantal correction given by eq. (58) is only important for $\vec{k}$ close to $\vec{k}_F$. If $k \gg k_F$, we can neglect the Pauli restriction $\theta(|\vec{k}-\vec{q}| - k_F)$ and the recoil contribution to the energy conservation law.

For $k > k_F$, we can use the small-ω expansion for the Lindhard dielectric function as discussed above, see eqs. (26) and (27). Then, one gets the following lifetime

$$\frac{1}{\tau} = \frac{\pi}{4} (k - k_F)^2. \tag{60}$$

This result shows that the inelastic collisions are drastically reduced as $k \to k_F$; the $(k-k_F)^2$ behavior follows the general trend given by Landau for the quasiparticles lifetime. Eq. (60) and the classical behavior given by (24) are collected in fig. 6 where $1/\tau$ is shown [26] for electrons as a function of k.

Before completing this section it is worth saying a few words about the different contributions appearing in eq. (58). One factor $(4\pi/q^2)^2$ represents the square of the interacting potential; $\frac{q^2}{4\pi} \text{Im}\left(-\frac{1}{\varepsilon(q,\omega)}\right)$ measures the probability of having an excitation of energy ω and momentum q in the electron gas, $\delta(\omega + \frac{1}{2}|\vec{k}-\vec{q}|^2 - \frac{1}{2}k^2)$ is the energy-conservation

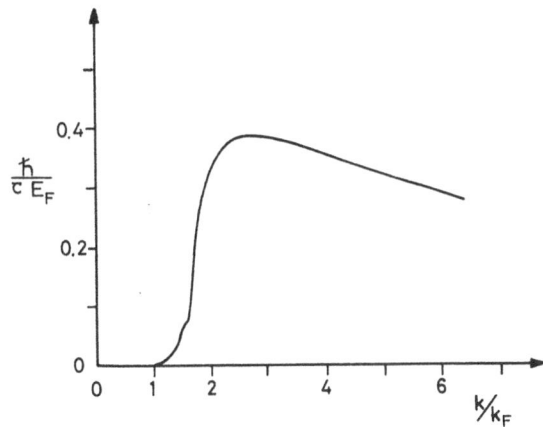

Figure 6. Qualitative behavior of $1/\tau$ as a function of k for electrons. For k near k_F, $1/\tau \approx (k-k_F)^2$, while for $k \gg k_F$, $1/\tau$ goes over to the classical limit.

law, and $\theta(|\vec{k}-\vec{q}| - k_F)$ represents the Pauli exclusion principle for electrons. Notice that the factor $\frac{q^2}{4\pi} \text{Im}\left(- \frac{1}{\varepsilon(q,\omega)}\right)$ measures the electron gas behavior in its interaction with the external charge. More about this will be commented below.

2.3 Bound Ion-Electron Composite Interacting with an Electron Gas. Auger Processes

In this section, we shall address the problem of an ion-electron composite [27,28] moving inside an electron gas. We begin by writing the Hamiltonian of the problem

$$H = H_0 + H_I + \sum_j \left(\frac{1}{|\vec{r}_j - \vec{r}_e|} - \frac{Z_1}{|\vec{r}_j - \vec{R}_I|} \right), \tag{61}$$

where H_0 is the Hamiltonian of the electron gas, $\vec{r}_j$, denotes the coordinates of the jth electron in the electron gas; $\vec{R}_I$ the coordinates of the ion and $\vec{r}_e$ denotes the coordinates of the electron in the ion-electron composite. H_I is the Hamiltonian of the ion-electron composite

$$H_I = - \frac{1}{2M} \nabla^2_{R_I} - \frac{1}{2} \nabla^2_{r_e} - \frac{Z_1}{|\vec{R}_I - \vec{r}_e|}. \tag{62}$$

The last term in eq.(61) describes the interaction between the electron gas and the ion-electron system. We shall be concerned mainly with one-electron ions. The eigenfunctions of H_I are given by $|i\rangle$

$$|i\rangle = e^{i\vec{k}_0 \vec{R}} u_0(\rho), \tag{63}$$

with

$$H_I|i\rangle = E_I|i\rangle \tag{64}$$

and

$$E_I = \frac{k_0^2}{2(M+1)} + \omega_0, \tag{65}$$

where $\vec{R} = (\vec{r} + M\vec{R})/(1 + M)$ represents the coordinates of the center of mass, $\vec{k}_0$ the total momentum of the composite, and $u_0(\rho)$ the wave function describing the relative motion of the electron in the composite with respect to the ion, i.e., $\vec{\rho} = \vec{r}_e - \vec{R}_I$, with energy ω_0. The ion-electron composite can lose its electron, then the new composite wave function is given by

20

$$|f\rangle = e^{i\vec{k}\cdot\vec{R}} e^{i\vec{k}_f\cdot\vec{\rho}}, \qquad (66)$$

where $\vec{k}$ is the final momentum of the composite and $\vec{k}_f$ is the momentum of the relative motion of the ion and the ejected electron. The corresponding eigenvalue is

$$E_f = \frac{1}{2(M+1)} k^2 + \frac{1}{2} \frac{M}{M+1} k_f^2. \qquad (67)$$

The Fermi golden rule gives us the transition probability per unit time of a transition from the initial state $|i\rangle|0\rangle$ of energy $E_I + E_0$ to the final state $|f\rangle|n\rangle$ of energy $E_f + E_n$; $|0\rangle$, $|n\rangle$, E_0 and E_n refer to many-body eigenvectors and eigenenergies of the homogeneous electron gas. Following the lines of previous analysis one finds

$$P^{loss} = \int dx \sum_{\vec{k}_f} \theta\left(|\vec{k}_f + \vec{v}| - k_F\right) \frac{16\pi^2}{q^2} \delta\left(\omega + \frac{k_f^2}{2} - \vec{q}\cdot\vec{v} - \omega_0\right)$$

$$\times \operatorname{Im}\left(\frac{-1}{\varepsilon(q,\omega)}\right) \left| \langle e^{i\vec{k}_f\cdot\vec{\rho}} | e^{i\vec{q}\cdot\vec{\rho}} | u_0(\rho)\rangle \right|^2. \qquad (68)$$

Several comments about this equation are worthwhile. First of all, $\vec{k}_f$ is the momentum of the ejected electron referred to the moving ion; thus the momentum with respect to the electron gas is $\vec{k}_f + \vec{v}$. In obtaining eq.(68) we have taken $\vec{v}$ to be a constant before and after the collision. Comparing eqs.(68) and (58), one should notice that $\vec{k}_f + \vec{v}$ has to be replaced by $\vec{k} - \vec{q}$ in (58). That this is the case is easily seen in eq.(68): if we choose the initial state $u_0(\rho)$ to be a plane wave $\exp\{i(\vec{k}-\vec{v})\vec{\rho}\}$, then $\vec{k}_f = \vec{k} - \vec{v} - \vec{q}$, the final state having a momentum $\vec{k} - \vec{q}$ referred to the electron gas. Finally, the energy conservation law, eq.(68), says that the difference between initial and final states energies ($\vec{k} = \vec{k}_0 - \vec{q}$)

$$\left(\frac{k_0^2}{2(M+1)} + \omega_0\right) - \left(\frac{(\vec{k}_0 - \vec{q})^2}{2(M+1)} + \frac{k_f^2}{2M/(M+1)}\right), \qquad (69)$$

goes to excitation of the electron gas. Notice that

$$\frac{k_0^2}{2(M+1)} - \frac{(\vec{k}_0 - \vec{q})^2}{2(M+1)} \qquad (70)$$

is replaced in eq.(68) by $\vec{q}\cdot\vec{v}$, neglecting recoil effects assuming $\vec{q}\cdot\vec{v} \gg \vec{q}^2/(M+1)$. We also assume $M/(M+1) \simeq 1$.

A similar argument can be used to calculate the capture probabilities. Instead of eq.(68) we find

$$P^{capture} = \int dx \sum_{\vec{k}_i} \theta\left(k_f - |\vec{k}_i + \vec{v}|\right) \frac{16\pi^2}{q^2} \delta\left(\omega - \frac{k_i^2}{2} - \vec{q} \cdot \vec{v} + \omega_0\right)$$

$$\times \operatorname{Im}\left(\frac{-1}{\varepsilon(q,\omega)}\right) \left| \langle u_0(\rho) | e^{-i\vec{q} \cdot \vec{\rho}} | e^{i\vec{k}_i \cdot \vec{\rho}} \rangle \right|^2. \tag{71}$$

Here, the ion captures an electron filling a state $\vec{k}_i + \vec{v}$ located below the electron-gas Fermi level.

Comparing eqs.(58) with eqs.(68) and (71) we find similar terms. To be specific, consider eq.(68); in this equation we find the same factor as in eq.(58) with an extra term $|\langle e^{i\vec{k} \cdot \vec{\rho}} | e^{i\vec{q} \cdot \vec{\rho}} | u_0 \rangle|^2$, which is associated with the probability for an electron of having a transition from the localized state u_0, to the plane wave $\exp(i\vec{k} \cdot \vec{\rho})$.

We note that the electronic states of the homogeneous electron gas are modified due to the presence of the localized state $u_0(\rho)$. A better

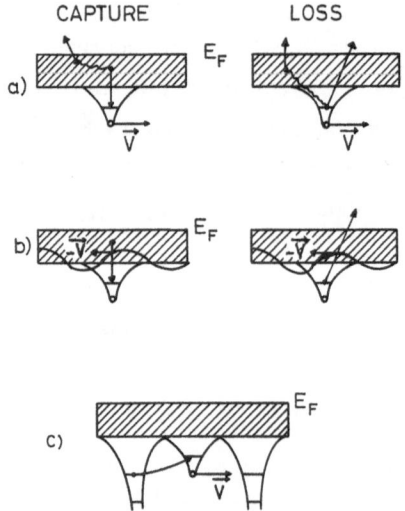

Figure 7. (a) An Auger process for an ion moving in a uniform electron gas; capture: a conduction electron jumps to a bound state on the ion; loss: an electron bound to the ion is excited to the conduction band. (b) Shows the coherent resonant process for an ion moving in a crystal: in a capture process, the crystal pseudopotential induces transitions between a conduction and a bound state; in a loss process, an electron bound to the ion is excited to a level of the conduction band. (c) Illustrates the shell process whereby an ion moving in a crystal captures one electron from an inner level of the target.

approximation to P^{loss} and $P^{capture}$ can be obtained if we choose the orthogonalized plane wave functions

$$|\vec{k}> = e^{i\vec{k}\cdot\vec{\rho}} - <u_0|e^{i\vec{k}\cdot\vec{\rho}}> u_0(\vec{\rho}), \qquad (72)$$

to describe the electrons of the conduction band.

2.4. Electronic Exchange Processes. Charge States of Ions in Solids

Several different mechanisms have been proposed as being responsible for electronic exchange processes of ions moving in condensed matter. The first mechanism plays an important role at low velocities. In the Auger process illustrated in fig.7a, an electron is captured (or lost) by the ion to (or from) a bound state assisted by a third body that may be a plasmon or an electron-hole pair. Condensed-matter effects are important here since electrons in valence-band states are involved. Other Auger processes associated with interactions between localized states of the target and bound states of the ion could contribute to the electronic exchange mechanism. These do not seem to play an important role here, however. Their corresponding cross sections are very small compared with those involving valence-band electrons in the region of velocities where Auger processes are significant.

A second mechanism leading to capture and loss is the coherent resonant interaction [29-31]. Electronic exchange processes are induced by the time-dependent crystal potential as seen from the fixed ion (fig.7b). The crystal atoms may be assumed to remain in their ground states in these interactions. For high ion velocities, the effective potential is mostly due to the core electrons; for low velocities, the interaction is mainly determined by the valence electrons. The coherent resonant interaction resembles a close atom-atom collision at high velocities. At low velocities condensed matter effects come into play.

The third mechanism we consider is the shell process, fig.(7c). In this case, an inner electron of a target atom can be captured by the moving ion [32-34]. Small impact-parameter collisions dominate in such interactions, thus the condensed matter properties of the medium are not very important here.

We discuss in the next sections the dynamic coherent resonant and the shell processes.

2.4.1. Coherent Resonant Processes. Auger processes are the only ones appearing in a uniform electron gas. New effects come from the crystal structure. Fig.(7b) illustrates the coherent resonant process we now discuss. The resonant processes are due to the potential seen by the moving ion. From the point of view of the ion, there appears a moving potential which gives rise to transitions between bound states of the composite and free electron states. An equivalent point of view is the one taken by Kaneko and Ohtsuki [29]. These authors analyzed the dynamical processes created by the electrostatic potential of a single crystal atom acting on the moving ion; this approach neglects, however, the interference effects associated with the periodicity of the crystal (see also the second article in ref.[29b]).

Coherent resonant processes are best understood by assuming the ion to be at rest; then, the effective pseudopotential seen by the electron of the composite can be written

$$V(\vec{r},t) = \sum_{\vec{G}} V(\vec{G}) \, e^{i\vec{G}\cdot(\vec{r}-\vec{v}t)}, \tag{73}$$

$\vec{G}$ being a reciprocal vector of the lattice.

The potential in eq.(73) may be considered to induce excitations between a bound state of the composite described by the orbital $u_0(\vec{\rho})$, and a valence-electron state described by an OPW wave function $|\vec{k}_i\rangle$. Notice that in the rest frame of the ion, the momentum $\vec{k}_f$ of the valence electron is given by $\vec{k}' - \vec{v}$, $\vec{k}'$ being the momentum in the laboratory frame.

Using standard perturbation theory, it is straightforward to calculate the probability amplitude, $a_{0\to f}$, for a loss process $u_0(\vec{\rho}) \to |\vec{k}_f\rangle$

$$a_{0\to f}(t) = -\sum_{\vec{G}} V(G) \, \langle k_f | e^{i\vec{G}\,\vec{\rho}} | u_0(\rho)\rangle \, \frac{e^{i(k_f^2 - \omega_0 - \vec{G}\,\vec{v})t/2}}{\frac{1}{2}k_f^2 - \omega_0 - \vec{G}\,\vec{v} - i\eta}. \tag{74}$$

The probability per unit time for the resonant loss process is given by

$$P_{0\to f}^L(t) = \lim_{t\to\infty} \frac{d}{dt} \, |a_{0\to f}(t)|^2. \tag{75}$$

Summing over all the empty final states, $\vec{k}_f$, we obtain the total probability, P_R, associated with the coherent resonant process

$$P_R^L = 2\pi \int \frac{d^3k_f}{(2\pi)^3} \, \theta(|\vec{k}_F + \vec{v}| - k_f)$$

$$\times \sum_G \sum_{\substack{\vec{G}'\vec{v}= \\ \vec{G}\vec{v}}} \left\{ V(\vec{G}) \; V(\vec{G}') \; \langle u_0(\rho)|e^{i\vec{G}\cdot\vec{\rho}}|\vec{k}_f\rangle\langle\vec{k}_f|e^{i\vec{G}'\cdot\vec{\rho}}|u_0(\vec{\rho})\rangle \right\}$$

$$\times \; \delta\left(\omega_0 - \frac{1}{2}\, k_f^2 + \vec{G}\cdot\vec{v}\right). \tag{76}$$

In the equation, we have interference effects [29] associated with those reciprocal vectors satisfying $\vec{G}\cdot\vec{v} = \vec{G}'\cdot\vec{v}$; for a direction such that this condition is never satisfied (irrational angles) we are left with the following result

$$P_R^L = 2\pi \int \frac{d^3 k_f}{(2\pi)^3} \; \theta(|\vec{k}_F + \vec{v}| - k_F) \sum_{\vec{G}} |V(\vec{G})|^2 \; |\langle u_0(\rho)|e^{i\vec{G}\cdot\vec{\rho}}|\vec{k}_f\rangle|^2$$

$$\times \; \delta\left(\omega_0 - \frac{1}{2}\, k_f^2 + \vec{G}\cdot\vec{v}\right) \tag{77}$$

The same result is obtained from eq.(76) if P_R^L is averaged over all the possible directions of the moving ion. eq.(76) includes effects associated with ion channeling that are reflected in the interference between different reciprocal vectors. If the ion charge states are not analyzed for specific directions of the fixed ion, eq.(77) seems to be more appropriate for discussing coherent resonant processes.

A similar argument yields the following result for the probability per unit time for a capture process due to resonant effects

$$P_R^C = 2\pi \int \frac{d^3 k_i}{(2\pi)^3} \; \theta\left(k_F - |\vec{k}_i + \vec{v}|\right) \sum_{\vec{G}} V(\vec{G})^2 \; |\langle u_0(\vec{\rho})|e^{-i\vec{G}\cdot\vec{\rho}}|\vec{k}_i\rangle|^2$$

$$\times \; \delta\left(\omega_0 - \frac{1}{2}\, k_i^2 - \vec{G}\cdot\vec{v}\right), \tag{78}$$

if we neglect interference effects.

It is of interest to compare eqs.(77) and (68). In eq.(77), $|V(\vec{G})|^2$ appears instead of the square of the Fourier transform of the Coulomb potential, $4\pi/q^2$; also, the dynamical factor $\int (d\omega/2\pi)(q^2/4\pi) \; \text{Im}(\varepsilon^{-1}(q,\omega))$ is replaced by 1, and inside the δ factor, ω is taken to be zero. The differences between the two results are easy to understand. The dynamical factor disappears in the resonant process due to the absence of any internal change in the structure of the lattice ions creating the moving potential. This also explains why ω must be taken to be zero inside the δ factor. Finally, $4\pi/q^2$ must obviously be replaced by $V(\vec{G})$.

Note that in Auger processes the momentum transfer may assume any value, while in resonant processes only reciprocal momenta $\vec{G}$ are involved. In the approximation used here this is the only effect of crystal structure. If instead of $\sum \vec{G}$ we employ $\int d^3q/(2\pi)^3$, we recover the results associated with the interaction with independent target atoms. The difference is only important when small momentum transfers are important. This occurs at low ion velocities. At high velocities the incoming ion suffers collision with only a single target atom at a time.

2.4.2 Shell processes. Fig.7c shows the interaction in which the moving ion captures an electron from the deep levels of the target. Cross [32] and Shevelko [33] have analyzed this process using an Oppenheimer-Brinkman-Kramers (OBK) approach. It is well-known that this method yields capture cross sections that are too large, but gives, at high enough velocities, a good description of the cross-section behavior as a function of the ion velocity.

The OBK approach [35] is based on the Born approximation, and different results may be obtained depending on the approximation used for the wave functions and the energy levels involved in the process. Assuming hydrogen wave functions and energy levels $E_0 = -Z_0^2/2n_0^2$ and $E_1 = -Z_1^2/2n_1^2$, for the electron in the incoming ion and in the target, respectively, the capture cross section for the transfer $1 \rightarrow 0$ is given by

$$\sigma_s = \frac{2^9}{5} \frac{\pi}{v^2} \frac{(Z_0 Z_1)^5}{(n_0 n_1)^3} \frac{1}{\left\{ \left(\frac{E_0 - E_1}{v} - \frac{v}{2} \right)^2 + \frac{Z_1^2}{n_1^2} \right\}^5}. \tag{79}$$

An improved cross section within the OBK approach can be obtained [36] by taking more appropriate values for E_0 and E_1.

Eq.(79) shows that outside the resonance condition the shell cross section decreases quickly with increasing n_1. In practice, only those capture processes associated with the deepest levels of the target are dominant.

At high velocities eq.(79) overestimates the shell cross sections; it is too large by a factor of 3. At intermediate velocities, this factor increases depending on the atomic number of the incoming ion; for a proton or He, that factor can be between 4 or 5 if $v > v_0$. This result is estimated from calculations of protons colliding with hydrogen. For heavier atoms, that factor increases to 10 or more for B, N, C, and O; these results have been obtained by Sols and Flores [37] who have

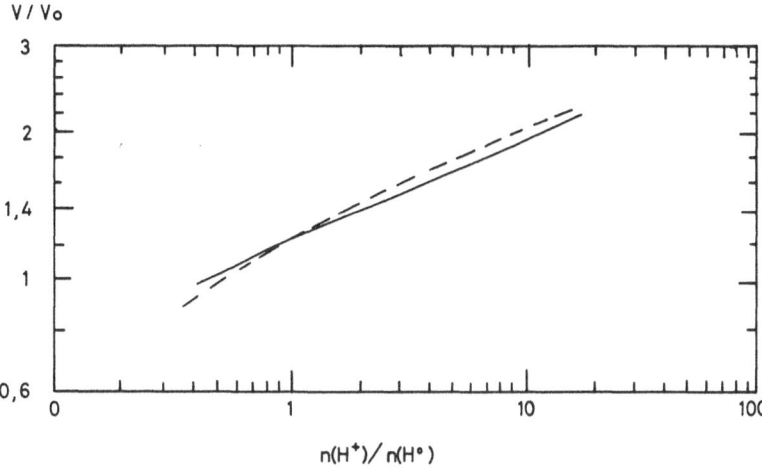

Figure 8. (Number of bare protons)/(number of neutral hydrogens) as a function of the atom velocity for an Al target. Experimental curve: dashed line.

calculated the corresponding shell cross sections (see below) using an eikonal method, as proposed by Chan and Eichler [38]. This should be a good approximation for the total cross sections.

2.5 Charge States

Since the pioneering work of Phillips [39] the charge states of moving ions have been experimentally measured over a wide range of energy [40-42].

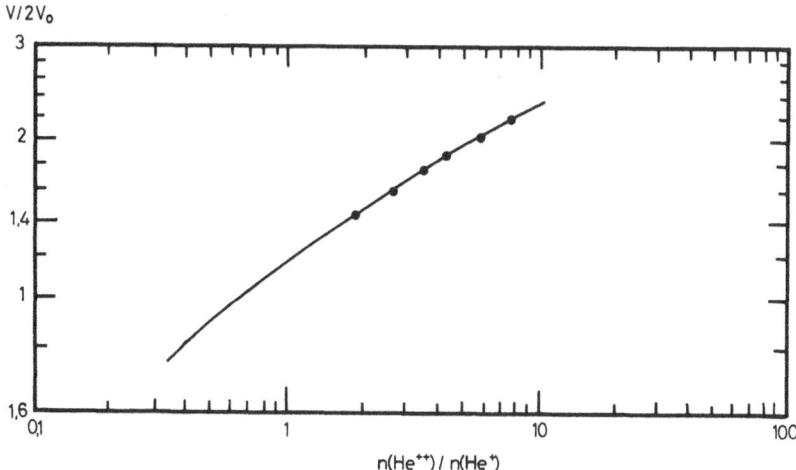

Figure 9. (Number of bare helium atoms)/(number of ions with one electron) as a function of the velocity for an Al target. Experimental results: dot points.

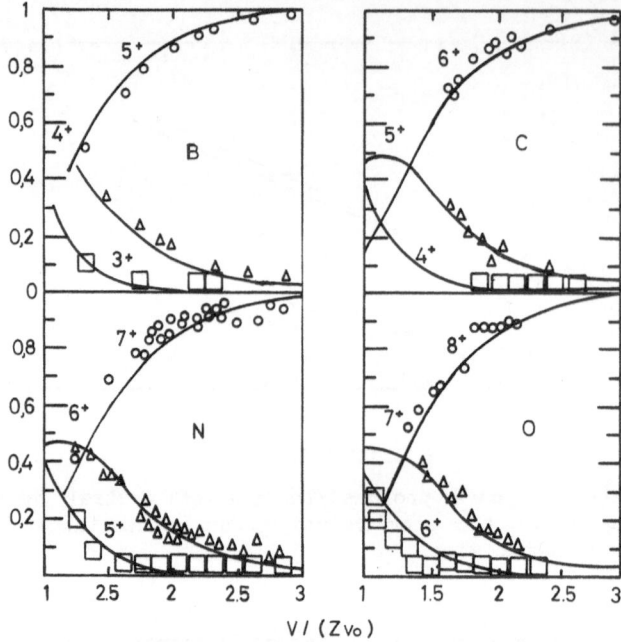

Figure 10. Fraction of charged ions (B, C, N and O) as a function of $v/Z_1 v_o$. Full lines: theoretical results.

Using the calculated probabilities for the exchange processes as discussed in sects.2,3 and 2.4, we can obtain the different charged ions inside the target, as a function of velocity. In figs.8-10 we show the charge fractions of different ions as a function of ion velocity. In figs.8 and 9, we show results for H and He [30-43], which are in good agreement with experiments [42]. In fig.10, we collect the results calculated [37] for B, C, N and O. For these cases a fair agreement with experiments [41] for $v > 1.5\ Z_1 v_0$ is also found.

3. STOPPING POWER FOR LIGHT IONS IN METALS

As mentioned in the introduction, in the limit of high velocities, $v > Z_1^{2/3} v_0$, the ion loses all its electrons and the ion-matter interaction can be described as discussed in sect.1, using linear-response theory. When $v \ll v_0$, the ion is dressed with quasiparticles which may strongly Üperturb electron states associated with the moving ion. Echenique deals with this problem in this volume using the local density functional method; in particular, it is shown there how the reaction of the medium to an ion for $v \ll v_o$ is similar to that for a static impurity.

The condition $v = Z_1^{2/3} v_0$ may be taken to define a regime of intermediate velocities. $Z_1^{2/3} v_0$ is the mean velocity of the electrons filling the levels of a neutral atom with nuclear charge Z_1; this result is obtained using Thomas-Fermi statistical theory [44]. The velocity $Z_1 v_o$ refers to the mean velocity of the electrons most strongly bound to the ion, while $Z_1^{2/3} v_o$ is the statistical mean velocity of all the electrons of the neutral atom. For velocities between v_o and $Z_1 v_o$, the ion exchanges electron with the medium, and the formal theory of stopping power should include the different charge states of the ion and the energy losses associated with the electron capture and loss mechanisms. This fundamental approach is a difficult task for heavy ions where many electrons participate in the exchange processes. In the following, we concentrate on discussing light ions, like He and H; we take the He case as the paradigm where to analyze in detail the fundamental theory of stopping power. In the last part of this section we also consider hydrogen, a case that presents some supplementary difficulties associated with its charge states in an electron gas.

3.1. Stopping Power for He. Charge States. Exchange Processes

Let us consider the He case, with the ion moving inside a metal with a given velocity, v. As discussed in sect.2, the ion suffers different electronic exchange processes with the medium and reaches some equilibrium charge states measured by the charge fractions $n(He^0)$, $n(He^+)$ and $n(He^{++})$. These charge fractions are obtained from the cross sections associated with the different capture and loss processes. In particular, if we consider the different processes of fig.11 and introduce the corresponding cross sections, $\sigma_{loss}(He^+)$, $\sigma_{capt}(He^+)$, $\sigma_{loss}(He^0)$ and $\sigma_{capt}(He^{++})$, we obtain the following equilibrium charge fractions

$$n(He^{++}) = \sigma_{loss}(He^+) \; \sigma_{loss}(He^0)/D, \qquad (80)$$

$$n(He^+) = \sigma_{loss}(He^0) \; \sigma_{capt}(He^{++})/D, \qquad (81)$$

and

$$n(He^0) = \sigma_{capt}(He^+) \; \sigma_{capt}(He^{++})/D, \qquad (82)$$

where

$$D = \sigma_{loss}(He^+) \; \sigma_{loss}(He^0) + \sigma_{loss}(He^0) \; \sigma_{capt}(He^{++})$$

$$+ \sigma_{capt}(He^+) \; \sigma_{capt}(He^{++}). \qquad (83)$$

The equilibrium charge fractions, $n(He^+)$ and $n(He^0)$, have been given in sect.2 in a logarithmic scale (see fig.9). In fig.12, we show [45] $n(He^0)$,

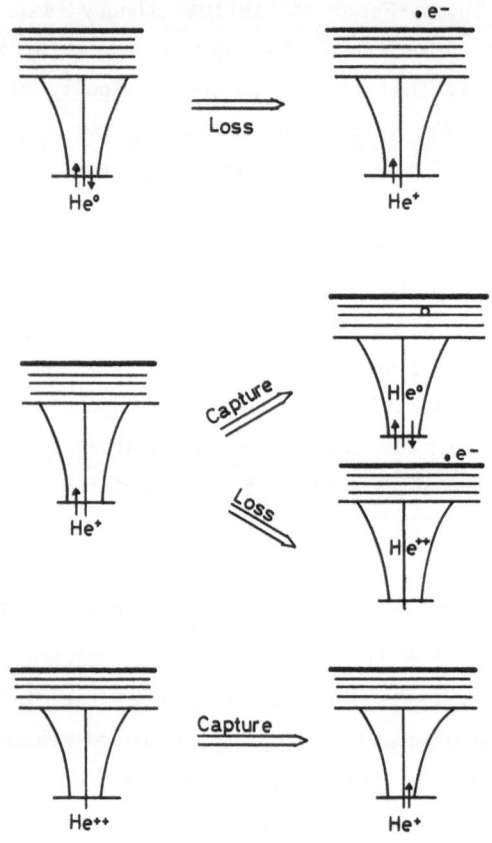

Figure 11. Different loss and capture processes for He^0, He^+ and He^{++}

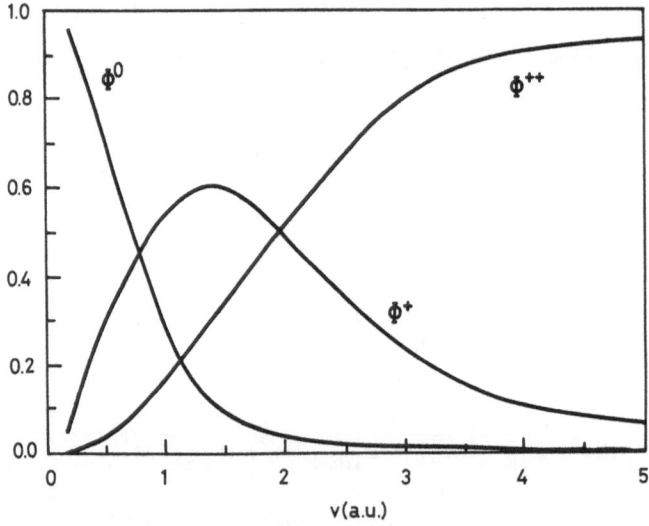

Figure 12. $n(He^0)$, $n(He^+)$ and $n(He^{++})$ as a function of the ion velocity.

$n(He^+)$ and $n(He^{++})$ as a function of the ion velocity. In the limit of low velocity, the only ion state is neutral He; for increasing velocities the charge fraction of He^+ increases and, eventually, in the limit of high velocities only He^{++} charge states survive.

In a very naive way, we would calculate the ion stopping power by considering each charge state and weighting its respective stopping power with its charge state fraction. In other words, if we call $\frac{dE}{dx}(He^0)$, $\frac{dE}{dx}(He^+)$, $\frac{dE}{dx}(He^{++})$ the stopping powers associated with the charge state fractions, $n(He^0)$, $n(He^+)$ and $n(He^{++})$, we would get in this approximation

$$\frac{dE}{dx} = n(He^0) \frac{dE}{dx}(He^0) + n(He^+) \frac{dE}{dx}(He^+) + n(He^{++}) \frac{dE}{dx}(He^{++}). \qquad (84)$$

In the limit of low and high velocities we have the following charge fractions

$$v \rightarrow 0 \begin{cases} n(He^0) \rightarrow 1 \\ \\ n(He^+) = n(He^{++}) = 0 \end{cases} \qquad (85)$$

$$v \rightarrow \infty \begin{cases} n(He^{++}) \rightarrow 1 \\ \\ n(He^0) = n(He^+) = 0 \end{cases} \qquad (86)$$

and eq.(84) yields the low and high velocity limits for the stopping power if we take for $n(He^0)$ the Local Density Formalism result [14], and for $\frac{dE}{dx}(He^{++})$ the linear-response limit discussed in sect.1.

In the actual calculation of the total stopping power, $\frac{dE}{dx}(He^{++})$ and $\frac{dE}{dx}(He^+)$ are obtained in linear theory while $\frac{dE}{dx}(He^0)$ is calculated using the Local Density Formalism results. This is a well justified approximation because of the different velocity regions in which $\frac{dE}{dx}(He^0)$, $\frac{dE}{dx}(He^+)$ and $\frac{dE}{dx}(He^{++})$ are relevant: $\frac{dE}{dx}(He^0)$ is dominant in the low-velocity limit, while $\frac{dE}{dx}(He^+)$ and $\frac{dE}{dx}(He^{++})$ are relevant only for higher velocities, where a linear theory analysis can be expected to be appropriate. It could be argued that this approach overestimates the stopping power of He^0 for intermediate velocities ($v_0 \leq v \leq 2v_0$), while at the same time it underestimates the stopping power of He^{++} and He^+. One could expect, however, that both effects tend to compensate as the results of the total stopping power (see below) suggest.

Eq.(84) is only a first approximation to the projectile stopping power, since in this equation all the energy losses associated with the charge-exchange processes have been neglected. These contributions are easily understood with the simple case shown in fig.13. Consider a He^0 atom moving inside the metal; due to a loss process, an electron of the 1s

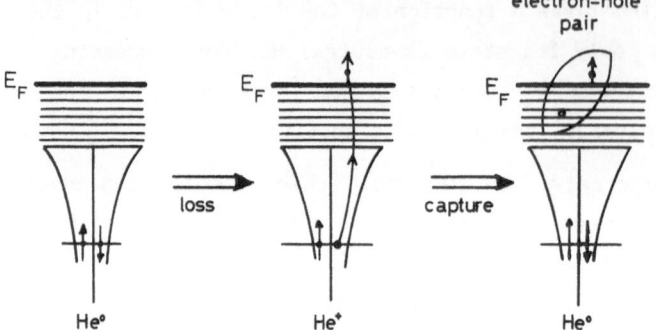

Figure 13. Successive loss and capture processes for a neutral He atom in an electron gas.

shell is transferred to an empty conduction band state; next, due to a capture process a conduction band electron is captured by the ion. The final state corresponds to the same initial ion charge state *plus an electron-hole pair excited in the metal.* The energies of these electron-hole pairs also contribute to the total ion stopping power.

Formally, we can add these contributions to the total stopping power, $\frac{dE}{dx}$, by means of the following terms

$$n(He^{++}) \frac{dE^C}{dx}(He^{++}) + n(He^+) \left\{ \frac{dE^L}{dx}(He^+) + \frac{dE^C}{dx}(He^+) \right\} + n(He^0) \frac{dE^L}{dx}(He^0),$$

(87)

where $\frac{dE^C}{dx}$ and $\frac{dE^L}{dx}$ represent the contribution to the total stopping power associated with the capture and loss processes.

In order to calculate the different contributions, $\frac{dE^C}{dx}$ and $\frac{dE^L}{dx}$, one has to take into account the different transition probabilities discussed in sect.2, and include there the energy transfer associated with the capture or loss processes. Proceeding in this way, one finds the following results for the shell processes

$$\frac{dE^C \text{shell}}{dx} = n_{at} \sum_n (E_n - E_0 + \frac{1}{2} v^2) \sigma_{n0},$$

(88)

where n_{at} is the target atomic density, E_n is the energy of the electron bound in the nth shell of the target atom, E_0 the binding energy of the captured electron, and σ_{n0} the cross section for the $n \rightarrow 0$ transition.

For the resonant capture and loss processes one gets

32

$$\frac{dE}{dx}^{C,L_{res}} = \frac{2\pi D_s}{v} \sum_{\vec{G}} \sum_{|\vec{k}+\vec{v}| \gtrless k_F} (\vec{G} - \vec{k}) \; \vec{v} \; |V(\vec{G})|^2$$

$$\times \; \left| <u_0|e^{i\vec{G} \; \vec{r}}|\vec{k} > \right|^2 \; \delta \left\{ E_{k0} \pm \vec{G} \; \vec{v} \right\}, \qquad (89)$$

where $E_{k0} = E_0 + \frac{k^2}{2}$, and u_0 is the wave function of the bound electron; D_s takes into account the spin degeneracy of the electron states: in the capture of an electron by the bare ion or in the loss by the neutral atom, $D_s = 2$, while in the capture and loss of an electron by He^+, $D_s = 1$.

For Auger capture and loss processes we get

$$\frac{dE}{dx}^{C,L_{Auger}} = \frac{2D_s}{v} \sum_{|\vec{k}+\vec{v}| \gtrless k_f} \int_0^\infty d\omega \int \frac{d^3q}{(2\pi)^3} \; (\vec{q} - \vec{k}) \; \vec{v} \; \frac{4\pi}{q^2}$$

$$\times \; \text{Im} \left\{ - \frac{1}{\varepsilon(\vec{q},\omega)} \right\} \; |M_{k,0}|^2 \; \delta(\omega \pm \Delta E), \qquad (90)$$

where $M_{k0} = <u_0|e^{i\vec{q} \; \vec{r}}|\vec{k}>$, and $\Delta E = \vec{q} \; \vec{v} + E_0 + \frac{k^2}{2}$.

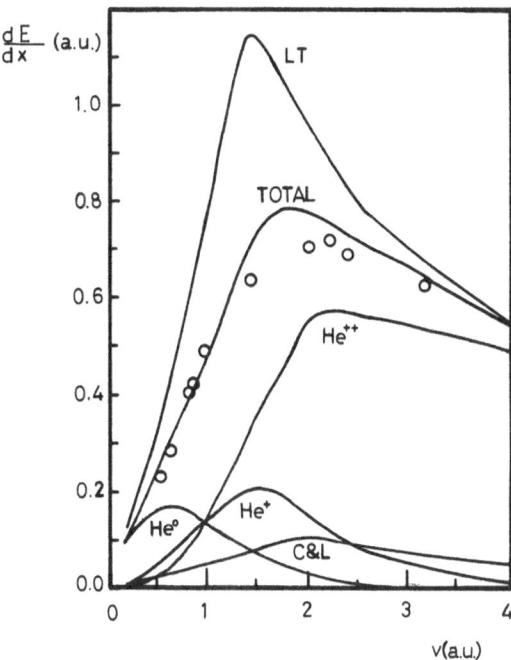

Figure 14. Total stopping power for He moving in Al. LT are calculations using linear theory and the RPA dielectric function. The total stopping power is separated in its different contributions (see text).

Fig.(14) shows the total stopping power (TOTAL) of the ions moving with velocity v in Al, including all the contributions discussed above. The curve labeled LT is the stopping power for bare ions calculated in linear theory using a RPA dielectric function. The different contributions to the curve labeled TOTAL have been separated to show the relevance of the various terms as a function of ion speed. When $v \simeq v_0$, the contributions from the neutral-atom fraction, from the singly-ionized fraction and the bare-ion fraction are comparable, each being about 30 % of the total, while that of charge-exchange processes (C and L) is about 10 %. The contribution from the inelastic processes of capture and loss is about 15 % for $v = 2v_0$. At high velocities, the stopping power of the bare ions dominates, as commented above; at low velocities, the neutral atom energy loss is the biggest contribution.

3.2 Stopping Power for H. Charge States

The case of H can be analyzed in a similar way to that of He. The main differences between both cases is due to the different ion charges. Thus, the main problem in the case of H is to define appropriately the different ion charge states: one can expect the ion to appear at least in two charge states, namely, H^+ and H^0; the main issue in this case is to know whether H^- may appear as a hydrogen charge state in the limit of low velocities; Vinter [45] and Guinea and Flores [46] have shown, using a selfenergy approach, the appearance of an H^- state in an electron gas, proving in this way that a proton can bind two electrons inside a metal. These results show that in the calculation of all the charge-exchange processes the proton can be assumed to have a well-defined bound level over the whole range of proton velocities. We also note that Norskov [47], using a local-density formalism, has concluded that the H^- configuration in aluminum has an energy 9 eV lower than the one for H^0, with a single electron bound to the proton. This result suggests that the H^- and H^0 configurations would be mixed for an ion velocity, v, such that $v^2/2 \simeq 9/27.2$ atomic units, namely, around $v = 0.8\ v_0$. The results shown below will confirm this analysis.

The equilibrium charge states, $n(H^+)$, $n(H^0)$ and $n(H^-)$, have been calculated in ref.[10] as discussed for He, and the total stopping power has been obtained [10] including all the capture and loss processes presented above.

Fig.15 shows the results for the equilibrium charge fractions as a function of the ion velocity. For very low velocities, only negative ions survive in the metal, while for high velocities only positive ions do. The

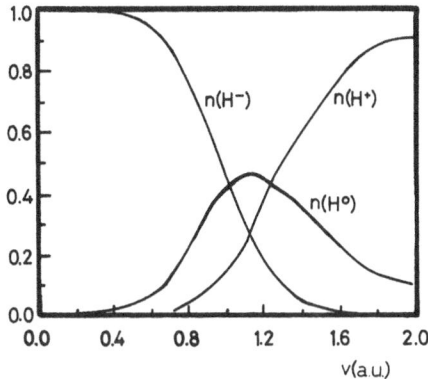

Figure 15. Equilibrium charge fractions $n(H^-)$, $n(H^0)$ and $n(H^+)$ as a function of the ion velocity.

intermediate region, $v \sim v_0$, is in good agreement with the rough estimate made above using Norskov's calculations [47].

The results [10] for the stopping power of protons moving with velocity v in Al are shown in fig.16 as a thick solid line (TOTAL). For comparison, it is shown in the same figure the stopping power for a bare proton calculated in linear theory, using a Lindhard dielectric function. The different contributions to the curve labeled TOTAL have been separated to show the relevance of the various terms as a function of the ion speed.

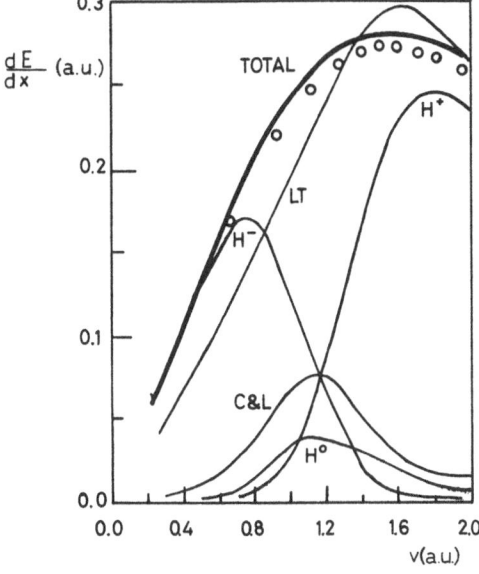

Figure 16. Total stopping power for H moving in Al (as for He in fig.14).

For $v = 1.2$ a.u., the contributions from the H^- and H^+ fractions are comparable each being 30 % of the total, while that of neutral H^0 is about 10 %. The inelastic processes of capture and loss contribute about 20 - 30 % in the range of velocities $v_0 < v < 1.4\ v_0$.

We conclude mentioning that the first-principle approach discussed in this section, shows at least a reasonable agreement with experiments, and offers a way to analyze the stopping power for ions moving in metals including their charge states. The results given have also shown that the capture and loss contributions to the stopping power are not negligible and should be included in a complete theory of the stopping power.

ACKNOWLEDGMENTS

The author thanks the Comisión Asesora de Investigación Científica Spain and Iberduero S.A. for their support.

REFERENCES

1. N. Bohr, Phil. Mag. *25* (1913) 16.
2. E. Fermi, Z.Phys. *29* (1927) 135.
3. E.J. Williams, Rev. Mod. Phys. *17* (1945) 217.
4. C.F.V. Weizsäcker, Z.Phys. *88* (1984) 612.
5. J. Lindhard, K. Dan, Vindensk. Selsk. Mat-Phys. Medd. *28* (1954) No.8.
6. N. Bohr, K. Dan. Vidensk. Selsk. Mat-Fys. Medd. *18* (1948) No.8.
7. W. Brandt, in *Atomic Collisions in Solids*, S. Datz, B.R. Appleton and C.D. Moak, eds., Plenum, New York (1975).
8. F. Guinea, F. Flores and P.M. Echenique, Phys. Rev. Lett. *47* (1981) 604.
9. H.D. Betz, Rev. Mod. Phys. *44* (1972) 465.
10. A. Arnau, P.M. Echenique, F. Flores and R.H. Ritchie; "Stopping power for Helium in Al", and "Stopping power for protons in Al for the whole range of velocities", to be published.
11. L. Hedin and S. Lundquist, Solid State Phys. *23* (1969) 1.
12. B.I. Lundquist, Phys. Konderns. Mater. *6* (1967) 206.
13. H.A. Bethe, Ann. Phys. (Leipzig) *5* (1930) 325.
14. P.M. Echenique in this volume.
15. R.H. Ritchie, Phys. Rev. *114* (1959) 644.
16. J.Lindhard and A.Winther, K.Dan. Vindensk. Selsk. Mat. Fys. Medd. *34* (1964) No.4.
17. H. Bichsel, Rev. Mod. Phys. *60* (1988) 663.
18. P. Sigmund, this volume.
19. F. García-Moliner and F. Flores, *Introduction to the Theory of Solid Surfaces*, Cambridge University Press, 1979.
20. M. Kato, J.Phys. Soc. Jpn *55* (1986) 1011.
21. P.M. Echenique, F. Flores and R.H. Ritchie, Solid State Physics *43* (1990) 235.
22. D. Pines, *Elementary Excitations in Solids*, Benjamin, New York (1964).
23. L. Van Hove, Phys. Rev. *95* (1954) 249.
24. J. Hubbard, Proc. R. Soc. (London) *A243* (1957) 336; *A240* (1957) 539.
25. R.H. Ritchie and J.C. Ashley, J.Phys. Chem. Solids *26* (1965) 1689.
26. B.I. Lundquist, Phys. Status Solidi *32* (1969) 273.
27. P.M. Echenique and R.H. Ritchie, Elhuyar 7 (1979) 1.
28. F. Guinea, F. Flores and P.M. Echenique, Phys. Rev. *B25* (1982) 6209.
29. T. Kaneto and Y.H. Ohtsuki, Phys. Stat. Solidi *B14* (1982) 491; Y.H. Othsuki, *Charged Beam Interactions with Solids*, Taylor (1983).

30. F. Sols and F. Flores, Phys. Rev. *B30* (1984) 4878.
31. J.D. Jackson and P.M. Platzman, Phys. Rev. *B22* (1979) 88.
32. M.C. Cross in *Inelastic Ion-Surface Collision*, N.H. Tolk, J.C. Tully, W. Heiland and C.W. White, eds., Academic Press, New York (1977).
33. V.P. Shevelko, Z. Phys. *A287* (1978) 19.
34. T. Kaneko, Nucl. Instrum. Methods *B2* (1984) 491.
35. H.C. Brinkman and H.A. BKramers, K. Wel. Amsterdam *33* (1930) 973.
36. F. Sols and F. Flores, Nucl. Instrum. Methods *13* (1986) 171.
37. F. Sols and F. Flores, Phys. Rev. *A37* (1988) 1469.
38. F.T. Chan and T. Eichler, Phys. Rev. Lett. *42* (1979) 58.
39. J.A. Phillips, Phys. Rev. *97* (1955) 404.
40. S.K. Allison, Rev. Mod. Phys. *30* (1958) 1137.
41. C.J. Sofield et al, Nucl. Instrumen. Methods *170* (1980) 257.
42. A. Itoh et al, Bull. Inst. Chem. Res. Kyoto University *60* (1983) 289.
43. P.M. Echenique and F. Flores, Phys. Rev. *B35* (1987) 8249.
44. L. Thomas, Proc. Cambridge Philos. Soc. *23* (1927) 542; E.Fermi, Z. Phys. *48* (1928) 73.
45. B. Vinter, Phys. Rev. *B17* (1978) 2729.
46. F. Guinea and F. Flores, J.Phys. *C13* (1980) 4137.
47. J.K. Norskov, Phys. Rev. *B20* (1978) 446.

DENSITY FUNCTIONAL THEORY OF STOPPING POWER

P.M. Echenique[1] and M.E. Uranga[2]

1. Dpto. de Física de Materiales, Facultad de Química
   Universidad del País Vasco/Euskal Herriko Unibertsitatea
   Apdo. 1072, E-20080 San Sebastián, Spain

2. Dpto. de Matemática Aplicada, E.T.S.I. Industriales
   Universidad del País Vasco/Euskal Herriko Unibertsitatea
   E-48007 Bilbao, Spain

1. DENSITY FUNCTIONAL FORMALISM

1.1 Theorem

The density-functional (DF) formalism provides a very useful tool to tackle many-body problems in a simplified manner. In this approach, the many-body problem is replaced by a Hartree-like equation, such as

$$\left[-\frac{\nabla^2}{2} + V_{eff}(\mathbf{r}) \right] \psi_i(\mathbf{r}) = \varepsilon_i \, \psi_i(\mathbf{r}). \tag{1}$$

(We use atomic units, $e^2 = \hbar = m = 1$ throughout except where otherwise indicated. The unit of energy is the Hartree (27.2 eV) and the Bohr's radius $a_0 = 0.529$ Å is the unit of length).

The crucial quantity in eq.(1) is $V_{eff}(\mathbf{r})$ which, in principle, contains all many-body effects. The DF formalism is based on two papers by Hohenberg and Kohn [1] and Kohn and Sham [2].

Our presentation here follows the one of Gunnarsson [3] in a NATO lecture course and relates to non-spin-polarized systems. The basic quantity in the density-functional approach is the electronic density to which many other ground-state properties are related. To see this, assume that N interacting electrons are moving in an external potential $v(\mathbf{r})$. The total Hamiltonian of the system is

$$H = T + U + V, \tag{2}$$

where T is the kinetic-energy operator, U is the electron-electron interaction term and V is the external potential operator corresponding to

Interaction of Charged Particles with Solids and Surfaces
Edited by A. Gras-Martí et al., Plenum Press, New York, 1991

the potential v(r). Then, at least in principle, the corresponding electron density n(r) could be calculated, which means that for a given external potential the density is uniquely determined. The converse statement is no so obvious, and it was proved by Hohenberg and Kohn [1] by *reductio ad absurdum*. Let us assume that the same density n(r) could be obtained for a different potential v'(r) differing from v(r) by more than a constant; assume, for simplicity, that the ground state is non-degenerate and ψ, ψ' is the ground state for v(r), v'(r). ψ and ψ' are different because they satisfy different Schrödinger equations. Since the expectation value of the Hamiltonian, under the above conditions, has its lowest value for the exact ground-state wave function, we can write, for the expectation value E',

$$E' = \langle\psi'|H'|\psi'\rangle < \langle\psi|H'|\psi\rangle = \langle\psi|H + V' - V|\psi\rangle$$

$$= \langle\psi|H|\psi\rangle + \langle\psi|V' - V|\psi\rangle = E + \int \{v'(r) - v(r)\}\, n(r)\, dr. \qquad (3)$$

However we could have started calculating the expectation value E,

$$E = \langle\psi|H|\psi\rangle < \langle\psi'|H|\psi'\rangle = \langle\psi'|H' + V - V'|\psi'\rangle$$

$$= E' + \langle\psi'|V - V'|\psi'\rangle = E' + \int \{v(r) - v'(r)\}\, n(r)\, dr. \qquad (4)$$

Adding (3) to (4) we obtain

$$E' + E < E + E', \qquad (5)$$

which disproves our assumption that there are two different potentials which give the same density n(r). Note that this does not mean that any given density n(r) can be obtained from an external potential v(r). What it means is that once the external potential v(r) is fixed, the density is univocally determined. The external potential fixes ψ, and therefore the density, and all the ground-state properties can be determined. We have established one of the basic results of the density-functional approach: *all ground-state properties are functionals of the density.*

1.2 Kinetic and Interaction Energy Functional

Two important examples of energy functionals are the kinetic and interaction energies. The sum of such energies is customarily denoted by F[n]

$$F[n] = \langle\psi|T + U|\psi\rangle. \qquad (6)$$

For a given potential $v(r)$, the total energy functional corresponding to an external potential $v(r)$, denoted by $E_v[n]$, is defined as

$$E_v[n] = \int v(r)\ n(r)\ dr + F[n].$$ (7)

Another basic result of the density-functional approach is precisely that *the ground state energy* **E**, *which is the functional* $E_v[n]$ *for the correct ground-state energy* **n(r)**, *is the lowest value that* $E_v[n']$ *can obtain for any density giving the total number of electrons* **N**,

$$\int n'(r)\ dr = N.$$ (8)

To proof this variational principle consider a state ψ', different from the ground state of H. Then

$$E_v[n] = \langle\psi|H|\psi\rangle < \langle\psi'|H|\psi'\rangle = \langle\psi'|T + V + U|\psi'\rangle$$

$$= \int n'(r)\ v(r)\ dr + F[n'] = E_v[n'].$$ (9)

From all the above we conclude that a general method to calculate ground-state properties of a system is to approximate the functional $F[n]$ and then minimize $E_v[n]$. The most known example of this approach is the Thomas-Fermi [4,5] method, in which it is assumed that the electrons are independent and that the kinetic energy density is the same as for a homogeneous medium. Then,

$$E_v[n] = \int v(r)n(r)\ dr + \frac{1}{2} \int \frac{n(r)n(r')}{|r - r'|}\ dr\ dr'$$

$$+ C \int [n(r)]^{5/3}\ dr,$$ (10)

where C is a numerical constant $C = (6\pi^2/5)(3/8\pi)^{2/3}$.

To minimize $E_v[n]$, we have

$$\delta \left\{ E_v[n] - \mu \int n(r)\ dr \right\} = 0,$$ (11)

where μ is a Lagrange multiplier. The minimum in $E_v[n]$ is now obtained with the density varied under the constraint

$$\int \delta n(r)\ dr = 0.$$ (12)

The corresponding Euler Lagrange's equation is

$$n(r) = C' \left[\mu - v(r) - \int \frac{n(r')dr'}{|r - r'|} \right]^{3/2},$$ (13)

with $C' = (3/5C)^{3/2}$, which is the well-known Thomas-Fermi equation. Note that eq.(13) gives the usual relation of the Thomas-Fermi theory, with its elegant scaling relations in atoms and ions, between the chemical potential μ and the electron density,

$$\mu = \frac{p_F^2(r)}{2} + V(r) = \frac{1}{2} (3\pi^2)^{2/3} [n(r)]^{2/3} + V(r), \tag{14}$$

where $p_F(r)$ is the Fermi energy of an homogeneous electron gas of density $n(r)$, and $V(r)$ is the external potential $v(r)$ plus the potential energy created at r by the electron cloud $n(r')$. That μ is the chemical potential can be seen [5] from equation (11) since

$$\mu = \frac{\partial E}{\partial N}. \tag{15}$$

1.3 Self Consistent Equations

The Thomas-Fermi approach, with exchange and correlation corrections incorporated, has been widely used in atomic and in solid-state calculations, but in this approximation the kinetic term is evaluated approximately and sometimes is too crude for detailed quantitative calculations. To find good approximations to the F[n] functional is not easy. Kohn and Sham realized that it is possible to separate numerically large contributions to F[n] which can be treated exactly, so that only the smaller remaining ones have to be treated approximately. Thus they used the partitioning

$$F[n] = T_s[n] + \frac{1}{2} \int \frac{n(r)n(r')drdr'}{|r - r'|} + E_{XC}[n]. \tag{16}$$

The term $T_s[n]$ is the kinetic energy of *non-interacting* electrons with density $n(r)$, the second term is the classical electrostatic interaction energy. Eq.(16) is a definition of the term $E_{XC}[n]$ which contains all the many-body effects.

The Euler equation is

$$\int \delta n(r) \left\{ v + \frac{\delta T_s}{\delta n} + \int \frac{n(r')dr'}{|r - r'|} + \frac{\delta E_{XC}}{\delta n} \right\} dr = 0, \tag{17}$$

($\delta/\delta n$ is a functional derivative) which leads to a set of Hartree-like one-body equations, like eq.(1),

$$\left[-\frac{\nabla^2}{2} + V_{eff}(r) \right] \psi_i(r) = \varepsilon_i \, \psi_i(r), \tag{18}$$

42

$$n(r) = \sum_{i=1}^{N} |\psi_i(r)|^2, \qquad (19)$$

$$V_{eff}(r) = v(r) + \int \frac{n(r')dr'}{|r - r'|} + \frac{\delta E_{XC}[n]}{\delta n(r)}. \qquad (20)$$

If the functional $E_{XC}[n]$ is known, eqs.(18-20) are solved self-consistently to obtain the ground-state density $n(r)$.

The main reason for the simplicity of the approach is twofold. On the one hand, the normal electron-electron interaction term does not enter explicitly; on the other hand, the effective potential is local in contrast to, for example, the Hartree-Fock potential, or more complicated self-energy treatments, in which

$$V_{eff}(r) \ \psi(r) = \int \Sigma(r,r',E) \ \psi(r') \ dr', \qquad (21)$$

where $\Sigma(r,r',E)$ is the complex non-local self-energy operator.

The simplicity, however, is only formal because the complexity of the many-body problem has been translated to the exchange and correlation functional $E_{XC}[n]$, but if the so-called (see below) local-density approximation is used for $E_{XC}[n]$, eqs.(18-20) can be, in many cases, quite easy to handle.

1.4 Local-Density Approximation

A very commonly used approximation is the local density (LD) approximation. In this approximation the local exchange-correlation energy $E_{XC}[n]$ is written in terms of the exchange-correlation energy of an electron in an homogeneous electron gas of density $n(r)$, ε_{XC}. Thus,

$$E_{XC}^L[n] = \int n(r) \ \varepsilon_{XC}(n(r)) \ dr. \qquad (22)$$

Then, the exchange-correlation contribution to the effective potential $V_{XC}[r]$ is

$$V_{XC}(r) = \frac{\delta E_{XC}[n]}{\delta n} = \frac{\partial}{\partial n(r)} \left[n(r) \ \varepsilon_{XC}(n(r)) \right], \qquad (23)$$

which is not only local in the sense of eq.(18), but depends on the density at the point r only.

Historically, besides the Thomas-Fermi equations, the first set of approximations to $V_{XC}(r)$ were found in a different way, from Dirac's solutions of the Hartree-Fock equations for an uniform electron gas [6].

The momentum-dependent exchange solution is given by

$$V_X(k) = - 4 \, f_{HF}(x) \left[\frac{3n_0}{8 \, \pi} \right]^{1/3}.$$ (24)

where n_0 is the electronic density, $x = k/k_f$ is the ratio of the momentum of the electron to its momentum at the Fermi energy, and $f_{HF}(x)$ is given by

$$f_{HF}(x) = \frac{1}{2} + \frac{1 - x^2}{4x} \, \ln \left| \frac{1 + x}{1 - x} \right|.$$ (25)

The approximation suggested by Slater [7] amounts to take the average value of $f_{HF}(x)$ under the Fermi level (i.e., 3/4) and replace n_0 by $n(r)$. Then

$$V_X^S(r) = - \frac{3}{2} \left[\frac{3n(r)}{\pi} \right]^{1/3}.$$ (26)

Note that when correlation effects are neglected, the variational principle of Kohn and Sham, following earlier work by Gaspar [8], will lead to

$$V_X^{KS}[r] = \frac{\partial}{\partial n(r)} \left[n(r) \varepsilon_X(r) \right]$$

$$= \frac{\partial}{\partial n(r)} \left[- \frac{3}{4} \, (\frac{3}{\pi})^{1/3} \, n(r)^{4/3} \right] = \frac{2}{3} \, V_X^S.$$ (27)

This final result is equivalent to the use of the value of $f_{HF}(x)$ at the Fermi level.

Many times a density parameter r_s, defined as

$$\frac{4}{3} \, \pi \, r_s^3 = \frac{1}{n(r)},$$ (28)

is used to characterize the density. In terms of r_s, the Kohn and Sham exchange potential, often written as μ_x, is

$$V_X^{KS} (n(r)) = \mu_X(r_s) = - \frac{0.61}{r_s}.$$ (29)

Results for ε_{XC} and V_{XC} including correlation effects have been given in the literature. Gunnarsson [3] writes that in the absence of spin polarization we can express [9]

$$V_{XC}(n(r)) = \beta(r_s)\mu_X(r_s),$$

$$\beta(r_s) = 1 + 0.0368 \, r_s \, \ln(1 + \frac{21}{r_s}).$$ (30)

The function $\beta(r_s)$ describes correlation effects. It has a fairly weak dependence on r_s. If correlation is neglected, $\beta = 1$, and one recovers the Kohn and Sham exchange; if $\beta = 3/2$, we recover Salter's exchange, and in the so-called Xα approximation $\beta = 3\alpha/2$, in which α is a fitting parameter.

A well-known approximation to the Wigner correlation energy may be written [3]

$$\varepsilon_C^W = - \frac{0.056\ n^{4/3}}{0.079 + n^{1/3}}. \tag{31}$$

Several analytical approximations to $\varepsilon_{XC}(n)$ have been devised, which are of comparable accuracy (about 1 - 2 %). Gunnarsson and Lundqvist [10] give

$$\varepsilon_{XC} = - \frac{0.458}{r_s} - 0.0666\ g(\frac{r_s}{11.4}) \tag{32}$$

where

$$g(x) = \frac{1}{2} \left[(1 + x^3)\ \ln(1 + x^{-1}) - x^2 + \frac{x}{2} - \frac{1}{3} \right], \tag{33}$$

The local-density approximation (LDA) assumes that the electron density variations are spatially slow, i.e., the density varies little over some typical distance, say a few inverse Fermi wave vectors; then one can assume that the electron "sees" an essentially homogeneous medium. The majority of calculations with the Kohn-Sham equations have been carried out with the use of LDA, which has yielded surprisingly good results even in cases where the density is not slowly varying. A reason of the success of the LDA might be found in the sum rule requiring that the exchange and correlation hole should contain exactly one electron [3]. This sum rule is fulfilled in the LDA, showing that the exchange and correlation E_{XC} might not be very sensitive to the shape and to the detailed form of the hole, if the sum rule is satisfied.

2. STOPPING OF A SLOW ION IN AN ELECTRON GAS

On the basis of the free-electron picture, with an additional assumption of independent, individual electron scattering, we can determine a formula in which the stopping power is written in terms of the scattering phase shifts of electrons scattered by an effective potential. This formula, by its definition, makes the description of the (spin independent) scattering correct. In the literature there are many ways to arrive at this formula. The techniques extend from theories of classical gas kinetics, through the Keldysh formalism for nonequilibrium

45

thermodynamics, to the general scattering description using the Møller operator. (We refer the reader to refs.[11-21]). Here we introduce the basic formula from classical considerations following Bonderup's lecture notes [11].

The average energy loss per unit length, $-dE/dx$ for an arbitrary ion of momentum $\mathbf{p}$ and mass M moving with velocity $\mathbf{v}$ through an electron gas of constant density n_0, is given by

$$- \frac{dE}{dx} = - \mathbf{v} \frac{d\mathbf{p}}{dx} = \frac{-dp_\parallel}{dt} \, , \tag{34}$$

where $p_\parallel$ is the momentum loss in the direction of $\mathbf{v}$. Since $1/M \ll 1$, we may consider the rest frame of the ion as a system of inertia in the calculation of the velocity changes for the electrons. As seen from the reference frame of the ion, the electrons are scattered by an effective screened Coulomb potential. The general structure of the formulae for energy loss can be obtained from simple considerations. Neglecting numerical factors we obtain, for the momentum transfer in the ion direction per unit time

$$\frac{dp_\parallel}{dt} = (k_F^2 \, v) \, v_F \, \sigma(v_F) \, v_F \, , \tag{35}$$

where the first factor is the number of electrons per unit volume, $\sigma(v_F)$ is the scattering cross section, and the last factor v_F is the momentum transfer. The product $(k_F^2 \, v)$ is the flux of electrons. Since $k_F^3 \sim n$, we finally obtain

$$\frac{dp_\parallel}{dt} = n \, \frac{v}{v_F} \, v_F \, \sigma(v_F) \, v_F = n \, v \, v_F \, \sigma(v_F). \tag{36}$$

Without going into any detail about the nature of the scattering, we have obtained the result that the stopping power is proportional to the ion velocity at low velocities

$$- \frac{dE}{dx} \sim n \, v \, v_F \, \sigma(v_F). \tag{37}$$

A more detailed understanding of this formula can be obtained as follows [11,12]. The distribution function in a moving gas is $f_0(\mathbf{v}+\mathbf{v}_e)$, where f_0 is the original distribution function. We suppose that, in each separate region of the gas, equilibrium is reached. Evidently, this approximation implies the neglect of all inner dissipative processes (viscosity, Debye-Onsager relaxation, off the energy-shell processes, etc.) in the system [12]. The required friction force F (the stopping power) can be calculated as the total momentum transferred, per unit time, to the heavy particle, by electrons colliding with it. The electrons carry momentum, v_e; after the collision, in which the electron momentum is turned through an angle θ, it carries away an average momentum $v_e \cos\theta$ (neglect the recoil of the heavy particle). The average momentum transferred to the rest particle in such an elastic collision is therefore $v_e(1-\cos\theta)$. Multiplying this by the flux of electrons with velocity v_e and by the differential scattering cross section for such a collision, and integrating, we obtain the total momentum transferred to the heavy particle,

$$F = \int f_0(v_e + v) \; |v_e| \; v_e \; \sigma_{tr}(|v_e|) \; dv_e, \tag{38}$$

where the transport cross-section is defined in terms of the scattering cross-section $\sigma(\theta)$ as

$$\sigma_{tr} = 2\pi \int_{-1}^{1} \sigma(\theta) \; (1 - \cos\theta) \; d(\cos\theta). \tag{39}$$

The scattering cross section $\sigma(\theta)$ for velocity v is given by

$$\sigma(\theta) = \frac{1}{v^2} \left[\sum_{L=0} (2L + 1) \; e^{i\delta_L} \; \sin(\delta_L) \; P_L(\cos\theta) \right]^2, \tag{40}$$

where $P_L(\mu)$ are the Legendre polynomials and δ_L are the scattering phase shifts at velocity v [12,15].

Note that the Pauli principle may be neglected in this evaluation. If the neglect of the Pauli principle leads to the inclusion of a non-allowed scattering from v_e to v_e', this is canceled by the scattering from v_e' to v_e because the cross sections are identical in both cases, and the momentum transfers connected with the two scattering processes are equal in magnitude and of opposite direction [22].

Changing the variable $v_e + v \rightarrow x$ we obtain, in eq.(38),

$$F = \int f_0(x) \; |x - v| (x - v) \; \sigma_{tr}(x - v) \; dx. \tag{41}$$

For small velocities ($v \ll v_F$), at zero temperature of the system, the opposite of the momentum transfer per unit time in the direction of motion, eq.(34), is given by

$$-\frac{dp_\parallel}{dt} = \int_{-1}^{1} 2\pi \; d\mu \int_{0}^{v_F} x^2 \; f_0(x) \; y(v - x\mu) \; \sigma(y) \; dx, \tag{42}$$

where $y = \sqrt{v^2 + x^2 - 2xv\mu}$. Expanding the right-hand-side of eq.(42) to first order in v,

$$- \frac{dp}{dt}^\parallel = \int_{-1}^{1} 2\pi d\mu \int_0^{v_F} x^2 f_0(x) \ (x - v\mu) \ (x\mu - v) \left[\sigma(x) - v\mu \frac{d\sigma(x)}{dx}\right] dx \qquad (43)$$

$$= \frac{4}{3} \pi v \int_0^{v_F} f_0(x) \ x^2 \left[4 \ x \ \sigma(x) + x^2 \frac{d\sigma}{dx}\right] dx \qquad (44)$$

$$= \frac{4}{3} \pi v \int_0^{v_F} \frac{2}{(2\pi)^3} \left[4 \ x^3 \ \sigma(x) + x^4 \frac{d\sigma}{dx}\right] dx, \qquad (45)$$

and integrating by parts the second term

$$= v \frac{4}{3} \pi \frac{2}{(2\pi)^3} \left[\int_0^{v_F} 4x^3 \ \sigma(x) \ dx + \left[x^4\sigma\right]_0^{v_F} - \int_0^{v_F} 4x^3 \ \sigma(x) \ dx\right], \qquad (46)$$

we obtain for the energy loss

$$- \frac{dE}{dx} = v \frac{4}{3} \pi \frac{2}{(2\pi)^3} v_F^4 \ \sigma(v_F) - \{v_F^4 \ \sigma(v)\}_{v \to 0}. \qquad (47)$$

For most potentials, $v^4 \ \sigma(v) \underset{v \to 0}{\to} 0$, and also $\frac{4}{3} \pi v_F^3 \frac{2}{(2\pi)^3} = n_0$. Then we have

$$- \frac{dE}{dx} = n_0 \ v \ v_F \ \sigma_{tr}(v_F). \qquad (48)$$

For a spherically-symmetric potential the transport cross section, and therefore the stopping power, can be written, using the properties of the Legendre polynomials, as

$$\sigma_{tr} = 2\pi \int_{-1}^{1} d(\cos\theta) \ (1-\cos\theta) \ \sigma(\theta) = \frac{4\pi}{k_F^2} \sum_{L=0}^{\infty} (L+1) \ \sin^2(\delta_L - \delta_{L+1}). \qquad (49)$$

The stopping power then is [13]

$$- \frac{dE}{dx} = n \ v \ v_F \ \sigma_{tr}(v_F) = \frac{3v}{k_F r_s^3} \sum_{L=0}^{\infty} (L+1) \ \sin^2\{\delta_L(E_F) - \delta_{L+1}(E_F)\} \ . \qquad (50)$$

$\delta_L(E_F)$ are the phase shifts at the Fermi energy for scattering of an electron off a spherically-symmetric potential. Since the ion is moving slowly compared with the electrons at the Fermi surface, we can use the results of a static calculation for such a potential.

The key question then remains to determine a good potential to describe the scattering of electrons at the Fermi level.

3. FRIEDEL SUM RULE

Suppose for simplicity that the solid has spherical shape and radius R, that will eventually be taken in the limit $R \rightarrow \infty$ [23]. In the absence of the potential due to the charge $Z = Z_1 - n_b$, where Z_1 is the external charge and n_b the number of bound states, the asymptotic form of the wave function is

$$\phi \rightarrow \frac{1}{k \, r} \sin(kr - \frac{L \, \pi}{2}), \qquad (51)$$

while, when the spherically-symmetric potential due to the charge Z is present, the asymptotic form of the wavefunction is ϕ

$$\phi \rightarrow \frac{1}{k \, r} \sin(kr + \delta_L - \frac{L \, \pi}{2}). \qquad (52)$$

Is order to count states in k-space we impose the condition that the wave function vanishes at R; this gives

$$k_n R + \delta_L(k_n) = (n + \frac{1}{2} L)\pi. \qquad (53)$$

The number of states between k and k + dk is then

$$\frac{dn}{dk} = \frac{R}{\pi} + \frac{1}{\pi} \frac{d\delta_L}{dk} . \qquad (54)$$

The first term is the free density of states, i.e., the density of states that one should have without the impurity so, the next quantity of the right

$$\frac{d}{dk} \left(\frac{\delta_L}{\pi} \right), \qquad (55)$$

is the change in the number of particle states. The Friedel sum rule [24-26] is obtained by integrating for all k and summing over all quantum numbers. Then the total charge of the impurity with charge Z is related to the δ_L as

$$Z = \frac{1}{\pi} \int_0^{k_F} dk \, \frac{d\delta_L}{dk} = \sum_{\substack{\text{all quantum} \\ \text{numbers}}} \frac{\delta_L(k_F)}{\pi} = \frac{2}{\pi} \sum_{L=0}^{\infty} (2L+1) \left\{ \delta_L(E_F) - \delta_L(0) \right\}. \qquad (56)$$

49

The Friedel sum rule can be written in a more familiar way by noticing that, from Levinson's theorem [27-29] the number of bound states is given by

$$n_b = \frac{2}{\pi} \sum_{L=0} (2L+1) \ \delta_L(0),$$ (57)

so the Friedel sum rule is

$$Z_1 = \frac{2}{\pi} \sum_L (2L+1) \ \delta_L(E_F).$$ (58)

The Friedel sum rule is a statement of charge neutrality: The change in electron charge around an impurity is exactly equal in magnitude, and opposite in sign, to the charge of the impurity. A neutral impurity would have a Friedel sum rule of zero. Note that this does not mean that all the phase shifts vanish. What the Friedel sum rule says is that the density fluctuations $\delta n(r) = n(r) - n_0$ should integrate to equal the impurity charge

$$Z_1 = 4\pi \int_0^\infty \delta n(r) \ r^2 \ dr.$$ (59)

This simple derivation of the Friedel sum rule does not take into account the fact that the electrons are interacting. Mahan [30] points out that the Friedel sum rule is believed to be exact in real systems. Langer and Ambegaokar [25] have shown it to be valid even in an interacting many-particle system. If one knows the exact impurity potential and the exact screening charge profile which it originated, one should find that the Friedel sum rule is satisfied.

Table 1. Ratios $(\Delta n(0)+n_0)/n_0$ of the total density to the mean electron density at the static proton position. The column labeled NL are the DF results, and the one labeled L the results of a linear calculation. The last column (H^0) shows the results for hydrogen.

r_s	NL	L	H^0
1	5.16	2.67	2.33
2	17.5	4.03	11.33
3	45.6	5.31	37.00
4	97.4	6.54	86.33
6	303.2	9.03	289.00

4. PROTON: POTENTIAL AND DENSITY FLUCTUATION RESULTS

The results of the density-functional calculation of induced charge densities and effective potentials differ substantially from the ones obtained from linear-response theory [12,31]. The density fluctuations and induced potentials of a static proton in an electron gas were evaluated by several authors [31]. The results for the nonlinear (density functional) density fluctuations can be seen together with the predictions of linear theory in table 1, which contains the ratio of the total density at the proton position to the mean electron density, for various values of the density parameter r_s. It is clear from this table that for low ion velocities the static linear response does not describe adequately the electron gas response and one has to go beyond it. The comparison between

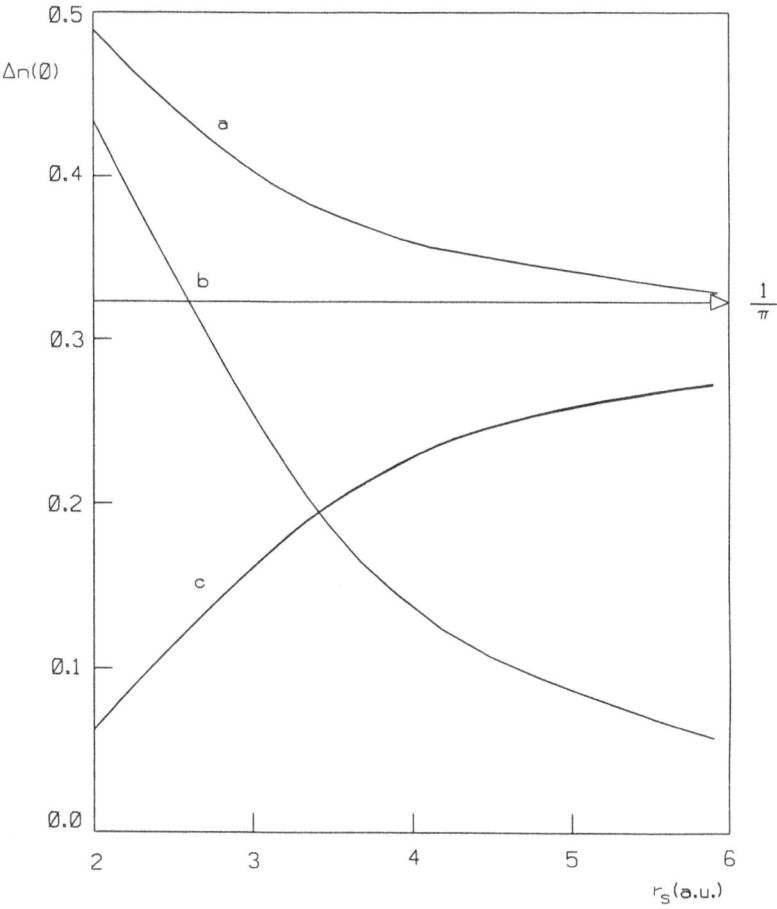

Figure 1. Induced electron density at the proton position, obtained in the DFA, as a function of r_s. The bound (c) and scattering (b) parts of the total density (a) are shown separately. From ref. [32].

the first column (NL) and the third (H^0) shows that for lower density of the host, the induced density fluctuation at the proton site becomes very similar to the atomic value.

In fig.1 we have plotted the induced charge density obtained at the proton site as a function of r_s. The bound (c) and scattering (b) parts of the total (a) density have been separated. This figure clearly shows that the bound-state character of the induced charge becomes more and more significant as the density decreases. A screening length λ can be defined as [31]

$$\frac{Z_1}{\lambda} - V_{Hartree}(\lambda) = \frac{1}{e}\frac{Z_1}{\lambda} \quad . \tag{60}$$

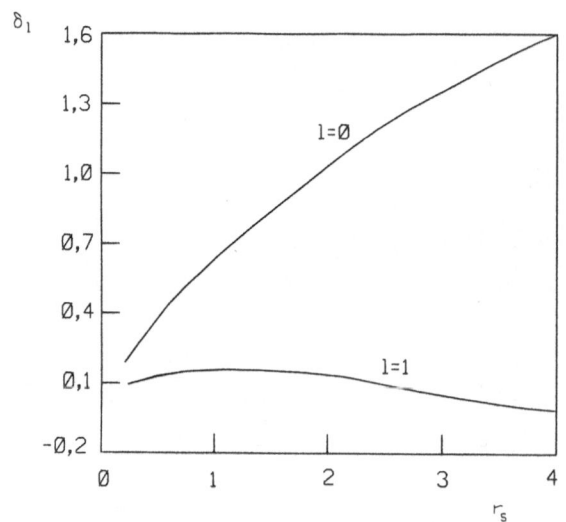

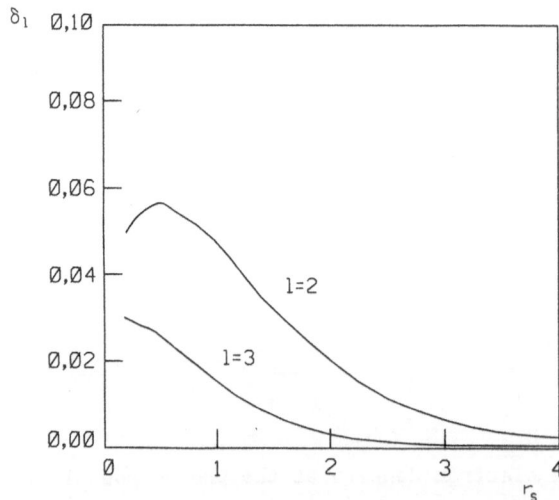

Figure 2. Phase shifts at the Fermi energy for electron scattering from a self-consistent density-functional potential around a proton.

The nonlinear results for the induced Hartree potential and the screening length are practically constant over the metallic range, $\lambda \sim 0.65$ and $v_H(0) \sim 1.2$. In contrast, the corresponding quantities from linear theory vary from 0.9 to 1.5 (λ), and from 0.8 to 0.5 ($V_H(0)$), when r_s varies from 2 to 6.

5. H AND He STOPPING

Echenique, Nieminen and Ritchie (ENR) [33] evaluated the phase shifts at the Fermi Level from a self-consistent density-functional potential around a proton and an alpha particle. The phase shifts were found to satisfy the Friedel sum rule to a 2 % accuracy. Puska and Nieminen [34] have reported the Fermi-level scattering phase shifts and the transport cross sections for atoms embedded in a homogeneous electron gas. The phase shifts at the Fermi energy, calculated from the self-consistent potential are shown in fig.2 as a function of the density parameter r_s. The effective number of phase shifts required scales as $1/\sqrt{r_s}$. In fig.3, taken from the original paper of ENR, we show the comparison between the results of the nonlinear theory of stopping, for protons and alpha particles, with the ones obtained by the method of Ferrell and Ritchie [35]. These authors have calculated the stopping power of an electron gas for slow, singly ionized He atoms, in linear response theory, they used a hydrogenic wave function, variationally determined by minimizing the total energy of the ion plus a bound electron in the electron gas.

As r_s decreases toward values much less than 1, the results of the DF calculation tend toward agreement with linear theory, i.e., proportional to Z_1^2 as they should, since at such high densities screening is so strong that bound states cannot exist. As r_s increases, the energy loss for both H and He decreases more rapidly than predicted by linear theory due to the fact that bound states of atomic character develop, thereby tending to screen out interactions with the electron gas. The energy loss of a He nucleus at large r_s is smaller than that of a proton at the same velocity. This is qualitatively different from any linear theory, in which the energy loss scales as the square of the ionic charge, and can be understood in terms of the atomic character of the scattering process in a very dilute electron gas. The Fermi energies at low densities are small, so the scattering is essentially that of a very low energy electron from a H and a He atom.

A comparison of experimental data for protons with the density functional predictions of ENR was made by Mann and Brandt [36]. They collected data on targets covering a wide range of atomic numbers, and

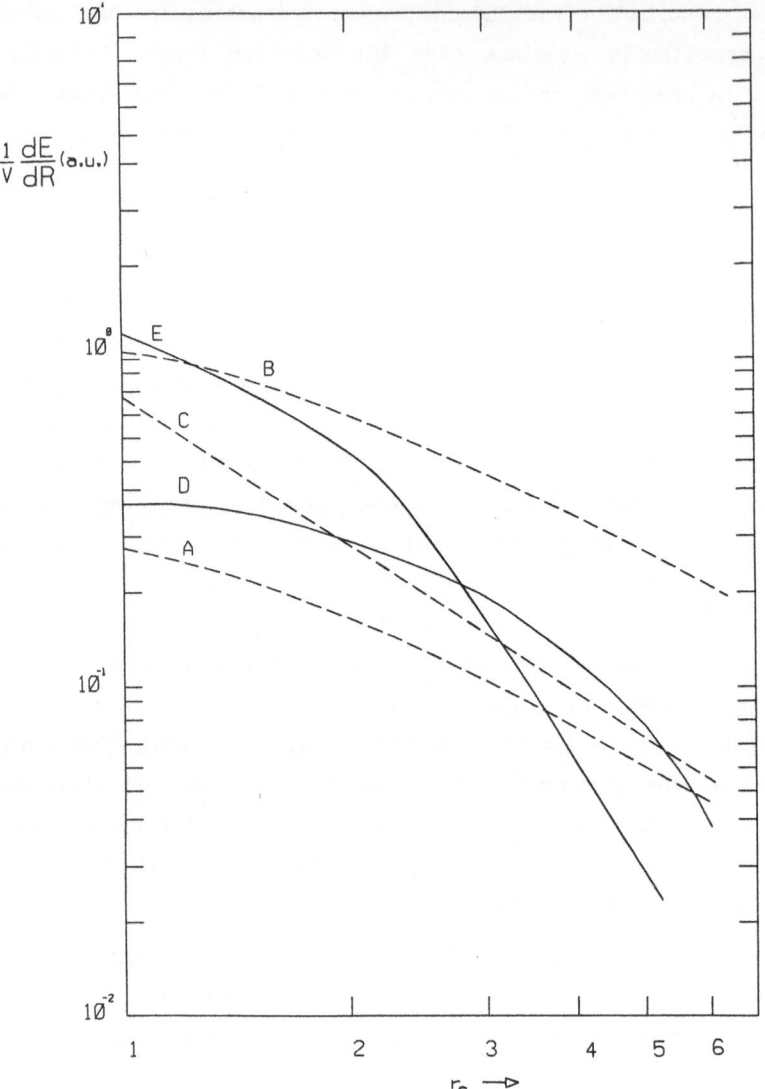

Figure 3. Low-velocity stopping power as a function of r_s. Curves A and B are calculated in linear-response theory for $Z_1 = 1$ and $Z_1 = 2$, respectively. Curve C is from ref.[35] for a slow, singly ionized He atom. Curves D and E are the density-functional results for a proton and a helium nucleus, respectively.

conclude from these comparisons that, within the uncertainties of the data,: a) the linear dependence on velocity holds up to $v \approx v_F$, and b) the density-functional predictions give good agreement with the data. We refer the reader to the original paper of Mann and Brandt or to the review of Echenique, Flores and Ritchie [37] (EFR) for further details.

A comparison between the ENR results and those obtained with the linear theory of Lindhard and Winther [38] is shown in EFR. For many solids used in experiments, the density-functional stopping powers show increases of around 66 % over those calculated from the linear theory. As stated above, though, the DF results decrease much more rapidly with increasing r_s that the corresponding linear-theory results. Ziegler, Biersack and Littmark [39] have used the ENR approach, with solid state charge distributions, to calculate the stopping of particles in solids at the Bohr velocity, 25 keV/amu. They compared the results to the empirical values for all solids. The agreement is quite reasonable, usually within 25 %. For several groups of targets, they used the ENR approach for guidance (for example for $Z_2 = 57 - 72$).

6. ANTIPROTON STOPPING

Nagy and co-workers have used the density-functional theory of stopping to evaluate the stopping power of a low-velocity antiproton moving in an electron gas [40]. Fig.4 shows the comparison of their calculated stopping power with the ones obtained for the proton-projectile

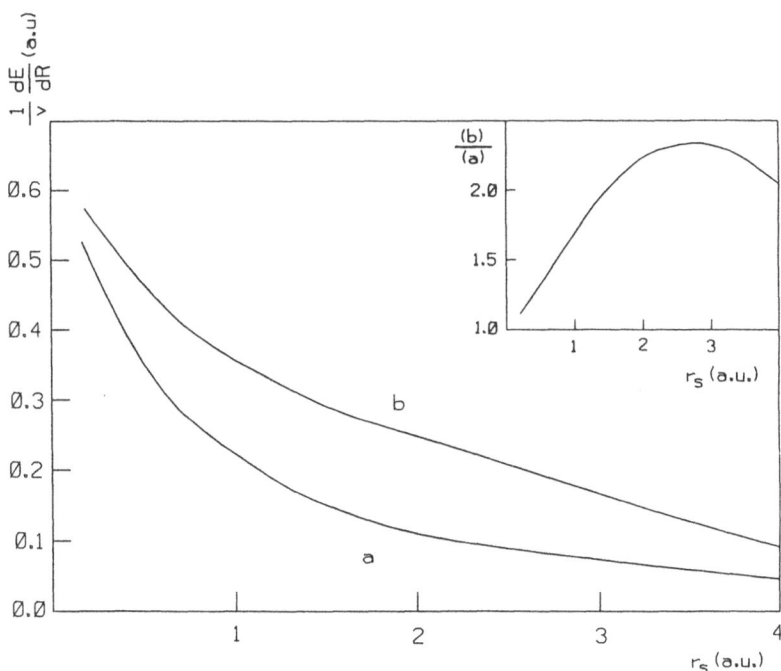

Figure 4. Stopping power of a homogeneous electron gas for slow antiprotons (curve a) and protons (curve b) as a function of the density parameter r_s. Both results are obtained with the same density-functional description. The curve in the inset shows the ratio (b)/(a) as a function of r_s.

case. Both results are obtained within the same DF description. At typical metallic densities, $v_F \propto 1$, the ratio of the proton and antiproton stopping power is roughly two, while at very high densities the two curves merge since the results tend to the linear-response result. This is illustrated in the inset of fig.4. Fig.5 shows the ratios (R) of the DF results for the stopping power of Nagy et al., with respect to those commonly obtained by using, for example, linearly-screened Thomas-Fermi [41] (Yukawa) or RPA potentials [12], treated in the first Born approximation. These latter descriptions are equivalent to linear dielectric-response theory, and are therefore insensitive to the sign of the projectile charge.

Another interesting feature is the contrasting behavior of the induced charge density for protons and antiprotons. As table 1 clearly shows, the static dielectric linear screening grossly underestimates the value of the induced charge density at the proton site $\Delta n(0)$. In DF

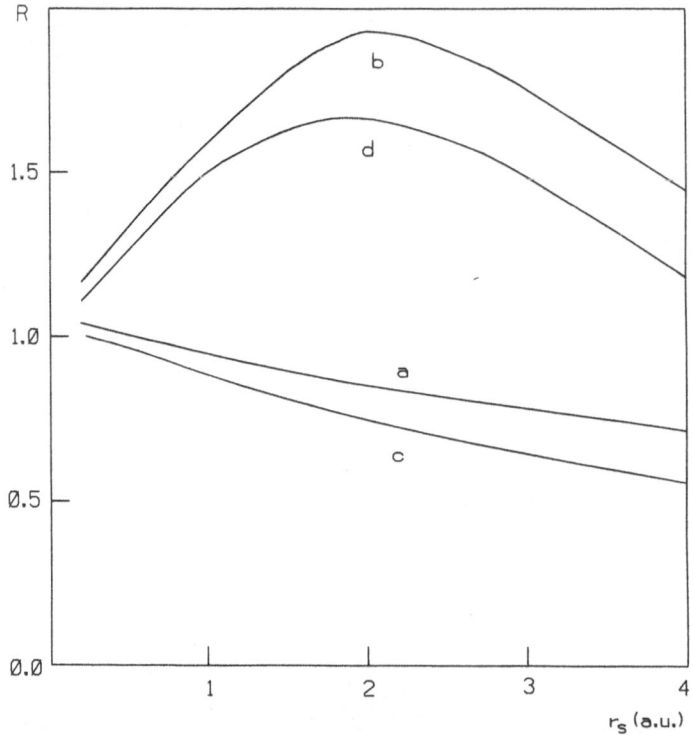

Figure 5. Ratios (R) of the stopping-power results based on DF calculations, with respect to those obtained in first Born approximation for the Thomas-Fermi (Yukawa) potential (curves a and b), and for the screened potential with RPA dielectric function (curves c and d). Curves a and c correspond to the antiproton while b and d correspond to the proton.

56

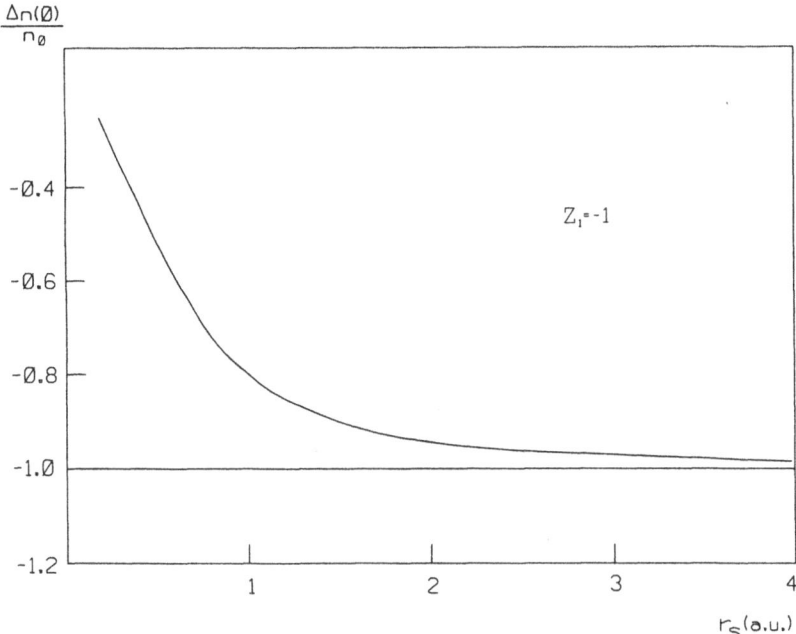

Figure 6. Ratio of the depletion-hole density $\Delta n(r)$, at the antiparticle site, to the background density n_0, as a function of the density parameter r_s.

calculations, the total screening density is very similar to that of atomic hydrogen. On the other hand, the electron gas is repelled by the antiproton and a depletion hole is created around the antiproton. An obvious bound for $|\Delta n(0)|$ is n_0, see fig.6. At metallic the electron-gas densities, the depletion is nearly complete, and remains constant roughly up to a distance $r \sim r_s$ at which point the antiproton is totally screened. The depletion charge density is essentially a "rigid" hole. For protons, the screening charge density is well localized, while for antiprotons it is more diffuse. In the latter case, the behavior of the radial density distribution is nearly parabolic at small r values, and reflects the almost constant depletion-charge density, fig.7.

Sørensen [42] and Nagy and Echenique [43] have presented analytical results for the higher-order correction to the stopping power, proportional to Z_1^3 [44-47], using the second Born approximation for the scattering amplitude to describe the scattering of a linearly-screened Thomas-Fermi (Yukawa) potential $-Z_1 e^{-\lambda r}/r$. The transport cross section, see eq.(49), is given by

$$\sigma_{tr}(v_F) = \frac{1}{2}\frac{4\pi}{v_F^4}(X_0 + Z_1 X_1)Z_1^2, \tag{61}$$

where

$$X_0 = \ln\left[1 + \frac{1}{g}\right] - \frac{1}{1 + g}, \tag{62}$$

$$X_1 = \frac{8g}{\lambda(3+4g)}\left\{4(1+g)\ln\left[1 + \frac{1}{4g(1+g)}\right] - \ln(1 + \frac{1}{g})\right\}, \tag{63}$$

with $\lambda = 3\alpha/\sqrt{r_s}$, $g = \alpha r_s/\pi$, and $\alpha = (4/9\pi)^{1/3}$. The values for the first-order term in eq.(61) are slightly smaller than those obtained in the dielectric description [45,42] (see fig.11 of EFR [37]). The correction due to X_1 is larger [43]. The results of the second-order Born approximation agree closely with those obtained by Hu and Zaremba [48] using the electron-gas quadratic density-response function. Fig.8 shows

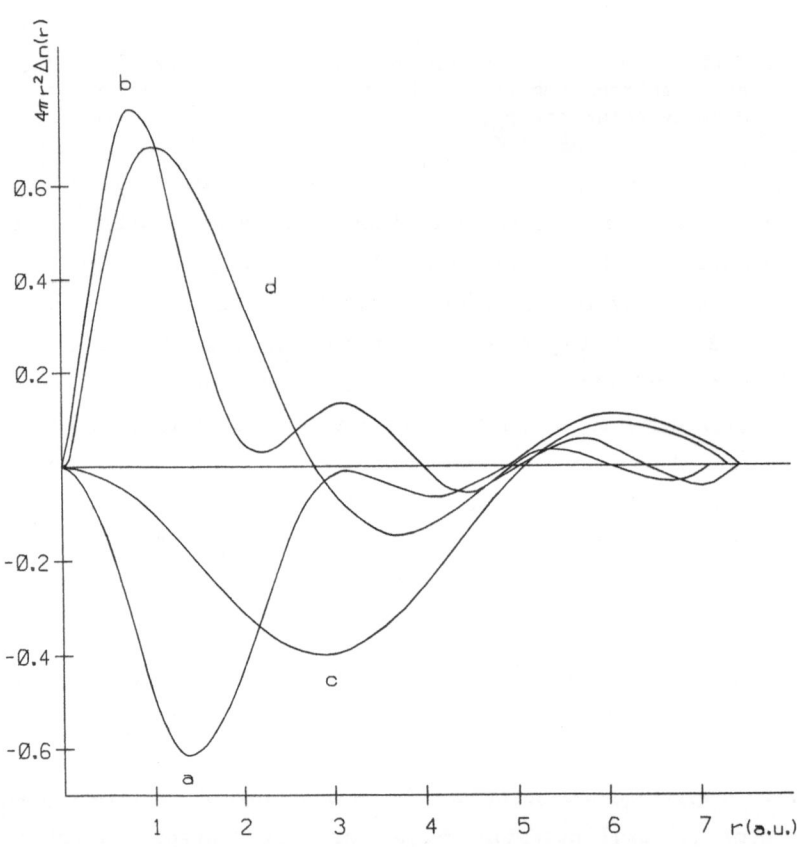

Figure 7. Radial density distributions, $4\pi r^2 \Delta n(r)$, as a function of the radial distance r. Curves b and d are the results for a proton, while curves a and c are for an antiproton. The density parameters are $r_s = 1.5$ (curves a and b) and $r_s = 3$ (curves c and d).

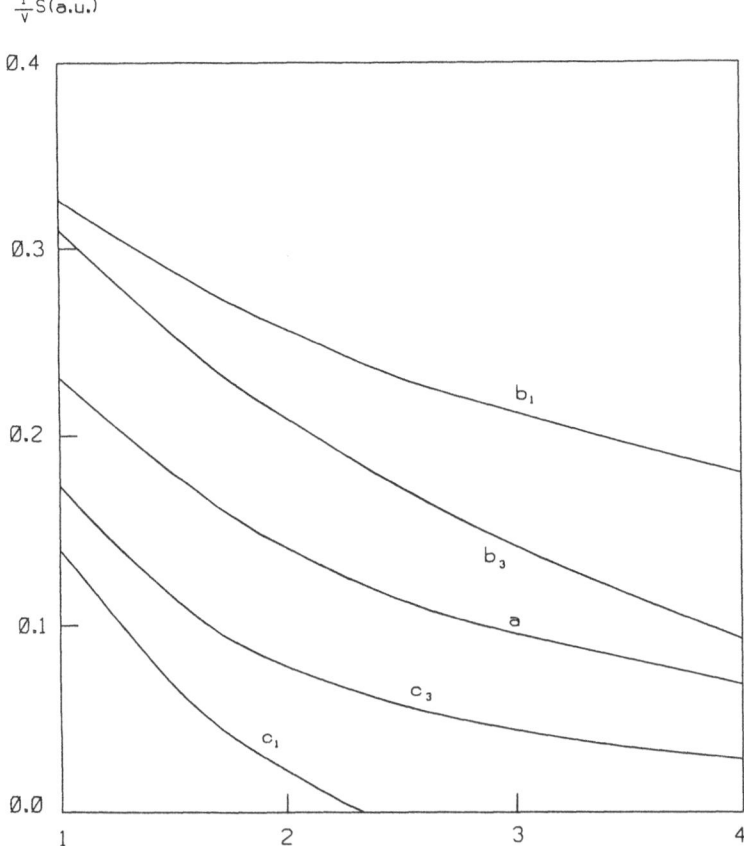

$\frac{1}{v}S$(a.u.)

r_s

Figure 8. Low-velocity stopping power for protons and antiprotons. Curve a
is the first-Born result for a Yukawa potential. Curves b_1 and
c_1 are obtained in the second Born approximation for protons and
antiprotons, respectively. Curves b_3 and c_3 are obtained from
eq.(50), by taking the coefficient in the exponent of the Yukawa
potential as a parameter, and then forcing the phase shifts to
satisfy the Friedel sum rule, eq.(58).

the results of the calculations of Nagy and Echenique. The curve a is the
first Born result, and curves b_1 and c_1 are obtained in the second Born
approximation for protons and antiprotons respectively. Curves b_3 and c_3
are obtained from eq.(50), taking λ as a parameter and then forcing the
phase shifts to satisfy the Friedel sum rule (see below). The results, for
the densities considered, of a self-consistent calculation starting from
the Yukawa potential and adding a local-density estimate for exchange and
correlation [49], performed by Sørensen [42], agree quite well with the
density-functional results of Nagy and co-workers [40].

7. MODEL POTENTIALS

The Friedel sum rule, eq.(58), is a strong condition on the scattering potential. Some authors have used it to determine the screening parameter in any one-parameter regular model potential. Several calculations have been done to treat the **proton** screening in an electron gas. Ferrell and Ritchie [35] used a Yukawa potential

$$V_{YK} = - \frac{e^{-\lambda r}}{r}, \tag{64}$$

whereas a Hulthén potential,

$$V_{CV}(r) = -\lambda \ [e^{\lambda r} - 1]^{-1}, \tag{65}$$

was used by Cherubini and Ventura [50]. Apagyi and Nagy [51], motivated by the fact that the static charge-density profiles obtained by the density functional method are close to hydrogenic 1s functions, used a hydrogen-type model potential

$$V_{AN} = - \frac{e^{-2\lambda r}}{r} \ (1 + \lambda r). \tag{66}$$

The screening parameter λ was approximated by an analytical formula, with high accuracy, in the $1 < r_s < 15$ range. The latter model was extended to a two parameter version to include the Kato's cusp condition [52]. For low electron density, all of these model potentials yield $\delta_0 \rightarrow \pi/2$ and $\delta_1 \rightarrow 0$. For *antiprotons*, we have already mentioned the calculations based on the Yukawa potential of Nagy and Echenique, and Sørensen. Nagy and co-workers [53] have suggested a simple parametric form of the dielectric function,

$$\varepsilon(q) = 1 + \frac{q_{TF}^2}{q^2} \ \frac{1}{1 + \lambda \ 3(\frac{q}{2v_F})^2}, \tag{67}$$

which leads to analytical expressions for the effective potential and induced hole density. The parameter is fixed via the cusp condition [54,55] for the total electron density $n(r)$ at the position of the probe charge,

$$\frac{n'(r)}{n(r)} \ \Big|_{r=0} = 2. \tag{68}$$

The results of Nagy et al. are in fair agreement with the ones obtained using the density-functional results for the phase shifts. The

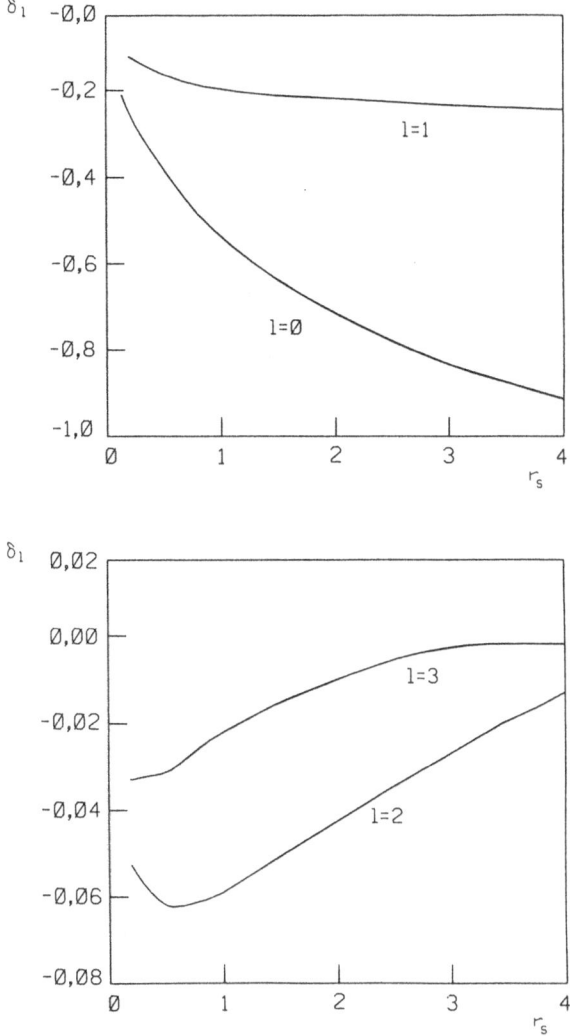

Figure 9. Phase shifts at the Fermi energy for electron scattering from a self-consistent density-functional potential around an antiproton.

density-functional results for the phase shifts at the Fermi level, for the case of the antiproton, are shown in fig.9 for completeness.

8. IONS WITH $Z_1 > 2$

Echenique, Nieminen, Ashley and Ritchie [56] (ENAR) extended the earlier work of ENR to ions with $Z_1 > 2$. Following Brandt and co-workers [57-59], who were able to condense a great amount of data introducing the concept of effective charge, ENAR define the effective charge, Z_1^*, in an operational manner, as

$$z_1^* = \left[\frac{\left(-\frac{dE}{dx} \right)_{z_1}}{\left(\frac{-dE}{dx} \right)_{z_1=1}} \right]^{1/2} . \tag{69}$$

For most charges and electron densities, self-consistency was achieved. The stopping power was then evaluated from q.(50). In fig.10 we show the effective charge, as a function of r_s, for a He nucleus. The effective charge, varies from $z_1^* = 2$ for $r_s \to 0$ to 0.46 for $r_s = 6$. It becomes less than 1 for $r_s > 2.7$.

The z_1 oscillations in the stopping power, and therefore in the effective charge, appear naturally in the self-consistent calculation since they are related to the appearance of new bound states, which are taken into account in a natural way in the formalism. A qualitative

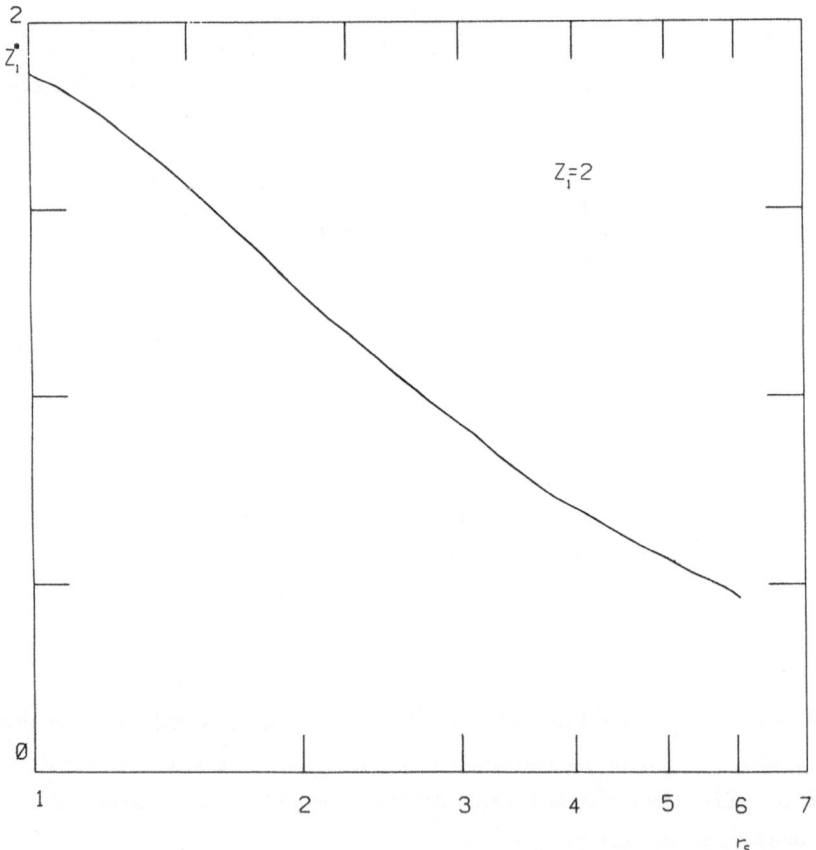

Figure 10. Effective charge, eq.(69), for a He nucleus, as a function of r_s, calculated using the density-functional approach.

understanding of the main feature of the oscillations can easily be achieved, in terms of the Friedel sum rule, scattering theory, and resonance levels in solids. If one of the phase shifts should dominate for all charges due to some resonance in the scattering, the Friedel sum rule leads to an oscillatory sinusoidal structure as a function of Z_1. In reality many phase shifts play a role depending on Z_1 and r_s, and the oscillatory structure is more complicated than the sinusoidal dependence. The Z_1 oscillations reflect the shell structure of the screening clouds around the ion, which is directly related to the formation of close shells. The maxima are related to the appearance of resonant states, for certain values of Z_1, in different shells; this causes the corresponding phase shifts δ to be close to $\pi/2$, and hence a maximum in the cross section and stopping power [34].

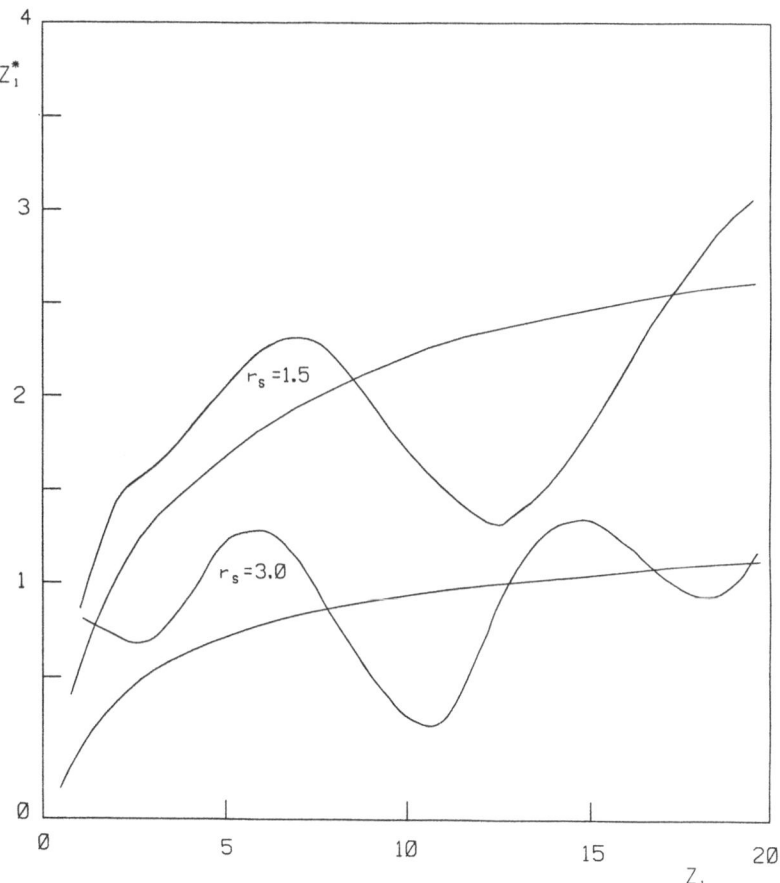

Figure 11. Effective charge, defined by eq.(69), as a function of the ion charge Z_1, for two different r_s values. The oscillating lines are the results of the DF. The results of Brandt and Kitagawa's model [59] are also shown.

In fig.11 we show the effective charge, defined in eq.(69), as a function of the ion charge, for two different values of the density parameter, r_s = 1.5 and 3. For a dilute electron gas, r_s = 3, the screening cloud approaches the free-atom electron structure, and minima appear at the formation of closed atomic shells. As the screening increases due to increasing electronic density, r_s = 1.5, this minimum shifts to higher ionic charges, since a stronger ionic potential is necessary to compensate the electronic screening and so have the strength to bind an extra electron. This is clearly shown in the graph as displacements of the minimum from the atomic value, Z_1 = 10, as r_s decreases. For very small r_s no bound states are formed, and linear-response theory should be valid, hence Z_1^* approaches Z_1. Comparison of experimental data with the theoretical predictions for stopping power of carbon, aluminum and silicon can be found in the literature [56,60,61]. In figs.12 and 13 we show the comparisons, taken from the original paper of ENAR, between Eisen's data [62,63] on the energy loss of "best channeled" ions in the <110> and <111> axial channels in Si. The electron density in a channel increases from a small value along the axis to values

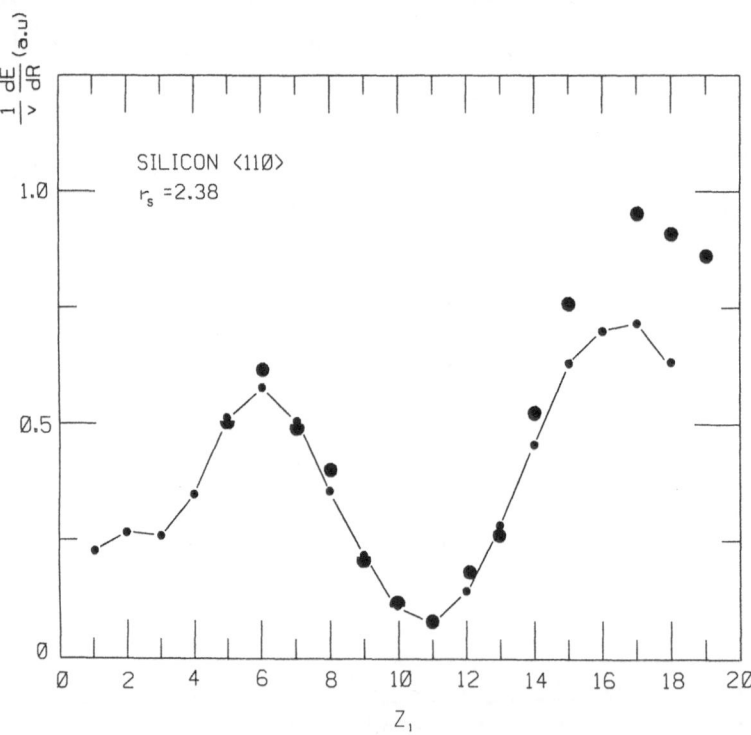

Figure 12. Stopping power for "well-channeled" ions in the <110> axial channel of silicon as a function of Z_1. The chained curve gives the theoretical predictions of ENAR for r_s = 2.38. The experimental points are from ref.[62].

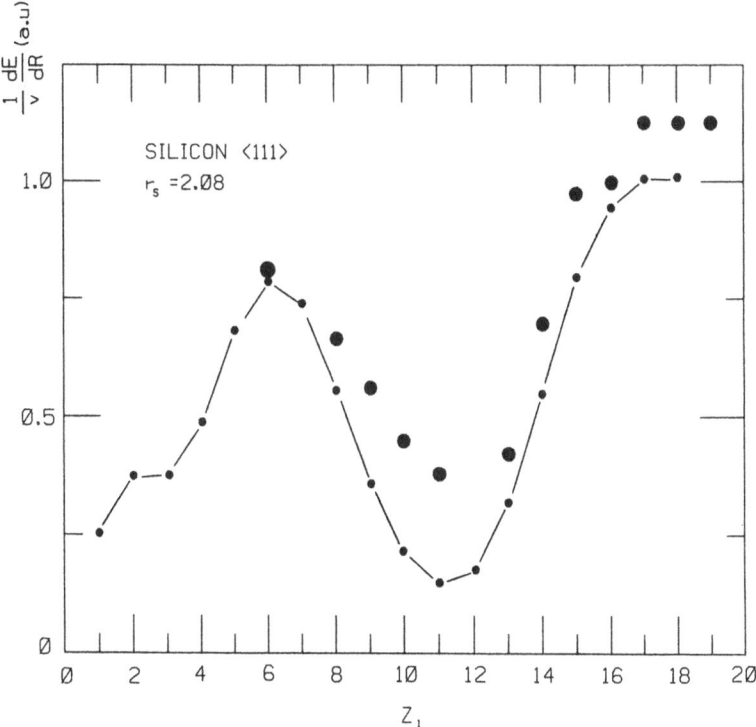

Figure 13. Stopping Power for "well-channeled" ions in the <111> axial channel of silicon as a function of Z_1. The chained curve gives the theoretical predictions for r_s = 2.08. The experimental points are from ref. [63].

an order of magnitude larger near the string of atoms defining the channel. This density variation is well characterized for the <110> axial channel in Si [64]. Since a range of electron densities (or impact parameters) is sampled in the energy-loss process, ENAR compared their theory with the data using an average electron gas density determined by making theoretical and experimental values equal at Z_1 = 5. For the <110> channel they took r_s = 2.38, and for the <111> channel, r_s = 2.08. As expected, these values correspond to lower densities than the average valence-electron density (corresponding to r_s = 2.0). In both figures the overall trends are closely reproduced by the theoretical calculations.

9. MEAN FREE PATH, STOPPING POWER, STRAGGLING

In linear theory it is easy to obtain the basic quantities that characterize the composite system of a charge Z_1 interacting with an electron gas described by a dielectric response function $\varepsilon(k,\omega)$ [41].

These are the width of the particle states Γ, (related to the inverse mean-free-path μ and to the imaginary part of the electron self-energy Σ_i), the stopping power and the straggling parameter W. The width of the particle states can be obtained, in a semiclassical way, from the probability $P(k,\omega)$ per unit time of losing energy ω and momentum k, thus creating a real excitation in the electron gas. For slow ions, this elementary process is the particle-hole generation. Γ is given [37] by

$$\Gamma = \frac{1}{\tau} = -2\,\Sigma_i = \int dx\, P(\mathbf{k},\omega), \tag{70}$$

where $\int dx \equiv \int dk/(2\pi)^3 \int_0^\infty d\omega/2\pi$. $P(k,\omega)$ can be written, in terms of the dielectric response function of the medium $\varepsilon(k,\omega)$, as

$$P(\mathbf{k},\omega) = \frac{16\pi^2 z_1^2}{k^2}\, \mathrm{Im}\left[\frac{-1}{\varepsilon(k,\omega)}\right]\, \delta(\omega - \mathbf{k}\cdot\mathbf{v}). \tag{71}$$

The stopping power and straggling parameter Ω^2 are then

$$-\frac{dE}{dx} = \frac{1}{v}\int dx\,\omega\,P(\mathbf{k},\omega), \tag{72}$$

$$\Omega^2 = n\,v\,W = \frac{1}{v}\int dx\,\omega^2\,P(\mathbf{k},\omega). \tag{73}$$

The low-velocity expression for Γ is given by [65]

$$\Gamma = \frac{v}{(2\pi)^3}\int_0^{2v_F} dk\, k^2 \left[\frac{4\pi Z_1}{k^2\varepsilon(k)}\right]^2. \tag{74}$$

So far, the elastic scattering between the screened ion and the electron is described in the first Born approximation. If we now interpret this result in terms of non-relativistic scattering theory, and make the substitution

$$\frac{4\pi Z_1}{k^2\varepsilon(k)} \rightarrow -2\pi\, f(\theta), \tag{75}$$

where $f(\theta)$ is the scattering amplitude, and $k = 2v_F \sin(\theta/2)$ is the momentum transfer, we obtain for the width of the states [65]

$$\Gamma = \frac{3}{2}\, n\, v\int_0^\pi d\sigma(\theta,v_F)\,\sin\left(\frac{\theta}{2}\right) \tag{76}$$

where $d\sigma = 2\pi\, |f(\theta)|^2 \sin\theta\, d\theta$.

Similarly, the well-known results for the stopping power and the straggling parameter [11,20] can be obtained,

$$\frac{dE}{dx} = 2 \, n \, v \, v_F \int_0^\pi d\sigma(\theta, v_F) \, \sin^2(\frac{\theta}{2}),$$ (77)

and

$$W = 3(v \, v_F)^2 \int_0^\pi d\sigma(\theta, v_F) \, \sin^3(\frac{\theta}{2}).$$ (78)

Eqs. (76-78) have been derived in a simple heuristic didactic way. From the general results of Müller-Hartman et al. [68], with the substitution $\omega = \mathbf{k}\mathbf{v} = kv \cos\phi = kvy$, and taking into account that for low ion velocities $k = 2v_F \sin(\theta/2)$, it is easy to obtain, neglecting the ion recoil, an expression $L(p)$ for the determination of the basic quantities,

$$L(p) = \frac{1}{2} \left[\frac{v_F}{\pi}\right]^2 \int_0^1 dy \int_0^\pi d\sigma(\theta, v_F) \left[2v \, v_F \sin(\frac{\theta}{2}) \, y\right]^p.$$ (79)

From eq. (79),

$$\Gamma = L(p=1),$$

$$-\frac{dE}{dx} = \frac{1}{v} \frac{dE}{dt} = \frac{1}{v} L(p=2),$$ (80)

$$W = \frac{1}{n \, v} \frac{d}{dt} \left[<(\Delta E - <\Delta E>)^2>\right] = \frac{1}{n \, v} L(p=3).$$

Eqs. (76-78) allow us to calculate the corresponding expressions, in terms of the phase shifts at the Fermi energy, for the elastic scattering of an electron by the self-consistent potential of the ion in the electron gas. From eq. (77) we recover the results of eq. (50). Expressions for Γ and W are given in refs. [65] and [66]. If we compare the density-functional results with the ones obtained using linear theory we find the same behavior that in the stopping power, that is, the DF results are about 50 % higher than those of linear theory when $1.5 \leq r_s \leq 3$. For higher r_s ($r_s \geq 4$), the non-linear results, as with the stopping power, are smaller than those of dielectric theory. For very small r_s ($r_s < Z_1^{-2}$) values, the two theories (linear and non linear) tend toward agreement independently of the charge of the projectile. The curves Γ and W, as a function of Z_1, show the characteristic Z_1 oscillations that can be explained in terms of the effective charge. Nagy et al [65] find that $\Gamma/v(Z_1^*)^2$ is about 0.31, within 30 % for $r_s = 1.5$, and for $r_s = 3.0$ it is about 0.28 within 15 %. That is, most of the oscillatory behavior in Γ is due to the variation of the ion effective charge. The same behavior, as the one described above for Γ, has been found for the straggling parameter [66,67].

10. INTERACTING ELECTRON GAS

Nagy and co-workers [69] have taken the theory one step further, by calculating the energy-width, stopping power and straggling for an interacting electron gas. They characterized the inner dissipative nature of the elementary electron-hole excitation by a complex local-field correction function, and used an appropriate low-frequency expansion of the imaginary part of the response function to modify the non-interacting electron gas results. Their results (we refer the reader to their paper for details) can be summarized as before in eqs.(79,80). The basic quantities are described by eq.(80), but now L(p) is given by

$$L(p) = \frac{1}{2} \left[\frac{v_F}{\pi}\right]^2 \int_0^1 dy \int_0^\pi d\sigma(\theta, v_F) \left[2v \; v_F \; \sin(\frac{\theta}{2}) \; y\right]^p$$

$$\times \left\{1 + C_1 \; \frac{12n}{\pi} \; a(n) \; f_G\left(\sin(\frac{\theta}{2})\right)\right\}, \tag{81}$$

where C_1 is a parameter; $C_1 = 1$ refers to an interacting electron gas, $C = 0$ refers to the non-interacting case discussed in the previous sections. $f_G(z)$ is given by

$$f_G(z) = z \left[f_1(z)\right]^2, \tag{82}$$

with

$$f_1(z) = \frac{1}{2} \left[1 + \frac{1 - z^2}{2z} \; \ln \left|\frac{z + 1}{z - 1}\right|\right], \tag{82'}$$

$a(n)$ is related to the imaginary part of the so-called dynamical local-field correction $G(k,\omega)$, which is defined in terms [69] of the collisional part $\phi(k,\omega)$ of the total effective interaction between density fluctuations: $4\pi/k^2 + \phi(k,\omega)$. There is a well-known connection [70] between $\phi(k,\omega)$ and $G(k,\omega)$,

$$\phi(k,\omega) = -\frac{4\pi}{k^2} \; G(\dot{k},\omega). \tag{83}$$

$G(k,\omega)$ represents the short-range correlations between electrons in the homogeneous electron-gas system and characterizes the *dynamic* electron-electron interaction. The determination of $Im[G(k,\omega)]$ is a tough problem that requires powerful many-body techniques [71,72]. Nagy et al. determined $Im [G(k,\omega)]$ according to the model of Gross and Kohn [73,74],

$$Im \; G(k,\omega) = a(n) \; \frac{k^2}{4\pi} \; \omega, \tag{84}$$

for small ω values; provided that the k dependence is valid in the range $0 \leq k \leq 2\, k_F$, Nagy et al. give

$$a(n) = \frac{\pi}{12n}\; 3.88\; r_s^{1/3}(\gamma_0 - \gamma_\infty)^{5/3}. \tag{85}$$

Here, the density dependent γ_0 and γ_∞ factors result from the compressibility and third-frequency-moment sum rule, respectively [75]. At metallic densities of the electron gas, these factors of the electron gas vary slowly and typical values are $\gamma_0 \cong 0.263$, $\gamma_\infty = 0.075$, for $r_s = 2$. The results of Nagy et al. show that, if we take into account the dissipative nature of the electron-hole excitations ($C_1 = 1$), we obtain, for Γ, dE/dx, and W, an increasing relative deviation with respect to the results based on the rigid electron-hole concept ($C_1 = 0$) as the density decreases. These deviations change from about 10 % ($r_s = 1.5$) to about 20 % ($r_s = 5$).

ACKNOWLEDGEMENTS

The authors thank Prof. I. Nagy for stimulating conversations, and E. Alvarellos for a critical reading of the manuscript. Many thanks are due to M. Peñalba and A. Arnau for their help. One of the authors (PME) gratefully acknowledges Iberduero S.A. for help and support.

REFERENCES

1. P. Hohenberg and W. Kohn, Phys. Rev. *136* (1964) B864.
2. W. Kohn and L.J. Sham, Phys. Rev. *140* (1965) A11113.
3. O. Gunnarsson, *The Density Functional Theory of Metallic Surfaces*, P. Phariseau, B.L. Györffy, and L. Scheire, eds., NATO Advance Institutes Series, vol. 41, 1, Plenum (1979).
4. L.H. Thomas, Proc. Cam. Philos. Soc. *23* (1926) 542; E. Fermi, Rc. Acc. Lincei, *6* (1927) 602.
5. N.H. March, *Theory of the Inhomogeneous Electron Gas*, S. Lundqvist and N.H. March, eds.,Plenum (1983) 7.
6. P.A.M. Dirac, Proc. Camb. Philos. Soc. *26* (1930) 376.
7. J.C. Salter, Phys. Rev. *81* (1951) 385.
8. R. Gaspar, Acta Phys. Hung. *3* (1954) 263.
9. L. Hedin and S. Lundqvist, *Solid State Physics*, H. Ehrenreich, F. Seitz, and D. Turnbull, eds., *23* (1969) 1; L. Hedin and B.I. Lundqvist, J. Phys. C 4 (1971) 2064.
10. O. Gunnarsson and B.I. Lundqvist, Phys. Rev. B *13* (1976) 4274.
11. E. Bonderup, Lecture notes on the *Penetration of Charged Particles Through Matter*, University of Aarhus, Denmark (1978), unpublished.
12. P.M. Echenique, I. Nagy, and A. Arnau, Int. J. of Quantum Chem. *23* (1989) 521.
13. B.A. Trubnikov and Yu N. Yavlinski, Sov. Phys. JETP *21* (1965) 167.
14. H.J. Davis and R. Dagonnier, J. Chem. Phys. *44* (1966) 4030.
15. J. Finneman, *M. Sc. Dissertation*, The Institute of Physics, Aarhus University, Denmark (1968), unpublished.
16. B.D. Josephson and J. Lekner, Phys. Rev. Lett. *23* (1969) 111.
17. E.G. d'Agliano, P. Kumar, W. Schaich, and M. Suhl, Phys. Rev. B *11* (1975) 2122.
18. A. Blandin, A. Nourtier, and D.W. Hone, J. Phys (Paris) *37* (1976) 369.
19. P. Minnhagen, J. Phys. C *15* (1982) 2293.
20. P. Sigmund, Phys. Rev A *26* (1982) 24-97.
21. K. Schönhammer, Phys. Rev. B *37* (1988) 7735.

22. L.D. Landau and E.M. Lifshitz, *Physical Kinetics*, Pergamon (1981) 41.
23. C. Kittel, *Quantum Theory of Solids*, Wiley, New York (1963).
24. J. Friedel, Phil. Mag. *43* (1952) 153; J. Friedel, Nuovo Cimento Suppl. 7 (1958) 287.
25. J.S. Langer and V. Ambegaokar, Phys. Rev. *121* (1961) 1090.
26. D.C. Langreth, Phys. Rev. *150* (1966) 516.
27. N. Levinson, Kgl. Dan. Videns K. Selsk. Mat.-Fys. Medd. *25* (1949) No.9.
28. A. Galindo and P. Pascual, *Mecánica Cuántica I, II*, Eudema Universidad, Madrid (1989).
29. L.D. Landau and E.M. Lifshitz, *Quantum Mechanism*, Pergamon (1977).
30. G.D. Mahan, *Many Particle Physics*, Plenum, New York (1981).
31. C.O. Almbladh, U. von Barth, Z.D. Popovic, and M.J. Stott, Phys. Rev. B *14* (1976) 2250; P. Jena and K.S. Singwi, Phys. Rev. B *17* (1978) 3518.
32. N. Barberan and P.M. Echenique, J. Phys. B *19* (1986) L81.
33. P.M. Echenique, R.M. Nieminen, and R.H. Ritchie, Solid State Commun. *37* (1981) 779.
34. M.J. Puska and R.M. Nieminen, Phys. Rev. B *27* (1983) 6121.
35. T.L. Ferrell and R.H. Ritchie, Phys. Rev. B *16* (1977) 115.
36. A. Mann and W. Brandt, Phys. Rev. B *24* (1981) 4999.
37. P.M. Echenique, F. Flores, and R.H. Ritchie, *Dynamic Screening of Ions in Condensed Matter*, Solid State Physics series, Erhenreich and Turnbull eds., Academic *43* (1990) 229.
38. J. Lindhard and A. Winther, K. Dan. Vidensk. Selsk, Mat. Pys. Medd. *34* (1964) No.4,
39. J. Ziegler, J.P. Biersack, and U. Littmark, *The Stopping and Range of Ions in Solids*, Pergamon (1985).
40. I. Nagy, A. Arnau, P.M. Echenique, and E. Zaremba, Phys. Rev. B *40* (1989) 11 983.
41. F. Flores, in this volumen.
42. A.H. Sørensen, Nucl. Instr. and Meth. B *48* (1990) 10.
43. I. Nagy and P.M. Echenique, *Proceedings of the 12th W. Brandt Symposium*, Donostia, Spain (1990).
44. W. Barkas, W. Birnbaum, and F.M. Smith, Phys. Rev. *101* (1956) 778.
45. W.H. Barkas, N.J. Dyer, and H.H. Heckman, Phys. Rev. Lett *11* (1963) 26.
46. H.H. Andersen, in *Semiclassical Descriptions of Atomic and Nuclear Collisions*, J. Bang and J. De Boer, eds.,Elsevier, Amsterdam, (1985) 409.
47. G. Basbas, Nucl. Instr. and Meth. B *4* (1984) 227.
48. C.D. Hu and E. Zaremba, Phys. Rev. B *37* (1988) 9268.
49. S.H. Vosko, L. Wilk, and M. Nusair, Can. J. Phys. *58* (1980) 1200.
50. A. Cherubini and A. Ventura, Lett. Nuovo Cimento *44* (1985) 503.
51. B. Apagyi and I Nagy, J. Phys. C *20* (1987) 1465.
52. B. Apagyi and I Nagy, J. Phys. C *21* (1988) 3845.
53. I. Nagy, in this book.
54. E. Steiner, J. Chem. Phys. *39* (1963) 2365.
55. A. Kallio, P. Pietiläinen and L. Lantto, Physica Scripta *25* (1982) 943.
56. P.M. Echenique, R.M. Nieminen, J.C. Ashley, and R.H. Ritchie, Phys. Rev. A *33* (1986) 897.
57. F. Shulz and W. Brandt, Phys. Rev. B *26* (1982) 4864.
58. W. Brandt, Nucl. Instr. and Meth. *194* (1982) 13.
59. W. Brandt and M. Kitagawa, Phys. Rev. B *25* (1982) 5631.
60. J.C. Ashley, R.H. Ritchie, P.M. Echenique, and R.M. Nieminen, Nucl. Instr. and Meth. B *15* (1986) 11.
61. P.M. Echenique, Nucl. Instr. and Meth. B *27* (1987) 256.
62. F.H. Eisen, Can. J. Phys. *46* (1968) 561.
63. J.S. Briggs and A. Pathak, J. Phys. C 7 (1974) 1929, display F.H. Eisen's previously unpublished data for the <111> channel in Si.

64. J.A. Golovchenko, D.E. Cox, and A.N. Goland, Phys. Rev. B *26* (1982) 2335.
65. I. Nagy, A. Arnau, and P.M. Echenique, Phys. Rev. B *38* (1988) 9191.
66. J.C. Ashley, A. Gras-Martí, and P.M. Echenique, Phys. Rev. A *34* (1986) 2495.
67. A. Arnau, P.M. Echenique, and R.H. Ritchie, Nucl. Instr. and Meth. B *33* (1988) 138.
68. E. Muller-Hartmann, T.V. Ramakrishnan, and G. Toulouse, Phys. Rev. B *3* (1971) 1102.
69. I. Nagy, A. Arnau, and P.M. Echenique, Phys. Rev. A *40* (1989) 987.
70. A.A. Kugler, J. St. Phys. *12* (1975) 35.
71. S. Ichimaru, Rev. Mod. Phys. *54* (1982) 1017.
72. R.G. Dandrea, N.W. Ashcroft, and A.E. Carlson, Phys. Rev. B *34* (1986) 2097.
73. E.K.U. Gross and W. Kohn, Phys. Rev. Lett. *55* (1985) 2850; *57* (1986) 923(E).
74. N. Iwamoto and E.K.U. Gross, Phys. Rev. B *35* (1987) 3003.
75. N. Iwamoto, Phys. Rev. A *30* (1984) 3289.

STATISTICS OF CHARGED-PARTICLE PENETRATION

P. Sigmund

Physics Department
Odense University
DK-5230 Odense M, Denmark

INTRODUCTION

Particle penetration phenomena are intrinsically statistical in nature. It is extremely unlikely that two particles penetrating a material medium at different times will undergo the same sequence of collision events. This applies even to ideal experiments with a well-collimated, monochromatic particle beam hitting a highly homogeneous target material with a uniform thickness. The energies and directions of motion of penetrating particles, as well as their charge and excitation states, will differ shortly after they hit the target. The same is true for the radiation effects they generate, such as excitation, ionization, and relocation of target atoms, and nuclear or chemical reactions. In brief, whatever quantity you find appropriate to measure, you will find fluctuations, and these fluctuations will not go to zero if you increase the number of particles in the beam.

A given experimental setup may not record a complete statistical distribution but only one quantity such as an average, a peak value, or a weighted integral. In that case it is important to know the precise connection of the measured quantity with the pertinent statistical distribution.

This series of lectures addresses sources of statistical fluctuation and ways to estimate pertinent distribution functions as well as averages and standard deviations, peak values and half-widths. Prime quantities will be beam parameters like energy loss (or stopping), angular deflection (multiple scattering), and charge state (capture and loss). The methods and results are highly relevant for the theory of radiation effects, but an actual discussion of that range of topics is beyond the scope of this chapter.

Coincidence techniques allow to measure several experimental parameters simultaneously. Such measurements may demonstrate some degree of correlation. Examples of correlated distributions will be mentioned, but the emphasis will be laid on the statistical distributions obeyed by one physical parameter at a time.

Sources of Fluctuation

Density fluctuations in the target. Fig.1 shows three particles penetrating through a dilute gas. Their trajectories are supposed to be well

Interaction of Charged Particles with Solids and Surfaces
Edited by A. Gras-Martí *et al.*, Plenum Press, New York, 1991

73

separated in space and time: they will hardly interfere. Particle 2 moves through a region which, at the time of penetration, is almost deserted of target atoms. Conversely, the path of particle 1 is unusually crowded. No doubt, particle 1 will undergo more encounters.

Details of the collision geometry. Since particle 1 suffers more collisions than particle 2, it is likely to be more heavily deflected and to lose more energy. Is it? Well, also the strength of the individual interactions matters. One violent, head-on collision may affect the trajectory more drastically than ten soft touches. This happens to particle 3 in fig. 1.

Quantal fluctuations. In fig.2, a particle in a well-defined state passes by a target atom at a distance (impact parameter) p. Let the projectile be a heavy particle such as an ion. Then, p can be specified with reasonable accuracy without violation of the uncertainty principle. Even then, the outcome of the collision is not determined uniquely: there is a spectrum of

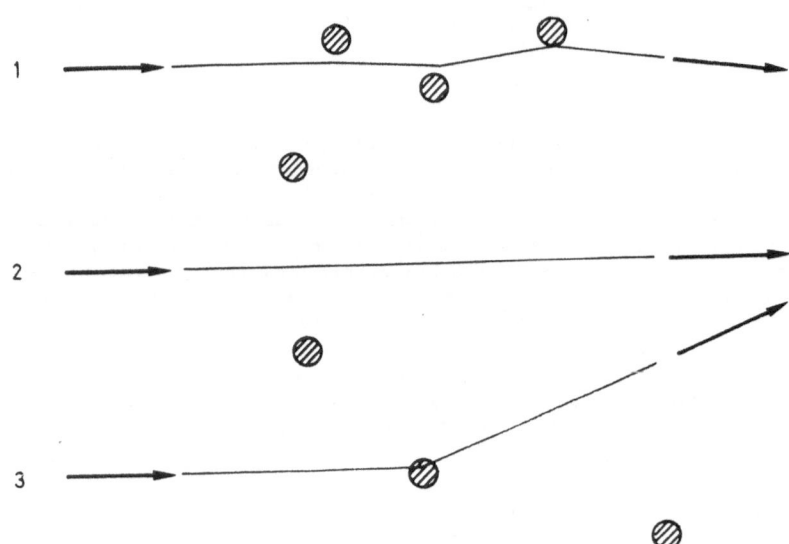

Fig.1. Three trajectories of fast particles in a gas target. See text

electronic excitations to a variety of states, each with its statistical weight determined by quantum mechanics.

Projectile state. In general, the charge and excitation state of the penetrating particle undergoes changes due to collisions, and the outcome of any subsequent collision must be more or less dependent on the projectile state.

Outline of the Lectures

The above processes will be considered in the order of increasing complexity. Lecture 1 considers processes in a dilute gas where collision events are well separated and statistically independent. Moreover, all cross sections are assumed constant throughout the target. The assumption of constant cross section is abandoned in lecture 2, which deals with true slowing-down and allows for charge-changing collisions. Lecture 3 considers processes in a dense medium where the assumption of independent collision events is not always justifiable. Lecture 4 addresses specific applications in the theory of energy loss and angular deflection.

I have tried to make most of this writeup reasonably self-contained and to apply a unified approach, but I assume that the reader has been pre-exposed to elementary penetration theory, e.g. by J.Lindhard's introductory lectures at the Alicante school, by chapter 13 of Jackson's (1975) text, or lecture notes such as those of Bonderup (1981) or myself (Sigmund,1975).

I am not aware of a comprehensive summary in the literature with the scope of the present lectures. There are early standard references on energy loss straggling (Symon,1948) and multiple scattering (Scott,1963), and there is N.Bohr's famous 1948 monograph which contains much enlightening physics.

The basic theoretical tools belong to the theory of probability (Feller, 1957,1966) and are in general use in the theory of stochastic processes in physics and chemistry (Chandrasekhar,1943; van Kampen,1981). The physical input like cross sections, the quantitative features of the results, and the terminology are specific to the field. In fact, particle penetration can well serve as a standard illustration of the general theory of stochastic proces-

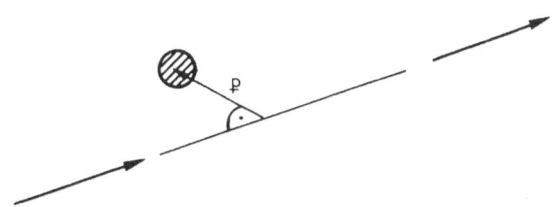

Fig.2. A collision at an impact parameter p

ses. This aspect has been explored in an illuminating study by Lindhard & Nielsen (1971) which I shall refer to repeatedly.

Lecture 1. THE POISSON LIMIT

The prototype of a theoretician's particle penetration experiment features a well-collimated, monochromatic beam of fast charged particles, ions, electrons, etc., at a reasonably high energy, interacting with a material target of homogeneous composition and density, and uniform thickness. Beam particles ("projectiles") may undergo collisions with target particles (mostly atoms, occasionally molecules or larger aggregates).

Within the scope of this school, explicit examples will deal mostly with ions at MeV and upper-keV energies, and occasionally with electrons of similar *velocities*.

A collision is any detectable event involving the projectile and one or more target particles. Collisions may be categorized by their outcome according to some label i = 1,2,... An i-event could typically be an excitation of a target atom to a specified state.

Initially, the following highly simplified situation will be considered,

i)   The beam is dilute and random,
ii) The target is dilute and random, i.e., an ideal gas,
iii) Beam particles do not undergo changes in charge or excitation state,
iv) The target is so thin that the energy lost by a projectile during passage through the target is small compared to its initial energy,
v)   Angular deflection of projectile particles is negligible, and
vi) The target is so thin that the total number of collisions is much smaller than the number of beam particles.

All these assumptions will be relaxed in due course from the bottom up. While v) and vi) will be abandoned shortly, assumptions i)-iv) will be kept throughout the first lecture. i) implies that neither the beam particles themselves nor the effects they generate in the target interfere with each other. ii) in conjunction with iii) has the central consequence that any two collision events are statistically independent. iv) in conjunction with iii) implies translational invariance across the target.

Although these restrictions seem severe, they turn out to be frequently justified in the real world. Of course, their limitations need to be identified.

1.1. Three Aspects of the Cross Section

Let a homogeneous beam, spread over an area A, hit a homogeneous target with a density of N [atoms/volume] and a thickness x. The mean number of collision events per unit time of type i will be

$$v_i = JANx\sigma_i, \qquad\qquad (1.1)$$

where J is the current density of the beam [particles/time/area]. The proportionality with J, A, N, and x is self-evident if the target is "almost transparent", i.e., if assumption vi) above is fulfilled. The remaining factor σ_i, which must have the dimension of an area, is called the cross section for an i-event. σ_i may be illustrated as the "black area per target particle" which, if hit by the beam, contributes exactly one count to the total signal v_i. In actual fact, σ_i is an *effective area*: Because of quantum mechanics, the chance for success is normally less than 1 whereever you hit. Conversely, there is a nonvanishing chance for success even if you hit outside a geometrical area of magnitude σ_i. Eq. (1.1) is the basis for an experimental determination of σ_i: In addition to the number v_i of events i, you need to measure the beam current J as well as target pressure and geometry.

It is not necessary that both the target and the beam be homogeneous. One is enough. Consider only one target atom so that ANx = 1, and let that particle be hit by a random (and *very dilute*) beam. Then, eq. (1.1) reduces to $\sigma_i = v_i/J$. This is the textbook prescription for quantal calculations of cross sections: σ_i is the mean number of events per unit time, divided by the incoming particle current density.

Alternatively, consider a random target but only one projectile. Dividing (1.1) by the total current JA, you find that

$$P_i = Nx\sigma_i \qquad\qquad (1.2)$$

must be the probability *per trajectory* for occurrence of the event i, i.e., the probability for the projectile to undergo an i- event while penetrating a path segment x in a dilute medium of density N. This property of the cross section is of prime importance in penetration phenomena.

1.2. Cumulative Quantities: Energy Loss

Energetic charged particles slow down by electronic and nuclear processes. Except at low speed ($v \ll e^2/\hbar$), electronic processes dominate. The terminology utilized in the present discussion refers to electronic processes. Most of the general results apply to nuclear processes as well, but those are usually accompanied by angular deflection which is still being ignored, cf. assumption v) above. A more specific discussion of energy loss processes is deferred to lecture 4.

Let an i-event now be a collision in which an energy quantum T_i is transferred from the projectile to a target atom. In a collision, the target atom is excited from its ground state E_0 to an eigenstate E_i, so that $T_i = E_i - E_0$. Assumption vi) will be abandoned. This implies that the target may be thick enough to allow for more than one collision, but still thin enough to make the *average energy loss per trajectory* small compared to the projectile energy E. This condition (iv) is easily fulfilled at not too low speed.

Let $F(\Delta E, x)d\Delta E$ be the probability distribution in energy loss ΔE. A central equation to be satisfied by this distribution can be found by considering a sandwich target with thicknesses y and x in series (fig.3). Let the energy loss of a specific beam particle exiting from the first layer (thickness y) be $\Delta E'$. In order to exit with an energy loss ΔE from the target, the projectile must have lost an energy $\Delta E - \Delta E'$ in the second layer (thickness x). Because of the statistical independence of the collision events, the following relation must be satisfied by F,

$$F(\Delta E, x+y) = \int d\Delta E' \ F(\Delta E', y) \ F(\Delta E - \Delta E', x) \ . \tag{1.3}$$

This is known as the Chapman-Kolmogorov equation in statistical physics. The convolution suggests going over to Fourier space,

$$F(\Delta E, x) = \frac{1}{2\pi} \int_{-\infty}^{\infty} dk \ e^{ik\Delta E} \ F(k, x), \tag{1.4}$$

so that

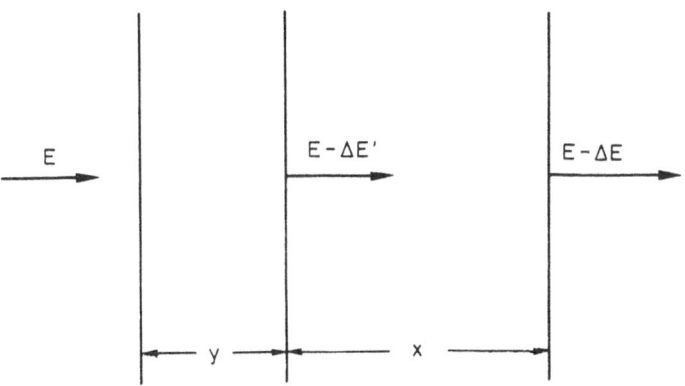

Fig.3. Energy loss in sandwich target. See text

$$F(k,x+y) = F(k,y)\, F(k,x)\ .\tag{1.5}$$

The solution of eq. (1.5) is an exponential in x,

$$F(k,x) = \exp[xC(k)]\tag{1.6}$$

with an arbitrary function C(k). That function depends on what actually can happen during the passage. Let x be so small that the projectile will undergo at most one collision. Then,

$$F(\Delta E,x) = [1 - \Sigma P_i]\ \delta(\Delta E) + \Sigma P_i\ \delta(\Delta E - T_i),\quad x\ \text{small},\tag{1.7}$$

where the first term on the right-hand side expresses the probability for no collision and, consequently, zero energy loss, while the second term collects all possibilities of exactly one collision and the associated energy loss. $\delta(\Delta E)$ is the Dirac function. Its Fourier transform,

$$\delta(\Delta E) = \frac{1}{2\pi}\int_{-\infty}^{\infty}dk\ e^{ik\Delta E}\ ,\tag{1.8}$$

will be utilized routinely in the following. Applying (1.2) and (1.8), you may write (1.7) in Fourier space,

$$F(k,x) = 1 - \Sigma\ Nx\sigma_i\ [1 - e^{-ikT_i}\ ],\qquad x\ \text{small}.$$

Comparing this with the small-x limit of (1.6) you find that

$$C(k) = -\ N\ \Sigma\ \sigma_i\ [1 - e^{-ikT_i}\ ] \equiv -\ N\ \sigma(k)\ .\tag{1.9}$$

This defines a quantity $\sigma(k)$ which occurs frequently and is called a *transport cross section*. From eqs. (1.4), (1.6), and (1.9), you finally obtain

$$F(\Delta E,x) = \frac{1}{2\pi}\int_{-\infty}^{\infty}dk\ e^{ik\Delta E - Nx\sigma(k)}\ .\tag{1.10}$$

This is the prototype of a rather general equation which, as we shall see, governs a wide range of particle penetration phenomena. I find it appropriate to call it the Bothe-Landau formula. The argument utilized in its derivation is quite general and applies to sums of many kinds of independent statistical variables. Applications include the statistical theory of errors, the fluctuations of the local electric field in a dielectric, the theory of pulse pile-up and noise, ion-beam induced mixing, and many others.

For simplicity, the derivation has been made for one type of target atom but allowing for a variety of target states. If there is a variety of target atoms or molecules, such as in atmospheric air, the respective densities N_μ, $\mu = 1,2,,\ldots$ are to be included in the statistical averaging, so that $N\Sigma\sigma_i$ is replaced by $\Sigma N_\mu \Sigma \sigma_{\mu i}$, where $\sigma_{\mu i}$ is the cross section for energy loss T_i to a target atom of type μ.

I have written eq. (1.10) as a Fourier transform rather than a Laplace transform, although the latter notation, used in Landau's 1944 paper, is more common. The Fourier notation is more comprehensive since it allows for both positive and negative values of the energy loss, i.e., energy loss and energy gain. When the cross section only allows for energy loss, the Fourier and Laplace formulation are strictly equivalent. In other applications like multiple scattering, the Fourier representation is superior.

1.3. Mean Energy Loss and Straggling

There are many ways to find the standard expressions for mean energy loss and fluctuation which you may be familiar with. Having a closed-form expression for the complete distribution, we may just as well use that. Multiplication of (1.10) by some power of the energy loss and integration over ΔE yields

$$\int_{-\infty}^{\infty} d\Delta E \ \Delta E^{\mu} \ F(\Delta E, x) = (i\frac{\partial}{\partial k})^{\mu} \ e^{-Nx\sigma(k)} \bigg|_{k=0} \tag{1.11}$$

for $\mu = 0,1,2...$ By means of eq. (1.9), the right-hand side is easily seen to reduce to 1 for $\mu = 0$, to $-iNxd\sigma(k)/dk$ for $\mu = 1$, and to $Nxd^2\sigma(k)/dk^2 - (Nxd\sigma(k)/dk)^2$ for $\mu = 2$, respectively, all taken at $k=0$.

The case $\mu=0$ expresses particle conservation,

$$\int d\Delta E \ F(\Delta E, x) = 1 \ . \tag{1.12a}$$

The first moment, $\mu = 1$, yields the mean energy loss

$$<\Delta E> = NxS \ , \tag{1.12b}$$

where the curled brackets indicate an average over many penetrating particles, and S is called the *stopping cross section*

$$S = \sum_i T_i \sigma_i \tag{1.13}$$

which is an atomic parameter. For $\mu = 2$, you find the variance, or *energy loss straggling*

$$<(\Delta E - <\Delta E>)^2> = NxW \tag{1.12c}$$

after subtracting the square of (1.12b). Here, the *straggling parameter* W,

$$W = \sum_i T_i^2 \sigma_i \tag{1.14}$$

has been introduced. It is likewise an atomic parameter independent of the density N.

If the single-event spectrum σ_i, $i = 1,2,3...$ is continuous rather than discrete, the energy transfer T_i is replaced by the continuous variable T, and σ_i by $d\sigma(T)$. Then, (1.13) and (1.14) read

$$S = \int T \ d\sigma(T) = \int dT \ T \ K(T) \ , \tag{1.13'}$$

$$W = \int T^2 d\sigma(T) = \int dT \ T^2 \ K(T) \ , \tag{1.14'}$$

where $K(T) = d\sigma(T)/dT$ is called the *differential energy loss cross section*. The limits of the integration are defined by conservation laws and may be thought of as being implicit in the definition of K(T).

1.4. The Number of Collisions

In order to demonstrate the power of the Bothe-Landau formula, let's leave the energy loss for a while and consider the probability that in a given passage, exactly n_j collisions of some type j will take place. Nothing has changed in the physics. All we need to do is changing the bookkeeping. Instead of adding all *energy losses* T_i for all excitation levels i, consider the *number* of collisions of type j only. In the new variable, eq. (1.10) reads

$$F(n_j,x) = \frac{1}{2\pi} \int dk \; e^{ikn_j} \; e^{-Nx\sigma(k)} \; ,$$

and after substitution of $T_i \rightarrow \delta_{ij}$, (1.9) takes on the simple form

$$\sigma(k) = \sigma_j(1 - e^{-ik}) \; .$$

It may be worthwhile to point out that the integration variable k is unrelated yet analogous to the k in eq. (1.4), and it has even a different dimension. The function $\sigma(k)$ defined above is likewise different from the $\sigma(k)$ defined by eq. (1.9), but both functions have the dimension of an area. Similar remarks apply in subsequent sections like 1.5 on multiple scattering.

After expansion of the double exponential $\exp(Nx\sigma_j e^{-ik})$ in powers of x, the integration over k can be carried out by means of eq. (1.8), so that

$$F(n_j,x) = e^{-Nx\sigma_j} \sum_{\nu=0}^{\infty} \frac{(Nx\sigma_j)^\nu}{\nu!} \delta(\nu-n_j) \; . \tag{1.15}$$

This shows that the number of j-events is distributed according to Poisson's law,

$$P(n) = \exp(-<n>) \; <n>^n/n! \tag{1.16}$$

around the mean value $<n> = <n_j> = Nx\sigma_j$. Nothing special has been assumed about the event j. Therefore, the number n of occurrences of *any* specific collision event is distributed according to eq. (1.16), provided that the assumptions i) - v) are satisfied.

In particular, the mean number of i-events is given by

$$<n_i> = Nx\sigma_i \; ,$$

and the variance by

$$<(n-<n_i>)^2> = <n_i> \; .$$

You could have started adopting the Poisson distribution via physical reasoning and from there have arrived at eqs. (1.12 -14). You may have seen this done in J.Lindhard's lectures.

In the limit of large $<n>$, the Poisson distribution is known to approach a gaussian, but in the opposite limit, major deviations occur. This has the important consequence that the energy loss spectrum will not be gaussian if the target is thin. A first example will be studied shortly.

1.5. Multiple Scattering

The theory of multiple scattering is very similar to the energy loss problem from a general point of view. Instead of focussing on the transferred energy, you need to consider momentum transfer and the accompanying angular deflection. While energy transfer is one-dimensional and, mostly, a one-way process going from some initial energy down toward zero, multiple scattering is two-dimensional, and the manifold of accessible directions of motion is bounded to the unit sphere.

Be sure that there are major differences between the physical processes leading to energy loss and multiple scattering: Energy loss is mainly electronic at high velocities, while multiple scattering is predominantly nuclear. This is, however, immaterial for the development of the statistical formulation.

1.5.1. Motion in a Plane. In order to establish a close analogy between energy loss and multiple scattering, consider first the academic case where all scattering takes place in a plane (fig.4). Let the direction of motion be

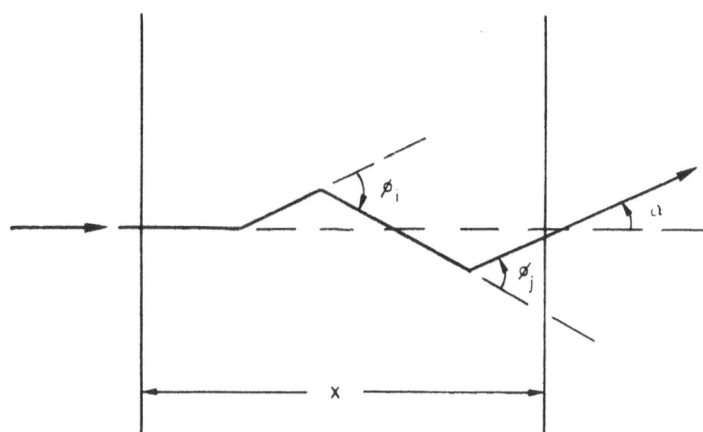

Fig.4. Multiple scattering in a plane

specified by an angle α to the x-axis. This angle may change in bits of magnitude ϕ_i, i = 1,2,3..., where $-\pi \leq \phi_i \leq \pi$. Collisions occur at random with a probability $P_i = Nx\sigma_i$, where σ_i now means the cross section for scattering into an angle ϕ_i. All energy loss will be ignored for clarity. In the presence of angular deflection, the travelled pathlength will exceed the target thickness. For definiteness, x will be taken to be the *path length* in connection with multiple scattering.

While the accumulated deflection may amount to many times 2π, the angle α is restricted to one periodicity interval, say $-\pi \leq \alpha \leq \pi$. Therefore, we may write a Fourier series instead of eq. (1.4),

$$F(\alpha,x) = \frac{1}{2\pi} \sum_{k=-\infty}^{\infty} F_k(x) e^{ik\alpha} \tag{1.17}$$

for the angular distribution $F(\alpha,x)d\alpha$ at pathlength x, where the k's are integers. Similar to (1.3), we must have

$$F(\alpha, x+y) = \int_{-\infty}^{\infty} d\alpha' \ F(\alpha', y) \ F(\alpha-\alpha', x) \ .$$

This reproduces the analog of eq. (1.5), from which you immediately obtain

$$F_k(x) = e^{xC_k} \tag{1.18}$$

with a set of unknown numbers C_k. For infinitesimal pathlength, similar to eq. (1.7), you deduce

$$F(\alpha, x) = (1 - \Sigma P_i) \ \delta(\alpha) + \Sigma P_i \ \delta(\alpha - \phi_i) \ . \tag{1.19}$$

Comparison of the Fourier components of (1.19) with the small-x limit of (1.18) yields

$$C_k = - N \ \Sigma \ \sigma_i \ (1 - e^{-ik\phi_i}) \equiv - N \ \sigma_k \ . \tag{1.20}$$

Fig.5. Tracks of alpha particles in a cloud
chamber. From Meitner & Freitag (1926)

Collecting (1.17) and (1.18) finally leads to

$$F(\alpha, x) = \frac{1}{2\pi} \ \sum_{k=-\infty}^{\infty} e^{ik\alpha - Nx\sigma_k} \tag{1.21}$$

with σ_k defined by eq. (1.20). Eqs. (1.21) and (1.20) determine the angular distribution. These expressions are very similar to the corresponding ones, (1.10) and (1.9), for the energy loss spectrum.

Most often in particle penetration, allowed angular deflections form a continuum. Then, (1.20) may be replaced by

$$\sigma_k = \int_{-\pi}^{\pi} d\sigma(\phi) \ (1 - e^{-ik\phi}) \ , \tag{1.20'}$$

where $d\boldsymbol{\sigma}(\phi)$ is the differential cross section for scattering by an angle $(\phi,d\phi)$.

1.5.2. Small—Angle Scattering in a Plane. Angular scattering may be a minor effect. An example is shown in fig. 5 which is typical for particles heavier than electrons at not too low energy. This feature is caused by the shape of the scattering cross section $d\boldsymbol{\sigma}(\phi)/d\phi$ which has a pronounced peak at small angles and decreases toward large angles, as you know from Rutherford's law.

For a narrow multiple scattering distribution, where small angles α dominate, it is mainly the short wavelengths, or large wave numbers k, that determine the sum (1.21). Then, the sum may be replaced by an integral,

$$\sum_{k=-\infty}^{\infty} = \sum_{k=-\infty}^{\infty} \Delta k \rightarrow \int_{-\infty}^{\infty} dk \ . \tag{1.22}$$

Moreover, since large scattering angles ϕ contribute very little to the integral (1.20'), one may as well extend the angular integration to infinity without any noticeable effect. Therefore,

$$\sigma(k) = \int_{-\infty}^{\infty} d\phi \ K(\phi) \ (1 - e^{-ik\phi}) \ . \tag{1.23}$$

It may look strange that infinite deflection angles should contribute to small-angle multiple scattering. The point is that they don't!

1.5.3. Small—Angle Scattering in Real Space. The main outcome of the previous paragraph was that with (1.21) written as an integral instead of a sum, and with an unlimited integration interval in (1.23), there is no formal difference between these integrals and the corresponding relations (1.9) and (1.10) for the energy loss. In other words, angular distributions in the limit of small angles could have been derived directly from eqs. (1.9) and (1.10) by changing nomenclature. This finding will be utilized now to determine multiple scattering distributions for motion in three-dimensional space.

All we have to do is finding suitable variables. You may wish to start by specifying a unit vector in the direction of motion $\underline{\Omega}$ in terms of a polar angle α and an azimuth ξ,

$$\underline{\Omega} = (\sin\alpha \ \cos\xi, \sin\alpha \ \sin\xi, \cos\alpha) \ . \tag{1.24}$$

Single-underlined letters will always denote vectors. If the angle α to the initial direction of motion, i.e., the z-axis, is small, (1.24) reduces to

$$\underline{\Omega} = (\alpha \ \cos\xi, \alpha \ \sin\xi, 1) \tag{1.25}$$

to first order in α. Note that all scatter takes place in the x-y plane. Thus, in small-angle scattering, attention is restricted to the lateral components of the velocity vector. Introducing a vector $\underline{\alpha}$ in the x-y plane with the components

$$\underline{\alpha} = (\alpha_x, \alpha_y) = (\alpha \ \cos\xi, \ \alpha \ \sin\xi) \ , \tag{1.26}$$

you may write down the distribution in $\underline{\alpha}$ as a straight generalization of eqs. (1.9) and (1.10) to two dimensions,

$$F(\underline{\alpha}, x) = \frac{1}{(2\pi)^2} \int d^2k \ e^{i\underline{k}\cdot\underline{\alpha} - Nx\sigma(\underline{k})} \tag{1.27}$$

and

$$\sigma(\underline{k}) = \int d^2\phi \ K(\underline{\phi}) \ (1 - e^{-i\underline{k}\cdot\underline{\phi}}) \ . \tag{1.28}$$

If you have doubts about the validity of the generalization, you may check that a convolution equation equivalent to (1.3) is satisfied, and that the correct limiting result is reached for small x. Not much more has been utilized in the derivation of the Bothe-Landau formula.

The quantity $K(\underline{\phi})d^2\phi$ introduced in (1.28) is the differential cross section for angular scattering by an increment $(\underline{\phi}, d^2\phi)$. $\underline{\phi}$ is a vectorial scattering angle in the x-y plane, and $d^2\phi$ is an element of solid angle around $\underline{\phi}$. The projection of the trajectory on the x-y plane is illustrated schematically in fig. 6. This illuminates the frequently used concept of

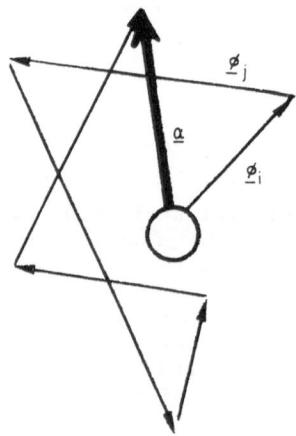

Fig.6. Small-angle multiple scattering in real space. Projection of a trajectory on a plane perpendicular to the initial direction of motion. Vectors $\underline{\alpha}$ and $\underline{\phi}_i$ are analogs to angles α and ϕ_i in fig.4

small-angle multiple scattering being a "random walk in the transverse plane".

When the target particles are atoms, it is safe to assume the differential cross section to exhibit azimuthal symmetry,

$$K(\underline{\phi}) \ d^2\phi = K(\phi) \ \phi d\phi d\chi \tag{1.29}$$

and to perform the integral in eq. (1.28) over the azimuth χ. This yields

$$\sigma(\underline{k}) = \int_0^\infty K(\phi) 2\pi\phi d\phi \ [1 - J_0(k\phi)] \equiv \sigma(k) \ , \tag{1.30}$$

where J_0 is a Bessel function of the first kind. Since $\sigma(k)$ depends only on the length of the vector k and not on its direction, you may also carry out the azimuthal part of the integration in eq. (1.27), with the result

$$F(\alpha, x) \ d^2\alpha = 2\pi\alpha d\alpha \ \frac{1}{(2\pi)^2} \ \int\limits_0^\infty 2\pi k dk \ J_0(k\alpha) \ e^{-Nx\sigma(k)} \ . \tag{1.31}$$

(Although all factors 2π drop out I have kept them for clarity). Eqs. (1.31) and (1.30) are valid under the assumptions of azimuthal symmetry and small scattering angle. They were first derived by Bothe (1921), who also gave explicit examples, one of which will be shown below.

1.5.4. Multiple Scattering at Small Angles: Moments. You may derive moments over the multiple scattering distributions from eqs. (1.27) and (1.28) by the procedure described in sect. 1.3 for the moments over the energy loss distribution. If you keep to cartesian coordinates, the extension to two dimensions is just a matter of notation, and you find the average (vectorial) scattering angle, expressed by curled brackets as the energy loss in previous paragraphs,

$$\langle \underline{\alpha} \rangle = Nx \int d^2\phi \ \underline{\phi} \ K(\underline{\phi}) \ , \tag{1.32}$$

which vanishes if the cross section shows azimuthal symmetry. The variance, of course, is nonvanishing, and given by

$$\langle (\underline{\alpha} - \langle \underline{\alpha} \rangle)^2 \rangle = Nx \int d^2\phi \ \phi^2 \ K(\phi) = Nx \int d\phi 2\pi\phi^3 \ K(\phi) . \tag{1.33}$$

This expression can be utilized to find rough estimates of beam broadening due to multiple scattering. However, the integrand $\phi^3 K(\phi)$ does usually not fall off sufficiently rapidly at large angels to make (1.33) useful as it stands. Some modification is necessary and will be mentioned in lecture 4.

1.5.5. Multiple Scattering in Real Space: Finite Angles. I shall briefly mention the generalization to two dimensions of the case of arbitrary scattering angles, the one-dimensional analog of which was discussed in sect. 1.5.1. Since the directional distribution must be periodic over the unit sphere, the Fourier series (1.17) needs to be replaced by an expansion in spherical harmonics. The convolution equation remains valid and can be reduced by means of the addition theorem of spherical harmonics. The resulting distribution is of interest mainly for the case of azimuthal symmetry of the cross section. Then, only the Legendre polynomials P_λ are left over of the spherical harmonics expansion, yielding

$$F(\underline{\Omega}, x) \ d^2\Omega = \frac{d^2\Omega}{4\pi} \ \sum_{\lambda=0}^\infty (2\lambda+1) \ P_\lambda(\cos\alpha) \ e^{-Nx\sigma_\lambda} \tag{1.34}$$

with

$$\sigma_\lambda = \int\limits_0^\pi 2\pi\phi d\phi \ K(\phi) \ [1 - P_\lambda(\cos\phi)] \ . \tag{1.35}$$

The similarity with the small-angle limit (1.30) and (1.31) is visible, and deriving those relations as a limiting case of (1.34) and (1.35) is indeed possible but will not be done here. The above results have been derived by Goudsmit and Saunderson (1940) with the aim of treating multiple scattering of electrons. A practical difficulty arises since the path length x cannot be identified with a layer thickness in case of substantial angular deflection. Therefore, the multiple scattering distribution needs to be considered jointly in real and velocity space. While this is possible, it complicates the matter to the point that Monte Carlo simulation (to be discussed in lecture 2) has been found to be a more powerful approach. A recent application of (1.34) to ion scattering will be mentioned in sect. 4.

1.6. Some Examples

It may now be instructive to look at a few simple explicit distributions. None of these will be the last word on the topic. All subjects will come again.

1.6.1. Bohr Straggling. Consider a stripped nucleus, i.e., a point charge with mass M_1 and charge Z_1e, moving with a speed v through a dilute gas of *free electrons* that have velocities << v. As a first approximation, consider the target electrons as initially at rest, i.e., treat the projectile-target interaction by Rutherford's law,

$$K(T) = 2\pi \frac{z_1^2 e^4}{mv^2} \frac{1}{T^2}, \qquad 0 \leq T \leq 2mv^2 \qquad (1.36)$$

where m and -e are the electron mass and charge. In specifying the upper limit of the accessible interval of transferred energy T, terms of order m/M_1 << 1 have been neglected.

Insertion of (1.36) into (1.14') and integration yields

$$W = 4\pi z_1^2 e^4 \qquad (1.37)$$

or, according to (1.12c),

$$\Omega^2 = <(\Delta E - <\Delta E>)^2> = Nx \; 4\pi z_1^2 z_2 e^4. \qquad (1.38)$$

This expression was first derived by N.Bohr (1915). In accordance with the usual notation, the target density N has been replaced by the electron density NZ_2, where Z_2 is the atomic number of the target nuclei and N the density of *atoms*. The implication is that *all* electrons in the target may be considered free. The presence of the nuclei and their possible effect on the energy loss spectrum has been ignored. Nevertheless, the above expression will be seen to be an excellent approximation to the straggling parameter at not too low projectile speed.

Eq.(1.38) may be compared to Bethe's (1930) formula for the mean energy loss,

$$<\Delta E> = Nx \frac{4\pi z_1^2 z_2 e^4}{mv^2} L , \qquad (1.39)$$

where $L = \log(2mv^2/I)$, and I is the mean excitation energy. The similarity of the two expressions is obvious and by no means accidental since both of them originate in the Coulomb interaction between the projectile and the electrons. However, while the straggling parameter is well defined even when all binding of the target electrons is neglected, the stopping cross section diverges if you set I = 0. This is due to the behavior of the integral over T near T = 0.

You may have recognized that all energy intervals of equal size from T = 0 to $T = 2mv^2$ contribute equal weights to W. This is very different from the stopping cross section where the various intervals contribute logarithmically, i.e., energy losses between 10 and 100 eV contribute as much as energy losses between 100 and 1000 eV etc., up to the maximum energy transfer. Thus, head-on collision events contribute with greater weight to straggling than to the mean energy loss.

1.6.2. An Integrable Energy Loss Spectrum. Lindhard and Nielsen (1971) have collected a number of examples where the Bothe-Landau formula can be evaluated analytically. An illuminating case is the spectrum for the following energy loss cross section,

$$K(T) = \frac{C}{T^{3/2}} e^{-\alpha T} , \quad 0 \leq T \leq \infty \tag{1.40}$$

(1.40) is similar to Rutherford's law, eq. (1.36), but T^{-2} has been replaced by $T^{-3/2}$, and the cutoff at $2mv^2$ has been replaced by an exponential with a parameter α which could be chosen to be $(2mv^2)^{-1}$.

For a continuum, eq. (1.9) reads

$$\sigma(k) = \int_0^\infty dT \; K(T) \; (1 - e^{-ikT}). \tag{1.41}$$

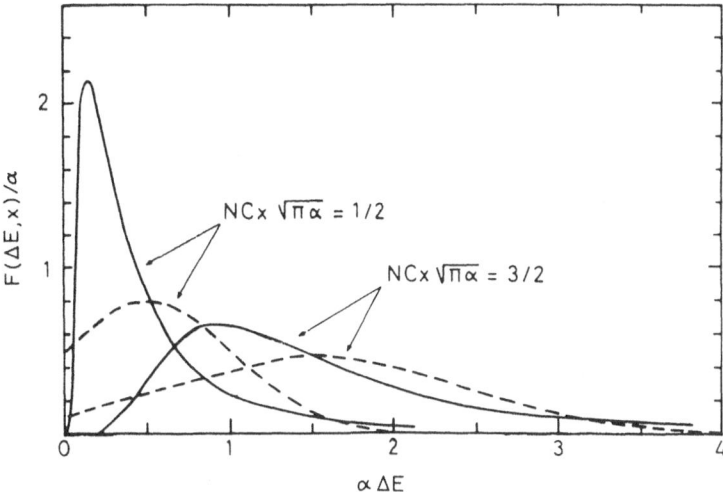

Fig.7.   Energy loss spectra calculated for model cross section (1.40), for two target thicknesses. Fulldrawn lines: Exact results, eq.(1.42); broken lines: Diffusion approximation, eq. (1.47). Redrawn after Lindhard & Nielsen (1971)

Integration with (1.40) yields

$$\sigma(k) = 2\pi^{1/2} C [(\alpha+ik)^{1/2} - \alpha^{1/2}]$$

and, after insertion into (1.10) and integration,

$$F(\Delta E, x) = \frac{NCx}{\Delta E^{3/2}} e^{-\alpha [\Delta E - NCx (\pi/\alpha)^{1/2}]^2/\Delta E} \tag{1.42}$$

Fig. 7 shows spectra for two target thicknesses. For the thinner target, the spectrum is very skew, much like the single-collision spectrum (1.40) which it approaches in the limit of vanishing thickness. The thicker target shows a spectrum that is intermediate between the single-collision case and a gaussian.

Moments over the energy loss spectrum can be determined either by integration over (1.42) or by means of the relations derived in sect 1.3. In addition to the normalization integral, one obtains

$$<\Delta E> = NCx \; (\pi/\alpha)^{1/2} \qquad\qquad (1.43)$$

and

$$<(\Delta E - <\Delta E>)^2> = NxC\pi^{1/2}/2\alpha^{3/2} \; . \qquad\qquad (1.44)$$

The mean energy loss and its standard deviation show different scaling properties with x. This is consistent with the varying shape of the distribution (1.42) with varying thickness.

The spectrum (1.42) has its maximum at an energy loss ΔE_p given by

$$\Delta E_p = -3/4\alpha + [(NCx)^2\pi/\alpha + (3/4\alpha)^2]^{1/2}$$

which, at large thicknesses, approaches

$$\Delta E_p = NCx \; (\pi/\alpha)^{1/2} - 3/4\alpha, \qquad\qquad (1.43')$$

This is less than the mean energy loss (1.43). The reason is that $<\Delta E>$ contains a contribution from the high-energy tail of the spectrum. We shall come back to this important finding in sect. 4.

1.6.3. Diffusion Approximation in Energy Loss. For thick targets, eq. (1.12c) predicts the energy loss spectrum to broaden. Hence, the Fourier integral should receive its main contributions from large wavelengths, i.e., small values of k. Expansion of eq. (1.9) in powers of k yields

$$\sigma(k) = - \sum_{\nu=1}^{\infty} \frac{1}{\nu!} \; (-ik)^\nu \int T^\nu d\sigma(T) = ikS + k^2W/2 - ik^3Q/6 \; ... \qquad\qquad (1.45)$$

where

$$Q = \sum_i T_i^3 \; \sigma_i = \int T^3 \; d\sigma(T) \qquad\qquad (1.46)$$

is a parameter analogous to S and W, eqs. (1.13) and (1.14). It can be shown to be related to the *skewness* of the energy loss spectrum.

If the series (1.45) is truncated at the term of second order in k, the integral (1.10) can be carried out and yields

$$F(\Delta E,x) = \frac{1}{(2\pi NxW)^{1/2}} \exp[- \frac{(\Delta E - NxS)^2}{2NxW}] \; , \qquad\qquad (1.47)$$

i.e., a gaussian centered around the mean energy loss NxS with the width given by the standard deviation $(NxW)^{1/2}$. While this is an appealing result, its derivation as well as the example shown in fig. 7 illustrate that it is only valid approximately, and only at large thicknesses. Bohr (1948) mentions as a sufficient criterion for the validity of (1.47) that the variance of the distribution, NxW, must exceed the square of the maximum energy transfer $2mv^2$ in a single encounter.

The procedure leading to eq. (1.47) is called diffusion approximation since it leads to a gaussian density of the type known from diffusion or heat conduction theory. One might conceive that inclusion of higher-order terms could improve the accuracy of this distribution. This is only partially true: Truncation of the series (1.45) at any finite term beyond $\nu = 2$ leads to *negative* values of the probability density F for some nonzero interval of the variable ΔE, regardless of the cross section (Lindhard and Nielsen, 1971). It is advisable, therefore, to base any improvements beyond the diffusion approximation upon the complete integral eq. (1.10) rather than the series expansion (1.45).

1.6.4. An Integrable Small–Angle Multiple–Scattering Distribution. Bothe (1921) gave an example of an integrable two-dimensional distribution. In the present notation it refers to the differential cross section

$$K(\phi) = \frac{C}{\phi^3} , \quad 0 \leq \phi \leq \infty \tag{1.48}$$

where C is a constant. Since Rutherford's law goes as ϕ^{-4} at small angles, (1.48) may be taken to model some kind of screened Coulomb interaction. Bothe's aim in applying (1.48) was actually to estimate the fluctuations in the electrostatic field in a dielectric, where the inverse cube stands for the dependence on distance of the field of a dipole. The relevance of (1.48) for multiple scattering was pointed out much later by Lindhard (1970). After insertion of (1.48) into (1.30) and integration, eq. (1.31) can be evaluated to yield

$$F(\alpha,x) = \frac{NCx}{[\alpha^2 + (2\pi NCx)^2]^{3/2}} . \tag{1.49}$$

The distribution is normalized as it must be, and it exhibits the single-scattering cross section (1.48) at large angles α, i.e., for $\alpha \gg 2\pi NCx$. Conversely, for $\alpha < 2\pi NCx$ the function is similar to a gaussian.

You may easily verify that the half-width of the distribution (1.49) is given by $\alpha_{1/2} = (2^{2/3}-1)^{1/2} 2\pi NCx$, i.e., strictly proportional to the target thickness x. On the other hand, eq. (1.33) predicts a r.m.s. scattering angle proportional to $x^{1/2}$. This has a similar origin as the difference between peak and mean energy loss found in the previous paragraph.

1.6.5. Diffusion Approximation in Small–Angle Scattering. The diffusion approximation in small-angle multiple scattering is found in analogy to eq. (1.45). It is most convenient to start at the vectorial form eq. (1.28) for $\sigma(k)$. For azimuthal symmetry in the cross section, the terms of first and third order in $\underline{k}$ vanish, so that

$$\sigma(k) = \frac{1}{4} k^2 \int_0^\infty \phi^2 \, 2\pi\phi d\phi \, K(\phi)... \tag{1.50}$$

After insertion of (1.50) into (1.27), integration is straightforward and yields

$$F(\alpha,x) = \frac{1}{4\pi Dx} \exp[-\frac{\phi^2}{4Dx}] , \tag{1.51}$$

where

$$D = \frac{1}{4} N\int\phi^2 \, 2\pi\phi d\phi \, K(\phi) \tag{1.52}$$

is a generalized diffusion coefficient with the dimension of an inverse
length. You may encounter problems when working with (1.52) that were mentioned already in sect. 1.5.4.

1.7. Discussion

The importance of Poisson statistics in particle penetration was recognized very early on. Poisson's original work from 1837 was rediscovered by
Smoluchowski (1904) and Bateman (1911) in connection with the theory of density fluctuations in an ideal gas and the laws of radioactive decay. Herzfeld
(1912) and Schweidler (1910) applied it to the statistics of ionization events, and Flamm (1914,1915) and Bohr (1913,1915) carried it on to the theory
of energy loss.

Eq. (1.12b) is presumably due to J.J.Thomson (1912) while (1.12c) dates
back to Bohr (1915). Eq. (1.10) is more recent. Although the general form of
the equation was established by Bothe (1921), and applied to the more complex
problem of multiple scattering, the specific application to energy loss was
made much later by Landau (1944) who followed a different route. The argument

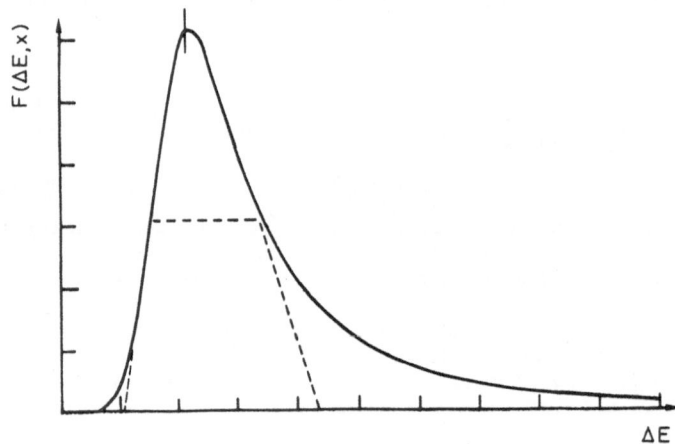

Fig.8.   Schematic illustration of the gaussian portion of the energy loss
           distribution and the single-collision tail. From Williams (1929)

applied here to derive the Bothe-Landau formula seems to go back to Moyal
(1955).

An application of the Bothe-Landau equation in the theory of radiation
effects was given some years ago (Sigmund & Gras-Marti,1981). It was shown
that the theory of charged particle stopping allows a one-to-one mapping on
the theory of atomic mixing by ion beams, with the energy loss spectrum corresponding to a relocation spectrum and the path length to bombarding-ion
fluence. Several available explicit solutions of the Bothe-Landau equation
(Lindhard & Nielsen,1971) were demonstrated to be relevant in a quantitative
sense.

The physics of energy loss straggling and multiple scattering was developed quite far by Williams (1929-1945) and Bohr (1948). The qualitative
picture will be explained for the energy loss but is very similar for multiple scattering. If the target is thin, the projectile undergoes very few collisions. Because of the strong peak of Rutherford's cross section at small T,
the most probable energy loss will be small, but because of the comparatively
slow, power-like decrease toward large T, there will be a noticeable tail

up to around the maximum energy transfer $2mv^2$ that is compatible with the conservation laws. With increasing thickness, the most probable energy loss increases, and so does the variance expressed by eq. (1.38). The maximum energy transfer in a single collision, on the other hand, does not increase with thickness. At thicknesses where the width Ω becomes comparable to or greater than $2mv^2$, the power-like tail at large T will gradually be overshadowed by a gaussian-like behavior from many low-T events (fig. 8).

The above picture is fully confirmed by fig. 7. The physical picture was utilized by Williams to construct energy loss spectra as a function of target thickness by dividing the accessible energy range into a single- and a multiple-scattering regime. With the advent of computers, one may refrain from going through the details of this matching procedure, but its qualitative content remains illuminating and will be mentioned repeatedly.

Lecture 2. VARYING PROJECTILE ENERGY; CHARGE-STATE STATISTICS; CORRELATED DISTRIBUTIONS

In this lecture, the central assumption of an ideal gas target will be kept, but noticeable energy loss will be allowed for, and the assumption of frozen projectile state (charge and/or excitation state) will be relaxed. Also an example of a correlated distribution of two statistical variables (energy loss and charge state) will be considered.

2.1. Allowing for Moderate Energy Variation

With increasing travelled pathlength a projectile will slow down, and the variation of the cross section along the trajectory will become noticeable. Since the energy loss spectrum will depend explicitly on the initial energy E, the convolution equation (1.3) must account for this and reads

$$F(E,\Delta E,x+y) = \int d\Delta E' \ F(E,\Delta E',y) \ F(E-\Delta E',\Delta E-\Delta E',x). \tag{2.1}$$

The rather innocent-looking additional variable $E-\Delta E'$ in the last factor complicates the matter considerably, in that it prevents eq. (2.1) from factorizing in Fourier space as was the case with eq. (1.5).

One way of getting around this complication is to replace the actual, stochastically distributed energy $E-\Delta E'$ by the *average* energy $f(E,y)$ attained by the beam after a penetrated pathlength y, starting from an energy E. This approximation appears appealing for moderately thick targets: For eq. (1.3) to be valid, the *average energy loss* had to be small; now it is only the *fluctuation* that should be negligible compared with the instantaneous energy. With this, eq. (2.1) reduces to

$$F(E,k,x+y) = F(E,k,y) \ F(f(E,y),k,x) \tag{2.2}$$

in Fourier space. This equation is satisfied by the expression

$$F(E,k,x) = \exp[-N \int_0^x dx' \sigma(f(E,x'),k)], \tag{2.3}$$

where

$$\sigma(E,k) = \Sigma \ T_i \ \sigma_i(E) \ [1 - e^{-ikT_i}], \tag{2.4}$$

and $\sigma_i(E)$ is the cross section for energy transfer T_i at a projectile energy E. The continuum equivalent to $\sigma_i(E)$ is $d\sigma(E,T) = K(E,T)dT$. In verifying (2.2) by insertion into (2.1), you will need the relation

$$f(f(E,y),x') = f(E,y+x'),$$

which is evident from the definition of $f(E,x)$. Eq. (2.3) expresses a convolution of energy loss spectra from successive layers of matter, each thin enough to justify the assumption of constant projectile energy.

The average energy degrades according to the *stopping power*,

$$dE/dx = - NS(E),\qquad\qquad\qquad\qquad (2.5)$$

where

$$S(E) = \Sigma\ T_i\ \sigma_i(E)\ \text{or}\ = \int dT\ T\ K(E,T).$$

The differential equation (2.5) can be integrated and yields

$$x = \int_{f}^{E} \frac{dE'}{NS(E')}.\qquad\qquad\qquad\qquad (2.6)$$

This relates the path length x to the average energy $f = f(E,x)$. The relation is not exact, but it is quite accurate for small and moderate straggling. A more precise evaluation will be given shortly.

2.1.1. Effective Thickness and Effective Energy. A spectrum calculated from eq. (2.3) may be approximated by a spectrum evaluated from the Bothe-Landau formula, if either the real target thickness x is replaced by an effective thickness x_{eff} or the initial energy E by an effective energy E_{eff}. These quantities may be defined according to eq. (2.3),

$$\int\ dx'\sigma(f(E,x'),k) = x_{eff}\ \sigma(E,k) = x\ \sigma(E_{eff},k).$$

Since $f(E,x) = E - \langle\Delta E\rangle \equiv E - NxS(E)$, you find

$$E_{eff} \cong E - \langle\Delta E\rangle/2 \qquad\qquad\qquad\qquad (2.7a)$$

and

$$x_{eff} \cong x\ [1 - \frac{\langle\Delta E\rangle}{2}\ \frac{\partial}{\partial E}\ \log\ \sigma(E,k)].\qquad\qquad (2.7b')$$

Eq. (2.7b') is useful when $\sigma(E,k)$ factorizes into an E-dependent and a k-dependent part. If the energy dependence goes as some power $E^{-\alpha}$, cf. eq. (1.36) where $\alpha = 1$, (2.7b') reduces to

$$x_{eff} = x\{1 + \alpha\langle\Delta E\rangle/2E\}.\qquad\qquad\qquad (2.7b)$$

It is advisable to apply either of these relations routinely in the analysis of stopping and multiple scattering measurements. Eq. (2.7a) is most suited for energy loss, while (2.7b) is superior for multiple scattering.

2.2. Transport Equations

With increasing energy loss, the scheme described in the previous paragraph must eventually break down. A rigorous scheme to fully incorporate the stochastic character of the energy loss may be found by going back to eq. (2.1), which was exact within the basic assumptions of a dilute target and independent events. When you allow for substantial energy variation, the

92

description becomes more transparent if the *final energy* E' rather than the *energy loss* ΔE is utilized as a variable. In this notation, eq. (2.1) reads

$$F(E,E',x+y) = \int dE'' \, F(E,E'',y) \, F(E'',E',x) .\qquad (2.1')$$

Now, assume y to be small enough so that the chance for more than one collision *in the first layer* is negligible. Then, $F(E,E'',y)$ is given by eq. (1.7), except that the projectile energy needs to be stated explicitly in $P_i(E) = Ny\sigma_i(E)$. Insert this into (2.1'), and keep only terms up to first order in y. Then, eq. (2.1') reads

$$-\frac{\partial}{\partial x} F(E,E',x) = N \sum_i \sigma_i(E) \, [F(E,E',x) - F(E-T_i,E',x)] .\qquad (2.8)$$

This is a differential equation in x and a difference equation in E. It allows you to determine the energy loss spectrum subject to the initial condition

$$F(E,E',0) = \delta(E - E'),\qquad (2.9)$$

provided that the loss quanta T_i and associated cross sections $\sigma_i(E)$ are known.

Before proceeding, let us do a similar calculation, starting from eq. (2.1') again, but assuming now that the second layer (thickness x) rather than the first one (thickness y) be infinitesimally thin. Applying the same steps, and afterwards renaming y to x you find

$$-\frac{\partial}{\partial x} F(E,E',x) = N \sum_i [\sigma_i(E') F(E,E',x) - \sigma_i(E'+T_i) F(E,E'+T_i,x)]. \qquad (2.10)$$

This looks very similar to (2.8). The most visible difference is in the energy variable: In eq. (2.8), it is the initial energy E that varies while the exit energy E' is a dummy variable; in eq. (2.10), the reverse is true.

Eq. (2.10) is called a *master equation* in statistical physics. In continuum notation, it reads

$$-\frac{\partial}{\partial x} F(E,E',x) = N\int dT \, [K(E',T) F(E,E',x) - K(E'+T,T) F(E,E'+T,x)], \qquad (2.11)$$

where $K(E,T)dT$ is the differential cross section in continuum notation. In this form, the equation is called a linear transport (or Boltzmann) equation of the *forward* type. Conversely, eq. (2.8), or its continuum version,

$$-\frac{\partial}{\partial x} F(E,E',x) = N \int dT \, K(E,T) \, [F(E,E',x) - F(E-T,E',x)] , \qquad (2.12)$$

is called a transport (or Boltzmann) equation of the *backward* type. In the forward equation, the *instantaneous* energy is the variable, while in the backward equation it is the *initial* energy.

For homogeneous media, the two formulations yield equivalent results. I find it more convenient to work with the backward equation when that is possible. A nice feature of the backward equation is the possibility to reduce it to an equation for a reaction yield: Assume you are interested in some function $I(E,x) = \int dE' F(E,E',x) \Sigma(E')$, where $\Sigma(E')$ expresses some reaction rate at energy E', a cross section for a nuclear reaction, an ionization cross section, or alike. All you need to do is multiplying (2.12) by $\Sigma(E')$ and integrate, and you have an equation that determines precisely the quantity you are interested in, no more, no less. This kind of shortcut has proved useful in the treatment of radiation effects, where solutions of the full transport equation are hard to obtain and would produce redundant information.

However, the backward equation is *nonlocal*, in that the energy variable appearing in it is not the energy at depth x. Therefore, the backward equation is also called a propagator equation (Lindhard & Nielsen,1971). Problems arise when you deal with an inhomogeneous medium such as a multilayer target.

Transport equations of forward or backward type, together with the Monte Carlo technique to be described below, have traditionally been the main tool in the statistics of particle penetration (Sigmund,1983; Huang et al.,1985). One of my goals in these lectures is to demonstrate that they are not the only powerful tool in the field.

2.2.1. An Application in Electron Spectroscopy. Transport equations have proved useful in the analysis of photoelectron and Auger spectra from solids (Tougaard & Sigmund,1982; Tougaard,1990a,b). As an example, look at the spectrum of electrons reflected from a semi-infinite solid. As was mentioned in Tougaard's lectures, electrons undergo a series of elastic (nuclear) scattering events before emerging from the target with energies (E',dE'), and their distribution in path length may be approximated by an exponential, $\exp(-x/L)d(x/L)$. The observed energy loss spectrum is then given by

$$G(E,E') = \frac{1}{L} \int_0^\infty dx \, e^{-x/L} F(E,E',x) \ .$$

From this and (2.8) you find an equation for $G(E,E')$ by integration,

$$G(E,E') = \frac{\lambda(E)}{L + \lambda(E)} [\delta(E'-E) + NL \sum_i \sigma_i(E) \, G(E-T_i,E')] \ , \qquad (2.13a)$$

where $\lambda = [N\Sigma\sigma_i]^{-1}$ is the mean free path. Alternatively, from (2.10) you find

$$G(E,E') = \frac{\lambda(E')}{L + \lambda(E')} [\delta(E'-E) + NL \sum_i \sigma_i(E'+T_i) \, G(E,E+T_i) \ . \qquad (2.13b)$$

These are generalizations of a relation derived by a different procedure in Tougaard's lectures. They allow to experimentally determine the parameters L, $\lambda(E)$, and $\sigma_i(E)$. The analysis becomes particularly transparent in the limit of small energy loss, $|E-E'| \ll E$ where λ and σ_i may be taken independent of E or E'. When that condition is not fulfilled, also the dependence of L on E or E' must be considered. Then, straight integration of (2.8) is not possible, and (2.13a) becomes invalid. The forward equation (2.13b), on the other hand, remains valid if L is replaced by L(E) (sic!).

2.3. Energy Moments

Analytic solutions of eq. (2.11) or (2.12) have been found for very special forms of the cross section only. General methods of attack are needed in order to arrive at solutions for realistic cross sections. One of them goes over the energy moments

$$M_n(E,x) = \int_{-\infty}^{\infty} dE'\ E'^n\ F(E,E',x).$$

Integration of the discrete backward equation, eq. (2.8), yields

$$- \frac{\partial}{\partial x} M_n(E,x) = N \sum_i \sigma_i(E)\ [M_n(E,x) - M_n(E-T_i,x)] \qquad (2.14)$$

with the initial condition (2.9)

$$M_n(E,0) = E^n . \qquad (2.15)$$

Eq. (2.14) can be solved for each n separately. The zero'th moment M_0 is the normalizing integral $M_0(E,x) = 1$ which expresses particle conservation. The first moment, $M_1(E,x)$, is the average energy $f(E,x)$ introduced above. With this notation, we may rewrite (2.14) for n = 1 in the form

$$- \frac{\partial}{\partial x} f(E,x) = N \sum_i \sigma_i(E)\ [f(E,x) - f(E-T_i,x)] . \qquad (2.16)$$

This equation provides a precise determination of the mean energy versus path length. The result derived in eq. (1.12b) emerges from (2.16) in the limit of negligible dependence of σ_i on E: Then, (2.16) is solved by $f(E,x) = E - NxS$, as it must be. Similar considerations apply to straggling.

2.4. Continuous–Slowing-Down Approximation

The simplest solution of eq. (2.8) is found by expanding the term $F(E-T_i,E',x)$ in powers of T_i and dropping all but the zero'th and first term. This so-called continuous-slowing-down approximation assumes that the energy loss comes in very small bits, $T_i \ll E$. The procedure is a simplification of the one utilized above to derive the diffusion approximation, cf. eq. (1.24); the latter includes one more term which allows for straggling.

In this approximation, eq. (2.8) reduces to

$$- \frac{\partial}{\partial x} F(E,E',x) = N\ S(E)\ \frac{\partial}{\partial E} F(E,E',x) \qquad (2.17)$$

with the solution

$$F(E,E',x) = \delta\left(x - \int_{E'}^{E} \frac{dE''}{NS(E'')}\right) \frac{1}{NS(E')} , \qquad (2.18)$$

as is verified by differentiation. The integral occurring in the argument of the Dirac function is identical with the one determining the energy versus path length in eq. (2.6), expressing the relation $E' = f(E,x)$ with $f(E,x)$ given in the approximation determined by eq. (2.6).

2.5. Varying Projectile State

I shall now go over to a seemingly very different subject. If the projectile is an ion, it may lose or capture electrons in collisions with the target atoms, and it may be excited or deexcited. One may distinguish between instantaneous processes determined by cross sections, such as excitation and capture, and delayed processes such as deexcitation and Auger cascades.

Consider projectile states $I = 0,1,2...$ and only instantaneous processes. Let σ_{IJ} be the cross section for a collision event in which the projectile state changes from I to J. Take a gas target thin enough to make the energy loss negligible compared to the projectile energy. This makes sense: Cross sections for charge exchange may be quite large, and a projectile may undergo many charge-changing collisions while losing only a small fraction of its energy.

Let the incoming beam particles all be in a well-defined state I and the probability for a projectile to emerge in state J after a penetrated thickness x be called $F_{IJ}(x)$. Then, similar to eq. (1.3), you may write down a convolution equation involving two layers of thickness y and x,

$$F_{IJ}(x+y) = \sum_L F_{IL}(y) \, F_{LJ}(x) \ . \tag{2.19}$$

This has the form of a matrix equation for the transfer matrix $\underline{\underline{F}}(x) = (F_{IJ}(x))$. Double-underlined quantities will always denote matrices. In matrix notation, (2.19) reads

$$\underline{\underline{F}}(x+y) = \underline{\underline{F}}(y) \, \underline{\underline{F}}(x) . \tag{2.20}$$

This is formally identical with eq.(1.5) except that $\underline{\underline{F}}$ is a matrix. The solution can be expressed by

$$\underline{\underline{F}}(x) = \exp(x\underline{\underline{C}}) \tag{2.21}$$

with some unknown matrix $\underline{\underline{C}}$. If you feel uncomfortable with a matrix in an exponent, just consider (2.21) as an abbreviation for an exponential series. This reduces the exponential to a sum of matrix products which are all well defined. The matrix $\underline{\underline{C}}$ may be determined from the thin-layer limit, just as the function $C(k)$ occurring in (1.6). Assume x small enough so that the probability for more than one change-of-state collision is negligible. Then,

$$F_{IJ}(x) = \delta_{IJ} [1 - Nx \sum_L \sigma_{IL}] + Nx \, \sigma_{IJ} \ , \qquad x \text{ small.} \tag{2.22}$$

Here, the first term on the right-hand side expresses the probability for the projectile to remain in the initial state I, and the second term expresses the probability for a transition from state I to state J; use has been made of eq. (1.2) applied to the cross sections σ_{IJ}. δ_{IJ} is the Kronecker symbol: $\delta_{IJ} = 1$ for I=J and $= 0$ for I≠J. Note that the unit matrix is $\underline{\underline{1}} = (\delta_{IJ})$.

Comparison of (2.22) with the limit of (2.21) for small x shows that

$$C_{IJ} = N(\sigma_{IJ} - \delta_{IJ} \sum_L \sigma_{IL}) \ . \tag{2.23}$$

Eq. (2.23) suggests introduction of a diagonal matrix $\underline{\underline{\rho}}$ with the elements

$$\rho_{IJ} = \delta_{IJ} \sum_L \sigma_{IL} \ . \tag{2.24}$$

With this, the transfer matrix $\underline{F}$ can finally be written in the form

$$F_{IJ} = (e^{Nx(\underline{\underline{\sigma}} - \underline{\underline{\rho}})})_{IJ} \, , \qquad\qquad (2.25)$$

where $\underline{\underline{\sigma}}$ is the matrix made up by the cross sections σ_{IJ}.

Eq. (2.25) comprises a solution in closed form of the charge-state distribution by instantaneous processes in the limit of negligible energy loss. In the literature, such distributions have been calculated by an alternative procedure which will be sketched briefly. Following the procedure applied in sect. 2.2, take the limit of small y in eq. (2.19) and insert eq. (2.22) for $F_{IL}(y)$. Then, you find a master equation of the backward type,

$$- \frac{\partial}{\partial x} F_{IJ}(x) = N \sum_{L} \sigma_{IL} \, [F_{IJ}(x) - F_{LJ}(x)] , \qquad\qquad (2.26)$$

with I running over all states and J fixed. The corresponding forward equation can be found by the same procedure but taking the limit of small x in eq. (2.19). After renaming y to x one finds

$$- \frac{\partial}{\partial x} F_{IJ}(x) = N \sum_{L} [F_{IJ} \, \sigma_{JL} - F_{IL} \, \sigma_{LJ}] , \qquad\qquad (2.27)$$

where now the coupling goes over the final charge state. With the initial state prescribed, (2.27) can readily be solved by Laplace transform or by straight numerical integration (Betz,1972).

Without going into details I like to mention that schemes to estimate *equilibrium* charge state distributions directly are available (Goscinski et al.,1982; Blomberg et al.,1986).

While the solution (2.25) is comprehensive, it does not appear particularly transparent. You may find it convenient to evaluate (2.25) by computer algebra, and some features may also be determined analytically. I am looking into this currently.

2.6. The Two-State Problem

It may be instructive to evaluate eq. (2.25) specifically for the two-state problem, where I takes on only the values 1 and 2, and $\underline{\underline{\sigma}}$ has only two nonvanishing elements σ_{12} and σ_{21}. Clearly, σ_{11} and σ_{22} vanish by definition: It does not make sense to introduce a cross section σ_{II} for an "event" that does not produce an observable change. Collecting elements of $\underline{\underline{\rho}}$ and $\underline{\underline{\sigma}}$ you find that

$$\underline{\underline{\sigma}} - \underline{\underline{\rho}} = \begin{bmatrix} -\sigma_{12} & \sigma_{12} \\ \sigma_{21} & -\sigma_{21} \end{bmatrix} . \qquad\qquad (2.28)$$

Multiplication of this matrix by itself leads to

$$(\underline{\underline{\sigma}} - \underline{\underline{\rho}})^2 = - (\sigma_{12} + \sigma_{21}) \, (\underline{\underline{\sigma}} - \underline{\underline{\rho}})$$

and hence,

$$(\underline{\underline{\sigma}} - \underline{\underline{\rho}})^{n+1} = (-)^n \, (\sigma_{12} + \sigma_{21})^n \, (\underline{\underline{\sigma}} - \underline{\underline{\rho}}) . \qquad\qquad (2.29)$$

Collecting the terms in the series for the exponential (2.25) yields

$$e^{Nx(\underline{\underline{\sigma}} - \underline{\underline{\rho}})} = \underline{\underline{1}} + \frac{1}{\sigma_{12}+\sigma_{21}} (\underline{\underline{\sigma}} - \underline{\underline{\rho}}) [1 - e^{-Nx(\sigma_{12}+\sigma_{21})}] . \qquad (2.30)$$

Finally, after collecting constant and exponential terms, respectively, you end up with

$$\begin{bmatrix} F_{11} & F_{12} \\ F_{21} & F_{22} \end{bmatrix} = \frac{1}{\sigma_{12}+\sigma_{21}} \left[\begin{bmatrix} \sigma_{21} & \sigma_{12} \\ \sigma_{21} & \sigma_{12} \end{bmatrix} + \begin{bmatrix} \sigma_{12} & -\sigma_{12} \\ -\sigma_{21} & \sigma_{21} \end{bmatrix} e^{-Nx(\sigma_{12}+\sigma_{21})} \right] , \qquad (2.31)$$

which is wellknown from the literature. The four functions F_{IJ} have been drawn up schematically in fig.9. It is seen that for large penetrated path-length x, the charge-state population becomes independent of the initial charge state.

2.7. Monte Carlo Schemes

The stochastic nature of the collision processes described so far makes particle penetration phenomena well suited for numerical treatment by Monte Carlo simulation. In fact, particle penetration has been one of the first major applications of computer simulation in physics. An early review was written by Berger (1963). This type of simulation is particularly useful as a tool in solving problems like some of those discussed in this lecture: Al-though analytical schemes are available, the numerical effort in the solution may become comparable to that of a straight simulation.

In principle, Monte Carlo schemes are supposed to simulate exactly the same physics as the analytical schemes discussed above. Whether or not this holds for any given simulation may be hard to verify, but a few characteri-stic features will be mentioned here.

Most Monte Carlo simulations make heavily use of the mean free path $\lambda(E)$, a concept that has barely been mentioned up till now. You ought to know the definition,

$$\lambda(E) = \frac{1}{N \sigma(E)} , \qquad (2.32)$$

where

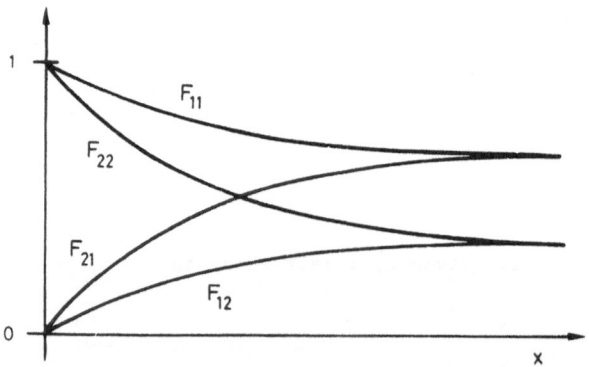

Fig.9   Transition probabilities for projectile states versus penetrated thickness; only 2 accessible projectile states, cf. eq. (2.31)

$$\sigma(E) = \sum_i \sigma_i(E) \qquad\qquad (2.33)$$

is the *total cross section*, i.e., the cross section for all possible collisions under consideration.

In a typical Monte Carlo simulation, a random number generator determines a free-flight path (x',dx') by the probability

$$e^{-x'/\lambda(E)}\, dx'/\lambda(E), \qquad\qquad (2.34)$$

which follows by application of eq. (1.16) for n = 0. Then the dice are thrown again, in order to determine what kind of collision is going to happen from the given variety. The probability for an event i is

$$\sigma_i(E)/\sigma(E) = N\lambda\sigma_i(E). \qquad\qquad (2.35)$$

With the choice of T_i having been made, the projectile is assigned a new energy $E-T_i$ and, more generally, a new direction of motion, state, etc. in accordance with the collision event i, and allowed to move freely until the next collision which is again governed by (2.34) and (2.35). This procedure is iterated as often as necessary, and repeated for a large number of projectiles -- and possibly target particles -- until adequate statistics is reached.

It is of interest to formulate the above procedure analytically. Consider the energy spectrum F(E,E',x) at a penetrated pathlength x. If the projectile makes its first collision at a pathlength x' > x, the energy E' at x will still be E. Conversely, if the first collision takes place at some pathlength x' < x, the particle will then have some energy $E - T_i$, the probability density for which is given by the collision probability (2.35). Therefore,

$$F(E,E',x) = e^{-x/\lambda(E)}\, \delta(E'-E) + \qquad\qquad (2.36)$$

$$+ \int_0^x \frac{dx'}{\lambda(E)}\, e^{-x'/\lambda(E)}\, \sum_i \frac{\sigma_i(E)}{\sigma(E)}\, F(E-T_i,E',x-x').$$

You may convince yourself that eq. (2.36) is equivalent to (2.8), i.e., the backward Boltzmann equation, by replacing $x' \to x-x'$ in the integral on the right-hand side, multiplying the whole equation by $\exp(x/\lambda(E))$, differentiating with respect to x and observing (2.32). Relations of the type of eq. (2.36) are alternative forms of transport equations which require reasonably large mean free paths in order to be of practical use. They have in particular been in use in neutron transport theory.

Note that unlike (2.8), eq. (2.36) contains *both* a summation over i and an integration over x. This is a manifestation of the fact that the random number generator is consulted *twice* in each step, first to find the travelled path length, and secondly to determine the type of collision. Simulation codes have also been developed that avoid a free-path distribution and, therefore, utilize the random number generator only once in each step, namely to determine the type of collision.

2.7.1. The Mean Free Path: Reality or Artefact? The significance of the total cross section and hence the mean free path in charged-particle penetration depends very much on the experimental situation. Because of the long range of the Coulomb interaction, the mean free path is generally very small. In fact, if free-Coulomb scattering were to apply universally, eq. (1.36) would imply $\lambda = 0$ at any projectile energy. This would hold for both energy loss and angular scattering. This feature is central and requires attention, both from a physics point of view and a computational one.

From a physics point of view, the problem is intimately related to experimental resolution. Consider the slowing-down of fast electrons such as in x-ray photoelectron or Auger spectroscopy (Tougaard,1990). The experimental resolution is in the 0.1 eV range, i.e., the main features of the electronic energy loss spectrum are readily resolved: By looking at a measured spectrum, you may be able to identify a fraction of electrons that have not suffered electronic energy loss. This implies that there is a well-defined mean free path, governed by electronic collisions. How is this possible for Coulomb scattering? Well, eq.(1.36) holds for *binary free-Coulomb* scattering: In a solid, electronic collisions are neither binary nor free.

Your experimental resolution may actually be in the meV region so that you may resolve energy losses to phonons (Lucas,1990). If so, the fraction of electrons that have not suffered energy loss will be smaller, and hence the mean free path shorter. Clearly, the total cross section must be the sum of the cross section for electronic and nuclear energy loss.

Thus, you arrive at the conclusion that in electron spectroscopy, the mean free path for energy loss is a meaningful concept. Its magnitude depends on the experimental resolution, but that dependence goes, so to speak, in steps.

For heavy charged particles, the energy resolution is generally poorer than in electron spectroscopy, but it depends on the energy and mass of the projectile and, of course, on the apparatus. For high-energy protons and alpha particles you may deal with resolutions in the keV range. Since the electronic energy loss spectrum in ordinary matter invariably has a pronounced peak in the lower eV region, the major portion of the events contributing to stopping are unresolved. In this situation, it does not make sense to distinguish between a projectile that has lost energy and one that has not. Mainly the gross features of the energy loss spectrum are important, mean value and straggling, peak energy loss and halfwidth. You may define a mean free path in terms of those energy losses which you may resolve, but this quantity has little physical meaning: If you improve the energy loss resolution of your apparatus by a factor of two, your mean free paths attain about half their original value, as follows from eq. (1.36) which is quite accurate for energy losses in the keV range.

Similar conclusions may be reached for low-speed heavy charged particles where the energy loss is dominated by nuclear stopping: Pertinent interaction forces have long ranges, and therefore the cross sections -- which follow from classical dynamics -- look similar to (1.36) with a pronounced divergence at $T = 0$. Discrete features enter in the meV range, i.e., near the threshold for phonon generation, which is usually several orders of magnitude below experimental resolution.

So much about physics. What about computation? Standard Monte Carlo techniques try to simulate eq. (2.36), and hence the existence of a nonzero mean free path is crucial. This is usually ensured by introduction of a cut-off energy T_C so that only energy losses $T \geq T_C$ enter. T_C is made small enough by trial and error to make the solutions insensitive to the precise choice of T_C.

You may actually prove that the solution of eq. (2.8) does not depend on T_c in the limit of small T_c. This is done by expanding the energy loss spectrum in powers of T_i, similar to what was done in eq. (1.45). Then, (2.8) reads

$$- \frac{\partial}{\partial x} F(E,E',x) = - N \sum_{\nu=1}^{\infty} \frac{(-)^{\nu}}{\nu!} Q_{\nu}(E) (\partial/\partial E)^{\nu} F(E,E',x) \qquad (2.37)$$

where

$$Q_{\nu}(E) = \sum T_i^{\nu} \sigma_i(E) \rightarrow \int T^{n} d\sigma(E,T) . \qquad (2.38)$$

You see that the zero'th moment Q_0, i.e., the total cross section, drops out altogether. No cut-off is necessary if only the first moment, i.e., the stopping cross section, is regular in $T = 0$.

This demonstrates a fundamental weakness of many Monte Carlo simulations in charged particle penetration: In order to get accurate results, you need to choose a small value of T_c. This implies that much computational power is spent in very soft collisions. Those are frequent but contribute to the energy loss spectrum mainly through the stopping cross section $Q_1(E) \equiv S(E)$, and less and less through the higher moments $Q_n(E)$.

The conclusion should be clear: A mean free path introduced by computational considerations only is an artefact that generates excessive computation times and should be avoided altogether.

A proper alternative is Monte Carlo computation simulating eq. (2.10). Here, the random number generator is used only once in each step, and the free path length is selected not randomly but in accordance with physical requirements: It should be smaller than the desired spatial resolution, and the probability for pertinent collision events should be $\ll 1$. You still need to introduce a cut-off T_c in order to define "pertinent collision events", i.e., events with $T_i > T_c$, but events with $T < T_c$ may now be included by way of a continuous friction-like term. Within these constraints, you have much freedom to vary the flight path as a function of energy and in accordance with target geometry: Just imagine the problems you encounter with a path length distribution like (2.34) when your target contains precipitates, bubbles, voids, or just a planar surface. These problems evaporate if you work with a flexible free path.

One of the more frequently used Monte Carlo codes (Biersack and Haggmark,1980) operates on the basis of a fixed free flight path Δx. That quantity is linked to the cutoff energy T_c by the requirement

$$N x \sigma_c(E) = \int_{T_c}^{\infty} K(E,T) dT = 1. \qquad (2.39)$$

Evidently, this model does not strictly simulate eq. (2.12) since energy losses $T < T_c$ are ignored, and since the collision probability on Δx is not $\ll 1$ as it should in a random medium. As a matter of fact, the code is supposed to simulate collisions in solids rather than in gaseous media.

With due changes in the wording, almost everything said about energy loss in this section also applies to angular scattering.

Why are truncation problems discussed in connection with the Monte Carlo scheme? Are they avoided in the analytical scheme? Well, yes and no. Consider eq. (1.7). Here the factor in square brackets is a complementary probability, i.e., the probability for no collision to happen during passage through a path segment x. If the total cross section is large, that probability may become negative. Note, however, that in an analytical calculation there is no problem to make x or, more precisely Nx, arbitrarily small: That probability serves only to define an initial condition, not a collision spectrum over a finite distance. Therefore, unless $\sigma(E)$ literally diverges, there should not be a problem. Physical cross sections are never literally divergent although they may be very large. If so, many-particle interactions must become significant, but that is another story.

2.8. Correlation of Energy Loss with Projectile State

Up till now we have looked into statistical distributions of one variable at a time, i.e., energy loss, angular deflection, or charge state. There are numerous kinds of correlated distributions in more than one statistical variable, some of which are conveniently discussed within the present framework that applies to a gas target.

Consider the energy loss spectrum of a projectile in the case where changes in projectile (charge) state are significant. In experiments with heavy ions at velocities where projectiles can be partially stripped, charge-changing collisions are frequent, and the energy loss must depend on the charge state.

Consider the function $F_{IJ}(\Delta E, x)\ d(\Delta E)$, which is the joint probability for a projectile in the initial state I to emerge from a target with thickness x in the final state J, *and* to have suffered an energy loss in the interval $(\Delta E, d(\Delta E))$. Applying the arguments leading to eqs. (1.3) and (2.19) simultaneously, you find that the following relation must be fulfilled

$$F_{IJ}(\Delta E, x+y) = \sum_L \int d(\Delta E')\ F_{IL}(\Delta E', y)\ F_{LJ}(\Delta E - \Delta E', x) \qquad (2.40)$$

for a sandwich target composed of two layers with thicknesses y and x. As in sect. 2.5, the energy loss is assumed small enough so that all cross sections may be taken independent of projectile energy.

In Fourier space, following eq. (1.5), eq. (2.40) reads

$$F_{IJ}(k, x+y) = \sum_L F_{IL}(k, y)\ F_{LJ}(k, x). \qquad (2.41)$$

This is a matrix equation with the solution

$$F_{IJ}(k, x) = (e^{x\underline{\underline{C}}(k)})_{IJ} \qquad (2.42)$$

for the matrix $\underline{\underline{F}}(k, x) = (F_{IJ}(k, x))$.

In order to determine the unknown matrix $\underline{\underline{C}}(k)$ we need a physical model for the collision events. You may operate with a cross section σ_{IJi} for collisions that lead to a change of state from I to J accompanied by some energy loss quantum T_i. Then, you find

$$F_{IJ}(\Delta E, x) = [1 - Nx \sum_L \sum_i \sigma_{ILi}]\ \delta_{IJ}\ \delta(\Delta E) + Nx \sum_i \sigma_{IJi}\ \delta(\Delta E - T_i) \qquad (2.43)$$

for small x by the usual argument, cf. eqs. (1.7) and (2.22). By comparison with the small-x limit of (2.42), this leads to

$$C_{IJ}(k) = - N \sum_i [\delta_{IJ} \sum_L \sigma_{ILi} - \sigma_{IJi} e^{-ikT_i}]. \tag{2.44}$$

The common picture (Vollmer,1974; Winterbon,1977) is to split up the collision events into two categories, charge-changing collisions with negligible energy loss, and energy loss events that do not affect the projectile charge. Adopting this picture you have

$$\sigma_{IJi} = \sigma_{IJ} \delta_{i0} + \delta_{IJ} \sigma_{Ii}, \tag{2.45}$$

where the transition $i = 0$ symbolizes zero energy loss, and σ_{Ti} is the cross section for energy loss T_i for a projectile in charge state I.

Straight application of the procedure leading to eqs. (1.9), (1.10), (2.24), and (2.25) generates the following joint distribution function

$$F_{IJ}(\Delta E, x) = \frac{1}{2\pi} \int dk\, e^{ik\Delta E} (e^{Nx(\underline{\underline{\sigma}} - \underline{\underline{\rho}} - \underline{\underline{I}})})_{IJ}, \tag{2.46}$$

where $\underline{\underline{\sigma}}$ and $\underline{\underline{\rho}}$ are the matrices introduced in eq. (2.25), and

$$\underline{\underline{I}} = \delta_{IJ} \sigma_I(k) = \delta_{IJ} \sum_i \sigma_{Ti} [1 - e^{-ikT_i}]. \tag{2.47}$$

Eq. (2.46), or the more general form based on (2.44) without the approximation (2.45), provides a closed-form solution of the problem of energy loss in gases in the presence of charge-changing collisions.

Numerical solutions, based on a somewhat different scheme and simplifying model assumptions on the dependence of energy loss on charge state are available (Vollmer,1974; Winterbon,1977). Significant pre-equilibrium effects in the stopping power have been reported (Clouvas et al.,1984). At larger thicknesses, mostly the straggling is affected. The latter point will be touched upon in lecture 4.

Lecture 3. SPATIALLY CORRELATED COLLISIONS

In this lecture, the assumption ii) mentioned in the beginning will be relaxed. Up till now, the penetrated target was assumed to be an ideal gas. This implies that successive collisions are statistically independent, as expressed by eq. (1.3) and its various modifications and generalizations. Even though this picture will show up to be justified to some degree for solids, we need to be able to estimate the magnitude of possible errors and to specify conditions under which errors may be pronounced.

In condensed matter, atoms are arranged in some densely packed structure. Ignore lattice vibrations for a moment. Then, the number of target atoms hit by a penetrating particle must be rather well defined once the impact point of the trajectory is given (fig.10). This is in contrast with the ideal gas where the trajectory is stochastic even when the impact point of the projectile is specified. It will turn out shortly that first-order averages like the mean energy loss are fairly insensitive to the structure while higher-order averages like straggling are not.

In the situation illustrated in fig.10, the most obvious cause of fluctuations is the varying impact point of the projectile in a broad beam, to which adds quantum mechanical uncertainty. Although inclusion of thermal and zero-point vibrations induces an element of randomness in the structure, there is no basis for an a priori assumption of complete statistical independence of successive collision events.

3.1. Impact Parameter

The central quantity in spatially correlated collisions is the impact parameter. Although it is a standard concept in classical and semiclassical collision theory, it has only been mentioned briefly in the introduction, but it was not utilized: Collisions in an ideal gas target are fully characterized by cross sections. The impact parameter may be useful in the computation of cross sections, but that is not a topic of the present set of lectures.

In classical collision theory, the impact parameter is the distance of an incoming straight-line trajectory from the target particle (fig.2) in a binary collision between free particles. It is a meaningful concept if a wave

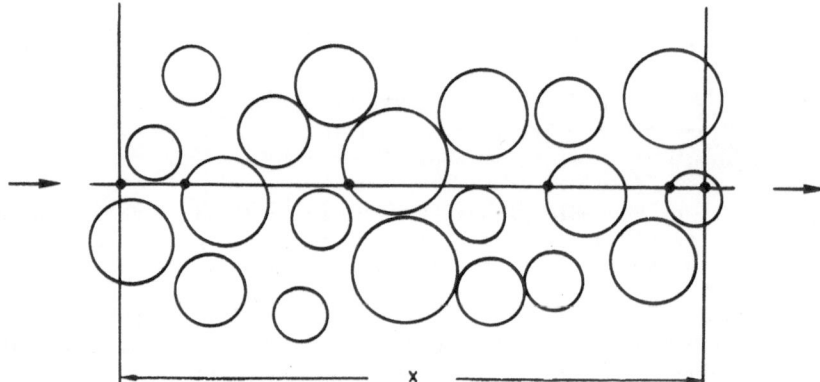

Fig.10. Trajectory of a fast particle in a solid. Angular deflection neglected. Redrawn after Dixmier et al. (1982)

packet can be constructed that follows the classical trajectory with an acceptable angular spread (fig. 11). The conditions for this to be possible have been analysed (Williams,1945; Bohr,1948; Lindhard,1965). For the specific case of free-Coulomb scattering, the central relation reads

$$\kappa = \frac{2|e_1 e_2|}{\hbar v} > 1 \, , \tag{3.1}$$

where e_1 and e_2 are the charges of the collision partners and v the relative velocity. This condition is most easily fulfilled in case of high charge of the target or the projectile, or both. Thus, the scattering trajectory of an ion colliding with a target *nucleus* may be well-defined classically, even though its interaction with the *electrons* of the target atom may be intrinsically quantal. This is the essence of semiclassical collision theory which has applications in many areas such as chemical reaction theory. In this lecture it will be assumed that the impact parameter to the target *nucleus* be well defined.

3.2. Collision Probability and Differential Cross Section

To get started, ignore all angular deflection of the projectile on the target nuclei, and assume the energy loss to be small compared to the incident energy, just as in sect. 1. Introduce a probability $Q_i(\underline{p})$ for energy transfer T_i to a target atom in a *single collision* at an impact parameter $\underline{p}$. The impact parameter is a *vector* in a plane perpendicular to the incoming trajectory (fig. 2), and T_i is an excitation energy of a target atom in the same notation as utilized in lectures 1 and 2.

From the definition of $Q_i(\underline{p})$ you find the cross section σ_i,

$$\sigma_i = \int d^2p \; Q_i(\underline{p}) \; . \tag{3.2}$$

This form is an explicit illustration of the fact that the cross section is a product of a geometric area and a collision probability, as was mentioned in the discussion of eq. (1.1).

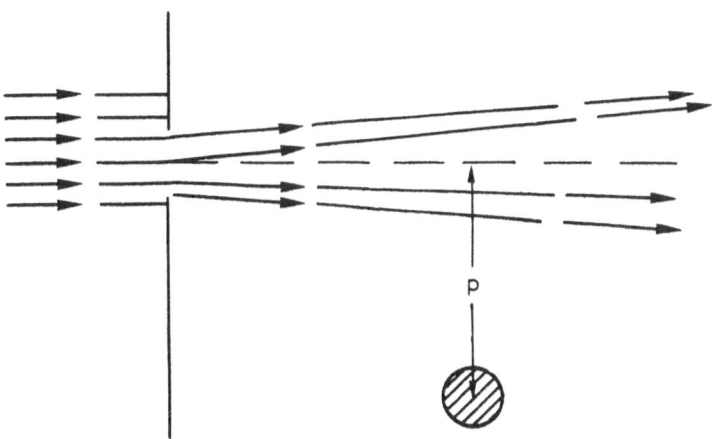

Fig.11.   The range of validity of the classical impact parameter picture is limited by diffraction effects at a virtual aperture defining a narrow beam

3.3. Stopping Cross Section and Straggling Parameter of an Atom

From (3.2) and eqs. (1.13) and (1.14), the stopping cross section and straggling parameter may be evaluated,

$$S = \int d^2p \; \sum_i T_i \; Q_i(\underline{p}) \; = \int 2\pi p dp \; <T(p)> \; , \tag{3.3a}$$

$$W = \int d^2p \; \sum_i T_i^2 \; Q_i(\underline{p}) \; = \int 2\pi p dp \; <T^2(p)> \; , \tag{3.3b}$$

Here, $<T(p)> = \sum_i T_i \; Q_i(\underline{p})$ is the mean energy transfer *in one collision* at impact parameter p, and correspondingly for $<T^2(p)>$. Azimuthal symmetry has been assumed by introducing the conventional impact parameter $p = |\underline{p}|$. In classical scattering on a central-force potential, the energy transfer is uniquely related to the impact parameter, $T = T(p)$; then, eqs. (3.3a,b) reduce to the wellknown relations

$$S = \int 2\pi p dp \; T(p); \qquad W = \int 2\pi p dp \; [T(p)]^2 \; , \qquad \text{(classical).} \tag{3.4}$$

105

The difference between eqs. (3.3) and (3.4) is determined by quantal uncertainty: Note that it is $<T^2(p)>$ and *not* $<T(p)>^2$ that enters into (3.3b).

3.4. Stopping Cross Section and Straggling Parameter of a Diatomic Molecule

A useful system for studying impact parameter correlation is an *ideal gas of diatomic molecules*. Such a medium can be considered in two ways from the present point of view. Either you regard the individual molecule as a target particle characterized by a set of collision cross sections, in which case collisions obey Poisson statistics. Alternatively, you may view the system as made up of *individual atoms that are pairwise correlated,* i.e., pairs of 1-atoms and 2-atoms oriented at random with a fixed interatomic distance d. This assumes that the only difference between a molecule and two separate atoms is the internuclear distance. This cannot be literally true, but the model contains the essentials of the effect we are looking at in the present lecture. Similar arguments have been applied in the theory of deuteron scattering (Glauber,1955). Most of the results reported here were found in previous work by the author (Sigmund,1976).

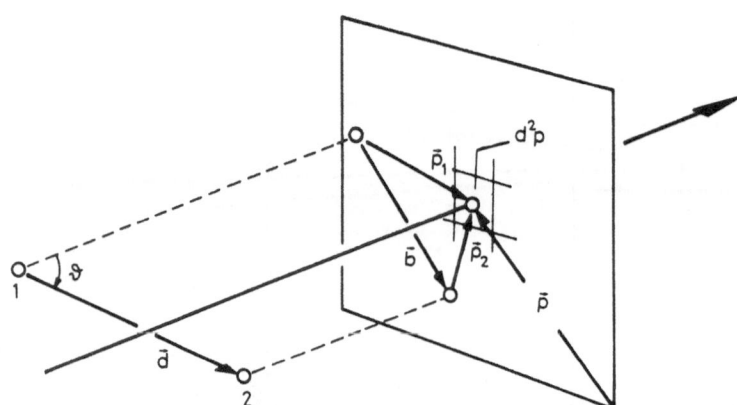

Fig.12.   Definition of vectorial impact parameters
characterizing ion-molecule collision

For simplicity, assume the energy transfer $T_\nu(p_\nu)$ to a target *atom,* ν = 1 or 2, to be uniquely related to the impact parameter $p_\nu = |\underline{p}_\nu|$. For a given orientation of the molecule (fig.12), the energy transfer for a specified geometry is given by

$$T(\underline{p}) = T_1(p_1) + T_2(p_2), \qquad\qquad (3.5)$$

where $\underline{p}$ specifies the intersection of the particle trajectory with the transverse plane, relative to some fixed point relative to the molecule. From (3.5), you find the stopping cross section of a target *molecule*

$$S = \int d^2p\ T(\underline{p}) = \int d^2p\ T_1(p_1) + \int d^2p\ T_2(p_2) . \qquad (3.6)$$

Since the integration over d^2p goes over the entire $\underline{p}$-plane, the integration variable can be replaced by $\underline{p}_1$ in the first and by $\underline{p}_2$ in the second integral. Then, you find the simple result that

$$S = S_1 + S_2 , \qquad\qquad (3.7)$$

106

i.e., the stopping cross section of a molecule is the sum of the stopping cross sections S_1 and S_2 of the constituent atoms. This is called Bragg's rule.

Before going on, let me mention some of the physics underlying this surprisingly simple result. Note first that the physical situation is that of an ideal molecular gas target penetrated by a homogeneous beam. Any given target _atom_ has a given distribution of hits by incoming projectiles, and within the model as specified above, that distribution is unaffected by the presence of another target atom in the molecule. Therefore, the mean energy loss to any given target _atom_, and hence the mean energy loss to the target as a whole, is unaffected by spatial correlation. Note also that the orientation of the molecule drops out in the transition from eq. (3.6) to (3.7). This is consistent with what has been said about the three aspects of the cross section after eq. (1.1).

Nevertheless, two main assumptions need to be fulfilled for (3.7) to hold. Firstly, the _trajectory_ of the projectile must be unaffected by the geometric configuration of the target. In quantitative terms, this means that the impact parameter p_2 must not change significantly by the angular deflection suffered in the collision with atom 1, or vice versa, dependent on which collision is first. For a beam penetrating with little angular deflection, this assumption should be very well justified. Secondly, the electronic energy transfer to a given atom at a given impact parameter must be unaffected by the presence of the other atom in the molecule. That assumption is a priori questionable, since electronic states in molecules differ by necessity from those of otherwise identical, but isolated atoms. Therefore, deviations from Bragg's rule, eq. (3.7), are expected, but they originate in the _electronic structure_ of the target molecule, rather than the _geometric configuration_. In practice, deviations from Bragg's rule are the more pronounced the greater the contribution from the valence electrons to the stopping cross section.

Consider now the straggling parameter of a molecule in the same model,

$$W = \int d^2p \; [T_1(p_1) + T_2(p_2)]^2 \; . \tag{3.8}$$

Going through the same procedure as above, you end up with

$$W = W_1 + W_2 + 2\Delta W \; , \tag{3.9}$$

where

$$\Delta W = < \int d^2p \; T_1(p_1) \; T_2(p_2) > \; . \tag{3.10}$$

Here the brackets indicate an average over all _orientations_ of the target molecules.

It is seen from eq. (3.10) that ΔW is nonnegative if $T_v(p) \geq 0$. This implies that the straggling parameter of a molecule is greater than the sum of the straggling parameters of the constituent atoms. This implies that a beam penetrating through a molecular gas experiences a greater fluctuation in energy loss than when penetrating an atomic gas consisting of otherwise similar _atoms_ at the same average _atom_ density fig. 13), even though the mean energy loss is the same. The physical origin of this is a bunching effect: With target atoms bundled in molecules, there is a good chance to hit two atoms at the same time once the projectile hits an atom at all. Conversely, the chance of penetrating without collision increases, too, since overlapping cross sections of the target atoms leave more empty space in between.

3.5. Rough Estimate of the Correlation in Straggling

In order to get an impression of the significance of ΔW we need some information on the impact parameter dependence of the energy loss to a target *atom*. For a rough estimate, it is sufficient to assume a gaussian dependence

$$T_\nu(p) = \frac{S_\nu}{\pi a_\nu^2} e^{-p^2/a_\nu^2} \qquad \nu = 1,2 , \qquad (3.11)$$

where a_ν is a characteristic distance defining the range of the electronic energy loss for a target atom ν. There has been provided a good theoretical basis for eq. (3.11) in recent years. The point will be discussed in lecture 4. At present, a recent result may illustrate the point (fig.14). At large impact parameters, the energy loss drops off exponentially rather than gauss-ian-like (Bohr,1913). Hence, (3.11) may slightly *underestimate* correlation effects.

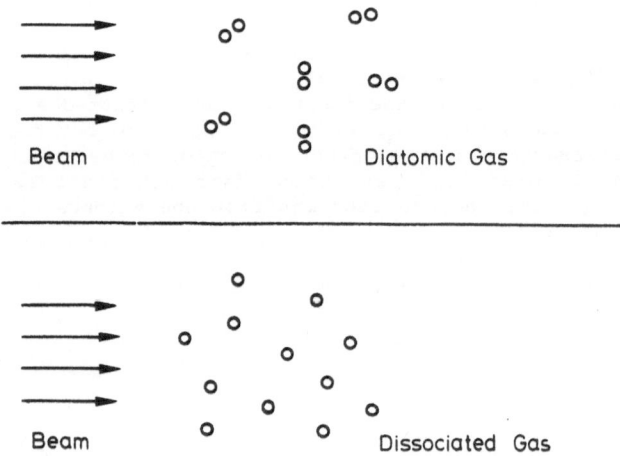

Fig.13. Penetration through an ideal diatomic gas (upper part) and an eqvivalent monoatomic gas of the same atom density

Insertion of (3.11) into (3.10) and integration yields

$$\Delta W = \frac{S_1 \, S_2}{\pi a^2} \langle e^{-b^2/a^2} \rangle , \qquad (3.12)$$

where

$$a^2 = a_1^2 + a_2^2 ; \qquad (3.13)$$

b is the projection of the molecular axis on the $\underline{p}$-plane,

$$b = d \sin\theta ; \qquad (3.14)$$

θ is the angle between the molecular axis and the beam direction (fig.12). After taking the angular average over the unit sphere, you find

$$\Delta W = \frac{S_1 S_2}{2\pi d^2} \, g(d/a) \tag{3.15}$$

with

$$g(d/a) = \pi^{1/2} \, \frac{d}{a} \, e^{-d^2/a^2} \quad \text{Im erfc}(-id/a) \ , \tag{3.16}$$

where erfc is the complementary error function (Abramowitz & Stegun,1965).

Before discussing this result, let us briefly confirm that it is more general than its derivation. In terms of the probability $Q_i(\underline{p})$, the differential cross section for energy transfer T to a diatomic 1-2 molecule in a given orientation reads

$$K(T) = \sum_i \sum_j \int d^2 p \, Q_{1i}(\underline{p}_1) \, Q_{2j}(\underline{p}_2) \, \delta(T - T_{1i} - T_{2j}) \tag{3.17}$$

as a generalization of eq.(3.2). This assumes again that the excitation processes for the two atoms in the model are independent of each other, except for the correlation via impact parameters. Eq.(3.17) may be inserted into

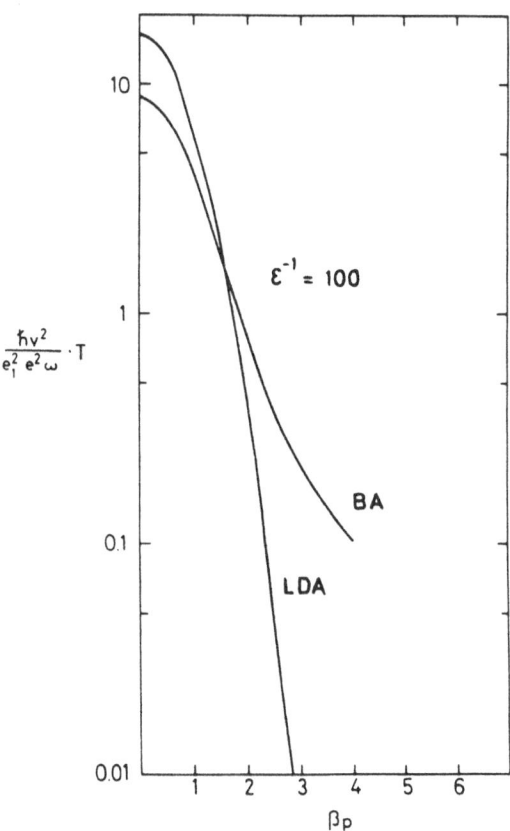

Fig.14. Calculated dependence of mean electronic energy transfer to a spherical harmonic oscillator, versus impact parameter p. e_1 = projectile charge; $\beta = (m\omega/\hbar)^{1/2}$; ω = oscillator frequency; $\varepsilon^{-1} = 2mv^2/\hbar\omega$. BA: First Born approximation; LDA: Local density approximation applied to Lindhard electron gas. From Mikkelsen & Sigmund (1987)

(1.13) or (1.14) to yield the stopping cross section S and straggling parameter W of a molecule. In the evaluation, you need the normalization condition for the collision probability

$$\sum_{i=0}^{\infty} Q_{\nu i}(\underline{p}) = 1 \; .$$

With this, you will be able to rederive eqs.(3.7), (3.9), and (3.10), with the correct atomic parameters S_ν and W_ν, (3a,b). Thus, eq. (3.15) remains valid when the proper quantum mechanical averages are inserted.

It is important that ΔW is proportional to the *product* of the two stopping cross sections. According to eq. (1.37), the atomic straggling parameter W_ν is independent of energy as a first approximation. Therefore, *the correlation effect in straggling must be most pronounced in the energy range of maximum electronic stopping power.*

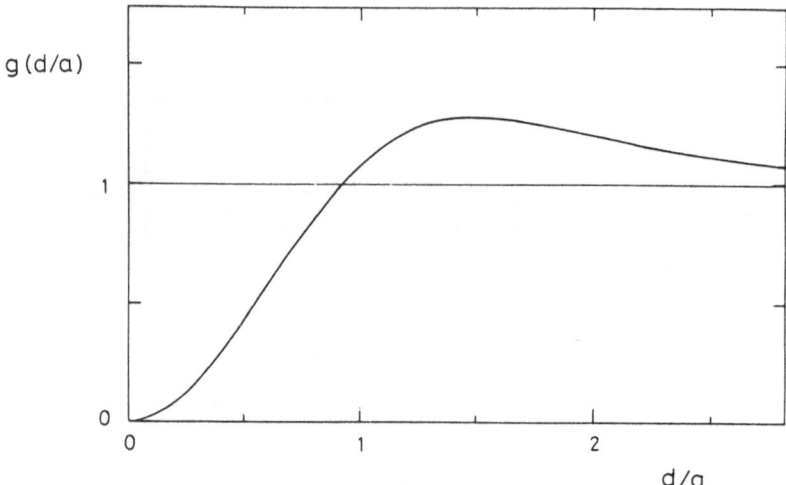

Fig.15. Function g(d/a), eq.(3.16)

The function g(d/a) has been plotted in fig. 15. It is seen to approach unity in the limit of d >> a. This result has been derived previously by a calculation that did not assume a specific impact parameter dependence (Sigmund,1976). More important, g appears to be quite close to unity down to d/a ~ 0.7, say.

Fig. 16 shows experimental results in comparison with the theoretical prediction, assuming g = 1. This supports the magnitude of this type of correlation effect.

3.5.1. An Excursion to Inner—Shell Ionization. I like to show you a different application of the same principles, mainly in order to clarify the underlying physics. Let us try to find the cross section σ_{++} for producing an inner-shell vacancy in *both* atoms 1 and 2 making up a diatomic molecule. Denoting the respective atomic excitation probabilities by $Q_1(\underline{p}_1)$ and $Q_2(\underline{p}_2)$, you find

$$\sigma_{++} = < \int d^2p \; Q_1(\underline{p}_1) \; Q_2(\underline{p}_2) >$$

if the processes in the excitation of the two atoms are statistically independent. The curled brackets denote an average over the orientation of the molecule.

With the notation indicated in fig. 12 you may rewrite this in the form

$$\sigma_{++} = \int d^2p_1 \int d^2p_2 \; Q_1(\underline{p}_1) \; Q_2(\underline{p}_2) \; <\delta(\underline{p}_1 - \underline{p}_2 - \underline{b})>,$$

where the orientation only enters through the vector $\underline{b}$ (fig.12). After carrying out the angular average you obtain

$$<\delta(\underline{p}_1 - \underline{p}_2 - \underline{b})> = \frac{1}{2\pi d^2} \; [1 - (\underline{p}_1 - \underline{p}_2)^2/d^2]^{-1/2}. \tag{3.18}$$

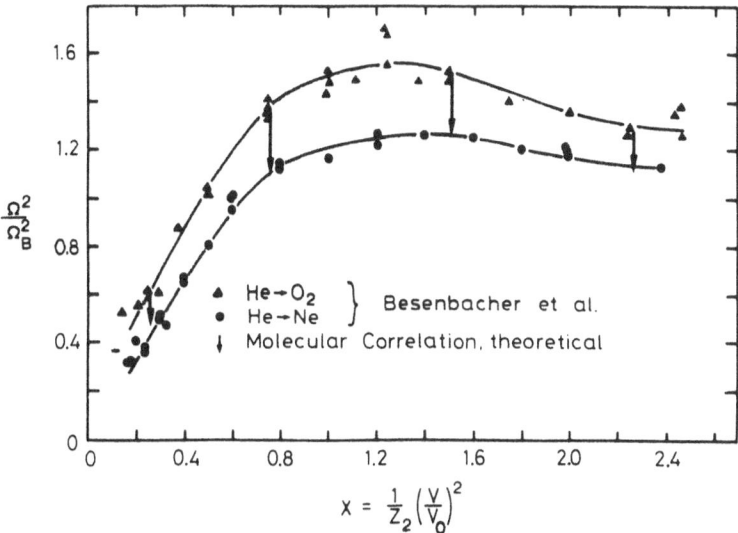

Fig.16. Measured energy loss straggling for He on O_2 and Ne versus scaled projectile energy. Ω_B is the Bohr value, eq. (1.38). Arrows indicate correlation correction, calculated from eq. (3.15) by use of measured stopping cross sections and setting g $\equiv$ 1. It is assumed that neon and atomic oxygen have very similar stopping parameters. From Besenbacher et al. (1981)

For inner-shell processes, impact parameters contributing to the integral are much smaller than the interatomic distance d. Hence, (3.18) reduces to $\sim (2\pi d^2)^{-1}$, and hence,

$$\sigma_{++} = \frac{\sigma_1 \; \sigma_2}{2\pi \; d^2},$$

where σ_1 is the cross section for production of an inner-shell vacancy in atom 1, and similar for σ_2.

You will recognize again the factor of $2\pi d^2$ in the denominator. This time its occurrence does not hinge on the gaussian impact dependence adopted in eq. (3.11). Estimating molecular effects approximating (3.18) by $(2\pi d^2)^{-1}$ works well for low-impact-parameter processes but breaks down for outer-shell effects (Sigmund,1976,1977).

3.6. Energy Loss in Solids

Dealing with solids we face the situation that the medium has some fixed structure so long as thermal vibrations are ignored. That structure may be regular or amorphous, but it is certainly not random: In a random medium, there is a possibility of two atoms lying on top of each other. If the average interatomic distance is large compared to the atomic radius, the error made by assuming randomness is negligible. This is why we do not have to worry about deviations from randomness in a gas.

Evidently, the fundamental eq.(1.3) is no longer applicable, so we need another starting point for the discussion of energy loss and straggling where the primary element of randomness is the impact point of the beam particle. The scheme reported here was developed by the author (Sigmund,1978).

Consider a solid target with z atoms in a piece of volume penetrated by the beam. These atoms, $\nu = 1,2,...z$ are located in positions $\underline{r}_\nu$, and an individual trajectory is initially assumed to be a straight line defined by a lateral position $\underline{p}$ (fig. 17). To get started, assume again a well-defined connection between impact parameter and energy loss, $T = T(\underline{p})$ in an individual collision. Then, the total energy loss in a given trajectory is given by

$$\Delta E = \sum_{\nu} T(\underline{p}-\underline{p}_\nu) , \qquad (3.19)$$

where $\underline{p}_\nu$ is the lateral component of the position vector $\underline{r}_\nu$.

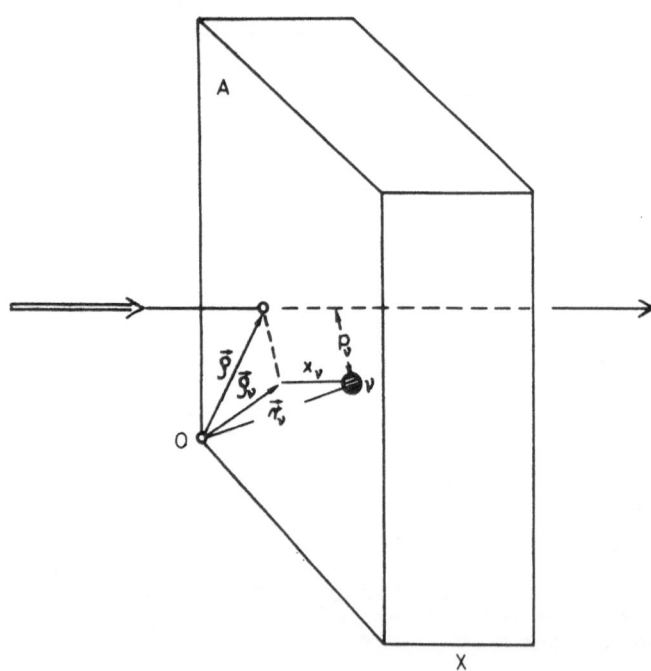

Fig.17.   Definition of geometric quantities determining
          energy loss in solids.

The average energy loss is found by randomizing the point of impact,

$$\langle \Delta E \rangle = \frac{1}{A} \int d^2\rho \, \Delta E \ , \tag{3.20}$$

where A is the cross sectional area of the beam. If A is a macroscopic area, it is much wider than the region where $T(\underline{\rho})$ is nonvanishing. Therefore, it is justified to extend the integration in (3.20) over the infinite plane, so that

$$\langle \Delta E \rangle = \frac{1}{A} \sum_{\nu} S = \frac{z}{A} S = NxS \tag{3.21}$$

by insertion of (3.19), where S is the atomic stopping cross section, $S = \int d^2\rho \, T(\underline{\rho})$, and $z = NxA$. As in the molecular gas, the mean energy loss is unaffected by the structure of the solid. The explanation given in sect. 3.4 applies also to the present case. Note, however, the assumption of a straight line trajectory throughout the target. That assumption will be relaxed shortly. A prominent situation where it is not valid is the case of channeling, where the motion is "governed" by the crystal structure (Lindhard,1965).

Consider now the mean-square energy loss,

$$\langle \Delta E^2 \rangle = \frac{1}{A} \int d^2\rho \, \sum_{\mu} \sum_{\nu} T(\underline{\rho}-\underline{\rho}_\mu) T(\underline{\rho}-\underline{\rho}_\nu) \ . \tag{3.22}$$

You may replace the integration variable $\underline{\rho}$ by $\underline{\rho}+\underline{\rho}_\mu$ in each individual term in the double sum (3.22), and split the sum over the terms $\mu=\nu$ and $\mu\neq\nu$, respectively. This yields

$$\langle \Delta E^2 \rangle = \frac{z}{A} W + \frac{z}{A} \sum_{\nu}' \int d^2\rho \, T(\underline{\rho}) \, T(\underline{\rho}+\underline{b}_\nu) \ , \tag{3.23}$$

where $\underline{b}_\nu$ indicates the lateral component of an internuclear distance vector, and the apostrophe indicates omission of the vector pointing from one target nucleus onto itself. W is the *atomic* straggling parameter, $W = \int 2\pi\rho d\rho \, [T(\rho)]^2$.

In the limiting case of a random medium, the sum over ν can be replaced by z times a random average over $\underline{b}_\nu$. Then, the second term on the right-hand side of (3.23) factorizes into $(zS/A)^2 = (NxS)^2 = \langle \Delta E \rangle^2$ as it must be. This recovers the random case eq. (1.12c).

In order to estimate the sign and magnitude of the deviation from the random case, take again the example of a gaussian impact parameter dependence (3.11). Then, integration of (3.23) yields

$$\langle \Delta E^2 \rangle = \frac{z}{A} W + \frac{z}{A} \sum_{\nu}' \frac{S^2}{2\pi a^2} e^{-b_\nu^2/2a^2} \ , \tag{3.24}$$

similar to eqs. (3.12) and (3.13). Here, a is the *atomic* screening constant corresponding to a_1 in (3.13), hence the factor 2 difference.

Eq. (3.24) implies an average over the target structure. This average can be expressed in compact form by introduction of the pair distribution function $g_2(\underline{r})$: $Ng_2(\underline{r})d^3r$ is the probability to find a target atom in a volume element d^3r at a vector distance $\underline{r}$ from a given target atom, averaged over all target atoms. Then,

$$\langle(\Delta E - \langle\Delta E\rangle)^2\rangle = NxW + Nx \frac{s^2}{2\pi a^2} N\int d^3r \, [g_2(\underline{r}) - 1] \, e^{-\rho^2/2a^2} , \qquad (3.25)$$

where $\langle\Delta E\rangle^2$ has been subtracted on both sides. Again, $\underline{\rho}$ is the lateral component of $\underline{r}$.

For a rotationally symmetric structure such as an amorphous target, where $g_2(\underline{r}) \equiv g_2(r)$, the angular integration can be carried out as in sect. 3.5, with the result that

$$\langle(\Delta E - \langle\Delta E\rangle)^2\rangle = Nx(W + \Delta W) \qquad (3.26)$$

with

$$\Delta W = 2Ns^2 \int_0^\infty dr \, [g_2(r) - 1] \, g\left(\frac{r}{a\sqrt{2}}\right) . \qquad (3.27)$$

Here, g is the function defined by eq. (3.16) and illustrated in fig.15.

As a test case, consider the pair distribution function of a diatomic molecular gas,

$$g_2(r) = 1 + \frac{\delta(r-d)}{4\pi Nd^2} , \qquad (3.28)$$

where d is the internuclear distance in the molecule and N the density of *atoms*. Here, (3.26) and (3.27) yield

$$\langle(\Delta E - \langle\Delta E\rangle)^2\rangle = NxW + Nx \frac{s^2}{2\pi d^2} \, g\left(\frac{d}{a\sqrt{2}}\right) . \qquad (3.29)$$

After consideration of the change of notation in both N and a, it is seen that eq. (3.28) completely agrees with the result derived in sect. 3.5, as it should.

For an amorphous solid, the *pair correlation function* g_2-1 looks like the one shown in fig. 18. It can be thought of as composed mainly of a region where $g_2-1 = -1$ at small distances and a peak at the nearest neighbor distance of the respective structure. Fig. 15 shows that the function $g(r/a\sqrt{2})$ is of the order of 1 for $r > a/\sqrt{2}$. a is of the order of the Thomas-Fermi radius, i.e., much smaller than the internuclear distance in the solid. Therefore it is justified to approximate $g = 1$ for all r that make up the integral (3.27), so that

$$\Delta W = s^2 \, 2N \int_0^\infty dr \, [g_2(r) - 1] . \qquad (3.30)$$

According to fig. 18, the integral over r is $= -2A$. The magnitude of the resulting straggling correction is very similar to the one found for molecules, but the prominent feature is the negative sign: Due to the dense packing of atoms in a solid, the number of hits made by a penetrating particle shows *less* fluctuation than in an ideal gas.

As in the end of the previous section, you may verify that the relations derived here for a well-defined energy loss function $T(\underline{p})$ remain valid also when quantum mechanical averages are evaluated. This means that it is the impact parameter dependent *mean energy loss* $\langle T(\underline{p}) \rangle$ that enters into ΔW, and hence the stopping cross section S that determines (3.30). This is important since the exponential eq. (3.13) explicitly applies to the mean energy loss $\langle T(p) \rangle$. The impact parameter dependence of the mean-square energy loss, $\langle T^2(p) \rangle$, has never been evaluated explicitly to the author's knowledge. The evaluation of W will be discussed in the following section.

You may ask about how realistically a model like this describes the stopping in a solid, and you may worry about two major simplifications, the straight-line trajectory and the neglect of collective effects.

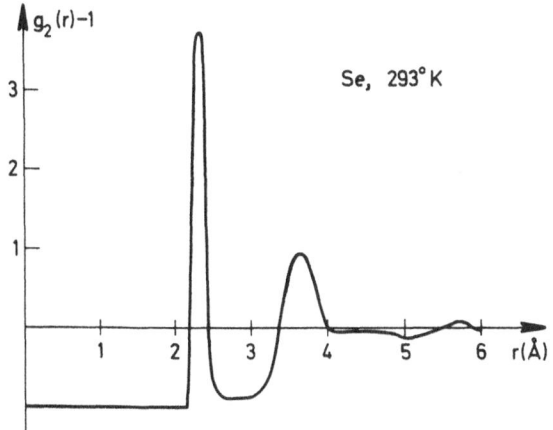

Fig.18.   Measured pair correlation function $g_2(r)$ for selenium. Evaluated by K. Carneiro from data by Hansen et al. (1975)

Collective effects may be important and will be touched briefly in sect. 4. Evidently, the present model does not apply to the portion of the energy loss spectrum which is due to conduction electrons.

Fig.18 shows that the pair correlation function vanishes at large distances. Hence, the integral over d^3r in (3.25) extends in practice only over a microscopic region about 4 A wide. The straight-line assumption needs to hold only over a pathlength of this magnitude, i.e., by no means across a macroscopic target thickness. Consider an experiment in which the total angular spread of the beam after penetration is 1°, and assume that this is caused by *one collision* per projectile. This means a change in impact parameter of = 0.07 A over a pathlength of = 4 A. A change in impact parameter by 0.07 A is a major effect when you consider inner-shell processes like K-shell excitation or wide-angle Rutherford scattering, but it is very minor when you look at electronic stopping and straggling. This is consistent with the fact that correlation effects vanish for small cross sections.

3.7. Trajectory Inversion

A puzzling experiment was reported about a decade ago (Pronko et al., 1979). A solid target was bombarded at normal incidence, and backscattered projectiles were observed very close to 180°. It turned out that in a narrow cone around normal reflection, the number and energy spectrum of reflected particles deviated markedly from the behavior off axis. It was soon recognized that this phenomenon was intimately connected to the fact that a particle reflected under 180° essentially collides with the same atoms at the same impact parameters on its way out as on its way in (fig.19).

Amsel (1982) pointed out a close relation between this experimental finding and the general theme discussed in this lecture. Obviously, the experimental finding cannot be explained in terms of random collision events of the type treated in sects. 1 and 2. Existing theoretical explanations are based on binary-collision computer simulation models involving a crystal lattice (Barrett et al.,1980) or simplifications thereof (Jakas & Baragiola, 1980; Crawford,1980). A clever transport calculation involving repeated pene-

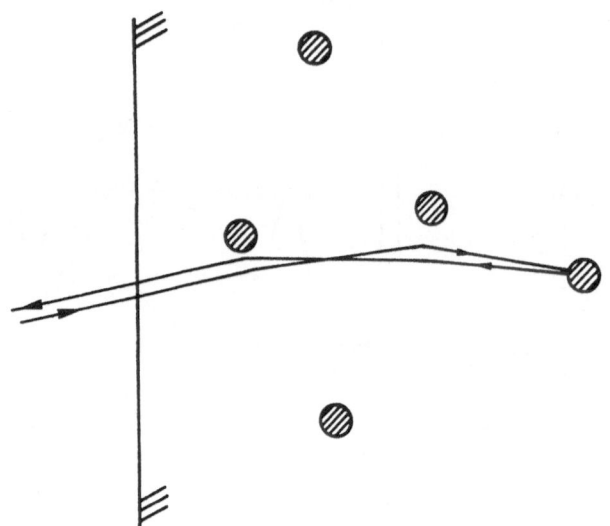

Fig.19. Trajectory inversion by scattering at 180°

tration of a random structure was reported more recently (Jakas and Ponce, 1984). An accurate analytical estimate seems still to be missing. The best agreement with experiment was achieved from the lattice simulation (Barrett et al.,1980). Simulating the medium as random -- even if this is applied only to the inward trajectory -- is probably inadequate for a subtle effect like this.

What I am going to do here is *not* meant to be an explanation of the measured 180° anomaly, but it deals with a closely related effect that should be measurable. Consider a particle penetrating a solid along a straight-line trajectory, and assume that it is reflected after some pathlength x under *exactly* 180°. The pathlength x is assumed short enough so that any motion of the target atoms (thermal or collision-induced) in the time span between the first and second passage of the projectile is negligible. The quantity to be estimated is the probability distribution in energy loss, and the question is how that distribution relates to the energy distribution of an "ordinary" projectile that has penetrated over a pathlength 2x (fig.19).

This matter can hardly be discussed without going into the physics of a "repeated" collision event. If a target atom has been excited or ionized on the way in, it may not have decayed before the return of the projectile, hence the interaction on the way out may be different even if the impact parameters are exactly the same as on the way in. If you restrict your attention to a light ion, there should be no major cause of concern: The excitation probability *per target electron* in an individual collision event will be small even though the probability for double excitation of an *atom* need not be negligible.

If you adopt this model, the second passage of the projectile will essentially *double up* the excitation probability *per target electron*. This is the same as doubling up the energy loss cross section of the atom. For random slowing-down, doubling the cross section has exactly the same effect as doubling the pathlength, since only the product of the two quantities enters into eq. (1.10). For a structured medium, this is not so, as follows from eqs. (3.26) and (3.27).

Doubling up the cross section in eq. (3.21) yields the average energy loss

$$\langle \Delta E \rangle = Nx \ 2S = 2x \ NS, \tag{3.31}$$

i.e., the same as the "ordinary" mean energy loss for a particle having penetrated a layer of thickness 2x. Doubling up the cross section in eq. (3.26) yields

$$\langle (\Delta E - \langle \Delta E \rangle)^2 \rangle = Nx \ 2W + Nx \ 4\Delta W = 2x \ NW + 2 \times 2x \ \Delta W \tag{3.32}$$

because of the factor S^2.

The additional factor of 2 in front of the correlation term in (3.32) causes the variance of the energy loss distribution to differ from that of a particle that has penetrated a pathlength 2x along a straight line in the same medium. The variance is smaller if ΔW is negative, i.e., for a solid, and greater if ΔW is positive, i.e., for a molecular gas. This effect should be measurable. Although its effect on the type of spectra observed in the experiment of Pronko et al. (1979) is in the right direction it can hardly explain the reported 180° anomaly which is generally assumed to be related to multiple scattering.

Lecture 4. APPLICATIONS

This lecture is supposed to give you an impression of how the theory can be applied to estimating energy loss spectra and multiple scattering distributions. In order to do this in a meaningful way, I need to give a brief summary on how and where to find pertinent cross sections and related quantities for elastic and inelastic scattering. At some point I shall switch gears, present fewer explicit derivations, and report and discuss various results with reference to sources in the literature.

4.1. Parameters Characterizing Energy Loss Spectra

For qualitative orientation, consider Rutherford's law, eq. (1.36), now written in the form

$$d\sigma(E,T) = \frac{C}{E} \frac{dT}{T^2} \ ; \qquad T_{min} \leq T \leq T_{max} \ , \tag{4.1}$$

where T_{max} is a kinematic upper limit. T_{min} is an effective lower limit, the origin of which depends on whether you deal with electronic or nuclear processes (Lindhard,1990). The same is true for the constant C. Consider projectile velocities large enough so that $T_{max} \gg T_{min}$. Then, the mean free path λ, as given by

$$1/\lambda = N\int d\sigma(E,T) = (NC/E) \ (1/T_{min} - 1/T_{max}) \ , \qquad (4.2)$$

depends sensitively on the minimum energy transfer T_{min}, as discussed in sect. 2.7.1. Consequently, calculations of mean free paths need to take proper account of the low-energy-loss portions of the cross section. In electronic stopping, this is mainly the region around the plasmon peak which is of particular interest in electron spectroscopy where such features can be resolved.

In the same approximation, eq. (4.1), the stopping power reads

$$-dE/dx = N\int Td\sigma(E,T) = (NC/E) \ \log(T_{max}/T_{min}) \ . \qquad (4.3)$$

This quantity depends equally much on T_{min} as on T_{max}, but either dependence is much weaker than the one on T_{min} in the mean free path. Thus, both large and small energy transfers contribute to the stopping power, but because of the logarithmic dependence, the stopping power is rather robust against minor modifications of the theoretical input. On the other hand, in applications the stopping power is needed with much greater accuracy than either mean free path or straggling (Feldman,1990).

The straggling in energy loss,

$$d\Omega^2/dx = N\int T^2 d\sigma(E,T) = (NC/E) \ (T_{max} - T_{min}) \ , \qquad (4.4)$$

shows essentially the opposite behavior from the mean free path in being most sensitive to the maximum energy transfer. It is easily verified that nuclear energy losses affect straggling, in particular so for light target atoms.

Higher moments are increasingly dependent on the high-T portions of the cross section. Collisions with near-maximum energy transfer are essentially free at moderate and high projectile velocities. This simplifies the computation of higher moments greatly.

4.2. Energy Loss Spectra versus Target Thickness

At target thicknesses $x \ll \lambda$, the energy loss spectrum is essentially the single collision spectrum with all structure present (although not necessarily resolved experimentally).

With x approaching the mean free path, the fine structure will start getting wiped out by plural scattering events (Bichsel & Saxon,1975; Tougaard,1990), and more global features gain importance.

4.2.1. Landau Formula. A very useful expression for the energy loss spectrum in the thin-layer limit was given by Landau (1944) on the basis of Rutherford's law, eq. (4.1). Integration of (1.9) after insertion of (4.1) yields

$$\sigma(k) = C'\{[1 - E_2(ikT_{min})]/T_{min} - [1 - E_2(T_{max})]/T_{max}\} \qquad (4.5)$$

with $C' = C/E$ in the notation of (4.1), where $E_2(z)$ is an exponential integral (Abramowitz & Stegun,1965). In the range of thicknesses under consideration, the main portion of the energy loss spectrum will lie in the region where

$$T_{min} \ll \Delta E \ll T_{max} \; . \tag{4.6}$$

Then, the integration over (1.10) is determined by the range of k-values for which $kT_{min} \ll 1$ and $kT_{max} \gg 1$. This allows Taylor expansions of the two terms on the right-hand side of eq. (4.5) in terms of kT_{min} and $1/kT_{max}$, respectively. The result is

$$\sigma(k) = C'\{ik[\psi(2) - \log(ikT_{min})] + k^2 T_{min}/2 \ldots - 1/T_{max} \ldots\}, \tag{4.7}$$

where $\psi(2) = 1-\gamma$ with $\gamma = 0.5772$ being Euler's constant.

Landau dropped all terms going to zero with either T_{min} or $1/T_{max}$ in eq. (4.7). With this, eq. (1.10) is easily written in the form

$$F(\Delta E,x) = \frac{1}{2\pi i} \frac{1}{NxC'} \int_{-i\infty}^{i\infty} du \; u^u \; e^{\Lambda u} \tag{4.8}$$

with $u = ikNxC'$ and $\Lambda = \Delta E/NxC' - \psi(2) - \log(NxC'/T_{min})$.

By direct integration or by applying eq. (1.11), you can verify that (4.8) obeys the normalization condition, but that both the mean energy loss and the straggling parameter diverge. Inclusion of the correction terms in eq. (4.7) does not improve on these particular features. However, it was shown recently by Lindhard (1985) that the Landau function characterizes the energy loss spectrum for thin penetrated layers even for more general cross sections, in terms of suitable scaling parameters determined by the cross section. Numerous tabulations of the Landau function are available (e.g. Kölbig & Schorr,1984).

Extensive comparisons with measured spectra have been performed (e.g. Baily et al.,1970). Experimental techniques allow to measure either the energy lost by the projectile or the energy deposited in the target. The two quantities are closely related but not necessarily identical: Energy may be carried away from the target by fast electrons and other forms of radiation (Bak et al.,1987; Bichsel,1988).

4.2.2. Transition to the Gaussian Regime. From the dependence of Λ, eq. (4.8), on x, you extract that the peak energy loss in the Landau limit scales as $NxC' \log(NxC' 2mv^2/I^2)$ for $T_{max} = 2mv^2$ and $T_{min} = I^2/2mv^2$. This reaches the mean energy loss given by the Bethe formula, $2NxC' \log(2mv^2/I)$, at a thickness x_0 given by

$$Nx_0 C'/T_{max} = 1 \; . \tag{4.9}$$

When (4.9) is fulfilled the target is sufficiently thick so that the probability for an event with near-maximum energy transfer approaches one. Eq. (4.9) is easily seen to be identical with Bohr's criterion, $NxW = T_{max}^2$, which was mentioned in sect. 1.6.3.

Presently, two approaches will be discussed which have been utilized to evaluate the shape of energy loss spectra for the transition region between the Landau and the Gaussian limit.

Vavilov (1957) rewrote the Bothe-Landau formula (1.10) in the form

$$F(\Delta E, x) = \frac{1}{2\pi} \int_{-\infty}^{\infty} dk \; e^{ik(\Delta E - NxS) - Nx\sigma_1(k)} \quad , \qquad (4.10)$$

where

$$\sigma_1(k) = \sigma(k) - ikS = \int dT (1 - T - e^{-ikT}) \; . \qquad (4.11)$$

He applied the free-Coulomb cross section (4.1) to $\sigma_1(k)$ only, i.e., to a term which is rather insensitive to the small-T portion of the cross section where binding of the target electrons is crucial. For the stopping cross section S, which depends sensitively on T_{min}, the most accurate available expression can be inserted.

Before passing on, you should note that with T_{max} occurring explicitly, it may be important to include relativistic effects if the projectile speed is high: For heavy projectiles, the thin-layer limit usually implies high projectile speed. Such corrections were included by Vavilov.

In the evaluation of the integral, Vavilov used a generalized diffusion approximation which follows from eq. (1.45). If the series in k is truncated after the term of third order in k you find, after integration, the following expression for the spectrum,

$$F(\Delta E, x) = \exp[(\Delta E - NxS - NxW^2/3Q)W/Q] \; (3NxQ/2)^{-1/3}$$

$$\times \; Ai[(\Delta E - NxS + NxW^2/2Q)/(3NxQ/2)^{-1/3}] \; , \qquad (4.12)$$

where $Ai(z)$ is Airy's function, and Q is the third moment defined by eq. (1.46). The range of applicability of this expression appears rather limited. In particular, inspection of the properties of the Airy function shows that the spectrum (4.12) turns negative for $\Delta E < NxS + NxW^2/Q - 2.338 \; (3NxQ/2)^{-1/3}$. This general feature of diffusion approximations going beyond second order was already mentioned in sect. 1.6.3.

An alternative procedure was proposed by Moyal (1955) and further developed recently (Sigmund & Winterbon, 1985). The exponent in (1.10) was expanded around $k = k_0$ instead of $k = 0$, where k_0 is the value of k at which the integrand has its maximum, i.e., where the function

$$f(k) = Nx\sigma(k) - ik\Delta E \qquad (4.13)$$

is stationary. Expansion up to second order in $k-k_0$ yields the wellknown saddle-point integral,

$$F(\Delta E, x) = \frac{\exp(-f(k_0))}{(2\pi f_0'')^{1/2}} \; , \qquad (4.14)$$

where $f_0'' = \partial^2 f(k)/\partial k^2$ at $k = k_0$. Moreover, $f_0' = 0$. Eliminating ΔE by means of (4.13), and expressing everything in terms of $\sigma(k)$ and its derivatives in $k = k_0$ you find

$$F(\Delta E, x) = \frac{\exp\{Nx(k_0\sigma_0' - \sigma_0)\}}{(2\pi Nx\sigma_0'')^{1/2}} \qquad (4.15)$$

and

120

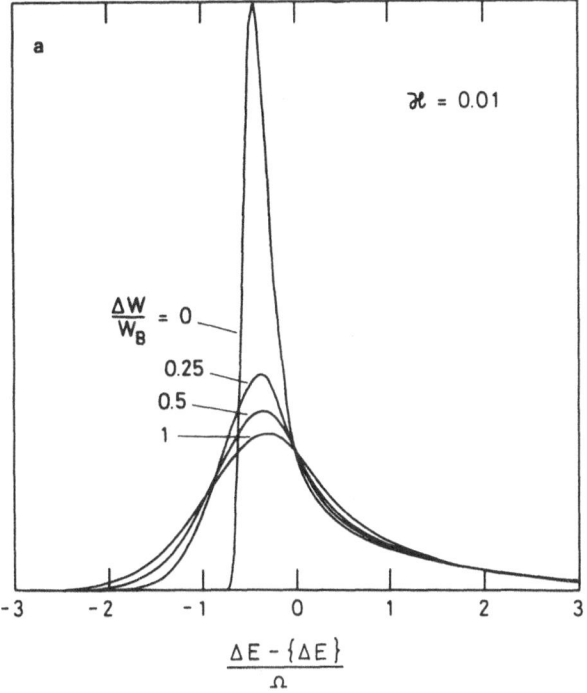

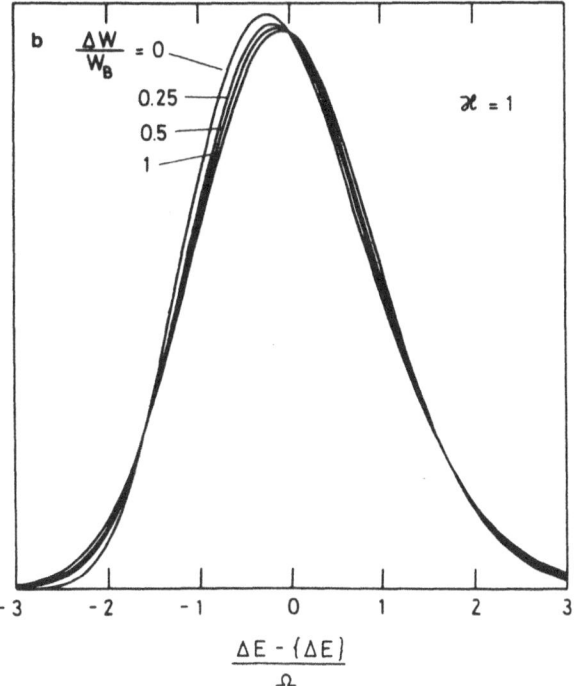

Fig.20. Energy loss spectra for modified free-Coulomb scattering, eq. (4.1), calculated from eqs. (4.15) and (4.16) for different target thicknesses. $\kappa = \Omega_B^2/T_{max}^2$. a) $\kappa = 0.01$; b) $\kappa = 1$. All spectra are normalized to 1. From Sigmund & Winterbon (1985)

$$\Delta E = -iNx\sigma_0'. \qquad\qquad (4.16)$$

Eqs. (4.15) and (4.16) form a parameter representation of F versus ΔE via k_0. This is very convenient for tabulating energy loss spectra.

The peak energy loss ΔE_p can be determined readily. For large thicknesses, it turned out to take the asymptotic form

$$\Delta E_p = NxS - Q/2W + \ldots \qquad\qquad (4.17)$$

where the neglected terms go to zero for large Nx. Here, Q is the third moment defined by eq. (1.46). You may easily convince yourself that the term Q/2W is identical with the expression $3/4\alpha$ in eq. (1.43') for the particular case of the cross section (1.40).

Eq. (4.17) demonstrates that the *relative* difference between peak and mean energy loss decreases with increasing target thickness while the *absolute* difference approaches a constant. For electronic stopping and free-Coulomb scattering, that constant is easily found to take on the value $Q/2W = mv^2/2$, where m is the electron mass. Note that these features are not predicted by the Vavilov formula (4.12). Experimental evidence supporting the existence of an intercept predicted by eq.(4.17) was referred to in the original paper and in a detailed investigation by Mertens (1986).

Fig. 20 shows energy loss spectra evaluated from (4.15) and (4.16) on the basis of the scattering law (4.1). Case a) corresponds to a thin target in the Landau limit. Case b) fulfills the Bohr criterion for a gaussian profile. A "shell correction" (or Blunck-Leisegang (1950) correction) ΔW of varying magnitude has been added to the straggling parameter W that is specified by eq. (4.4). The origin of this correction will be discussed below. It is seen that the shape of the spectra depends critically on the magnitude of ΔW at target thicknesses where the spectrum is not close to gaussian.

4.2.3. Thicker Targets. When the relative energy loss gets sizable, the spectrum will skew again. The direction of the skewness depends on whether the energy loss increases or decreases with decreasing energy. For high-speed particles, the stopping power increases with decreasing velocity. Hence, the peak energy loss will become greater than the mean energy loss. The topic has been treated in classical papers by Symon (1948) and Tschalär (1968).

4.3. Survey of Models for Calculating Electronic Stopping Parameters

Accurate estimates need to be based on cross sectional input that goes beyond eq. (4.1). For electronic collisions, this implies taking properly into account the effects of binding and orbital motion of the electrons involved in the interaction, as well as possible collective effects. For nuclear collisions, it implies proper care of the screening of the Coulomb potential of the projectile and target nuclei. This section is supposed to give you a brief survey of where to find theoretical input. This survey is not complete, and I have included this material only since I feel that it ought to appear *somewhere* in this book. Most of it is more recent than pertinent texts and reviews (Fano,1963; Inokuti,1971; Sigmund,1975; Bonderup,1981).

4.3.1. "Exact" Calculations. There will still go some time until the first accurate *ab initio* calculations of stopping power and straggling are available that allow for all important items, including many-electron effects, charge exchange by the projectile, nonuniform trajectories and, where necessary, relativistic effects. However, standard procedures from atomic collision theory may be applied to evaluate stopping parameters in a way to at least incorporate all those effects, like shell corrections and higher-order Born terms, which can be treated on the basis of more conventional stopping models. Stopping cross sections of inert-gas atoms for a point charge have been reported recently (Schiwietz,1990). Such calculations seem suitable as benchmark tests for analytical procedures, and vice versa.

4.3.2. Free-Electron Gas. The free-electron gas in Lindhard's (1954) description has been utilized successfully in stopping theory, both as a model to study processes qualitatively and for quantitative evaluation of stopping parameters. "Shell corrections" emerge readily from this model, i.e., deviations from the Bethe logarithm of the stopping number L defined by eq.(1.39). These are caused by the orbital motion of the target electrons (Lindhard and Winther,1964). Higher-order Born terms in the stopping power have been calculated (Esbensen & Sigmund,1990). The scheme has also been applied to straggling (Bonderup & Hvelplund,1971; Sigmund & Fu,1982) and to higher moments (Capuj et al.,1987).

The free-electron gas has been used extensively to describe low-velocity stopping in solids where only the most weakly bound electrons contribute. Explicit calculations of stopping power and straggling have been performed either within the original linear theory or within a nonlinear extension (Echenique,1990). Caution appears indicated when the electron gas model is applied to insulators.

Stopping parameters have been evaluated also for atoms, molecules and solids by averaging electron-gas stopping parameters over a given charge distribution in accordance with the local-density approximation. Early attempts within this scheme (Lindhard & Scharff,1953; Bonderup,1967; Chu and Powers,1972; Chu,1976) gave very useful guidance. Before going into similar computations now, you should ponder whether the degree of sophistication of your available electron densities and the necessary amount of computation are compatible with the simplicity of the underlying physical model.

The local density approximation has also been applied to estimate the impact parameter dependence of electronic energy loss (Winterbon,1983; Gras-Marti,1985; Ascolani & Arista,1986). Such estimates have been demonstrated to be only qualitative (Mikkelsen & Sigmund,1987). One obvious shortcoming is the dependence at large distances which is known to be essentially exponential (Bohr,1913; Bloch,1933) with the slope depending on projectile velocity. In the local density approximation, this dependence follows that of the electron density which is independent of the projectile speed. The quantitative difference may be significant, as is seen from fig. 14.

4.3.3. Harmonic Oscillator. While the electron gas model may occasionally have been overdone, the harmonic oscillator was probably underrated for some time. The *classical* oscillator was the standard of all stopping theory from Bohr (1913) to Bethe (1930). It was utilized later in the investigation of polarization phenomena at high projectile speeds (A.Bohr,1948) and, notably, higher-order perturbation corrections to the stopping power (Ashley et al.,1972; Lindhard,1976).

Quantal treatments of charged-particle stopping by a harmonic oscillator have become available recently, both for the first (Sigmund & Haagerup,1986), second, (Mikkelsen & Sigmund,1989) and third (Mikkelsen,1990) order in the Born series. The strength of the oscillator model is the explicit inclusion of binding forces and the possibility to study the dependence of the energy loss on the impact parameter (Mikkelsen & Sigmund,1987). It was demonstrated that accurate estimates could be found of atomic stopping cross sections by replacing a target atom by an ensemble of oscillators as in classical dispersion theory. Successful applications of this have been provided recently. An example is shown in fig. 21.

123

4.3.4. Dense Gas of Harmonic Oscillators. In order to include collective effects, a model has been designed recently which allows to study the modification of atomic stopping parameters when the atom is embedded into a dense medium. It consists of an assembly of quantal harmonic oscillators, with the force centers distributed at random in space. The mutual electrostatic interaction of the electrons is included selfconsistently, but the Pauli principle has been ignored. The system has been treated both to leading order in the projectile charge (Belkacem & Sigmund,1990) and to the next higher order (Esbensen & Sigmund,1990).

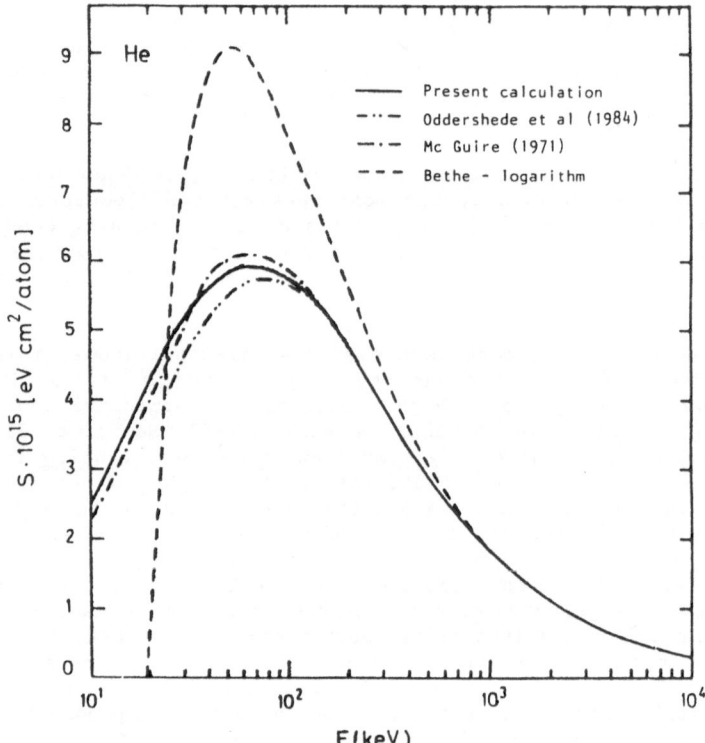

Fig.21.   Stopping cross section of helium for protons, calculated within the first Born approximation. Solid line: Target atom represented by an ensemble of harmonic oscillators, weighted by dipole oscillator strengths; Dot-dashed line: Numerical result (McGuire,1971); Double-dot-dashed line: Kinetic theory (Oddershede & Sabin,1984); Dashed line: Bethe logarithm. From Mikkelsen & Mortensen (1990)

Not unexpectedly, it was found that the characteristic resonance frequency of the system is the wellknown expression

$$\Omega = (\omega_0{}^2 + \omega_p{}^2)^{1/2} \qquad\qquad (4.18)$$

from classical dispersion theory. Here, ω_0 is the resonance frequency of an isolated oscillator, and ω_p is the plasma frequency. With Ω as a parameter, many results obtained for an isolated oscillator can be scaled to a dense medium. Even at a high electron density, corresponding to $\omega_p{}^2/\omega_0{}^2 = 10$, the scaled stopping number resembles much more that of an isolated oscillator than that of a free electron gas (Fig.22). This is probably related to the observation that at the same high density, the discrete features of the excitation spectrum of the harmonic oscillators have not yet been lost.

4.3.5. Binary—Encounter Model and Kinetic Theory. Binary-encounter models are based on binary-collision kinematics for free particles but take explicit account of the orbital motion of the target electrons. These models are relevant when the projectile speed is not large compared to characteristic electron speeds in the target atoms.

They come in three essentially different versions. The *classical* binary-encounter model (Gryzinski,1965; Gerjuoy,1966; Harberger et al.,1974) assumes free-Coulomb scattering between moving collision partners and a lower cutoff energy transfer T_c which may be used as an adjustable parameter. The model has little predictive power but seems to fit stopping powers near the maximum (Kührt & Wedell,1981).

The classical binary-encounter model fails by a factor of two in predicting the asymptotic behavior at high velocities. This is due to the ne-

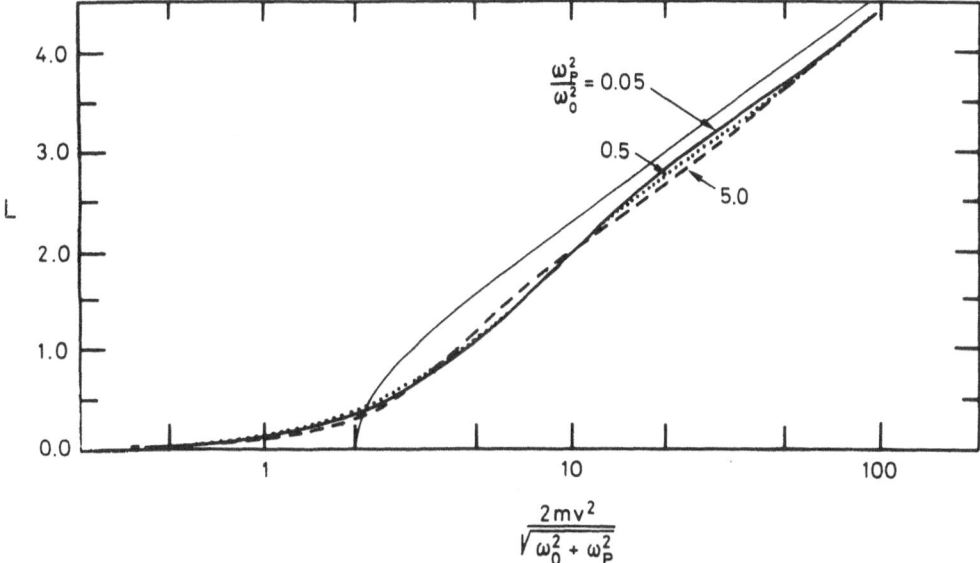

Fig.22.   Stopping number L of a dense medium of harmonic oscillators. L is related to the stopping cross section by eq. (1.39). The three similar curves correspond to electron densities varying over a factor of 100. The thin solid line characterizes a free electron gas ignoring Fermi motion. $\omega_p^2 = 4\pi ne^2/m$; ω_0 = oscillator frequency. From Belkacem & Sigmund (1990)

glect of resonance excitations. This shortcoming has been circumvented in the kinetic theory (Sigmund,1982) which incorporates kinematic corrections into the Bethe formula and avoids adjustable parameters. An example is included in fig. 21. The scheme predicts the leading shell corrections in both stopping power and straggling, but it led to erroneous results for shell corrections in higher-order Born terms, as demonstrated recently (Mikkelsen & Sigmund,1989). The kinetic theory predicts velocity-proportional stopping at low projectile speed for any target atom. Although this is a result that is common to most other stopping models that are supposed to be valid at low speed, it is most likely not universally true.

The classical binary-encounter model has been used successfully to evaluate higher than second-order moments over the cross section, i.e., moments that determine the skewness, curtosis etc. of the energy loss spectrum (Kührt & Wedell,1983; Sigmund & Johannessen,1985).

A third version of the binary encounter model has been applied to low-velocity stopping in connection with linear and nonlinear dielectric theory. This aspect has been reviewed by Echenique (1990).

4.3.6. Firsov Model. Firsov's (1959) model has been frequently used in rough estimates of low-velocity stopping. It is geared toward heavier ions and is mostly a continuum model of charge exchange (Winterbon,1968). The popularity of the model stems from the fact that it provided an explicit estimate of the impact-parameter dependence of the mean energy loss in an individual collision (Robinson & Torrens,1975). Also fluctuations have been estimated on this basis (Hvelplund,1971,1975), even though they must be underestimated within the model as it stands.

4.3.7. Heavy—Ion Stopping. Much less attention has been devoted to the theory of stopping and straggling of swift heavy ions than that of light ions. While the range of validity of the Born series at the one end (low Z_1) and the classical approximation at the other end (high Z_1) are reasonably well established via Bloch's (1933) treatment, much less is known about the values of the pertinent target parameters, about impact parameter dependences, etc. In addition, the notoric problem of effective projectile charges has not yet found a generally accepted solution. An interesting development is the use of classical-orbital simulation models in heavy-ion stopping which incorporates multiple-excitation effects (Olson,1989).

4.4. Straggling Calculations

It is convenient to relate the straggling parameter per atom, $W = \Omega^2/Nx$, to Bohr's value, $W_B = 4\pi Z_1{}^2 Z_2 e^4$, cf. eq. (1.38), which was derived on the basis of free-Coulomb scattering between a heavy projectile and the target electrons. From Bethe's quantum theory of stopping, Williams (1931b) derived a logarithmic correction term which, for high velocities, can be brought into the following form (Livingston & Bethe,1937; Fano,1963)

$$W/W_B = 1 + \frac{2}{3} \frac{<v_e{}^2>}{v^2} \log(2mv^2/I_1) ,$$ (4.19)

where $<v_e{}^2>$ is the mean square velocity of the target electrons in the initial state. I_1 is a mean excitation energy defined by

$$\log I_1 = \frac{\Sigma f_i T_i \log T_i}{\Sigma f_i T_i} ,$$ (4.20)

and f_i is the dipole oscillator strength for the transition to an excitation level T_i above the initial state. Eq. (4.19) is to be understood as an asymptotic expansion, similar to the shell correction expansion in the stopping power. An additional numerical constant under the logarithm in eq. (4.19) was omitted in Fano's paper, as pointed out subsequently (Hvelplund,1968; Inokuti et al.,1978).

The quantity I_1, eq.(4.20), differs from the more common mean excitation energy I that enters the stopping power,

$$\log I = \Sigma f_i \log T_i ,$$ (4.21)

in that higher excitation levels have a higher weight in (4.20) than in (4.21). Tabulations of I_1, based on calculations from Hartree atomic wave functions have become available (Inokuti et al.,1981).

126

Explicit numerical evaluations of the straggling parameter (1.14), based on the first Born approximation and avoiding the asymptotic expansion (4.19), were performed for hydrogenic wave functions (Bichsel,1970) and for the harmonic oscillator (Sigmund & Haagerup,1986). Straggling calculations for the free electron gas (Bonderup & Hvelplund,1971; Sigmund & Fu,1982) confirmed the general asymptotic form (4.19) and, in addition, allowed predictions of the low-velocity behavior. The range of validity of kinetic theory (Sigmund,- 1982) is much greater in straggling than in stopping power, in view of the increasing weight of excitations into the continuum.

Fig. 23 shows a calculated straggling parameter for bare protons in atomic hydrogen. One may notice the wide range of applicability of Bohr's expression down to the velocity range where the stopping power has its maximum.

Higher than second-order moments are conveniently and accurately evaluated from the classical binary-encounter model (Sigmund & Johannessen,1985).

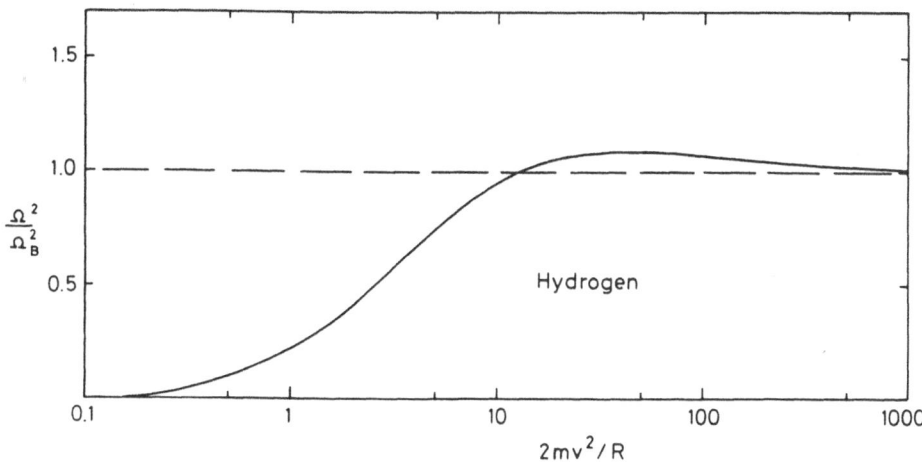

Fig.23. Straggling parameter calculated from atomic oscillator model (cf. text to fig. 21) for atomic hydrogen. R is the Rydberg energy 13.6 eV. From Sigmund & Haagerup (1987)

Because of the dominance of large energy transfers, electronic binding forces are unimportant while proper account of the initial motion of the target electrons is crucial. Excellent agreement was found in a comparison with numerical calculations within the first Born approximation (Bichsel,1970). An example is shown in fig. 24.

4.5. Straggling Measurements

As a general rule of thumb, experimental errors tend to more seriously affect measurements of fluctuations than those of mean values. For this reason, the accuracy of straggling data will in general not be comparable with that of stopping power measurements. On the other hand, the requirements on the accuracy of straggling data are less severe than those on stopping powers.

4.5.1. Measurements on Gas Targets.
From the point of view of testing theoretical predictions, gas targets are to be preferred. Targets may be made extremely thin (in terms of areal density Nx) and reasonably homogeneous. A very thorough study was performed by Besenbacher et al. (1981), where also

references to earlier work may be found. Their data were analysed in terms of existing theory at the time, mostly electron gas calculations combined with the local density approximation (Besenbacher et al.,1980). A systematic comparison of this set of measurements with more recent calculations would be worthwhile in my opinion.

4.5.2. Measurements on Solids. In applications of ion beams, straggling data are of interest mainly for solids. The main source of error in such measurements is nonuniform target thickness. This holds for both transmission and reflection measurements.

For heavier ions, charge exchange straggling appears to be a major contribution to the width of the energy loss spectrum. Since the stopping power depends on the charge state, fluctuating charge states generate fluctuations in energy loss. The fundamental mathematics was discussed in sect. 2.8, and more explicit estimates are available in the literature (Vollmer,1974; Winterbon,1977; Cowern et al.,1979, Sofield et al.,1981). However, there are fundamental problems regarding charge states of ions in solids and their relation to energy loss which are unresolved. Charge states of swift heavy ions emerging from solids are generally higher than after penetration of gas targets (Lassen,1951; Bohr & Lindhard,1954), while there does not seem to be a substantial difference in stopping power (Moak & Brown,1966). The problem has been discussed extensively (Betz,1972,1990), but a coherent picture has not yet emerged. Therefore, caution is indicated regarding theoretical predictions of charge exchange straggling.

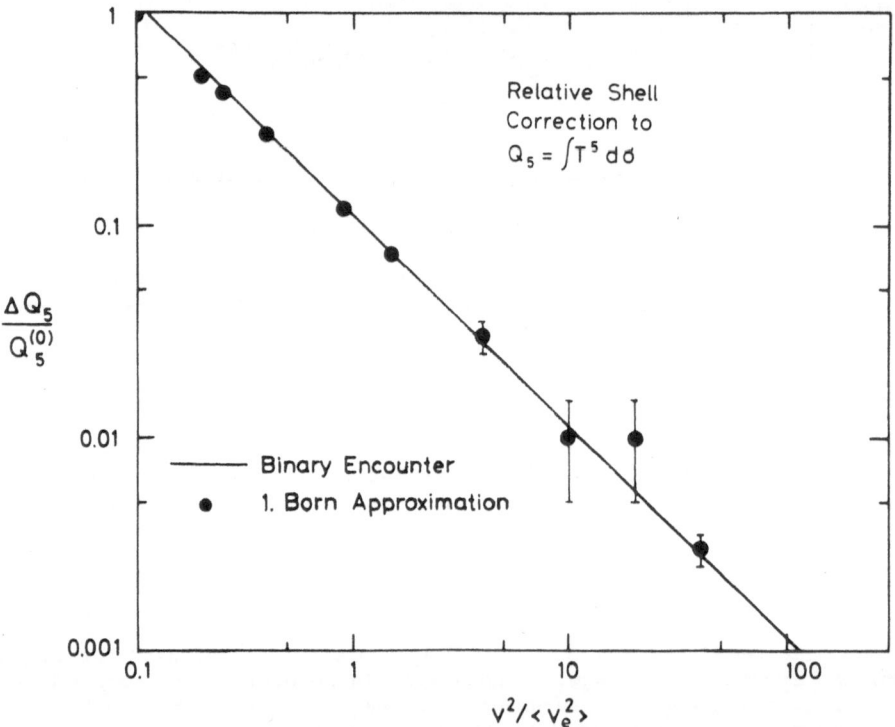

Fig.24. Deviation of fifth moment over the energy loss cross section from the free-Coulomb value $Q_5^{(0)}$ for an initially stationary electron. Solid line: Classical binary-encounter calculation (Sigmund & Johannessen,1985). Dots: First Born approximation (Bichsel,1970). Error bars reflect the number of decimals given in the published tables. $\langle v_e^2 \rangle$ is the mean square electron speed of the target electron

Fig. 25 shows two examples of measured straggling data for light and heavy ions, respectively. Fig.25a refers to protons and deuterons and should mainly demonstrate the wide range of applicability of the Bohr straggling formula. The authors were concerned about the uniformity of their target foils and tried to eliminate this source of error. Fig.25b confirms the significance of charge-exchange straggling for penetrating heavy ions.

In low-velocity heavy-ion straggling, inner-shell excitations have been demonstrated to be more prominent than in the corresponding stopping power (Lennard et al.,1986).

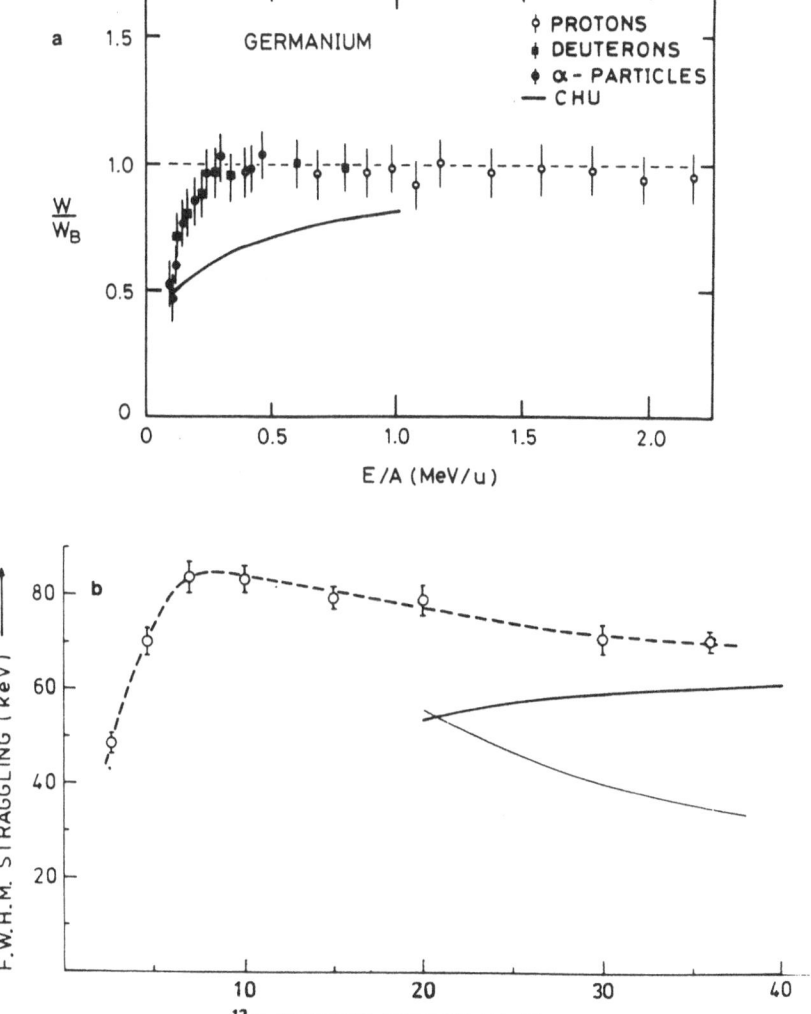

Fig.25.   Straggling measured on thin solid targets in transmission: a) Protons and deuterons on germanium, compared with local density approximation (Chu,1976). From Malherbe & Alberts (1982). b) Carbon in aluminium. Thick solid line: Livingston & Bethe (1937); thin solid line: Estimated contribution from charge-exchange straggling. The dotted line is the sum of the contributions. Note that the vertical axis represents a half-width. From Cowern et al. (1979)

4.6. Multiple Scattering

The basic equations governing multiple scattering of charged particles have been derived in sect. 1.5, in particular paragraphs 1.5.3 and 1.5.5. The remainder of this lecture serves to apply the general theory mainly to penetrating ions and to discuss a few related phenomena. As discussed by Lindhard (1990), angular deflections of penetrating charged particles are governed primarily by collisions with the *nuclei* of the target, in contrast to energy loss which is predominantly *electronic* except at low projectile speeds.

4.6.1. Parameters Characterizing Multiple Scattering Distributions. Usually, angular distributions of penetrating charged particles have a maximum at zero deflection angle and fall off monotonically with increasing angle, cf. eq.(1.49). Exceptions can be observed, e.g., in the angular distributions of ions emerging in a specific charge state (Kanter,1983). Such more complicated cases will be disregarded here. Similar to what was discussed for energy loss spectra, angular distributions consist of a central portion which

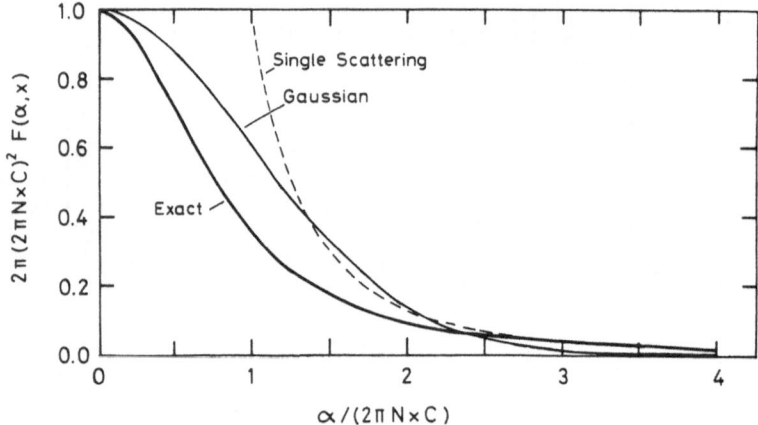

Fig.26.   Multiple scattering distribution for model cross section (1.48). See text

is gaussian-like and determined by a number of small-angle scattering events, and a tail at large angles that is due mainly to a single deflection. Just as in case of the energy loss spectrum, the single-collision tail is shifted toward larger and larger angles when the target thickness increases. However, because of the different magnitude of the cross sections for energy loss and angular deflection, *you may well observe a pronounced single-collision tail in the angular distribution, even though the energy loss spectrum is close to gaussian.*

As mentioned in sect. 1.6.4, it is necessary to carefully distinguish between the half-angle and the r.m.s. angle of an angular distribution, since the two in general show different dependences on target thickness.

4.6.2. Analytical Estimates. It is instructive to inspect a simple procedure developed by Williams (1939,1940). Similar to energy loss (cf. fig.8), separate the angular distribution into two parts, a small-angle portion composed of multiple scatterings at angles $\phi < \phi^*$, and a large-angle portion made up by a single scattering at an angle $\phi > \phi^*$. Assume the two regimes to contribute equal portions to the scattered intensity, so that the probability for a deflection $\phi > \phi^*$ becomes

$$P^* = Nx \int_{\phi^*}^{\infty} d\sigma(\phi) = 1/2, \qquad (4.22)$$

and describe the multiple-scattering portion by the diffusion approximation, eq.(1.51), with a diffusion coefficient $D = D^*$, eq.(1.52) given by

$$D^* = \frac{1}{4} N \int_0^{\phi^*} \phi^2 2\pi\phi d\phi \, K(\phi). \qquad (4.23)$$

You may test this scheme on the cross section (1.48) where the rigorous solution is known. From (4.22) you find $\phi^* = 4\pi NxC$ for this particular cross section. Fig.26 shows a comparison of the two limiting distributions with the exact one. The overall agreement is reasonable. The exact fit of the gaussian portion at $\alpha = 0$ is, however, a specific feature of the cross section (1.48).

4.6.3. Scattering Cross Sections. Cross sections for angular deflection in ion-atom collisions can most often be calculated from classical dynamics. The reason is two-fold. Firstly, according to Bohr (1948), trajectories of scattered particles may be sufficiently well localized if eq. (3.1) is fulfilled. For high atomic numbers and not too high velocities, this is easily satisfied. Conversely, for low atomic numbers and high velocities, the Coulomb interaction is essentially unscreened, and Rutherford's law emerges from both classical and quantal dynamics.

Screening introduces changes in the cross section which do not affect the validity of the classical orbital picture at low velocities (Lindhard et al.,1968). However, at high speed, in the range of validity of the Born approximation ($z_1 z_2 e^2 / \hbar v < 1$), substantial changes occur in the cross section at small angles which do affect the multiple scattering distribution at small angles.

Molière (1947,1948) evaluated angular distributions on the basis of a Thomas-Fermi screened-Coulomb potential,

$$V(R) = \frac{z_1 z_2 e^2}{R} \, \Phi(R/a) \, , \qquad (4.24)$$

where R is the distance between the colliding nuclei. a is a screening radius which may take on various forms, amongst those the one adopted by Lindhard et al. (1963,1968)

$$a = 0.8853 \, a_0 \, (z_1^{2/3} + z_2^{2/3})^{-1/2} \, . \qquad (4.25)$$

The screening function $\Phi(R/a)$ is usually set equal to the one found from the Thomas-Fermi description of the *atom*, either in the straight version ("Thomas-Fermi") or in the Lenz-Jensen form. This can be rationalized but not proven, and the main justification lies in comparisons with measured scattering cross sections (Loftager et al.,1979).

4.6.4. Scaling Properties. The differential scattering cross section belonging to the interaction potential (4.24) can be cast into a universal form in terms of suitable variables (Lindhard et al.,1968). For small-angle scattering, it reads

$$d\sigma = \pi a^2 \frac{d\bar{\phi}}{\bar{\phi}^2} f(\bar{\phi}), \qquad (4.26)$$

where $\bar{\phi}$ is a scaled scattering angle,

$$\bar{\phi} = \frac{Ea}{2z_1z_2e^2} \phi ,$$ (4.27)

and $f(\bar{\phi})$ a universal function that is determined by the screening function $\Phi(R/a)$ in eq. (4.24). (4.26) allows to write the multiple scattering distribution in the scaled form

$$F(\alpha,x)d^2\alpha = \tau \frac{d\bar{\alpha}}{\bar{\alpha}^2} f(\bar{\alpha},\tau),$$ (4.28)

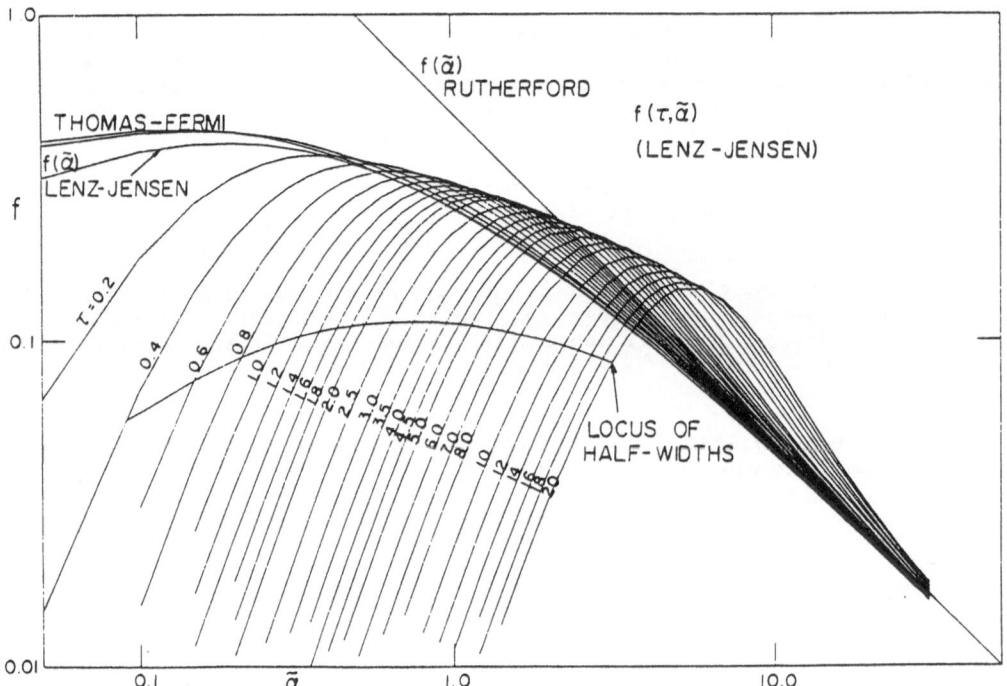

Fig.27. Calculated multiple scattering distributions in Thomas-Fermi scaling variables (4.29). The plot is such that the model cross section (1.48) (dashed line in fig.26) would become a horizontal straight line. For details see text. From Sigmund & Winterbon (1974)

where τ and $\bar{\alpha}$ are the scaled target thickness and scattering angle, respectively,

$$\tau = \pi a^2 Nx; \qquad \bar{\alpha} = \frac{Ea}{2z_1z_2e^2} \alpha .$$ (4.29)

The function $f(\bar{\alpha},\tau)$ has been evaluated numerically and is shown in fig.27, together with the function $f(\bar{\alpha}) \equiv f(\bar{\alpha},0)$ which characterizes the single scattering distribution. Apart from scale, fig.27 reflects the qualitative features brought about in fig.26.

Scaling variables of the present type were introduced by Molière (1948). The specific form (4.26) was utilized by Meyer (1971). Fig.27 is taken from Sigmund & Winterbon (1974,1975), which differs from Meyer's evaluation in two features. Firstly, Meyer's numerical accuracy was not sufficient to ensure that $f(\overline{a}, \tau)$ approached the single-scattering law in the limit of zero thickness. Secondly, Meyer made an attempt to correct for the non-Poisson character of collisions in solids by introducing a second function $f_2(\overline{a}, \tau)$ which depended on a somewhat arbitrarily introduced cutoff radius.

Also the lateral spreading of a collimated ion beam has been evaluated and was found to show very similar scaling properties as the angular distribution (Marwick & Sigmund,1975). Also the combined angular and lateral spread has been calculated (Sigmund et al.,1978).

Extensive experimental tests have been performed on the scaling properties for different elements, the predicted dependence of the halfwidth $\overline{a}_{1/2}$ on τ, and the specific shape of the distribution (Andersen et al.,1974). Fig. 28 shows a recent test on the halfwidth over a very wide range of the reduced thickness τ (Anne et al.,1988). A thorough test of the shape of the distribution was carried out on amorphous Al_2O_3, where the simple scaling properties do not apply (Schmaus & L'Hoir,1982). The multiple scattering distribution was evaluated numerically, following the above principles but without applying scaling properties. There was achieved excellent agreement over several decades (fig.29), The effective target thickness (cf. sect. 2.1.1) was used as a fitting parameter. Related work by Schmaus et al. (1982) demonstrated that multiple scattering profiles also allow to extract information about the *fluctuations* in target thickness.

Two cases of lacking agreement between theory and experiment were pointed out more recently. In measurements involving heavy projectiles on light targets (Geissel et al.,1985), discrepancies were found in the tails.   These

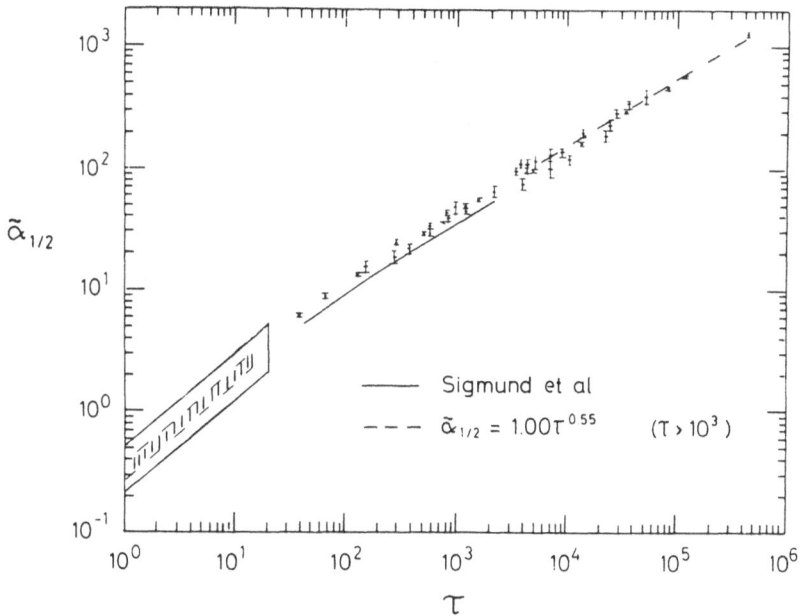

Fig.28. Measured half-width of multiple-scattering distributions for high-energy (20-90 MeV/u) O, Ar, Kr, and Mo ions versus scaled target thickness. From Anne et al. (1988)

could be attributed to a breakdown of the small-angle approximation and were corrected (Winterbon,1987,1989). This feature is characteristic of heavy projectiles incident on light targets where the maximum single-event scattering angle is small in absolute terms.

Secondly, in measurements involving high-energy hydrogen ions, disparities were located in the plot of half-width versus target thickness which were correctly ascribed to the breakdown of the classical scattering cross section at the very small scattering angles involved (Kumbartzki et al., 1985).

4.6.5. Non—Poisson Effects. Measurements of multiple scattering distributions on molecular gas by Sidenius revealed minor differences from the scaling properties predicted for atomic gases. These measurements initiated my effort to study the effects of correlated impact parameters on the statistics of particle penetration, as discussed in lecture 3 on the case of energy loss straggling.

The case of molecular gases is well described (fig. 30 and papers by Sidenius et al.,1976 and Besenbacher et al.,1978), but in practical terms, the correlation effect is insignificant except for extremely thin targets.

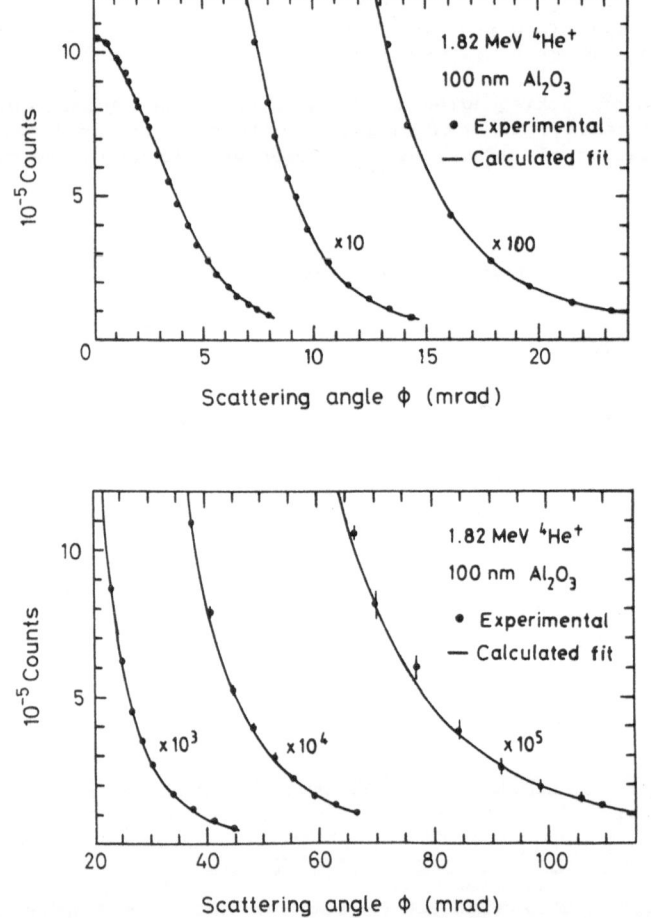

Fig.29. Measured multiple scattering distribution for 1.82 MeV He on Al_2O_3. See text. From Schmaus & L'Hoir (1982)

Unlike in the case of energy loss, the theory is not readily generalizable to solid targets. The reason is that multiple scattering distributions are not well characterized by a second moment but need explicit determination of a *distribution*, cf. the remarks made in the end of sect. 1.6.4. I have not yet managed to properly calculate a distribution for non-Poisson collision statistics -- either for energy loss or for angular scattering. A new effort in that direction could shed more light on effects of trajectory inversion. This effect was discussed in sect. 3.7 with regard to energy loss straggling, even though it is generally believed that multiple angular deflection effects should be dominating.

4.7. Energy Loss versus Angle

4.7.1. Detector Geometries.
The precise result of an energy loss measurement depends in general on the detector geometry. Three characteristically

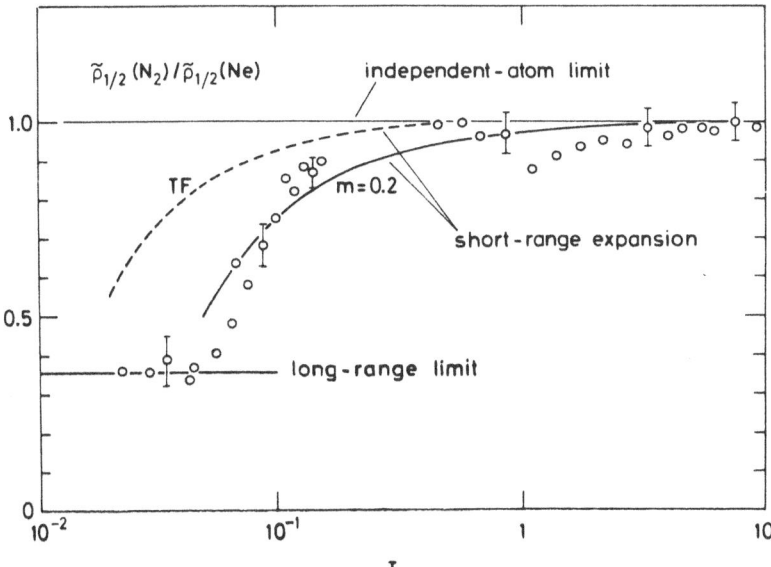

Fig.30. Ratio of multiple-scattering half-widths measured for heavy keV ions incident on nitrogen and neon gas targets. The measured quantity is not the half-angle but the half-width $\rho_{1/2}$ of the lateral spread. Measurements from Copenhagen and Aarhus Universities. Details in the original paper. From Sigmund (1977)

different cases, all referring to transmission measurements, are illustrated in fig. 31.

In the upper case, essentially all transmitted particles are collected, and hence, all energy loss including nuclear stopping will be recorded.

In the center graph, only particles transmitted with negligible angular deflection will be recorded, and hence, nuclear energy loss will be more or less suppressed since such collision events are accompanied by angular deflection.

In the lower case, the energy loss can be determined as a function of scattering angle. For angles exceeding the multiple scattering half-width, the mean energy loss may then be written in the form

$$\langle \Delta E \rangle = NxS + T(\alpha), \qquad\qquad\qquad (4.30)$$

where $T(\alpha)$ is the energy loss (nuclear and/or electronic) in a single scattering event at an angle α. The dependence on thickness and scattering angle complicates the analysis, but it also allows to extract additional information, e.g., on the impact parameter dependence of energy loss. Pertinent generalizations of multiple scattering theory have been provided (Meyer et al., 1977; Jakas et al.,1983,1984; Gras-Marti,1985; Sigmund & Winterbon,1985).

4.7.2. Measurements. The situation has been analysed by Geissel et al. (1984) for slow heavy ions and by Lantschner et al. (1987) for protons at energies around the stopping power maximum. In both cases, there was achieved reasonable qualitative understanding in terms of known stopping mechanisms.

Measurements with high-energy hydrogen ions (5-7 MeV) yielded rather dramatic effects shown in fig. 32 (Ishiwari et al.,1982). From eq. (4.30), you deduce single-collision losses of $T(\alpha)$ = 10 and 6 keV for Be and Ag, respectively. These values are much larger than recoil losses at the scattering angles in question, and they are significantly larger than mean electron-

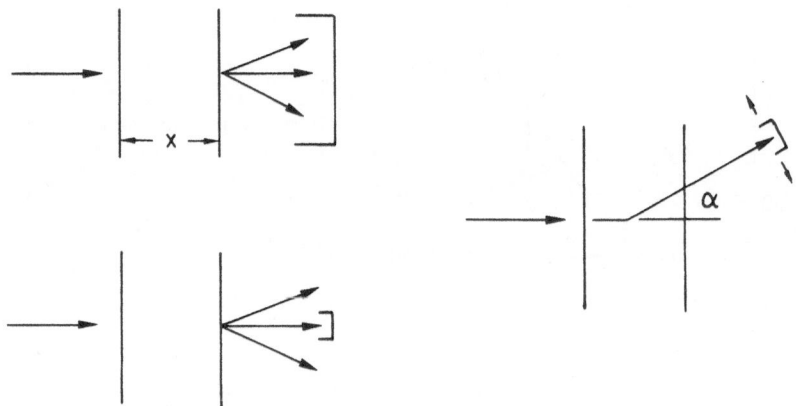

Fig.31. Detector geometries in stopping measurements

ic losses at the impact parameters involved. In particular for beryllium, the quoted value is almost two orders of magnitude greater than the excitation energy of the K-shell. These considerations were quantified by Gras-Marti (1985).

4.7.3. Nonuniform Target Thickness. Qualitative considerations by Lindhard (1984) and numerical estimates by Mertens & Krist (1986) revealed the importance of nonuniform foil thickness for measurements of this type. This is illustrated in fig. 33. Recent data taken on gold foils with controlled uniformity show a significantly reduced effect (Ishiwari et al., 1990). It is unlikely, however, that the large disparity in the beryllium data can be explained by foil nonuniformity alone.

These measurements were taken in an energy range where the fundamental process of energy loss is supposed to be well understood. A measurement on foils of controlled quality of beryllium or another light target material, done at another laboratory, ought to be performed in order to give a hint as to whether the existing data are due to an experimental artefact or whether some fundamental aspect has been left out of consideration.

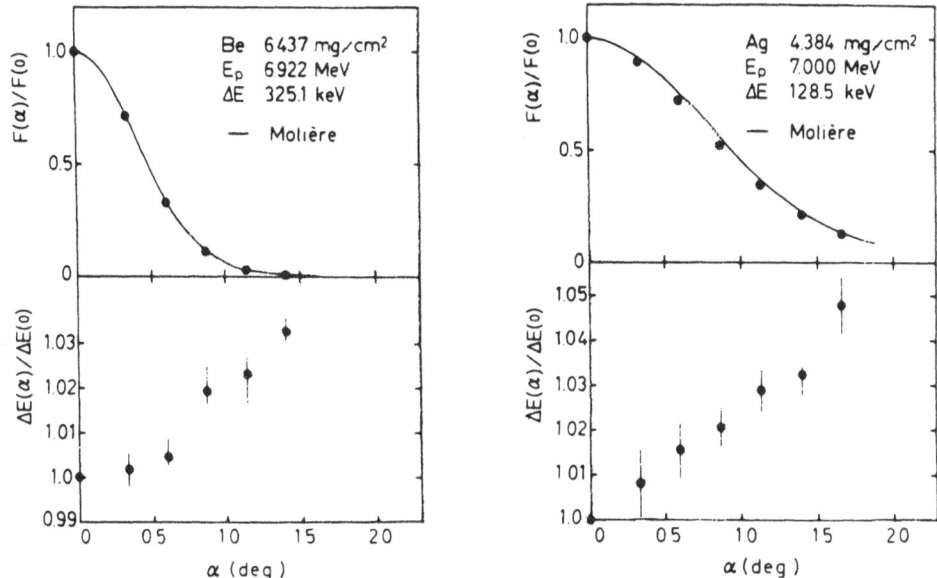

Fig.32. Mean energy loss versus emergence angle of 7 MeV protons in Be and Ag foils. Also the multiple scattering distributions are shown. From Ishiwari et al. (1972)

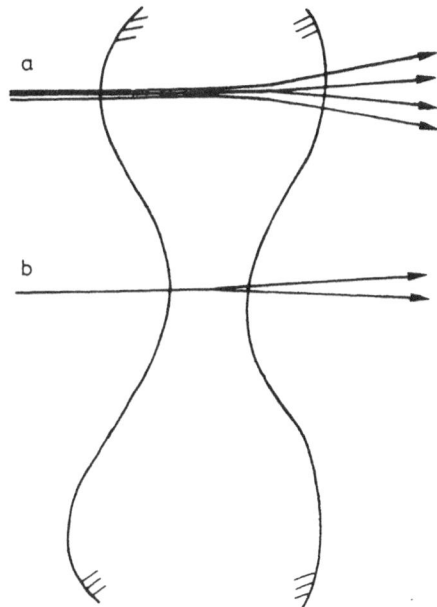

Fig.33. Schematic illustration of the fact that for a nonuniform target, high energy loss tends to correlate with high multiple-scattering angles

4.8. Multiple Scattering of Diatomic Molecules

I should like to conclude these lectures by reporting some results of my recent work that has not yet been finished. This gives me the opportunity to mention a situation where assumption i) from lecture 1 is not any longer valid in some sense.

Molecules penetrating through solid matter will in general dissociate rapidly, and the fragments will move away from each other due to their mutual Coulomb repulsion. A few years ago, it was pointed out that collecting molecular fragments emerging from a very thin foil on a position-sensitive detector may reveal information on the initial structure of the molecule (Gaillard et al.,1978). This has led to the construction of sophisticated detecting devices (Faibis et al.,1986). This equipment has been utilized successfully to describe both the average structure and the thermal vibrations in real space of some simple molecules.

Data of this type are affected by multiple scattering of the molecular fragments. Evidently, the broadening effect of multiple scattering must compete with that of the thermal vibrations in the molecule. An important question is whether multiple scattering and Coulomb explosion superimpose linearly so long as the pertinent angles are small (which is what they are).

Fig. 34 shows the initial configuration of a diatomic molecule starting to Coulomb-explode, i.e., at time t = 0. The two trajectories are curved, and they are affected by a scattering event taking place at some time t > 0. After this event, Coulomb explosion continues, starting from different velocities of the projectile atoms. If only one scattering event occurs, you may write the following expression for the asymptotic relative velocity $\underline{v} = \underline{v}_1 - \underline{v}_2$,

$$\underline{v}(\infty) = \underline{v}_0(\infty) + P(t)\ \underline{\Delta v}\ , \tag{4.31}$$

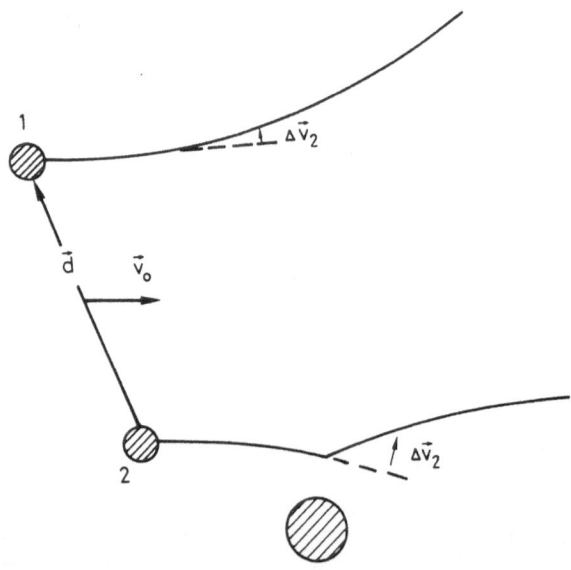

Fig.34. Interaction of a Coulomb-exploding molecule with a target atom. Unlike in ordinary multiple scattering, trajectories are curved in between collisions due to the mutual Coulomb force between the atoms

138

where $\underline{v}_0(\infty)$ is the asymptotic relative velocity due to Coulomb explosion in the absence of a scattering event, $\underline{\Delta v}$ the change in relative velocity due to the scattering event, and P(t) a factor that depends on the time at which the scattering event occurs. Eq. (4.31) is the leading portion of an expansion in powers of $\underline{\Delta v}$, i.e., it assumes multiple scattering to be a small correction. Note that it is the relative velocity that is of interest when molecular fragments are observed in a position-sensitive detector.

Fig. 35 shows the function P(t) for the two limiting cases, P_1 and P_t, where $\underline{\Delta v}$ is directed parallel and perpendicular, respectively, to the molecular axis. It is seen that both P_1 and P_t approach 1 at large t. This was to be expected since at large t, the fragments move independently, so multiple scattering will superimpose. For small t, i.e., scattering in the early stage of Coulomb explosion, a transverse kick is enhanced while a longitudinal kick is suppressed according to fig. 35. In particular the second feature appears surprising at first sight. You will accept it by noticing that the main effect of a kick in the longitudinal direction cannot depend on the sign of Δv, i.e., must be of second order.

The main conclusion is illustrated in fig. 36 for two limiting cases. Small-angle deflections occur by a kick given perpendicular to the direction of motion. If the moving molecule is oriented perpendicular to the beam direction, the asymptotic velocity (4.31) will be governed by P_1 in fig. 35, i.e., the effect of multiple scattering will be less than in the absence of Coulomb explosion. Conversely, if the molecule is oriented parallel to the beam, multiple scattering will be governed by P_t, i.e., will be enhanced. However, the two atoms will be scattered at the same impact parameter. Therefore, the *absolute* momentum transfer will be large, but the increment $\underline{\Delta v}$ in *relative* velocity will be small.

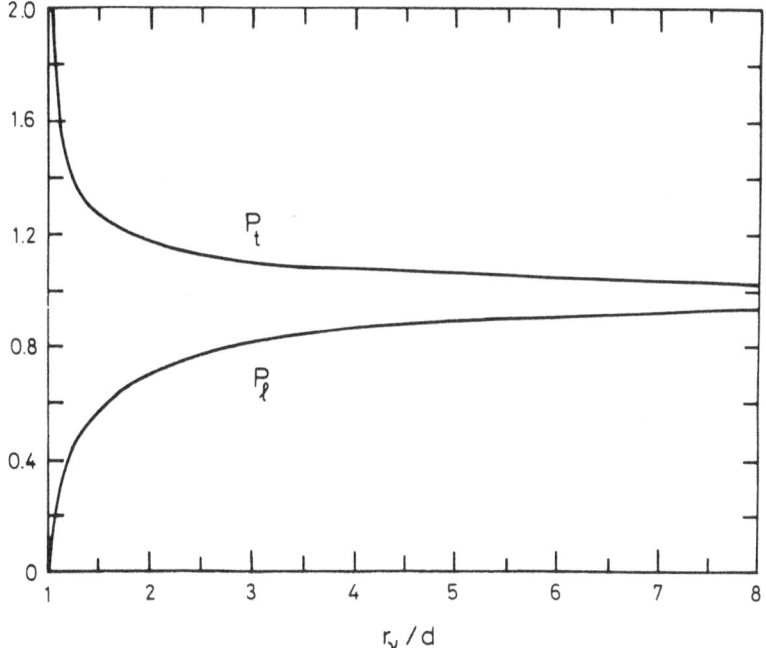

Fig.35.   Function P(t) entering eq. (4.31) as a function of the time t at which a Coulomb-exploding molecule gets a kick. P_t: Kick perpendicular to the instantaneous molecular axis: P_1: Kick along the axis. r_v is the internuclear distance at the time at which a collision occurs

As a result, one may conclude that the image of a Coulomb-exploding molecule on a detector will be *less* smeared by multiple scattering, that the smearing will be strongly anisotropic, and in some directions less pronounced than what you would expect on the basis of straight superposition, but more pronounced in others.

CONCLUDING REMARKS

I should like to conclude by making a few remarks on the scope of the material presented and on my views on the need for future activity.

I have addressed mostly penetrating ions. The reason for this, apart from my own interest and that of the majority of the participants in the school, is the wide variety of interesting phenomena and important applications. However, all the central concepts apply to all kinds of *penetrating particles,* and all the calculational details apply with little modification to *penetrating charged particles*.

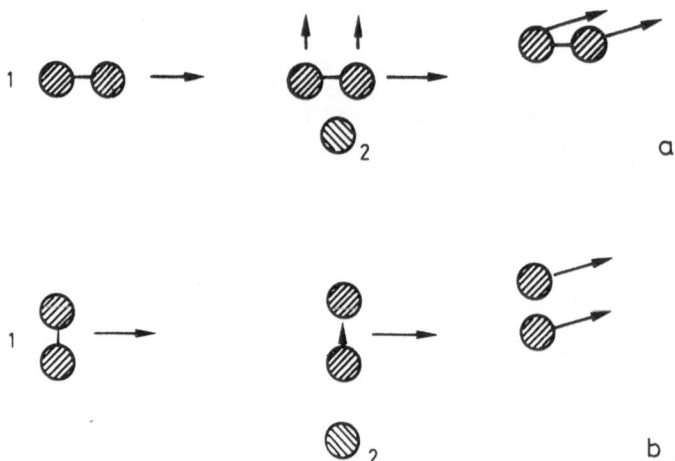

Fig.36.   Small-angle scattering of Coulomb-exploding molecule (schematic).
a) Oriented parallel to direction of motion;
b) Oriented perpendicular to direction of motion

There is essentially no limitation on the velocity range in which the concepts and tools apply. I have concentrated on particles penetrating foils and, hence, have deemphasized phenomena associated with true slowing-down. For very slow particles, penetration in thick targets and reflection from surfaces are the dominating events, neither of which has been addressed in any detail. However, the basic statistical concepts apply to those aspects in unmodified form.

I have emphasized binary-collision statistics and mentioned collective aspects only very briefly. Describing particle penetration in terms of polarization (or "wake") fields has been found to lead to interesting phenomena that are investigated experimentally by studying penetration properties of molecular beams and spectra of emitted secondary electrons. The main reason why I disregarded this area here is the fact that most theoretical studies in the area deal with mean fields and ignore statistical fluctuations.

There are many interesting statistical problems that arise with crystal lattice effects in particle penetration. This topic is covered in other lectures at this school.

Radiation effects have hardly been mentioned at all. They are caused by secondary electrons and/or recoiling atoms and are frequently dominated by low-energy particles, i.e., are associated with near-complete slowing down. Considerable difficulties are encountered with low-energy particles, such as increasing importance of impact parameter correlation, breakdown of the binary-collision approximation and the mean-free-path concept, and postcollisional relaxation phenomena.

I have already indicated on various occasions where I see a need for future theoretical activity. There are the classical areas, such as improved estimates of atomic stopping, straggling, and scattering parameters. At least the regime of heavier ions offers many challenges. I find that the study of non-Poisson penetration statistics is only in its beginning. The study of charge-changing collisions has experienced major progress recently with regard to the atomic physics aspects. Implications on particle penetration should be looked into, not the least for solid targets. I mentioned above the need to look into fluctuations in collective phenomena like wake fields, and I gave a brief discussion of molecule penetration which is an active field on the experimental side which might take beenefit from fresh theoretical input.

Acknowledgements

I should like to thank the organizers of this school, especially Alberto Gras-Marti and Herbert Urbassek, for providing the challenge to write a summary of this field at a time which, from the point of view of my being available for doing the job, could hardly have been chosen more adversely. Thanks are due to Arne L. Tofterup for checking a preliminary manuscript, and to Herbert Urbassek for very carefully going through the final writeup. Part of my work in this area was supported by a NATO Collaborative Research Grant.

REFERENCES

Abramowitz, M. & Stegun, I.A , 1965, *Handbook of Mathematical Functions*, Dover Publ.

Amsel, G., 1982, Nucl. Instrum. Methods **194**: 1.

Andersen, H.H., Bøttiger, J., Knudsen, H., Møller Petersen, P., & Wohlenberg, T., 1974, Phys. Rev. A **10**: 1568.

Anne, R., Herault, J., Bimbot, R., Gauvion, H., Bastin, G., & Hubert, F., 1988, Nucl. Instrum. Methods **B34**: 295.

Ascolani, H. & Arista, N.R., 1986, Phys. Rev. A **33**: 2352.

Ashley, J.C., Ritchie, R.H., & Brandt, W., 1972, Phys. Rev. B **5**: 2393.

Baily, N.A., Steigerwaldt, J.E., & Hilbert, J.W., 1970, Phys. Rev. B **2**: 577.

Bak, J.F., Burenkov, A., Petersen, J.B.B., Uggerhøj, E., Møller, S.P., & Siffert, P., 1987, Nucl. Phys. **B288**: 681.

Barrett, J.H., Appleton, B.R., & Holland, O.W., 1980, Phys. Rev. B **22**: 4180.

Bateman, H., 1911, Philos. Mag. **21**: 745.

Belkacem, A., & Sigmund, P., 1990, Nucl. Instrum. Methods B **48**: 29.

Berger, M.J., 1963, Meth. Comp. Phys. **1**: 135.

Besenbacher, F., Heinemeier, J., Hvelplund, P., & Knudsen, H., 1977, 1978, Phys. Lett. **61A**: 75; Phys. Rev. A **18**: 2470.

Besenbacher, F., Andersen, J.U., & Bonderup, E., 1980, Nucl. Instrum. Methods **168**: 1.

Besenbacher, F., Andersen, H.H., Hvelplund, P., & Knudsen, H., 1981, Mat. Fys. Medd. Dan. Vid. Selsk. **40**: no. 9.

Bethe, H. 1930, Ann. Physik **5**: 325.

Betz, H.D., 1972, Rev. Mod. Phys. **44**: 465.

Betz, H.D., 1990, this school.

Bichsel, H., 1970, Phys. Rev. B **1**: 2854.

Bichsel, H., 1988, Rev. Mod. Phys. **60**: 663.

Bichsel, H. & Saxon, R. P., 1975, Phys. Rev. **11**: 1286.

Biersack, J.P. & Haggmark, L.G., 1980, Nucl. Instrum. Methods **174**: 257.

Bloch, F., 1933, Ann. Physik 16: 285.

Blomberg, A., Åberg, T., & Goscinski, O., 1986, J. Phys. B **19**: 1063.

Blunck, O. & Leisegang, S., 1950, Z. Physik **128**: 500.

Bohr, A, 1948, Mat. Fys. Medd. Dan. Vid. Selsk. **24**: no. 19.

Bohr, N., 1913, Philos. Mag. **25**: 10.

Bohr, N., 1915, Philos. Mag. **30**: 581.

Bohr, N., 1948, Mat. Fys. Medd. Dan. Vid. Selsk. **18**: no. 8.

Bohr, N. & Lindhard, J., 1954, Mat. Fys. Medd. Dan. Vid. Selsk. **28**: no. 7.

Bonderup, E., 1967, Mat. Fys. Medd. Dan. Vid. Selsk. **35**: no. 17.

Bonderup, E., 1981, *Penetration of Charged Particles through Matter*, Aarhus University.

Bonderup, E., & Hvelplund, P., 1971, Phys. Rev. A **4**: 562.

Bothe, W., 1921, Z. Physik **4**, 161, 300; **5**, 63.

Capuj, N.E., Eckardt, J.C., Lantschner, G.H., & Arista, N.R., 1987, Phys. Rev. **36**: 3715.

Chandrasekhar, S., 1943, Rev. Mod. Phys. **15**: 1.

Chu, W.K., 1976, Phys. Rev. A **13**: 2057.

Chu & Powers, 1972, Phys. Lett. **40a**: 23.

Clouvas, A., Gaillard, M.J., dePinho, A.G., Poizat, J.C., & Remillieux, J., 1984, Phys. Lett. **103A**: 419.

Cowern, N.E.B., Sofield, C.J., Freeman, J.M., & Mason, J.P., 1979, Phys. Rev. A **19**: 111.

Crawford, O.H., 1980, Phys. Rev. Lett. **44**: 185.

Dixmier, M., L'Hoir, A., & Amsel, G., 1982, Nucl. Instrum. Methods **197**: 537.

Echenique, P., 1990, this school.

Esbensen, H. & Sigmund, P., 1990, Ann. Phys. (New York), July issue.

Fano, U., 1963, Ann. Rev. Nucl. Sci. **13**: 1.

Faibis, A., Koenig, W., Kanter, E.P., Vager, Z., 1986, Nucl. Instrum. Methods B **13**: 673.

Feldman, L.C., 1990, this school.

Feller, W., 1957, 1966, *An Introduction to Probability Theory and its Applications*, Vols. I & II, Wiley, New York.

Firsov, O.B., 1959, Zh. Eksp. Teor. Fiz. **36**: 1517; Sov. Phys. JETP **9**: 1076.

Flamm, L., 1914, 1915, Sitzungsber. Akad. Wiss. Wien, Math. Nat. Wiss. Kl. **123**: 1393; **124**: 597.

Gaillard, M.J., Gemmell, D.S., Goldring, G., Levine, I., Pietsch, W.J., Poizat, J.C., Ratkowski, A.J., Remillieux, J., Vager, Z., Zabransky, B.J., 1978, Phys. Rev. A **17**: 1797.

Geissel, H., Lennard, W.N., Andrews, H.R., Ward, D., & Phillips, D., 1984, Phys. Lett. **106A**: 371.

Geissel, H., Lennard, W.N., Andrews, H.R., Jackson, D.P., Mitchell, I.V., Philips, D., & Ward, D., 1985, Nucl. Instrum. Methods B **12**: 38.

Gerjuoy, E., 1966, Phys. Rev. **148**: 54.

Glauber, R.R., 1955, Phys. Rev. **100**: 242.

Goscinski, O., Åberg, T., & Peltonen, M., 1982, Phys. Lett. **90A**: 464.

Goudsmit, S. & Saunderson, J.L., 1940, Phys. Rev. **84**: 1092.

Gras-Marti, A., 1985, Nucl. Instrum. Methods **B9**: 1.

Gryzinski, M., 1965, Phys. Rev. **138**: A305, A322, A366.

Hansen, F.Y., Knudsen, T.S., & Carneiro, K., 1975, J.Chem. Phys. **62**: 1556.

Harberger, J.H., Johnson, R.E., & Boring, J.W., 1974, Phys. Rev. A **9**: 1161, 1172.

Herzfeld, K.F., 1912, Phys. Z. **13**: 347.

Huang, W., Urbassek, H.M., & Sigmund, P., 1985, Philos. Mag. A **52**: 753.

Hvelplund, 1968, prize essay, Aarhus University.

Hvelplund, P., 1975, Phys. Rev. A **11**: 1921.

Inokuti, M., 1971, Rev. Mod. Phys. **43**: 297.

Inokuti, M., Itikawa, Y., & Turner, J.E., 1978, Rev. Mod. Phys. **50**: 23.

Inokuti, M., Dehmer, J.L., Baer, T., & Hansen, J.D., 1981, Phys. Rev. A **23**: 95.

142

Ishiwari, R., Shiomi, N., & Sakamoto, N., 1982, 1983, 1984, Phys. Rev. A **25**: 2524; **27**: 810; **30**: 82.

Ishiwari, R., Shiomi-Tsuda, N., Sakamoto, N. & Ogawa, H., 1990, Nucl. Instrum. Methods B **48**: 65.

Jackson, J.D., 1975, *Classical Electrodynamics*, Wiley, New York.

Jakas, M.M. & Baragiola, R.A., 1980, Phys. Rev. Lett. **44**: 425.

Jakas, M.M., Lantschner, G.H., Eckardt, J.C., & Ponce, V.H., 1983, 1984, phys. stat. sol. (b) **117**: K131; Phys. Rev. A **29**: 1838.

Jakas, M.M. & Ponce, V.H., 1984, J. Phys. D **17**: 1303.

Kampen, N.G.v., 1981, *Stochastic Processes in Physics and Chemistry*, North Holland, Amsterdam

Kanter, E.P., 1983, Phys. Rev. A **28**: 1401.

Kölbig, K.S., & Schorr, B., 1984, Comp. Phys. Commun. **31**: 97.

Kumbartzki, G.J., Neuburger, H., & Polster, W., 1985, Nucl. Instrum. Methods B **12**: 315.

Kührt, E., & Wedell, R., 1981, 1983, Phys. Lett. **86A**: 54; **96A**: 347.

Landau, L., 1944, J.Phys. USSR **8**: 201.

Lantschner, G.H., Eckardt, J.C., Jakas, M.M., Capuj, N.E., & Ascolani, H., 1987, Phys. Rev. A **36**: 4667.

Lassen, N.O., 1951, Mat. Fys. Medd. Dan. Vid. Selsk. **26**: no. 5; no. 12.

Lennard, W.N., Geissel, H., Phillips, D., & Jackson, D.P., 1986, Phys. Rev. Lett. **57**: 318.

Lindhard, J., 1954, Mat. Fys. Medd. Dan. Vid. Selsk. **28**: no. 8.

Lindhard, J., 1965, Mat. Fys. Medd. Dan. Vid. Selsk. **34**: no.14.

Lindhard, J., 1970 in *Atomic Collision Phenomena in Solids*, D.W.Palmer et al., eds., North Holland, Amsterdam, p. 675.

Lindhard, 1976, Nucl. Instrum. Methods **132**: 1.

Lindhard, J.,1984, unpublished private communication.

Lindhard, 1985, Phys. Scr. **32**: 72.

Lindhard, J., 1990, this school; no written version has been submitted to the proceedings; the elementary parts of J.Lindhard's lectures have been written up by Bonderup (1981).

Lindhard, J., Scharff, M., & Schiøtt, H.E., 1963, Mat. Fys. Medd. Dan. Vid. Selsk. **33**: no. 14.

Lindhard, J. & Nielsen, V., 1971, Mat. Fys. Medd. Dan. Vid. Selsk. **38**: no.9.

Lindhard, J., Nielsen, V., & Scharff, M., 1968, Mat. Fys. Medd. Dan. Vid. Selsk. **34**: no. 14.

Lindhard, J. & Scharff, M., 1953, Mat. Fys. Medd. Dan. Vid. Selsk. **27**: no. 15.

Lindhard, J. & Winther, A., 1964, Mat. Fys. Medd. Dan. Vid. Selsk. **34**: no. 4.

Livingston, M.S. & Bethe, H.A., 1937, Rev. Mod. Phys. **9**: 245.

Loftager, P., Besenbacher, F., Jensen, O.S., & Sørensen, V.S., 1979, Phys. Rev. A **20**: 1443.

Lucas, A., 1990. this school.

Malherbe, J.B. & Alberts, H.W., 1982, Nucl. Instrum. Methods **192**: 559.

Marwick, A.D. & Sigmund, P., 1975, Nucl. Instrum. Methods **126**: 317.

McGuire, E.J., 1971, Phys. Rev. A **28**: 2096.

Meitner, L., & Freitag, K., 1926, Z. Physik **37**: 481.

Mertens, P., 1986, Nucl. Instrum. Methods **B13**: 91.

Mertens, P. & Krist, Th., 1986, Nucl. Instrum. Methods B **13**: 95.

Meyer, L., 1971, phys. stat. sol. (b) **44**: 253.

Meyer, L., Klein, M., & Wedell, R., 1977, phys. stat. sol. (b) **83**: 451.

Mikkelsen, H.H., 1990, to appear in Nucl. Instrum. Methods B.

Mikkelsen, H.H., & Mortensen, E.H., 1990, Nucl. Instrum. Methods B **48**: 39.

Mikkelsen, H.H. & Sigmund, P., 1989, Phys. Rev. A **40**: 101.

Mikkelsen, H.H. & Sigmund, P., 1987, Nucl. Instrum. Methods **B27**: 266.

Moak, C.D., & Brown, M.D., 1966, Phys. Rev. **149**: 244.

Molière, G., 1947, 1948, Z. Naturforsch. **2a**: 133; **3a**: 78.

Moyal, J.E., 1955, Philos. Mag. **46**, 263.

Oddershede, J., & Sabin, J.R., 1984, Atomic Data & Nucl. Data Tables **31**: 275.

Olson, R.E., 1989, Phys. Rev. A **39**: 5572; Radiat. Eff. **110**: 1.

Poisson, D.S., 1837, *Recherches sur la probabilité des jugements en matière criminelle et en matière civile*, Bachelier, Paris.

Pronko, P.P., Appleton, B.R., Holland, O.W., & Wilson, S.R., 1979, Phys. Rev. Lett. **43**: 779.

Robinson, M.T., & Torrens, I.M., 1974, Phys. Rev. B **9**: 5008.

Schiwietz, G., 1990, this school.

Schmaus, D. & L'Hoir, A., 1982, Nucl. Instrum. Methods **194**: 75.

Schmaus, D., L'Hoir, A., & Cohen, C., 1982, Nucl. Instrum. Methods **194**: 81.

Schweidler, E.v., 1910, Phys. Z. 11, 614.

Scott, W.T., 1963, Rev. Mod. Phys. **35**: 231.

Sidenius, G., Andersen, N., Sigmund, P., Besenbacher, F., Heinemeier, J., Hvelplund, P., Knudsen, H., 1976, Nucl. Instrum. Methods **134**: 597.

Sigmund, P., 1975, in *Radiation Damage Processes in Materials*, ed. by C.H.S.Dupuy, Noordhof, Leiden, p.3.

Sigmund, P., 1976, Phys. Rev. A **14**: 996.

Sigmund, P., 1977, Mat. Fys. Medd. Dan. Vid. Selsk. **39**: no. 11.

Sigmund, P., 1978, Mat. Fys. Medd. Dan. Vid. Selsk. **40**: no. 5.

Sigmund, P., 1982, Phys. Rev. A **26**: 2497.

Sigmund, P., 1983, Phys. Scr. 28, 257 (1983)

Sigmund, P., 1989, presented at the *International Conference on Atomic Collisions in Solids*, Aarhus, unpublished.

Sigmund, P. & Fu, D.J., 1982, Phys. Rev. A **25**: 1450.

Sigmund, P. & Gras-Marti, A., 1981, Nucl. Instrum. Methods **182/183**: 25.

Sigmund, P. & Haagerup, 1986, Phys. Rev. A **34**: 892.

Sigmund, P., Heinemeier, J., Besenbacher, F., Hvelplund, P., & Knudsen, H., Nucl. Instrum. Methods **150**: 221.

Sigmund, P. & Johannessen, K., 1985, Nucl. Instrum. Methods B **6**: 486.

Sigmund, P. & Winterbon, K.B., 1974, 1975, Nucl. Instrum. Methods **119**: 541; **125**: 491.

Sigmund, P. & Winterbon, K.B., 1985, Nucl. Instrum. Methods B **12**: 1.

Smoluchowski, M.v., 1904, *Festschrift Ludwig Boltzmann*, p. 626.

Sofield, C.J., Cowern, N.E.B., & Freeman, J.M., 1981, Nucl. Instrum. Methods **191**: 462.

Symon, K.R., 1948, Thesis, Harvard Univ., unpublished.

Thomson, J.J., 1912, Philos. Mag. **23**: 449.

Tougaard, S., 1990a, this school.

Tougaard, S., 1990b, J. Elect. Spect. Related Phen. **52**: 243.

Tougaard, S. & Sigmund, P., 1982, Phys. Rev. B **25**: 4452.

Tschalär, C., 1968, Nucl. Instrum. Methods **61**: 141; **64**: 237.

Vavilov, P.V., 1957, Zh. Eksp. Teor. Fiz. **32**: 920; Sov. Phys. JETP **5**: 749.

Vollmer, O., 1974, Nucl. Instrum. Methods **121**: 373.

Williams, E.J., 1929, 1931a, 1931b, 1933, 1939, Proc. Roy. Soc. A **125**: 420; **130**: 328; **135**: 108; **139**: 163; **169**: 531.

Williams, E.J., 1940, Phys. Rev. **58**: 292.

Williams, E.J., 1945, Rev. Mod. Phys. **17**: 217.

Winterbon, K.B., 1968, Can. J. Phys. **46**: 2429.

Winterbon, K.B., 1977, Nucl. Instrum. Methods **144**: 311.

Winterbon, K.B., 1983, Radiat. Eff. **79**: 251.

Winterbon, K.B., 1987, 1989, Nucl. Instrum. Methods B **21**: 1; **43**: 146.

ACCELERATORS AND STOPPING POWER EXPERIMENTS

H.H. Andersen

Physics Laboratory, H.C. Ørsted Institute
Universitetsparken 5
DK-2100 Copenhagen Ø, Denmark

ABSTRACT

In a series of four lectures, organized in three chapters, the accelerators used for energy-loss measurements are described. The transport and diagnostics of the ion beams are discussed, and experimental details of stopping-power experiments are treated. A number of accelerator topics of less direct connection to stopping powers like storage rings, cooling, microbeams and accelerator mass spectrometry are also touched upon.

1. INTRODUCTION

The present set of lecture notes is aiming mainly at students who have not worked with accelerators themselves but are interested in knowing how energy-loss data are obtained and which pitfalls to look out for in experimental publications. They may, hopefully, also serve as a simple introduction for students wanting to enter the experimental field themselves. The notes are at the tutorial level. That means they are not provided with footnotes and references like original scientific publications, but each chapter has been provided with a brief list of suggestions for further reading.

To give some sense of completeness to the presentation, a number of hot topics from recent accelerator developments are included. These developments either go into the direction of improving the quality of the provided ion beams, which is certainly of interest for energy-loss measurements, or for improved analytical accelerator facilities, which will also be of interest for most students of the present school.

Stopping power measurements have been performed over 11 orders of magnitude in energy from 1 eV to 10^{11} eV. With the coming Pb beam at the CERN SPS machine this range may hopefully be extended by two orders of magnitude in the upward direction. Beams of these projectiles are nearly

Interaction of Charged Particles with Solids and Surfaces
Edited by A. Gras-Martí *et al.*, Plenum Press, New York, 1991

145

always obtained from accelerators. To be useful for stopping power the measurements beams should be well defined with respect to species, energy and direction. Most of these lectures are dealing with how to produce such beams and how to ensure they remain in such a condition.

In a few cases beams produced by non-accelerator techniques may be utilized, e.g., beams produced from sources of natural radioactivity, mainly alpha particles (~ 5 MeV) and fission fragments but also alpha recoils and beams produced through thermal neutron reactions in light elements.

The topic of particle detectors, also of importance for stopping power measurements, will not be treated here. Some coverage of this topic will be found in the lectures and references of Feldman at the present school.

A number of subsections of the present notes like accelerator developments, ion sources, beam diagnostics, storage rings and accelerator mass spectrometry have been the topics of full summer schools. Nearly all subsections have at least been covered as full lectures at summer schools. These facts imply that by necessity the present lectures are rather superficial in their coverage at many places. As the author is no expert in all fields covered, another reason for this superficiality may at some places be all too apparent.

2 ACCELERATORS

Two basically different types of "standard" accelerators are shown in fig.1. In the upper example the charged particles are produced in an ion source kept at high potential and accelerated over a number of DC high-voltage gaps to ground potential, where the beam is analyzed with regard to mass, charge state and energy before impinging on the target. This is the standard DC accelerator, a variation of which allowing the ion source to be kept close to ground potential, viz. the tandem accelerator, will be discussed below. In the standard AC accelerator, also shown in fig.1, the ion source may be kept close to ground potential, and acceleration is achieved through use of an AC voltage over an acceleration gap. To have effective beam utilization the beam must be bunched to ensure that injection into the structure will take place at a phase angle that will give rise to acceleration. As discussed below, the accelerator may be circular allowing the use again and again of the same acceleration gap or linear, utilizing a series of acceleration gaps often in connection with a travelling RF wave.

I) DC-Accelerators

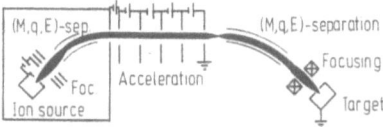

II) AC-Accelerators

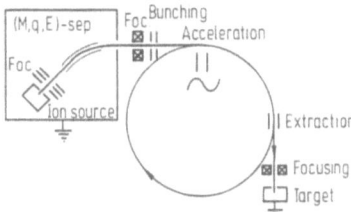

Figure 1. Standard AC and DC accelerators showing ion source, accelerating stage, beam analysis and target. The acceleration stage only discerns between the AC and DC systems. (M, q, E means mass, charge state and energy).

We shall not here be treating electron accelerators but they do not differ much from ion accelerators. Only the ion source is replaced by a hot cathode or filament supplying the electrons.

2.1 Ion Sources

For energy loss measurements we are interested in beams of all elements in the periodic system. For tandem accelerators we need negative ions, and to get heavy ions to very high energies it is very useful to have highly charged positive ions. We do hence need systems that produce negative, positive and multicharged ions of all elements. For a few elements, stable negative ions do not exist but in most of these cases a compound ion may be accelerated instead (e.g., NH_3^- to obtain nitrogen beams from a tandem accelerator).

Positive and negative ion beams may be produced by a large number of methods. The most important ones are through extraction from an appropriate plasma. These and a number of other methods are summarized below in a very condensed style. The list is certainly not complete but does contain the most important methods.

A) Plasma composition. The plasma is created by a discharge in a carrier gas, most often a noble gas. The wanted element may be introduced into the discharge: as a gas (if gaseous, e.g., noble gases, hydrogen, oxygen, nitrogen ...), as a volatile or gaseous compound (H_2S, BF_3, CO_2, CuCl, $NiCl_2$ ...), or by evaporation (S, Cs, Bi, Hg ...).

B) Positive or negative ions from a plasma. Most plasmas contain negative as well as positive ions to some extent spatially separated. Extraction is done at the right place (axially on axis, axially off axis, radially) to obtain the wanted ion. It is difficult to extract negative ions from a plasma without loading high-voltage supplies with a heavy electron current.

C) Multicharged ions. Make the plasma *very* hot and achieve a long lifetime in the plasma to enhance production of multicharged ions through a multistep process.

D) Sputtering. A high fraction of sputtered ions may be charged. Positive ions may be produced directly by sputtering from a high work-function surface, negative ions by sputtering from a low work-function surface. By partial surface coverage with oxygen or cesium, most solid materials may be prepared in either state. O or Cs may be used as the sputter beam.

Heavy ion sputtering may be used to enhance production of cluster ions if such species are wanted.

E) Surface ionization. Let a beam of neutral atoms impinge on a hot surface. Conditions are as for sputtering but it is difficult to make a low work-function hot surface without having excessive electron emission. The method is used extensively for on-line accelerator systems (ISOLDE at CERN).

F) Field emission. The high field at a very sharp tip is used to extract ions from a molten metal seeping through the tip. Clusters and higher charge states are obtained abundantly.

G) Neutral beam ionization. A very intense electron or photon beam is used to ionize the atoms. The cross section for e-beam ionization is very low. Resonance absorption may be used for lasers, often as a two- or three-photon process.

H) Charge exchange plus to minus. The positive beam is sent through an alkali vapor at low energy. Most often lithium is used. This method is necessary for production of He^-.

I) Charge exchange minus to plus and plus to multiplus. This electron stripping process may be utilized at high energies in a gas or a solid foil (cf. lectures by Lindhard and Betz at this school).

J) Laser impact. A high-intensity laser pulse impacting on a solid produces a burst of highly-charged ions. Their energy distribution is very broad.

148

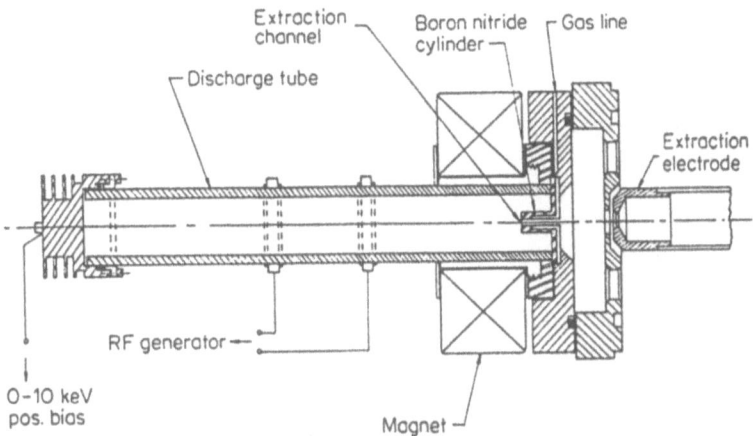

Figure 2. RF ion source with capacitive coupling of the radiofrequency power. Also inductively-coupled systems exist.

Table 1. Ion sources. LMIS: Liquid Metal Ion Source. ANIS: Aarhus Negative Ion Source. EBIS: Electron Beam Ion Source. ECRIS: Electron Cyclotron Resonance Ion Source.

Name	Elements	Currents	Charge state	Comments
RF	Gases, Li	1mA	+	
Duoplasmatron	Gases	100mA	+	
Penning	Gases, C	5mA	+	
Sputter Sources	Most solids	10–100µA	+	
Nielsen-Almen	Gases, vapor, chlorinated compounds (CCl_4)	100µA	+, (2+)	
Freeman	Solids	10mA	+, (2+)	
Sidenius (high temperature)	Most elements	100µA	+, (2+)	
LMIS	Ga, In, Sn, Au Pb, Tl, Pb, Bi	µA	+ ... 4+	Clusters. Extremely high brightness.
Duoplasmatron (off axis)	H,O	100µA	–	
Penning (radial)	C^-, NH^-, Halogens	100µA	–	
Middleton	Most solids	50µA	–	
ANIS	Most solids	50µA	–	
EBIS, ECRIS	Most elements	30µA	+ ... 40+	
Bucket source	Gases	100mA–1A	+	

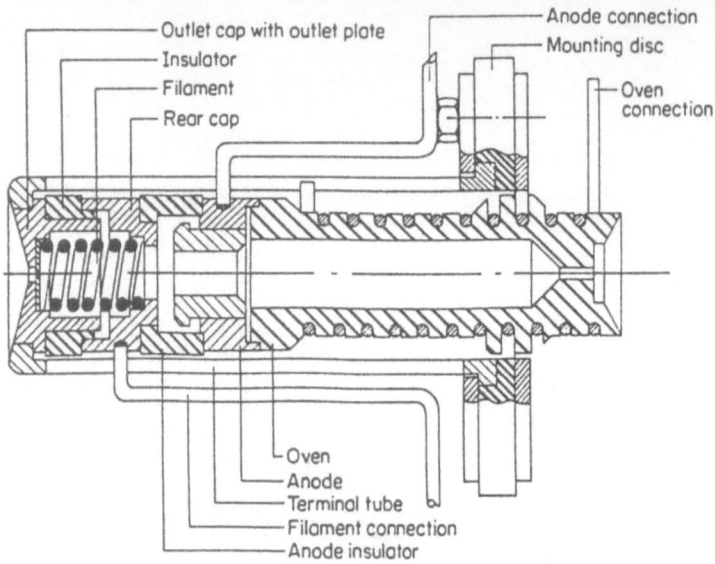

Figure 3. The Sidenius high-temperature ion source as produced by Danfysik.

Virtually all these principles have been used in accelerator systems and quite a number appears in commercial ion sources. A number of ion sources are listed in table 1 below under their nicknames together with some of the ions produced.

A few of these ion sources will be described in somewhat more detail. Together with the duoplasmatron source, the RF source is the oldest work horse of ion acceleration. In the example shown in fig.2, the RF power used for heating the plasma is coupled capacitively, but also inductively-coupled systems exist.

The Sidenius high-temperature source works for most elements and yields relative large currents, as seen in the table. The trick is to make the discharge chamber into an oven totally surrounded by a heater, as seen in fig.3. With the whole system at very high temperature, it is an absolute necessity to have a very good vacuum to avoid corrosion of the source parts by oxygen and nitrogen. The source is particularly well suited to produce beams of refractory transition metals.

A relatively new development is the liquid-metal ion source (LMIS) also called the field ion emission source (FIE). The principle is shown in fig.4. A liquid metal is fed through a porous tungsten cone to a very sharp (R ~ 50 nm) tip. The field lines from the extraction plate are strongly concentrated at the tip, giving rise to field ionization. The

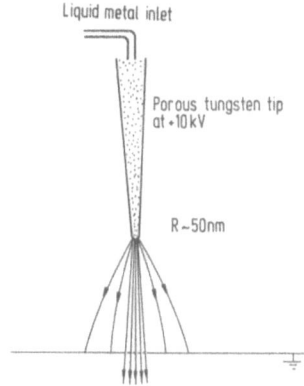

Figure 4. The field-emission principle utilized in the liquid metal ion source.

field must not be raised to such high values that the tip itself is eroded through field evaporation. The principle is the same as that utilized in the field ion microscope. As the emitting surface is extremely small, the extracted beam will have a very high brightness (see below) and be well suited for microbeam work.

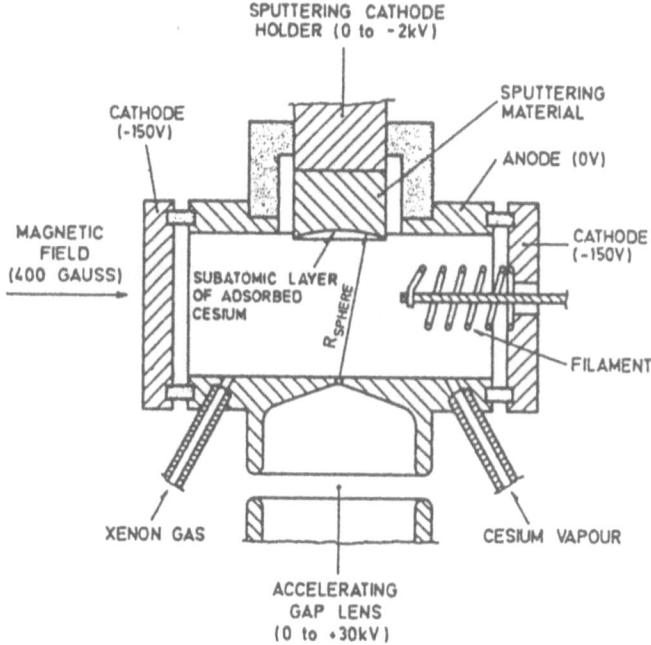

Figure 5. The ANIS (Aarhus Negative Ion Source) source. Sputtering is from a spherical cathode with center at the outlet hole. The continuous cesium bombardment of the cathode maintains the cesium coverage necessary for the creation of negative ions.

The development of tandem accelerators for heavy ions in the early 1970's created a big need for sources of negative ions. Most of these utilized cesium sputtering to create the negative ions. The Middleton source has an external cesium beam, while the cesium sputtering in the ANIS source is achieved through introduction of cesium vapor in the discharge plasma, where xenon is used as a carrier gas. The principle is shown in fig.5. The sputter cathode, from which the negative beam is obtained, is machined as a hollow part of a sphere with center in the outlet opening. Negative ions are accelerated over the plasma sheath in the direction of the hole, and exit without use of an extraction voltage. The electron loading is hence negligible. The source is now available with a rapid exchange system for the cathode from Danfysik.

Much more complicated systems for ion production and transport may be used if high-energy heavy ions are needed. An ambitious plan is to produce 33 TeV lead ions in the CERN SPS machine as shown in fig.6. The system starts with an ECR ion source delivering 30 μA (electric current, i.e., 6×10^{12} ions/s) Pb^{+30}. After being accelerated through a linac to 850 MeV, stripping is done to charge state +53. In the PS accelerator the beam is accelerated to 650 GeV and fully stripped before injection in the SPS accelerator. The final beam is of much higher intensity than needed for energy loss measurements.

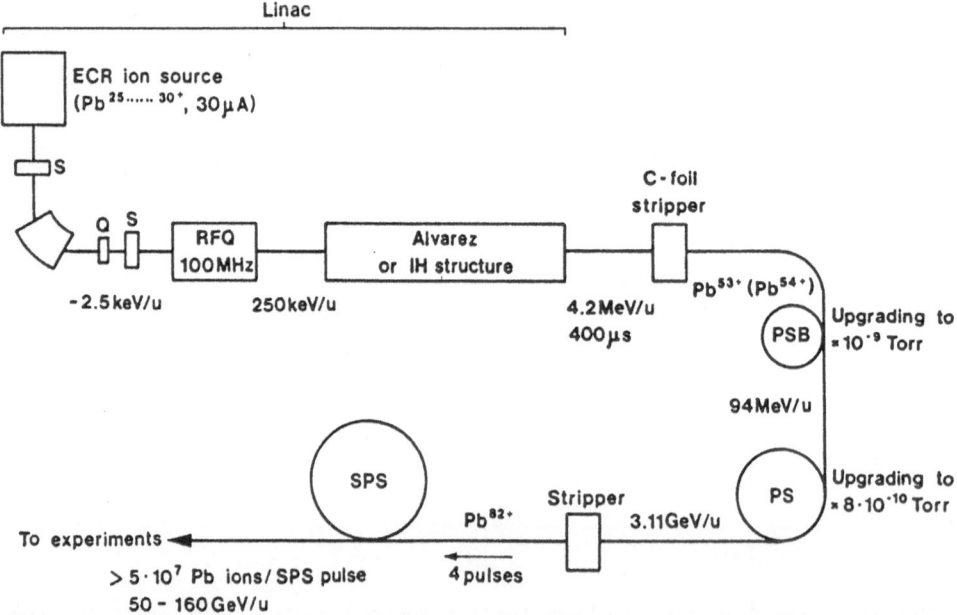

Figure 6. The planned 33 TeV Pb^{+82} facility at CERN utilizing most of the CERN accelerator complex. (From R. Billinge et. al., CERN 90-01 (1990)).

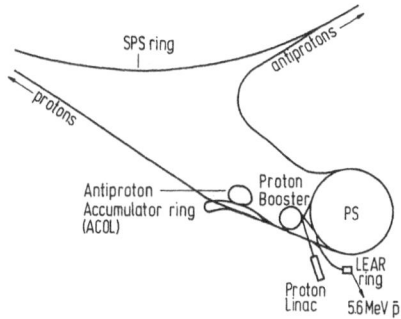

Figure 7. The CERN accelerator complex as utilized for production of low-energy p̄ beams.

The most exotic beams presently used for stopping-power measurements are low-energy antiprotons. Here, the total CERN machinery (except LEP) may be looked upon as the ion source, as shown in fig.7. Protons are accelerated through a linac, the proton booster and the PS, before impinging on a solid target to produce antiprotons that are captured and accumulated in the ACOL ring. They are then decelerated in the PS and injected into LEAR (Low Energy Antiproton Ring). Here, the beam quality is improved by cooling (see below) as was also the case in ACOL. After further deceleration, the beam for stopping measurements may be extracted.

Fortunately, many energy-loss data have been obtained with less complicated equipment. But the simpler the source, the more complex is usually the emerging beam. Ions of carrier-gas atoms, impurities, atmospheric residuals exit from the source in a lot of charge states. After DC acceleration to somewhere between a few keV and 100 keV, a preliminary purification may be done before injection into larger structures.

As clearly seen from fig.7, the same elements tend to repeat over and over again as the energy is increased: Accelerator complexes are fractal structures. We have hence until now described not only ion sources but also miniature DC accelerators.

2.2 DC Accelerators

DC machines with acceleration voltages below 500 kV are nearly always built with a solid-state electronic power supply to provide the high voltage. This is mostly obtained by some variation over the old Cockcroft-Walton rectifier voltage multiplication principle. Machines working at voltages up to 500 kV may further conveniently be built in open air without too large a risk of sparking except under very humid conditions. For higher voltages the structure is built into a pressure vessel filled with an insulating gas (N_2, CO_2 and expensive but most spark-preventing, SF_6) up to pressures of 1 MPa. According to the discussion in the previous section, the ion source in a single-stage DC machine must be placed on the high-voltage terminal, which makes the source difficult to access. Particularly in pressurized machines simple RF gas-fed sources are hence mostly used. Machines with a maximum voltage of 3 MV supplied electronically have been marketed, but, to the best of the lecturer's knowledge, never reached this voltage with any customer. Apparently, no DC machines with more than 2.4 MV supplied electronically are being marketed today.

DC accelerators for megavolt voltages are usually utilizing the van-de-Graaff principle. Charge is transported either on an insulating belt or a string of metal beads or ladder steps insulated from each other, to the high voltage terminal. Such machines were developed commercially in the 1950's for experimental nuclear physics and are still sold in substantial numbers as the low MeV range is the most important one for ion beam analysis of materials. The principle is shown in fig.8.

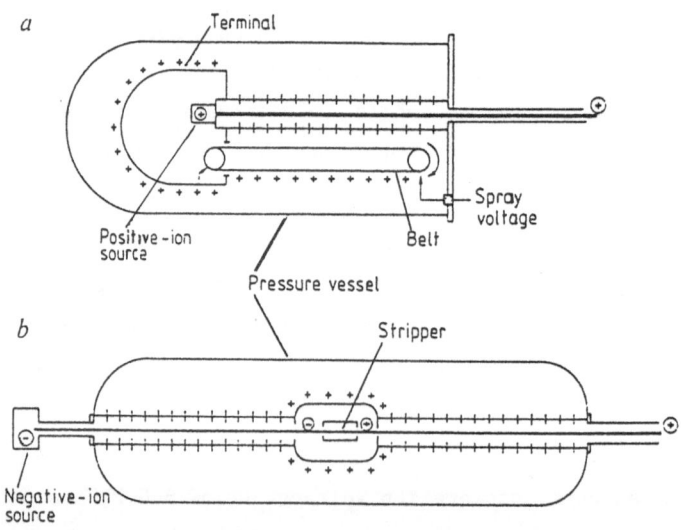

Figure 8. Schematic diagrams of a) single-ended and b) tandem electrostatic accelerators built according to the van de Graaff principle, where charge is transported to the high-voltage terminal on an insulating conveyor. The machines shown are operating within a pressure vessel containing insulating gases and intended for MV voltages.

Belts are very difficult to make homogeneous and the belt revolution frequency is hence often seen as a ripple on the acceleration voltage. With good belts or ladders, the ripple in well-maintained machines may be brought down to 100 V, or even below, out of 5 MV. The momentum spread in the accelerated beam will hence reach approximately $\Delta p/p = 10^{-5}$.

The high voltage is stabilized through several feedback loops. High-frequency variations are measured with a capacitive pickup of a signal from the terminal. Lower-frequency variations may be measured by a rotating voltmeter (a device that measures the size of the electrical field within the pressure vessel) and by a double slit pickup signal. For the latter case, the accelerated beam is energy analyzed in a bending magnet. At the magnet exit a pair of slits may be hit by the beam. An asymmetric current loading indicates that the beam energy is off the mark. All these error signals are used to regulate the current through a corona discharge from the high-voltage terminal, and hence regulate the voltage. The slit stabilization will of course only work when a sufficient beam current is available. It may hence not be used for extremely weak beams, as used in accelerator mass spectrometry.

In the single-ended pressure-vessel-contained machines the high-voltage platform is particularly difficult to access. The tandem principle as illustrated in fig.8b is particularly useful in facilitating the introduction of more complicated heavy-ion sources. Here, negative ions are produced in an ion source placed at 50 to 100 kV. The negative ions are injected into the machine and accelerated to the positive high-voltage terminal where they are stripped to a positive charge state compatible with their velocity in a foil or gas stripper. They are now accelerated once more down to the ground potential. If the acceleration voltage is U (in MV) and the charge state after stripping +n, the energy after passing through the machine will be (n+1)U MeV. The energy of the emerging beam may be further increased by inserting yet another stripper station approximately a third of the way down the beam tube, to drive the charge state even higher. Such extra stripper stations are very difficult to service and their use gives rise to an abundance of different charge states and energies in the emerging beams. Tandem accelerators are these days running with terminal voltages well above 20 MV in Oak Ridge and Daresbury. The Strassbourg Vivitron, which is under construction, is aiming for 40 MV.

An accelerator facility needs a substantial amount of ancillary facilities. We have already mentioned the energy analysis for beam stabilization (and purification). We shall also need devices for beam diagnostics and focusing. A number of such gadgets will be mentioned below. Further, the entire acceleration and beam transport must take place in a high vacuum to avoid scattering and charge exchange through collisions with background gas molecules. The better the vacuum, the cleaner the emerging beam.

Finally, it should be mentioned that the accelerating tube itself is also a focusing device. Depending on the shaping of the field at the exit of the tube, the focal point may be placed within or after the tube.

2.3 AC Accelerators

Plans now exist to accelerate protons to energies 6 orders of magnitude larger than what may be achieved by the largest tandem accelerators. It is of course out of the question to achieve such a goal by DC acceleration. Cosmic distances would be needed to support such voltage differences. Fortunately, acceleration may also result from the application of AC voltages. AC accelerators may be divided according to two different taxonomies. Along one dividing line, we talk about cyclic vs. linear accelerators. Along another, about acceleration in a standing wave field, versus wake-riding on a travelling wave.

Let us first very briefly discuss cyclic accelerators. The most well known is the cyclotron. The ion beam follows a nearly circular path in a constant magnetic field. The period of revolution is independent of velocity, depending only on particle mass and charge and the strength of the magnetic field. Every time an acceleration gap is passed an RF oscillator ensures that the particle is accelerated. As higher energies are reached, the cyclotron frequency decreases due to the relativistic mass increase. The cyclotron is hence driven with a variable frequency; we have the synchrocyclotron. A serious problem both with cyclotrons and synchrocyclotrons is the stability of the orbits. It may be shown rather easily that vertical focusing obtains only provided the magnetic field decreases with radius. If it decreases too fast, however, the circular orbits will become unstable in the horizontal plane. We must hence have a weakly decreasing field to ensure stable orbits, in mathematical terms $- 1 < \partial(\log B_z)/\partial(\log r) < 0$, where the logarithmic derivative is called the field index, B_z is the axial field and r the radius. Note that since the field between the poles is rotation free, $\partial B_z/\partial r = \partial B_r/\partial z$. The decrease of the field with increasing radius also calls for a synchrotron-like variation of the frequency.

For proton energies larger than approximately 800 MeV the magnets get too bulky. It is more convenient to work with a fixed radius and increase the magnetic field with acceleration. Such a machine is called a synchrotron. It is obvious that in such a structure, the particles must be accelerated in bunches, although bunches may be accelerated stably over a considerable phase angle. The structure is tuned to accept beams being accelerated at some intermediate voltage while the voltage is increasing. Particles having achieved too high an energy will hence arrive too early and suffer less acceleration, and vice versa for particles of too low an energy: We have phase stabilization, but particles may oscillate to and fro in phase. For not heavily relativistic beams this will give an uncertainty in the beam energy roughly corresponding to half the energy gain per stage. A rather large number of acceleration steps are hence required to give a well-defined energy, or the phase-angle accepted by the machine must be made very small resulting in a substantial loss in intensity.

Synchrotrons were first built according to the orbit stability criteria outlined above. This so-called weak focusing resulted in rather wide beams. Courant, Snyder and Livingston did therefore in 1952 propose to separate the bending and focusing. This is the so-called strong focusing principle. We have a lattice with alternating dipole magnets and quadrupole magnets (see below) for focusing. This allows the beams to be made very small with corresponding savings in the vacuum chamber and magnet materials. Today large circular accelerators like SPS and LEP at CERN, HERA at DESY and the Tevatron at Fermilab are built according to this principle which will also be used in the huge proposed SSC in Texas.

It is difficult to get a particle started in a circular accelerator. Hence a linac is usually employed as injector. Such a machine may be of the Wiederöe or Alvarez type. In the Wiederöe structure, the particle beams pass through a structure of circular electrodes connected to an RF oscillator. The electrodes increase in length along the structure to ensure that the transmission time through the electrodes stays constant while the particles are accelerated. In the gap between the electrodes, the particles are not only accelerated if they are in phase, they are also focused by the fringing fields.

In the Alvarez structure the particles travel along with a wave in a waveguide. As it is difficult to make large variations in the propagation velocity of the wave, such a structure is best suited for relativistic particles. Also here, focusing is achieved simultaneously with acceleration.

A new development is in the direction of very high current low-energy machines mainly for materials science purposes. One may, e.g., contemplate mA 1 MeV nitrogen beams for treatment of metal surfaces. Such beams may be accelerated and steered either in the radio-frequency quadrupole (RFQ) accelerator, where the electrodes are shaped to give quadrupole focusing fields, or in the MEQUALAC shown in fig.9, where the Wiederöe structure is used for acceleration but focusing is made with separate quadrupoles. These may be arranged in a square lattice to give a large number of parallel acceleration channels. Such a machine will be able to convert a very large fraction (~ 60 %) of the RF power into beam energy, but the beam will not be very well-defined in energy.

Briefly to return to the question of circular accelerators, parts of the CERN accelerator complexes were shown in figs.6 and 7. They are nearly all synchrotrons. If circular machines are used at very high energies ($\gamma \sim 10^5$) synchrotron radiation will be a serious problem. To get even

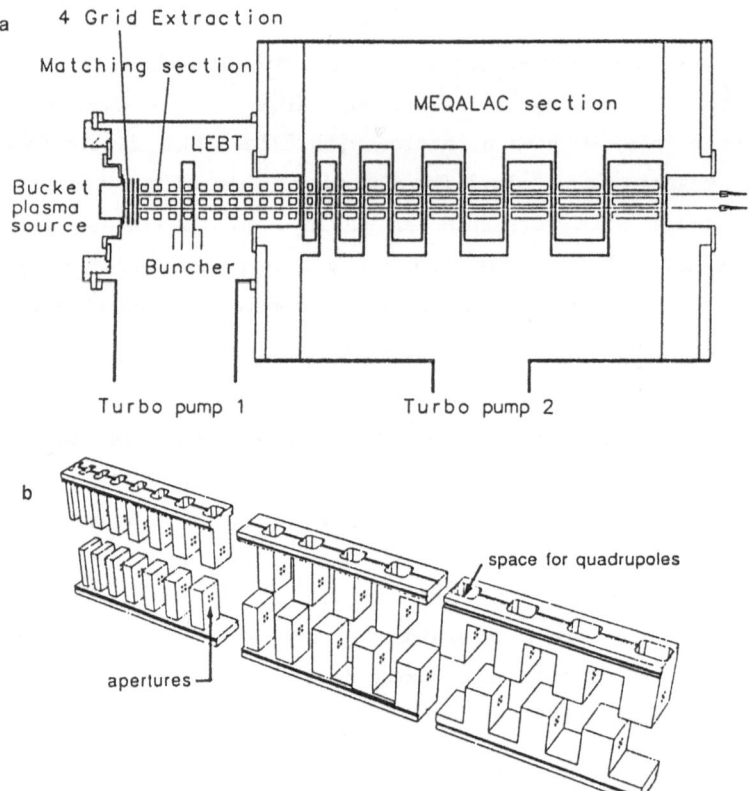

Figure 9. Schematic drawing of the MEQUALAC experimental setup: a) shows the plasma bucket source, the low-energy beam transport section with the buncher and the acceleration section (in practice there are many more acceleration gaps than drawn); b) is an exploded view of a 4-beam acceleration structure. (From W. Urbanus, Thesis, Utrecht 1990).

Table 2. Accelerators

Energy/velocity	Accelerator type
1 keV - 3 MeV	DC. Solid-state electronics HV. $\Delta E = 20 - 50$ eV.
0.5 MeV - 20 MeV/u	van de Graaff. $\Delta E \simeq 100$ eV.
$v_0 - 0.1$ c	Linacs, Wiederöe structure.
	Cyclotrons
$1.01 < \gamma < 10^5$	Synchrotrons.
$1.01 < \gamma < \ldots$	Linacs, Alvarez structure.
$\gamma < \ldots$	New principles.

higher in energy, linear structures must be used. Today it is difficult to get the average field over a linear structure higher than 5 MV/m. This must be increased considerably to make a 1 TeV linear accelerator feasible.

Table 2 gives a brief overview of the accelerator principles discussed.

2.4 Ion Optics

Charged-particle optics is a sophisticated topic of modern applied physics. Several textbooks exist in the field (some are mentioned in the literature list below) and literally millions of lines of computer programs have been written to allow numerical solutions of complicated problems. Here is not the space - nor the intention - to go into any depth. Rather a few simple elements: the magnetic dipole, the electrostatic dipole, their combination in the Wien velocity filter, the Einzel lens (the electrostatic acceleration–deceleration lens) and the magnetic quadrupole, shall be mentioned and some of the elements treated superficially.

Let us first look at the magnetic dipole, fig.10. The radius, R, of the circular orbit in a homogeneous magnet is given by

$$\frac{2ME}{q^2} = B^2 R^2, \tag{1}$$

where M is the particle mass, E its energy, q its charge state, and B the magnetic field strength. (All in appropriate practical units!). We do immediately see that we need three parameters to characterize the beam, viz. mass, energy and charge state. Fortunately, q varies strictly as whole numbers and M nearly so. We do hence not need three different analysis systems fully to determine our beam species, but many

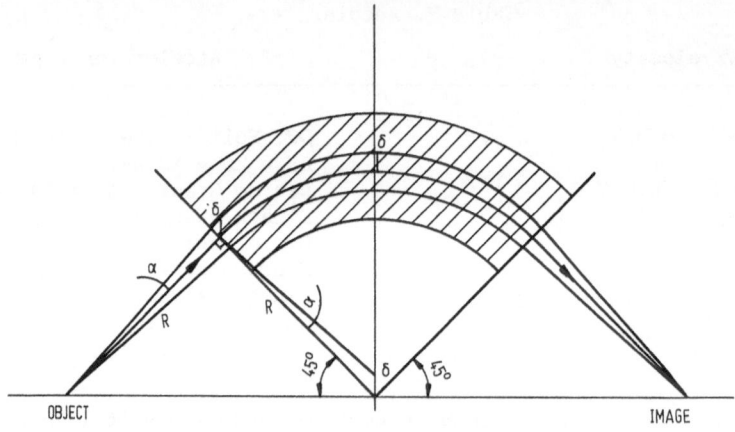

Figure 10. Orbits in a symmetrical 90° magnetic dipole. The orbits have identical radii within the field. From the two triangles with side δ and acute angle α it is seen that the orbits all have their center on the line of symmetry. They will hence be focused on the image plane. The magnification is one.

combinations may get through one system. Note further that we have only focusing in the plane shown. As with the cyclotron, we do need a field decreasing in the z-direction to give a doubly focusing system. This may be achieved by proper shaping of the pole pieces.

The other obvious possibility is the electrostatic dipole. With reference to fig.11, we see that the transit time of a particle with velocity v through an electric field ε covering a beam path of length a is

$$t = a/v = a \left(\frac{M}{2qV_{acc}} \right)^{1/2} , \qquad (2)$$

where V_{acc} is the acceleration voltage. For small deflection angles the beam is, within the field-covered region, deflected the distance

$$d = \frac{1}{2} (q\epsilon/M) \left(\frac{\delta\ a^2\ M}{2qV_{acc}} \right) , \qquad (3)$$

and the deflection angle is

$$\nu = d/a = a\epsilon/V_{acc}. \qquad (4)$$

First, you note that for particles not breaking up or suffering charge exchange during acceleration, the deflection will depend on the ratio between analyzer field and voltage only: You get through what you feed in. But intruders may be sorted out. Secondly you will need voltages over the analyzer comparable to the acceleration voltage. This is obviously

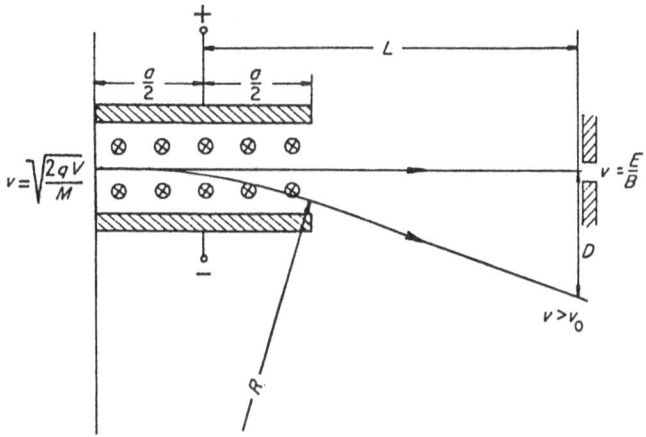

Figure 11. The Wien velocity filter.

inconvenient as soon as the energy is larger than, say, 100 keV. Hence, you will often instead use a combination of crossed electric and magnetic field called a Wien filter. As seen from fig.11, it is a velocity filter. To avoid neutral particles of any energy passing through, a small electrostatic deflection of the beam is often introduced.

The third optical element to be discussed is the magnetic quadrupole. As mentioned in connection with the discussion of strong focusing, this is the main focusing element in cyclic accelerators. It should also be mentioned that electrostatic quadrupoles are often used as mass filters at low energies. Fig.12 demonstrates that the quadrupole focuses in one plane

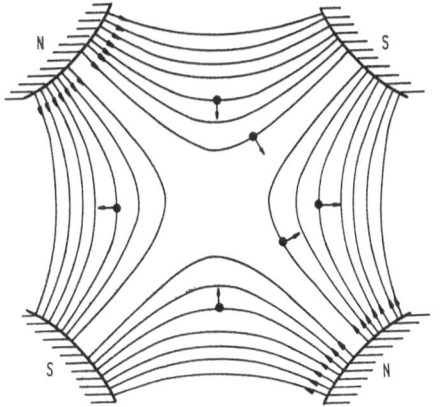

Figure 12. Field lines in a magnetic quadrupole. The arrows show the direction of the force acting on a positively charged particle travelling into the plane of the paper.

only but defocuses in the orthogonal plane. The quadrupole must hence be followed by a second one, where the field is turned 90°. In practice it is difficult with a quadrupole doublet to achieve the same focus in both planes; you get an astigmatic image. Hence an L-2L-L triplet is usually chosen, where L is the length of the individual quadrupoles.

It would be convenient if a magnetic solenoid lens could be used but this is only possible at rather low energies or for very light particles such as electrons in the electron microscope. High field superconducting magnets may change that situation in the future.

After having introduced a number of optical elements, we may go on to speculate what we may achieve using these elements on an ion beam. Here we shall need the very central term of emittance. This has the origin in Liouville's theorem. The theorem applies strictly to N particles moving in 6N-dimensional phase space, but we shall here confine ourselves to a 6-dimensional phase space (and shortly go on to split that into three 2-dimensional spaces). In the 6-dimensional space the beam will behave like an incompressible fluid as long as the particle moves only under the influence of Hamiltonian ("conservative") forces. Hence, if the trajectories of all particles in the beam are given in this 6-dimensional space, we cannot change the phase space volume by optical manipulations.

To be more precise let us define a z-coordinate along the "central beam" of our system and the x,y-coordinates perpendicular to this line. Further, let

$$x' = dx/dz, \qquad y' = dy/dz, \tag{5}$$

i.e., for $(x', y') \ll 1$, x', y' are the angles to the central beam. Let us assume that the beam intensity drops off sharply outside a certain contour in the (x,y,x',y') space. We then define the emittance

$$E_{xy} = \int_{\Omega xy} dx dx' dy dy'. \tag{6}$$

We may further often make the reduction to 2-dimensions mentioned above, and define

$$E_y = \int_{\Omega y} dy dy', \tag{7}$$

and analogously for x, see fig.13. Obviously, if we accelerate the beam, we increase v_z in proportion to $T^{1/2}$, but keep v_y and v_x constant (T is here the kinetic energy). Hence, for pure acceleration, the normalized emittance

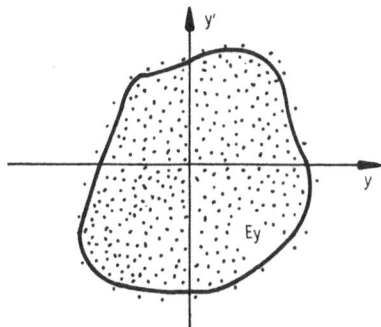

Figure 13. Definition of the emittance E.

$$\varepsilon_x = \pi E_x \ T^{1/2} \tag{8}$$

stays constant. The factor π is traditionally included. Similarly,

$$\varepsilon_{xy} = \pi^2 \ E_{xy} \ T \tag{9}$$

is a constant of motion.

Eqs.(8) and (9) represent the phase-space volumes mentioned above. The emittance will hence remain constant under the influence of Hamiltonian forces. This statement does also include the average action of the other particles in the beam ("space charges") that has to be taken into account in a self-consistent manner. But scattering on individual ions ("beam scattering") is not included and will lead to a growth of normalized emittance.

We further define the *brightness* of a beam with an intensity of I particles per second as

$$\text{Brightness} = I/\varepsilon_{xy}. \tag{10}$$

If the beam has a broad energy spectrum, as for synchrotron radiation, we define further.

$$\text{Brilliance} = d(I/\varepsilon_{xy})/dE. \tag{11}$$

When the beam, as is often the case, does not drop off sharply at a certain contour, we present instead isobrightness contours in phase space.

Let us, as very simple examples, mention that a beam focused nearly to a spot will be depicted as a vertical line in fig.13 (as will also the beam emerging from the liquid metal ion source, fig.4). In contrast, a "parallel" beam will be depicted as a horizontal line. The basic lesson is that we may only focus at the expense of angular divergence.

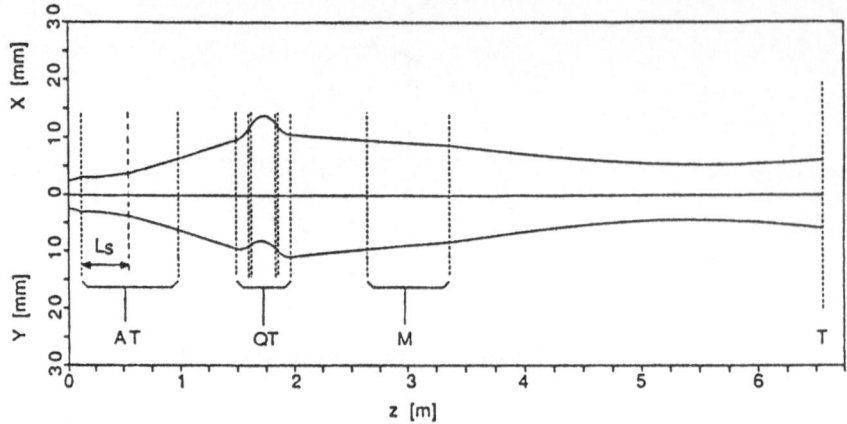

Figure 14. The rms beam envelope in a 1 MV ion-implantation facility in the x-z and y-z plane for a 300 μA Xe$^+$ beam accelerated from 30 to 200 keV. Space charge is taken into account. The dotted blocks indicate the acceleration tube (AT), the quadrupole triplet lens (QT) and the 30° switching magnet (M). (From W. Urbanus, Thesis, Utrecht 1990).

Equipped with the optical characteristics of our beam focusing elements and of the accelerator together with the emittance conservation theorem, we may calculate the transport of our beam. This may be done by a matrix formulation if we only worry about beams close to the axis and disregard space charge. If space charge is taken into account we need a more complicated transport equation (or an iterative approach) and, finally, if we want to handle wider beams and take spherical aberrations into account, we shall most conveniently use a ray-tracing program. An example of a beam calculation is shown in fig.14, while actual emittance measurements are treated under beam diagnostics in sect.3.

2.6 Microbeams

A rapidly developing and interesting application of high-quality high-brightness ion beams is for use in ion microprobes. The elements of such a microprobe are shown in fig.15. The similarity to the elements of a light microscope is obvious. The beam from an accelerator - typically 3 MeV protons for PIXE analysis - is focused on a pair of object slits.

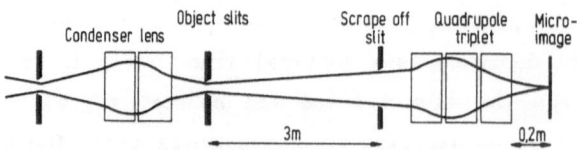

Figure 15. The optics of an ion microprobe.

These will typically be 30 × 30 μm^2 and be reached with a beam of an angular divergence of a few mrad. Note that the scale perpendicular to the beam axis in fig.15 is strongly exaggerated. The acceptance of the object slits will be considerably smaller than the emittance of the beam, which means that most of the beam is thrown away. This has two consequences. First, it might be difficult to get a sufficient current through the object slits to get a useful beam for analytical purposes, and second, there will be lots of slit scattering from the object slits. These are not made any larger because it is difficult to design a final quadrupole system that demagnifies by a factor larger than 50. A final image smaller than 1 × 1 μm^2 will be aimed at. The scattering means that a pair of scrape-off apertures will be necessary to get rid of most of the slit-scattered particles. The scrape-off aperture should be made sufficiently small to eliminate most of the scattered particles but sufficiently large to prevent the main beam from hitting the jaws, as this would cause further slit scattering. This design aim is not always met in practical systems and may not be easy to achieve. It may be necessary to make the object slit a double one, to limit the angular opening of the impinging beam.

The final focusing is made with a quadrupole lens, usually a magnetic one. Both doublets, triplets and quartets are known. The state-of-art is to obtain 0.5 × 0.5 μm^2 beam of sufficient intensity to perform PIXE (i.e., 100 pA). One may wonder why a two-stage demagnification has not been attempted. This should allow the object slits to be made

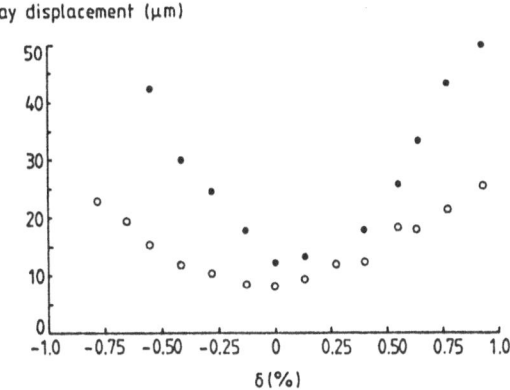

Figure 16. The effect of chromatic aberration measured in the Oxford microprobe final triplet quadrupole. The points show the broadening of the image observed visually through a microscope as the mean momentum of the beam is varied. Difficulties in measuring spot sizes less than 10 μm contribute to the results at low δ values. (O: x-direction, ●: y-direction). (From G.W. Grime, Nucl. Instr. Methods *197* (1982) 97).

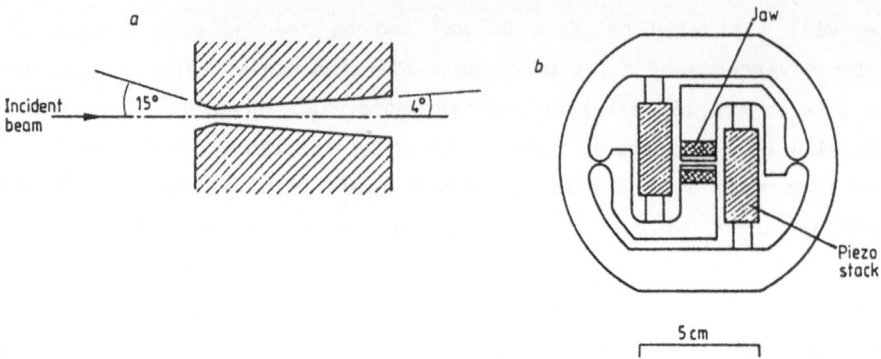

Figure 17. The micro-slit assembly of the old Heidelberg microprobe. (a)
The profile of the jaws, which is calculated to minimize
small-angle scattering. (b) The method for adjusting the size
and position of the slit aperture using high-voltage
piezoelectric transducers. (From R. Nobiling et al., Nucl.
Instr. Methods *130* (1975) 325).

substantially larger and hence allow a more intense beam at the final
focus.

One of the problems of a final lens system is the rather strong
chromatic aberration, as illustrated in fig.16. From the slopes of the
lines, it is seen that to obtain a 1 μm focus, the ripple of the beam must
be smaller than $\Delta p/p = 5 \times 10^{-4}$. It appears that the stability of the
accelerator may play a substantial role in limiting the resolution of
nuclear microprobes. Further, particles slit scattered over very small
angles but having suffered some energy loss will also contribute to the
blurring of the image. Perfect alignment of the focusing elements is very
important, but difficult to obtain.

An interesting question is how to design the object aperture. Fig.17
shows the micro-slits used in the older Heidelberg microprobe. The
tapering of the slits is calculated to minimize small-angle scattering,
but without taking full account of multiple lateral scattering. Further,
it will be very difficult to machine the edge, where the two tapered
sections meet, to such a perfection that small irregularities that may be
penetrated by the beam are not left. Such irregularities will give rise to
energy loss but little scattering. Finally, as the demagnification of the
Heidelberg final doublet was 1:4.5 and 1:25 in two perpendicular planes,
it is seen that one pair of jaws should certainly not be more than 10 μm
apart to give an acceptable image. 3 MeV protons have a range of
approximately 40 μm in stainless steel. Elementary geometrical
calculations on the design of fig.17 show that the beam will penetrate the
slit to a distance of approximately 2.5 μm away from the edge. As this

happens on both sides, 30 % of the transmitted beam will have penetrated the slits! This shows the importance of using large slits and perhaps a two-stage demagnification.

For the Heidelberg microprobe, a change has later been made to cylindrical slits. They are easier to machine and align. They should, however, not be made from too thin cylinders. If the projectile range is R, the slit opening a and the radius of the jaws ρ, it may be shown that the slit-penetrating fraction of the beam is $R^2/(4a\rho)$. If ρ is made large, there will be lots of small-angle lateral scattering taking place in the jaws. Penetration and small-angle scattering must be balanced against each other. There appears to be a scope for optimization using modern multiple-scattering theory and good energy-loss data.

A microprobe with a resolution substantially better than 1 μm will be very interesting for energy-loss measurements. Even with a very weak beam, such an instrument will be useful both for estimating thickness irregularities of thin foils and for making good straggling measurements independent of intermediate-scale thickness variations.

3. BEAM DIAGNOSTICS

3.1 Emittance Measurements

According to the definitions in the previous chapter (sect.2.4), a full characterization of beam emittance demands the measurement of a quadruple differential distribution. Such a measurement may be achieved with a device shown in fig.18.

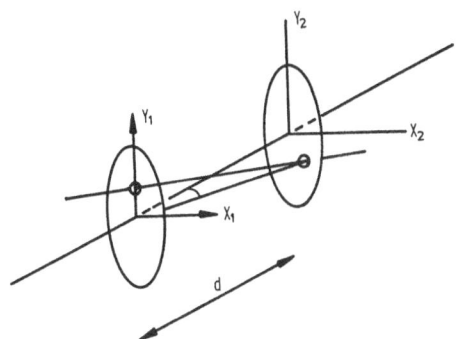

Figure 18. Emittance measuring device. The first aperture defines (x_1, y_1) and the second aperture defines (x', y') through $x' = (x_2 - x_1)/d$, $y' = (y_2 - y_1)/d$.

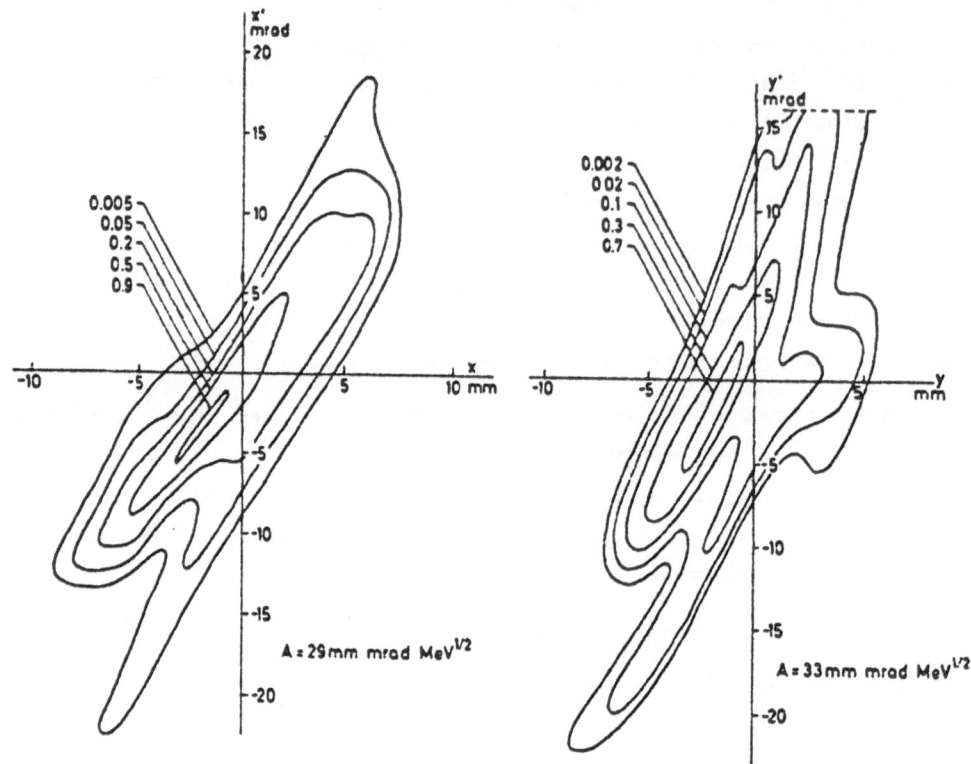

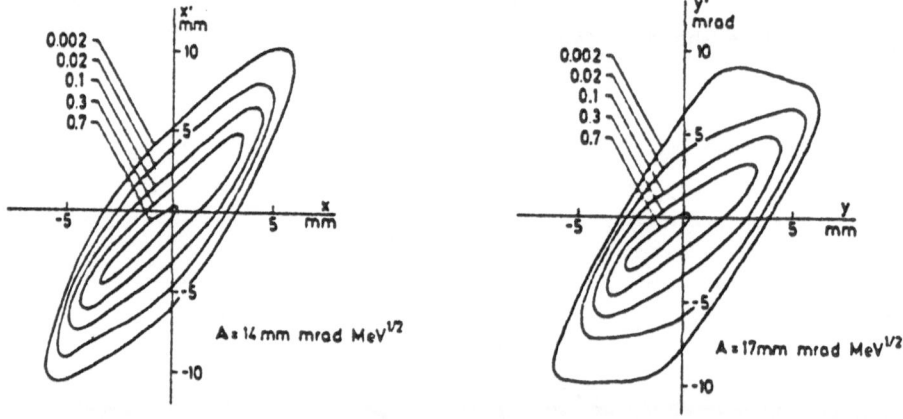

Figure 19. Emittance diagrams in two planes for gold beams from the ANIS source run under different plasma conditions. Each brightness contour is labelled with its fraction of the peak brightness. (From P. Tykesson, H.H. Andersen, and J. Heinemeir, IEEE Trans. Nucl. Sci. *NS23* (1976) 1104. © 1976 IEEE).

A first aperture is used to define the beam position (x_1, y_1). For each position, the total transverse beam plane is scanned with a second aperture to determine the beam intensity $I(x_1, y_1, x_2, y_2)$. The measurements are rather tedious, but may easily be computerized. They demand nevertheless a rather long-time stability of the beam to make sense. Substantially simpler measurements of $I(x_1, x_2)$ may be performed by a pair of slits but do not give nearly as detailed a picture.

An example of a full set of brightness contour data for the ANIS source running in two different modes with a Au beam is shown in fig.19. It is worth noting that emittance in the (x, x') and (y, y')-plane may easily be coupled. This happens most simply through application of a longitudinal magnetic field.

For control and optimization of the beam from a running accelerator, a full emittance measuring device is clearly too complicated. Most often simple control devices are inserted to ensure that the beam is focused and optimized at the predetermined (x, y, z) positions. This may most simply be done by inserting a fluorescent disc in the beam. Somewhat better are current measurement devices that may be inserted automatically in the beam. Particularly to control the output of injectors and ion implanters, a beam profile monitor is very useful. A vibrating needle scans the profile and is driven synchronously with an oscilloscope. The current measured to the needle is fed to the y-axis of the scope to picture the profile directly. Particularly to control the astigmatism of quadrupole

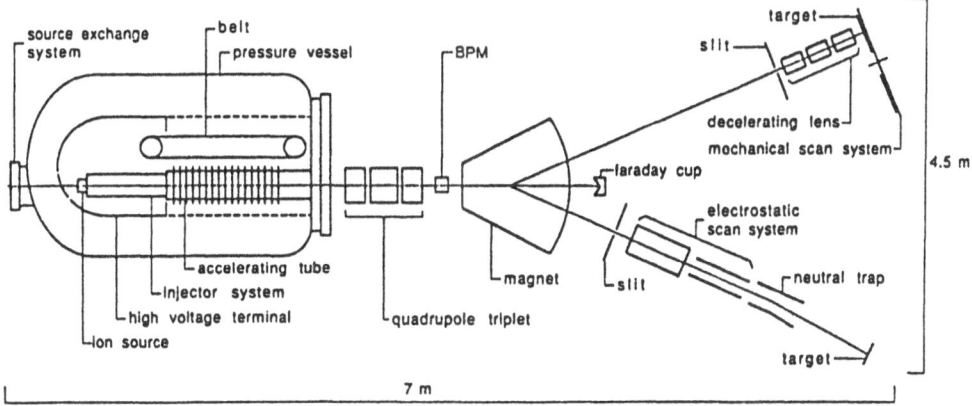

Figure 20. Schematic layout of a 1 MV ion-implantation facility. The position of the beam profile monitor (BPM) and Faraday cup in the direct beam line are shown. Individual beam lines will each be provided with a Faraday cup. (From W. Urbanus, Thesis, Utrecht 1990).

focusing devices, simultaneous scans in the x and y-directions are most useful.

Fig.20 shows a typical layout of a 1 MV electrostatic accelerator with beam-diagnostic devices. Individual beam lines will also be provided with Faraday cups and perhaps even profile monitors.

3.2 Current Measurements

The standard current-measuring device for charged-particle beams is the Faraday cup. A simple version is shown in fig.21. The purpose for the negatively biased suppression ring is to prevent secondary-electron emission from influencing the current measurements. The folklore tells you that the aspect ratio of the cup, L/d, should be larger than three. In some versions it is even claimed that the end of the cup should be shaped like a hollow cone. The former claim is somewhat more scientifically based than the latter, but a factor-of-three aspect ratio certainly does not guarantee accuracy. It appears much more important that both cup and electrodes are made from high-work-function materials to minimize secondary electron emission. Further, to minimize backscattering from the end of the cup, this part should preferable be made from a low-Z material. Such a choice may conflict with attempts to minimize γ or neutron emission from the cup, in case of light-particle beams of higher energy. Anyway, such an optimization should be attempted rather than relying on a conical bottom of the cup!

The design of fig.21 will allow current measurements at the 5 % - 10 % accuracy level, but neglects a number of secondary and tertiary

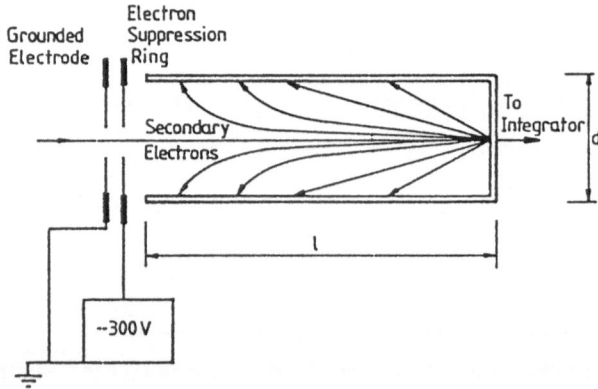

Figure 21. Simple Faraday cup for beam current measurement and integration. The cup is provided with a grounded aperture and a negatively biased electron suppression electrode.

processes. Let us mention a few: the scattered beam from the main aperture may generate very low-energy electrons that may end up in the cup. Backscattered positively charged particles from the cup will go directly to the suppression ring and falsify the current measurements as will positively charged sputtered species. Neutral backscattered and sputtered species may generate secondary electrons on the bias ring. These will be driven to the cup and falsify current measurements. (This is a tertiary, but important source of error). Similar things may happen for particles hitting the electron suppressor ring directly or after scattering at the main aperture.

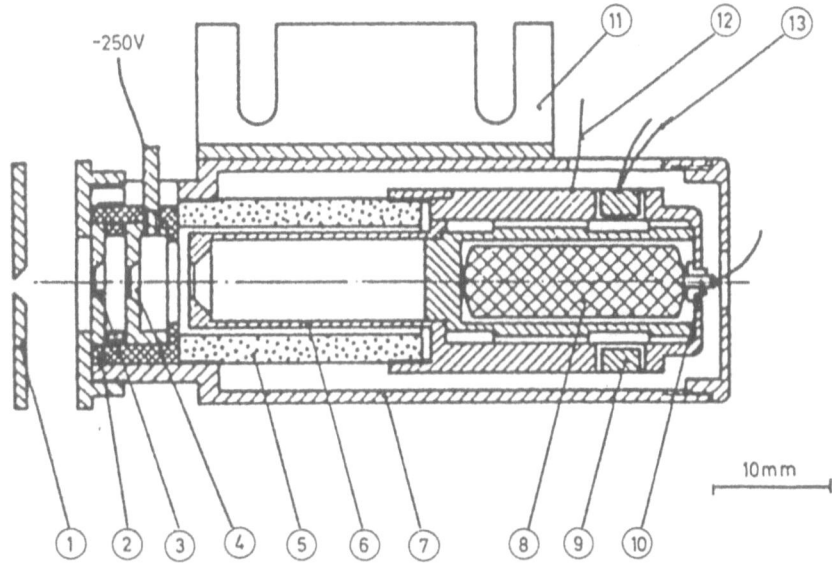

Figure 22. Details of a complicated Faraday cup with option for calorimetric checks on particle charge state. The numbers in the figure refer to the following construction details: (1) beam-limiting grounded aperture, (2) ceramic insulator ring, (3) grounded aperture connected to the shield surrounding the cup [this aperture is necessary to prevent particles slit-scattered at (1) from hitting (4), the secondary-electron suppression ring], (5) ceramic insulator ring supporting main cup, (6) Faraday cup (all similarly hatched parts touching each other are in electrical and thermal contact), (7) grounded shield (preventing low-energy secondaries in the target chamber from striking the cup), (8) heater for calibration of cup used as calorimeter, (9) copper ring electrically insulated from cup by thin mica strips, (10) feed-through for heater power, (11) mechanical mounting support, (12) current feed-through, (13) thermocouple for calorimeter. The figure does not illustrate construction details aimed at ensuring adequate vacuum performance. (From H.H. Andersen and H.L. Bay, p. 145 in R. Behrisch, ed., *Sputtering by particle bombardment*, Springer, 1977).

Fig.22 shows a more complicated design aiming at circumventing most of the above mentioned difficulties.

This device may further be used to control whether the beam contains the expected charge state of the particles through use as a calorimeter. It is simple to control whether the beam contains neutrals by use of electrostatic deflection of the charged part of the beam in front of the cup. Such a control may be made with high precision as the remainder of the beam is measured directly. Less precise is the control of the charge state of a heavy-ion beam. Here, a consistency check is made between the beam power and the beam current.

Generally, the device may measure currents with an accuracy of 1 %. As a general advice to theoreticians reading experimental papers, it may be stated that if experimentalists claim such an accuracy and do not claim to have made a very special effort to reach it, they should mostly not be trusted.

For an intense beam all devices inserted in the beam must be constructed to withstand the heat load. Examples are known of beams actually melting their way through misconstructed beam stops.

It should finally be mentioned that bunched beams may conveniently be measured by the pick up signal in a coil surrounding the beam tube. Rather sophisticated so-called beam transformers are marketed commercially.

3.3 Analysis of Energy, Mass, and Charge Distribution

We have already in the previous subsection touched upon the question of charge states of the accelerated particles. There, it was mentioned that consistency between current measurements and beam calorimetry could be used for confirming the assumed charge state. Such a check is, however, not possible if the beam energy is not known with better precision than the accuracy of the check. But for stopping-power measurements, particularly in regions where the stopping power varies rapidly, it is necessary to know the beam energy with substantially higher accuracy. In fact, as mentioned in the next chapter, it is often so that the main uncertainty in stopping-power measurements stems from the energy calibration.

We must hence attempt a high accuracy in our energy calibration. Absolute energy calibrations may be attempted in a number of ways. The field of a magnetic analyzer may be mapped in all detail, including the fringing fields. Velocities may be measured by time-of-flight or, at moderately relativistic velocities, by the Cerenkov radiation angle of the

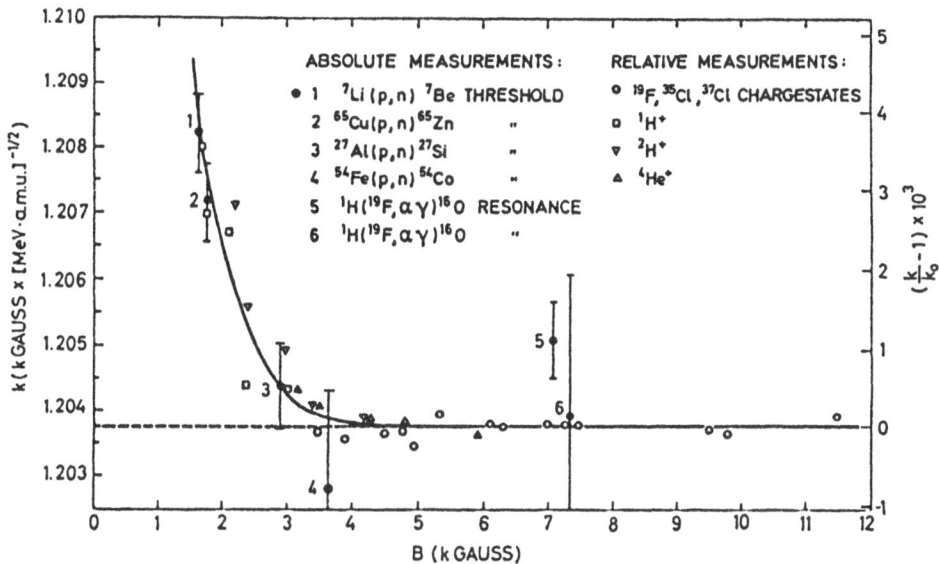

Figure 23. Non-linearities in the energy-analyzing magnet of the Aarhus tandem accelerator as mapped by several threshold and resonance reactions together with consistency checks with heavy ions in different charge states. (From H.H. Andersen, P. Hornshøj, L. Højholt-Poulsen, H. Knudsen, B.R. Nielsen and R. Stensgaard, Nucl. Instr. Methods *136* (1976) 119).

beam in an appropriate detector gas. But very often it is possible to perform a relative energy determination in a magnetic analyzer that is calibrated by known resonance or threshold reactions. Differential hysteresis in the analyzer magnet does, however, mean that the strict proportionality between the calculated $B\rho$ value of the beam and the field measured somewhere in the magnet does not hold. It is necessary to map the field in some detail as indicated in fig.23.

If we are using heavy ions from a single-ended van-de-Graaff, or if we investigate the beam of an injector to a tandem accelerator, we may find a most confusing mixture of particles, energies and charge states. If the wanted beam is difficult to achieve, it may not even be the most abundant one. Let us, for the sake of illustration, imagine that we want to run an 1 MeV Ar^+ beam from a facility like the one shown in fig.20. On one of the beam lines we measure the signal as a function of the analyzer magnetic field. The measured spectrum may look like fig.24. The peaks shown may not all be there and all the peaks may not be identifiable, depending mainly on the cleanliness of the ion source and the quality of the beam-line vacuum. Apart from Ar^{3+} only single and doubly charged species are shown. We have included common atomic and molecular species

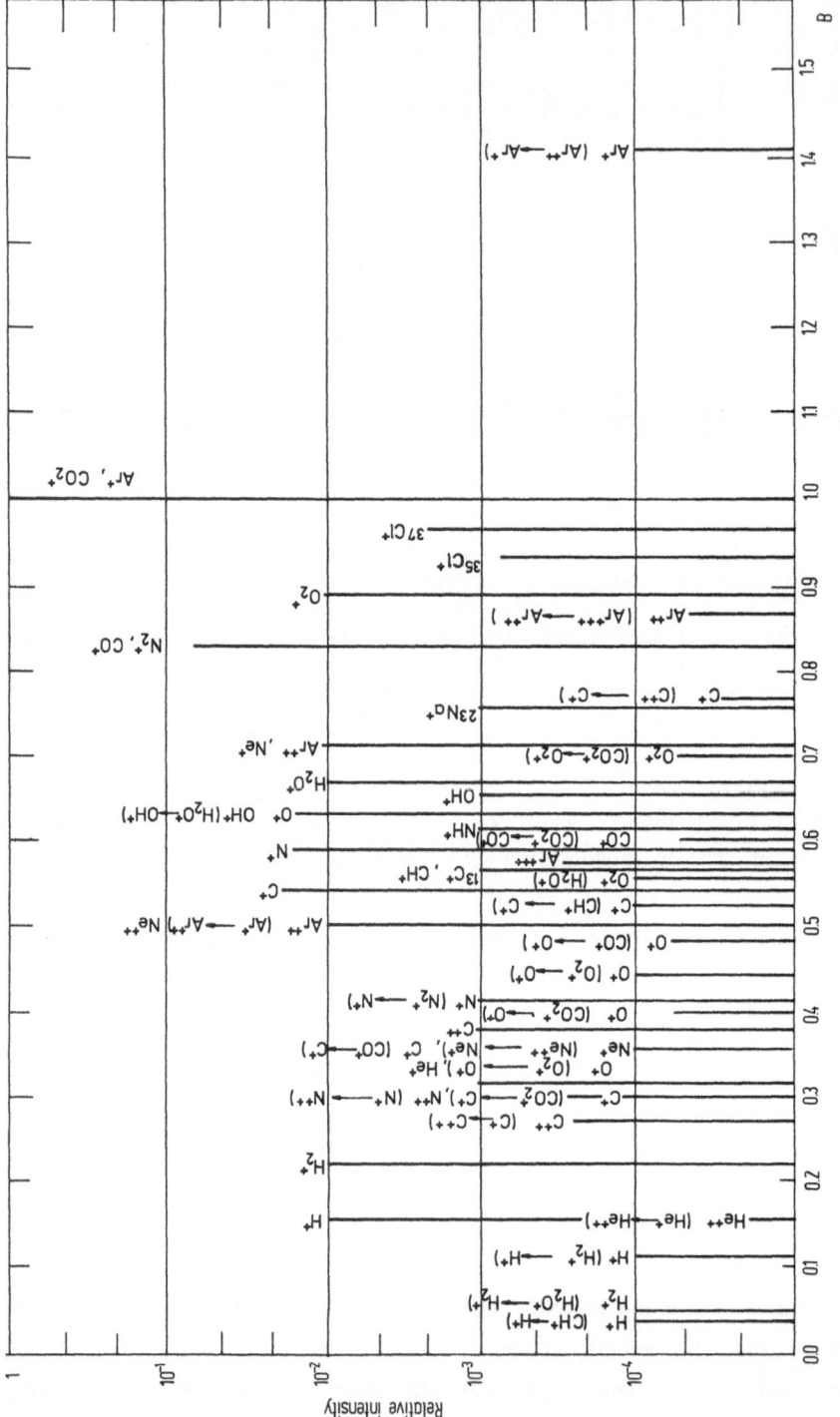

Figure 24. Fictitious magnetic analyzer spectrum for an Ar^+ main beam. The magnetic field is given relative to the one transmitting the main beam.

emerging from the source as singly and doubly charged ions and taken account of charge exchange and molecular breakup through background-gas collisions between the acceleration tube and the analyzer.

Some cases are troublesome. An O^+ beam that changes to O^{2+} between acceleration and the analyzer magnet will emerge together with a He^+ beam that has undergone no charge exchange. Confusion between H_2^+ and D^+ is an obvious complication. The author has experienced an even more confusing mix-up. A copper beam was run from an isotope separator through running CCl_4 over hot copper filings. Volatile $CuCl$ was obtained and fed into the iron source. Now, the compound $COCl^+$ will also appear. Through the two stable chlorine isotopes (mass 35 and 37), this compound has the same two masses as copper (63 and 65) and nearly the same intensity ratio. The beam scanner had to be watched carefully for the compounds containing carbon-13 to make sure that the contamination was negligible.

A continuous background will always be found in the analyzer spectrum. This is due to particles suffering charge exchange during acceleration. The background is usually stronger in tandem accelerators than in single-ended machines, particularly if the terminal stripper is a gas.

These complications serve to underline the point mentioned in sect.2.4: Analysis by one system does not give full beam characterization. The author has found it very useful for heavy ion beams of above say 400 keV to be able to insert a scattering foil in the magnetically analyzed beam. The scattered beam is then watched with a surface-barrier detector. If more than one peak appears, the presence of molecular species is indicated. The position of the peak(s) in the spectrum gives the energies of the species involved, which in connection with the magnetic analysis most often will allow unambiguous identification. The width of the peak gives an indication of the approximate mass. For energies above say 5 MeV, a $(\Delta E, E)$ gas-counter telescope may be used as seen in the next section. There are hence very good possibilities for unambiguous identification of beam contaminants, but do not take it for granted that they are not there.

3.4 Accelerator Mass Spectrometry

In view of all the complications of the previous subsection it may sound absolutely overambitious to attempt to measure a beam 14 orders of magnitude weaker than the main beam. Before discussing how this is actually done, we may first ask who would be interested in reaching such a seemingly unachievable goal. The answer to that question is very broad: archaeologists, geophysicists, planetary physicists, geologists, etc.

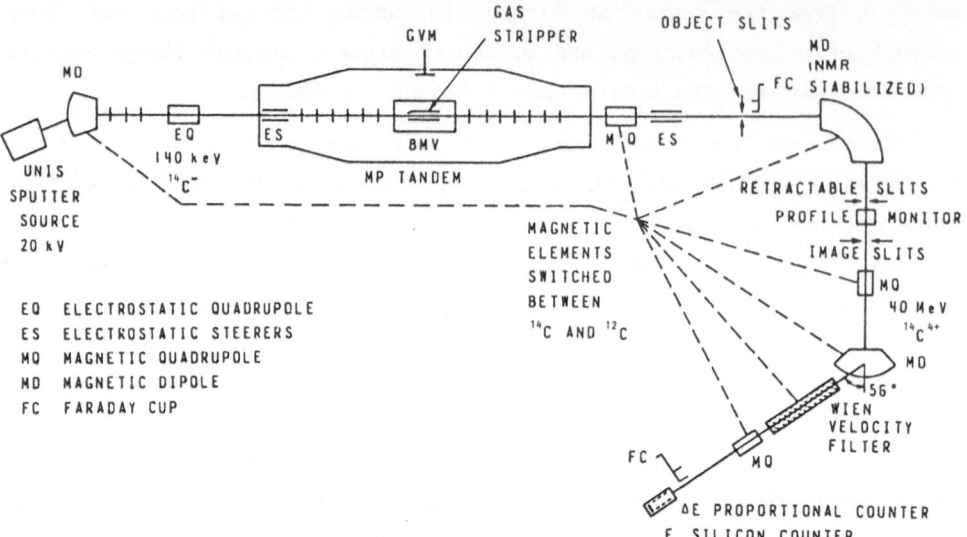

Figure 25. Schematic layout of the Chalk River accelerator mass
spectrometry system. (From H.R. Andrews, G.C. Ball, R.M. Brown,
N. Burn, W.G. Davies, Y. Imahori, J.C.D. Milton, and W.
Workman, Nucl. Instr. Methods *B5* (1984) 134).

Isotopes of interest are for example ^{10}Be, ^{14}C, ^{28}Al, ^{32}Si, ^{36}Cl, etc. All cosmogenic isotopes of intermediate lifetimes. For carbon-14 it does sound much more precise to count the more than 10^{10} carbon-14 atoms in one gram of recent, organic carbon than to limit yourself to the 15 beta disintegrations per minute from the same sample.

In practice the job is done with a setup like the one shown in fig.25. After magnetic analysis, the beam is first sent through a Wien filter (sect.2.4) and then a (ΔE, E)-telescope (sect.3.3). This triple analysis allows unambiguous identification, but absolute values may be achieved only through intermediate counting of standards, as the transmission through the system is not well known.

Systems like the one shown are now in use several places in the world. Milligram samples of carbon are routinely dated up till ages between 35 and 40 ka, and a number of the other isotopes mentioned are also being studied.

4. STOPPING POWER AND STRAGGLING MEASUREMENTS

The stopping power is defined as the limit of $\overline{\Delta E}/\Delta x$ for $\Delta x \to 0$. Here Δx is the target thickness and $\overline{\Delta E}$ the average energy loss of a large number of particles penetrating the target. We shall often write ΔE

instead of $\overline{\Delta E}$. For projectiles heavier than protons the limit may not be well-defined experimentally. It takes a certain target thickness, δx, to establish charge-state equilibrium within the target. If a target is bombarded with heavy ions of a charge state rather far away from their equilibrium one, substantial transients in the stopping powers have been observed.

Quite often we talk euphemistically about investigations of energy-loss processes when we mean investigations of stopping powers. To make a good stopping-power measurement, it is, however, equally important to measure the average energy loss and the target thickness. A few methods exist, where a direct measurement of target thickness is not necessary like measurements of absolute heights of Rutherford Backscattering Spectra (RBS) and Doppler shifted γ-spectra from decay in flight. Here the depth scale is obtained indirectly. Both methods will be discussed in more detail below.

4.1 Target Characterization

The main characteristics of our targets are their average thicknesses, but their homogeneity both with respect to thickness, composition and structure should also be checked. The main methods for

Table 3. Thickness determinations.

Method	Unit for "dE/dx"	Accuracy (2σ)
Interferometry, Tallystep Ruler (!). (Solids)	keV/μm	3 %
Light absorption (Solids)	keV/μm	5 %
Pressure, cell length and temperature (Gases)	eV/(atom/cm^2) (N^{-1} dE/dx)	1 %
Light absorption (Gases)	eV/(atom/cm^2)	5 %
Atom counting by scattering, etc. (Solids)	eV/(atom/cm^2)	1.5 %
Weighing known area	keV/(mg/cm^2) (ρ^{-1} dE/dx)	0.1 %
Energy loss (for relative stopping powers)		Accuracy of used dE/dx

thickness determination are mentioned in table 3. Note that the weighing method is by far the most accurate but that it demands a rather large area. If the average energy loss is not measured over exactly the same area, the accuracy may be irrelevant. Note also that thickness determination through scattering experiments demands an absolute current integration involving all the complications discussed in sect.3.2. On top of that comes the knowledge of cross sections and detector geometry.

We have already mentioned in connection with weight determination of the target thickness that thickness homogeneity may be a serious issue. For dE/dx measurements this difficulty may be circumvented by use of the calorimetric energy-loss method or by mechanical target scanning. Both methods are discussed below. If straggling information is wanted, it is, however, a much more serious business to ensure thickness homogeneity. This is the reason that nearly all good straggling measurements are made on gaseous targets. Microbeams of really high lateral resolution (0.1 μm) may here turn out to become very useful (sect.2.6).

The composition (purity) of the target may conveniently be ensured by one or several ion-beam analysis methods when accelerators are used anyway. Fortunately trace impurities will have no decisive influence on stopping powers. The structure of solid targets may be a more serious business. Most measurements are performed on polycrystalline targets. Such targets are often textured and channeling may hence influence the energy loss processes. An RBS check may disclose problems, but errors of the order of a few per cent may occur.

4.2 Energy Calibration

Methods for energy calibration have already been discussed in sect.3.3. Here we shall look on their importance for stopping-power measurements. If the accuracy of the energy calibration is σ_E and if dE/dx is proportional to E^n, then the accuracy of the stopping power is

$$\sigma = |n| \; \sigma_E. \tag{12}$$

On all parts of the stopping curve, $|n| < 1$. If we measure relative energy losses, like in the calorimetric method,

$$\sigma = 2 \; |n| \; \sigma_E. \tag{13}$$

It is self evident that it does not help to make a good stopping-power experiment if it is not known to which energies the data refer. Nevertheless, lots of stopping powers are published without due consideration to the energy calibration. Using such data, one should always check whether a calibration accuracy consistent with the claimed σ of the data has been properly documented. Often, the energy calibration turns out to be the main source of error.

As measurements are performed for finite thicknesses, a first order correction should always be made to the measuring energy. Data points should be attributed to $E_0 - \Delta E/2$ not to E_0. Because of the curvature of the dE/dx vs. E curve, a second-order correction will be important for large ΔE. As a rule of thumb, we may obtain 2 % accuracy for a 5 % energy loss without the first-order correction. If the correction is included, we may obtain the same accuracy for a 25 % energy loss. We should, however, remember that for such thick targets another second-order effect gains in importance: The distribution in energy losses (straggling) together with the energy dependence of the stopping power gives rise to an error. The error goes in the same direction as the second-order correction for finite ΔE, i.e., if the dE/dx vs. E curve curves upwards too large stopping powers are measured; if it curves downwards, too small a dE/dx will be found.

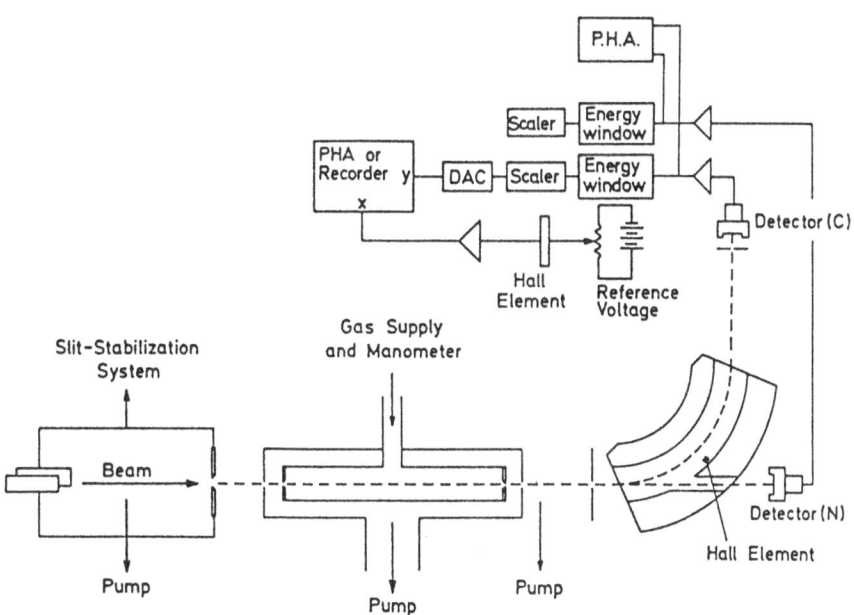

Figure 26. Experimental setup for magnetic analysis of energy-loss distributions at a 2 MV van-de-Graaff accelerator. Measurements are performed on a gas target. Note the normalization to the neutral beam. (From F. Besenbacher, H.H. Andersen, P. Hvelplund and H. Knudsen, Kgl. Danske Videnskab. Selsk. Mat. Fys. Medd. *40* No. 3, 1979).

4.3 Energy Analysis After the Target

The oldest method for performing stopping-power experiments is probably to shoot a beam of known energy through the target and analyze the emerging energy distribution with a magnetic spectrometer. A typical setup is shown in fig.26 as used for gas-stopping measurements. With magnetic analysis it is often necessary to go through the energy spectrum stepwise and hence some sort of beam normalization will be necessary. For solid targets such a normalization will most often be to particles backscattered from the target, but here the neutralized component of the beam is used.

A much more complicated spectrometer, as used by the GSI and GANIL group for measuring stopping power of solids for a 3 GeV ^{36}Ar beam, is shown in fig.27. An example of an energy spectrum with and without stopping foil is shown in fig.28. The main problem in achieving high accuracy using a method based on energy analysis before and after the target is that of target homogeneity. The better an energy resolution is aimed at, the smaller a source area must be used for the spectrometer. If 1 % is aimed at for ΔE, a 4×10^{-4} relative energy resolution must be achieved for the spectrometer. (Remember that the entrance spectrometer of fig.27 enters twice in a correlated fashion, both for determining the absolute energy and for the starting point for the ΔE measurement). Such a resolution is possible but as mentioned above, the target area must be

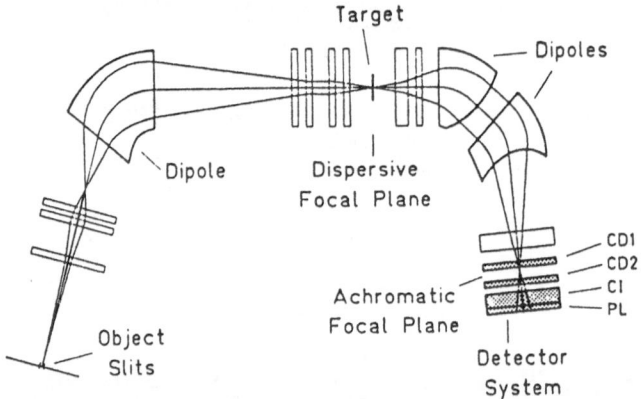

Figure 27. The energy loss spectrometer SPEG and the detector system used for energy-loss measurements of a 3 GeV ^{36}Ar beam in solid targets at GANIC. (From Th. Schwab, H. Geissel, P. Armbruster, A. Gillibert, W. Mittig, R.E. Olson, K.B. Winterbon, H. Wollnik, and G. Münzenberg, Nucl. Instr. Methods *B48* (1990) 69).

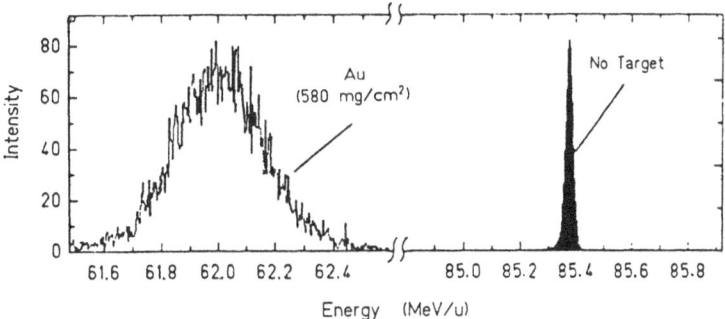

Figure 28. Energy distribution of a 85.38 MeV/amu ^{36}Ar beam with and without passage through a 580 mg/cm^2 gold target. (From Th. Schwab et al., loc. cit. fig.27)

made very small. For solids, particularly for foils thinner than 0.1 mm and hence for measurements at lower energies, foil inhomogeneities will ruin the results and a 2σ level below 1.5 % is extremely difficult to achieve. Similar considerations will hold for electrostatic analysis, like in the setup shown in fig.29.

Fig.30 shows how the data acquisition system may simultaneously be used to step the electrostatic analyzer through the distribution.

Also time-of-flight techniques may be used for the energy analysis. Fig.31 shows an example from multi-GeV proton measurements in single crystals.

The energy loss will in a number of circumstances depend on the scattering angle of the particles. These phenomena have been extensively

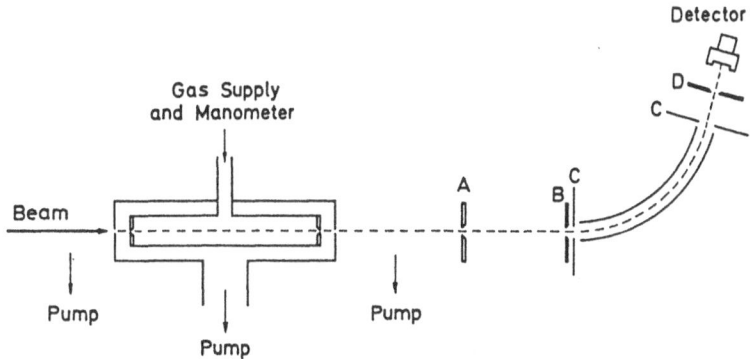

Figure 29. Energy-loss measurements on a gas target with electrostatic analysis after target passage. (From F. Besenbacher, H.H. Andersen, P. Hvelplund and K. Knudsen, Kgl. Danske Videnskab. Selsk. Mat. Fys. Medd. *40* No 3, 1979).

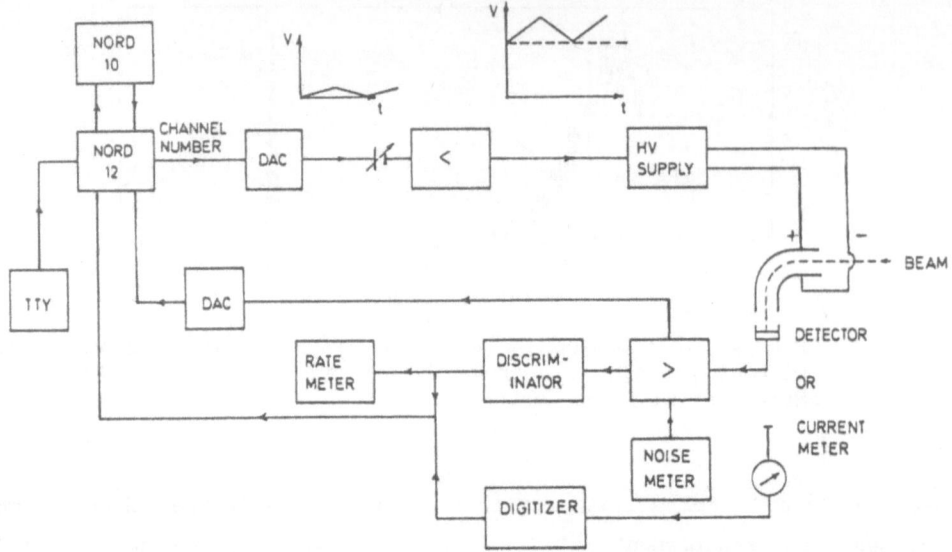

Figure 30. Data acquisition and analyzer stepping system for an electrostatic energy analysis system. (From H.H. Andersen, F. Besenbacher, and P. Goddiksen, Nucl. Instr. Methods *168* (1980) 75).

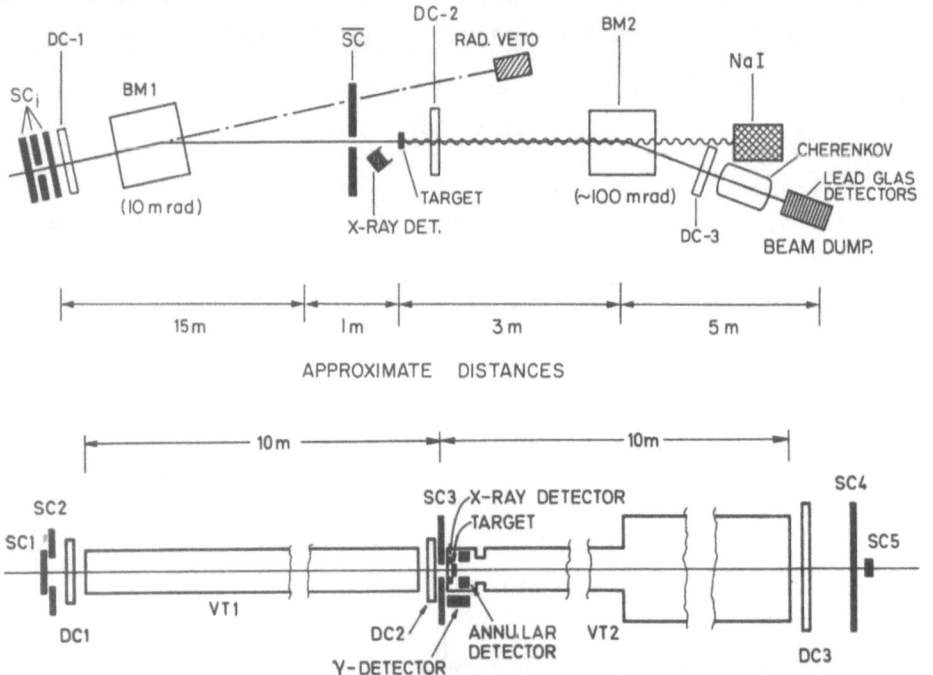

Figure 31. Time-of-flight and angular scattering setup for stopping-power measurements of multi-GeV protons channeled through silicon single crystals. (From E. Uggerhøj, Thesis, University of Aarhus, 1988).

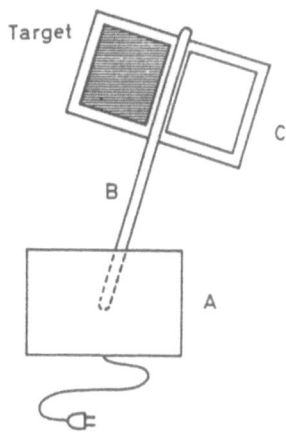

Figure 32. Mechanical target-scanning system. (From R. Ishiwari, N. Shiomi-Tsuda, N. Sakamoto and H. Ogawa, Nucl. Instr. Methods *B47* (1990) 111).

investigated using time-of-flight techniques, but are discussed in the lectures of Lennard of this school and shall hence not be treated here in any detail.

In principle, the question of target thickness homogeneity plagues all spectrometric ΔE measurements but Ishiwari et al. introduced a device to get around the difficulty, as shown in fig.32. The target is mechanically scanned through the beam spot, at least in one direction. Even with a surface barrier detector as spectrometer, $2\sigma = 0.5$ % for dE/dx is claimed for MeV protons. The energy calibration for these measurements is, however, not documented in detail, but an impressive set of data have been published over the years.

A similar scanning system might be used with a spectrometer with higher resolution. It is, in this connection, important that the data acquisition system is switched off in synchronization whenever the beam might hit the frame of the foil support system (as is indeed the case for the Ishiwari et al. measurements). For all high-precision stopping data obtained from full energy-loss spectra, attention should further be given to the measured energy-loss distributions. Either the peak value theoretically corrected for asymmetries should be used for the average energy loss or great care should be exercised in measuring the wings of the distributions. Bad statistics in the wings combined with asymmetries in the peaks means that standard statistics may not be used in estimating the uncertainties in the average energy losses.

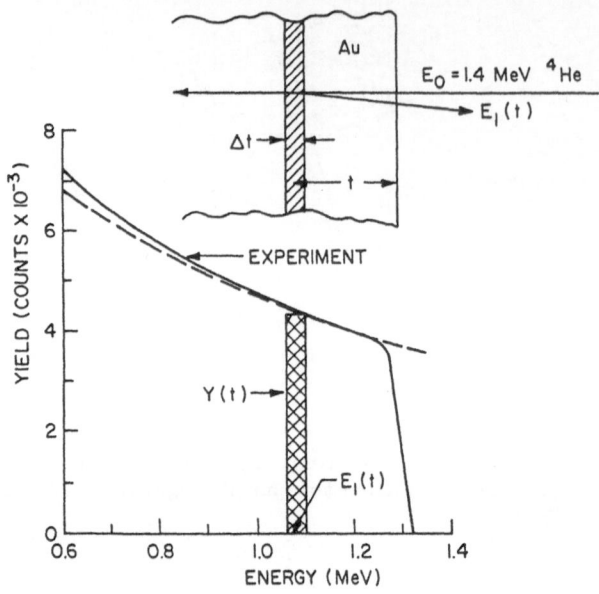

Figure 33. Energy loss relations for a beam backscattered at depth t below a surface. (From L.C. Feldman and J.W. Mayer, *Fundamentals of Surface and Thin Film Analysis*, North-Holland, 1986).

4.4 Backscattering Measurements

The basic energy-loss relations for a backscattered beam at depth t are shown in fig.33. In concert with the discussion in the previous section, it would be better to use the stopping power at $E_0 - \Delta E_{in}/2$ and at $E_1 - \Delta E_{out}/2$ rather than at E_0 and E_1. Whenever the method is inverted, i.e., to allow deductions of stopping powers rather than of scattering depths, the main problem is that stopping powers at two often rather widely separated energies enter the expression. Mostly, some energy dependence of the stopping power is assumed and a parameter optimization is performed to deduce stopping powers.

By depositing a thin layer on a lighter substrate, we may use RBS as a standard method for ΔE measurements on targets that are not self supporting. The thickness of the deposited film may be measured either by a quartz-oscillator microbalance during deposition or through absolute backscattering-yield measurements. In the former case, the quartz crystal itself should be used as the substrate, which may be difficult, or we shall depend on absolute current measurements and knowledge of detector geometry. In both cases it has proven difficult to press the 2σ level below 3 %.

184

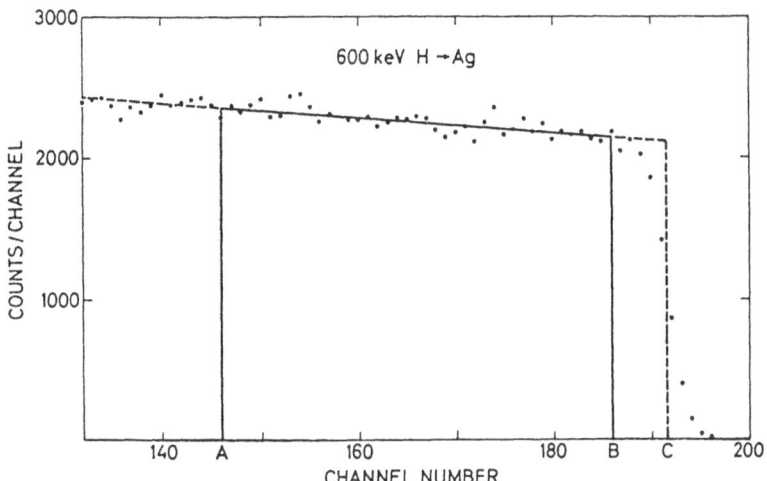

Figure 34. Example of extrapolation of a backscattering spectrum to the edge. Section A-B is used to establish the height of the spectrum. The edge position is determined through fitting an error function to the spectrum from position B and upwards. In the present case the energy resolution is determined to be 10.5 keV fwhm. (From H.H. Andersen, H. Knudsen, and V. Martini, Nucl. Instr. Methods *149* (1978) 137).

From fig.33 it is further seen that the number of target atoms passed by the projectile while being slowed down from E to E – dE is inversely proportional to the stopping power. Hence the number of counts per unit energy interval in the backscattered spectrum must also be inversely proportional to dE/dx. If the integrated current and the absolute detector geometry is known, we may extract dE/dx at E = E_0, provided the scattering cross section is known with sufficient accuracy. On top of previously mentioned difficulties, we shall, however, determine as well spectral height as position of the edge (fig.34). It is difficult to get below 2σ = 5 %. The scatter in published data indicate that 10 % – 15 % may be a more realistic value in spite of the fact that a 1 % accuracy is often claimed. This may have to do with the previously mentioned unwillingness to realize the difficulty of performing absolute current measurements.

It has been attempted to get around the problem of absolute measurements through standardization to a well known material of not too different an atomic number. It is not possible to make self-supporting thin foils from rare-earth metals as they will immediately oxidize. Hence, backscattering targets were prepared through vacuum evaporation and covered in situ with a thin Au protective layer. As secondary emissions

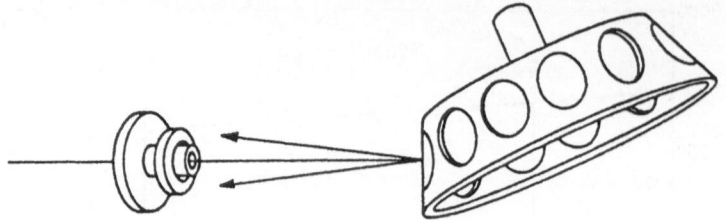

Figure 35. Target-changing setup for backscattering experiments. (From H.H. Andersen, H. Knudsen and V. Martini, Nucl. Instr. Methods *149* (1978) 137).

are different from different surfaces, current measurements were totally avoided through rapid random switching between a number of target materials. An annular detector was used and the targets mounted on a conically-shaped ring as shown in fig.35 to allow rapid switching. Counting time for each position was 2 seconds and the total cycle was repeated 100 times. Great care has to be exercised to optimize the performance of the annular detector. The result will be spectra like the one shown in fig.36. Note that in this case, the target was not

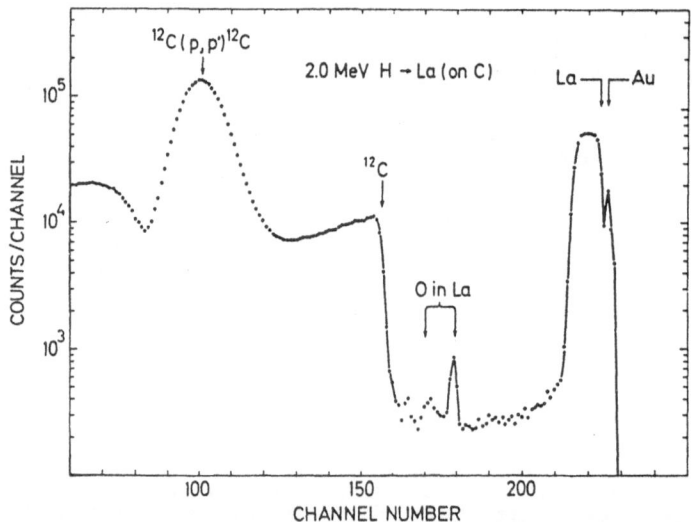

Figure 36. Energy spectrum of 2.0 MeV protons backscattered from a thin La target deposited onto a C backing. The La target was covered by a very thin protective layer of Au but did nevertheless contain substantial amounts of oxygen. (From H.H. Andersen, H. Knudsen and V. Martini, Nucl.Instr.Methods *149* (1978) 137. See also H. Knudsen, H.H. Andersen and V. Martini, Nucl. Instr. Methods *178* (1980) 41).

sufficiently thick to allow a precise determination of the height of the spectrum. Note also that in spite of the precautions, a substantial amount of oxygen was present in the target.

4.5 The Calorimetric Technique

For materials from which self-supporting foils of good heat conductivity may be made, stopping-power measurements for light MeV projectiles may be performed by the calorimetric technique shown in fig.37. The technique works at liquid-helium temperatures and capitalizes on the enormous sensitivity of semiconductor thermometers at these temperatures together with the vanishing thermal radiation coupling between systems. The beam is passed through the stopping foil and into a beam stop. The power deposited by the beam is calibrated by the use of electrical heaters and the relative energy loss is given by

$$\frac{\Delta E}{E} = \frac{P_F}{P_F + P_B}. \tag{12}$$

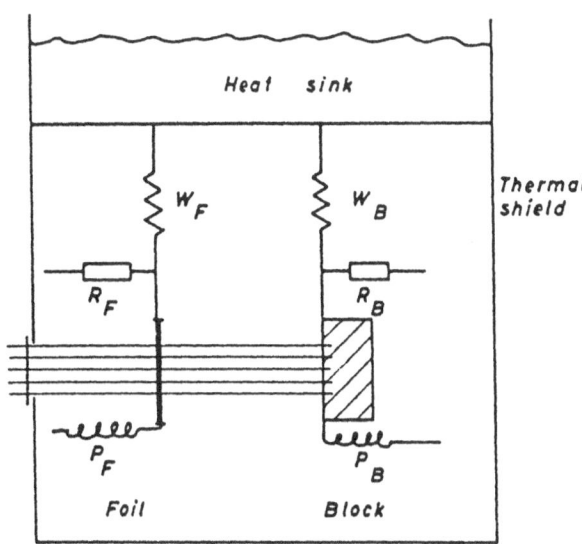

Figure 37. Principle of the calorimetric energy-loss measuring technique. W_F and W_B are thermal resistors, R_F and R_B thermometers, and P_F and P_B electrical heaters used for calibrating the powers deposited by the beam. (From H.H. Andersen, *A Low-temperature technique for measurement of heavy-particle stopping powers of metals*, Risø Report No 93 (1965). See also, e.g., H.H. Andersen, C.C. Hanke, H. Sørensen, and P. Vajda, Phys. Rev. *153* (1967) 338 and H.H. Andersen, J.F. Bak, H. Knudsen, and B.R. Nielsen, Phys. Rev. *A16* (1977) 1929).

The main advantage, apart from the sensitivity of the temperature and power measuring system, is that a large foil area (1 cm^2) may be homogeneously irradiated and exactly the same area have its thickness determined through weighing. Disadvantages are twofold. First, we perform relative measurements and are hence, as previously discussed, very sensitive to the energy calibration. Some checks of the relative energy calibration may be performed. Protons and deuterons, and ^{3}He and ^{4}He must pairwise yield identical stopping powers at identical velocities. Second, whatever junk comes along with the beam is measured. The main problem is slit-scattered particles and it takes a very careful beam-line design to avoid a serious influence.

Many years of measurements have established that a precision level of $2\sigma = 0.5$ % may be reached by the calorimetric method. Straggling information may, unfortunately, not be obtained and the method may not be used for very weak beams like antiprotons from LEAR.

4.6 Doppler-shift techniques

An ingenious method for stopping measurements is shown in fig.38. A boron-containing substance is irradiated with thermal neutrons. Through an (n, α) reaction, 800 keV Li recoils are generated in random directions. The Li nuclei are in an exited state that decay during slowing down. The Doppler-shifted γ-spectrum, shown in fig.39, will hence reflect the energy versus time spectrum of the Li nuclei. The main advantage is that the stopping target may be in any shape or phase if only doped with boron. Disadvantages are mainly that the method works only for Li ions with energies below 800 keV. Similar methods may be used for recoils from accelerator-induced nuclear reactions. Here, the recoil has its initial velocity vector within a narrow cone and complicated corrections for angular scattering must be applied. A minor problem is that the density of

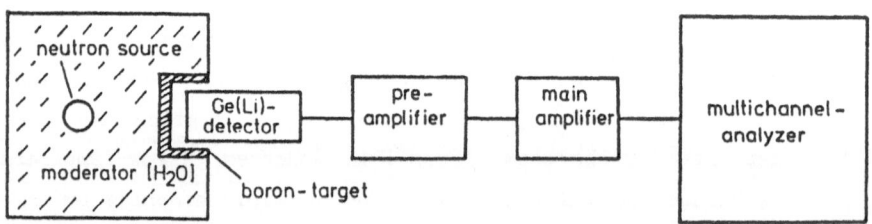

Figure 38. Experimental set-up for Doppler-shift measurements of stopping powers of lithium 7. (From W. Neuwirth, U. Hauser and E. Kühn, Z. Physik *220* (1969) 241).

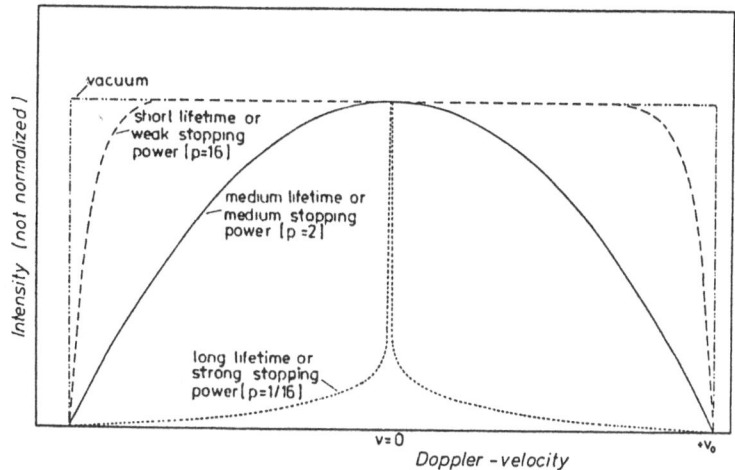

Figure 39. Examples of shapes of Doppler-shifted γ-spectra for different lifetimes or different stopping powers. (From W. Neuwirth et al., loc. cit.).

the target must be known. Most serious is that in evaluation of the data a velocity-proportionality is usually assumed to simplify the data analysis. Where this has been proven correct (e.g., B_4C), a high accuracy (2 %) may be reached. In other cases the published results appear to be rather far off the mark.

Table 4. Accuracy of the dE/dx measurements.

Method		$2\sigma_{min}$
ΔE,	Magnetic	1.5 %
	Electrostatic	2 %
	Time-of-flight	2 %
	Surface barrier	2 %
	Any method, but target averaging	0.5 %
RBS		3 %
RBS,	Spectral height	5 %
RBS,	Relative height, REE targets	5 %
RBS,	Relative height, well-behaved targets	3 %
Calorimetric		0.4 %
Doppler shift, ^{7}Li		(2 %)

4.7 Accuracy of stopping measurements

The conclusions of this chapter are given in table 4. It is worthwhile once again to emphasize that the energy calibration of the beam must be sufficiently good to warrant these accuracies. And, also once again, that very often the accuracy of published stopping data turns out to be not nearly so good as claimed by the authors of the data.

5. CONCLUSIONS

Recent years have seen a number of improvements in the quality of charged particle beams. They are obtainable with higher brightness, better purity and higher precision of their energy than previously. Important prerequisites for such improvements are new devices (e.g., the liquid metal ion source) or new techniques (e.g., beam cooling). New systems like nuclear microprobes, heavy-ion storage rings and accelerator-mass spectrometers have been built utilizing these advances, but apart from the antiproton stopping-power measurements at the LEAR ring at CERN, the advances have not been used for energy-loss measurements. The extremely precise energy definition in storage rings as well as the very small area sampled by a microbeam ought to be particularly useful for measurements of energy straggling.

6. SUGGESTIONS FOR FURTHER READING

2.1 Ion sources

G.D. Alton, *The Sputter Generation of Ion Beams*, Nucl. Instr. Meth. *B37/38* (1989) 45-55.

G.D. Alton and M.L. Mallory, eds., *Proc. Int. Conf. on Heavy Ion Sources*, IEEE Trans. Nucl. Sci. *NS23* (1976) 885-1183.

H.H. Andersen, *Formation and Stability of Sputtered Clusters*, Vacuum *39* (1989) 1095-99.

R. Billinge et al., *Concept for a Lead-Ion Accelerating Facility at CERN*, CERN 90-01 (1990) 33 p.

S. Bliman, ed., *The Physics of Multiply Charged Ions and ECR Ion Sources*, J. de Physique *50* (1989) C1, 904 p.

I.G. Brown, ed., *The Physics and Technology of Ion Sources*, John Wiley, Chichester (1989) 444 p.

R. Geller, *Sources of Highly Charged Ions*, in Proc. European Particle Accelerator Conference, S. Tazzari, ed., World Scientific, Singapore (1989) 88-92.

G. Sidenius, *Ion Sources for Low Energy Ion Accelerators*, Nucl. Instr. Meth. *151* (1978) 349-62.

T. Tagaki, ed., *Ion Implantation Technology*, Nucl. Instr. Meth. *B37/38* (1989) 1-994.

P. Tykesson, H.H. Andersen and J. Heinemeir, *Further Investigations of ANIS. (The Aarhus Negative Ion Source)*, IEEE Trans. Nucl. Sci. *NS-23* (1976) 1104-08.

2.2, 2.3 Accelerators

Richard Fornow, *Introduction to Experimental Particle Physics*, Cambridge Univ. Press, Cambridge, (1989). Paperback 413 p.

G. Frick and E. Jaeschke, eds., *Electrostatic Accelerators and Associated Boosters*, North-Holland, Amsterdam, (1990). Nucl. Instr. Meth., *A287* (1990) 336 p.

W.M. Gibson, *Nuclear Reactions*, Penguin, Harmondsworth (1971). Chapter 2. Paperback.

H. Ryssel and I. Ruge, *Ion Implantation*, John Wiley, Chichester (1986) 457 p.

W. Scharf, *Particle Accelerators and Their Uses*, Harwood, London and New York (1986), vols. 1 and 2. 998 p.

T. Tagaki, see section 3.1.

S. Tazzari, ed., *Proc. European Particle Accelerator Conference*, World Scientific, Singapore (1989). 1-1530

S. Turner, ed., and W. Bryant and S. Turner, eds., *Proc. CERN Accelerator Courses*. Several CERN reports during the 1980's.

W. Urbanus, *DC and RF Ion Accelerators for MeV Energies*. (Thesis, Utrecht, 1990) 130 p.

2.4 Ion optics

D.C. Carey, *Optics of Charged Particles*, Academic Press, New York (1987) 291 p.

P. Dahl, *Introduction to Electron and Ion Optics*, Academic Press, New York and London (1973) 147 p.

M. Month and S. Turner, eds., *Frontiers of Particle Beams*, Springer, Berlin (1988) 700 p.

A. Septier, ed., *Focusing of Charged Particles*, Academic Press, New York and London (1967), vols. 1 and 2. 509 + 471 p.

M. Szilagyi, *Electron and Ion Optics*, Plenum, New York (1988) 539 p.

A. Sørensen, *Liouville's Theorem and Emittance*, p. 18-36 in S. Turner, ed., Proc. Third General Accelerator Physics Course, CERN 89-05 (1989).

W. Urbanus, *DC and RF Accelerators for MeV Energies*, Thesis, Utrecht (1990) 136 p.

H. Wollnik, *Optics of Charged Particles*, Academic Press, New York (1987) 291 p.

2.5 Cooling, Storage Rings

S.P. Møller, *Cooling Concepts*, p. 1-17 in S. Turner, ed., Proc. 3rd general accelerator physics course, CERN 89-05 (1989).

2.6 Microbeams

G.W. Grime and F. Watt, eds., *Nuclear Microprobe Technology and Applications*, North-Holland, Amsterdam (1988). Nucl. Instr. Meth., *B30* (1988) 227-506.

P.W. Hawkes, ed., *Aspects of Charged Particle Beams*, Academic Press, London and New York (1989).

G. Legge, ed., *Proc. Second int. Conf. on Nuclear Microprobe Technology and Applications*, North-Holland, Amsterdam (1990). Nucl. Instr. Meth. *B52* (1990) No 2.

K. Traxel, *Nuclear Microbeams: Realization and Use as a Scientific Tool*, Nucl. Instr. Meth. *B50* (1990) 177-88.

F. Watt and G.W. Grime, *Principles and Applications of High-Energy Microbeams*, Adam Hilger, Bristol (1987) 339 p.

3. Beam Diagnostics

M. Month and S. Turner, eds., *Frontiers of Particle Beams; Observation, Diagnosis and Correction*, Springer, Berlin-Heidelberg (1989) 509 p.

Most books mentioned for sects. 3.2, 3.3 and 3.4.

3.4 Accelerator Mass Spectrometry

H.E. Gove, A.E. Litherland and D. Elmore, eds., *Accelerator Mass Spectrometry*, Proc. AMS '87, North-Holland, Amsterdam (1987) 445 p. Nucl. Instr. Meth. *B29* (1987) No 1, 2.

W. Wölfli, H.A. Polach and H.H. Andersen, eds., *Accelerator Mass Spectrometry*, Proc. AMS '84, North-Holland, Amsterdam (1984) 357 p. Nucl. Instr. Meth. *B4* (1984) No 2.

Proceeding from AMS '90: Nucl. Instr. Methods *52* (1990) No.3-4.

4. Stopping Power and Straggling

H.H. Andersen, *Studies of Atomic Collisions in Solids by Means of Calorimetric Techniques*, Thesis, Univ. of Aarhus (1974) 279 p.

H.H. Andersen and J.F. Ziegler, *Hydrogen Stopping Powers and Ranges in all Elements*, Pergamon, New York (1977) 317 p.

D. Powers, *Description of Methods for Stopping Power Measurements in Stopping Powers for Protons and Alpha Particles*, ICRU-Report (to be published, 1990).

See also the numerous research papers quoted in the figure captions.

ACKNOWLEDGEMENT

Numerous colleagues are thanked for permission to reproduce their figures in the present notes. Similarly, North-Holland Physics Publishing, The Institute of Electrical and Electronic Engineers, Springer Verlag and Det Kongelige Danske Videnskabernes Selskab are thanked for permission to reproduce material originally published by them.

CHARGED PARTICLE-SURFACE INTERACTIONS AND SPECTROSCOPIES

A.A. Lucas

Institute for Studies in Interface Sciences
Facultés N.-D. de la Paix
B-5000 Namur, Belgium

In a set of lectures, we treat the interaction of a charged particle with a solid when the particle travels outside of the solid. This situation is relevant to a number of experimental techniques used to investigate solid surfaces, such as UPS and XPS (ultraviolet and X-ray photoelectron spectroscopy), EELS (electron energy loss spectroscopy), several ion spectroscopies, electron tunneling, etc. There is a vast body of old and recent literature on the classical and quantum behavior of a charge near a solid surface in the different energy regimes from the static to the fast-particle cases. Some aspects of this work will be reviewed in the present School. Irrespective of the nature of the charge (electron, positron, proton, complex ion, etc.), its interaction with the solid always includes and, at large distances, is dominated by the ubiquitous "image potential". A description of the image potential, particularly of its quantum manifestation in EELS, is the main subject of the present lectures. The single unifying concept of an effective dielectric function will be introduced and used throughout the lectures to construct the static and dynamic potential of the charge to which the solid responds.

The first lecture will begin at the elementary level of solving the electrostatic Poisson equation for the potential distribution of a static point charge in the presence of one or several parallel interfaces separating homogeneous media of different dielectric constants. Although a well known textbook problem, it will be given a new and computationally powerful solution. It turns out that it is indeed possible to define a single surface impedance function which serves as the effective dielectric constant for characterizing the long-wavelength electrostatic response of a plane-stratified medium of arbitrary structure to the presence of the

Interaction of Charged Particles with Solids and Surfaces
Edited by A. Gras-Martí *et al.*, Plenum Press, New York, 1991

193

static charge. The classical image potential energy of the charge outside the stratified medium will be given by the usual image construction in a single interface provided that the dielectric constant of the normally homogeneous substrate be replaced by the effective dielectric function of the multilayer structure. The infinitely many images of the multiple interfaces are automatically incorporated in the continued fraction expression of the effective dielectric constant. Extension to the case where the source charge is embedded inside the multilayer substrate will also be given. A particularly elegant, closed-form result will be shown to apply to the case of a periodic superlattice. Illustration will be provided for Si/Ge and GaAs/AlAs superlattices.

The quantization of the Fourier components of the potential amplitude leads to a quantum mechanical version of the static image charge concept. In this version, the image potential energy is viewed as a shift in zero-point energy of the quantized polarization modes subjected to a stretching of their harmonic amplitudes when the source charge is brought from infinity to its actual position near the interface.

Besides its computational convenience the great formal advantage of the central concept of an effective dielectric function is to be applicable to the response of layered materials to dynamical probes as well as to static ones. This situation is encountered in charge particle spectroscopies such as EELS as well as in several light spectroscopies such as Infrared Absorption (IA) and Attenuated Total Reflection (ATR).

In the second part of these lectures, the nature of polaritons in multilayer materials will be briefly explained. It will be shown that their dispersion relations and their role in the optical properties such as observed in IR and ATR can be formulated in terms of retarded, effective dielectric functions again calculable as continued fractions.

In the third chapter a detailed study of EELS will be presented in the reflection geometry. Only the dipole mechanism of scattering by surface and interface phonons and plasmons of long wavelengths will be considered. This mechanism is again expressible in terms of an effective, dynamical dielectric function. The concept of coherent excitation of the collective modes will be presented and shown to serve as the starting point for constructing a semiclassical theory of EELS. The quantitative success of this theory will be illustrated with several experimental EELS spectra of both homogeneous and layered substrates.

The fourth and last part of the lectures will deal with constructing a model barrier for electron tunneling through a Metal-Vacuum-Metal junction such as occurs in Scanning Tunneling Microscopy. A short review of tunneling theories will be given and a recent formulation of multidimensional tunneling in terms of scattering theoretic Green functions will be expounded.

The following list of references provides the background information which will be followed in the oral lectures and which can be consulted by the students as an introduction to the fields.

ACKNOWLEDGEMENTS

The author is grateful to Th. Laloyaux, A. Dereux, I. Derycke, J.P. Vigneron, Ph. Lambin, M. Tamborelli for discussions and comments. Part of the work was developed under the Belgian Interuniversity Research programs of the Prime Minister's Office, Science Policy Programming.

REFERENCES

1. J.D. Jackson, *Classical Electrodynamics*, J. Wiley and Sons, N.Y. (1975), p. 147.
2. A.A. Lucas, J.P. Vigneron, P.H. Cutler, T.E. Feuchtwang, R.H. Good Jr., and Z. Huang, J. Phys. Colloq. *45* (1984) C9-125.
3. A.A. Lucas, E. Kartheuser, R. Brado, Phys. Rev. *B2* (1970) 2488. M. Sunjic, A.A. Lucas, Phys. Rev. *B3* (1971) 719. A.A. Lucas, Phys. Rev. *B4* (1971) 2939. A.A. Lucas, M. Sunjic, Surface Sci. *32* (1972) 439.
4. A.A. Lucas and E. Kartheuser, Phys. Rev. *B1* (1970) 3588. M.S. Thomas et al., Phys. Rev. *B9* (1974) 1489.
5. A. Dereux, J.P. Vigneron, Ph. Lambin, A.A. Lucas, Phys. Rev. *B38* (1988) 5438. Ph. Lambin, J.P. Vigneron, A.A. Lucas, A. Dereux, Phys. Scripta *35* (1987) 343.
6. D. ter Haar, *Selected Problems in Quantum Mechanics,* Infosearch, (London) 1964, p.162. I.I. Goldman *et al., Problems in Quantum Mechanics*, Infosearch, (London) 1960, p.103.
7. R.J. Glauber, Phys. Rev. *131*, (1963) 2766; in *Quantum Optics*, Proc. Int. School "Enrico Fermi", Varenna, Academic Press, (New York) 1969, p.15.
8. A.A. Lucas and M. Sunjic, Phys. Rev. Lett. *26* (1971) 229. M. Sunjic and A.A. Lucas, Phys. Rev. *B3* (1971) 719. M. Sunjic and A.A. Lucas, Prog. Surface. Sci. *2* (1972) 75.
9. Ph. Lambin, J.P. Vigneron, A.A. Lucas, Comput. Phys. Commun., to be published.
10. E. Evans and D.L. Mills, Phys. Rev. *B5* (1972) 4126.
11. Ph. Lambin, J.P. Vigneron, A.A. Lucas, Phys. Rev. *B32* (1985) 8203.
12. A.A. Lucas, J.P. Vigneron, Solid State Commun. *49* (1984) 327.
13. R. Strümpler and H. Lüth, Surface Sci., to appear.
14. B. Voigtländer, D. Bruchman, S. Lehnald, H. Ibach, Surface Sci., to appear.
15. C.J. Powell, Phys. Rev. *175* (1968) 972.
16. A.A. Lucas, Phys. Rev. *B20* (1979) 4990.
17. A. Närmann, R. Monreal, P.M. Echenique, F. Flores, W. Heiland, S. Schubert, Phys. Rev. Lett. *64* (1990) 1601.
18. J.P. Vigneron, Ph. Lambin, J. Phys. *A13* (1980) 1335; *A14* (1981) 1815.
19. J. Bardeen, Phys. Rev. Lett. *6* (1961) 57.
20. C.B. Duke, *Tunneling in Solids*, Academic Press, New York (1969).
21. J. Tersoff and D. Hamann, Phys. Rev. *B31* (1985) 805.
22. N. Lang, Phys. Rev. Lett. *55* (1985) 230.

23. A.A. Lucas, Europhys. News *21* (1990) 61.
24. A.A. Lucas, H. Morawitz, G.R. Henry, J.P. Vigneron, Ph. Lambin, P.H. Cutler, T.W. Feuchtwang, Phys. Rev. *B37* (1988) 10708.

ELECTRON SPECTRA IN SOLIDS

R.H. Ritchie[1,2], R.N. Hamm[1], J.C. Ashley[1]
and P.M. Echenique[3]

1. Health and Safety Research Division
   Oak Ridge National Laboratory, Oak Ridge
   TN 37831-6123, U.S.A.

2. Department of Physics, University of Tennessee
   Knoxville, TN 37996, U.S.A.

3. Departamento de Física, Facultad de Química
   Apartado 1072, San Sebastián
   E-20080, Spain

ABSTRACT

These lectures cover special topics about the interaction of swift electrons with condensed matter. The focus is on inelastic processes via the linear response function of the medium and sum rules that it satisfies. The spin dependence of low-energy electron interactions and exchange scattering in an electron gas, and the low-energy end of electron-hole cascades in metals are discussed. Aspects of the localization of initially unlocalized, coherent excitations are entered into briefly.

1. INTRODUCTION

The charge to participants in this Summer School is to emphasize research areas where further effort is needed. We will adhere to this in attempting to cover selected topics that present challenges to current understanding.

These lectures are intended to give a general introduction to some of the concepts encountered in the study of swift electron interactions in condensed matter. We do not attempt to be exhaustive in our coverage of relevant topics but will emphasize fundamentals rather than applications, special topics and putatively neglected areas rather than conventional ones. In this lecture we will review briefly some aspects of the dielectric function of solids, with emphasis on metals. Since Tougaard's lectures cover many aspects of electron transport, we will focus on our state of knowledge about the cross sections for energy loss by swift

Interaction of Charged Particles with Solids and Surfaces
Edited by A. Gras-Martí *et al.*, Plenum Press, New York, 1991

electrons in condensed materials, and their representation for use in transport calculations.

Good survey papers about the subject of electron interactions in solids are available. We mention Kumakhov and Komarov [1], Ohtsuki [2], Schattschneider [3], Schou [4], Tougaard [5], Echenique et al.[6], and Seiler [7].

2. THE INTERACTION OF A SWIFT ELECTRON WITH CONDENSED MATTER

A formal expression [8] involving the linear response function of the N electrons in an isotropic system may be found by equating the transition rate of the electrons in the system under a given perturbation to the rate obtained assuming that the response is proportional to the inverse dielectric function $\varepsilon_{k,\omega}^{-1}$. One finds,

$$\text{Im}\left[\frac{-1}{\varepsilon_{k,\omega}}\right] = \frac{4\pi^2 e^2}{\hbar k^2 \Omega} \sum_n |<n|\rho_{\vec{k}}|0>|^2$$

$$\times \left[\delta(\omega + \omega_o - \omega_n) - \delta(\omega_n - \omega_0 + \omega)\right], \tag{1}$$

where $(\hbar\omega_n, |n>)$ is the (eigenenergy, eigenvector) of the nth excited state of the unperturbed system, $\rho_{\vec{k}} = \sum_1^N e^{i\vec{k}\cdot\vec{r}_i}$ is the density operator and Ω is the normalization volume of the system. Sum rules have been used to constrain approximate representations of the response function of matter in many applications. For example, Bethe [9] in his pioneering theory of charged particle energy loss, was the first to prove and use a sum rule generalized from a dipole rule employed in early optical response theory. Placzek [10] later generalized the Bethe approach, while Fano and Turner [11] made specific application of these ideas to atomic systems.

Sum rule results may be found for general response functions using methods similar to those employed for atomic systems. Perhaps the most useful of these is the pair

$$\frac{m}{2\pi^2 ne^2} \int_0^\infty \omega \, \text{Im}\left[\frac{-1}{\varepsilon_{k,\omega}}\right] d\omega = 1, \tag{2}$$

$$\frac{m}{2\pi^2 ne^2} \int_0^\infty \omega \, \text{Im}\left[\varepsilon_{k,\omega}\right] d\omega = 1, \tag{3}$$

where m is the electron mass and n is the mean density of all electrons in the system. Higher-order sum rules defined as [12]

$$\sum_j (k) = \frac{m}{2\pi^2 ne^2} \int_0^\infty \omega^j \, \text{Im}\left[\frac{-1}{\varepsilon_{k,\omega}}\right] d\omega, \tag{4}$$

may be of value. In particular,

$$\sum_2 (k) = \frac{\hbar k^2}{2m} + \frac{4}{3} T/\hbar + 0_2, \tag{5}$$

where T is the mean kinetic energy per electron in the ground state of the system and 0_2 represents terms originating in two electron correlations, which go rapidly to zero for large k. The expression for $\sum_3$ is quite long, and $\sum_4$ diverges.

In the most widely used model, the inverse mean free path (IMFP), $\lambda^{-1}(E_o)$, for the excitation, by a swift electron having energy E_o, of electronic transitions in the medium, is written in terms of $\varepsilon_{k,\omega}^{-1}$ as

$$\lambda^{-1}(E_o) = \frac{2e^2}{\pi\hbar v^2} \int_0^{E_o} d\omega \int_{k_-}^{k_+} \frac{dk}{k} \, \text{Im}\left[\frac{-1}{\omega_{k,\omega}}\right], \tag{6}$$

in first Born approximation, and neglecting exchange. Also, $k_\pm = (2m/\hbar^2)^{1/2} [E_o^{1/2} \pm (E_o - \hbar\omega)^{1/2}]$. The differential IMFP (DIMFP) for energy loss $\hbar\omega$ is taken to be the integrand of the ω integration. In these applications $\varepsilon_{k,\omega}$ is assumed to depend on the magnitude of $\vec{k}$ and not on its direction.

2.1 The Electron Gas

Here we review some aspects of the dielectric function of the electron gas, since it has been used extensively, as well as empirically, in representing the interaction of energetic electrons with many different kinds of solids. We emphasize the properties of the dynamic function $\varepsilon_{k,\omega}$ rather than those of the static dielectric function $\varepsilon_{k,o}$. The latter has been studied extensively in the context of many body theory [13] and is fairly well understood by now.

The theory of the interaction of charged particles with solids was advanced by Lindhard [14] through his discovery of an analytical expression for the quantal dielectric function of an electron gas. This dielectric function, variously termed the Lindhard or RPA function, has had extensive application to metals and semiconductors in the ensuing years. As useful as this advance has been to solid state physics in the technology of band state and pseudopotential theory, it is still more so in leading to qualitative ideas about the general response function of condensed matter.

A systematic development of interactions in the electron gas may be carried out using a diagrammatic approach. Lindhard [14] employed a Hartree self-consistent approach in his original work. We follow standard notation [15-17]. The Green function of an electron in a noninteracting system may be written

$$G_p^0 = \left[p_0 - \Omega_{\vec{p}} + i\delta \, \text{sgn}\left(|\vec{p}| - p_F \right) \right],$$ (7)

where $p = (\vec{p}, p_0)$, $\Omega_{\vec{p}} = \hbar^2 \vec{p}^2 / 2m$, p_F is the Fermi wave number and δ is a positive infinitesimal.

The Green function for the interacting, translationally invariant system is

$$G_p = \left[p_0 - \Omega_{\vec{p}} - \sum_p \right]^{-1},$$ (8)

where $\sum_p$ is the exact self-energy of the fully interacting electron gas, as indicated by the Feynman diagram of fig.1a. In the G-W

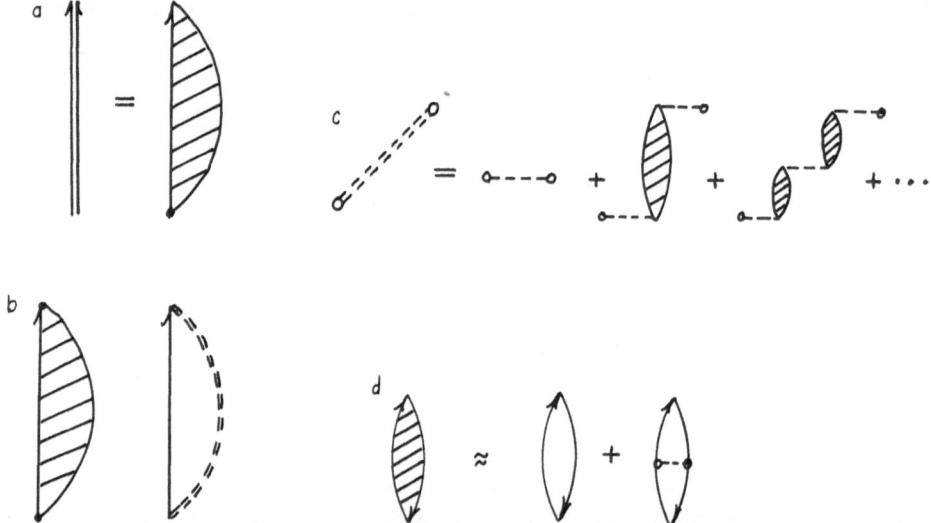

Figure 1. Feynman diagrams representing various quantities of importance in the theory of the dielectric function of the electron gas: a) The exact Green function; b) illustration of the G-W approximation to the self-energy operator; c) representation of the bubble approximation to the effective interaction; d) bubble diagrams, showing the empty one yielding the RPA dielectric function and the crossed bubble considered by Hubbard [22].

approximation [13] $\sum_p$ is given in terms of the Green function itself and a polarization propagator (or effective interaction) as in fig.1b, where the latter, in turn, may be approximated by the infinite series of terms depicted in fig.1c. Here, the shaded bubble represents the proper polarization subdiagram sum $\Lambda_{k,\omega}$.

The dielectric function $\varepsilon_{k,\omega}$ is related to the effective interaction, $W_{k,\omega}$, by

$$W_{k,\omega} = \frac{v_k}{\varepsilon_{k,\omega}}, \tag{9}$$

where $v_k = 4\pi e^2/k^2$ and $\varepsilon_{k,\omega} = 1 - v_k \Lambda_{k,\omega}$. If the shaded bubble in fig.1c is approximated by an empty bubble one obtains the RPA dielectric function, which is nearly the same as the Lindhard function [14] except that in this time-ordered theory, $\varepsilon_{k,\omega} = \varepsilon_{k,-\omega}$, whereas Lindhard's function is the retarded version of this, in which $\varepsilon_{k,\omega} = \varepsilon^*_{k,-\omega}$

In many applications there is essentially no difference between the results obtained using these two forms. In this case the proper polarization subdiagram sum yields

$$\Lambda^o_{k,\omega} = 2 \int \frac{d^4p}{(2\pi)^4} \, G^o_{\vec{p},p_o} \, G^o_{\vec{p}-\vec{k},p_o-\omega}, \tag{10}$$

in the RPA. We omit reference to spin in this limit.

The simple analytical form of the RPA-Lindhard function has expedited many numerical studies of the response of solids to swift charged particles [18-20,5] as well as many attempts to modify it to account for spin effects, local field and higher-order interactions, etc., in an approximate way. It has also been useful in the evaluation of the exchange and correlation energy in metals [20], and in the study of pseudopotentials in solids [21]. Among these modifications is the early introduction by Hubbard [22] of a correction for exchange processes in the primitive bubble diagram, such as shown in fig.1d. He suggested replacing $\Lambda^o_{k,\omega}$ by

$$\Lambda^{(1)}_{k,\omega} = \Lambda^o_{k,\omega} / (1 + f_k \Lambda^o_{k,\omega}) \tag{11}$$

where $f_k = k/2(k^2 + k_F^2)$ and k_F is the Fermi wave number. Much work has been done in this same spirit [23-27]. Mermin [28] has proposed a number-conserving approximation to $\varepsilon_{k,\omega}$ that includes damping in an empirical way.

No systematic comparisons of theoretical electron transport parameters such as stopping power and inelastic mean free path have been made using the various approximations to the electron gas dielectric function (see, however, [29]. The indications are that differences found using the various approximations to $\varepsilon_{k,\omega}$ are probably less than the uncertainties that arise in attempting to represent the response function of a real solid by employing the electron gas model.

2.1.1 Exchange Effects. An interesting point arises when one attempts to compute energy losses by low-energy electrons to solids. A fundamental requirement of quantum mechanics is that the total wave function of an assembly of identical particles must be antisymmetric under interchange of both spin and space coordinates. This leads not only to the operation of the Pauli exclusion principle in the scattering process, but in general affects the cross section for scattering to allowed final states.

The simplest case that illustrates this is the scattering of a swift, non-relativistic electron, on a free, stationary electron. The differential cross section for this process is [30]

$$d\sigma = \frac{2\pi e^4}{E} \left[\frac{1}{(E-E')^2} + \frac{1}{E'^2} - \frac{1}{E'(E-E')} \right],$$ (12)

where E is the energy of the incident electron, E' is its energy after the collision and $E_s = E - E'$ is the energy of the secondary, $E_s \leq E'$.

For scattering of a low-energy electron in an electron gas, such that $E - E_F \ll E_F$, it may be shown [31,32] that

$$d^2 \lambda^{-1} = \lambda_o^{-1} \, dE' dE_s,$$ (13)

where the constant

$$\lambda_o^{-1} = \left\{ \tan^{-1}\left(\frac{1}{\gamma^2}\right) + \frac{\gamma}{1+\gamma^2} - \frac{\gamma}{\sqrt{1+2\gamma^2}} \tan^{-1}\left[\frac{\sqrt{1+2\gamma^2}}{\gamma^2}\right] \right\} / 8\gamma a_o E,$$ (14)

and $\gamma = \left(\frac{32}{9\pi^4}\right)^{1/6} r_s^{1/2}$. This expression is subject to the usual condition that the more energetic of the electrons after the collision is regarded as the secondary, and energy conservation requires that $E_s - 1 \leq E - E'$, where now all energies are measured in units of the Fermi energy E_F. The one-electron radius in the electron gas is $r_s = a_o k_F [4/(9\pi^2)]^{1/3}$, and a_o is the first Bohr radius.

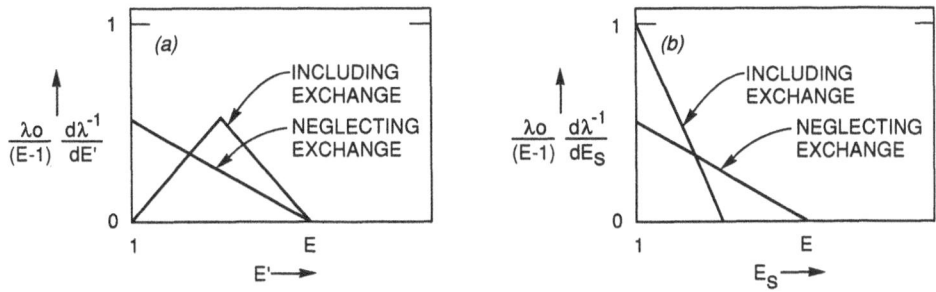

Figure 2. The distribution in energy of primary and secondary electrons after a collision of a low-energy electron in an electron gas. These calculations were made with and without exchange (Ritchie and Ashley, [32]).

Fig.2 shows the distribution of energies following a scattering event, from eq.(13), comparing the results obtained both including and neglecting the antisymmetry in parallel spin interactions. These results were found by calculating the matrix elements corresponding to the Feynman diagrams of fig.3, in which this antisymmetry is displayed explicitly. With this approach, and employing a screened, spin-independent Thomas-Fermi interaction it was found [32] that the mean free path at low energies was appreciably longer than that calculated neglecting the antisymmetry requirement [33], and than experiment would indicate, although the interpretation of available data is difficult at these low energies. Subsequently Kleinman [25] introduced a different approximation for the exchange-corrected electron-electron (e-e) interaction and found little change from the Quinn results [33]. However, the Kleinman e-e interaction is not antisymmetrized for parallel spins.

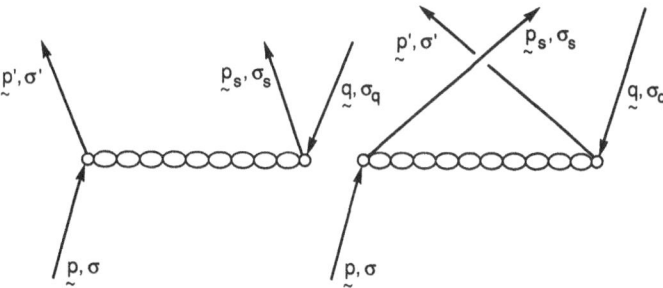

Figure 3. Feynman diagrams representing direct and exchange interactions of a swift electron in an electron gas. The momentum and spin of the interacting electrons are indicated by p and σ, respectively.

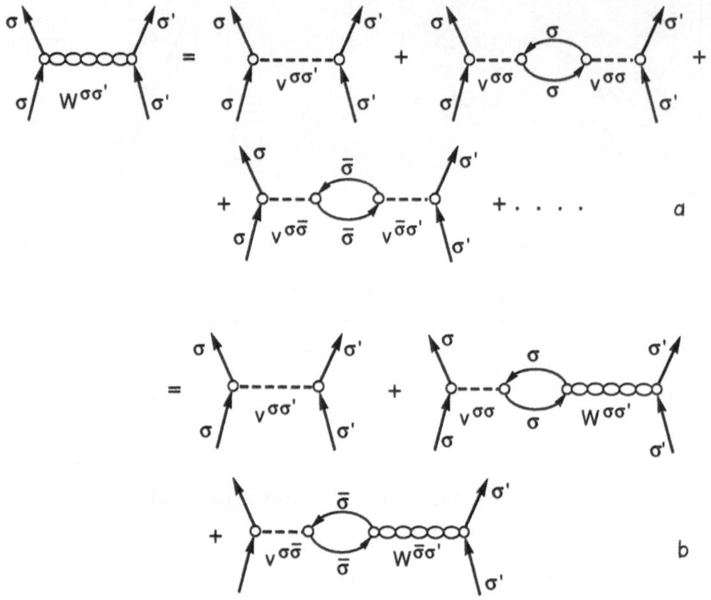

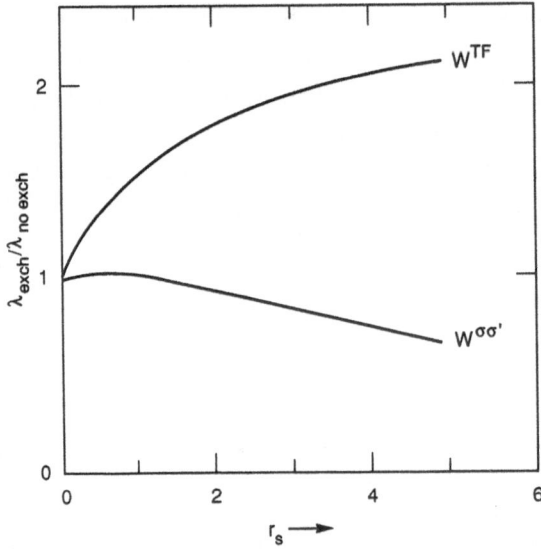

Figure 4. Diagrams representing the sum of possible spin-dependent
interactions in the electron gas: a) Detailed representation of
electron-hole pair excitation and deexcitation;
b) representation of the integral equations for the effective
interaction in the electron gas.

Figure 5. The ratio of the mean free path of low-energy electrons in an
electron gas including exchange to that neglecting exchange. The
top curve, labeled W_{TF} was calculated using a spin-independent
interaction, while the curve labeled $W^{\sigma\sigma'}$ corresponds to using a
spin-dependent interaction (Penn, [37]).

Since these early papers much work has been done on the effects of exchange and correlation on the effective e-e interaction [34-36,27]. Penn [37] has used some results of Kukkonen and Overhauser [34] that allow for an interaction $v^{\sigma}n^{\sigma}$ between parallel spin electrons and a different interaction $v^{\sigma\bar{\sigma}}$ between antiparallel spin electrons. Fig.4a indicates the infinite sum of interactions that must be considered under these conditions, while fig.4b represents the integral equations that may be solved to find the effective spin-dependent interaction $W^{\sigma\sigma'}$.

Using data from reference [34] on $v^{\sigma\sigma}$ and $v^{\sigma\bar{\sigma}}$ Penn [34] found results for the IMFP as a function of the density parameter r_s of the electron gas shown in fig.5. It is somewhat surprising that at $r_s = 2$ there is little difference between his results and those found taking $W^{\sigma\sigma} = W^{\sigma\sigma'} = W_{TF}$, where W_{TF} is the Thomas-Fermi form of the screened interaction, and neglecting antisymmetry. However one sees that for $r_s = 5$ the spin dependent interaction yields a quite different IMFP than is found assuming spin independence as in the Thomas-Fermi form.

Subsequent study of the spin-dependent interactions [35,36,27] have involved approximate diagrammatic evaluations of $v^{\sigma\sigma}$ and $v^{\sigma\bar{\sigma}}$, accounting for local field effects and vertex corrections via many-body fields, using

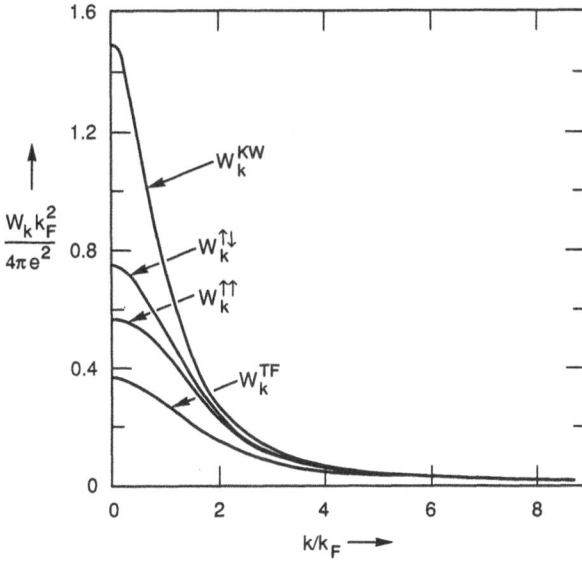

Figure 6. Static, effective, spin-dependent interactions in an electron gas for $r_s = 4$ (Vignale and Singwi, [36]). The label on the curves are evident, except that W^{KW} is the Kukkonen-Wilkins form [34].

sum rules in some cases. Fig.6 compares the approximate static effective interactions in momentum representation, from the work of Vignale and Singwi [36], with the spin-independent W_{TF} for r_s = 4. Also shown is the Kukkonen-Wilkins approximation to the effective interaction according to [36].

Experimental study of the IMFP for this case would be quite useful in selecting between the various theoretical predictions for the spin-dependence of the static interaction. For comparison, additional evaluations of the IMFP using the several alternative approximations to the effective spin-dependent interactions would be very useful.

2.1.2. The Effect of the Exclusion Principle in Intermediate States. The most common approach to evaluation of the mean free path (MFP) for low-energy electrons in condensed matter is to assume that the interaction is in some sense small, thus to employ the first Born approximation to calculate the transition rate of the electron in interaction with the medium, using the linear response function of the medium from electron gas theory or, e.g., from a semi-empirical variation of this theory (see sect.2.2 below). Lindhard [14] has argued that screening of the incident electron by electrons in the medium has the effect of strongly reducing interactions in the system, thus tending to validate the use of perturbation theory even at low incident energies.

As one sees from the extensive literature cited in the previous section, even at low relative energies the interaction is far from simple and disagreement exists between some workers as to the proper exchange and correlation potential for use in calculating such scattering. Given the complex exchange and correlation interactions between electrons in an electron gas, it is interesting to evaluate the magnitude of some effects such as the operation of the exclusion principle in intermediate states, when a low-energy electron is added to such a system. Hamm et al.[38] have

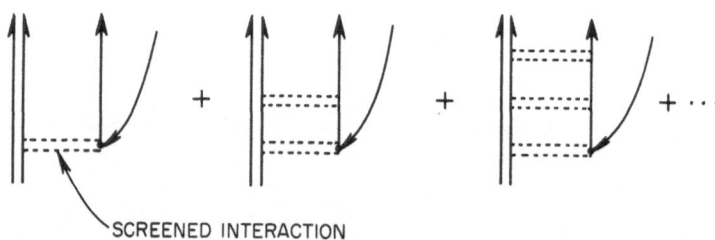

SCREENED INTERACTION

Figure 7. Diagrams showing the infinite sum of repeated interactions in the electron gas that correspond to the Bethe-Goldstone [39] equation.

carried out such a calculation using the Bethe-Goldstone [39] equation for the scattering of two Fermions immersed in a Fermi sea of identical particles and interacting through a static potential. This equation was originally used by Bethe and Goldstone to evaluate the MFP of nucleons in an infinitely large nucleus. More recently, a similar approach has been used [40] to evaluate the energy loss rate of positrons in a metal.

Fig.7 shows Feynman diagrams representing the repeated scattering of a low-energy electron on a particular electron, excited from the Fermi sea, with both constrained to remain outside of the excluded part of momentum space. Following Bethe and Goldstone [39], we write a wave function for the pair in momentum space as

$$2\left[E(\vec{k}_o) - E(\vec{k})\right]\psi_{\vec{P},\vec{k}_o}(\vec{k}) = \int \frac{d^3k'}{(2\pi)^3} W_{\vec{k}-\vec{k}'} \psi_{\vec{P},\vec{k}_o}(\vec{k}') \theta(\vec{P}+\vec{k},\vec{P}-\vec{k}), \qquad (15)$$

where $\vec{P}$ is the center-of-mass momentum, $\vec{k}_o$ is the initial relative momentum, and $\theta(\vec{p},\vec{p}') = H(|\vec{p}| - k_F) H(|\vec{p}'| - k_F)$, where $H(x)$ is the

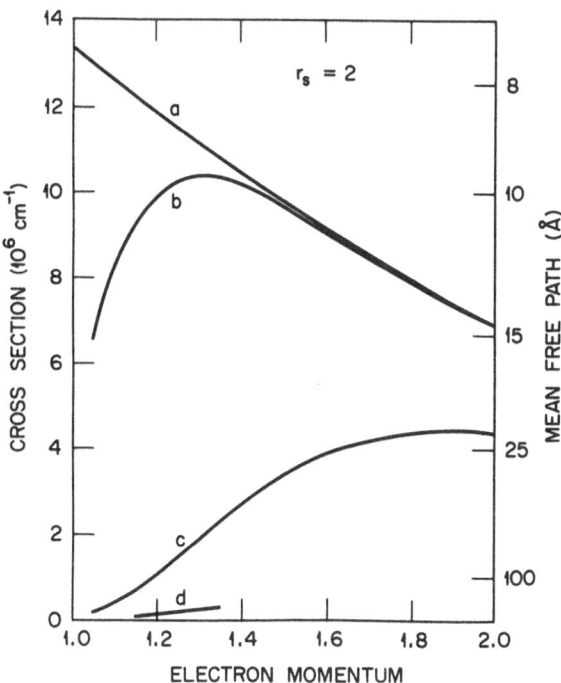

Figure 8. Cross section for electron-electron scattering for $r_s = 2$ calculated (a) without intermediate states corrections, (b) with intermediate states corrections, (c) with final states correction; (d) experimental data for Al (see [38]). The electron momentum is measured in units of Fermi momentum.

Heaviside step function; W_k is the Fourier transform of the effective two-body interaction.

This equation was solved numerically for several different electron gas densities using the Thomas-Fermi static interaction potential for W. Some of the results for the inelastic cross section are presented in figs.8 and 9. One sees that the effect of the Pauli exclusion principle on this quantity is important at energies up to several times the Fermi energy, but that its operation in intermediate states is relatively small for energies greater than twice E. One must emphasize that this treatment neglects vertex corrections and other, more complicated diagrams. However, the conclusions drawn here are likely to be qualitatively correct.

2.1.3 Electron-Hole Cascades in the Electron Gas. The emission of electrons from solids is a fundamental phenomenon that may take place when energetic charged particles or photons interact near a surface of the material. Interest in this phenomenon has continued unabated since its discovery by Herz in the last century. It is of critical importance in

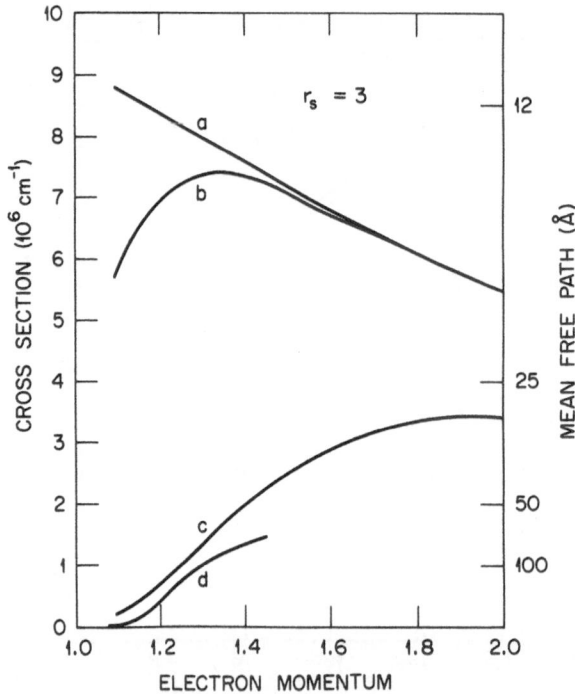

Figure 9. Cross section for electron-electron scattering for $r_s = 3$ calculated (a) without intermediate states corrections, (b) with intermediate states corrections, (c) with final states correction; (d) experimental data for Au (see [38]). The electron momentum is measured in units of Fermi momentum.

scanning electron microscopy, in particle multipliers, in x-ray and Auger microprobes, positron beam studies, and in many other applications.

A great many studies of this phenomenon have been carried out. Recent reviews of secondary electron emission [4,5,7,41,42] are available. Theoretical treatments have used Monte Carlo simulations of individual particle trajectories and focus on the energy and angular distributions of emitted electrons. Typically, one employs cross sections for elastic scattering on ion cores, excitation of plasmons and inner shell levels, as well as valence levels, and decay of plasmons by single-electron excitation. In none of the simulations that we are aware of are Auger cascades in the valence electron sea accounted for.

Some time ago, we developed the theory of coupled electron-hole (e-h) cascades in the electron gas [43]. Taking advantage of the analytically convenient form of the DIMFP's for electron and hole creation at low energies, the Boltzmann energy balance equations were written and solved to yield the following expressions for the coupled electron-hole cascade due to a uniformly distributed monoenergetic source

$$u(x) = \frac{6(x+1)}{5\mu_o x_o^3} \left\{ \left(\frac{x_o}{x}\right)^4 - \frac{x}{x_o} + \frac{5}{9} \left[\left(\frac{x_o}{x}\right)^3 - 1 \right] \right\} + \frac{2(x+1)}{\mu_o x_o^2} \delta(x-x_o)$$

(16)

$$v(x) = \frac{6}{5\mu_o x_o^3} \left\{ \left(\frac{x_o}{y}\right)^4 - \frac{y}{x_o} - \frac{5}{9} \left[\left(\frac{x_o}{y}\right)^3 - 1 \right] \right\}$$

(17)

Here $u(x)$ is the fluence [electrons/(unit area) (unit energy interval)] of electrons with energy $x = (E-E_F)/E_F$, due to a delta function source with energy x_o. All energies are measured in units of E_F from the Fermi energy. Similarly, $v(y)$ is the fluence of holes with energy $y = (E_F -E)/E_F$. The quantity μ_o depends on the electron gas density [43] and is not important for the present discussion. The most striking feature of this result is the inverse fourth-power dependence of the fluxes on the energy variables in these equations. Fig.10 shows the electron and hole fluxes separately, due to a monoenergetic source, where energies now are scaled to the source energy x_o.

Definitive experimental confirmation of the validity of the above result has been obtained by Gadzuk and Plummer [44] on hot e-h cascades in field emission of electrons from metals, and by Donder and Lee [45] in observation of cascades in photofield emission.

In the past several years, many different groups have constructed

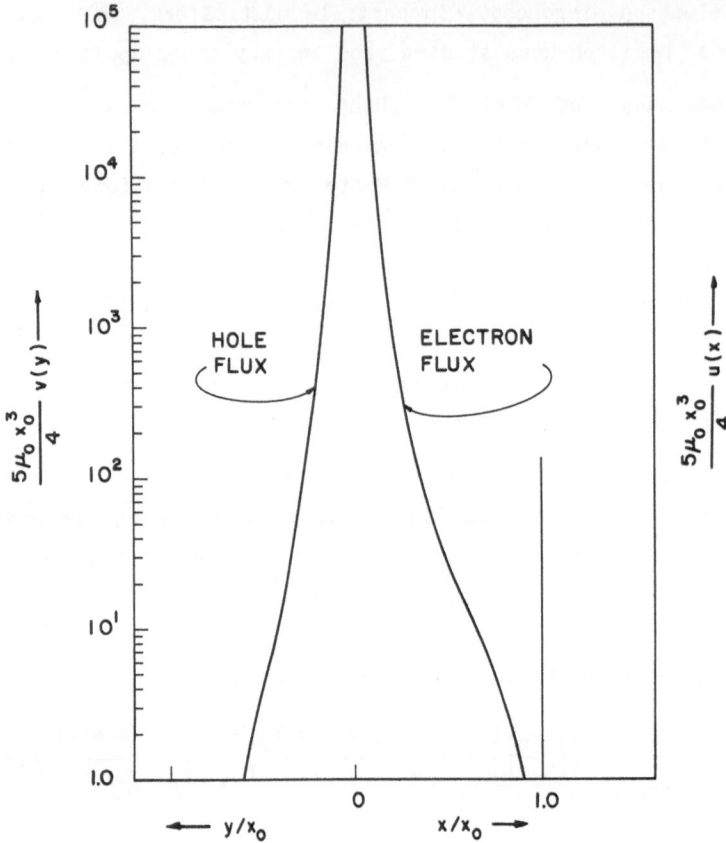

Figure 10. The fluence of electrons and holes in an electron gas due to a coupled e-h cascade generated by a monoenergetic source with energy $E = x_0 E_F$. All energies are measured from the Fermi level and in units of the Fermi energy. The fluxes are scaled as shown, where the symbols are defined in the text (Ritchie [43]).

Monte Carlo codes to simulate penetration, backscatter, and secondary emission from metals and semiconductors bombarded by charged particles. Detailed cross sections have been used (sometimes quite schematic ones) for losses to inner shells, single-particle and plasmon generation as well as plasmon decay, with detailed simulation of electron cascades. Luo and Joy [41], especially, have emphasized the importance of representing plasmon decay into single-particle states.

However, none of these codes have attempted to simulate e-h cascades in the valence electron sea. It turns out that the simple solution given above for the flux in an e-h cascade may be used to represent very effectively the low-energy end of the spectra of electrons emerging from a number of different metals, bombarded by electrons under a variety of

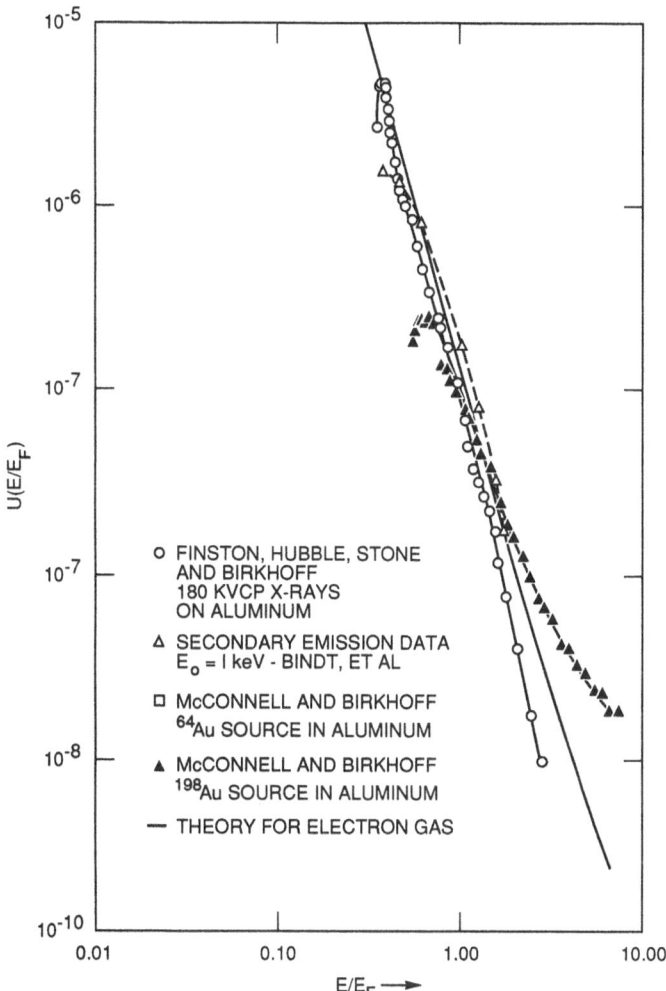

Figure 11. The low-energy end of electron slowing-down flux spectra in Al metal. The ordinate shows the fluence divided by the mean energy of the sources, while the abscissa shows electron energy measured from the Fermi level and in units of E_F. The different sets of experimental data are described on the figure and in ref.[46], while the theoretical curve is a plot of eq.(16) in the text.

different conditions. We may term this a "Pedestrian Approach" to the theory of secondary electron emission from metals.

The key to this representation is the fact that at low energies, the fluence spectrum described by eq.(16), when divided by x_o, the source electron energy, is independent of x_o. This suggests that if one plots the ratio of the measured fluences in the low-energy region, to the energy per unit volume liberated in the medium by the source electrons, the results should be comparable with those predicted by eq.(16). The rationale for

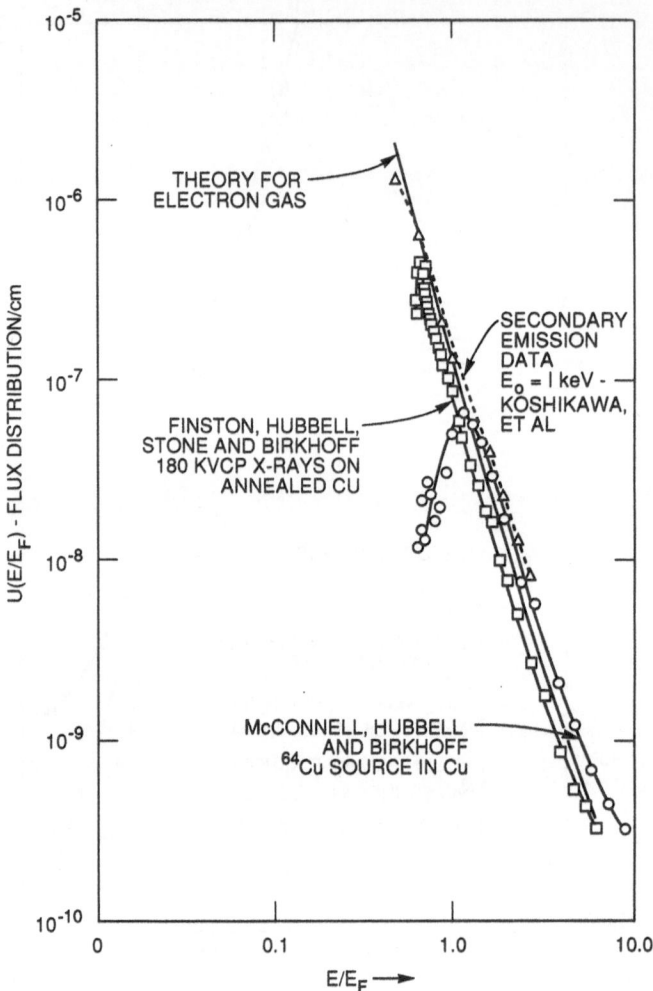

Figure 12. The low-energy end of electron slowing-down flux spectra in Cu metal. The energy is measured from the Fermi level. See the figure and ref.[46] for a description of the data shown.

this is that one expects that all high-energy excitation processes should give rise finally to e-h cascade processes in the valence electron sea, and that the scale of these cascades should be set by the total energy liberated in the medium per unit volume.

In figs.11-13 we have plotted, in this manner, experimental data from quite different experimental configurations. "Keplertron" data obtained by Birkhoff and collaborators consisted of electron emission from metals loaded uniformly with beta-emitting radioisotopes [46]. Other emission data were gotten from experiments with metal electrodes, separated by a vacuum gap and irradiated by x-rays. Curves of current flowing between the electrodes, as a function of the voltage applied between them, were

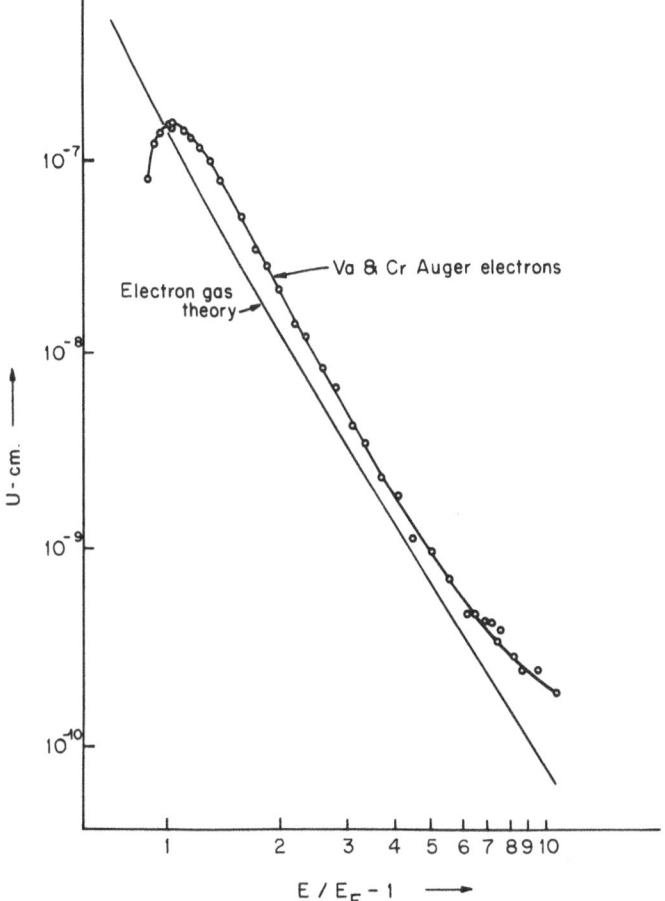

Figure 13. The low-energy end of electron slowing-down flux spectrum in Cr metal. See ref.[46] for a description of the data.

analyzed to yield the flux of electrons in the bulk of the metal. In both kinds of experiments the energy of the source electrons was found in a straightforward manner. For the data on secondary emission due to keV electrons incident on a surface [42,47], the source energy density was found from a knowledge of the stopping power for the source electrons at the entrant surface of the medium. Under the rather good assumption that the electron fluence at the surface is isotropic in direction, $\phi(E)/\varepsilon$, the electron fluence per unit energy liberated in the medium, for such secondary emission experiments, is taken as

$$\frac{\phi(E)}{\varepsilon} = \frac{\delta(E_o)\ 2j(E_o)}{\frac{dE}{dx} \int_o^\infty j(E)dE \left\{ 1 - \sqrt{\frac{E_F + \phi_w}{E}} \right\}} \tag{18}$$

213

Here $j(E)$ is the electron current [electrons/(unit area) (unit energy)] leaving the surface, $\delta(E_o)$ in the secondary emission coefficient at the source energy E_o, and the factor of 2 comes from the assumption of isotropic directional distribution. ε is the energy liberated in the medium per unit volume by the incident ionizing radiation, and dE/dx is the stopping power of the medium for electrons, as they enter.

As one sees, the absolute agreement between experimental data plotted in this way and our theory of the coupled e-h cascade is quite reasonable, in view of the usual uncertainty about surface conditions, work functions, etc., in such experiments. Our results here suggest that similar plots of the many existing sets of experimental data would be useful in delineating possible band structure and other solid-state effects that may affect such low-energy spectra.

2.2 Electron Interaction in Real Solids

Representation of the response function of an arbitrary solid is a special task for each material. Considerable work has gone into approximating such functions for ideal crystals [48]. A general expression for the inverse dielectric function of a solid that is characterized by the set of reciprocal lattice vectors $\{\vec{G}\}$ may be written [49]

$$\text{Im}\left\{\varepsilon^{-1}_{\vec{G},\vec{k},\omega}\right\} = \frac{4\pi^2 e^2}{\hbar k^2 \Omega} \sum_n \langle 0|\rho_{\vec{k}}|n\rangle \langle n|\rho_{\vec{k}+\vec{G}}|0\rangle \left[\delta(\omega + \omega_{no}) - \delta(\omega - \omega_{no})\right],$$
(19)

where $\omega_{no} = \omega_n - \omega_o$. The zero-order sum rule corresponding to that given in eq. (2) is [50], [49]

$$\frac{m}{2\pi^2 ne^2} \int_o^\infty \omega\, \text{Im}\left\{-\varepsilon^{-1}_{\vec{G},\vec{k},\omega}\right\} d\omega = \frac{\vec{k}(\vec{k}+\vec{G})}{k^2} \langle 0|\delta_{\vec{G}}|0\rangle,$$
(20)

again neglecting two-electron correlations. Higher-order sum rules are quite lengthy and will not be discussed here.

Consider now the variation of the so-called energy loss function $\text{Im}(-\varepsilon^{-1}_{k,\omega})$ in the k,ω plane. If one employs the RPA dielectric function as modified by Mermin [28] to allow for damping of the final states, one finds the surface shown in fig. 14. The parameters used in the calculation are shown in the figure caption. The prominent peak at small wave number k is seen to spread out, become smaller and move to larger values of energy transfer $\hbar\omega$ with increasing k. The peak for small k corresponds to the existence of plasmons in the medium, while at large k the "Bethe ridge" is centered around the line $\omega = \hbar k^2/2m$. Fig. 15 shows a logarithmic plot of

214

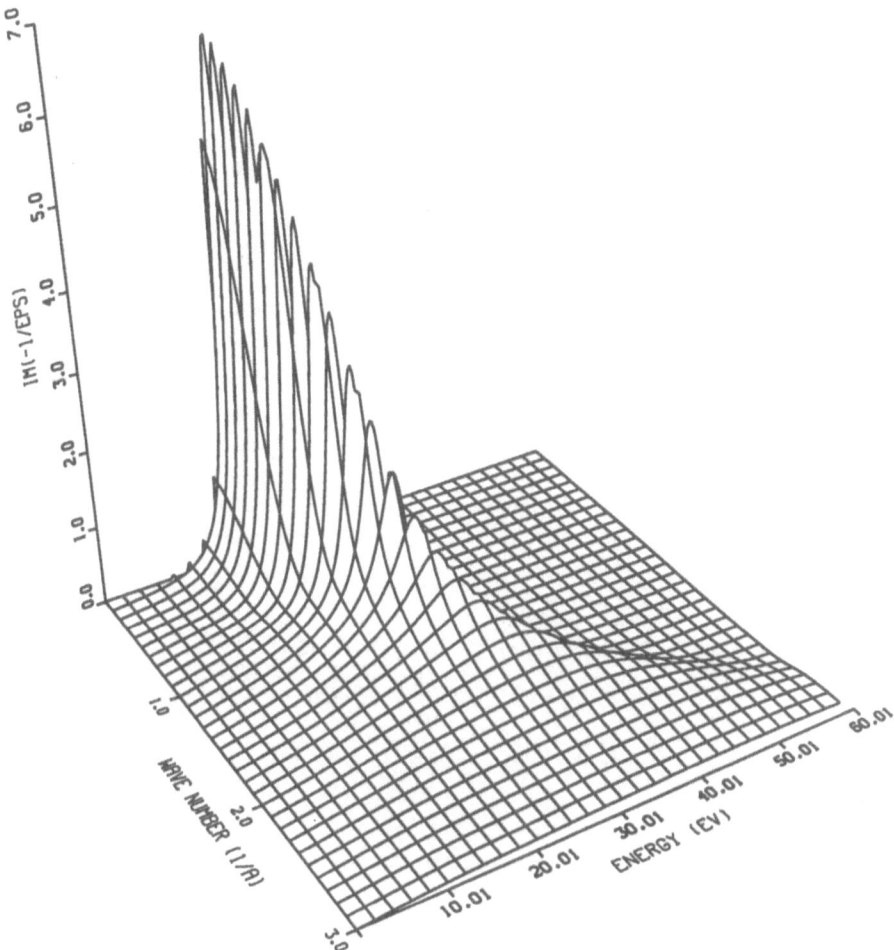

Figure 14. A contour plot of the energy-loss function of a damped electron gas, as it depends on wavenumber k (in Å^{-1}), and on energy transfer E = ℏω (in eV). The damping constant was taken to be 3 eV and the plasma energy 15.4 eV.

$Im(-\varepsilon_{k,\omega}^{-1})$ from a different viewpoint, in order to display clearly the parabolic region of single-particle excitations as k increases. This region is also coincident with the Bethe ridge.

In eq.(4), the Bethe sum rule is $\sum_1 = 1$, as it must, for the function displayed in figs.14 and 15. The second-order sum rule is $\sum_2 = \omega_p$, the plasma frequency of the electron gas, as k → 0. When k → ∞,

$$\sum_1 (k) \rightarrow \frac{\hbar k^2}{2m} + \frac{4}{5} T/\hbar,$$

for the RPA loss function. In the neighborhood of the Bethe ridge, when k

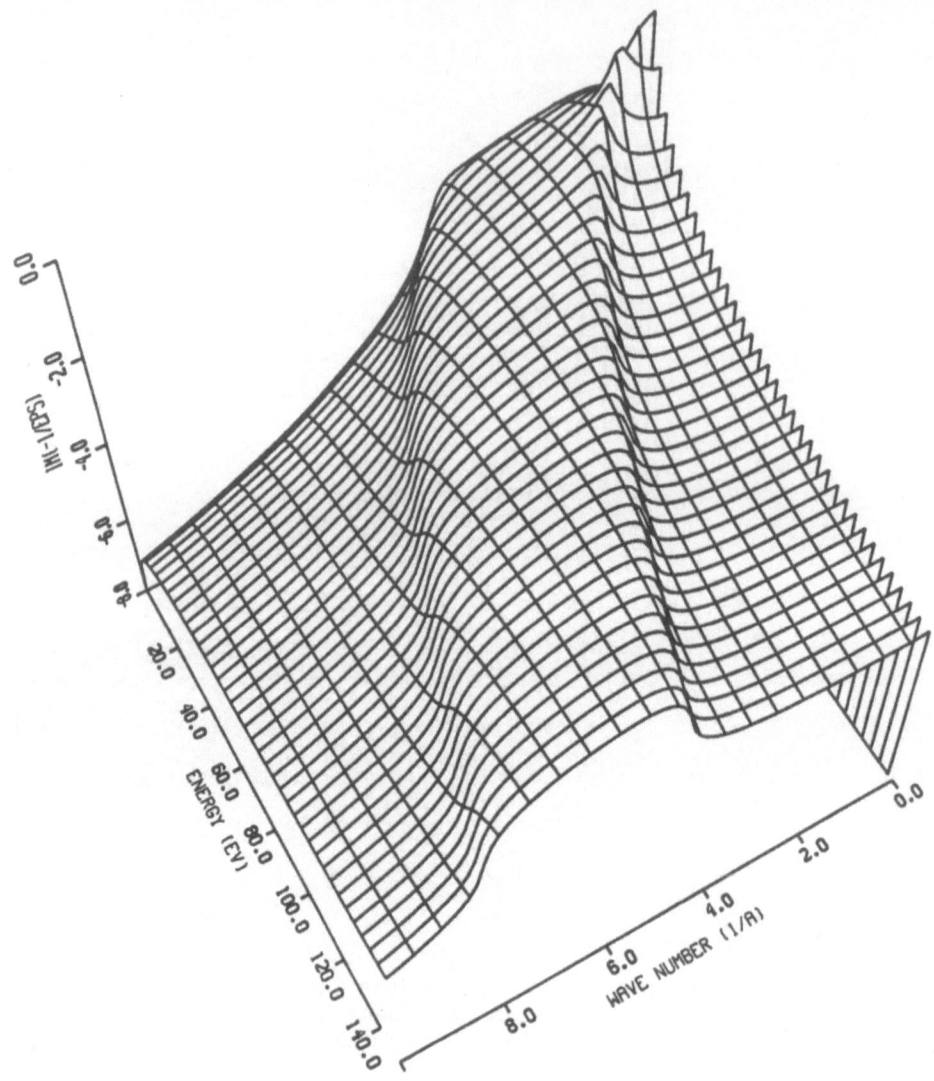

Figure 15. The same as fig.14, except that the loss function is plotted on a logarithmic scale in order to emphasize the region of single-particle losses.

is large, the full width of the Bethe ridge at half height, for fixed and large k, is $2\sqrt{2E_oE_F}$, where $E_o = \dfrac{\hbar^2k^2}{2m}$ is the energy corresponding to the center of the ridge. Thus there is a doppler-broadening of the ridge, corresponding to the ejection with large momenta of electrons from the whole Fermi sea. This is expected to be a general property of $Im(-\varepsilon^{-1})$ for any solid; that is, the width of the large-momentum part of the Bethe ridge corresponding to a given energy band, should be proportional to the square root of the width in energy of that band [51]. This general feature may be incorporated into schemes for extrapolating measurements of $\varepsilon_{o,\omega}$ for a given solid into the k - ω plane [12,46,52].

Most electron transport experiments have been done using polycrystalline or amorphous solids. Estimates of transport parameters such as DIMFP's, stopping powers, straggling parameters, etc., are sometimes based on schematic or empirical formulas [53]. Present solid-sates theory does not permit detailed construction of dielectric functions for real solids without prohibitive amounts of numerical computation. Approximate representations of the dielectric function of semiconductors [54] have been made, mostly based on that for the electron gas in the RPA.

The author and his colleagues have represented the dielectric function for an amorphous insulator using an Orthogonalized Plane Wave (OPW) basis set [55,46]. The OPW approach has had extensive use in solid-state physics. Applications to obtain transport parameters in many different solids has been made [46,56]. DIMFP's for energy loss in liquid water have been derived from a model for $\varepsilon_{k,\omega}$ based on measurements of $\varepsilon_{o,\omega}$, the optical dielectric function, and a sum-rule-constrained extrapolation of these measurements into the k,ω plane [57]. This general idea has been used to obtain transport parameters for other solids as well [58,59] and to determine the energy loss of an ion channeled in a crystalline solid [49]. A statistical model based on the electron-gas response function has also used to compute DIMFP's for several different solids [60].

Ritchie and Howie [61] have suggested that experimental data on $\varepsilon_{o,\omega}$ may be extrapolated usefully into the k - ω plane by fitting the data with sums of functions containing parameters that give dispersion of the loss surface into a Bethe-type ridge, with a position and width commensurate with sum rules for the solid under consideration. This procedure was also used in reference [57] to fit data on liquid H_2O.

The considerations in section 2.1.1 above, of exchange in the electron gas for low electron energies, are important for a full understanding of this effect in a model system. A theory of exchange corrections to DIMFP's of electrons at arbitrary energy is not available. An empirical procedure based on the relativistic version of eq.(12) has been used to correct DIMFP data inferred from optical data or from theory neglecting exchange [57,55].

3. THE IMPACT PARAMETER REPRESENTATION OF CHARGED PARTICLE ENERGY LOSSES IN CONDENSED MATTER

In quantal collision theories, momentum and energy are usually taken to be good quantal variables [9]. Classical collision theory, on the other hand, uses position and time to describe interactions between a probe and a target [62]. In modern physics, one may wish to express quantal theories in terms of space-like variables. For example, experiments are now common in which one measures, by means of a narrowly focused beam of swift electrons, the distribution in energy of losses experienced in a very small region of space. Also, in experiments with channeled ions, and in microdosimetry, one is interested in the spatial coherence of unlocalized excitations created by swift ions and electrons, and their ultimate localization through transfer of energy to, e.g., single-particle excitations.

The problem of visualizing quantal collisions in the space (and perhaps time) variable has been faced by several workers over many years [62]. Fano [63] has pointed out that in condensed media, excitations may be coherent over distances comparable with v/ω_{res}, the Bohr cutoff impact parameter, and thus may involve the collective motion of $\sim 10^6$ atoms. Here ω_{res} is the principal resonant frequency of the valence electrons in the particular medium. He noted the lack of a comprehensive theory of the impact parameter representation of collisions in condensed matter, and has described a preliminary approach to such a theory. Fano's [63] approach to this problem involves the assumption that atoms or molecules in the system are weakly interacting with one another.

Here we represent the interaction probability in terms of the dielectric response function of matter and the usual momentum transfer variable, transformed into a space-like variable, by several alternative methods. The resulting transforms are compared.

3.1 The Inverse Mean Free Path

We write the standard expression for the differential inverse mean free path (DIMFP) for energy transfer $\hbar\omega$, and for momentum transfer $\hbar\vec{K}$ to a dielectric medium by a particle with charge Ze traveling with velocity $\vec{v}$ in the medium as,

$$\frac{d^3\lambda^{-1}}{d\omega \, d^2K} = \frac{Z^2e^2}{\pi^2\hbar v^2} \frac{1}{K^2+\omega^2/v^2} \, \text{Im}\left(\frac{-1}{\varepsilon_{k,\omega}}\right). \tag{21}$$

Here the vector $\vec{K}$ is understood to be in a direction perpendicular to $\vec{v}$, and, $\hbar k$, the magnitude of the total momentum, is $\hbar k = \hbar(K^2 + \omega^2/v^2)^{1/2}$. The goal is to express the DIMFP in terms of a spatial variable that we will interpret as an impact parameter. This formula is applicable to swift electrons with speed $v \gg e^2/\hbar$, as well as to ions.

3.1.1 The "Van Hove" Transform. Van Hove [64] introduced a pair distribution function $G(\vec{r},t)$ that is obtainable directly from the response function of the medium. It gives the quantal average probability of finding a particle at $(\vec{r},t)$ if a particle is located at the point $\vec{r} = 0$ when $t = 0$. It is a complex function because it reflects the quantal properties of the system and has been applied extensively in analyzing the properties of dense gases and liquids as inferred from neutron scattering experiments.

Pines and Nozieres [65] state that the comparable quantity in a polarizable medium characterized by a dielectric function is to be interpreted as describing correlations between density fluctuations at different space-time points. Rather than dealing with the correlation function G as defined by Van Hove, we proceed to compute the two-dimensional Fourier transform of the DIMFP, in order to compare directly with other forms described below. Thus

$$\lambda_{VH}^{-1} = \frac{2 \, Z^2 e^2}{\pi^2 \hbar \, v^2} \int_0^\infty d\omega \int d^2K \, J_0(bK) \, \frac{\mathrm{Im}(-1/\varepsilon_{k,\omega})}{K^2 + \omega^2/v^2}, \tag{22}$$

where we have integrated over ω in order to obtain an easily surveyed result. A simple form is obtained if the plasmon is approximated as a dispersionless resonance at $\omega = \omega_p$, and if the integral over K is extended to infinity. Then

$$\lambda_{VH}^{-1} \sim \frac{2}{\pi} \, \frac{Z^2 e^2 \omega_p}{\hbar v^2} \, K_0\!\left(\frac{\omega_p b}{v}\right), \tag{23}$$

where K_0 is the modified Bessel function of the second kind and order zero. Although the appearance of the Bohr impact parameter, v/ω_p, in the argument of the Bessel function seems quite reasonable, we give a more realistic distribution below.

3.1.2 The Chang-Raman transform. An approach to the problem of expressing quantal probabilities in terms of an impact parameter-like variable has been advocated by Fano [63], based on a method introduced into high-energy physics by Chang and Raman [66]. To apply this method to the DIMFP of eq.(21) we notice that this equation is positive definite and may thus be written

$$\frac{d^3\lambda^{-1}}{d\omega\ d^2K} = |\sigma(\vec{K})|^2, \tag{24}$$

where the $\vec{K}$ dependence is not written explicitly. We now seek to eliminate $\vec{K}$ in favor of a spatial variable, that we will interpret as an impact parameter. Thus

$$\frac{d\lambda^{-1}_{CR}}{d\omega} = \int d^2K\ |\sigma(K)|^2 = \int d^2K \int d^2K'\ \sigma(K)\ \sigma^*(K')\ \delta^2(\vec{K}-\vec{K}')$$

$$= \frac{1}{(2\pi)^2} \int d^2b \int d^2K \int d^2K'\ \sigma(\vec{K})\ \sigma(\vec{K}')\ e^{i\vec{b}(\vec{k}-\vec{k}')}. \tag{25}$$

The integrand of eq.(25) is now set equal to the DIMFP in impact parameter space, viz.,

$$\frac{d^3\lambda^{-1'}_{CR}}{d\omega\ d^2b} = \frac{z^2e^2}{4\pi^4\hbar v^2} \left| \int d^2k\ e^{i\vec{k}\cdot\vec{b}}\ \frac{1}{k} \left[\mathrm{Im}\left(\frac{-1}{\varepsilon_{k,\omega}}\right) \right]^{1/2} \right|^2. \tag{26}$$

Although seemingly reasonable, this form does not appear to be useful when the DIMFP has a narrow resonance in the $k - \omega$ plane, as when plasma oscillations may occur, since eq.(26) gives an indeterminate result in this case. In addition, the dependence on b does not seem as realistic as for the transform discussed next.

3.1.3 The Energy Transfer Transform.
Here we follow an approach similar to that used to get the Chang–Raman transform [66], but now do not require that the distribution in b be positive definite. Writing

$$\frac{d\lambda^{-1}_{ET}}{d\omega} = \frac{z^2e^2}{\pi^2\hbar v^2} \int \frac{d^2K}{k^2} \int \frac{d^2K'}{k'^2}\ \vec{k}\cdot\vec{k}'\ \mathrm{Im}\left[\frac{-1}{\varepsilon_{k,\omega}}\right] \delta^2(\vec{K}-\vec{K}'), \tag{27}$$

where $\vec{k} = \vec{K} + \frac{\omega}{v^2}\ \vec{v}$, and using the integral representation of the two–dimensional delta function as before, we find

$$\frac{d^3\lambda^{-1}_{ET}}{d\omega\ d^2b} = \frac{z^2e^2\omega}{\pi^2\hbar v^3} \int_0^\infty \frac{K\ dK}{K^2+\omega^2/v^2}$$

$$\times \left[\frac{\omega}{v}\ K_0\left(\frac{b\omega}{v}\right)\ J_0(Kb)\ +\ K\ K_1\left(\frac{b\omega}{v}\right)\ J_1(Kb) \right] \mathrm{Im}\left(\frac{-1}{\varepsilon_{k,\omega}}\right), \tag{28}$$

interpreting the integrand as before. We term this the "energy transfer transform" since it agrees precisely with the formula found by computing

220

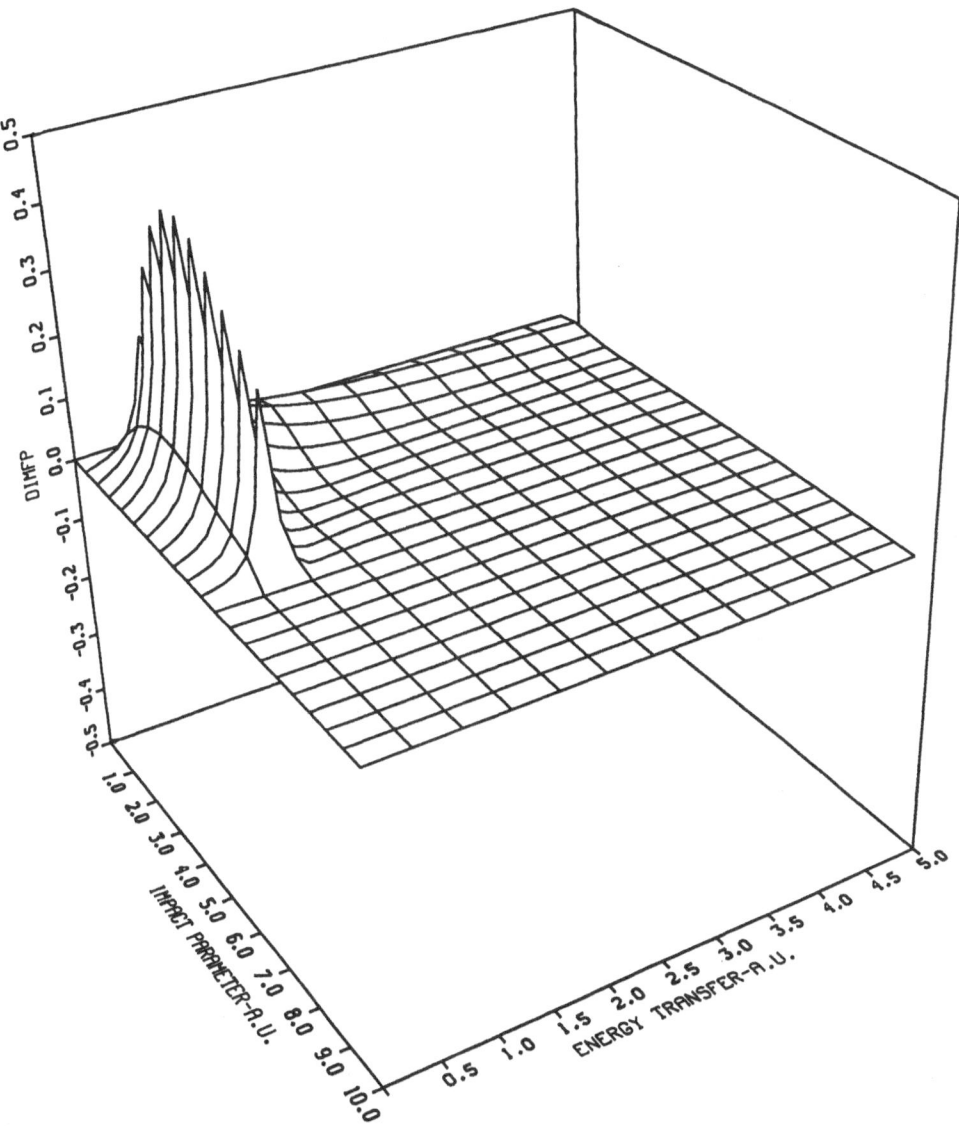

Figure 16. The DIMFP of a swift particle in an electron gas, plotted as a function of impact parameter b and energy transfer ħω. The DIMFP is normalized arbitrarily and b and ħω are expressed in a.u. An approximate expression for the dielectric function has been used for analytical convenience. The particle is assumed to have a speed of 4 a.u. and $\hbar\omega_p$ is taken to be 15.4 eV.

the energy transferred to the medium at fixed impact parameter, using semi-classical dielectric theory, and then dividing the integrand by ħ.

We are now in a position to express the DIMFP in b − ω space rather than as in the usual k − ω variables. Applying now the Energy Transfer transform to the DIMFP we find the result plotted in fig.16 in the ω − b;

plane but using a simplified form for $\varepsilon_{k,\omega}$ to expedite the calculation. Here the prominent region begins for b equal to a few Å near the plasma frequency, and continues to larger ω but at even smaller values of b. This may be termed the "Bohr Ridge", in analogy with the so-called Bethe ridge in the conventional plot of the generalized oscillator strength. The latter corresponds to the ejection of electrons with large momenta and energy, while the Bohr Ridge represents the ejection of electrons with large energies that originate from regions of space close to the track of the charged particle.

3.2 Secondary Electron Generation Processes and Their Degree of Localization

An interesting application of the Impact Parameter Representation (IPR) is in the characterization of the secondary electron (SE) signal in scanning electron microscopy (SEM). The SE signal has long been recognized in SEM as the most useful indicator of surface topography. Recent experimental work in scanning transmission electron microscopy has shown [67,68] that it is possible to obtain SE signals with 1 nm spatial resolution and 1 eV energy resolution, and that reflection SE images of oxidized Cu show oxide islands and details of their interaction with surface steps.

The relative importance of the processes by which secondary electrons are generated in the complex slowing-down and cascade interactions in solids has been considered by many workers, but less attention has been paid to assessing their degree of localization, i.e., the relevant impact parameter or distance from the electron beam where secondaries are generated. Monte Carlo calculations by Luo and Joy [41], among others, indicate that the majority of secondaries originate from plasmon decay. Others [67,68] question whether plasmon decay is sufficiently well-localized to explain their measured high spatial resolution.

We address this question using the Energy Transfer Transform of eq.(28) to obtain the IPR for energy deposition in the conduction band of aluminum metal. Fig.17 shows the IPR distribution calculated from this equation, after integration over ω, using a simple representation for the plasmon contribution to the inverse dielectric function. The computation was done for three different bombarding electron energies. The somewhat surprising result is that each of the curves decreases strongly as b increases, going asymptotically as $\exp(-2\omega_p b/v)$ as $b \to \infty$. It turns out that for each of these primary energies the mean value of b, averaged over

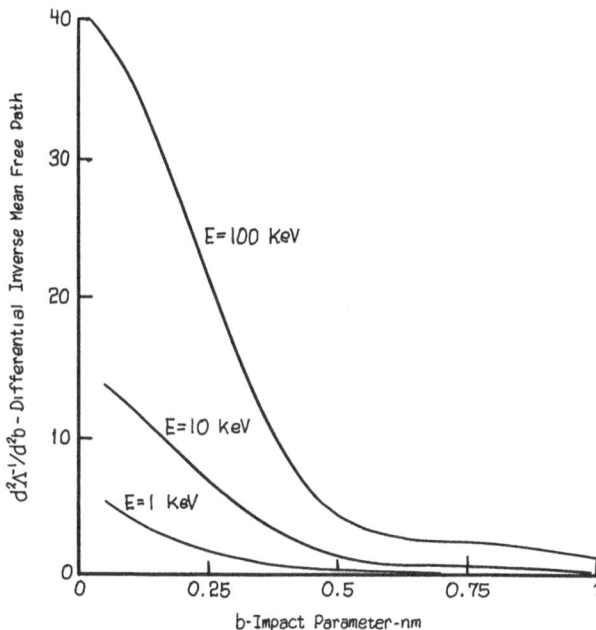

Figure 17. A plot of the DIMFP, differential in impact parameter b only, corresponding to plasmon excitation in an electron gas. Single-particle excitation is not included here. The electron energies correspond to those of a narrow beam, such as those used in scanning transmission electron microscopy. For convenience in plotting, the curves have been scaled by multiplying the results from eq. (28) by the factor $2\pi\hbar v^3/e^2\omega_p$, where v is the electron speed.

these distributions, is less than 1 nm. For emphasis in plotting, the calculated DIMFP values have been multiplied by $2\pi\hbar v^3/e^2\omega_p^2$. The small fluctuations in these curves correspond to quantal effects in the plasmon field. Note that propagation and decay of the plasmon are described in detail in this treatment. It appears that the narrow spatial resolution of these IPR distributions is due to the presence of relatively large momentum components in the interaction spectrum of the swift electron and the electron gas.

The conclusion is that if plasmons make an important contribution to energy transfer to the solid in SEM, then good spatial resolution is expected because of the intrinsic nature of the spatial deposition of energy by the plasmon in metals. Of course, it may be that in some situations, plasmon excitation does not play as important a role as single-particle excitations. In this case it is expected from Monte Carlo studies that lateral spread of secondary electrons is not inconsistent with the good spatial resolution found in SE.

4. SUMMARY

These lectures have emphasized special but basic topics that originate in the interaction of electrons with matter. Important questions about spin-dependent effective interactions, exchange scattering, the role of the exclusion principle when applied consistently in low-energy electron scattering in metals, electron-hole cascades in the electron gas, sum rules, and aspects of localization of coherent, initially unlocalized excitations in matter have been considered.

ACKNOWLEDGEMENTS

The author thanks his colleagues V.E. Anderson, A. Howie, F. Flores, A. Gras-Marti, and G.J. Basbas for their kind cooperation and friendship over many years of pleasant association. This research was sponsored by the Office of Health and Environmental Research, U.S. Department of Energy under contract DE-AC05-84OR21400 with Martin Marietta Energy Systems, Inc., and the U.S.-Japan Cooperative Science Foundation Joint Research Project No.87-16311/MPCR-168

REFERENCES

1. M.A. Kumaklov and F.F. Komarov, *Energy Loss and Ion Ranges in Solids*, Gordon and Breach, New York (1981).
2. Y.H. Ohtsuki, *Charged Beam Interactions with Solids,* Taylor and Francis, New York (1983).
3. P. Schattschneider, *Fundamentals of Inelastic Electron Scattering*, Springer Verlag, New York (1986).
4. J. Schou Scanning Microsc. *2* (1988) 607.
5. S. Tougaard, Surface and Interface Analysis, vol.12, Wiley, New York (1988).
6. P.M. Echenique, F. Flores and R.H. Ritchie, Solid State Phys. *43* (1990) 229.
7. H. Seiler, in *Electron Beam Interactions with Solids,* SEM, Inc, Chicago (1984) p.33.
8. P. Nozieres and D. Pines, Il Nuovo Cimento (X) *9* (1958) 470.
9. H.A. Bethe, Ann. Physik *5* (1930) 325; Z. Physik *76* (1932) 293.
10. G. Placzek, Phys. Rev. *86* (1952) 377.
11. U. Fano and J.E. Turner in NAS-NRC Nuclear Science Series, Report 39, Committee on Nuclear Science, Publication 1133 (1964) p.49.
12. R.H. Ritchie, R.N. Hamm, J.E. Turner and H.A. Wright, Proc. 6th Symp. Microdos., EUR6064 DE-EN-FR (1978) p.345.
13. K.S. Singwi and M.P. Tosi, Solid State Physics *36* (1981) 177.
14. J. Lindhard, K. Dan. Vidensk. Sel.-Mat.-Fys. Medd. *28* (1954) no.8.
15. T.D. Schultz, *Quantum Field Theory and the Many-body Problem*, Gordon and Breach, New York (1964).
16. A.L. Fetter and J.O. Walecka, *Quantum Theory of Many-Particle Systems,* McGraw Hill, New York (1971).
17. G.D. Mahan, *Many-Particle Physics,* Plenum Press, New York (1984).
18. R.H. Ritchie, Phys. Rev. *106* (1957) 874.
19. R.H. Ritchie, Phys. Rev. *114* (1959) 644.
20. L. Hedin and S. Lundqvist, Solid State Phys. *23* (1969) 1.
21. V. Heine and I.V. Abarenkov, Phil. Mag. *9* (1964) 451.
22. J. Hubbard, Proc. Roy Soc. Lond. *A240* 539; ibidem *A243* (1957) 336.
23. L. Kleinman, Phys. Rev. *160* (1967) 585.
24. L. Kleinman, Phys. Rev. *172* (1968) 383.
25. L. Kleinman, Phys. Rev. *B3* (1971) 2982.

26. D.C. Langreth, Phys. Rev. *181* (1969) 753.
27. S. Yarlagadda and G.F. Giuliani, Solid State Comm. *69* (1989) 677.
28. N.D. Mermin, Phys. Rev. *B1* (1970) 2362.
29. A. Pathak and J. Yussef, Phys. Status Solidi *649* (1973) 431.
30. C. Moller, Z. Phys. *70* (1931) 786; Ann. Phys. *14* (1932) 531.
31. W.R. Ferrell, R.H. Ritchie and T.L. Ferrell, Am. J. Phys. *52* (1984) 915.
32. R.H. Ritchie and J.C. Ashley, J. Phys. Chem. Solids *36* (1965) 1689.
33. J.J. Quinn, Appl. Phys. Lett. *2* (1963) 167.
34. C.A. Kukkonen and A.W. Overhauser, Phys. Rev. *B20* (1979) 550.
35. N. Iwamoto and D. Pines, Phys. Rev *29* (1984) 3924.
36. G. Vignale and K.S. Singwi, Phys. Rev. *32* (1985) 2156.
37. D. Penn, Phys. Rev *B22* (1980) 2677.
38. R.N. Hamm, R.H. Ritchie and J.C. Ashley, ORNL-TM-2072 (1968).
39. H.A. Bethe and J. Goldstone, Proc. Roy. Soc. (London) *A238* (1957) 551.
40. A. Zhang and P. Platzman, Phys. Rev. *B37* (1988) 7326.
41. S. Luo and D. Joy, Scanning Microsc. *2* (1988) 1901.
42. R. Bindi, H. Lanteri, P. Rostaing, J. Phys. *D13* (1980) 461.
43. R.H. Ritchie, J. Appl. Phys. *37* (1966) 2276.
44. J.W. Gadzuk and E.W. Plummer, Phys. Rev. Lett. *26* (1971) 92.
45. P.J. Donders and M.J.G. Lee, Phys Rev *B41* (1990) 1781.
46. See R.H. Ritchie, C.J. Tung, V.E. Anderson and J.C. Ashley, Rad. Research *64* (1975) 181 for the original references; R.H. Ritchie and V.E. Anderson, IEEE, NS-18(6) (1971) 141.
47. T. Koshikawa and R. Shimizu, J.Phys. *D27* (1974) 1303.
48. See ref.[1], chapter 3, and the references given therein.
49. O.H. Crawford and C.W. Nestor, Jr., Phys. Rev. *A28* (1983) 1260.
50. D.L. Johnson, Phys. Rev. *B9* (1974) 4478.
51. A. Gras-Marti and R.H. Ritchie, (to be published).
52. R.H. Ritchie, to be published.
53. See, e.g., D. Liljequist, J. Appl. Phys. *57* (1985) 657 and F. Salvat, J.D. Martinez, R.Mayol and J. Parellada, J. Phys. *D18* (1985) 299.
54. See L.C. Emerson, R.D. Birkhoff, V.E. Anderson and R.H. Ritchie, Phys. Rev. *B7* (1973) 1798 and references therein to the Callaway-Tosatti model. A more recent version of this model has been given by Z.H. Levine and S.G. Louie, Phys. Rev. *B25* (1982) 6310.
55. C.J. Tung, J.C. Ashley, R.D. Birkhoff, R.H. Ritchie, L.C. Emerson and V.E. Anderson, Phys. Rev. *B16* (1977) 3049.
56. J.C. Ashley and V.E. Anderson, J. Electron Spectrosc. Rel. Phenom. *24* (1981) 127.
57. R.H. Ritchie, R.N. Hamm, J.E. Turner and H.A. Wright, Proc. 6th Symp. Microdos., EUR 6064 DE-EN-FR,(1978) 345.
58. D. Penn, Phys. Rev. *B35* (1987) 482.
59. J.C. Ashley, J. Appl. Phys. *63* (1988) 4620.
60. J.C. Ashley, C.J. Tung and R.H. Ritchie, Surf. Sci. *81* (1979) 409.
61. R.H. Ritchie R H and A. Howie, Phil. Mag. *36* (1977) 463.
62. See, e.g., N. Bohr, Kgl. Danske Vid. Sels. Mat.-Fys. Medd. *18* (1948); E.J. Williams, Rev. Mod. Phys. *17* (1945) 217; J. Neufeld, Proc. Phys. Soc. London *A66* (1953) 489.
63. U. Fano, in *Charged Particle Tracks in Solids and Liquids,* Inst. of Phys., Conf. Series No. 8, London, (1970) p.1.
64. L. Van Hove, Phys. Rev. *95* (1954) 249.
65. D. Pines and P. Nozieres, *The Theory of Quantum Liquids,* W.A. Benjamin, New York (1966).
66. N.P. Chang and K. Raman, Phys. Rev. *181* (1969) 2048.
67. A.L.Bleloch, A. Howie, R.H. Milne, and M.G. Walls, Ultramicrosc. *29* (1989) 175.
68. D. Imeson, R.H. Milne, S.D. Berger, and D. McMullan, Ultramicrosc. *17* (1985) 243.

LOW ENERGY ION PENETRATION AND COLLISION CASCADES IN SOLIDS

H.M. Urbassek

Institut für Theoretische Physik
Technische Universität
D-3300 Braunschweig, FRG

1. INTRODUCTION

Effects of the interaction of fast particles with solids and surfaces form the general subject of this Advanced Study Institute. The present lecture specifically addresses the phenomena occurring under what is called low-energy ion bombardment, i.e. where nuclear stopping dominates more or less over electronic stopping. In this regime, the slowing down of the bombarding ion and its range distribution will be studied. Furthermore, the effects on the irradiated target, in particular the formation of a collision cascade and the sputtering of the target, will be analyzed. It is hereby attempted to give an introduction to the field, stressing the relevant concepts and outlining the analytical and computational tools available. Introductory and review articles covering in greater detail several aspects of the topics addressed here are available [1-3]. The present lecture attempts to describe both analytical and simulational approaches, in order to allow for a fair comparison of the virtues and drawbacks of the respective methods.

In order to understand the interaction of charged particles with solids, evidently the relevant interaction forces must be presented first (sect.2). Then, the slowing down and transport of particles in the target are described on the basis of transport equations (sect.3) as well as by computer simulation schemes (sect.4). Applications on range distributions and collision cascades end the presentation.

2. INTERACTION FORCES

Problems of ion penetration of materials are determined by the interaction potentials between the projectile and the target particles. For very close encounters between the projectile and a target atom

Interaction of Charged Particles with Solids and Surfaces
Edited by A. Gras-Martí *et al.*, Plenum Press, New York, 1991

nucleus, a Coulomb interaction can be assumed to hold. At larger distances, the electron clouds surrounding the projectile and the target atom shield the Coulomb field. At very low interaction energies finally, the electronic binding of the target material may become important and must be taken into account in the form of many-body potentials.

A typical energy E_{scr} which separates the regimes of strongly and weakly screened Coulomb interaction between the projectile (atomic charge Z_1) and a target atom (Z_2) is given by the Coulomb energy at the so-called *screening distance* a. In Gaussian units, it is

$$E_{scr} = \frac{Z_1 Z_2 e^2}{a}.$$ (1)

The screening length a can be calculated using similarity considerations within the Thomas-Fermi theory of the atom [4]

$$a = 0.8853a_0 \left[Z_1^{2/3} + Z_2^{2/3} \right]^{-1/2},$$ (2)

where a_0 = 0.529 Å is the Bohr radius. The energy E of the projectile of mass M_1 when colliding with a target atom of mass M_2 is often expressed in dimensionless units

$$\varepsilon = \frac{E_{rel}}{E_{scr}} = \frac{M_2}{M_1 + M_2} \frac{aE}{Z_1 Z_2 e^2},$$ (3)

with the relative energy $E_{rel} = M_2 E / (M_1 + M_2)$.

The interaction potential between the projectile and a target atom at a distance r is usually expressed as

$$V(r) = \frac{Z_1 Z_2 e^2}{r} \Phi\left[\frac{r}{a}\right],$$ (4)

where $\Phi(r/a)$ is called the screening function. If Φ does not depend on the specific projectile-target system, this equation predicts that the screening of all projectile-target combinations can be expressed by a single screening function Φ, and that the material parameters enter here only via the single screening length a. It is clear that such a formula is most adequate for relatively high energies ε, where the interaction stems from the overlap of the core electrons of the collision partners, and that it loses its foundation at too low collision energies in the 10 eV region or so. Here valence electron overlap is important, which is not well described by Thomas-Fermi screening.

A typical screening function Φ, the so-called Kr-C function, is displayed in fig.1. It is universal in the sense that it depends only via its argument x = r/a on the projectile-target system. It has been derived as an analytical fit function through a multitude of data obtained for many diverse projectile-target atom combinations. The data for each individual system have been derived from the classical electrostatic repulsion of the atomic charge densities, including the kinetic energy increase of the overlapping electron clouds [5].

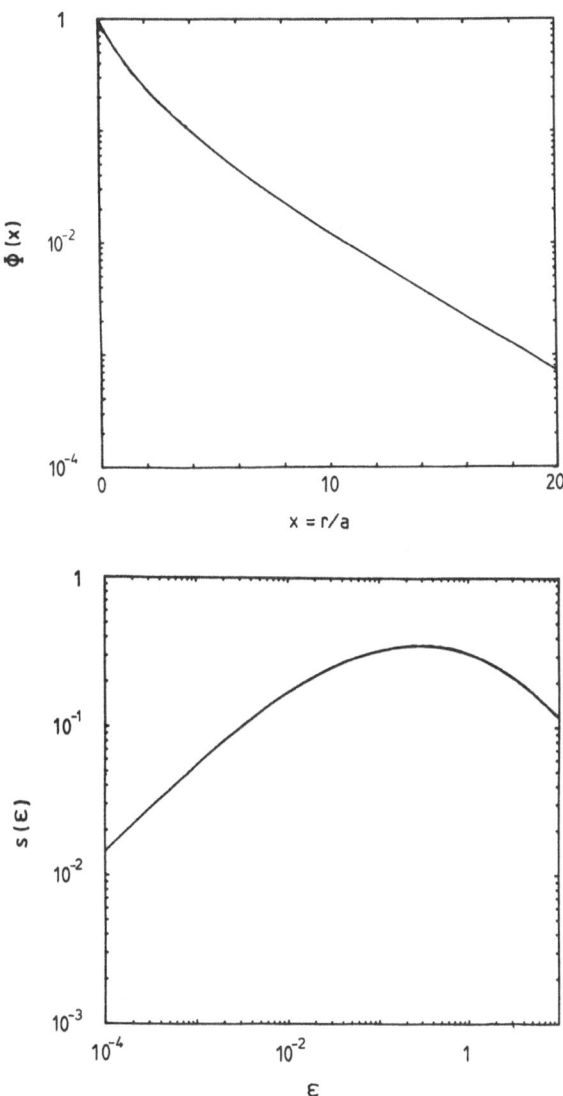

Figure 1. Screening function Φ versus scaled interatomic distance x=r/a, and dimensionless stopping power s versus reduced energy ε, for Kr-C potential.

At lower energies, in the range of the electronic band energy, the concept of binary interaction potentials between particles tends to break down. In this region, the use of many-body potentials has been advocated. For metals, the total potential energy of an ensemble of atoms may be written as

$$E_{tot} = \frac{1}{2} \sum_{i \neq j} V(r_{ij}) + \sum_{i} F(\rho_i),$$ (5)

where i, j count the atoms, V is a binary potential depending on the distance between atoms i and j, and F is an embedding energy, which takes account of the energy needed to embed atom i in an electron gas of density ρ_i. This electron density is calculated by summing up the contributions of the atomic electron densities ρ of all surrounding atoms at the location of atom i,

$$\rho_i = \sum_{j \neq i} \rho(r_{ij}).$$ (6)

If the tabulated electron densities of atoms in the gas phase are used for $\rho(r)$, the calculation of the embedding energy F is actually quite straightforward to use in computer simulation schemes. This constitutes the basis of the so-called *embedded atom method* [6]. It appears that up to now no systematic study of the influence of realistic many-body potentials on ion penetration phenomena has been performed.

For the description of the directional forces of covalent bonding in semiconducting and insulating materials, many-body potentials appear to be even more important. For these systems the modeling of many-body potentials is, however, more complicated than for metallic systems.

While moving through the target, the projectile loses some of its energy to electrons. Such processes are discussed in great detail elsewhere in these Proceedings. It may suffice here to note that in the range of energies considered here, one usually assumes that the energy loss occurs continuously along the path of the projectile. It is proportional to the velocity v of the projectile, and its magnitude is approximately [7]

$$\left[\frac{dE}{dx}\right]_e = N \, S_e = 8\pi \, e^2 \, a_0 \, Z_1^{1/6} \, N \, \frac{Z_1 Z_2}{(Z_1^{2/3} + Z_2^{2/3})^{3/2}} \frac{v}{e^2/\hbar}.$$ (7)

Low-energy electronic stopping is only approximately described by this formula. Deviations from the velocity proportionality, as well as so-called Z_1-oscillations (and also oscillations in the target atomic

230

number Z_2) have been measured. These deviations are discussed in the paper by Lennard and Geissel in these Proceedings.

2.1 Cross Section and Stopping Power

When describing the collisions of two atoms, one is often not interested in the details of the trajectory, but rather in the statistics of energy transfer. This is conveniently given by the *cross section*. It may be defined as follows: If a particle of energy E moves a small distance ΔR in a random medium of density N consisting of particles at rest, the probability dP of undergoing a collision with energy transfer between T and T+dT is

$$dP = N \ \Delta R \ \sigma(E,T) \ dT. \tag{8}$$

The cross section σ can be calculated from the potential V by well known rules [8]. For qualitative orientation, Lindhard et al. recommend the so-called power-law cross section

$$\sigma(E,T)dT = C_m \ E^{-m} \ T^{-1-m} \ dT. \tag{9}$$

The cross section constant C_m is given in terms of the masses and the atomic charges of the colliding atoms as

$$C_m = \frac{\pi}{2} \ \lambda_m a^2 \left[\frac{M_1}{M_2}\right]^m \left[\frac{2Z_1 Z_2 e^2}{a}\right]^{2m}. \tag{10}$$

It describes elastic nuclear scattering using classical mechanics. This cross section approximates the scattering in a power potential $V(r) \propto r^{-1/m}$. By choosing the power exponent m and the constant λ_m suitably, this cross section can be used to model realistic interatomic cross sections over restricted regions of energy. For very high energies $\varepsilon \gg 1$, m = 1 is valid. For $0.1 \leq \varepsilon \leq 2$, m = 1/2 has been advocated. For $\varepsilon \leq 10^{-3}$, a value of m with $0 \leq m \leq 1/4$ should be characteristic.

For general cross sections describing screened Coulomb interaction, Lindhard et al. [4] found an approximate scaling relation. In terms of the variable $t = \varepsilon^2 T/T_m$, where T_m is the maximum energy transfer in a single nuclear collision, these authors express the cross section as

$$d\sigma = \sigma(E,t) \ dt = \pi a^2 \ \frac{dt}{2t^{3/2}} \ f(t^{1/2}). \tag{9a}$$

In this approximation, $d\sigma$ depends on only one variable t instead of E and T. The function f has to be determined from the interaction potential used.

An important information to be obtained from the cross section is the average energy loss ΔE a particle of energy E suffers when travelling a path length ΔR. With the probability dP of eq.(8) it is

$$\Delta E = \int T \, dP = N \, \Delta R \int T \, \sigma(E,T) \, dT. \tag{11}$$

For infinitesimal ΔR, this leads to the concept of *stopping power*

$$\frac{dE}{dR} = - N \, S_n(E), \tag{12}$$

with the stopping cross section

$$S_n(E) = \int T \, \sigma(E,T) \, dT. \tag{13}$$

For power law cross sections (9) it is

$$S_n(E) = \frac{1}{1 - m} \, C_m \, E^{1-2m}, \tag{14}$$

for $M_1 = M_2$. Hence, for a hard interaction $m = 0$, the stopping power increases linearly with E, whereas for $m = 1/2$, i.e. around the maximum of the stopping power, it is constant.

In fig.1 the stopping power pertaining to the Kr-C potential is plotted. Here the dimensionless function $s(\varepsilon)$ has been introduced which is related to $S_n(E)$ by

$$S_n(E) = 4\pi a \, Z_1 \, Z_2 \, e^2 \, \frac{M_1}{M_1 + M_2} \, s(\varepsilon). \tag{15}$$

3. TRANSPORT EQUATIONS

Analytical reasoning in this field usually takes its starting point with so-called transport equations. These are linear integro-differential equations, which may be derived from the well known (nonlinear) Boltzmann equation in the limit that every particle collides only with particles at rest. In the following, I shall state the conditions under which they are valid, then sketch their derivation, and discuss the respective usefulness of the so-called forward and backward equations. Finally I shall give some hints on how they may be solved analytically. I shall concentrate on the equation describing the slowing down and transport of a particle in matter. Similar equations can be derived for the particles set in motion in the target, which will be presented shortly.

3.1 Equations

Following Lindhard et al.[9], let us find the equation describing the time evolution of the phase space density $f(\mathbf{x}, \mathbf{v}, t)$ of a particle in a target. It is defined such that $f(\mathbf{x}, \mathbf{v}, t) \, d^3x \, d^3v$ gives us the probability to find the particle in an infinitesimal volume element d^3x around point $\mathbf{x}$ and in an infinitesimal velocity element d^3v around velocity $\mathbf{v}$, at time t. An equation for f is readily derived, if the following conditions hold:

1. The interaction is binary, i.e. every particle interacts with at most one other particle at the same time. This is valid at not too low interaction energies.

2. The target may be treated as structureless. Again, this is a good approximation for particle motion in matter at not too low interaction energies, with the notable exception of channeling phenomena.

3. The interaction is local, such that all details of the dynamics during the scattering events are neglected. This allows collisions to be described in terms of a collision cross section. Again this should be a valid assumption at not too low bombarding energies.

4. One of the collision partners is initially at rest. Exceptions are found in the case of a heavy ion stopped in a light target, with evidently only small implications [10].

We thus see that the conditions under which the transport equations hold restrict their application essentially to high energies. It is not clear from the above what this means quantitatively, but it has been found that these assumptions hold in some cases down to quite low energies (in the 50 or 100 eV range), as long as information is sought for energy, angle or space-averaged quantities [11]. For sputtering applications they are even used successfully in the eV range.

Under these conditions, an equation can be derived for f. Let a particle start at time $t = 0$ at $\mathbf{x} = \mathbf{x}_0$ with velocity $\mathbf{v} = \mathbf{v}_0$. The probability $f(\mathbf{x}, \mathbf{v}, t+dt) \, d^3x \, d^3v$ of finding the particle at $(\mathbf{x}, d^3x)$ with $(\mathbf{v}, d^3v)$ at time t+dt has two additive contributions from the probability $f(\mathbf{x}, \mathbf{v}, t)$ at time t:

1. If there was no collision in the time interval dt, there is a contribution proportional to $f(\mathbf{x}-\mathbf{v}t, \mathbf{v}, t) d^3x \, d^3v$. The proportionality factor is the probability of no collision during dt; this is $[1-N\int d^3v' K(\mathbf{v} \rightarrow \mathbf{v}')vdt]$, where $K(\mathbf{v} \rightarrow \mathbf{v}') \, d^3v'$ denotes the collision cross section of a particle with velocity $\mathbf{v}$ to scatter into velocity $(\mathbf{v}', d^3v')$, $v = |\mathbf{v}|$, and N is the target atom number density.

2. If there was a collision in dt, the contribution to $f(\mathbf{x}, \mathbf{v}, t+dt)$ is $N\int d^3v' K(\mathbf{v}' \rightarrow \mathbf{v})v'dt \, f(\mathbf{x}, \mathbf{v}', t)$. Here terms of order $(dt)^2$ or higher are neglected.

Balancing these expressions we obtain

$$f(\mathbf{x}, \mathbf{v}, t + dt) = \left[1 - N \int d^3 v' \ K(\mathbf{v} \rightarrow \mathbf{v}') \ v \ dt \right] f(\mathbf{x} - \mathbf{v}t, \mathbf{v}, t)$$

$$+ \ N \int d^3 v' \ K(\mathbf{v}' \rightarrow \mathbf{v}) \ v' \ f(\mathbf{x}, \mathbf{v}', t), \tag{16}$$

and by collecting terms linear in dt,

$$\left[\frac{\partial}{\partial t} + \mathbf{v} \cdot \frac{\partial}{\partial \mathbf{x}} \right] f(\mathbf{x}, \mathbf{v}, t) = N \int d^3 v' \left[K(\mathbf{v}' \rightarrow \mathbf{v}) \ v' \ f(\mathbf{x}, \mathbf{v}', t) \right.$$

$$\left. - \ K(\mathbf{v} \rightarrow \mathbf{v}') \ v \ f(\mathbf{x}, \mathbf{v}, t) \right]. \tag{17}$$

In case electronic stopping is operative, a term

$$+ \ N \frac{\partial}{\partial E} \left[S_e(E) \ f(\mathbf{x}, \mathbf{v}, t) \right] \tag{18}$$

has to be added on the right hand side.

Eq. (17) is known as a *forward* transport equation. The time evolution of f with the initial condition

$$f(\mathbf{x}, \mathbf{v}, t = 0) = \delta(\mathbf{x} - \mathbf{x}_0) \ \delta(\mathbf{v} - \mathbf{v}_0) \tag{19}$$

can equally be described by a so-called backward equation. This is a common feature of - so-called Markovian - linear evolution equations. Backward equations have been extensively used for range, defect and sputtering calculations, since they allow to formulate equations for quantities in which part or all of the end variables $\mathbf{x}$ and $\mathbf{v}$ have been integrated away, cf. sect.5. But, in principle, the very same information is present in the forward as in the backward type of equations for the distribution f with initial condition (19). In the present section, I shall concentrate on the presentation of the forward equation (17).

The evolution of the phase space density f_{rec} of target atoms can be described quite similarly as in eq. (17) above. Let us consider, for the sake of simplicity, the case of self-bombardment,

$$f_{rec}(\mathbf{x}, \mathbf{v}, t = 0) = \delta(\mathbf{x} - \mathbf{x}_0) \ \delta(\mathbf{v} - \mathbf{v}_0). \tag{20}$$

Then the evolution equation reads

$$\left[\frac{\partial}{\partial t} + \mathbf{v} \cdot \frac{\partial}{\partial \mathbf{x}} \right] f_{rec}(\mathbf{x}, \mathbf{v}, t) =$$

$$N \int d^3 v' \left[K(\mathbf{v}' \rightarrow \mathbf{v}) \ v' \ f_{rec}(\mathbf{x}, \mathbf{v}', t) + K(\mathbf{v}' \rightarrow ., \mathbf{v}) \ v' \ f_{rec}(\mathbf{x}, \mathbf{v}', t) \right.$$

$$\left. - \ K(\mathbf{v} \rightarrow \mathbf{v}') \ v \ f_{rec}(\mathbf{x}, \mathbf{v}, t) \right]. \tag{21}$$

234

This equation is in complete analogy with the projectile equation (17). The term added on the right hand side stems from the creation of a recoil atom in a collision. The expression $K(\mathbf{v}' \rightarrow ., \mathbf{v}) \, d^3v$ denotes the collision cross section of a particle with velocity $\mathbf{v}'$ to scatter at a target atom at rest, and to impart to it a velocity $(\mathbf{v}, d^3v)$.

3.2 Methods of Solution

The solution of eq.(17) or (21) is not trivial. Usually, however, one is not interested in the full information present in the distribution $f(\mathbf{x}, \mathbf{v}, t)$. Often, for example, the detailed time structure of the slowing down and transport process is irrelevant, and one may restrict attention to time-integrated quantities. Furthermore, if a target with a plane surface is irradiated with a fixed incident polar angle, in many applications one is not interested in information on the lateral space variables, or the azimuth of the velocity vector around the surface normal. In such a case, the information sought is contained in the distribution $f(z, E, \mu)$, which depends only on the depth z into the target, the energy E of the particle and the cosine of the polar angle ϑ of the velocity vector towards the surface normal, $\mu = \cos\vartheta$.

It is straightforward to derive an equation for $f(z, E, \mu)$ from eq.(17). Let us first express the scattering cross section K in energy and angle variables,

$$K(\mathbf{v} \rightarrow \mathbf{v}') \, d^3v' = K(E, \mathbf{e} \rightarrow E', \mathbf{e}') \, dE' \, d^2e'$$

$$= \sigma(E, T) \, dT \, \frac{1}{2\pi} \, \delta(\mathbf{e} \cdot \mathbf{e}' - \hat{\mu}) \, d^2e' . \qquad (22)$$

Here, $\mathbf{e}$ is the direction of motion of a particle with velocity $\mathbf{v}$, and $\mathbf{e}'$ analogously; $T = E - E'$ is the transferred energy and $\hat{\mu}$ is the cosine of the projectile deflection angle, expressed as a function of E and T; $\sigma(E, T) \, dT$ is the cross section for energy transfer T in its usual notation [4].

For a stationary and homogeneous beam of current density ψ [particles per unit time and area], we may introduce the projectile flux by

$$\Phi(z, E, \mu) = \frac{1}{\psi} \, v \, f(\mathbf{x}, E, \mathbf{e}) . \qquad (23)$$

Expanding Φ into Legendre polynomials

$$\Phi(z,E,\mu) = \sum_{\ell=0}^{\infty} (2\ell + 1) \; P_\ell(\mu) \; \Phi_\ell(z,E),$$

(24)

$$\Phi_\ell(z,E) = \frac{1}{2} \int_{-1}^{+1} d\mu \; P_\ell(\mu) \; \Phi(z,E,\mu),$$

eq.(17) becomes (for details see ref.[12]):

$$\frac{\partial}{\partial z}[\mu\Phi(z,E,\mu)]_\ell = N \int dE' \left\{ \sigma(E',E'-E) \; P_\ell(\hat{\mu}) \; \Phi_\ell(z,E') \right.$$

$$\left. - \; \sigma(E,E-E') \; \Phi_\ell(z,E) \right\} + Q_\ell(z,E),$$

(25)

where $[F(\mu)]_\ell$ denotes the ℓ-th Legendre moment of an arbitrary function $F(\mu)$, in accordance with eq.(24), and Q_ℓ denotes the ℓ-th Legendre moment of the source $Q(z,E,\mu) = \delta(z) \; \delta(E-E_0) \; \delta(\mu-\mu_0)$.

If the slowing down and transport of the recoil atoms in the collision cascade is studied, the recoil flux $\Psi(z,E,\mu)$ is introduced, in complete analogy to the projectile flux Φ. It obeys the equation

$$\frac{\partial}{\partial z}[\mu\Psi(z,E,\mu)]_\ell = N \int dE' \left\{ \sigma(E',E'-E) \; P_\ell(\hat{\mu}) \; \Psi_\ell(z,E') \right.$$

$$\left. + \; \sigma(E',E) \; P_\ell(\tilde{\mu}) \; \Psi_\ell(z,E') \; - \; \sigma(E,E-E') \; \Psi_\ell(z,E) \right\} + Q_\ell(z,E).$$

(26)

Here, $\tilde{\mu}$ denotes the cosine of the angle between the velocity of the scattering particle before the collision and the recoiling particle.

Even the solution of eq.(25) (or (26), respectively) is rather an intricate task and is possible only with the help of approximations. These may be introduced either in the cross sections (*synthetic cross sections* [13]), or in the distribution function. Explicit solutions are obtained, though, for the case where the spatial distribution of f is irrelevant. These are of interest since they characterize the energy distribution of recoil atoms in a collision cascade, and may be used to calculate the energy spectrum of sputtered particles, cf. sect.5.2. By integrating for example eq.(26) over all space and all angles, one obtains

$$N \int dT \left\{ \left[\sigma(E + T,T) + \sigma(E + T,E)\right] \Psi(E + T) - \sigma(E,T) \; \Psi(E) \right\}$$

$$+ \; \delta(E - E_0) = 0.$$

(27)

236

In order to render the notation not too complicated, we denote this angle and energy integrated quantity again by Ψ. The derivation of eq. (27) above is possible of course only if the cross section does not depend on space. In particular, it must be assumed that the slowing down and transport process occurs in a target which stretches out homogeneously in all directions (the *infinite medium*, in which the plane $z = 0$, where the projectile is injected into the target, constitutes only a reference plane).

Eq. (27) can be solved rigorously for power cross sections (9), with methods detailed in [1,14]. Here we shall be interested however only in the leading term of this solution for small detector energies $E \ll E_0$, which can readily be obtained for general forms of the cross section [15].

To this end, substitute $t = T/(E+T)$ in the gain terms of eq. (27), and $t = T/E$ in the loss term. We obtain

$$
N \int dt \left\{ E' \left[\sigma(E',t) + \sigma(E',1-t) \right] \Psi(E') - E \, \sigma(E,t) \, \Psi(E) \right\}
$$

$$
+ \delta(E - E_0) = 0. \tag{28}
$$

It is now advantageous to introduce the lethargy

$$
u = \ln \frac{E_0}{E}. \tag{29}
$$

For convenience, we introduce

$$
\tilde{\Psi}(u) = (E_0 e^{-u})^2 \, \Psi(E = E_0 e^{-u}), \tag{30}
$$

and obtain

$$
N \int dt \left\{ e^{u'} \left[\sigma(u',t) + \sigma(u',1-t) \right] \tilde{\Psi}(u') \right.
$$

$$
\left. - e^u \, \sigma(u,t) \, \tilde{\Psi}(u) \right\} + \delta(u) = 0. \tag{31}
$$

Taylor expansion in u' around u yields

$$
N \, \tilde{\Psi}(u) \int dt \left\{ e^{u'} \left[\sigma(u,t) + \sigma(u,1-t) \right] - e^u \, \sigma(u,t) \right\}
$$

$$
+ e^u \frac{\partial}{\partial u} \tilde{\Psi}(u) N \int dt (u' - u) e^{u'-u} \left[\sigma(u,t) + \sigma(u,1-t) \right]
$$

$$
+ \delta(u) = 0. \tag{32}
$$

Note that (u'-u) is a function of t only. The first integral vanishes due to energy conservation. The second term can be written in a compact way by introducing a certain moment of the scattering cross section, the *slowing down cross section* $\tilde{\sigma}$

$$\tilde{\sigma}(u) = - \int dt \ (1 - t) \ \ln(1 - t) \ \Big[\sigma(u,t) + \sigma(u,1 - t)\Big], \tag{33}$$

$$\tilde{\sigma}(E) = - \int dT \ \sigma(E,T) \ \left\{\Big[1 - \frac{T}{E}\Big] \ln\Big[1 - \frac{T}{E}\Big] + \frac{T}{E} \ln \frac{T}{E} \right\}. \tag{34}$$

Hence eq.(32) finally reads

$$\frac{\partial}{\partial u} N \ \tilde{\sigma}(u) \ \tilde{\Psi}(u) = \delta(u). \tag{35}$$

This equation is readily solved. In the original variables, we obtain

$$\Psi(E) = \frac{E_0}{N \ E^2 \ \tilde{\sigma}(E)}. \tag{36}$$

This constitutes the asymptotic solution for the energy dependence of the recoil flux Ψ, valid for $E_0 \gg E$. The essential idea of this solution scheme is to introduce convenient variables. In these variables, a Taylor expansion is particularly effective, since $N \ \tilde{\sigma}(u) \ \tilde{\Psi}(u)$ depends only slightly on u, cf. eq.(35).

For Lindhard cross sections (9), it is

$$\tilde{\sigma}(E) = \frac{\psi(1) - \psi(1 - m)}{m(1 - m)} \ C_m \ E^{-2m} = \frac{1}{\Gamma_m} \frac{S_n(E)}{E}, \tag{37}$$

where $\psi(x)$ denotes the digamma function [16], $\Gamma_m = m/[\psi(1)-\psi(1-m)]$, and $S_n(E)$ is the stopping cross section (13). For this special case, the result (36) reduces to the well known expression [17,14]

$$\Psi(E) = \Gamma_m \frac{E_0}{N \ S_n(E) \ E}. \tag{38}$$

Inclusion of the space dependence of the phase space distribution is of course necessary for many studies. In this case, the angular dependence of the distribution has to be included as well, since both these features are invariably linked in the transport process via the term $\mathbf{v} \cdot \frac{\partial}{\partial \mathbf{x}} f$. A rigorous analytical solution of this space-dependent problem has been achieved only in cases where the energy dependence of the distribution could be ignored; this is the traditional topic of transport theory [18,19]. In the cases of interest to us here, the energy dependence of the flux is crucial, and one therefore has to make use of approximation methods. The spatial moment method seeks equations for the moments

$$\Psi_\ell^n(E) = \int_{-\infty}^{\infty} dz\ z^n \int_{-1}^{+1} d\mu\ P_\ell(\mu)\ \Psi(z,E,\mu),\tag{39}$$

and attempts to reconstruct the full distribution Ψ from the calculated moments [20,12]. The calculation of the spatial moments is straightforward only in the case of an infinite medium. Corrections for half-space distributions have been derived [21].

Other methods have been employed to investigate the space-dependent slowing down and particle transport in solids. A viable approach consists in expanding the distribution function $\Psi(z,E,\mu)$ in a set of functions $g_\ell(\mu)$ of the direction cosine μ. In view of the structure of eq.(25) or (26), an expansion in Legendre polynomials appears most natural. Truncation of the Legendre-expansion at the L-th term, i.e., setting Ψ_{L+1} = 0, constitutes the so-called P_L-approximation [19,22,23]. In certain situations, other expansions also prove valuable. Close to the target surface, for example, most particles stream in the direction out of the target, and at the very surface itself, *all* particles (except the impinging projectile) stream outwards. To describe such a strongly anisotropic situation, an expansion in two sets of functions, one containing particles flying outwards, the other containing particles flying inwards, prove useful [24], and has been applied to particle emission problems [25,26].

4. COMPUTER SIMULATION

Analytical reasoning of the sort described above gives valuable information on the general trends of particle slowing down and transport in matter, insight into the dependence of the phenomena encountered on the parameters of the system - ion and target atom species and mass, cross sections, bombarding energy, etc. - and allows to derive quantitative predictions in good agreement with observation; the linear cascade theory of sputtering is an example hereof (cf. sect.5). There exist cases, however, where information beyond that supplied by analytical theory is needed; examples are the study of slowing down and transport under conditions where the assumptions underlying analytical theory are violated, or the evaluation of slowing down quantities in unusual geometries.

Computer simulation has proven to be a useful tool for such and similar studies. An ever increasing number of simulation schemes have been devised for such purposes. Some of these are quite widespread and are being used in parallel to experiments in order to help with their interpretation. In the following, I wish to describe shortly the two basic types of simulations which exist in this field, Monte Carlo and molecular dynamics simulations. [As a sort of intermediate, a third type of algorithm is in use [27], which is deterministic, but treats all interactions as binary collisions]. I shall point out how Monte Carlo simulations can be used to obtain a solution of the linear transport equations of sect.3, which features are used to render the simulations more realistic by departing from the basic assumptions underlying transport theory, and describe some of the principles to be observed in molecular dynamics studies.

4.1 Monte Carlo Solution of the Linear Transport Equation

The phase-space density f, whose time evolution is described in eq.(17), denotes the ensemble average over many trajectories of an energetic particle incident on a structureless target. One may view each trajectory as being determined microscopically by the actual locations of the target atoms with which it collides. These locations form an intrinsically stochastic element in the slowing down and transport process of the bombarding particle. A second stochastic element enters the description: At each collision, the impact parameter and the pertinent azimuth angle are random. Due to fluctuations in these quantities, each projectile trajectory looks different. As in experiments, the quantities of interest are the mean values obtained after averaging over many projectile trajectories.

The simplest scheme to obtain a solution of the linear transport equation is to simulate the effect of an energetic projectile on a structureless target. By performing averages over many particle trajectories, an approximate solution to the transport equation is obtained. This method is termed a Monte Carlo method, because of the stochastic nature with which random numbers are used in the simulation process. Of course, in this way, also information on the fluctuations around the average quantities can be obtained.

The simulation algorithm for a single projectile is obviously the following: When the projectile enters the target medium, it travels on a straight line until the first collision occurs. The free flight path length *l* must be sampled from an exponential distribution,

$$F(l) = \frac{1}{\lambda} e^{-l/\lambda}, \qquad\qquad (40)$$

with mean $\lambda = 1/(N\sigma_{tot})$. Here, σ_{tot} denotes the total collision cross section. In case σ_{tot} is infinite, the cross section is usually cut off at small energy transfers; this issue is discussed in some detail in the lectures by Sigmund (these Proceedings). Eq.(40) is easily proven as follows. The probability that the projectile collides on a small path element Δl is $N\sigma_{tot}\Delta l = \Delta l/\lambda$. The probability that it does *not* collide while travelling a finite path length l, which is a large multiple of Δl, $l = n\ \Delta l$, then is

$$\lim_{n \to \infty} \left[1 - \frac{\Delta l}{\lambda} \right]^n = \lim_{n \to \infty} \left[1 - \frac{l}{n\lambda} \right]^n = e^{-l/\lambda}. \qquad\qquad (41)$$

The probability of actually making its first collision after flying the path length l is then $N\sigma_{tot}dl\ e^{-l/\lambda}$, i.e. eq.(40). Note that when electronic stopping is included, the sampling procedure for l has to be modified.

If a collision occurs, its outcome is stochastic. The impact parameter p, which is related by kinematics to the scattering angle and the energy transfer, has to be sampled, and so must the azimuth φ of the collision in order to describe the trajectory of the scattered and the recoiling particles in three-dimensional space. This sampling is most easily performed by noting that each area element p dp dφ is equiprobable, i.e., φ and p^2 are sampled from uniform distributions.

The free flight and collision processes are then repeated indefinitely, or - in practice - until some stopping criterion is met. For range or reflection calculations, for example, as soon as the projectile energy drops below some predetermined threshold, the simulation of its trajectory is stopped. Since all the degrees of freedom of the projectile are known at all times, detectors for whatever quantity is to be measured may be built, and the information stored for further processing.

In practice, the collision and free flight routines are straightforward to implement, whereas statistics considerations (variance reduction techniques [28]), and the design of detectors require more thought. The algorithm described here is implemented in this or a similar form in several existing Monte Carlo codes. The code BEST fits this description exactly [29]; it was designed to provide a numerical solution of the linear transport equations (17), or (25) and (26).

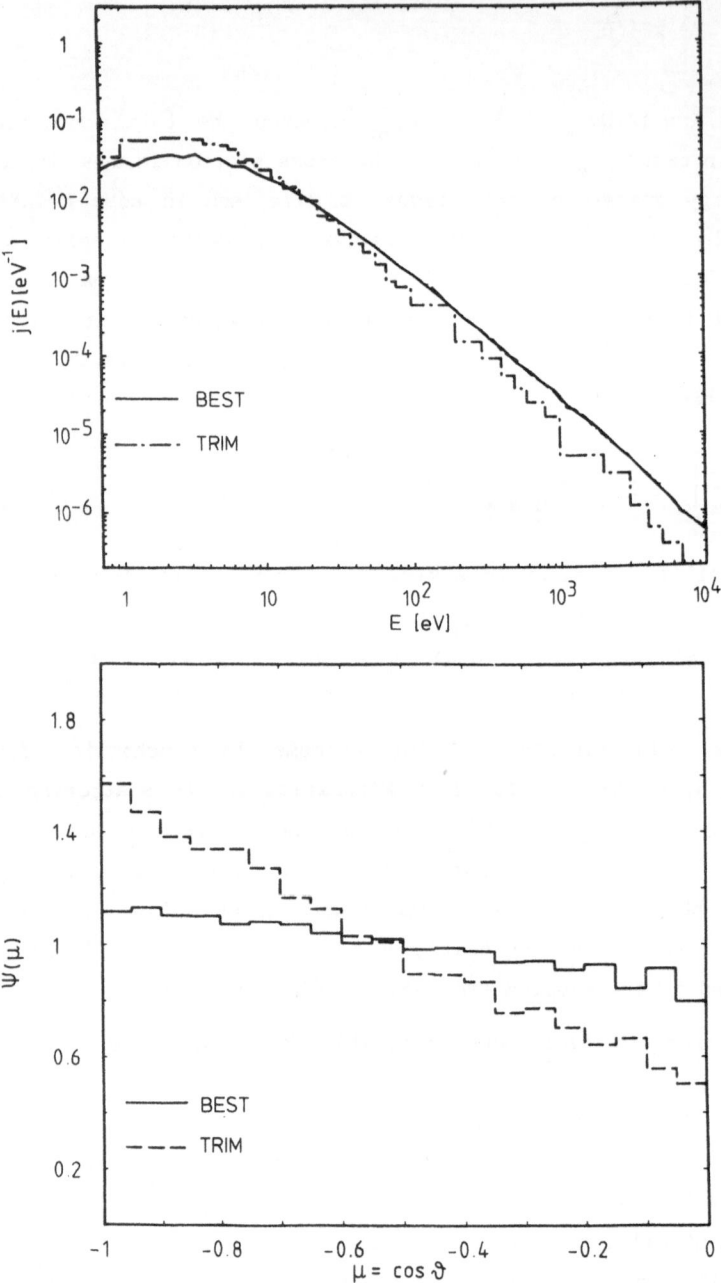

Figure 2. Comparison between the energy and angular distributions of sputtered particles calculated by the TRIM code and by a simulational solution of the transport equation with the code BEST. Comparison adapted from ref.[29]. The Kr-C interaction potential has been assumed. TRIM takes electronic stopping into account, whereas BEST neglects it. TRIM data taken from ref.[30]. Top: Energy spectrum $j(E)$ of Ni atoms sputtered by 100 keV Ne^+ ions at normal incidence. Bottom: Angular distribution $\Psi(\mu)$ of Ni atoms sputtered by 5 keV Ne^+ ions at normal incidence. A cosine-distributed emission current corresponds to a horizontal line $\Psi(\mu) = 1$ in this plot, where $\mu = \cos\vartheta$, and ϑ is the polar angle of the emitted particle with respect to the inward surface normal.

4.2 More General Monte Carlo Procedures

There exist quite a number of different Monte Carlo procedures, most of which do not attempt to solve the linear transport equations, but rather try to model the slowing down and transport of particles in matter as realistically as possible. Usually computational speed is also a design criterion. Thus these codes deviate in one or several points from the assumptions underlying the analytical theory. The most prominent changes are the following:

1. The target is not assumed structureless, but rather some sort of near-neighbor order (as in amorphous matter) is adopted. Thus departures from the exponential distribution of free flight paths (40) are implemented.

2. Scattering information is not derived from the cross sections, but from the potential itself. Thus the dynamics during the collision can be taken into account. As a result, for fixed impact parameter, the outcome of the collision may deviate from the asymptotic trajectories calculated with the help of the collision cross section.

3. An attempt is made to mimic part of the binding and the low-energy many-body forces in realistic materials. These include the displacement threshold, bulk binding energies, and alike.

The most prominent example of such a Monte Carlo code is probably TRIM [2]. It is widely used to help in the interpretation of experimental data.

It is clear from this list that differences between analytical theory - or the numerical solution of the transport equation - and the results of these general Monte Carlo schemes have to be expected. Two examples of a comparison between the data obtained by TRIM and the solution of the transport equations are shown in fig.2. In particular, in the angular distribution the differences are obvious. A more broadly based comparison between different Monte Carlo codes exists [11].

4.3 Molecular Dynamics

A more basic approach to particle-solid interaction phenomena is represented by molecular dynamics computer simulation. Here, the locations and initial velocities of the projectile and all target atoms are taken as input. With the help of Newton's equations of motion, and taking into account the forces between all interacting particles, the time evolution of the entire projectile-target system can - in principle - be obtained. Usually, it is necessary to perform several runs with different initial conditions, e.g. in order to vary the initial position of the projectile on the target surface, or to simulate thermal targets. In the same way as described above for Monte Carlo calculations, detectors are built and store the desired information.

Molecular dynamics is clearly the method of choice when complex problems are to be investigated, in which the binary interaction picture is surmised to break down. For the problems of ion penetration and sputtering, it is still disputed whether molecular dynamics results differ significantly from the predictions of analytical or Monte Carlo calculations [31]. This is in part due to the different interaction potentials used in the molecular dynamics simulations, and in the Monte Carlo and analytical calculations, and in part due to the various parameters which can to some degree be adjusted in many Monte Carlo simulations. Phenomena which are difficult to investigate properly without molecular dynamics type of calculations include the melting of materials in the late phase of a collision cascade [32] or processes occurring under cluster bombardment [33]. Since these phenomena are based on collisions between moving particles, they are outside the range of the Monte Carlo algorithms in the simple form as discussed above.

Molecular dynamics simulations are usually quite expensive in terms of computation time. From an algorithmical point of view, the coding is essentially straightforward. The choice of the ordinary differential equation solver used, and of whether a nearest neighbor list is to be kept (in order to keep control of the interaction partners of any particle), are the main decisions to be taken in designing a code. However, a number of questions pertaining to the physics of the simulated volume of matter need to be settled:

1. Which interaction potential is to be used?

2. Which is the initial state of the system? Are thermal effects to be included, and how?

3. What is to happen if particles or energy arrives at the boundaries of the simulated region? Often, for example, visco-elastic forces act on the outermost atoms to dampen atomic motion there.

4. How many runs need be performed? Here it is important to understand to what extent the simulation results depend on the exact bombardment point of the projectile on the target surface.

5. Can electronically inelastic effects safely be ignored? If not, how to include them?

Often, the information needed to settle these issues is not evident.

Many of these questions need also be solved when employing analytical theory or Monte Carlo simulations. I chose to put this list here in order to make clear that however fundamental the molecular dynamics approach appears to be, it resides on quite a number of - sometimes debatable - physical input.

5. APPLICATIONS TO ION RANGES AND SPUTTERING

5.1 Ranges

One of the first and best studied questions in ion penetration pertains to the end location of the projectile in the material. Consider a beam of projectiles of energy E_0 impinging with direction e_0 on the surface $z = 0$ of a target. Since different projectiles will stop at different depths, one may inquire into the probability $R(E_0, e_0, z)$ dz that the projectile will come to rest at a depth between z and z+dz inside the target. $R(E_0, e_0, z)$ is called the (projected) range distribution of the projectile. A (backward) equation can be established [34], which determines $R(E_0, e_0, z)$ completely, once the electronic stopping and the nuclear collision cross section have been specified.

The probability $R(E_0, e_0, z)$ dz can be calculated as well by considering the projectile after it traveled a path length Δz into the solid. It then continues its motion from a location $\Delta z \cos\vartheta_0$, where ϑ_0 denotes the angle between e_0 and the surface normal. As in the balance leading to eq.(16), there may have occurred a collision on the path Δz. Denoting the collision cross section with $K(E_0, e_0 \rightarrow E', e')$ dE' d^2e', cf. eq.(22), the probability for such a collision is

$$N \, \Delta z \int dE' \; d^2e' \; K(E_0, e_0 \rightarrow E', e'), \tag{42}$$

and the contribution to the probability that the projectile is stopped at depth (z,dz) is

$$N \, \Delta z \int dE' \; d^2e' \; K(E_0, e_0 \rightarrow E', e')$$
$$\times R(E_0 - T - NS_e \, \Delta z, e', z - \Delta z \cos\vartheta_0), \tag{43}$$

since the energy of the projectile and its direction have changed due to the collision; here, electronic stopping has been taken into account.

Conversely, if there occurred no collision during the path length Δz, the contribution to the probability of being stopped at depth (z,dz) is

$$\left[1 - N\Delta z \int dE' \; d^2e' \; K(E_0, e_0 \rightarrow E', e') \right] R(E_0 - NS_e\Delta z, e_0, z - \Delta z \cos\vartheta_0). \tag{44}$$

The sum of these two contributions, eqs.(43) and (44), must equal $R(E_0,e_0,z)$. Expanding this balance in terms linear in Δz, we obtain

$$- \cos\vartheta_0 \frac{\partial}{\partial z} R(E_0,e_0,z) =$$

$$= N \int dE' \ d^2e' \ K(E_0,e_0 \rightarrow E',e') \left\{ R(E_0,e_0,z) - R(E_0 - T,e',z) \right\}$$

$$+ N \ S_e(E_0) \frac{\partial}{\partial E_0} R(E_0,e_0,z). \qquad (45)$$

This constitutes the well known range equation [34,1]. The solution of this equation is generally no simple task. Of most interest, however, is often the calculation of spatial averages of the range distribution, such as the *mean range*

$$\bar{z}(E_0,e_0) = \int dz \ z \ R(E_0,e_0,z). \qquad (46)$$

An equation for $\bar{z}$ can be obtained directly from the general range equation. By multiplying eq.(45) with z and integrating over all z, we obtain

$$\cos\vartheta_0 = N \int dE' \ d^2e' \ K(E_0,e_0 \rightarrow E',e') \left\{ \bar{z}(E_0,e_0) - \bar{z}(E_0 - T,e') \right\}$$

$$+ N \ S_e(E_0) \frac{\partial}{\partial E_0} \bar{z}(E_0,e_0). \qquad (47)$$

Here we integrated by parts on the left-hand-side assuming slowing down to take place in an infinite medium such that

$$\int dz \ R(E_0,e_0,z) = 1. \qquad (48)$$

If we insert the cross section, eq.(22), and expand the equation for the mean range, eq.(47) in Legendre moments, we obtain in analogy to eq.(26)

$$\delta_{\ell 1} = N \int dT \ \sigma(E_0,T) \left\{ \bar{z}_\ell(E_0) - P_\ell(\hat{\mu}) \ \bar{z}_\ell(E_0-T) \right\}$$

$$+ NS_e(E_0) \frac{\partial}{\partial E_0} \bar{z}_\ell(E_0), \qquad (49)$$

where $\mu_0 = \cos\vartheta_0$, $\hat{\mu}$ is the scattering angle of the projectile in the laboratory frame expressed in terms of T and E_0, $\delta_{\ell 1} = 1$ for $\ell = 1$ and 0 otherwise, and, cf. eq.(24),

$$\bar{z}(E_0,\mu_0) = \sum (2\ell+1) \ P_\ell(\mu_0) \ \bar{z}_\ell(E_0). \qquad (50)$$

From (49) we immediately see that all Legendre moments $\ell \neq 1$ of $\bar{z}$ vanish identically such that $\bar{z}$ is proportional to $\mu_0 = \cos\vartheta_0$,

$$\bar{z}(E_0, \mu_0) = \mu_0 R_p(E_0). \tag{51}$$

R_p is called the *projected range*. From eq.(49) we see that it obeys

$$1 = N \int dT \ \sigma(E_0, T) \left\{ R_p(E_0) - \hat{\mu} R_p(E_0 - T) \right\}$$

$$+ NS_e(E_0) \ \frac{\partial}{\partial E_0} R_p(E_0). \tag{52}$$

In the general case, this equation has to be solved numerically, and a variety of techniques have been devised to this end, cf. ref.[2]. An approximate analytical solution may be obtained, though, in the case of a heavy projectile slowing down in a light medium. In this case, the projected range is essentially equal to the path length of the projectile, since the heavy projectile will be deflected only negligibly from its initial direction. We may hence set $\hat{\mu} = 1$ in eq.(52). Furthermore, since the energy transfer T in a single collision is small, we are allowed to perform a Taylor expansion in this variable, and obtain from eq.(52)

$$1 = N \int dT \ \sigma(E_0, T) \ T \ \frac{\partial}{\partial E_0} R_p(E_0) + NS_e(E_0) \ \frac{\partial}{\partial E_0} R_p(E_0), \tag{53}$$

or

$$R_p(E) = \int_0^{E_0} \frac{dE'}{N \ [S_n(E') + S_e(E')]}, \tag{54}$$

where the nuclear stopping cross section $S_n(E)$, eq.(13), was employed. This constitutes an approximate analytical expression for the projected range.

As stated above, usually numerical schemes will be used to solve the projected range equation (52). The same applies to the analogous equations which can be set up for the higher spatial moments of the range distribution. Nowadays, often computer simulation is used to calculate range distributions and their spatial averages.

5.2 Sputtering

Due to the impact of an energetic projectile on a target medium, a number of target atoms are set in motion; these collide in their turn with other target atoms, etc. In this way a *collision cascade* develops, which brings about a number of measurable phenomena:

1. Target atoms are knocked off their initial location, and come to rest elsewhere. Thus, vacancies and interstitials are created.

2. Atoms close to the surface may obtain enough momentum to be emitted out of the target surface; they are said to be sputtered.

3. In multi-component targets, the elemental concentrations are changed, and the material is atomically mixed.

In the following, I shall concentrate on the description of sputtering phenomena. Already this topic is - due to its many applications - quite broad and rich in physical aspects, such that I have to concentrate here on the exposition of some basic facts. The reader who is interested in further details should consult recent reviews [3,35,36].

The basic quantity needed to evaluate sputtering phenomena is the target atom flux close to the surface, $\Psi(z=0,E,\mu)$. In eq.(36), we evaluated the spatially-integrated flux, $\Psi(E)$. It can be shown that in an infinite medium - that is, in a medium in which the plane at which projectiles enter the target only forms a reference plane, target atoms extending on both sides of that plane - the particle flux at the surface retains the same energy dependence as displayed by $\Psi(E)$; this holds for small energies $E \ll E_0$. Then, furthermore, the flux is isotropic, since all low-energy recoil atoms lost the memory of the initial projectile direction. The magnitude of the flux is determined by the so-called deposited energy distribution at the surface, $F_D(z = 0)$. That quantity is defined such that $F_D(z)$ dz denotes the amount of energy deposited in low-energy recoil motion in the target between depth z and z+dz. This distribution describes the spatial extension of the collision cascade. It has been calculated in the infinite medium approximation using the method of spatial moments, applied on the backward transport equation [12]. The flux thus reads

$$\Psi(z,E,\mu) = \frac{1}{2} \frac{F_D(z)}{E^2 N\tilde{\sigma}(E)} , \tag{55}$$

where the factor 1/2 is necessary to take account of the angular normalization of the distribution. Due to the predominance of very low energy particles in the flux shown in eq.(55), a low-energy cross section may be used in the determination of $\tilde{\sigma}$. Using Lindhard's power cross section, eq.(9), with m = 0, the low-energy flux can be expressed as

$$\Psi(z,E,\mu) = \Gamma \frac{F_D(z)}{2 \; N \; C \; E^2},\tag{56}$$

Here the slowing down cross section $\tilde{\sigma}$ has been calculated using eq.(37). It is $\Gamma = \Gamma_0 \cong 0.61$, and $C = C_0 \cong 2 \; \overset{\circ}{A}{}^2$.

Only those particles are sputtered which can leave the target; to do so, they have to overcome a surface barrier. Here, often a planar surface barrier is assumed, which acts only on the velocity component normal to the surface. Thus, a surface barrier of height U is overcome only by those particles with energy E and direction cosine μ which obey $E \; \mu^2 > U$. If the surface was a continuum, this picture would be exact. It is easily shown [17] that the particle *current* $j(E,\vartheta) = \mu \; \Psi(z{=}0,E,\mu)$, after passage of the surface barrier, reads

$$j(E,\vartheta) = \Gamma \frac{F_D(z = 0)}{2 \; N \; C} \; \frac{E}{(E + U)^3} \; \cos\vartheta,\tag{57}$$

where ϑ is the polar angle of a sputtered particle with respect to the outward surface normal. E and ϑ denote here quantities *outside* the target.

From (57) we observe that the particle current is cosine distributed. Its energy spectrum follows the so-called Thompson distribution $E/(E{+}U)^3$, with a maximum at U/2 and an E^{-2}-tail at large energies. The total number of sputtered particles per incoming ion, the sputter yield, is given by

$$Y = \int_{0}^{E_0 - U} dE \int_{0^o}^{90^o} d\vartheta \; \sin\vartheta \; j(E,\vartheta).\tag{58}$$

Evaluation, using eq.(57), yields (for $E_0 \gg U$)

$$Y = \frac{1}{8} \frac{\Gamma \; F_D(z = 0)}{N \; C \; U}.\tag{59}$$

This represents Sigmund's well known sputter yield formula [37,17]. It describes sputter phenomena excellently in the regime of collision cascade sputtering. This regime is defined by essentially two requirements: (i) The collision cascade must be dilute, i.e. the fraction of moving atoms in the cascade volume must be small, and (ii) the asymptotic energy distribution (55) or (56) must hold. The first requirement is usually satisfied in strongly bonded materials like metals or semiconductors, with the exception of heavy ion bombardment of heavy materials with energies around the nuclear stopping power maximum [17,35]; here, the energy density in the material becomes so high that virtually all atoms in the collision cascade volume are set in motion. The second requirement fails, in particular, at low energy bombardment, below 1 or 2 keV, say.

It should be added that particles may be emitted during energetic ion bombardment of materials by other mechanisms than the one presented here. In particular, electronic excitations may lead to particle emission from insulating solids. Furthermore, there exists evidence that thermal or collective flow phenomena may contribute to sputtering in the case of high energy densities and/or low material binding energies. The investigation of these mechanisms is still in progress [35].

6. CONCLUSION

In these lectures on low-energy ion penetration and collision cascades in solids, I tried to provide the student with an introduction to this vast field, emphasizing theoretical concepts, methods and results. A number of issues had to be left out. For these, I should like to refer the reader to ref.[2] on stopping powers and ranges of ions in matter, refs.[3,35] on sputtering, and [38] on special questions of computer simulation of stopping and sputtering. Finally, the topic of ion reflection from solids - not covered in these notes - is reviewed in ref.[39].

REFERENCES

1. P. Sigmund, Rev. Roum. Phys. *17* (1972) 823; 969; 1079.
2. J.F. Ziegler, J.P. Biersack, U. Littmark, *Stopping Powers and Ranges of Ions in Matter*, vol 1, J.F. Ziegler, ed., Pergamon, New York (1985).
3. R. Behrisch, ed., *Sputtering by Particle Bombardment*, vols. 1 and 2, Springer, Berlin (1981,1983).
4. J. Lindhard, V. Nielsen, M. Scharff, Mat. Fys. Medd. Dan. Vid. Selsk. *36*, no.10 (1968).
5. W.D. Wilson, L.G. Haggmark, J.P. Biersack, Phys. Rev. *B15* (1977) 2458.
6. A.E. Carlsson, in Solid State Physics 43, H. Ehrenreich, D. Turnbull, eds., (1990) p 1.
7. J. Lindhard, M. Scharff, Phys. Rev. *124* (1961) 128.
8. L.D. Landau, E.M. Lifshitz, *Mechanics*, Pergamon, Oxford (1960); *Quantum Mechanics*, Pergamon, Oxford (1958).
9. J. Lindhard, V. Nielsen, M. Scharff, P.V. Thomsen, Mat. Fys. Medd. Dan. Vid. Selsk. *33*, no.10 (1968).
10. V.I. Shulga, M. Vicanek, P. Sigmund, Phys. Rev. *A39* (1989) 3360.
11. P. Sigmund, M.T. Robinson, M.I. Baskes, M. Hautala, F.Z. Cui, W. Eckstein, Y. Yamamura, S. Hosaka, T. Ishitani, V.I. Shulga, D.E. Harrison, Jr., I.R. Chakarov, D.S. Karpuzov, E. Kawatoh, R. Shimizu, S. Valkealahti, R.M. Nieminen, G. Betz, W. Husinsky, M.H. Shapiro, M. Vicanek, H.M. Urbassek, Nucl. Instr. Meth. *B36* (1989) 110.
12. K.B. Winterbon, P. Sigmund, J.B. Sanders, K Dan. Vidensk. Selsk. Mat. Fys. Medd. *37*, no.14 (1970).
13. M.M.R. Williams, Prog. Nucl. Energy. *3* (1979) 1.
14. W. Huang, H.M. Urbassek, P. Sigmund, Phil. Mag. *A52* (1985) 753.

15. M. Vicanek, H.M. Urbassek, in preparation.
16. M. Abramowitz, I.A. Stegun, eds., *Handbook of Mathematical Functions*, Natl. Bureau Standards, Washington DC (1965).
17. P. Sigmund, in *Sputtering by Particle Bombardment*, vol.1, R. Behrisch, ed., Springer, Berlin (1981), p.9.
18. J.J. Duderstadt, W.R. Martin, *Transport Theory*, Wiley, New York (1979).
19. K.M. Case, P.F. Zweifel, *Linear Transport Theory*, Addison-Wesley, Reading (1967).
20. J.B. Sanders, Thesis, Univ. Leiden, 1968.
21. J. Bottiger, J.A. Davies, P. Sigmund, K.B. Winterbon, Radiat. Eff. *11* (1971) 69.
22. H.M. Urbassek, Nucl. Instr. Meth. *B4* (1984) 356; *B6* (1985) 585.
23. H.M. Urbassek, M. Vicanek, Phys. Rev. *B37* (1988) 7256.
24. J. Yvon, J. Nucl. Energy *14* (1957) 305.
25. K.T. Waldeer, H.M. Urbassek, Nucl. Instr. Meth. *B18* (1987) 518.
26. K.T. Waldeer, H.M. Urbassek, Appl. Phys. *A45* (1988) 207.
27. M.T. Robinson, I.M. Torrens, Phys. Rev. *B9* (1974) 5008; M.T. Robinson, in ref.[3], vol.1, p.73.
28. A. Alcouffe, R. Dautray, A. Forster, G. Ledanois, B. Mercier, eds., *Monte Carlo Methods and Applications in Neutronics, Photonics and Statistical Physics*, Lecture Notes in Physics *240*, Springer, Berlin (1985).
29. M. Vicanek, H.M. Urbassek, Nucl. Instr. Meth. *B30* (1988) 507.
30. J.P. Biersack, W. Eckstein, Appl. Phys. *A34* (1984) 73.
31. H.H. Andersen, Nucl. Instr. Meth. *B18* (1987) 321.
32. R.S. Averback, T. Diaz de la Rubia, R. Benedek, Nucl. Instr. Meth. *B33* (1988) 693.
33. V.I. Shulga, P. Sigmund, Nucl. Instr. Meth., in press.
34. J. Lindhard, M. Scharff, H.E. Schiott, Mat. Fys. Medd. Dan. Vid. Selsk. *33*, no.14 (1963).
35. P. Sigmund, Nucl. Instr. Meth. *B27* (1987) 1.
36. P.C. Zalm, Surf. Interface Anal. *11* (1988) 1.
37. P. Sigmund, Phys. Rev. *184* (1969) 383; *187* (1969) 768.
38. W. Möller, p.151 in *Materials Modification by High-Fluence Ion Beams*, NATO ASI Series E 155, R. Kelly and M.F da Silva eds., Kluwer, Dordrecht (1989).
39. E.S. Mashkova, V.A. Molchanov, *Medium-Energy Ion Reflection from Solids*, North-Holland, Amsterdam (1985).

INTERACTION OF LOW-ENERGY IONS, ATOMS AND MOLECULES WITH SURFACES

W. Heiland[1]

Arizona State University
Tempe, USA

1. Permanent address: Universitat Osnabruck
D-4500 Osnabruck, FRG

1. INTRODUCTION

The study of the interaction of swift particles with matter dates back to the first years of this century. The literature of those days is decorated with many Nobel prize winners, like Curie, Bohr, Stark, Rutherford. World War I ended the first era of particle-solid interaction for reasons nobody cared to think about so far. The early work was inspired by the detection of radioactivity, the α and β rays, and by the experiments made possible by the detection of the "Kanalstrahlen" by Goldstein [1]. Even though the Kanalstrahlen are based on a long history of gas discharge experiments, there was no connection made between the sputtering effect observed much earlier (Grove [2]) and the observation of the energy loss of swift particles in matter.

It is worth noting that Bohr [3,4] discussed the matter for electrons (β rays) and He^{++} (α rays) within the same model. Based on Rutherford's concept of the atom, i.e., a positively charged, small but heavy nucleus surrounded by a cluster of electrons, Bohr states: *there is no reason to discriminate materially between the collisions of an atom with an α or β particle, apart of course from the differences due to the difference in their charge and mass.* The basis was also laid to discuss: i) the energy loss as a function of velocity (not energy), ii) the probability distribution of the loss of velocity, iii) relativistic effects, and iv) ionization produced by α and β rays.

Experimentally there are the results using foils or "sheets of matter" to measure stopping power or the range of α-particles in matter (Bragg [5]), not to forget Rutherford [6]. There was also a series of

Interaction of Charged Particles with Solids and Surfaces
Edited by A. Gras-Martí *et al.*, Plenum Press, New York, 1991

253

experiments by Fuchtbauer [7] and Stark et al.[8,9] where projectile excitation, channeling effects and blistering, i.e., severe radiation damage, were observed or at least predicted (channeling). Stark fell to Nazism, the experimental results fell into the subconsciousness of the scientific community for almost 60 years, whereas the particle-foil interaction became a standard procedure for many different experiments. The years around 1930 became the next major era of particle-solid interaction studies spirited by the advances of quantum mechanics, the detection of nuclear reactions and the development of accelerators. Again we find many famous scientists in the publication lists: Bohr, Bethe, Bloch, Fermi, Teller, Lamb, Lawrence, Davisson and Germer, Stern and Estermann.

At that time we certainly find a tendency to separate the low-energy particle world from high-energy particle physics. The magic of unification is not observed, even though the two important experiments to prove the wave nature of matter were a success of low energy particle-surface interaction experiments: LEED (Low Energy Electron Diffraction) [10] and diffraction of atoms [11]. The neglect of surface physics was purely due to impure surfaces: vacuum technology was totally insufficient to do reproducible surface experiments with any given materials. World War II was certainly not the time to do LEED or atomic diffraction experiments: no relevance to the weaponry and other type of hardware thought to be important for mankind. The third era of particle-solid interaction may be related to the fact that when nuclear physicists moved on to do particle physics at higher and higher energies, accelerators in the range up to a few MeV became accessible to solid-state physics. The surface-physics world started into existence at around the same time (1960) due to the development of the ultra high vacuum (UHV) technology. Since then we find the ever increasing list of acronyms labeling an increasing number of experimental tools to study surfaces, their structure, reactions on surfaces, thermal vibrations of surface species and the electronic properties of surfaces (see appendix).

From the point of view of the present paper the key publications are by H.D. Hagstrum [12] and by D.P. Smith [13]. H.D. Hagstrum started a long lasting study of the charge exchange between ions and metastable atoms with clean and adsorbate covered surface, which laid the basis for our present understanding of these processes. D.P. Smith was the first to reproducibly perform low energy ion-surface scattering experiments, paving the way to a new tool to study the chemistry and the structure of the surfaces. He recognized the extreme surface sensitivity and the effect of

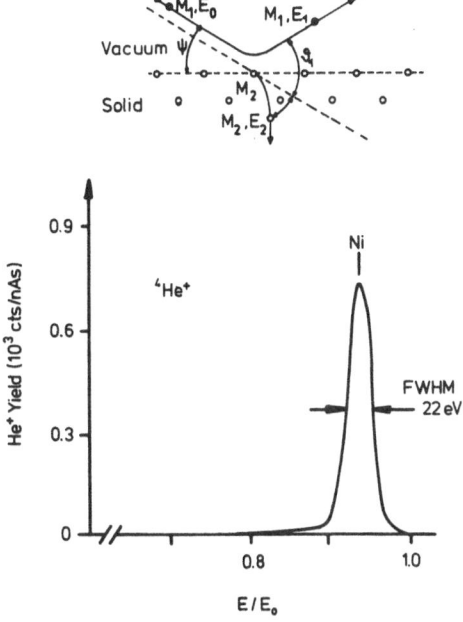

Figure 1. Definition of the laboratory scattering angles θ_1 of the projectile and θ_2 of the recoil in a binary collision. M_1 and M_2 are the masses of the projectile and recoil, E_0 is the primary energy, E_1 the projectile energy after the collision and E_2 the recoil energy. The impact angle ψ is measured against the surface (glancing angle). The ^{4}He energy spectrum is for $E_0 = 600$ eV, $\theta = 60^\circ$, $\psi = 30^\circ$, the target is a Ni(110) single crystal surface. The plane of scattering is at an angle of 3° with respect to the [110] surface direction (Taglauer et al. [15]).

shadowing. The latter effect has been called "Rutherford's classical shadow" by J. Lindhard [14] in a paper usually connected with channeling.

2. LOW ENERGY ION SURFACE SCATTERING

2.1 Experimental Results

The basic result of a low-energy ion scattering experiment is the observation of a single peak in the energy spectrum, fig.1 [15]. The energetic position E_1 of the peak in the spectrum is in good agreement with the following result

$$E_1 = K(\theta_1)\, E_0, \tag{1}$$

where E_0 is the primary energy of the projectile (ion), $K(\theta_1)$ the kinematic factor and θ_1 the laboratory scattering angle, fig.1. $K(\theta_1)$ is

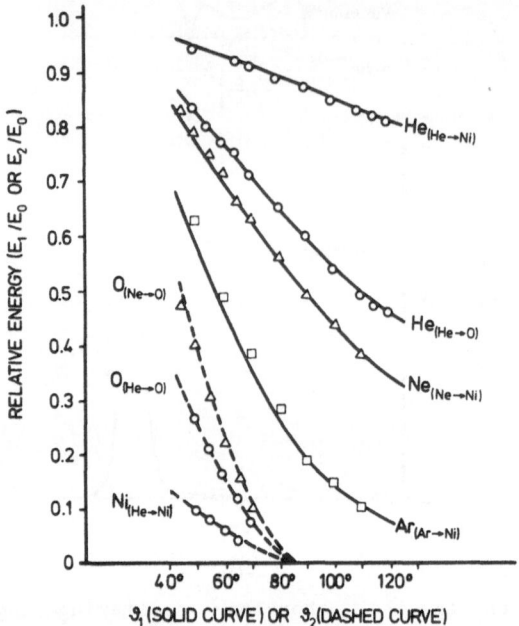

Figure 2. Angular dependence of the secondary energy of backscattered ions. The curves are calculated assuming single binary collisions (Smith [13]).

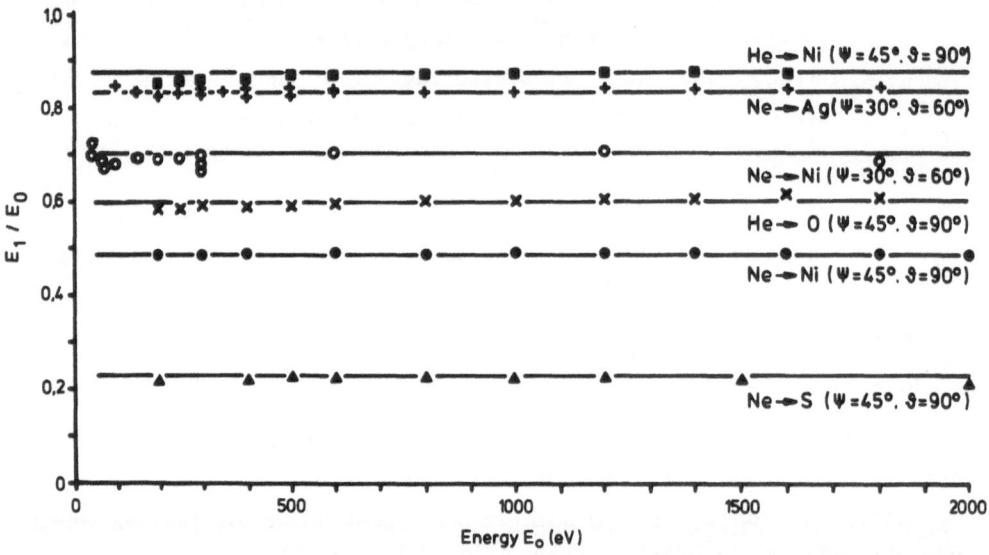

Figure 3. Energy dependence of E_1 of backscattered ions. The straight lines are the theoretical values for single binary collisions (Heiland and Taglauer [16]).

given by

$$K(\theta_1) = \left(\frac{M_1}{M_1 + M_2}\right)^2 \left[\cos \theta_1 \pm \left\{(M_2/M_1)^2 - \sin^2\theta_1\right\}^{1/2}\right]^2. \qquad (2)$$

M_1 is the mass of the ion, M_2 is the mass of the target atom. The validity of eq.(1) can be tested by measuring the angular dependence [13], fig.2, and the energy dependence [16], fig.3. Also the angular dependence of the recoiling atoms is shown in fig.2, which follows the equation

$$E_2 = \frac{4 M_1 M_2}{(M_1 + M_2)^2} E_0 \cos^2\theta_2, \qquad (3)$$

or

$$E_2/E_0 = T = \gamma_{12} \cos^2\theta_2, \qquad (3a)$$

where θ_2 is the recoil angle of the atom in the laboratory system. E_2/E_0 is often called T, the energy transfer from the ion to the atom.

A comparison of the He$^+$ backscattering spectrum, fig.1, with a Li$^+$ backscattering spectrum [15] fig.4, gives one clue for the simplicity of the results shown in figs.1-3: a comparison of the intensity scales shows a much higher yield of Li$^+$ compared to He$^+$. The Li$^+$ peak is broadened and obviously contains ions which are due to scattering events not in agreement with eq.(1). A computer-aided analysis of the He$^+$ and Li$^+$ data shows that, in fact, the He$^+$ peak is due to single scattering events as

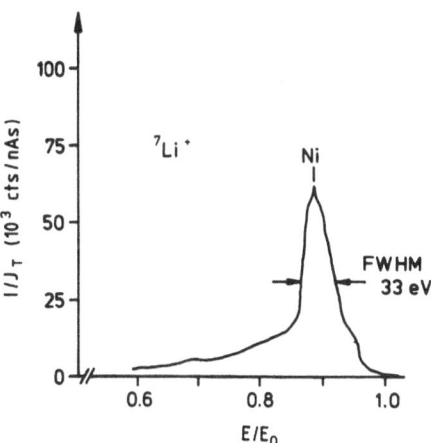

Figure 4. ^{7}Li$^+$ scattering from Ni(110) under identical condition as ^{4}He$^+$ in fig.1. Note the difference in peak shape and scale (Taglauer et al.[15]).

postulated by eq.(1), whereas the Li[+] peaks contains approximately 90% multiple scattering events and 10% single scattering events [17]. The hypothesis, which explains the data, is that He[+] ions involved in multiple scattering events are neutralized and cannot be measured, if the experiment does not afford the detection of neutrals, as it was the case in the experiments discussed so far. The neutralization of Li[+] is negligible in comparison with He[+], which was also proven by independent experiments [18].

As stated already, the application of eqs.(1-3) implies, beside the conservation of energy and momentum, the binary collision approximation (BCA), i.e., each ion scatters from a given target atom as if they were free point masses. Furthermore, in the calculations mentioned [15,17] the particle trajectories are approximated by the asymptotes. Fig.5 shows as an example the trajectory of a K[+] scattered from an Ir(110) surface along a [110] chain [19]. The deviation of the real trajectory from the asymptote between the atoms A and B is negligible, (<.01 Å), compared to the distance of closest approach of 0.9 Å for the second collision in fig.5. It is obvious from fig.5 that with decreasing energy the accuracy of the asymptotic approximation will decrease, because for the same scattering angle the distance of closest approach will increase with decreasing energy.

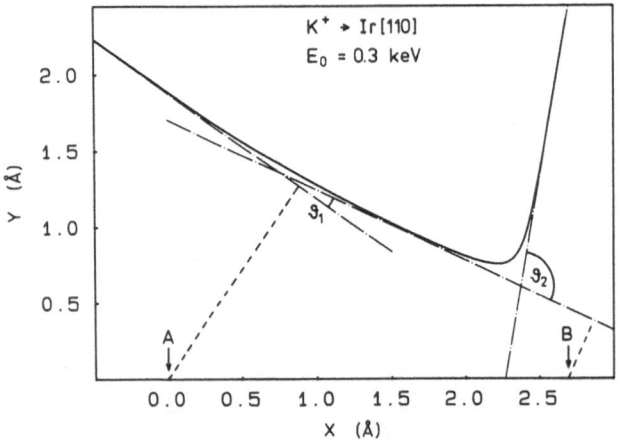

Figure 5. Classical trajectory (solid line) of a 300 eV K[+] ion scattered by two Ir atoms located at positions A and B. The total scattering angle is 115°. The asymptotes to the trajectory are the dash-dotted lines. The impact parameters for the collisions are indicated by dashed lines (Hetterich [19]).

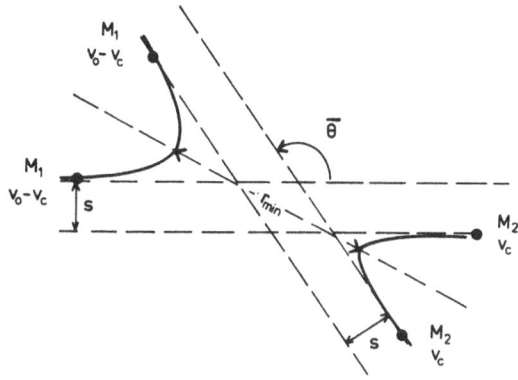

Figure 6. Scheme of a binary collision in the center of mass system
(C.O.M.). The masses M_1 and M_2 are the projectile and target
respectively. The velocities are v_0 for the projectile and v_c
for the center of gravity (the target is at rest before the
collision). $\bar{\theta}$ is the scattering angle, s is the impact parameter
and r_{min} the distance of closest approach.

2.2 Classical Scattering Theory

The theory is usually given in the center of mass (C.O.M.) system,
i.e., all velocities are relative to the velocity v_c of the center of mass
which is

$$v_c = v_0 \frac{M_c}{M_2}, \tag{4}$$

where v_0 is the initial velocity of the projectile ($E_0 = M_1 v_0^2/2$), and the
reduced mass M_c is given by

$$M_c = \frac{M_1 M_2}{M_1 + M_2}. \tag{5}$$

The advantages of the C.O.M. system are that the total linear momentum
will be zero, the velocities scale with $1/M$ and $E_c = M_c v_0^2/2$ is constant,
as is the angular momentum $J_c = M_c v_0 s$, where s is the impact parameter
(see fig.6). The angular momentum is constant also in the laboratory
system, of course, a fact generally known as the second law of Kepler. The
angles transform from the lab to the C.O.M. system, when $v_2^0 = 0$ (initial
velocity of the target), with

$$\bar{\theta}_2 = 2\theta_2 = (\pi - \theta_1), \tag{6}$$

see fig.6, and

259

$$\tan \theta_1 = \frac{M_2}{M_1 + M_2} \frac{\sin \bar{\theta}_1}{\cos \bar{\theta}_1}, \tag{7}$$

for the recoil $\bar{\theta}_2$ and scattering angle $\bar{\theta}_1$ respectively. In a head-on collision, or $\theta_1 = 180^\circ$, the particles approach each other to a distance defined as the collision diameter d which is given by the equality of kinetic and potential energy. For a Coulomb interaction potential this is

$$\frac{1}{2} M_c v_0^2 = \frac{2Z_1 Z_2 e^2}{d}. \tag{8}$$

For other scattering angles the distance of closest approach is given by

$$r_{min} = \frac{d}{2} \left(\text{cosec} \frac{\bar{\theta}}{2} \pm 1 \right). \tag{9}$$

The motion of the particles in the C.O.M. system is described by the equation of motion, which is derived from the conservation of angular momentum and energy respectively,

$$J_c = M_c r^2 \dot{\bar{\theta}} = m_c v_0 s, \tag{10}$$

and

$$E_c = \frac{1}{2} M_c (\dot{r} + r^2 \dot{\bar{\theta}}^2) + V(r), \tag{11}$$

where $\dot{\bar{\theta}} = d\bar{\theta}/dt$ and $\dot{r} = dr/dt$; $V(r)$ is the interaction potential, which is assumed to be static and having spherical symmetry, hence no transverse forces act on the particles and they will move in a plane. Eqs.(11) and (12) lead to

$$\dot{r} = v_0 \left(1 - \frac{V(r)}{E_c} - (\frac{s}{r})^2 \right)^{1/2}, \tag{12}$$

and

$$\dot{\bar{\theta}} = \frac{v_0 \, s}{r^2}, \tag{13}$$

and since $d\bar{\theta}/dr = (d\bar{\theta}/dt) \cdot (dt/dr)$ one obtains

$$\dot{\bar{\theta}} = \pi - \int_{-\infty}^{+\infty} \frac{s \, dr}{r^2 \left(1 - \frac{V(r)}{E_c} - \frac{s^2}{r^2} \right)^{1/2}}, \tag{14}$$

which is called the "classical scattering integral". This integral has analytical solutions for special cases of the potential only.

For the energy range of interest here we list five examples of screened Coulomb potentials,

$$V(r) = \frac{z_1 z_2 e^2}{r} \phi \left(\frac{r}{a}\right), \tag{15}$$

where ϕ is called the screening function and a the screening parameter. Bohr [20] gave the following screening function

$$\phi_B(x) = \exp(-x), \tag{16}$$

with $r/a = x$ and

$$a_B = 0.8853 \ a_0 \ (z_1^{2/3} + z_2^{2/3})^{-1/2}, \tag{17}$$

with $a_0 = 0.529$ Å. Another choice of a is given by Firsov [21]

$$a_F = 0.8853 \ a_0 \ (z_1^{1/2} + z_2^{1/2})^{-2/3}, \tag{18}$$

based on the Thomas-Fermi model of atoms. The screening parameter of Ziegler, Biersack and Littmark [22]

$$a_{ZBL} = 0.8854 \ a_0 \ (z_1^{0.23} + z_2^{0.23})^{-1}, \tag{19}$$

is based on a numerical evaluation of theoretically calculated potentials. This procedure led also to the so called "universal potential", which is an analytical approximation to the numerically calculated potentials. This procedure yields a screening function

$$\phi_{ZBL}(x) = 0.1818 \ e^{-3.2x} + 0.5099 \ e^{-0.9423x} +$$

$$0.2802 \ e^{-0.4028x} + 0.02817 \ e^{-0.2016x}. \tag{20}$$

A previous study of experimental scattering cross sections, with various interatomic potentials [23], has led to the average Lenz-Jensen [24,25] screening function

$$\phi_{LJ}(x) = (1 + 0.9389 \ y + 0.4272 \ y^2 + 0.01150 \ y^3 + 0.01288 \ y^4) \ e^{-y}, \tag{21}$$

with $y = (9.67 \ x)^{1/2}$. The Bohr screening length is used. The potential obtained from eq.(20) has been approximated by Lindhard et al.[26] by the so called "power potential" screening which has the form

$$\phi_{pp}(\frac{r}{a}) = \frac{C}{1/m} (\frac{r}{a})^{1-1/m}.$$ (22)

The parameter m can be used to fit the power potential to any given potential, depending on the atoms and energies involved. The advantage of the power potential is the analytical form obtained with it for, e.g., differential cross sections for the energy transfer T

$$d\sigma = C E^{-m} T^{-1-m} dT,$$ (23)

(see the original papers for details).

Later on we will make comparisons also to the Moliére [27] screening function

$$\phi_m(x) = 0.35 e^{-0.3x} + 5.5 e^{-1.2x} + 0.1 e^{-6x},$$ (24)

which has been found to give good agreement with different experimental results.

The physical basis for the "universal potential" are the Coulomb interactions between the nuclei, the nuclei and the electrons, the electrons and the electrons respectively, the Pauli excitation of the electrons and the exchange potentials [28].

In reality also the classical image potential influences ion trajectories above a surface. At very low energies it affects the trajectories indeed [29,30], whereas at higher energies the influence of the image potential can be neglected for most applications [31].

2.4 Cross Section

The classical differential scattering cross section in a binary collision is defined by

$$d\sigma = \pi |\frac{\delta s^2}{\delta f}| df,$$ (25)

where $f(\theta) = \sin^2(\theta/2)$; θ is the scattering angle in the C.O.M. system. Eq.(14) gives the connection of θ with the scattering potential. The scattering cross section is the decisive quantity which controls the intensity in any scattering experiment. The intensity in the lab system is given by

$$I(E_0,\theta_1) = A N_s \frac{d\sigma}{d\Omega} P_s \Delta\Omega,$$ (25a)

with $d\Omega = 2\pi \sin\theta_1 d\theta_1$, where A is the primary intensity (particles/s),

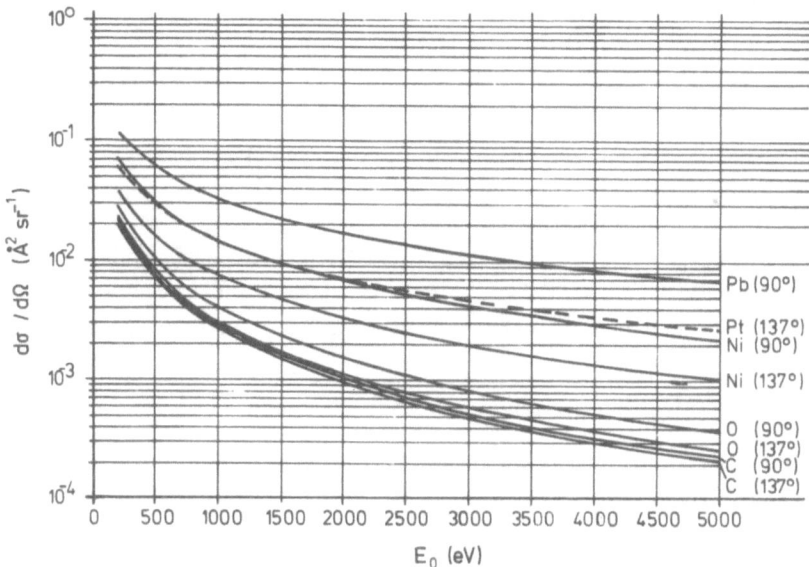

Figure 7. Differential scattering cross section as a function of the
projectile energy, calculated using a Thomas-Fermi interatomic
potential, for different atom-atom pairs and laboratory
scattering angles of 90° and 137°. The projectiles are He atoms
(Heiland and Taglauer [76]).

N_s the number density of particles (particles/cm^2), $d\sigma/d\Omega$ is the
differential scattering cross section (cm^2/sr), and $d\Omega$ is the solid angle
of acceptance of the detector (sr) sitting at the angle θ_1. In A also the
transmission of the instrument and the detector sensitivity is factorized.
P_s is the ion survival probability. Fig.7 shows some calculated
differential cross sections in the energy range of interest. The Firsov
screening length [21] and the Moliére screening function [27] have been
used in the calculation.

2.5 The Binary Collision Approximation (BCA)

There are four aspects with the BCA to be considered: the neglect of
the thermal motion of the target atoms, the neglect of the binding of the
target atoms, the neglect of inelastic losses and the assumption that
quantum effects can be neglected. The velocity of a target atom of mass 58
(e.g., Ni) at room temperature is below 10^5 cm/s, which is small compared
to the velocity of a He$^+$ at 1 keV or a heavy K$^+$ at the same energy which
move with $2 \cdot 10^7$ cm/s and $7 \cdot 10^6$ cm/s respectively. On the same basis we can
argue with respect to the binding between surface atoms. The velocity of
the atoms is a measure for the thermal vibration frequency. Since the
amplitudes are of the order of 0.1 Å, the frequencies are of the order of

10^{13} s^{-1}. The collision time measured, e.g., by the velocity and the collision diameter, 1.7 Å for $He^+ \rightarrow Ni$, or 25 Å for $K^+ \rightarrow Ni$ at 1 keV is of the order of 10^{-15} and $3 \cdot 10^{-14}$ s, respectively. Hence, the lattice can not respond to the collision, i.e., coupling to the phonon system is not possible within the collision time.

We know that diffraction of atoms occurs at thermal energies [11], which can be expected from an estimate of the de Broglie wavelength and a comparison of this number with the lattice constants of solids. For 1 eV He we have λ_B = h/mv = 0.15 Å, which is already small compared to 2 Å, which in turn is a small planar distance in solids.

A thorough quantum-mechanical approach can be found in textbooks, i.e., Landau and Lifshitz [32]. The elastic differential scattering cross section is given by the square of the "scattering amplitude" $f(\theta)$

$$d\sigma = |f(\theta)|^2 \, d\Omega, \qquad (26)$$

with $d\Omega = 2\pi \sin\theta \, d\theta$ in the C.O.M. system. The scattering amplitude is defined by the wave function of the scattered particle, which is the sum of the initial, planar wave function and the scattered, spherical wave function

$$\psi = e^{ikz} + f(\theta) \frac{e^{ikr}}{r}, \qquad (27)$$

where z is the coordinate in the direction of the incoming wave packet and k its wave-vector. Small changes of θ will be proportional to the change of linear momentum p, i.e., $\Delta\theta \sim \Delta p/p$, with $\Delta\theta \ll \theta$. The uncertainty in momentum p and on the impact parameter s will be $\Delta p \cdot \Delta s \sim h$, and hence $\Delta\theta \gg h/(sp)$, since $\Delta p \gg h/s$. Now $s \cdot p$ is the angular momentum, which is quantized to hl, therefore the condition for classical behavior is $\theta l \gg 1$ or $l \gg 1$. That is, at small θ, deviations from classical behavior are most likely to occur.

The inelastic losses which have do be dealt with quantum mechanically in principle can be expected to be small. We will discuss the effects in sect.4. But an extrapolation of energy loss data at high energies [33] yields losses of the order of 10 eV at energies of 1 keV. So, as long as large-angle scattering events are concerned, where the elastic losses are large, the trajectories and the interaction times with the surface are short, and then inelastic losses may be neglected.

2.7 Recoil Sputtering and Ion-Induced Desorption

In general there are three processes which cause the removal of

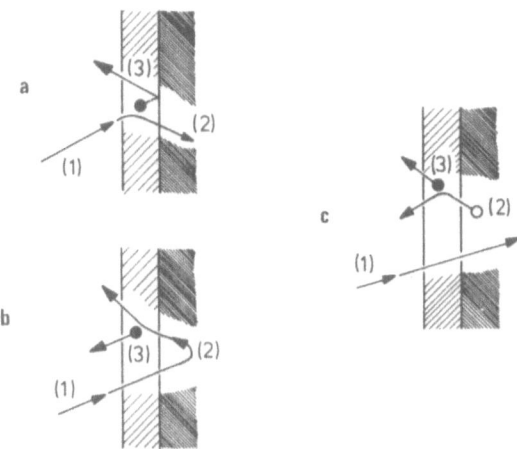

Figure 8. The three postulated sputtering mechanisms, schematically. a: direct knock-off, b: reflected ion and c: the cascade sputtered atom contribution (Winters and Sigmund [34]).

surface atoms by an ion beam: i) the direct recoil or direct knock-off process, ii) the reflected ion contribution, and iii) the cascade sputtering process (Winters et al.[34,35] (fig.8). At small perpendicular energies the cascade sputtering can become negligible and only the processes i) and ii) prevail. As shown by Eckstein [36], the multiple-scattering energy losses decrease also with decreasing perpendicular energy such that the distinction between direct knockoff and reflected ion contribution becomes futile, since the primary ion will not penetrate the surface nor will it loose an appreciable amount of energy before hitting the then recoiling atom. Further aspects and applications of direct recoil spectroscopy have been reviewed recently by Rabalais [37].

Also, in this case, the shadow cone effect can be used to study adsorbed atom positions (O'Connor et al.[38]). Very useful is the detection of negative ions, which allows studies of practically all possible adatoms (beside N). Especially the detection of H^- is to noted, since H evades practically all other surface analytical tools.

In the general case, i.e., when the primary ion beam penetrates the solid, the reflected ion beam flux will have an energy distribution and this "internal" flux will knock off surface atoms. Furthermore, the collision cascade within the substrate will also cause the desorption of surface atoms. Within the Sigmund-Winters theory, which is also a classical collision theory, the knock off cross section is given by

$$\sigma_1 = \frac{1}{\cos \beta} \int_E^{\gamma_{13}E_0} d\sigma_{13} \; (E_0 \; , E'),$$ (28)

where β is the angle between the surface normal and the incident beam, E_0 is the ion energy, E' is a threshold energy for the desorption of an adatom with mass M_3 by the ion with mass M_1, and

$$\gamma_{13} = 4 \; \frac{M_1 + M_3}{(M_1 + M_3)^2}.$$ (29)

The contribution of the reflected ions to the desorption is given by

$$\sigma_2 = \sigma_1 \; R_{12} \left[4 - \frac{4 \; \ln x}{3(x^{1/3} - 1)} \right] \cos\beta,$$ (30)

where σ_1 is the total cross section of the knock off contribution, eq.(28), R_{12} is the particle reflection coefficient for the ions with mass M_1 from the substrate atoms with mass M_2, and $x = \gamma_{13} \; E_0/U_3$ is the ratio of the maximum energy transfer in a 1-3 collision, to the surface binding energy of the adatom to the substrate. Finally, the substrate sputtering contribution can be estimated using Sigmund's sputtering theory [39]

$$\sigma_3 = 4 \; \gamma_{23} \; C_{23} \; \frac{U_2}{U_3} \; S_{12} \; \frac{1 - [(1+\ln y)/y]}{(1 - \gamma_{23}/y)^2},$$ (31)

where C_{23} is estimated from the theory

$$C_{23} = \frac{1}{2} \; \pi\lambda_m \; a^2 \left[\frac{M_2}{M_3} \right]^m \left[\frac{2 \; Z_2 \; Z_3 \; e^2}{a} \right]^{2m},$$ (32)

m is the parameter of the inverse power potential (Sigmund [39], Lindhard et al.[26]), λ_m is a constant of the order of 1, and a is the Bohr screening length, eq.(17). U_2 is the sublimation energy of the substrate. The parameter $y = \gamma_{12} \; \gamma_{23} \; E/U_3$. The sputter yield is calculated to be

$$S_{12} = 0.042 \; \alpha \; \frac{C_{23}}{1 - m} \; \gamma_{12}^{1-m} \; \frac{E^{1-2m}}{U_2}.$$ (33)

Typical numbers for m and λ are m = 0.25 and λ = 1.62; for lighter ions like (He), m = 0.40 and $\lambda \cong 0.5$ have been estimated (Winters and Taglauer [35]). The parameter α is a function of m and M_1/M_2 (Sigmund [39]).

Cross sections for ion impact desorption calculated by the Winters-Sigmund theory agree with a variety of experimental results [34,35,40-43]. Ion desorption is an important process in fusion

266

experiments, and it plays probably a role in the space shuttle glow and related effects. Little use is made of the available data in every day "crystal cleaning" in UHV surface experiments, where very often unnecessary "thick" surface layers are sputtered away instead of just desorbing the impurity atoms off the surface. The sputtering causes unwanted radiation damage and leads eventually to surface structures which cannot be annealed thermally.

2.8 The Classical Shadow Cone, Rainbow Scattering and Channeling

In his paper of 1965 Lindhard [14] proposed an "idealized experiment" to discuss the phenomenon of "the shadow behind a repulsive scattering in an external, parallel beam of atoms. The scattering is supposed to be classical, and we suppose that there is a screen perpendicular to the beam, at a distance d behind the scattering center. This idealized experiment may be said to represent a pair of atoms, the scattering center being one atom, the second atom being placed in the screen, so that we ask for the probability of hitting the second atom". This describes exactly

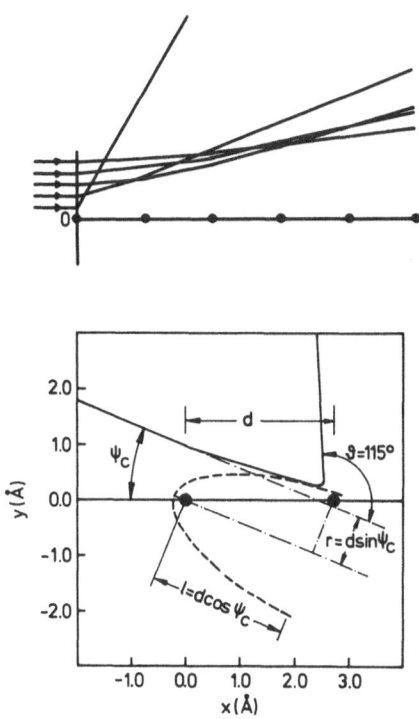

Figure 9. Scheme of a model shadow cone by Bogh [44], and scheme of the shadow cone for K^+ scattering from Ir (110) along a [110] direction, E_0 = 5.9 keV (Hetterich [19]).

267

the situation shown in fig.9. Further experimental details will be discussed in sect.3.2. Lindhard did not consider surface scattering.

For a beam parallel to the z-axis, the position r of a scattered particle on the screen is given for small angles by

$$r = s + \frac{b}{s} d,$$ (34)

where s is the impact parameter and $b = Z_1 Z_2 e^2/E$ (for Rutherford scattering). It follows immediately that there is a minimum distance r_{min} which limits an area into which no particles will be scattered

$$r_{min} = 2 \sqrt{b} \ d.,$$ (35)

for $s = \sqrt{b} \cdot d$. The form of the shadow as a function of d is parabolic, fig.9. When the beam direction is tilted with a glancing angle Ψ with respect to z_1, at a critical angle $\Psi_c = 2 \sqrt{b}/d$, the second atom will be hit (fig.9). A sketch of a shadow cone published by Bogh [44] in 1973 illustrates how the individual trajectories form, as an envelope, the parabolic shadow cone (fig.9). The intensity distribution of the scattered particles obtained along a line on the screen will be zero up to r_{min} for a fixed atom (no thermal motion). Outside it will fall off with

$$f(r) = \frac{1}{2} \left[\left(1 - \frac{r_{min}^2}{r^2} \right)^{1/2} + \left(1 - \frac{r_{min}^2}{r^2} \right)^{-1/2} \right].$$ (36)

The thermal motion reduces the steepness of the edge (see sect.3). It is worth mentioning that Lindhard predicts quantal effects at the edge of the shadow cone.

More recently Jackson [45] and Oen [46] gave a more universal account of the shadow cone. In agreement with Lindhard, it is shown that for small scattering angles or, more accurately, in the impulse approximation (i.e., $\theta_1 \cong \tan \theta_1$), the shadow cone radius at a distance d behind an atom is given by $r_c = s_c + d \theta_c$. For a Moliére potential instead of the unscreened Coulomb potential, the shadow cone radius can be expressed as a function of a single dimensionless parameter bd/a^2, where $b = Z_1 Z_2 \ e^2/E$ as above, and a is the screening length in the Moliére potential. With an accuracy better than 1 % , the ratio of the Moliére critical shadow cone radius r_c to the Coulomb shadow cone radius $2 \sqrt{b} d$ is given by Oen

$$\frac{r_c}{2\sqrt{b} \ d} = \begin{cases} 1.0 - 0.12 \ \alpha - 0.01 \ \alpha^2, & 0 \leq \alpha \leq 4.5, \\ \\ 0.924 - 0.182 \ \ln\alpha - 0.0008 \ \alpha, & 4.5 \leq \alpha \leq 100, \end{cases}$$ (37)

and, accordingly, for the critical impact parameter

$$\frac{s_c}{\sqrt{b} \, d} = \begin{cases} 1.0 & \alpha \le 0.6, \\ 1.03 - 0.04 \, \alpha + 6 \cdot 10^{-6} \, \alpha^4 & 0.6 \le \alpha \le 100, \\ 1.093 - 0.1785 \, \ln\alpha, & 10 \le \alpha \le 100, \end{cases} \qquad (38)$$

where the parameter $\alpha = 2\sqrt{b}d/a$.

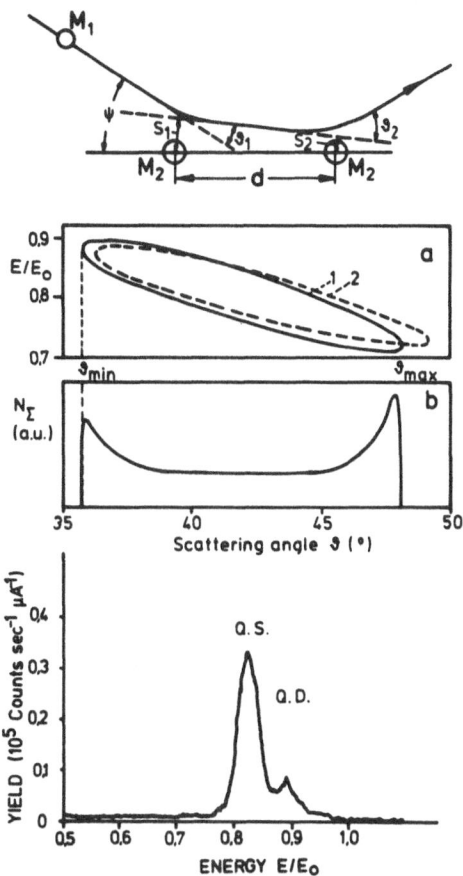

Figure 10. (a) Scheme of a double collision sequence (Parilis [58]). s_1 and s_2 are the impact parameters, and θ_1 and θ_2 the scattering angles of the two collisions, respectively. (b) The loops are the theoretical expectations for a rigid [110] chain in Cu (110) (solid line 1) and for two [110] chains (dashed line 2). The particles are Ar incident with E_0 = 1 keV and an impact angle Ψ = 20°. (c) N_Σ is the total reflected intensity for one chain (Yurasova et al. [7]). (d) The energy spectrum is for Ne$^+$ scattered from Ag (110) at E_0 = 600 eV, Ψ = 30° and θ = 60°. Q.S. and Q.D. are the theoretically expected single scattering and double scattering peaks (Heiland and Taglauer [76]).

269

It should be noted that, for usual distances in surfaces, the shadow cone can be approximated by the trajectory of a particle scattered by ~ 2° off the first atom at energies around 1 keV Heiland and Taglauer [47], Godfrey and Woodfruff [48], Niehus and Comsa [49]). Further examples of calculated low-energy cones are found in the literature [50-52].

When in an experiment like that shown schematically in fig.9 the detector is set not at a scattering angle of ~ 180° but a an angle of ~ 60°, multiple scattering is observed [53-57]. If zig-zag collisions are neglected the situation can be treated as proposed first by Parilis and Bitenski [58] (fig.10). As can be seen, the energy of the backscattered particle will be a function of the lattice constant of the string and of the impact angle. Consequently, there are two peaks in the energy spectra corresponding to "quasi-single" and "quasi-double" scattering. Measuring the two peaks as function of the scattering angle results in a E_1/E_0 versus θ_1 plot in characteristic "loops", the size of which depends also on the impact angle. The extension of the loop is determined at the low θ side by the shadow cone effects described above, and at the high θ side by the corresponding blocking effect. Experimentally, only the low θ side shows a sharp well-defined intensity maximum, whereas the blocking effect is much more blurred by thermal and 3D trajectories [16,59]. Nevertheless, the fact of two energies for one scattering angle establishes the case for a classical rainbow. At the end of the loop, the intensity variation $dI/d\theta \rightarrow \infty$ for a completely rigid chain. The peaks are indeed found in classical calculations (Yurasova [57], fig.10). Recently there has been a number of studies with respect to rainbow effects at lower energies (< 50 eV) (Horn et al.[60], Hulpke [61]).

If now the impact angle is lowered beyond Ψ_c, defined by the shadow cone, the experiment approaches the conditions for surface channeling. Channeling was found in computer studies of ions travelling in solids [62]. The first physical model for the effect was given by Lindhard [14,63]. For a string of atoms the critical angle for channeling to occur is given by

$$\Psi_1 = \frac{2\, z_1 z_2\, e^2}{E\, d}\,, \tag{39}$$

for a Coulombic string potential, where d is the distance between atoms in the string. The key point for both axial channeling, eq.(39), and planar channeling is the conservation of the energy perpendicular to the string or plane respectively. For further details see, e.g., the monograph by Morgan [64].

Surface channeling experiments were proposed by Oen [65] and the effect was demonstrated by Thompson et al.[66]. It found widespread use for the study of surface structures [67,68], charge exchange effects [69], study of excited states [70-72], and surface magnetism [73-75].

3. INSTRUMENTATION AND APPLICATIONS OF ION SCATTERING

3.1. Instrumentation

The apparatus for studies of low energy particle-surface interactions has to be a UHV system. The details of the parts concerned with the ion beam, i.e., the primary beam and the detection of scattered or recoiled particles, are governed by eqs.(1) and (3). Since there are essentially 3 parameters only -energy, mass ratio and scattering angle- this imposes the need for a collimated, monoenergetic, mass analyzed primary beam achieved by an ion source and a magnetic sector field or a Wien-filter (for details see, e.g., ref.[76]). If time of flight (TOF) measurement is chosen for the energy analysis of the secondary particles, the primary beam has to be pulsed. With TOF usually the variation of the scattering angle is traded off, because flight tubes of a length of about 1 m are cumbersome to move around the target. (For an exemption see, e.g., ref.[37]). Variability of the scattering angle is achieved with electrostatic energy analyzers which are made to rotate around the target. The trade off is then the inability to detect neutral particles.

The energetic and angular resolution is estimated by differentiation of eqs.(1) and (3), which results in

$$\frac{\Delta M_1}{M_1} = \frac{\Delta E_1}{E_1} \frac{A + 1}{2A} \frac{A^2 - \sin^2\theta_1 + \cos\theta_1 \, (A^2 - \sin^2\theta_1)^{1/2}}{A^2 + \sin^2\theta_1 - \cos\theta_1 \, (A^2 - \sin^2\theta_1)^{1/2}}, \tag{40}$$

for the scattered particles, $(A = M_2/M_1)$, and for the recoils

$$\frac{\Delta M_2}{M_2} = \frac{1 + A}{1 - A} \left(\frac{\Delta E_2}{E_2} + 2 \, \tan\theta_2 \, \Delta\theta_2 \right). \tag{41}$$

For many applications an energy resolution of 1 % is sufficient. The angular resolution which has to be matched to the energy resolution is then of the order of 0.5 to 1°. A collimator in front of the electrostatic analyzer or the flight tube is a useful device for that purpose. The same care has to be taken for the angular divergence of the primary beam. Most ion sources in use afford an energy divergence below 1 %. Stability of the source is an important aspect for smooth long time experiments.

A good, reproducible working goniometer for the target is essential;

x, y, z movement and 2 axes of rotation are very important. The preparation of the target, i.e., ion beam polishing and/or cleaning, has to be done over the whole target surface, not just the beam spot, in many cases. Also a surface chemical analysis has to be repeated from different spots on the target. For these purposes the x, y, z motion is needed. For all studies of the surface structure or the position of adsorbates, impact and azimuthal angle variation are essential.

Care has to be taken to allow the combination with other surface analytical tools, e.g., Auger-Electron Spectroscopy (AES), Secondary-Ion Mass Spectrometry (SIMS) or LEED.

3.2 Applications

3.2.1 Chemical Analysis.
Chemical surface analysis by ion scattering (ISS) [13,76], or recoil spectroscopy (ERD, for Energetic Recoil Detection, see, e.g., refs.[36,37]), is the direct application of eqs.(1-3). At a fixed primary energy and a fixed detection angle the energy of the scattered ions or recoils is measured. For the ISS case a scattering angle of 90° is most convenient since eq.(1) reduces to

$$\frac{E}{E_0} = \frac{M_2 - M_1}{M_2 + M_1}, \tag{42}$$

i.e., the estimate of the masses found in the spectra is extremely simple. An example is shown in fig.11, albeit for an angle of 137°.

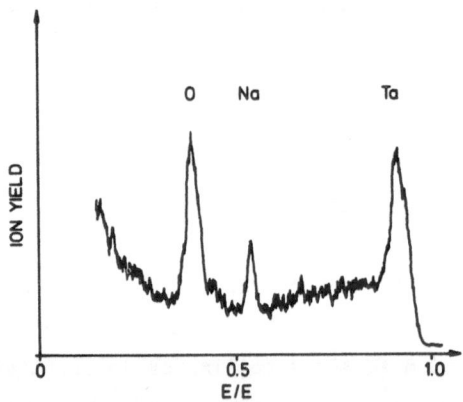

Figure 11. Backscattering spectrum of $^4\mathrm{He}^+$ from a "dirty" Ta_2O_5 surface. E_0 is 1 keV. The analysis is made with a cylindrical mirror analyzer, hence $\Psi = 90^{\circ}$ and $\theta = 137^{\circ}$ (Heiland and Taglauer [76]).

It is worth noting that in most cases the substrate signal decreases with increasing coverage with other species, such that the yield from the substrate can be described by [76,77]

$$I_s = I_0 \, P_s \, (N_s - \alpha N_a) \, \frac{d\sigma_s}{d\Omega} \, \Delta\Omega, \qquad (43)$$

where N_s and N_a are the number densities (cm^{-2}) of substrate and adsorbate atoms, P_s is the ion survival probability and α the shadowing coefficient. Eq.(43) is a direct consequence of the classical shadow cone. It allows an immediate, at least qualitative statement whether any species is adsorbed on the surface or is incorporated into deeper layers. The effect makes ISS an unique tool for the study of catalyst surfaces [78] and alloy surfaces [79-82].

3.2.2 **Application of the Shadow Cone.** The first intuitive application of the shadow cone was the observation [13] that He^+ scattering off CO adsorbed on Mo gave only rise to an O signal. This is in agreement with the Blyholder model [83] of CO adsorption on transition metals, i.e., CO adsorbs linearly with the C end bound to the metal. Consequently, in ISS only an O signal is observed. This finding has been corroborated for the adsorption of CO on W and Ni [84,85]. In the latter study, also the variation of the impact angle Ψ was used to gain further insight into the height of the adsorbate, i.e., its bond length and its position. The technique has been used successfully in numerous cases [86] mostly in

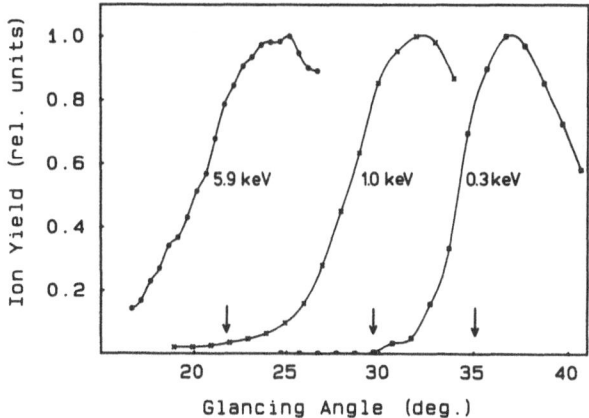

Figure 12. Intensity versus angle of incidence (glancing angle) dependence of K^+ scattered from Ir (110) for three ion energies. The lines are drawn to guide the eye. The arrows mark the critical angles at 0.8 of the peak intensity. All data are taken at a target temperature of 370 K (Hetterich et al.[19]).

agreement with results of comparably complicated techniques like LEED or SEXAFS (Surface Extended X-ray Absorption Fine Structure).

The application of the shadow cone is simplified if the total scattering angle θ is chosen close or equal to 180°, and the backscattering intensity is measured for a given θ as a function of the impact angle Ψ, fig.12 [87]. The technique has been named ICISS [88] (Impact Collision ISS), ALICISS (Alkali ICISS [49,89]) or NICISS (Neutral ICISS [90,91]), which are self explaining acronyms; NICISS implies a TOF technique. The merit of the large angle is that the impact parameter in the second collision of the double collision event is close to zero, hence the analysis of the experimental results is straightforward. What is actually done in this type of experiments is the probing of the shadow cone of one atom by another atom, a nifty application of Lindhard's Gedanken experiment [14].

As pointed out by Aono [88] the technique is self calibrating. Since in most cases some lattice parameters of a given surface are known, the shadow cone can be measured at a few points at least and the shadow cone can be fitted to these points.

These findings give way for further applications beside the measurement of structural parameters: the estimate of potential parameters (fig.13 [87]) and an estimate of the r.m.s. thermal amplitudes of surface atoms [87,92]. Fig.14 [87] shows the influence of the thermal vibrations on the intensity distribution of the shadow cone. It is not necessary to

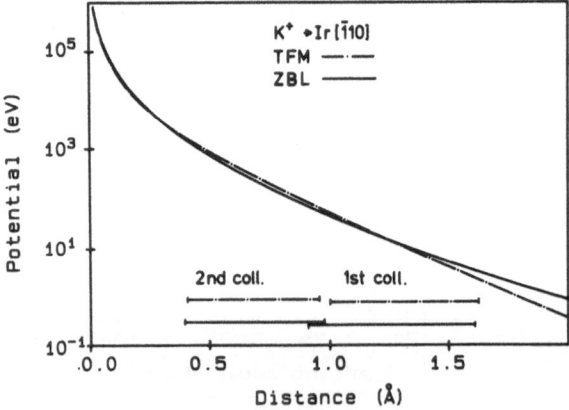

Figure 13. Comparison of the TFM and ZBL potential for K-Ir. The parallel lines (first collision and second collision) indicate the range covered by the two collisions in the experiment. Potential differences are 300 eV at 0.4 Å, 47 eV at 0.9 Å and 5 eV at 1.6 Å (Hetterich et al. [19]).

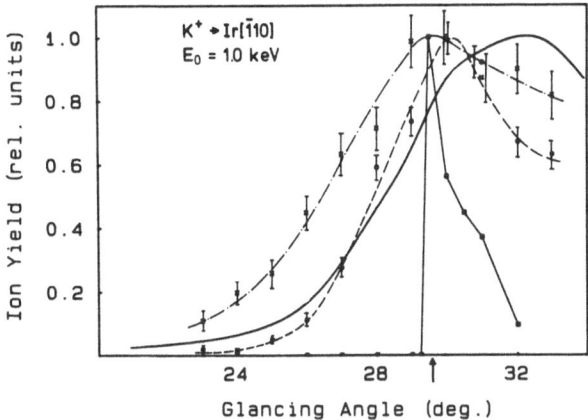

Figure 14. Comparison of experimental I versus Ψ data with MARLOWE results
for "T = 0 K" and T = 370 K, using a target Debye temperature
of 150 K and 280 K. The solid line is the experimental result;
squares: the case of no vibrations. (●) for θ_D = 280 K and (×)
for θ_D = 150 K. The lines are drawn to guide the eye. The bars
are the statistical error (Hetterich et al.[19]).

use a full three dimensional analysis of the data. Shadow cone experiments
can be analyzed by 2D approximations [88-91]. The technique has been
applied successfully to the study of reconstructed clean metal
surfaces [89,93-95], of carbide surfaces [88], of a number of metallic
overlayers on metals [96] and on Si, e.g., Ag on Si(111) [97-100] and Au
on Si(111) [101], adsorbate systems [102,103], and NiAl single crystal
surfaces [104].

3.2.3 Multiple Scattering and Surface Channeling. Multiple scattering
was first reported at rather high energies (50 keV Ar^+ → Cu) [53]. At low
energies it is an obvious effect when changing the primary ion He^+ to Ne^+
or Ar^+ [47]. An excellent review of the "regularities" of these effects is
given by Mashkova and Molchanov [105]. A typical low energy result [106]
is shown in fig.15, i.e., K^+ scattering from Au (110) (1x2) at an energy
of 600 eV, an impact angle of 30° and a scattering angle of 70°. The
experimental data are compared with calculated results using the computer
code MARLOWE [107,108]. The agreement between experiment and calculation
of these triply differential spectra is excellent. The fit includes the
structural data [109] of the two top layers of the surface, the surface
Debye temperature and the scattering potential. The scattering potential
is approximated by the Moliére screening with a reduced Firsov screening
length a = 0.8 a_F. The potential can be checked independently of the

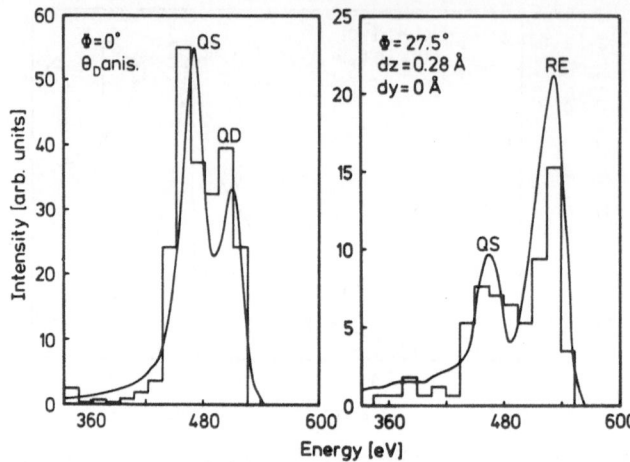

Figure 15. Comparison of experimental spectra for K^+ scattered from Au (110) with MARLOWE calculated spectra. The surface is the (1x2) phase of Au (110). The structure parameters used in the calculation are from Moritz and Wolf [109]. The primary ion is K^+ at E_0 = 600 eV, the impact angle Ψ = 30°, the scattering angle θ_1 = 70°. The spectrum (left) is for scattering along [110], the spectrum (right) is taken 27.5° off this direction (Overbury et al.[106], Hemme et al.[107]).

structure. The surface Debye temperature $\theta_D \cong 110$ K is in good agreement with the theoretical estimates of Jackson [110]. The structural parameters agree with the data from a LEED analysis [109], which were disputed at the time on the basis of electron microscopy [111] and x-ray diffraction results [112].

Other examples of this combination of multiple scattering and computer-aided analysis are found in the literature [15,17,59,113,114]. An important aspect is the finding that single-scattering events and also in-plane scattering sequences ("string" or "chain" effect) have in general a very low probability. Whenever alkali ions or TOF techniques are used such that all particles are detected, the majority of the particles found have rather complicated 3D trajectories.

Another experimental approach to study the structure of single crystal surfaces is surface channeling. There are two approaches in fact, the one is making use of a position-sensitive detector (PSD) (fig.16a [115]), and the other measures the dechanneled intensity at an angle off the specular angle (fig.16b [116]). It is, of course, also feasible to scan a detector over the area covered by the PSD instead [117,118]. The PSD replaces simply the photographic plate used in the very early

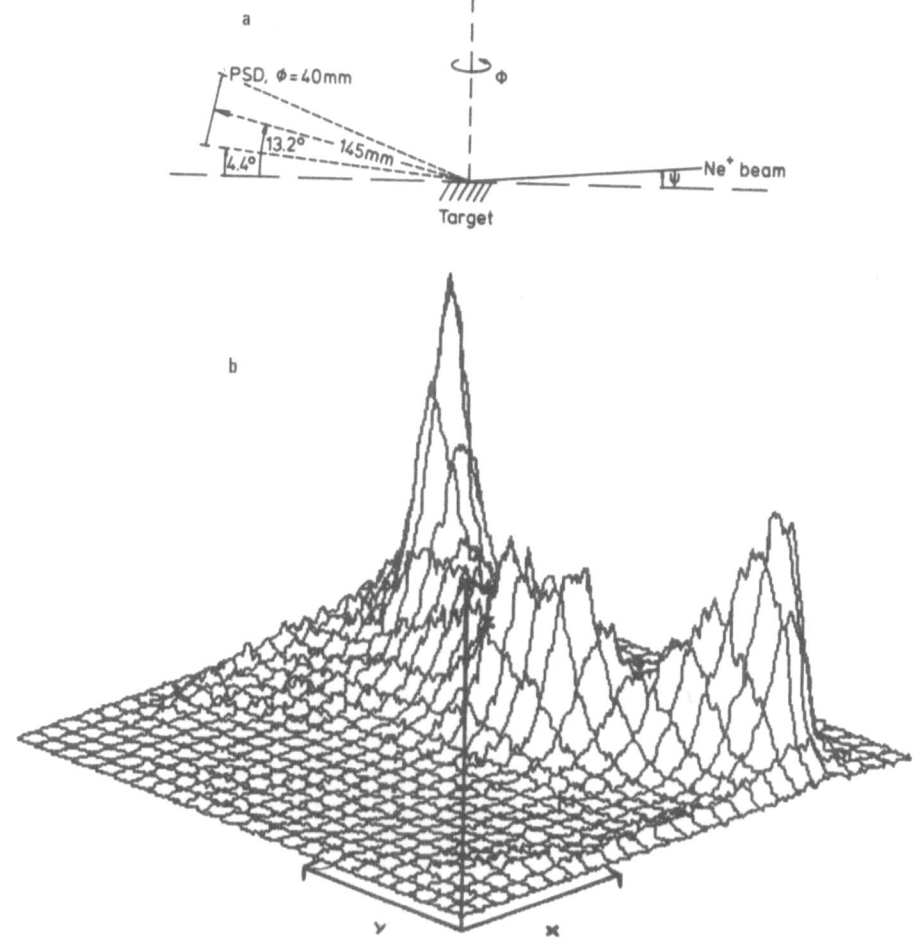

Figure 16 a. Scheme of a surface channeling experiment using a position
sensitive detector (Niehof and Heiland [115]).

b. Surface channeling of Ne$^+$ at 4 KeV, $\Psi = 7^o$ from Ir (110), in
the (1x3) phase (Hetterich and Niehof).

experiments [66]. The pattern observed (fig.16) are "Lambert projections"
of the surface structure [119]. It is obvious to "read" qualitative
information from these patterns. The computer-aided analysis in this type
of experiment benefits from the fact that no particles are lost, i.e., the
"economy" of the code used is better. An energy analysis is not necessary.
The energy spectra are narrow, but slightly asymmetric due to energy loss
straggling (see sect.4).

Even though in the second approach (fig.17), the "surface blocking"
experiment, most of the scattered flux may miss the detector, the results

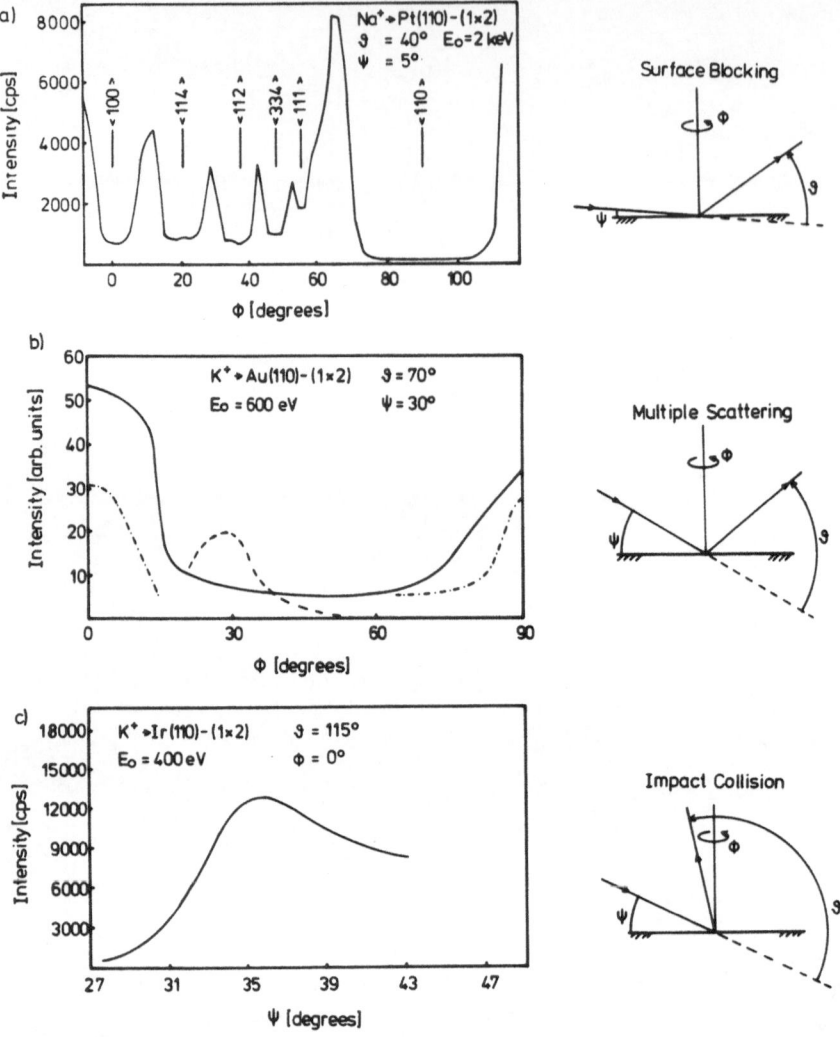

Figure 17A. Schematic representation of a surface blocking experiment (a),
multiple scattering (b), and impact collision experiment (c)
(Heiland [86]).

are qualitatively simple (fig.17). The yield versus azimuthal angle plots
are allow the direct "reading" of the surface periodicity and symmetry, as
does the eyeball analysis of a LEED pattern. A combination of a surface
channeling experiment with an ICISS procedure is, hence, a very powerful
technique to study surface structure, r.m.s. amplitudes of thermal
vibrations, and surface defects. The channeling experiments serve to
characterize the structure qualitatively and to orient the target
quantitatively (!). Any misalignement will cause, e.g., the maxima at each
side of a major surface channel to have different heights. Defects make
the channeling minima shallower. The ICISS can then be used to acquire
quantitative data.

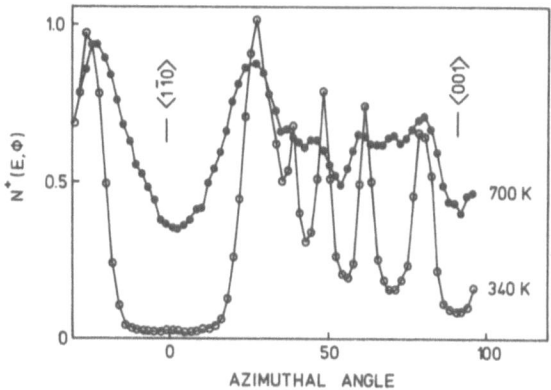

Figure 17B. Surface blocking experiment on Au (110) in the (1x2) phase (open circle) and in the (1x1) phase (filled circles). The ion is Na^+, E_0 = 2 keV, Ψ = 5°, θ_1 = 40° (Derks et al.[116]).

The effect of point defects is demonstrated in fig.18, for the [110] chains of Au (110) (1x2) [120]. With increasing temperature starting at a threshold at 510 K, increasingly more single vacancies are found in the [110] chains. This process initiates the (1x2) = (1x1) phase transition of the Au (110) surface [121].

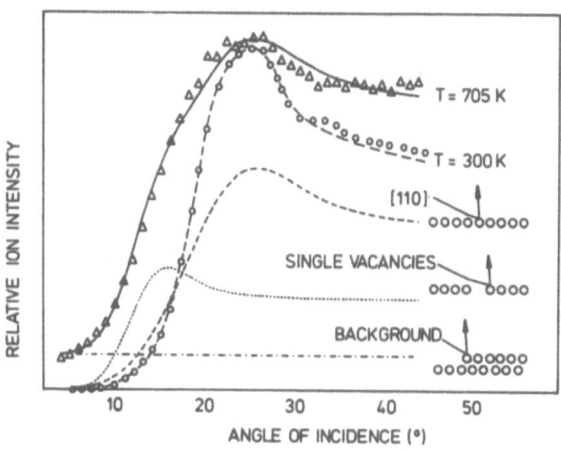

Figure 18. Ion impact collision experiment using 4.6 keV Ne^+ on Au (110) in the (1x2) phase (circles) and the (1x1) phase (triangles). The dashed line is a two-atom model calculation for a perfect [110] chain, the dotted line for a chain with a single vacancy, the dash-dotted line for a step. The (1x2) experiment is fitted with the perfect chain, the (1x1) experiment is fitted with the sum (solid line) of the three contributions, as shown (van de Riet et al.[95]).

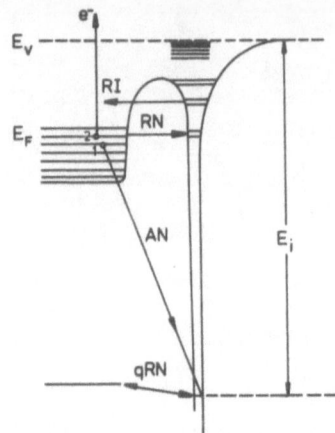

Figure 19. Hagstrum model for the charge exchange between a solid (Fermi energy E_F) and an atom (ionisation energy E_i). The bands in the upper part of the atomic potential well indicate excited states which are broadened near the surface. RI = resonance ionisation (electron loss). RN = resonance neutralization (electron capture). AN = Auger neutralization. qRN = quasiresonant neutralization between core states of the solid state atoms and empty states in the projectile.

4. INELASTIC ION-SURFACE COLLISIONS

4.1 Charge Exchange Processes and Inelastic Energy Loss - a Survey

Charge exchange processes of the primary particle with the surface atoms are the most import inelastic effects in low-energy particle-surface interaction. Most of the pioneering work was done by Hagstrum [12,122], and his terminology will be used to discuss, in a first chapter, the phenomena on a qualitative basis. fig.19 shows schematically the connection between a metal, characterized by its band structure, and a free atom characterized by its electronic energy levels in a Coulomb well. When the atom is close enough to the surface, a variety of processes can occur depending on the initial state of the atom.

If the atom comes in as a singly charged ion, e.g., He$^+$ with a hole at - 25.4 eV, there are three processes possible: i) Auger neutralization (AN), whereby two electrons in the conduction band of the solid are excited by the Coulomb field of the ion, such that one electron falls into the hole of the ion, and the second electron leaves the band with a kinetic energy approximately equal to the potential energy minus its actual binding energy. These electrons, when they are emitted into the vacuum, carry information about the density of states and the band structure of the solid. The information can be extracted from the energy spectra of the electrons. The technique is called ion neutralization spectroscopy (INS). ii) resonance neutralization (RN), whereby one electron tunnels through the barrier between the solid and the ion, into an excited state of the atom. The capture may be followed by an Auger process within the atomic system, which is called Auger deexcitation (AD). The electron can also tunnel back, which is then a resonance ionization (RI), giving a new chance for either AN or RN. The excited state may also survive through the collision and give rise to line emission, when decaying according to the rules of atomic physics. iii) the core hole in the approaching ion may also be filled by a resonant process involving core states of a surface atom. The process has been found in surface physics in He^+ - Pb scattering by Erickson and Smith [123]. It is basically a Stueckelberg-type charge exchange process. The role of the valence-band electrons in this exchange is not yet understood. The process is called quasi-resonant neutralization (qRN).

If the atom is a metastable neutral atom, the charge exchange process may start with a RI process or with an AN process, where however, in this case, the electron will be emitted from the atom. After the initial RI process the particle is a singly charged ion, as discussed above, and any sequence of events will start as discussed. The energy analysis of the emitted electrons allows also the evaluation of electronic surface properties (MDS = metastable deexcitation spectroscopy). It is a useful technique to study adsorbates, because the interaction with the substrate band electrons is reduced.

The processes discussed so far can be treated in the "adiabatic limit", which implies, in a first approximation, very low energies or velocities. In that case, the charge exchange processes are dominated by matrix elements describing the tunneling process, i.e., exponentials. The ion survival probability is then given by

$$P_i = \exp(- v_c/v_\perp),\tag{44}$$

where v_c is a constant of the order to 10^7 cm/s, and $v_\perp$ the velocity component of the ion perpendicular to the surface. More sophisticated theoretical treatments take into account the level shifts and broadenings of the atomic level due to the image potential and/or the screening by the solid-state electrons. The important result is that, up into the keV region, the Hagstrum formula is in some cases a good approximation.

In general, however, with increasing ion velocity dynamic effects, and effects due to the increasing overlap of the electronic shells, become more important and hence the adiabatic treatment insufficient. Clear evidence for the onset of such processes is the observation of kinetic electron emission [124]. Electrons emitted at low ion energies are due to the potential energy of the ion only, as discussed in connection with INS and MSD. With increasing energy, processes known from atomic collision physics become important, i.e., Pauli excitation or ionization, and also dynamic resonant and Auger processes. The dynamic interaction of the Coulomb field with the periodic potential of the solid is seen as forces acting on the electrons leading to excitation and ionization, or simply electron-hole pair creation. The moving charge couples also to the eigen-frequencies of the solid-state plasma, thereby exciting plasmons.

The problem of the collisional excitation or ionization can still be handled within the framework of the Hagstrum model. By introducing a ionization probability P_{coll}, the final probability may be written as

$$P = P_{in} \, P_{coll} \, P_{out}, \tag{45}$$

where P_{in} and P_{out} are terms of the form of eq. 44. By an analogous procedure, approximate approaches to tackle the problem in multiple scattering are occasionally successful in describing the overall ion survival probability [48,125-128], i.e.,

$$P = P_1 \, P_2 \, \cdots, \tag{46}$$

where P_i is now evaluated as a function of the distance, which is known in principle from the trajectory. The model deviates from the original Hagstrum model where the surfaces has no structure in x-y direction, if z is normal to the surface. In eq. (46) it is tacitly assumed that the charge exchange occurs at the distance of closest approach between the ion and a surface atom, i.e., the problem is handled more like the charge exchange in the scanning tunneling microscope (STM). The dependence on z is already found in Hagstrum's papers, in the ansatz for the transition rate for the electron

$$R(z) = A \exp(-az), \tag{47}$$

where A is the maximum transition rate and a parameter with the character of a screening length (inverse of). Note that $A/a = v_c$ in eq. 44. From eq. (47), the ion survival probability is

$$P = \exp\left\{ -\frac{A}{a} \cdot \frac{1}{v} \exp(-az_0) \right\} \tag{48}$$

where $z_0 = r_{min}$ is the distance of closest approach.

The next useful - in some cases - correction to the Hagstrum model is the consideration of the parallel velocity component of the ion; in fact, for studies at grazing incidence the correction by the Galilean transformation is simply a must [129-131]. However, whenever the parallel velocity component is large enough to ask for the Galileo transformation, it may be also large enough to cause enough electronic excitation such that the adiabatic concept underlying the Hagstrum model breaks down.

A way to check the limits of the adiabatic approximation is to use the Massey parameter, which provides an estimate. If there are two electronic states n and m with an energy difference ΔU_{nm}, the relative velocity of motion of the nuclei to which the states n and m belong is v, and if there is a characteristic interaction length a, e.g., the collision diameter, then

$$\frac{\Delta U_{nm} \, a}{hv} \gg 1 \tag{49}$$

is the condition for a high survival probability of the initial state m (Nikitin [132]). The energy terms are adiabatic and so are the wave functions entering in this estimate. Also, of course, the basic assumption is made that the electronic motion (fast) and the nuclear motion (slow) can be dealt with separately (Born-Oppenheimer approximation). Considering the case of electron-hole pair excitations at the Fermi edge of metal, eq.(49) shows that not very high velocities are needed to make the Massey parameter approach unity.

In a recent application of energy loss and charge exchange theories [133-136] it was shown that the neutralization of, e.g., He^+ at grazing scattering condition, can be described by

$$\frac{d}{dt} \, n(He^+) = - \frac{1}{\tau^A} \, n(He^+) + \frac{1}{\tau^R} \left[1 - n(He^+) \right] \tag{50}$$

where n is the number of particles in the beam, τ^A and τ^R are the Auger and Resonance lifetimes respectively, i.e., $1/\tau^A$ is the probability per unit time for an Auger process to happen. At low energy the resonance process is rather unlikely (Hagstrum), such that the neutralization can be estimated from the Auger process only

$$\frac{1}{\tau^A} = \int_{-\infty}^{+a} \frac{d^3 q}{(2\pi)^3} \int_0^\infty \frac{d\omega}{(2\pi)} \sum_k \frac{16 \, \pi^2}{q^2} \, \delta(\omega + E_k - q \, v - E_0)$$

$$\times \, Im(- \frac{1}{\varepsilon(q,\omega)}) \, \left| <k|\exp(-iqr)|u_0(r)> \right|^2, \tag{51}$$

283

(in atomic units), where ω and k are the energy and momentum of the excitation in the solid, E_k and $|k>$ the energy and wavefunction of the metal electron filling the atomic level E_0, and u_0 is the wavefunction of the atomic level. It is worth noting that the Galileo transformation is

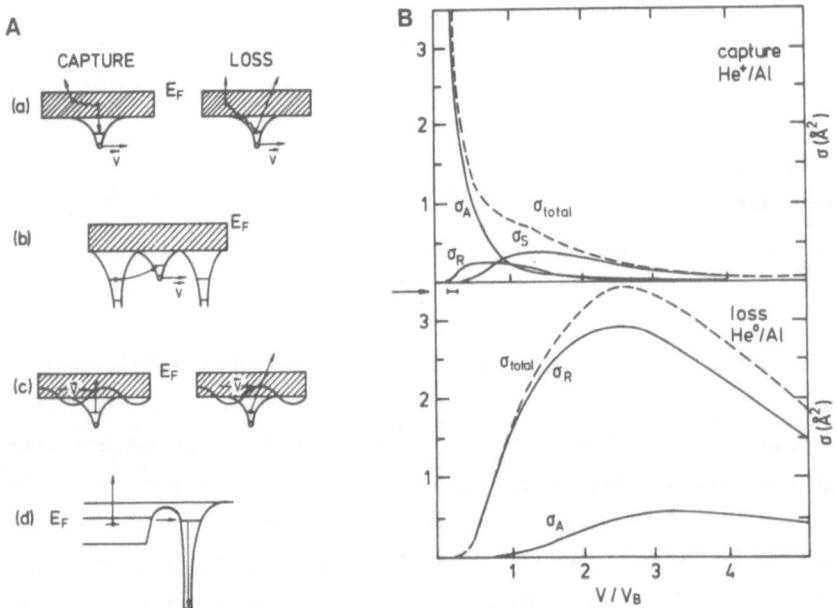

Figure 20A. Schematic representation of the different capture and loss processes considered. a) Auger processes, b) inner shell capture, c) dynamic resonant processes: the crystal potential is represented by the sinusoidal line; d) resonant neutralization followed by Auger deexcitation. (Sols and Närmann [137,154]).

B. Comparison of the cross sections of the different capture and loss processes as a function of velocity. σ_A: Auger process. σ_R: Dynamic resonant process. σ_S: Inner shell capture. σ_{total}: Sum of all cross sections.

included. More important is the truly dynamic description of the process, which allows to make use of the most advanced theory of the electronic stopping which includes the mechanism discussed here. The relative importance of the Auger process, compared to other capture and loss mechanisms, can be seen in fig.20 [134,137].

An energy spectrum of neutralized particles is then given by

$$\frac{d}{dQ}\, n(\text{He}^0) \propto \exp\left\{-\frac{Q - Q_0}{(\gamma_s^+ - \gamma_s^0)v\,\lambda^A}\right\}\left\{1 + \frac{2}{\sqrt{\pi}}\int_0^y e^{-t^2}\,dt\right\} \tag{52}$$

where γ_s^+ and γ_s^0 are the surface friction coefficients of the ion and the neutral, respectively, Q_0 is the energy loss after the whole scattering process, and $\lambda^A = v\,\tau^A$. The second term describes the energy loss straggling [138], with

$$y = \frac{1}{\sqrt{2}}\left(\frac{Q - Q_0}{\Omega} - \frac{\Omega}{(\gamma_s^+ - \gamma_s^0)\,v\,\lambda^A}\right), \tag{52a}$$

where Ω/v^2 is the straggling parameter. The energy loss Q_0 is defined by

$$Q_0(x) = \int_x^L \gamma_s^0\, v\, dl, \tag{53}$$

for a total trajectory length L. Correspondingly,

$$Q_+(x) = \int_x^L \gamma_s^+\, v\, dl. \tag{54}$$

Q in eq.(51) is $Q \in [Q_0,\ Q_+]$, and hence Q_0 in eq.(51) is the loss of a

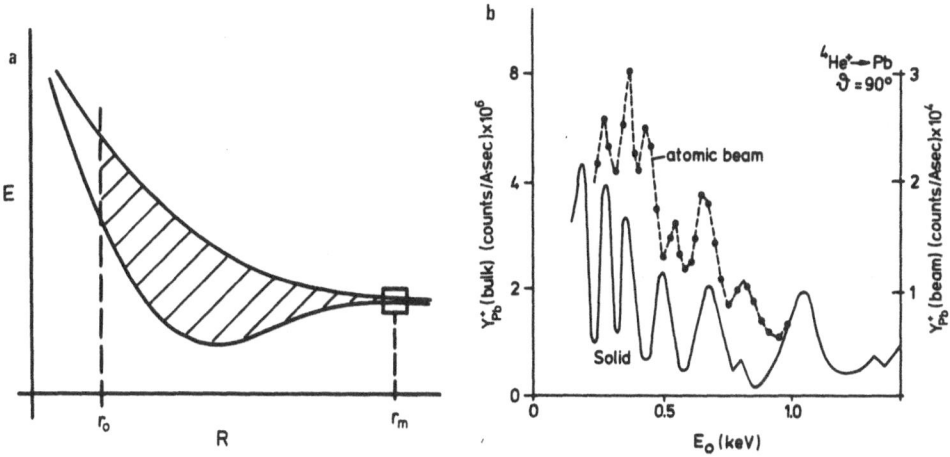

Figure 21a. Schematic representation of the interaction potential in case
of a quasi resonant charge exchange between two atoms. r_0 is
the distance of closest approach, r_m the mixing distance for
the electronic wave functions involved in the collision.

b. He$^+$ scattering from polycrystalline Pb and an atomic Pb beam,
as a function of the primary energy (Zartner et al.[140]).

particle being neutralized at the beginning of the trajectory and that is never re-ionized. Recent results from Xu and Sullivan support the theory discussed here [139].

Depending on the type of atom and its velocity, for the study of excited states the resonant capture and loss processes are not negligible. At velocities above the Bohr velocity also the ionization of "core states" starts to contribute to the loss processes, fig.20 (σ^A in the loss world).

The qRN channel, fig.19, can be handled in a good approximation as a pure ion-atom collision process [140-142]. The charge exchange occurs between two atomic terms, e.g., He^+ - Pb, between the He 1s and the Pb 5d levels. The phenomenon is well known in atomic collision physics and is understood to be caused by a phase interference of the electron wavefunction which develops either on the He^+ - Pb or on the He - Pb^+ potential, fig.21a. The signature of the effect are oscillations with frequency ω of the ion yield, as a function of the velocity described by the ion yield, fig 21b.

$$P \propto A_1 + A_2 \sin^2(A_3), \tag{55}$$

where A_1 and A_2 are monotonous functions of v. A_1 is for the RN + AD neutralization channel and A_2 is for the AN-channel.

$$A_3 = \frac{1}{hv} \int_{r_0}^{\infty} \Delta E(R)\ dR, \tag{56}$$

where ΔE is the energy difference between the two potential curves, see fig.21. r_0 is the distance of closest approach. eq.(55) neglects the possible influences of the angular momenta (quantum number l) and of the m substates. Of interest are the selection rules for m, which are $\Delta m = 0$ for radial coupling and $\Delta m = \pm 1$ for rotational coupling [142]. Radial coupling is occurring in collisions at large impact parameters, whereas rotational coupling comes into play when during the collisions appreciable angular momentum is at work.

4.2.1 Charge Exchange Experiments. Out of the multitude of ion yield measurements, we show in fig.22 double differential ion yields of He^+ and Li^+ backscattered from Ni (110) [15]. The Li^+ data follow exactly the differential scattering cross section calculated from a Thomas-Fermi-Moliére potential. If the He^+ data are divided by $\exp(-v_c/v_\perp)$, with $v_c = 1.78 \cdot 10^7$ m/s, they fit to the $d\sigma/d\Omega$ values. There are a number of ion-surface combination showing the AN behavior, as listed in table 1. There are also some combinations which show a non-monotonous

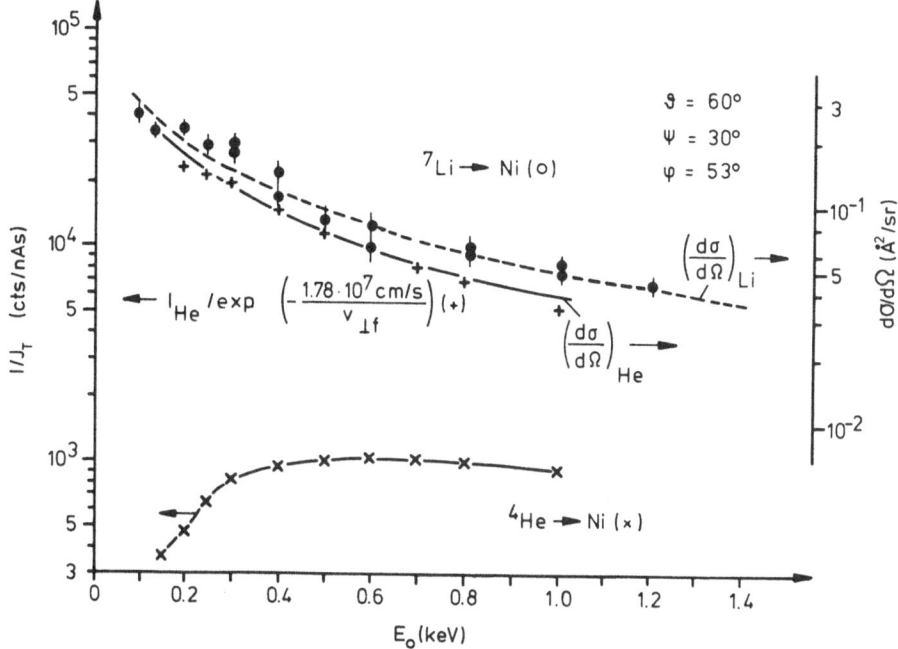

Figure 22. Energy dependence of the He^+ and Li^+ intensities (left ordinate), compared with the calculated differential scattering cross sections for single scattering (right ordinate). The exponential yields P = 0.11 for He^+ at 600 eV, independent of the azimuthal angle. $v_\perp$ is the normal velocity component of the scattered He^+ ions (Taglauer et al.[15]).

v-dependence, and finally the group of nine materials in table 1 show clear evidence for the qRN charge exchange [143,144].

Fig.23 shows the oscillation in the ion yield, for different directions on two Pb single-crystal surfaces [145]. The latter results can be understood in terms of the selection rules for m, in case of rotational coupling, and in terms of the orientation of the t_{2g} and e_g orbitals (Henck and West [146], Wedler [147]) on a single-crystal surface. Since $\Delta m = \pm 1$, the m = 2 substates of the Pb 5d states can not contribute to the charge exchange, that is, whatever the orientation of e_g or t_{2g} might be, the oscillation must show an azimuthal dependence. For the two cases shown in fig.23, the orientations of the orbitals is such for Pb (110) and [001] that only t_{2g} is seen by the He, it is the one t_{2g} orbital perpendicular to the surface. For Pb (111) and [121], the He particles pass through one e_g and one t_{2g} orbital. These orbitals are oriented 35.3^o and 54.7^o, respectively, with respect to the (111) plane. The trajectory is 45° to the plane, on the way in and out. It should be noted that the

Table 1. Schematic classification of ion-neutralization behavior.

Class I: $p \sim \exp(-v_c/v_\perp)$ is a good approximation for the ion survival

He$^+$	Ne$^+$	Ar$^+$
Al, Si, Ni	Zn, Sb, Te	
Cu, Zn, Zr	W	
Nb, Pd, Ag		
Cd, Ta, Pb, Au		

Class II: $P \sim A_1 + A_2 \sin^3 A_3$ is a good approximation

He$^+$	Ne$^+$	Ar$^+$
Ga, Ge, As	Ga	
In, Sn, Sb		
Tl, Pb, Bi		

Class III: others

He$^+$	Ne$^+$	Ar$^+$
S, Sc, Ti		
Nb, Mo, Te	Ti, Cd, La	In, Tl, Pb
La, Ce, Nd	Ce, Sm	Ag, Cd, Su
Sm, Gd, Dy	Gd, Dy, Er	Sb, Te, La
Er, Yb, Hf	Yb, Hf, Ta	Ce, Nd, Sm
Ta, W	W, Tl	Gd, Dy, Er, Yb, Hf, Pt, Au

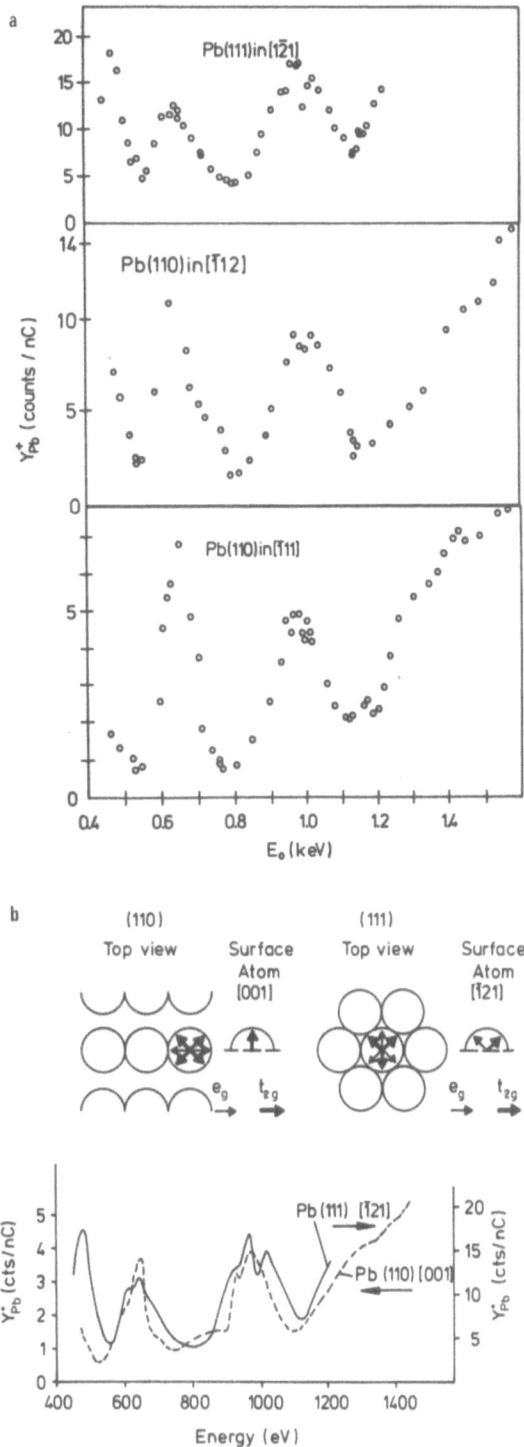

Figure 23a. Ion yield of He[+] backscattered from Pb (111) and Pb (110), for three orientations of the plane of scattering, i.e., parallel to [121] and parallel to [111] and [112]. The abscissa is the primary kinetic energy of the ions (Oelschig [145]).

b. The data for [121] and [111] as in (a) (Oelschig [145]).

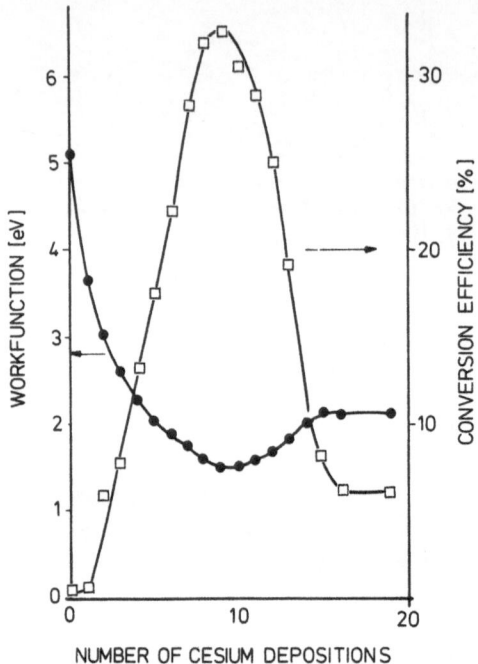

Figure 24. Work function change of W covered with Cs, and the change of the H^- yield, for H^+ incident at 400 eV with $\Psi = 8°$. One layer of Cs deposition is about 0.35×10^{14} atoms/cm^2. (The conversion efficiency is the total reflected H^- current divided by the incident H^+ current) (van Wunnik et al.[130,131]).

ion yield is, for all azimuthal orientation, higher from (111) than from (110). It seems plausible to assume that the one t_{2g} orbital perpendicular to (110) is the cause. There is no orbital perpendicular to (111), nor is any orbital oriented under 45° to the surface.

Clear evidence of resonance capture has been found, e.g., for negative H formation [129-131], and for the neutralization of alkali ions [148], in both cases at grazing incidence. By changing the work function of the surface, it is possible to "tune" the position of the Fermi edge such that it matches the energy level of the atomic state in question. Under the circumstances of the experiments, the Galileo transformation has to be taken into account. Fig.24 gives an example for work function dependence of the H^- formation [130,131].

4.2.2. Charge Exchange and Inelastic Loss. The inelastic loss in low-energy scattering experiments has been studied for large-angle scattering experiments [149-153] and more recently in greater detail for grazing scattering [154-157]. As mentioned above, the losses are small in

290

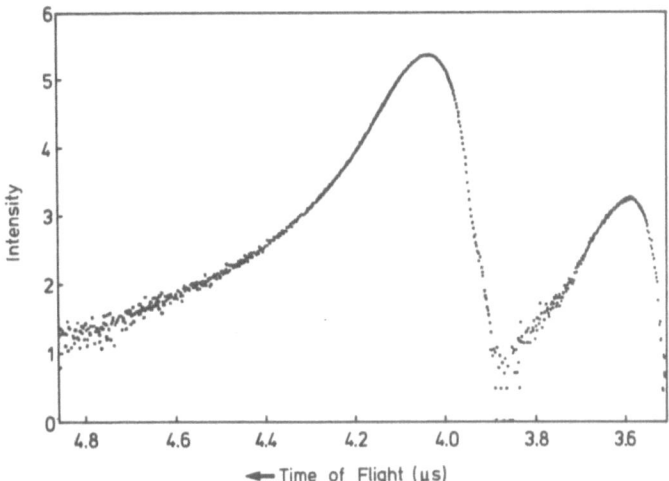

Figure 25. He^0 and He^+ time of flight spectrum for He^+ incident on Ni (110) with E_0 = 2 KeV, Ψ = 5°, θ_1 = 10° (Närmann [154]).

general and only in a few cases distinct losses have been observed being due to projectile excitation [149,153]. The grazing incidence experiments afford better comparison with theory because the elastic losses can be neglected. For example, for 2 keV He scattered into a total scattering angle of 10° from a metal, the elastic loss estimated from eq.(1) is 4 eV only in a single scattering. However at an impact angle of 5° channeling

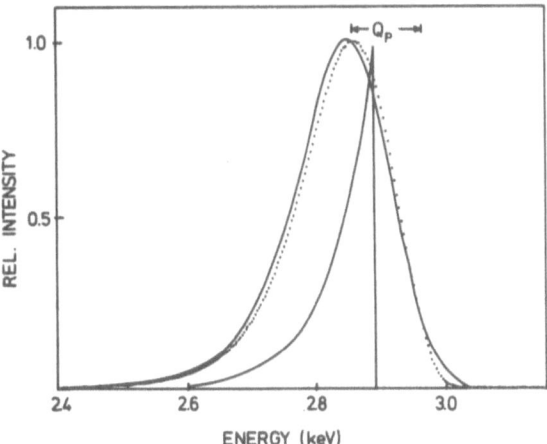

Figure 26. Comparison of a 3 keV He^0 energy spectrum (dots) (Ψ = 5°, θ = 10°) for scattering from Ni (110), with theoretical estimates without straggling (peaked solid line) and with "gaussian" straggling (solid line) (Monreal et al.[135]).

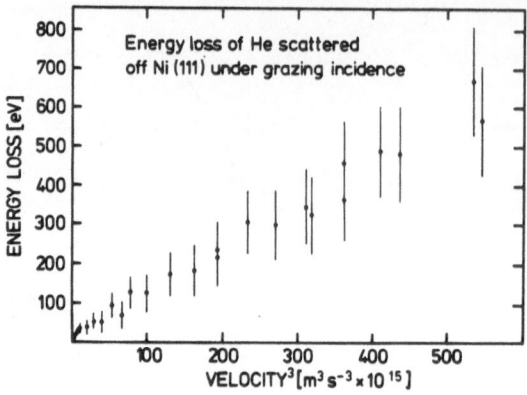

Figure 27. The peak energy loss for He^0 plotted versus v^3 (Närmann et al.[157]).

conditions prevail and the loss estimated from eq.(1) is an upper limit. Fig.25 shows TOF spectra from He scattered from Ni under channeling conditions. The majority of the scattered particles is indeed neutral, i.e., the ion fraction is very low, in agreement with estimates from eq.(51). Fig.26 shows the comparison of the experimental energy loss spectrum with calculated spectra using eq.(52), and a gaussian approximation for the straggling. Further predictions of the theory are proven by the energy loss dependence on the energy, fig.27. The experiment shows that $Q \propto v^3$, which is based on the finding that the trajectory length L increases with v, and that the friction coefficient γ_s is

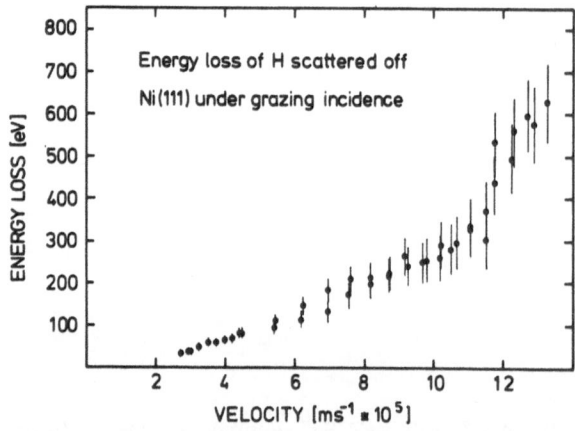

Figure 28. The peak energy loss for H^0 plotted versus v (Närmann et al.[157]).

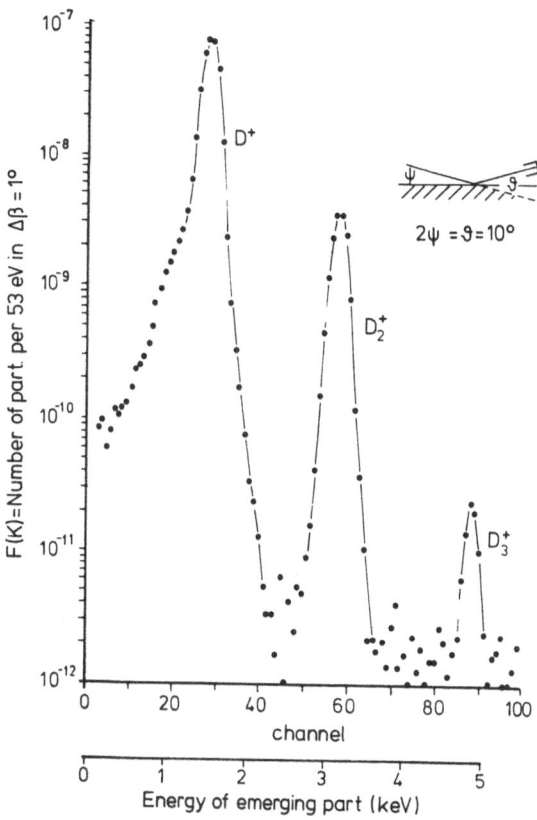

Figure 29. Energy spectra of D_3^+ and its dissociation products D_2^+ and D^+, incident at $E_0 = 4.92$ keV on Au, $\Psi = 5^o$, $\theta = 10^o$. $\Delta\beta$ is the angle of acceptance of the detector (Eckstein et al. [159]).

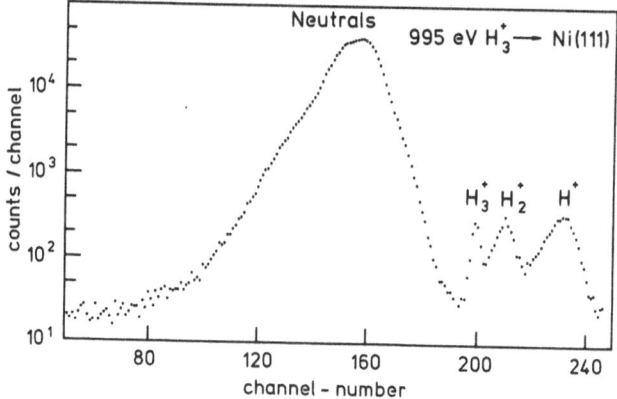

Figure 30. Time of flight spectra of He^+ and of its ionic and neutral dissociation products, $E_0 = 495$ eV, $\Psi = 5^o$, $\theta = 10^o$ (Willerding et al. [160]).

apparently velocity dependent, i.e., $Q = v\gamma_s L$. The velocity dependence of γ_s can be read from eq.(52), i.e., with increasing velocity the survival of He^+ along the surface will increase, thereby γ_s^+ contributes more and more to the total loss.

Using hydrogen, a different behavior is observed, i.e., $Q \sim v$ (fig.28). In this case, the dominant charge state at low energy is not the neutral state but H^-. With increasing velocity H^o takes over, and the total loss increases much less than in the He case, because the neutrals experience the lower energy losses. At about 8 keV, the surface channeling condition is no longer valid, and the particle are scattered increasingly through the bulk.

5. THE INTERACTION OF MOLECULAR IONS WITH SURFACES

5.1 A Survey

After some preliminary experiments by Panin [158] and Smith [13], the first experiment casting some light on the physics involved in the fast molecule-surface interaction was performed by Eckstein et al.[159] (fig.29). The binding energy of D_3^+ and D_2^+ is much smaller than the energy loss of the surviving D_2^+ and D_3^+, but the loss is mainly inelastic. The question arises whence the dissociation comes about: charge exchange or vibrational-rotational excitation?. As Eckstein et al pointed out - based on the lore of charge fraction experiments - there has to be neutralization. The next experiments with H_3^+ gave indeed ample evidence for neutralization, fig.30 [160]. The ion peaks are "pulled out", in this TOF experiment, by post acceleration. The problem with the neutral peak is that it probably contains H_2 and H (no H_3 because it is not stable). Without re-ionization, the two neutral species can not be told apart in a TOF experiment.

The problem can be overcome by using diatomic molecules. With H_2^+ at sufficiently low energy, the ion yield becomes very low and the neutral spectrum shows two components (fig.31 [161]), i.e., a narrow distribution peaking somewhat below the primary energy, and a broad distribution extending above the primary energy. The narrow peak can be identified as neutral H_2 by its energy position, but also by comparison with H scattering at half the energy. The broad distribution is dissociated H, and the extension above the primary energy is due to the transformation of the center of mass to the laboratory system,

$$E = \frac{1}{2} E_0 + \frac{1}{2} E_D \pm \sqrt{E_0 E_D} \cos \alpha, \qquad (57)$$

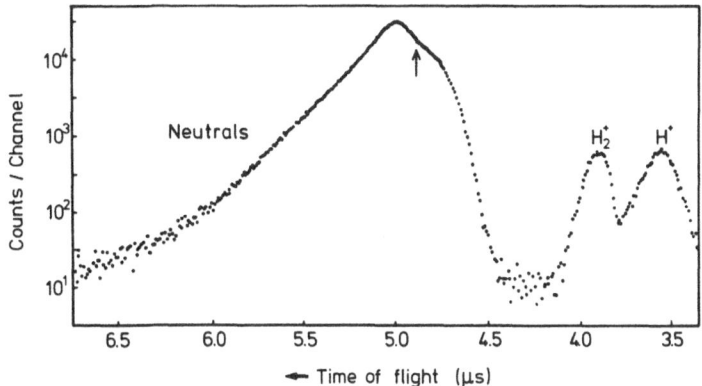

Figure 31. Time of flight spectra of H_2^+ and of its dissociation products from Al (111), for $E_0 = 200$ eV (Imke [161]).

where E_D is the dissociation energy and α the angle of the molecular axis with respect to the beam direction. Calculations using the MARLOWE code, gave good agreement with the assumption that the dissociation occurs with

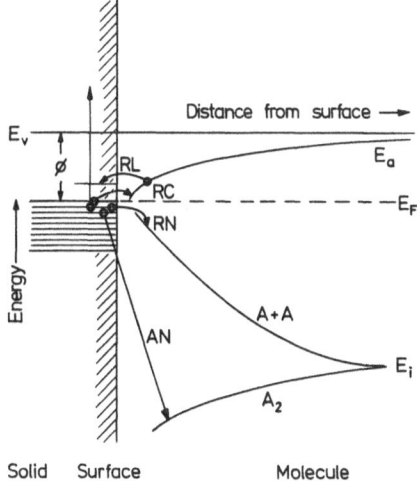

Figure 32. Hagstrum model for the charge exchange between a solid and a diatomic molecule A_2 (compare with fig.19). The capture (RC) and loss (RL) processes involving the affinity level E_a, i.e., the formation of A_2^- by resonant capture (RC), are formally distinguished from RN, which is also a capture process. RN will in most cases lead to dissociation, AN, to ground-state neutral molecules.

a distribution of dissociation energies corresponding to a repulsive interaction potential [162]. The electron capture would cause a Franck-Condon transition from the initial ionic state into either the ground state or the antibonding state of the molecule, fig.32. This figure is the basic scheme of the dissociation by electron capture.

Parilis and Bitenski [163], Jakas and Harrison [164] and van den Hoek et al.[165,166], made calculations for collisional dissociation. The model is based on classical scattering, and dissociation occurs when the relative kinetic energy between the two atoms of a diatomic molecule

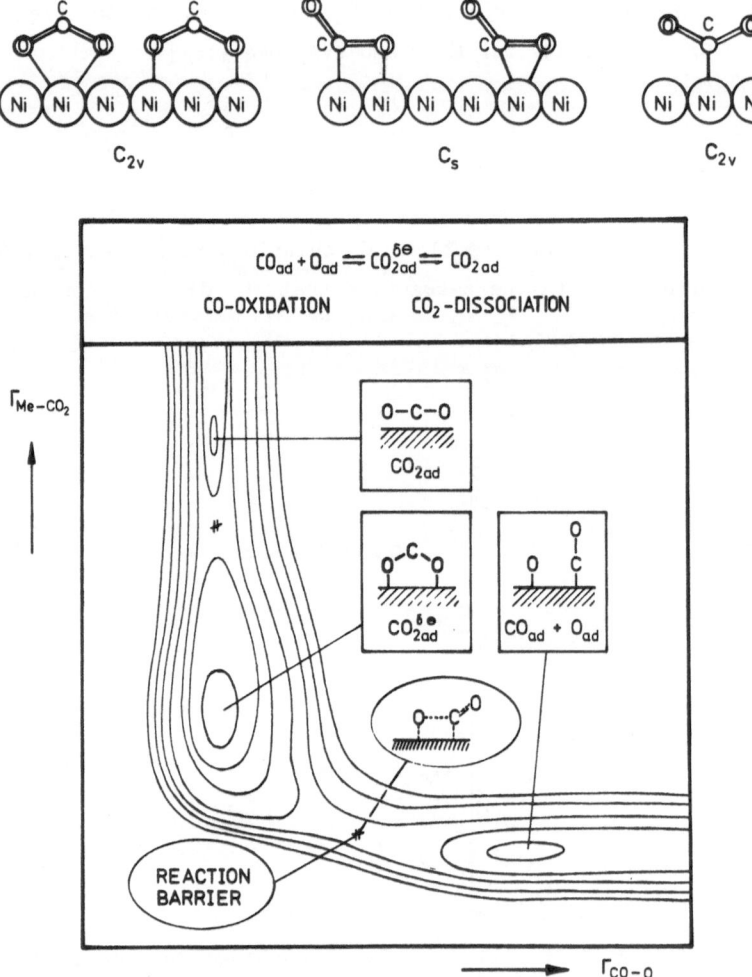

Figure 33. Possible coordination sites of $CO_2^{\delta-}$, together with the corresponding point group symmetries (top). Schematic two-dimensional potential energy diagram for CO_2 metal interaction (vertical axis), and CO-O dissociation (horizontal axis) (Bartos et al.[175]).

becomes larger than their binding energy. This model gives satisfactory agreement with dissociation experiments in the 20 keV region of a Russian group (see ref.[105]).

The case of the electronic dissociation has been further developed by Imke et al.[167,168], for the case of hydrogen and oxygen. By taking into account the total symmetry of the system it can be shown that the orientation of the molecular axis plays a role in the capture process. These symmetry rules have been elaborated previously by Dunn [169], for gas-phase electron capture into molecules. In surface experiments, the validity of these rules have been demonstrated by comparing the scattering of N_2 and O_2 (Schubert et al.[170]), and by experiments concerning the formation of C^- and O^- (van Pinxteren et al.[171]).

The formation of negative molecules has been shown to play a major role in molecule-surface interactions experimentally [172-180]. In comparison with corresponding gas-phase experiments, the term "harpooning" is used to name this type of reaction. First experimental evidence for the reaction to occur in molecule-surface interaction was found by Haochang et

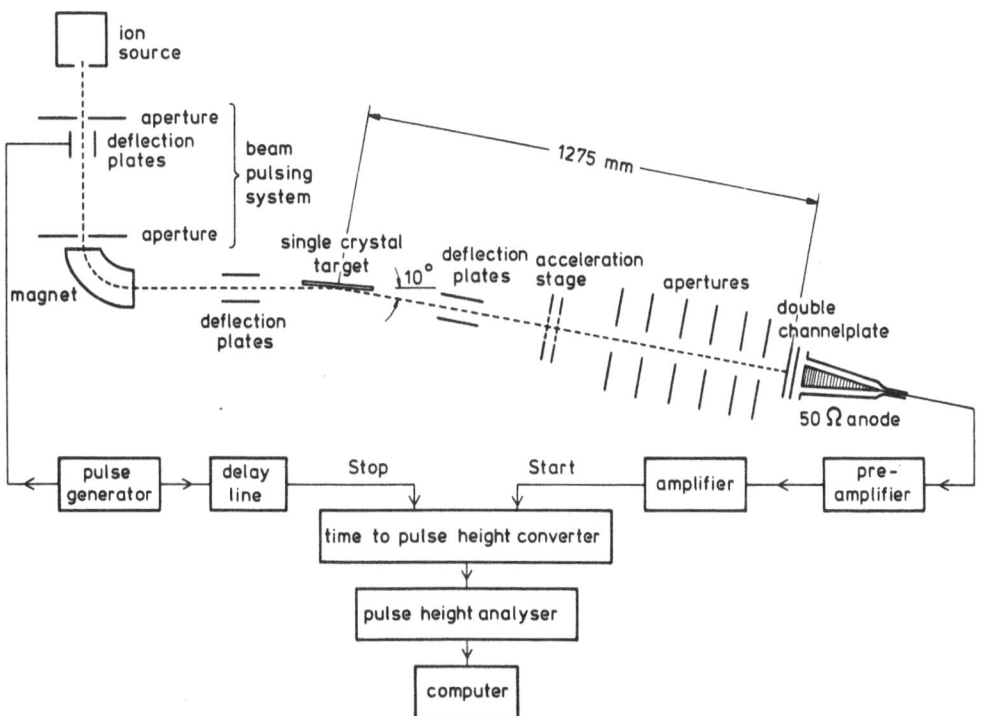

Figure 34. Scheme of a TOF ion scattering system (Willerding et al.[188]).

al.[186] ($O_2^+ \rightarrow$ Ag), and by Dannon and Amirov [187] ($I_2 \rightarrow$ diamond).

The relevance of the negative state may be estimated from fig.33, which shows that in the "classical" reaction $CO + O = CO_2$ on a surface CO_2^- is the all important intermediate step for the reaction [174]. A similar role is played by O_2^- in the dissociation of O_2, after the physisorption of O_2 on metals like Ag and Ni, which then leads to chemisorption and oxide formation.

5.2 Experimental Results from Molecule-Surface Interactions

Fig.34 shows the experimental setup for the Osnabruck molecule-surface interaction experiments [188]. It is a UHV TOF system, which has been used for the energy loss experiments too (see sect.4.2.4). Fig.35 is a collection of spectra for the primary ions N_2^+, CO_2^+, O_2^+ and CO_2^+. Experimentally, the O_2^+ is the worst ion to work with. The cathode in the Wittmaack ion source used in the experiment has a few hours lifetime only with O_2 as a gas, in comparison to several weeks with rare gases or even N_2. At the primary energy used for the data [189] of fig.35, i.e.,

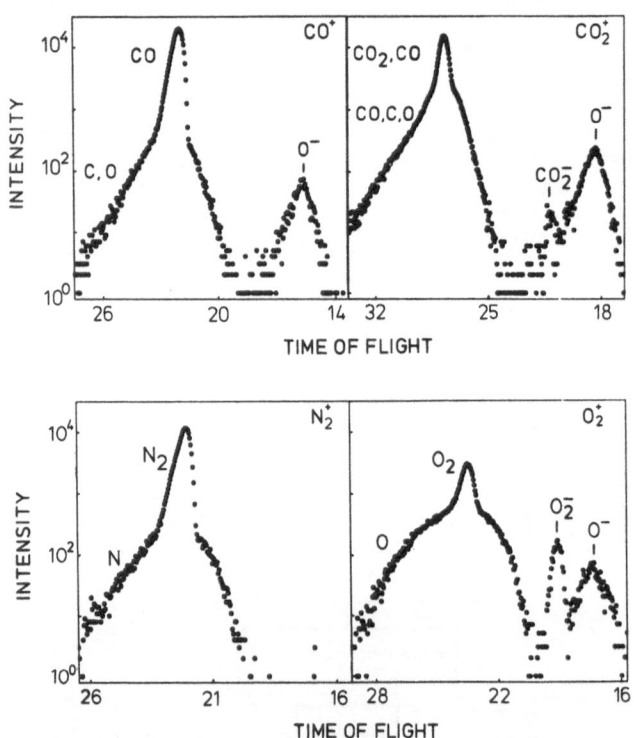

Figure 35. TOF spectra for CO^+, CO_2^+, N_2^+ and O_2^+ incident on Ni (111), $E_0 =$ 400 eV, $\Psi = 5^\circ$, $\theta = 10^\circ$ (Schubert et al.[189]).

480 eV, there are no positive ions, that is, the positive ion yield is certainly below 10^{-4}. The most interesting features are the negative ions, which can partly be identified from their energy (or time) position in the spectra. In case of the dissociation of C and O-containing molecules, where both C^- and O^- can occur, the time resolution is not sufficient to separate the two species. In case of CO and CO_2 , the neutral "broad" peak may also contain different species which cannot be separated as in the case of H_3^+ (fig.30).

The observation of any O_2^- or CO_2^- ions is already enough evidence to establish the usefulness of these beam experiment to study intermediate states of surface chemical reactions [186,188,189]. Further proof was obtained in a series of experiments, where the work function of the surface was varied by K adsorption (fig.36).

To discuss the results we use the O_2^+ case and distinguish two cases: Auger Neutralization (AN) is the fastest and dominating process, and b) resonant neutralization into a dissociative triplet state occurs as the first ion-surface interaction. Hence we have the following set of possible reactions,

a.1) $O_2^+ + e^-$ (AN) $\to O_2$

a.2) $O_2 + e^-$ (RC) $\to O_2$ (resonant capture)

a.3) $O_2^{\delta-} - e^-$ (RL) $\to O_2$ (resonant loss)

a.4) $O_2^{\delta-}$ diss $O^- + O$

a.5) $O^- - e^-$ (RL) $\to O$

a.6) $O + e^-$ (RC) $\to O^-$.

The possibility of charge exchange processes (a.5,a.6), and of the dissociation (a.4), are a complication in the analysis of the experimental results. In case of a resonant transition into a dissociative state we have three possible "reactions",

b.1) $O_2^+ + e^-$ (RN) $\to O + O$

b.2) $O + e^-$ (RC) $\to O^-$

b.3) $O^- - e^-$ (RL) $\to O$.

If the work function is changed, the negative ion yield is expected to increase, as it occurs with atomic ions (see sect.4.2.1.). This is indeed the case for oxygen, as shown in fig.30. It is seen that all O^- signals increase the same way, whether the original particles are O^+, CO^+, CO_2^+ or O_2^+. This means that the formation of O^- is decided very late in the

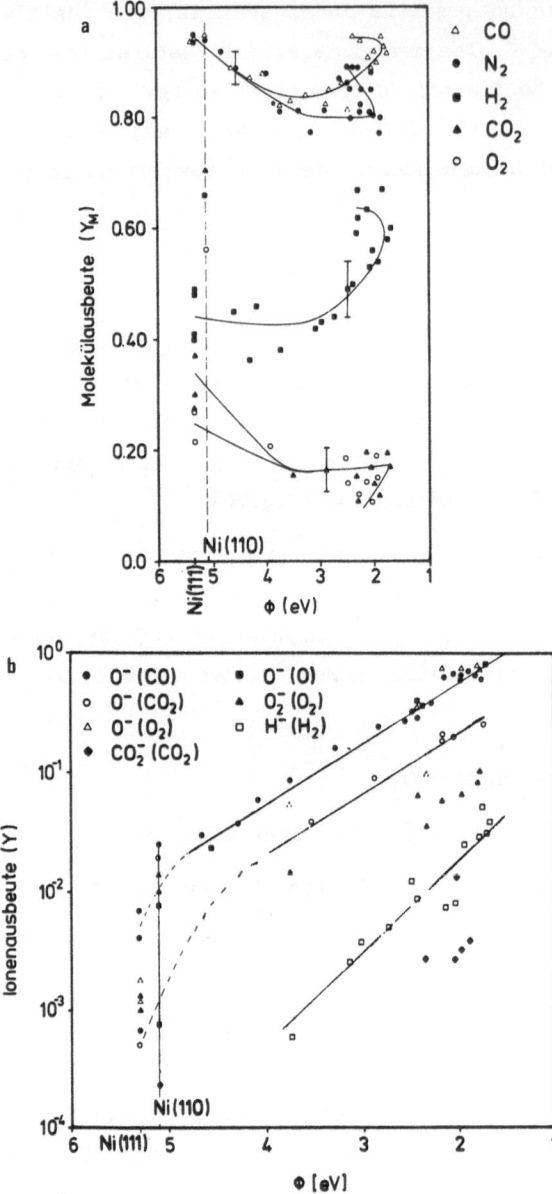

Figure 36a. Survival probability of different molecular ions interacting
with clean Ni (110), Ni (111) and Ni (111) covered with K. The
work function change monitors the K coverage (Schubert et
al. [189]).

b. Negative ion yields for the same molecular ions interacting
with clean Ni (110), Ni (111) and Ni (111) + K (Schubert et
al. [189]).

collision, in agreement with, e.g., the findings for other atomic
particles [129-131,148]. However, the molecular survival depends in a
quite variable manner on the work function and the molecular species
involved (fig.36). CO and N_2 show little dependence at all, and

essentially the same behavior, in agreement with the isoelectronic character of these two species. For H_2 we observe an increase of the molecular survival, which is in qualitative agreement with the model for the hydrogen dissociation given previously [167]. The H_2 ground state is of the order of 10 eV below the Fermi energy, for the clean surface already. A decrease of the work function has therefore no great influence on the Auger process. However, the resonant transition moves out of the Frank-Condon region (fig.32), when the work function decreases, i.e., Auger wins by default.

Exactly the same qualitative pattern should be observed if the main source of dissociation in case of O_2 and CO_2 would be the RN into a triplet state. The experiment, however, shows a strong decrease of the molecular survival for O_2 and CO_2, i.e., we conclude that the reactions a.1 + a.2, followed by a.4, are the dominating steps. Qualitatively, we expect also from the atomic scattering results capture into the affinity levels with increasing work function.

Experiments using O_2 as primary particle, $O_2 \rightarrow$ Ag (111), by Kleyn et al.[190] have resulted in O_2^- ions coming out of the surface interaction, which is an essential prove that the initial charge state is not very important, and that the harpooning reaction prevails.

Since we know the velocity of the scattered particles, and since we can estimate the interaction length with the surface, it is easy to find that the time for, e.g., the $O_2^{\delta-}$ or the $CO_2^{\delta-}$ to dissociate at the surface, is of the order of 10^{-13} s. Obviously this time is long enough for at least some of the molecules to dissociate. In the case of a surface with a very low work function, practically all particles dissociate within a short time, and static chemisorption experiments have no chance to identify these intermediate negative species. Another aspect of the beam experiments is the zero coverage, i.e., each scattered particle is scattered from a surface where no other molecules of the same type are adsorbed. In the CO adsorption case on K covered surface, the question was discussed (Bonzel [191]) whether the CO dissociates on such surfaces, because experiments using isotopes have shown that atomic exchange occurs, but only observable at high coverage. Our results obviously support Bonzel's interpretation that the high coverage experiments show not necessarily dissociation but intermolecular exchange only, because the high K coverage surface (CO free!) shows no CO dissociation (fig.36). Another important field connected with charge exchange processes and molecular interactions are the methods of film growth using ion beams [192].

ACKNOWLEDGEMENT

I am grateful for the hospitality of the Physics Department of the Arizona State University, Tempe, during the early months of 1990. I thank I.S.T. Tsong for many helpful discussions, and his efforts to make my visit to Tempe possible and fruitful.

REFERENCES

1. E. Goldstein, Monatsberichte der Königl. Preuss. Akademie der Wissenschaften zu Berlin 33 (1880) 82 and Wiedemann'sche Annalen der Physik u. Chemie 54 (1898) 38.
2. W.R. Grove, Philos. Mag. 5 (1853) 203.
3. N. Bohr, Philos. Mag. 25 (1913) 10.
4. N. Bohr, Philos. Mag. 30 (1915) 581.
5. W.H. Bragg, Jahrb. d. Rad. und El. 2 (1905) 4.
6. E. Rutherford, Phil. Mag. 10 (1905) 163, 193 and Phil. Mag. 12 (1906) 154.
7. Chr. Füchtbauer, Physik. Zeitschr. 7 (1906) 748.
8. J. Stark and W. Steubing, Ann. d. Phys. 28 (1909) 974.
9. J. Stark and G. Wendt, Ann. d. Phys. 38 (1912) 669, 690 and 921.
10. L.C. Davisson and L.H. Germer, Phys. Rev. 30 (1927) 705.
11. I. Estermann and O. Stern, Z. Phys. 61 (1930) 95.
12. H.D. Hagstrum, Phys. Rev. 96 (1954) 336.
13. D.P. Smith, J. Appl. Phys. 18 (1967) 340 and Surf. Sci. 25 (1971) 171.
14. J. Lindhard, Mat. Fys. Medd. Kong. Dansk Videnskab. Selskab 34 (1965) No.14.
15. E. Taglauer, W. Englert, W. Heiland and D.P. Jackson, Phys. Rev. Letters 45 (1980) 740.
16. W. Heiland and E. Taglauer, Nucl. Instr. Meth. 132 (1976) 535.
17. D.P. Jackson, W. Heiland and E. Taglauer, Phys. Rev. B24 (1981) 4198.
18. A.L. Boers, Nucl. Instr. Meth. B2 (1984) 353.
19. W. Hetterich, H. Derks and W. Heiland, Nucl. Instr. Meth. B33 (1988) 401.
20. N. Bohr, Phys. Rev. 58 (1940) 654 and Phys. Rev. 59 (1941) 270.
21. O.B. Firsov, JETP 7 (1958) 308.
22. J.F. Ziegler, J.P. Biersack and U. Littmark, *The Stopping and Range of Ions in Solids,* Vol. 1 *of The Stopping and Ranges of Ions in Matter,* ed. by J.F. Ziegler, Pergamon Press, N.Y. (1985).
23. P. Loftager, F. Besenbacher, O.S. Jensen and V.S. Sorensen, Phys. Rev. A20 (1979) 1143.
24. W. Lenz, Z. Phys. 77 (1932) 713.
25. H. Jensen, Z. Phys. 77 (1932) 722.
26. J. Lindhard, V. Nielsen and M. Scharff, Mat. Fys. Medd. Kong. Dansk Vidensk. Selskab 36 (1968) No.10.
27. G. Moliére, Z. f. Naturf. A2 (1947) 133.
28. W.D. Wilson, L.G. Haggmark and J.P. Biersack, Phys. Rev. 15 (1977) 2458.
29. E. Hulpke and K. Mann, Surf. Sci. 133 (1983) 171.
30. A.-D. Tenner, K. T. Gillen, T.C.M. Horn, J. Los and A. W. Kleyn, Phys. Rev Lett. 52 (1984) 2183.
31. M. Kato, R.S. Williams and M. Aono, Nucl. Intr. Meth. B33 (1988) 462.
32. L.D. Landau and E.M. Lifschitz, *Quantum Mechanics*, Pergamon Press, N.Y. (1974), p. 219ff.
33. H.H. Andersen, *The Stopping and Ranges in Matter,* Vol. 2, J.F. Ziegler, ed., Pergamon Press, N.Y. (1977).
34. H.F. Winters and P. Sigmund, J. Appl. Phys. 45 (1974) 4760.

35. H.F. Winters and E. Taglauer, Phys. Rev. *B35* (1987) 2174.
36. W. Eckstein, Nucl. Instr. Meth. *B27* (1987) 78.
37. J.W. Rabalais, CRC Critical Reviews in Solid State Materials Sciences *14* (1988) 319.
38. D.J. O'Connor, R.J. MacDonald, W. Eckstein and P.R. Higginbottom, Nucl. Instr. Meth. *B13* (1986) 235.
39. P. Sigmund, Phys. Rev. *B184* (1969) 383.
40. E. Taglauer, W. Heiland and J. Onsgaard, Nucl. Instr. Meth. *168* (1980) 571.
41. E. Taglauer, G. Marin, W. Heiland and U. Beitat, Surf. Sci. *63* (1977) 507.
42. E. Taglauer and W. Heiland, J. Nucl. Mat. *93* (1980) 823.
43. W. Heiland, J. Kraus, S. Leung and N.H. Tolk, Surf. Sci. *67* (1977) 437.
44. E. Bogh, in *Channeling*, D.M. Morgan, ed., J. Wiley, London (1973).
45. D.P. Jackson, Nucl. Instr. Meth. *132* (1976) 603.
46. O.S. Oen, Surf. Sci. *131* (1983) L470.
47. W. Heiland and E. Taglauer, J. Vac. Sci. Tech. *9* (1972) 620.
48. D.J. Godfrey and D.P. Woodruff, Surf. Sci. *89* (1979) 76.
49. H. Niehus and G. Comsa, Surf. Sci. *10* (1984) 18.
50. M. Aono, Nucl. Instr. Meth. *B2* (1984) 374.
51. C.S. Chang, U. Knipping and I..S.T. Tsong, Nucl. Instr. Meth. *B18* (1986) 11.
52. C.S. Chang, U. Knipping and I.S. T. Tsong, Nucl. Instr. Meth. *B28* (1987) 493.
53. S. Datz and C. Snoek, Phys. Rev. *134* (1964) A347.
54. E. Taglauer and W. Heiland, Surf. Sci. *33* (1972) 27.
55. E.S. Mashkova and V.A. Molchanov, Rad. Eff. *13* (1972) 183.
56. S.H.A. Begeman and A.L. Boers, Surf. Sci. *30* (1972) 134.
57. V.E. Yurasova, V.I. Shulga and D.S. Karpuzov, Can. J. Physics *46* (1968) 759.
58. E.S. Parilis, N.Y. Turaev and V.M. Kivilis, Int. Conf. Phen. Ion. Gases, Belgrade (167) p.47
59. W. Heiland, E. Taglauer and M.T. Robinson, Nucl. Instr. Meth. *132* (1976) 603.
60. E. Hulpke, Surf. Sci. *52* (1975) 615.
61. T.C.M. Horn, Pan Haochang, P.J. van den Hoek and A.W. Kleyn, Surf. Sci. *201* (1988) 573.
62. M.T. Robinson and O.S. Oen, Appl. Phys. Lett. *2* (1963) 30.
63. J. Lindhard, Physics Letters *12* (1964) 126.
64. *Channeling*, ed. by D.V. Morgan, J. Wiley, London (1973).
65. O.S. Oen, private communication.
66. W.M. Thompson and H.J. Pabst, Rad. Eff. *37* (1978) 105.
67. R. Sizmann and C. Varelas, Nucl. Instr. Meth. *132* (1976) 633.
68. R. Schiffner, K. Goltz and C. Varelas, Vakuum-Technik *26* (1977) 3.
69. H. Winter, Proc. of the XIII Werner Brand Symposium, San Sebastián, Spain (1989); R. Zimny, H. Nienhaus and H. Winter, Nucl. Instr. Meth. *B48* (1990) 361.
70.· H.J. Andrä, Phys. Lett. *54A* (1975) 315.
71. N.H. Tolk, J.C. Tully, J.S. Kraus, W. Heiland and S.H. Neff, Phys. Rev. Lett. *41* (1978) 643.
72. H. Obermeyer, K. Snowdon, H. Hemme and W. Heiland, Z. Phys. *B61* (1985) 187.
73. H. Winter, H. Hagedorn, R. Zimny, H. Niehaus and J. Kirschner, Phys. Rev. Lett. *62* (1989) 296.
74. C. Rau, Appl. Phys. *A49* (1989) 579.
75. M. Schleberger, C. Huber, A. Närmann and J. Kirschner, to be published.
76. W. Heiland and E. Taglauer in: *Methods of Experimental Physics 22* (1985) 299, M.G. Lagally and R.L. Park, eds.
77. R.E. Honig and W.L. Harrington, Thin Solid Films *19* (1973) 43.

78. E. Taglauer, Appl. Phys. *A38* (1985) 161.
79. J. Du Plessis, G.N. van Wyk and E. Taglauer, Surf. Sci. *220* (1984) 381.
80. P. Varga and G. Hetzendorf, Surf. Sci. *162* (1985) 584.
81. T.M. Buck in: *Chemistry and Physics of Solid Surfaces IV*, R. Vanselove and J.C. Campuzzano, eds.
82. S. Tang and N.Q. Lam, Surf. Sci. *223* (1989) 179.
83. G. Blyholder, J. Phys. Chem. *68* (1964) 2772.
84. W. Heiland, W. Englert and E. Taglauer, J. Vac. Sci. Tech. *15* (1978) 419.
85. W. Englert, W. Heiland, E. Taglauer and D. Menzel, Surf. Sci. *83* (1979) 243.
86. W. Heiland, Vacuum *39* (1987) 367.
87. W. Hetterich, H. Derks and W. Heiland, Appl. Phys. Letters *52* (1988) 371.
88. M. Aono, Nucl. Instr. Meth. *B2* (1984) 374.
89. H. Niehus, Surf. Sci. *166* (1986) L107.
90. H. Niehus and G. Comsa, Surf. Sci. *151* (1985) L171.
91. H. Niehus, Surf. Sci. *130* (1983) 41.
92. G. Engelmann, E. Taglauer and D.P. Jackson, Nucl. Instr. Meth. *B13* (1986) 240.
93. H. Niehus, Surf. Sci. *145* (1984) 407.
94. J. Möller, K.J. Snowdon, W. Heiland and H. Niehus, Surf. Sci. *178* (1986) 475.
95. E. van de Riet, H. Derks and W. Heiland, Surf. Sci., in press.
96. S.H. Overbury and D.R. Mullins, J. Vac. Sci. Tech. *A7* (1984) 1942.
97. C.S. Chang, T.L. Porter and I.S.T. Tsong, J. Vac. Sci. Tech. *A7* (1984) 1406.
98. C.S. Chang, T.L. Porter and I.S.T. Tsong, Vacuum *39* (1989) 1195.
99. T.L. Porter, C.S. Chang and I.S.T. Tsong, Phys. Rev. Letters *60* (1988) 1739.
100. M. Aono, R. Souda, C. Oshima and Y. Ishizawa, Surf. Sci. *168* (1986) 713.
101. K. Oura, M. Katayama, F. Shoji and T. Hanawa, Phy.s Rev. Lett. *55* (1985) 1486.
102. Th. Fauster, Vacuum *38* (1988) 129.
103. D.R. Mullins and S.H. Overbury, Surf. Sci. *210* (1989) 481.
104. D.R. Mullins and S.H. Overbury, Surf. Sci. *199* (1988) 141.
105. E.S. Mashkova and V.A. Molchanov, *Medium Energy Ion Reflection from Solids*, North Holland, Amsterdam (1985).
106. S.H. Overbury, W. Heiland, D.M. Zehner, S. Datz and R.S. Thoe, Surf. Sci. *109* (1981) 239.
107. H. Hemme and W. Heiland, Nucl. Instr. Meth. *B9* (1985) 41; H. Derks, H. Hemme, W. Heiland and S.H. Overbury, Nucl. Instr. Meth. *B23* (1987) 374.
108. M.T. Robinson and I.M. Torrens, Phys. Rev. *B9* (1974) 5008.
109. W. Moritz and D. Wolf, Surf. Sci. *88* (1979) L29.
110. D.P. Jackson, Surf. Sci. *43* (1974) 431.
111. L.D. Marks, Phys. Rev. Letters *51* (1983) 1000; D.J. Smith and L.D. Marks, Ultramicroscopy *16* (1985) 101.
112. I.K. Robinson, Phys. Rev. Letters *50* (1983) 1145.
113. S.H. Overbury, Nucl. Instr. Meth. *B27* (1987) 65.
114. R.L. McEachern, D.M. Goodstein and B.H. Cooper, Phys. Rev. *B39* (1989) 10503.
115. A. Niehof and W. Heiland, Nucl. Instr. Methods *B48* (1990) 306.
116. H. Derks, W. Hetterich, E. van de Riet, H. Niehus and W. Heiland, Nucl. Instr. Meth. *B48* (1990) 315.
117. A.J. de Wit, R.P.N. Bronckers, T.M. Hupkens and J.M. Fluit, Surf. Sci. *82* (1979) 177.
118. E. van de Riet, J.B.J. Smeets, J.M. Fluit and A. Niehaus, Surf. Sci. *214* (1989) 111.

119. H. Niehus and E. Preuss, Surf. Sci. *119* (1982) 349; E. Preuss, Rad. Eff. *38* (1978) 151 and Surf. Sci. *110* (1981) 287.

120. E. van de Riet, Thesis, Utrecht 1990.

121. J.C. Campuzzano, M.S. Foster, C. Jennings, R.F. Willis and U. Unertl, Phys. Rev. Letters *54* (1985) 2684.

122. H.D. Hagstrum, P. Petrie and E.E. Chaban, Phys. Rev. *B38* (1988) 10264.

123. R.L. Erickson and D.P. Smith, Phys. Rev. Letters *3* (1975) 297.

124. G. Falcone and Z. Sroubek, Phys. Rev. *B39* (1989) 1999.

125. W. Englert, E. Taglauer and W. Heiland, Surf. Sci. *117* (1982) 124.

126. D.P. Woodruff, Nucl. Instr. Meth. *194* (1982) 639.

127. R.J. MacDonald, D.J. O'Connor, J. Wilson and Y.G. Shen, Nucl. Instr. Meth *B33* (1988) 446.

128. J.M. Beuken, E. Pierson and P. Bertrand, Surf. Sci. *223* (1989) 201.

129. W. Eckstein, H. Verbeek and R. S. Battacharya, Surf. Sci. *99* (1980) 356.

130. J.N.M. van Wunnik, J.J.C. Geerlings and J. Los, Surf. Sci. *131* (1983) 1.

131. J.N.M. van Wunnik, J.J.C. Geerlings, E.H.A. Granneman and J. Los, Surf. Sci. *131* (1983) 17.

132. E.E. Nikitin, *Theory of Elementary Atomic and Molecular Processes in Gases,* Clarendon Press, Oxford (1974).

133. P.M. Echenique, R.M. Niemienen, J.C. Ashley and R.H. Ritchie, Phys. Rev. *A33* (1986) 897.

134. F. Sols and F. Flores, Phys. Rev. *B30* (1984) 4878.

135. R. Monreal, E.C. Goldberg, F. Flores, A. Närmann, H. Derks and W. Heiland, Surf. Sci. *211/212* (1989) 271.

136. H. Derks, A. Närmann and W. Heiland, Nucl. Instr. Meth. *B44* (1989) 125.

137. F. Sols, Thesis Madrid, 1985.

138. M.A. Kumakhov and F.F. Komarov, *Energy Loss and Ion Ranges in Solids,* Gordon and Breach (1981).

139. N.S. Xu and J.L. Sullivan, Vacuum *39* (1989) 1201.

140. A. Zartner, E. Taglauer and W. Heiland, Phys. Rev. Letters *40* (1978) 1259.

141. G.J. Lockwood and E. Everhart, Phys. Rev. *125* (1962) 657; E. Everhardt, Phys. Rev. *132* (1963) 2083; H.F. Helbig and E. Everhart, Phys. Rev. *136* (1964) A675; W. Lichten, Phys. Rev. *139* (1965) A27.

142. R.J. Janev and H. Winter, Physics Reports *117* (1985) 265.

143. E. Taglauer and W. Heiland, Proc. Int. Vac. Congress & Int. Conf. Sol. Surfaces (Vienna, 1977) p.2495

144. T.W. Rusch and R.L. Erickson in: *Inelastic Ion-Surface Collisions,* N.H. Tolk, J.C. Tully, W. Heiland and C.W. White, eds., Academic Press (1977) 73.

145. S. Oelschig, Diplomarbeit Osnabruck 1988.

146. L.L. Henck and J.K. West, *Principles of Electronic Dynamics,* J. Wiley (1989).

147. H. Wedler, *Adsorption,* Verlag Chemie, Weinheim (1970).

148. J. Andrä, R. Zimny, H. Winter and H. Hagedorn, Nucl. Instr. Meth. *B9* (1985) 572.

149. W. Heiland and E. Taglauer in *Inelastic Ion-Surface Collisions* Academic Press, N.Y. (1977) p.27

150. W. Eckstein, V.A. Molchanov and H. Verbeek, Nucl. Instr. Meth. *149* (1978) 599.

151. P. Bertrand and H. Ghalim, Physica Scripta, *T6* (1983) 168.

152. F. Shoji and T. Hanawa, Nucl. Instr. Meth. *B2* (1984) 401.

153. F. Shoji, Y. Nakayama, K. Oura and T. Hanawa, Nucl. Instr. Meth. *B33* (1988) 420.

154. A. Närmann, Thesis Osnabruck 1989; A. Närmann, R. Monreal, P.M. Echenique, F. Flores, W. Heiland and S. Schubert, Phys. Rev. Lett. *64* (1990) 1601.

155. K.J. Snowdon, D.J. O'Connor and R.J. MacDonald, Phys. Rev. Lett. *61* (1988) 1760; K.J. Snowdon, D.J. O'Connor and R.J. MacDonald, Surf. Sci. *221* (1989) 465.
156. M. Kato, I. Iitaka and Y.H. Ohtsuki, Nucl. Instr. Meth. *B33* (1988) 432.
157. A. Närmann, K. Schmidt, W. Heiland, R. Monreal, F. Flores and P.M. Echenique, Nucl. Instr. Meth. *B48* (1990) 378.
158. B.V. Panin, Sov. Phys. JETP *15* (1962) 215.
159. W. Eckstein, H. Verbeek and S. Datz, Appl. Phys. Lett. *27* (1975) 527.
160. B. Willerding, K. Snowdon and W. Heiland, Z. f. Physik *B59* (1985) 435.
161. U. Imke, Thesis Osnabruck 1987.
162. W. Heiland, U. Beitat and E. Taglauer, Phys. Rev. *B19* (1977) 1677.
163. I.G. Bitenskii and E.S. Parilis, Nucl. Instr. Meth. *B2* (1984) 384.
164. M. Jakas and D. Harrison, Surf. Sci. *149* (1985) 500.
165. P.J. van den Hoek and E.J. Baerends, Surf. Sci. *221* (1989) L791.
166. P.H.F. Reijnen, P.J. van den Hoek, A.W. Kleyn, U. Imke and K.J. Snowdon, Surf. Sci. *221* (1989) 427.
167. U. Imke, K.J. Snowdon and W. Heiland, Phys. Rev. *B34* (1986) 41 and 49.
168. U. Imke, H. Rechtien and P.H.F. Reijnen, Surf. Sci. *221* (1989) 454.
169. G.H. Dunn, Phys. Rev. Lett. *8* (1962) 62.
170. S. Schubert, J. Neumann, U. Imke, K.J. Snowdon, P. Varga and W. Heiland, Surf. Sci. *17.1* (1986) L375.
171. H.M. van Pinxteren, C.F.A. van Os, R.M.A. Heeren, R. Rodink, J.J.C. Geerlin and J. Los, Europhys. Lett. *10* (1989) 715.
172. C.T. Campbell, Surf. Sci. 157 (1985) 43 and Surf. Sci. *173* (1986) L641.
173. D. Andersson, B. Kasemo and L. Wallden, Surf. Sci. *152/153* (1985) 576.
174. B. Bartos, H.-J. Freund, H. Kuhlenbeck, M. Neumann, H. Lindner and K. Müller, Surf. Sci. *179* (1987) 59.
175. B. Bartos, H.-J. Freund, H. Kuhlenbeck and M. Neumann, Springer Series in Surf. Sci. *8* (1987) 164.
176. G. Illing, D. Heskett, E.W. Plummer, H.-J. Freund, J. Somers, Th. Lindner, A.M. Bradshaw, V. Buskotte, M. Neumann, U. Starke, K. Heinz, P.L. de Andres, D. Saldin and J.B. Pendry, Surf. Sci. *206* (1988) 1.
177. J. Wambach, G. Odorfer, H.-J. Freund, H. Kuhlenbeck and M. Neumann, Surf. Sci. *209* (1989) 159.
178. A.F. Carley, D.E. Gallagher and M.W. Roberts, Surf. Sci. *183* (1987) L263.
179. A.F. Carley, D.E. Gallagher and M.W. Roberts, Spectrochimica Acta *43A* (1987) 1447.
180. R.G. Coperthwaite, P.R. Davies, M.A. Morris, A.W. Roberts and R.A. Ryder, Catalysis Letters *1* (1988) 11.
181. H.-J. Freund and P. Messmer, Surf. Sci. *172* (1986) 1.
182. P. Feibelmann, Surf. Sci. *160* (1985) 139.
183. J..W. Gadzuk and J. Nøfrskov, J. Chem. Phys. *81* (1984) 2828.
184. J.W. Gadzuk, Comm. At. Mol. Phys. *16* (1985) 219.
185. S. Holloway, J. Vac. Sci. Tech. *A5* (1987) 476.
186. Pan Haochang, T.C.M. Horn and A.W. Kleyn, Phys. Rev. Lett. *57* (1986) 3035.
187. A. Danon and A. Amirav, Phys. Rev. Lett. *61* (1988) 2961.
188. B. Willerding, H. Steininger, K.J. Snowdon and W. Heiland, Nucl. Instr. Meth. *B2* (1984) 453.
189. S. Schubert, U. Imke and W. Heiland, Surf. Sci. *219* (1989) L 576.
190. A.W. Kleyn, Proc. ICPEAC 16, New York (1989).
191. H.P. Bonzel, Surf. Sci. Reports *8* (1988) 43.
192. S.R. Kasi, H. Kang, C.S. Sass and J.W. Rabalais, Surf.Sci. Report *10* (1989) 1.

Appendix: List of acronyms of surface analysis techniques

1. Elastic scattering techniques

Acronym	Technique	Probe particle	det.particle
EELS	Electron energy loss spectr.	Electron	Electron
HREELS	High resolution EELS	"	"
SEM	Scanning electron microscopy	"	"
ISS (LEIS)	(Low energy) ion scat. spectr.	Ion	Ion
MEIS	Medium energy ion scattering	"	"
RBS	Rutherford back scattering	"	"
ICISS	Impact collision ISS	"	"
NICISS	Neutral ICISS	"	Atom
ERD	Elastic recoil detection	"	Ion
FIM	Field ion microscopy	Atom	Ion

2. Diffraction methods

LEED	Low energy electron diff.	Electron	Electron
RHEED	Reflection high-energy electron diffraction	"	"
HEAD	Helium atom diffraction	Atom	Atom

3. Ionisation methods

AES	Auger electron spectrocopy	Electron	Electron
PES	Photoelectron spectroscopy	Photon	Electron
ESCA	Electron spectroscopy for chemical analysis = PES	Photon	Electron
UPS	Ultraviolet photoelectron spectroscopy	Photon	Electron
ARUPS	Angular resolved UPS	"	"
XPS	X-ray photoelectron spectr.	"	"
SEXAFS	Surface extended X-ray absorption fine structure	"	"
NEXAFS	Near edge X-ray absorption fine structure	"	"
FEM	Field electron microscopy	(el. Field)	Electron

4. Bond breaking

PSD	Photon stimulated desorption	Photon	Ion
ESD	Electron stimulated desorption	Electron	Ion
ESDIAD	ESD ion angular distribution	"	"
SIMS	Secondary ion mass spectrom.	Ion	Ion
SNMS	Secondary neutral mass spectr.	"	Atom
FABS	Fast atom bombardment spec.	Atom	Atom
ERD	Elastic recoil detection	Ion	Ion

5. Charge exchange methods

INS	Ion neutralisation spectr.	Ion	Electron
MDS	Metastabe de-excitation spec.	Atom	Electron
STM	Scanning tunneling microscopy	(tip)	Electron

APPLICATIONS OF ION BEAMS TO MATERIALS SCIENCE

L.C. Feldman

AT&T Bell Laboratories
Murray Hill, NJ 07974-2070
U.S.A.

Modern materials technology is largely based on the modification and control of the surface and interface regions of solids. Thin-film deposition has reached extraordinary levels where control on the monolayer scale is readily achievable. Lasers, ion beams, and electron beams are all used to modify the near-surface region of solids. Ceramic processing is strongly controlled by the interfacial phenomena involved in sintering, and the metal/polymer interface is crucial to the formation of material composites. This materials development would not be possible without the advent of analysis techniques which can examine the surface and interface regions of solids. This talk discussed the use of energetic ion beams in materials science, with special emphasis on scattering as a surface and interface analysis tool.

Materials scientists make use of ion beams throughout the energy range accessible in the laboratory. The lowest energies (1 eV - 10 keV) are used in processes such as sputtering, reactive ion etching, surface analysis and various lithography schemes. The 100 keV regime is often associated with ion implantation, which has had its biggest impact in electronic devices. This region also represents the lower limit for Rutherford backscattering and applications to near-surface analysis. The region of 1 MeV and above is primarily associated with near-surface analysis, although there has been recent interest in high-energy ion implantation in this energy regime.

The strength of ion-beam analysis is the quantitative, elemental analysis of the near-surface composition of a solid. The most common use of high-energy ion beams is elastic scattering, and has assumed the name Rutherford backscattering spectrometry (RBS). Given the apparatus for RBS, it is relatively simple to supplement materials analysis capabilities with

Interaction of Charged Particles with Solids and Surfaces
Edited by A. Gras-Martí *et al.*, Plenum Press, New York, 1991

309

channeling for near-surface crystallography, nuclear reaction analysis (NRA) for quantitative light-element detection, particle induced x-ray emission (PIXE) for elemental analysis, and elastic recoil detection (ERD) for convenient depth profiling of hydrogen and its isotopes.

Information on the crystalline structure of a solid comes from the combination of ion scattering and channeling. Under channeling conditions, the beam is aligned with a major crystalline direction of a solid. In the case of an ideal single crystal, the beam can scatter from the first monolayer(s) of the solid only; scattering from atoms within the crystal is decreased (almost turned off) because of the shadowing effect of the first atoms. Atoms near the surface or at an interface, which are not on lattice sites, also provide a non-reduced scattering yield. Such atoms might be the silicon and oxygen within an amorphous oxide. Under channeling conditions, the scattering spectrum is dominated by atoms off lattice sites, and the intrinsic surface peak of the single crystal.

Interface studies using ion-scattering techniques can be divided by the thickness of the overlayer, those corresponding to 10 monolayers or less being "surface-physics" oriented and those more than 10 monolayers corresponding to "thin-film" studies. Of course, there is no sharp transition; most scientists in this field are consistently interested in bridging the gap between these two regimes.

Epitaxial structures (particularly heteroepitaxial) represent a class of thin-film systems susceptible to ion channeling techniques. The combination of RBS and channeling provides a mass-dispersive and depth-dispersive crystallography over a useful depth range. Some of the more novel examples will be discussed.

Analysis of strained-layer semiconductor structures has been a recent area of interest using channeling studies. In these systems, there is a relatively large lattice mismatch (> 1 %) between the semiconductor film and the semiconductor substrate. Pseudomorphic growth of the film results in a tetragonal distortion, which is measured by channeling techniques or x-ray diffraction.

Studies in the monolayer regime, (surface regime), are usually directed at establishing the detailed properties of the interface. Surface studies test theories of epitaxial growth which often predict critical thicknesses smaller than a few layers.

The series of lectures will cover the basic particle-solid interactions used in analysis, and illustrate their applicability to understanding solids at the forefront of materials science. A list of comprehensive references is included.

REFERENCES

W-K. Chu, J.W. Mayer, and M-A. Nicolet, *Backscattering Spectrometry*. Academic Press, New York (1978).

L.C. Feldman and J.W. Mayer, *Fundamentals of Surface and Thin Film Analysis*, North-Holland, New York (1986).

J.F. Ziegler, *Helium Stopping Power and Ranges in All Elements*; Pergamon, Oxford, U.K. (1977).

J.W. Mayer and E. Rimini, eds., *Ion Beam Handbook for Materials Analysis*, Academic Press, New York (1977).

J.R. Bird and J.S. Williams eds., *Ion Beams for Materials Analysis*, Academic Press, Australia (1989).

L.C. Feldman, J.W. Mayer and S.T. Picraux, *Materials Analysis by Ion Channeling*, Academic Press, New York (1982).

L.C. Feldman, *Critical Rev. Solid State Mater. Sci. 10* (1981) 143.

L.C. Feldman and J.M. Poate, *Ann. Rev. Mater. Sci. 12* (1982) 149.

J.F. van der Veen, *Surf. Sci. Rep.* (1985) p.199.

J.A. Davies in *Materials Characterization Using Ion Beams*, J.P. Thomas and A. Cachard, eds., Plenum Press, London (1978) p.405.

J. Lindhard, *Mat. Fys. Medd. Dan. Vid. Selsk. 34* (1965) p. 1.

J.H. Barret, *Phys. Rev. B 3* (1971) 1527.

DESORPTION INDUCED BY ELECTRONIC TRANSITIONS:
BASIC PRINCIPLES AND MECHANISMS

R.A. Baragiola[1] and T.E. Madey

Department of Physics and Astronomy
and Laboratory for Surface Modification
Rutgers University, Piscataway, NJ 08855, U.S.A.

1. Present address: Dept. Nucl. Engn. and Engn. Physics
University of Virginia, Charlottesville
VA 22901, U.S.A.

1. INTRODUCTION

In DIET, Desorption Induced by Electronic Transitions, neutral and ionized atoms and molecules are ejected from solids by electronic excitations of the surface bonds induced by incident electrons, photons or heavy particles. By focusing in electronic transitions, we exclude from DIET desorption induced by direct momentum transfer (sputtering) or by thermal agitation. The term DIET includes electron stimulated desorption (ESD), photon stimulated desorption (PSD) and various forms of heavy-particle induced desorption.

Interest in this field originates from positive and negative influences that desorption has in different areas. For instance, desorption produces unwanted effects complicating the measurements of ultra low pressures, altering samples in surface analysis, shortening the lives of electron multipliers and vacuum tubes, and contaminating nuclear fusion devices and particle storage rings. On the other hand, desorption can be used to desirable ends in electron beam lithography, in mass spectrometry, in gas discharge cleaning of surfaces, and in the determination of the structure of adsorbed molecules. Furthermore, its occurrence is of importance in the evolution of surfaces of planetary bodies. From the point of view of basic science, the study of DIET is one of the richest test grounds for our understanding of fundamental processes in atomic physics, and of the evolution of excited states in solids on the femtosecond scale.

Interaction of Charged Particles with Solids and Surfaces
Edited by A. Gras-Martí *et al.*, Plenum Press, New York, 1991

Our aim here is to introduce this richness to the reader; for recent reviews of the field, and historical relations, the reader is referred to refs.[1-16]. We attempt to explain the current understanding of the main physical processes involved, how they are related to other aspects of particle-solid interactions, and how they can be applied to understand experimental observations. We start by presenting the main concepts of DIET in sect.2. This is followed by the description of the repulsive state leading to desorption sect.3, the formation of this repulsive state, sect.4 and its evolution, sect.5. We introduce the special case of DIET by heavy particles in sect.6, and end up in sect.7 by identifying the main unknowns and the stumbling blocks for further understanding.

2. GENERAL PRINCIPLES

2.1 Electronic Excitations. Phenomenological Description

Most DIET observations refer to light desorbed ions (H^+, F^+, O^+, etc.), due to large signals and ease of detection. Usual experiments employ incident electrons and produce data in the form of mass spectra, yields, and angular and energy distributions of desorbed species. Typical findings are as follows: 1) the immense majority of desorbed particles are neutral; 2) yields rise from a threshold in the range 5 - 40 eV, pass through a maximum in the 10^2 eV range, and then decay with energy; 3) neutrals are ejected with energies below 1 eV, ions with several eV; 4) emission generally is most intense in the direction of the normal to the surface, and is suppressed in off-normal directions; 5) desorption from single crystals form angular patterns, which reflect initial bond geometries.

Although a large number of observations follow these "typical" lines, no regularities or scaling laws have been found, contrary to the case of sputtering of neutrals, and even of ions. Furthermore, wide variations from these typical results have been reported as research has been broadened to include many materials, and to include more sophisticated detectors. Among those results are desorption at eV incident energies, thresholds in yield-energy curves of hundreds of eV, ion kinetic energies above 20 eV, observation of multiply-charged ions and of large clusters, etc. In many cases, workers have monitored changes in the target and inferred decomposition yields, not necessarily related to desorption. Other DIET probes have included light from UV sources and synchrotrons, light ions and fission fragments.

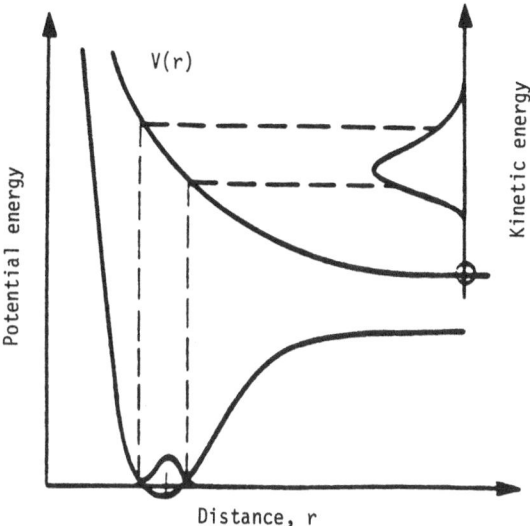

Figure 1. Schematic potential-energy curves for the interaction between a surface and an adsorbate atom. The kinetic-energy distribution results from mapping the initial distribution of interatomic distances, by means of the repulsive potential-energy curve.

In DIET, atomic motion results from the release of potential energy from a repulsive electronic state. This can occur because of one-electron excitations of bonding levels, or by more complex multiple excitations or ionizations. One-electron mechanisms are conceptually the same as those occurring in photon or electron dissociation of free molecules, which result after an initial Franck-Condon transition to repulsive molecular states [17]. This is illustrated in fig.1 for the case of a diatomic molecule. A Frank-Condon, or vertical, transition occurs so suddenly (~ 0.1 fs), that the nuclei in the molecule do not have time to adjust to the change in potential produced by the change of the electron distribution [18]. Therefore, the evolution of the repulsive state starts with the nuclei having the same instantaneous positions and velocities they had in the initial state.

The nuclei will then evolve along the repulsive potential-energy curve (or energy surface, for polyatomic molecules). In the absence of other electronic transitions, the molecule will dissociate if the initial repulsive energy is larger than the dissociation limit of the molecule. In this simple picture, the energy distribution of the dissociation products is determined by mapping the initial distribution of interatomic distances, by means of the potential-energy curve of the repulsive state. This description is known as the "reflection approximation", and is shown in fig.1.

Let us apply this simple idea to desorption from surfaces, to see how far it can take us. We compare production of H^+ by electrons from free H_2, to the desorption of H^+ from atomic hydrogen adsorbed on a metal surface. The peak cross section in the gas phase is 6×10^{-16} cm^2 at 100 eV, while H^+ desorption cross sections for hydrogen adsorbed on metals are typically about 10^{-23} cm^2 [19]. We have to conclude that there is an enormous difference between the observations and what can be predicted on the basis of a simple dissociation model. The difference is not limited to the case of H on metals. Typical maximum values of gas-phase dissociation cross sections, for electrons on diatomic molecules, are of the order of 10^{-16} cm^2, while desorption cross sections are in the range 10^{-18} - 10^{-25} cm^2 or

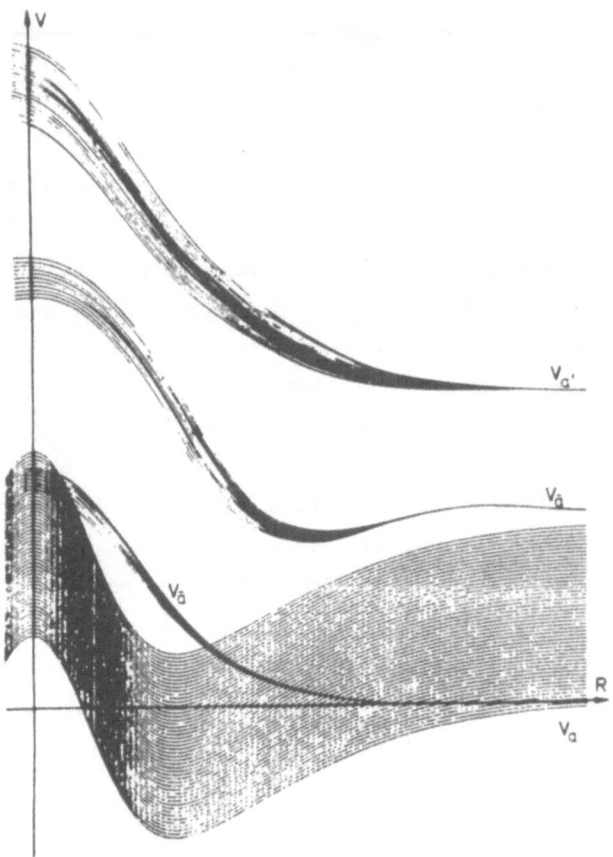

Figure 2. Excitation spectrum of a metal with an adsorbed atom. A continuum of excited particle-hole states of the solid is indicated above the ground-state potential curve V_a. An excited antibonding state V_a is shown embedded in the continuum. Also shown are another excited state V_a and an ionic state V'_a. The widths of the excited curves result from the coupling to the continuum, which is not indicated explicitly as is done for the ground state. (From ref.[20]).

less for neutrals, and smaller by one to more than three orders of magnitude for ions. Differences are also found in the distributions in charge, mass, and energy of the fragmentation products, and in the dependence of cross sections on electron energy. The general rule is that cross sections for rupture of an internal molecular bond are high, and may be comparable to gas-phase processes, but those to break the molecule-metal bond are much lower.

2.2 Quenching of the Excitations: Reneutralization

Our present understanding of the above differences between gas phase and surface cross sections, is that processes at surfaces are governed by the existence of a large density of other final states with the same energy as the dissociation channel, and which result from the presence of the solid. Transitions to those states can quench the initial excitation, and "heal" the ruptured bond, as illustrated in fig.2 [20]. Here we assume weak coupling between electrons in the molecule and those in the solid, and have added to the picture the states corresponding to the molecule in its ground state and one excited electron of the solid. Transitions can now occur at the crossings of the repulsive state with the continuum, which lead to a relaxed molecule and an excited conduction electron.

Many alternative descriptions exist for the quenching of the initial excitation. For instance, in the one-electron energy diagram of fig.3, many transitions are possible. For typical adsorption distances ($\sim$ 2 Å), transition rates for one-electron, resonant neutralization (RN) are usually much faster (0.1 - 1 fs) than those for two-electron, Auger neutralization (AN) (3 - 100 fs) [21]. The most likely sequence is that the excited electron in the molecule first tunnels to an empty state in the solid; then, the ground state of the molecule is repopulated by an Auger transition, in which an electron tunnels to the molecule and another is excited in the metal. We see here that the final state is one with two excited electrons in the solid, and the molecule in its ground electronic state. The two excited electrons will cascade down to the Fermi level generating many excited electrons. The molecule could remain in an excited vibrational state, if enough energy were gained during its stay in the repulsive state. All this energy will eventually be converted into thermal agitation of the solid/adsorbate system.

Since electronic transition times are much shorter than the time it takes for an atom in the repulsive energy surface to be ejected from the solid (100 - 1000 fs), practically all excitations will be quenched before significant motion occurs, frustrating desorption.

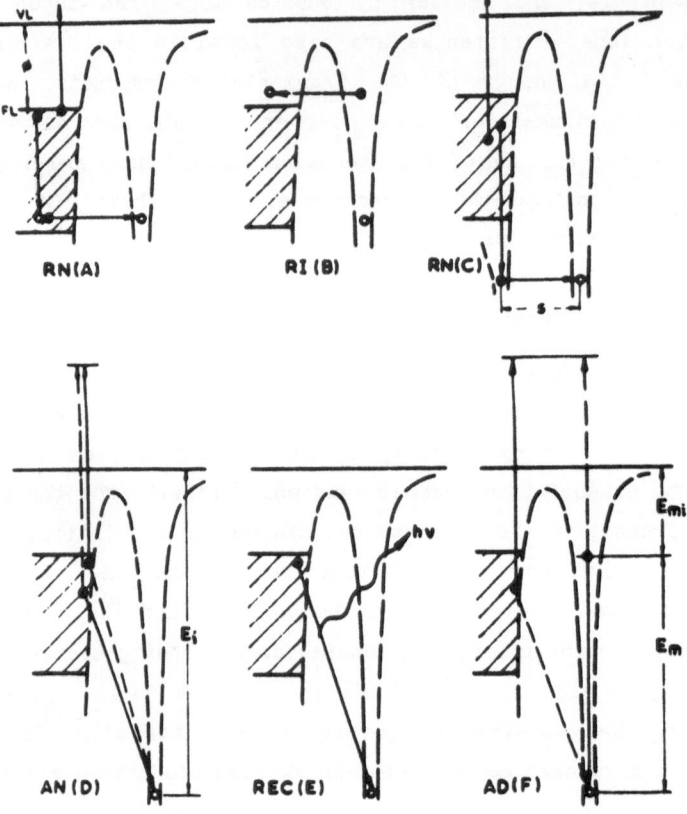

Figure 3. Electron-energy diagrams for: (A) resonance neutralization (RN); (B) resonance ionization (RI); (C) RN from core level; (D) Auger neutralization (AN); (E) radiative electron capture (REC); (F) Auger deexcitation (AD). VL: vacuum level, FL: Fermi level, ϕ: work function. Electronic transitions are indicated by arrows, initially filled (empty circles) electron states in atoms or solid by filled (open circles). Light arrows indicate secondary transitions following RN, and dashed lines in AN and AD, exchange transitions. (From ref.[21]).

In some cases, though, the molecule will stay in the repulsive state a time sufficiently long for dissociation to occur. The cases more likely to lead to desorption are those in which the molecular fragment spends less time close to the surface, since then electron exchange with the solid will be less likely.

The population N(s) of the excited repulsive state decays at a rate

$$dN(s)/dt = - N(s) R(s),\qquad (1)$$

where R(s) is the electronic transition rate at the distance s between the surface and the receding species. Therefore, the desorption probability is given by

$$P_d = e^{-\int \{R(s)/v(s)\} \, ds}, \tag{2}$$

where the velocity of the desorbing particle increases from a low value at s_0, due to the force $F_r = dU/dr$ created by the repulsive potential $U(r)$ [22] along the direction of motion.

Due to the exponential fall of the electron density outside atoms and solids, the expected form of R is

$$R(s) = A \, e^{-as}, \tag{3}$$

at large s, with a saturation at small distances. This dependence persists roughly in the more elaborate theories [23], from which we can infer that values of a are of the order of 1.5 Å^{-1} for metals. The pre-exponential factor A ranges between 10^{15} s^{-1} and $6 \times 10^{17} \text{ s}^{-1}$, for metals, and is expected to depend strongly on the work function of the surface and the energy and extent of the atomic orbitals. Critical velocities $v_c = A/a$ thus fall in the range $6 \times 10^7 - 4 \times 10^9$ cm/s, covering the range of estimates derived from other types of experiments [24].

The integrand in (2) decays extremely fast from its initial value at s_0, over distances of the order of $1/a$. Neglecting the zero point velocity, we get a mean normal velocity

$$v_m \sim \sqrt{F_r(s_0)/aM} \, \cos\theta, \tag{4}$$

in the range of interest, where M is the mass of the desorbing particle and θ the angle between the initial velocity vector and the surface normal. With this approximation, the desorption probability becomes

$$P_d \sim \exp\left\{- M \, A \, \sqrt{a} \, e^{-as_0} \, |F_r(s_0)|^{1/2} \, \cos\theta\right\}, \tag{5}$$

For the special case of the Coulomb potential, $|F_r|^{1/2}$ takes the simple form $(Z_1 Z_2)^{1/2}/s$, where Z_1 and Z_2 are the charges of the two repelling ions.

Expression (5) can be used now to predict the dependence of desorption on different parameters. Due to the very strong exponential dependences, and our lack of good estimates of A, we cannot obtain reliable values of P_d. We can still, however, make a large number of predictions for the dependence of P_d on the following parameters, (where k_M, k_s and k_θ are lumped constants):

M : From (5) we obtain that $P_d = \exp\{- k_M \sqrt{M}\}$, which points to a strong isotope effect. *Ceteris paribus*, the lighter isotope should desorb more readily, since it is faster at the same translational energy. This is in agreement with experimental observations [25]. This mass dependence is also probably the main factor why H^+ emission is normally much stronger than O^+ and OH^+, for the ubiquitous OH adsorption on surfaces [26].

s_o: This is the strongest dependence, since $P_d = \exp\{-k_s e^{-as}o\}$, and since as_o is typically quite larger than one. This means that desorption is very strongly favored for species which desorb far from the surface. This explains why it is easier to break interatomic bonds of adsorbed molecules, than metal-adsorbate bonds.

θ : From (5), $P_d = \exp\{- k_\theta / \cos\theta\}$, and therefore desorption is suppressed in off-normal directions.

A : Desorption probabilities decrease exponentially with electronic transition rates. The pre-factor A is very high for resonance processes. Therefore, significant desorption of ions from metals is only expected for very few atomic species, like H, O, and F, which have a high ionization potential, and from alkali ions, whose ionization potentials are smaller than the work function of most materials. This behavior is observed experimentally. Since wide band-gap materials do not have occupied states with the same energy as valence states of atoms, electron transfer rates should be small, and larger desorption yields are expected for adsorbed gases in these materials, as observed [27]. Transition probabilities are expected to be larger for multiply-coordinated species (i.e., bonded to several atoms), which explains why desorption generally occurs more readily from atop sites than from bridge-bonded sites [28].

Due to the nature of the exponential function, all dependences mentioned above will be stronger the higher the exponent in (5), i.e., the lower the desorption probability.

2.3 Complementary Description of Quenching

Sometimes it is useful to describe the same processes we have just discussed in terms of time. One can write the probability of desorption as

$$P_d = \exp\left(- \frac{t_n}{t_e}\right),$$

where t_e is an average lifetime of the repulsive electronic state, and t_n is the time taken by the nucleus to escape from the quenching effect of the surface. In this framework, desorption requires a metastable repulsive configuration, i.e., with lifetime of the order of the ejection times. Ejection times to distances of 3 Å are t_e(fs) ~ 40 $\sqrt{M/E}$, where E (in eV) is the final energy of the ejected species, and M its mass in amu. Most desorption studies are described by desorption times ranging from 10 fs (20 eV H) to 200 fs (1 eV CO).

Electronic lifetimes vary very widely with the electronic configuration. In metals, they are 0.1 fs, or lower, for valence and conduction states. These are so much shorter than t_n , that desorption of metal atoms cannot be observed from clean metals [29]. Longer lifetimes are expected in insulators, where a hole may take more than 1 fs to move to an adjacent site. Particularly long times can result for the decay of self-trapped excitons in non-metals. For adsorbed gases, t_e will increase exponentially with distance to the surface, t_e ~ $1/R(s) = t_0 \, e^{as}$, in correspondence with eq.(2) above. An important property for desorption is that t_e is much larger for two holes, a point to which we will return in sect.3.3. Long lifetimes can also occur for inner-shell holes. Unlike the case of metals, where these holes will be screened very fast (~ 0.1 fs), in insulators the core holes will exert an essentially unscreened Coulomb force in the lattice, which can survive for times of the order of desorption times.

We end this section with the conclusion that DIET is driven by an initial excitation to a repulsive state, but the probability that this state leads to desorption is governed by the probability to survive bond reformation and neutralization by fast electron exchange with the surface. We now turn to the discussion of the nature of the repulsive state, its formation and evolution.

3. THE NATURE OF THE REPULSIVE STATE

3.1 Background

In general, the nature of the excited state is not known and cannot be predicted with certainty. But we can learn about it by studying the threshold energy for excitation. This procedure is simple for photo-desorption since the photon has a definite energy. Electrons, on the other hand, will be accelerated by the surface potential. The available energy for excitation increases from the value far outside the surface to inside the solid, as the electrons drop in to the bottom of the conduction band (the Fermi level, for metals), acquiring additional energy (the work function, for metals). Furthermore, in the case of an ionized repulsive state, different thresholds could be obtained, depending on whether the

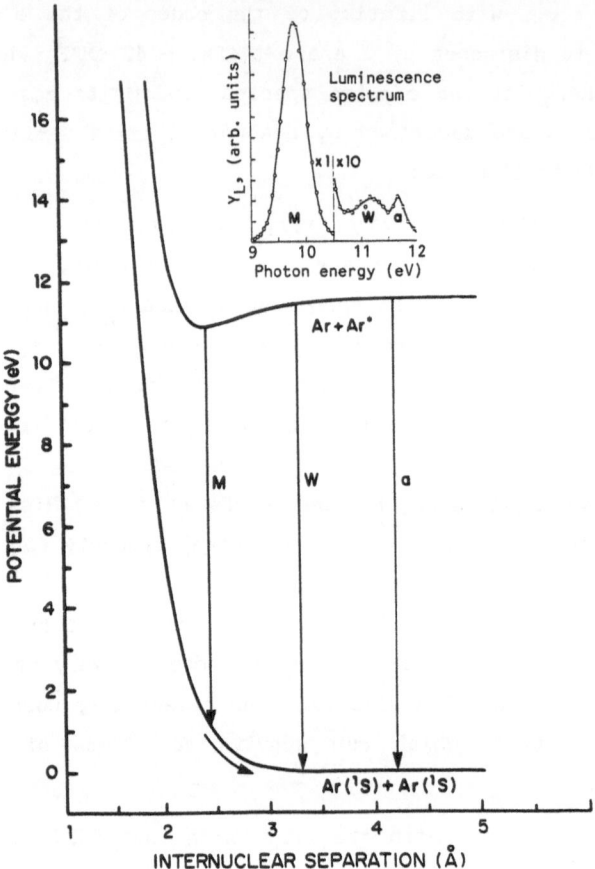

Figure 4. Interaction potential for two isolated Ar atoms, together with transitions corresponding to the luminescence spectrum shown in the inset. *M* and *W* are dimer decays and **a** is a group of unresolved atomic-like decays. (From ref.[31]).

electron is ionized to the vacuum level or to the bottom of the conduction band. It thus appears, that threshold energies may depend on the location of the excitation event and, in most cases, can be uncertain by a few eV.

3.2 Valence Excitations

The lowest thresholds for desorption are usually found to be in the range of typical valence excitations, i.e., below 15 eV for most adsorbed gases. A particular case of DIET by single valence excitations, where the potential-energy curves are known, is the desorption of condensed rare gases. Here, excitation leads to the formation of a molecular exciton, which decays to the repulsive part of the ground-state potential-energy

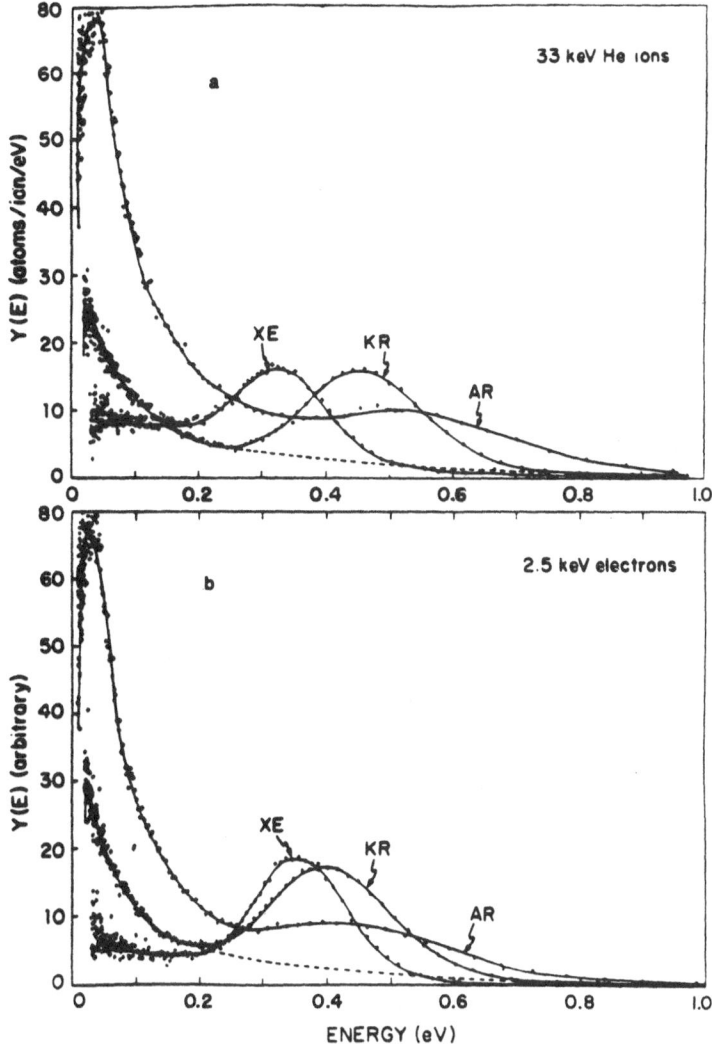

Figure 5. Energy-distribution curves of desorbed atoms for a) ion, and b) electron bombardment, of Ar, Kr, and Xe. (From ref.[31]).

curve [30], as shown in fig.4, giving desorbed atoms of characteristic energies. This has been confirmed experimentally [31], as shown in fig.5.

Another example of DIET by valence excitation is provided by the emission of negative ions for very low electron energy. In dissociative attachment (DA), low energy electrons attach to a molecule (A-B) to form a transient negative ion (A-B)$^-$, which later dissociates into A + B$^-$,

$$e^- + AB \Rightarrow AB^{-*} \Rightarrow A^- + B^*,$$

where * indicates an electronically-excited species.

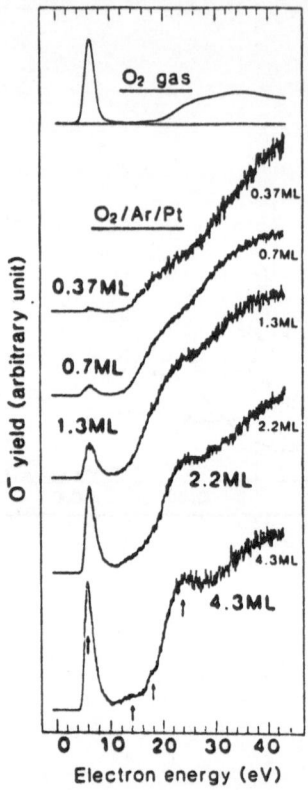

Figure 6. O$^-$ desorption yields, as a function of electron energy, for the gas phase and for 0.1 monolayer (ML) of O$_2$ on Pt. As the thickness of the Ar isolating layer increases, O$^-$ emission resembles more gas-phase processes. (From ref.[32]).

Threshold energies for DA are usually less than 5 eV, smaller than those for production of negative ions via dipolar dissociation,

$$e^- + AB \Rightarrow AB^* \Rightarrow A^- + B^+ + e^-$$

An example of DIET by DA is shown in fig.6, for O$_2$ on Pt covered with insulating spacer layers of condensed Ar [32]. For very thin Ar layers, the substrate quenches DIET by DA, and desorption of negative ions is mainly given by dipolar dissociation. As the thickness of the spacer layer increases, the interaction with the metal becomes weaker, and emission tends to resemble gas-phase DA.

A valence excitation mechanism was the first one proposed for DIET by both Redhead [33] and Menzel and Gomer [34]. It now appears to be relatively unimportant for ion desorption in most cases, as can be deduced by the appearance of sharp onsets in desorption yields, related to the

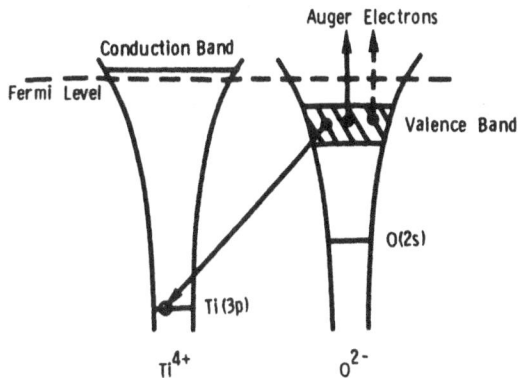

Figure 7. Auger stimulated-desorption mechanism in the maximal-valency oxide TiO_2. Formation of a *3p* core hole is followed by an interatomic Auger decay leading to the formation of an O^+ ion. (From ref.[35]).

excitation of core levels. Correlations between desorption and core excitations point to an Auger mechanism, where bond braking is produced by two final valence excitations or holes resulting from the Auger decay of the core hole. This particular form of DIET is called ASD, for Auger Stimulated Desorption, and is usually more efficient than one-electron excitations.

3.3 Two Hole Final States: Auger Stimulated Desorption (ASD)

The proposal by Knotek and Feibelman (KF) in 1978 [35], that Auger processes constitute a general mechanism for decomposition of oxides, provided a landmark in the study of DIET processes. KF propose that interatomic Auger decay of cation core holes in oxides cause DIET, when the stoichiometry of the oxide is that of "maximal valency", and when the Pauling electronegativity difference is > 1.7. Maximal valency means that the cation is ionized to the noble gas configuration. The importance of this property is that after a core hole is created in either the cation or the anion, the Auger decay will ionize the valence electrons, which are localized in the anion. Thus, in TiO_2 (fig.7), a core hole can decay by a high-order Auger process transforming O^{--} into O^+. This reversal of the Madelung potential at the anion (O) site, produces a strong repulsion from the neighboring cations (Ti^{4+}), which can drive desorption of O^+.

A close look shows that the predictions of the KF model are not accurate. For example, the model predicts that desorption should occur from MgO, but this oxide is known to be very hard to decompose under electron bombardment [36]. The model also predicts no ESD from NiO, because its valence band has both O2*p* and Ni3*p* character, so a core hole created in Ni decays primarily by an intra-atomic Auger process, producing a more positive cation (Ni^{4+}). However, ESD from NiO has been observed [37].

Some of the problems of this model are explained by the concurrent work by Ramaker et al.[38], who recognized that ASD can also be applied to non-ionic solids, where the two-hole (2h) final states were expected to be short lived. The modified ASD model proposes that desorption in covalent systems (in their example, SiO$_2$), starts by a core-hole excitation. Decay of the core hole by an Auger process may result in localization of two holes in a bonding orbital. If the lifetime of the 2h state is long enough, it will relax by Coulomb repulsion into atomic displacements. Two-hole lifetimes can be much larger than one-electron lifetimes, if the ionized state drops below the valence band (this is not possible in a one-electron picture). This is equivalent to the condition that the hole-hole repulsion energy U be larger than the band width W. Holes located below the band edge cannot be resonantly neutralized, but can be filled by slower Auger transitions if the width E_g of the band gap is not too large. Otherwise, they will stay ionized until filled by a radiative transition, in times much larger than desorption times. The important

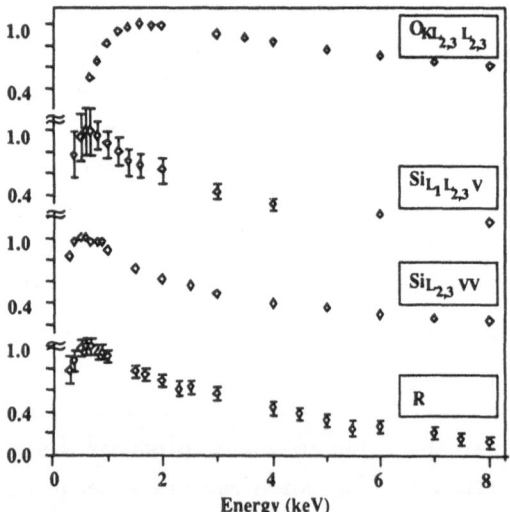

Figure 8. Energy dependence of the ratio R of Auger signals from damaged to undamaged SiO$_2$, and of Auger intensities from different core levels. (From ref.[39])

parameters which determine the likelihood of this 2h mechanism, are the relation between the energies U, W, and E_g.

This extended ASD model was applied to explain why SiO_2 loses oxygen under electron bombardment, and successfully predicted that the energy dependence of the O loss, fig.8, should essentially be that of the cross section for excitation of the *2p* level of Si [39]. The model can also be applied to covalent bonds of atoms to metal surfaces. In this case, the important parameter determining the lifetime of the 2h states is the tunneling rate for electrons in the valence band of the metal, which was described above.

It should be pointed out that double ionization does not require a core hole, so desorption from 2h final states can occur by valence excitations, below the threshold for core excitation, though with a reduced probability.

3.4 ASD by Multiple Electron Excitations

The analysis above shows that, in general, 2h final state are more effective than 1h final states, for producing desorption. This inspires the question: are even more complex excitations important, or at least observable?

The answer comes from close inspection of the predictions of the

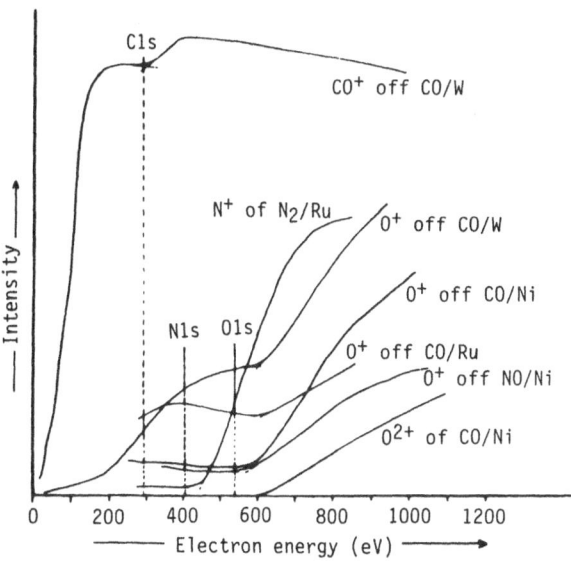

Figure 9. Compilation of ion yields versus electron energy, for desorption from metals. Notice how desorption onsets are related, but delayed, with respect to *1s* thresholds. (From ref.[40]).

2h-ASD model. Many times, sharp rises do not occur at the precise threshold energy for excitation of core electrons, but they are displaced to higher energies by tens of eV, suggesting the occurrence of additional excitations.

This is shown clearly in the compilation of fig.9 [40]. For example, in ESD from adsorbed CO and NO, O^+ and O^{++} onsets are delayed tens of eV from the $O1s$ core threshold at ~ 535 eV. In a more detailed study using synchrotron radiation [41], it was seen that photon-stimulated ion yields from CO/Ni(100) do not follow the absorption coefficient above the $O1s$ level. The delay of the onset, by more than 20 eV, can be attributed to simultaneous excitation of two electrons from the 1σ and 3σ orbital of CO.

Multielectron excitations in adsorbed gases are an order of magnitude more efficient for desorption than ASD with only one electron excitation. They can be even more important in ion emission from insulators, where normal ASD models predict no emission of cations. In SiO_2, the emission of single and multiply charged Si ions has an onset delayed ~ 30 eV with respect to the $Si2p$ level, (fig.10) [42]. To see why multiple electron excitations are needed here, notice that atoms in SiO_2 are in the configuration $Si^{1.6+}$ and $O^{0.8-}$ [43]. Therefore, while Auger decay of a $Si2p$ hole can produce Si^{3+}, this ion will not be emitted since it will be even more attracted to the adjacent O^- ions in the lattice. Ejection is possible if we create additional ionization of O^- to O^+. The particular case of Si^{3+} emission, thus requires the excitation of no less than four valence electrons.

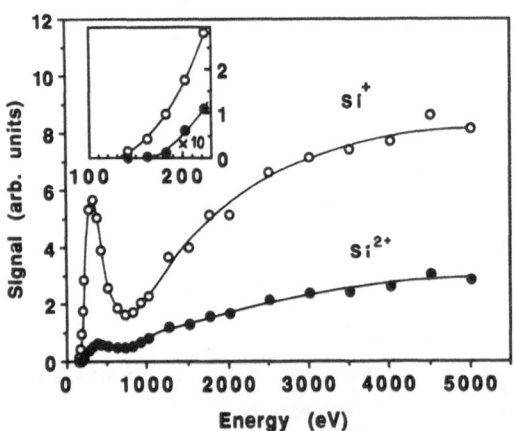

Figure 10. Ion yields from a 65 Å SiO_2 film versus electron energy. Notice that emission is delayed with respect to the $Si2p$ threshold, and that Si^{2+} yields rise more slowly with energy (From ref.[57]). The dip at ~ 800 eV depends on sample treatment.

The need for multielectron excitations can also explain why the energy dependence for desorption of ions, fig.10, does not follow the inner shell excitation or the total damage cross sections, fig.8.

4. THE EXCITATION OF THE REPULSIVE STATE

4.1 Primary Excitations

For photons, or for particles with energies high enough that bulk attenuation and large-angle deflection effects are minimal, the production rate of the initial excitation can be obtained readily from known cross sections. Primary electrons can backscatter and pass through the surface layer again, thus increasing the ionization produced at the surface. The effect will be more pronounced for large incidence angles, since then only soft collisions are required to turn the electron back to the surface.

4.2 Secondary Excitations

Excitations can also occur by secondary electrons produced in the ionization cascade generated by the projectile in the solid [44]. The question is: when are secondary effects important? A determining parameter here is the threshold energy for the production of the repulsive state. If it is high, the effect of secondary electrons should be small, since most secondaries have energies of only a few eV, and desorption yields should be proportional to the excitation cross sections. If threshold energies are low, the number of excitations are expected to be proportional to the electronic stopping power S_e, the total amount of electronic energy deposited by the projectile per unit path length. This has been found to be the case for desorption of ices by electrons and light ions [2]; (at high values of deposited energy, cooperative effects between single excitations make yields proportional to S_e^2 or even S_e^3 [45]).

The investigation of the role of secondary electrons in DIET, can also be done by comparing the shape of photon-stimulated desorption yields with that of photoelectron emission. For instance, the application of this method to desorption from $H_2O/Si(100)$ at the $Si2p$ threshold, has led to the conclusion that desorption in this case is not caused by secondary electrons [46], but is due to direct excitations.

It is possible, however, that the excited state is formed by a cooperation between concurrent inner and outer-shell excitations. A high degree of ionization can increase lifetimes by pulling hole states below the valence band, as mentioned in sect.3.3, and also by reducing the overlap with neighboring atoms, by the reduced size of the ionic orbitals.

4.3 Atomic Motion Prior to Auger Decay

During the lifetime of the core hole, atomic motion may occur due to the change in electronic configuration of the bonding orbitals. That is, atoms can be pulled closer or moved apart depending on the shape of the potential. The amount of motion will depend on the strength of this potential and on the lifetime of the core hole. In the case of free SiH_4, the decay of a $2p$-excited molecule can proceed by a fast dissociation into $(Si2p)H_3 + H$ followed by autoionization of the core excited SiH_3^* fragment [47], made possible by the long lifetime (> 10 ps) of the $Si2p$ hole. A similar mechanism was proposed for the dissociation of core excited HBr [48].

4.4 Lifetimes of Core Holes

Auger rates depend on the type of atom, the level, and the environment. Values of K-shell lifetimes, important for many ASD processes, increase from 0.1 fs to 1.7 fs, in going from Be to Si. These atomic lifetimes are reduced in metals, since the core hole attracts itinerant electrons. In insulators, the lifetime of core holes in cations should be longer than in free atoms, due to the decrease in the valence electron density at the core site. The opposite effect should occur for anions, which present a negative ion configuration. Auger decay in ionic insulators can be described principally by "interatomic" transitions, since the participating valence orbitals are "centered" on different atoms [49]. While interatomic Auger lifetimes are short (1 - 10 fs), intraatomic lifetimes are much longer, due to the decreased orbital overlap. Calculations for the $2p$ levels of Mg/O and Na/O yield long lifetimes, 2400 fs [50], though estimates for the Na $2p$ level in NaF are surprisingly shorter, at 5 fs [51].

4.5. Multiple Auger Decay and Shake-off

The decay of core holes produces mainly double ionization, but a small fraction (~ 10 %) of higher-charged ions is expected, due to multiple Auger decay [52]. Additional ionization may result from shake-off, if the energy above threshold is higher than the binding energy of the outer electron of the inner-shell ionized atom. For CO, the shake-off contribution to ionization is of the order of 20 % [53]. Holes in deep core levels may decay by an Auger cascade [54], producing a Coulomb explosion from a high concentration of holes. The average amount of ionization produced by each core hole increases with the binding energy of the hole, (and with the atomic number Z of the atom). Although many of

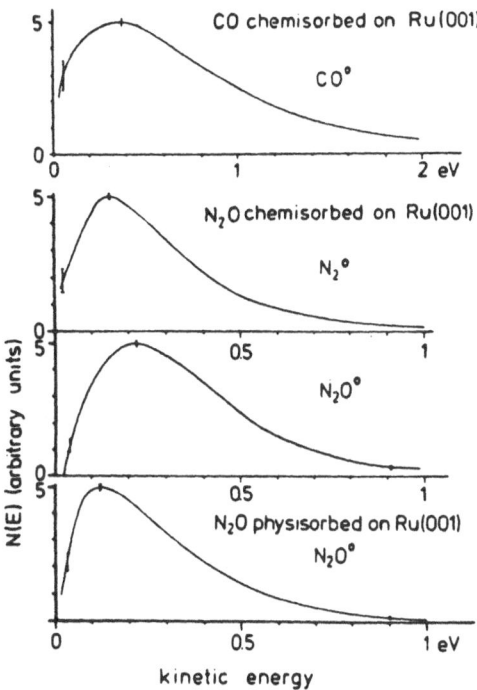

Figure 11. Energy distributions of neutral species desorbed by 250 eV electrons. (From ref.[55]).

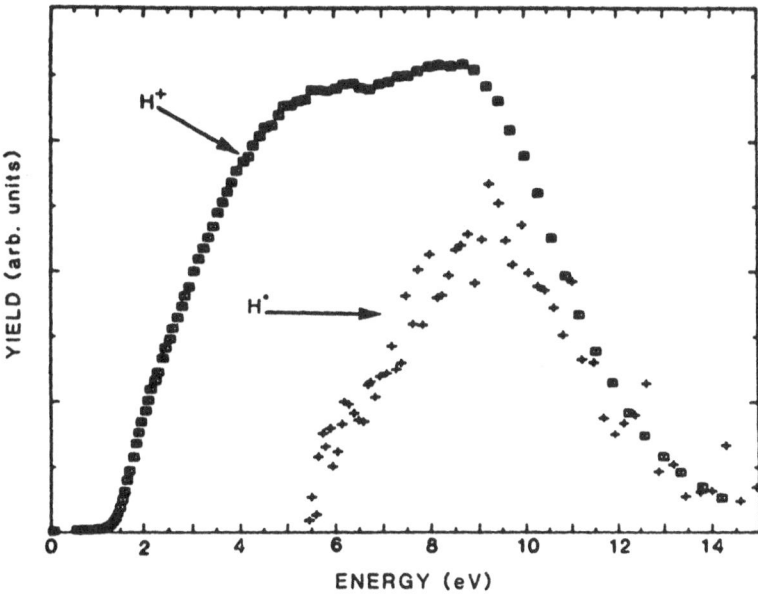

Figure 12. Energy distributions of H^+ and metastable H^* from H_2O ice at 150 K, under 45 eV electron bombardment. The vertical scale is different for the two curves. (From ref.[56]).

these processes may seem unlikely, it must be remembered that in
desorption they can be important, even dominant, through the increase in
hole lifetimes due to multiple ionization.

5. THE EVOLUTION OF THE REPULSIVE STATE

5.1 Energy Distributions

From the discussion in sect.2.1, it can be expected that by measuring
energy distribution curves (EDCs) we can gain information about the shape
of the potential-energy surface which leads to desorption. EDCs are
determined by the interatomic distances at the time of excitation, and by
the shape of the repulsive surface, but also by relaxation effects, and by
energy-dependent electron-transfer effects. Fig.11 shows that EDCs for
neutral species peak at a fraction of an eV, and are very narrow [55].
Ions and excited neutrals have larger energies; EDCs usually peak at a few
eV, as shown in fig.12. Higher energies for ions result from preferential
neutralization of low-velocity particles. This is, presumably, the main
reason why H^+ usually have relatively high energies. Metastable H emitted
from condensed H_2O have larger mean energies than emitted H^+, suggesting
than slow metastables are quenched by resonance ionization to the
solid [56]. Coulomb repulsions involving multiply-charged ions can produce
EDCs with long tails extending to tens of eV, as observed in SiO_2 [57],
and shown in fig.13.

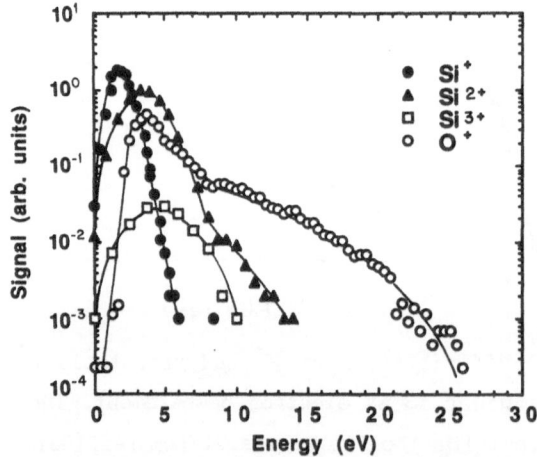

Figure 13. Energy-distribution curves for ions ejected from a 65 Å SiO_2
film by 2 keV electrons. (From ref.[57]).

5.2 Maximum Energy from Coulomb Repulsion

For two isolated ions of charges z and Z, separated by a distance r, the Coulomb repulsion energy is

$$E_c(eV) = 14.39 \; z \; Z \; /r(\text{Å}). \tag{6}$$

This maximum energy is lowered in a solid, due to screening of the Coulomb interaction by the valence electrons.

The repulsion energy is shared between a pair of atoms, and is inversely proportional to their masses, in order to conserve momentum. This tends to make light ions, like H^+, more energetic. However, the trend shown by measurements is that energies are not inversely proportional to mass, but rather, they are of similar magnitude. This points in the direction of "recoilless" desorption [58], where the surface as a whole absorbs all momentum. For Coulomb repulsion, a spread dr in the separations gives an energy spread of (E_c/r) dr, and so these two quantities are related, i.e., the larger the mean energy, the larger the energy spread of the emitted ions.

The value of r relevant for the maximum energy of the desorbed ions, will depend on whether the ionized state of the molecule previous to the Auger decay is repulsive or attractive, as mentioned above. If it is attractive, the final energy of the fragments will be larger than that expected for the equilibrium separation of the initial state, r_0. This effect is expected to be important for Auger lifetimes > 10 fs.

5.3 Lattice Relaxation During Desorption

So far we have considered only pair potentials. In reality, when an atom is being ejected, the other lattice atoms adjust their positions adiabatically to their final, relaxed configuration, thereby slowing down the ejected ion. As shown in fig.14, this rearrangement will be particularly severe for ionic or partially ionic lattices, due to the long-range nature of the interatomic potentials [59]. This effect is considered to be responsible for the apparent absence of direct desorption from alkali halides.

Another possible desorption process involving the motion of lattice atoms, is one in which an initial motion inside the material propagates through collision cascades and reaches the surface, where it can eject an atom which has not participated in the primary excitation event. This phenomenon, similar to the case of collisional sputtering, is very likely present in the desorption of weakly bound, rare-gas solids.

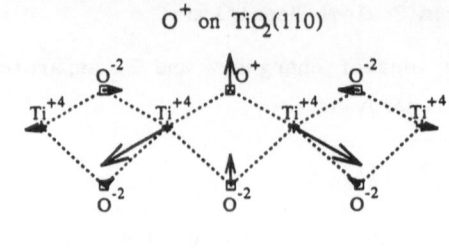

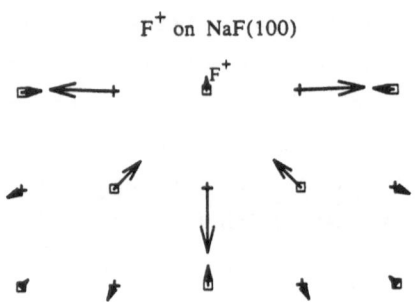

Figure 14. The forces acting after a positive ion is formed at an anion site, for a) O^+ on TiO_2(110), and b) F^+ on NaF(100). In a), large forces act on O^+ in the direction of the surface normal, due to a favorable geometry, so desorption occurs. In b), the net force on F^+ is much smaller, and the lattice distortion prevents F^+ desorption. (From ref.[16]).

A conclusion of sects.5.1 - 5.3 is that the promise of obtaining information on the repulsive state from the EDCs cannot in general be realized, due to the utmost importance of the energy-dependent electron-transfer rates and, in some cases, of energy losses of desorbed atoms in the lattice relaxation accompanying desorption. A further complication is caused by experimental problems, since low ion energies are usually hard to measure, due to contact potential differences, stray fields and poorly known transmission characteristics of mass spectrometers.

5.4 Angular Distributions

The observation of spot patterns due to desorption of alkali-halides [60], showed that the desorbed flux is not isotropic in single crystals. A similar phenomenon occurs in DIET from adsorbed gases, where angular patterns carry information on the bond directions in the adsorbed molecules. This property is the basis of the technique called ESDIAD [61,62], an acronym for *Electron Stimulated Desorption Ion Angular*

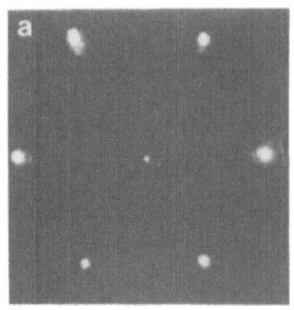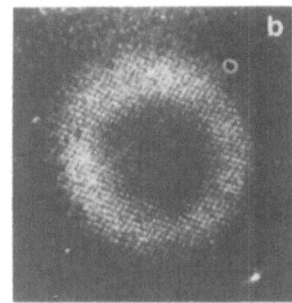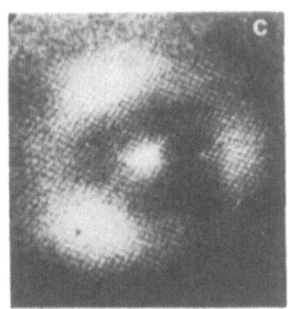

Figure 15. Low Energy Electron Diffraction (LEED) and ESDIAD patterns, for ammonia adsorbed on Ni(111). a) clean LEED at 110 eV; b) H^+ ESDIAD of NH_3 on Ni(111); c) H^+ ESDIAD of NH_3 on oxygen predosed Ni(111). The ESDIAD patterns were taken with 300 eV electrons. (From ref.[65]).

Distributions. In ESDIAD, the desorbed positive ions are accelerated into a channelplate electron multiplier. The shower of electrons issuing from the channelplate then forms the image by striking a phosphor screen, or other suitable position-sensitive detector. Recently, the technique has been extended to the detection of metastable atoms [63] and negative ions [64].

What determines the angular patterns in ESDIAD? For bonds which are singly coordinated, or binary, a transition to an antibonding state should produce a repulsive force directed, by reasons of symmetry, along the axis of the two atoms. In general, we do not observe sharp spots, but actually broad distributions of ions. The beam widths are influenced by bending vibrational modes in the surface molecules.

In practice, ESDIAD has proved to be a valuable technique for the study of structures of adsorbed layers. Unlike LEED, the information is direct, and since the imaging forces are local, long-range order is not required for directed ion emission. An example of ESDIAD is shown in fig.15, for H^+ from NH_3 adsorbed on Ni(111) [65]. A halo is observed, showing that molecules are bonded via the N atom with H atoms, pointing away from the surface. The H atoms have random azimuthal orientations, probably due to free rotations around the Ni-N bond. When a small quantity of oxygen is adsorbed, the O-H interactions lock the ammonia molecule to a fixed orientation, as in fig.16, producing a pattern with three peaks.

Since ions are observed in DIET, other effects may distort the picture. In interpreting ESDIAD patterns, one has to take into account

Model for Orientation of NH₃   **Model for Orientation of H₂O**

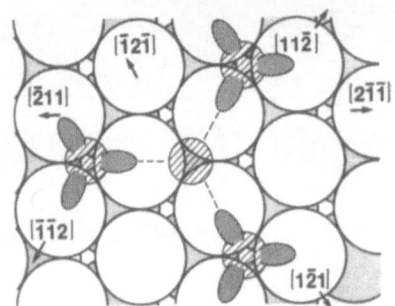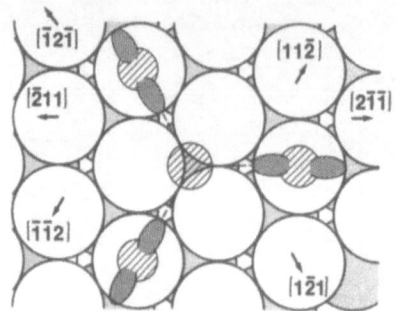

Figure 16. Model of NH$_3$ adsorption on oxygen-predosed Ni(111). (From
ref.[65]).

that the dominant effect of electron transfer will favor emission of
singly-coordinated species, of atoms at edges or steps, of atoms more
distant from the surface. Electron transfer will deplete the intensity for
emission at small angles to the surface. A further complication is caused
by long-range interactions, which deflect the ions from the initial bond
direction [66]. The simplest one is that caused by the image charge
attraction, which tends to curve the trajectories away from the surface
normal. More complicated deflections can occur in the case of other atoms
coadsorbed in an ionic state [67]. Other important factors to consider in
interpreting ESDIAD patterns, are the effect of initial vibrations and
rotations in beam widths, and the existence of possible atomic relaxation
effects between the moments of excitation and desorption events
(cf.sect.4.3).

Strong directional effects, such as those observed in DIET, are of
importance in all DIET measurements made using a detector with limited
aperture [68], and should be considered when evaluating published results.

5.5 Charge Distributions of Desorbed Particles

As mentioned in sects.1 and 2.2, the vast majority of desorbed
species are neutral, due to the importance of electron capture by RN or
AN, also responsible for determining the charge distribution in
collisional sputtering. In insulators, neutralization rates are greatly
reduced, since there are usually no occupied states in the solid opposite
empty states in the desorbing ion. This is due more to the localization of
the occupied valence electrons in oxygen, rather than to the bulk band
structure [69]. A recent study [70] shows interesting correlations between

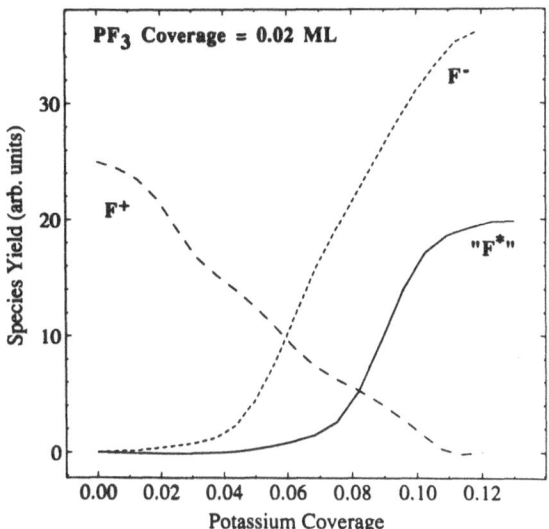

Figure 17. Electron-stimulated desorption yields for F^+, F^- and F^* as a
function of K coverage on Ru(0001) at 90 K, under bombardment
with 200 eV electrons. The coverage of PF_3 is 0.02 ML
(monolayers). All yields are plotted in the same scale. (From
ref.[70]).

the yields of F^+, F^-, and metastable F^* from PF_3 versus work-function
changes induced by co-adsorption of K, fig.17. These data are interpreted
in terms of a model which considers the positions of the affinity level
(F^-), the metastable level (F^*), and the ionization level (F^+), with
respect to the Fermi level, due to K-induced changes in work function.

During electron bombardment of SiO_2, ions are emitted with charges up
to 3+, as shown in fig.18 (also see sect.4.4). The analysis of the charge
distribution is very complex in this case, since it is likely caused by a
convolution of initial charge states following Auger decay and AN during
ejection. AN of multiply charged ions [71], is believed to proceed via a
series of single Auger transitions causing observable electron emission;
it should occur at a lower rate than for singly-charged ions: the orbitals
shrink as a result of the smaller screening, causing less overlap with
neighboring atoms, and therefore a lower probability of hole hopping and
inter-atomic Auger decay (AN) [72]. Electron capture from neutral atoms
leads to the formation of negative ions, even in insulators, as shown by
the recent observation of emission of O^- and Si^- from SiO_2 [73].

Neutralization processes are among the least known steps in DIET.
Some surprising findings which have not been explained, are the general
absence of N^+ emission [74], and the very strong impurity F^+ signals

observed from most surfaces. In fact, F⁺ signals are observed from
surfaces which show no detectable F peak in Auger electron spectroscopy,
and which have, therefore, surface concentrations less than 1 % . Relative
positive ion yields along the second row from C to F, do not correlate
with the ionization potentials. A correlation exists with the long-range
parts of the potential of the neutral atom, (or with the electron
affinities, which in the case of N is zero), but this seems to be
accidental.

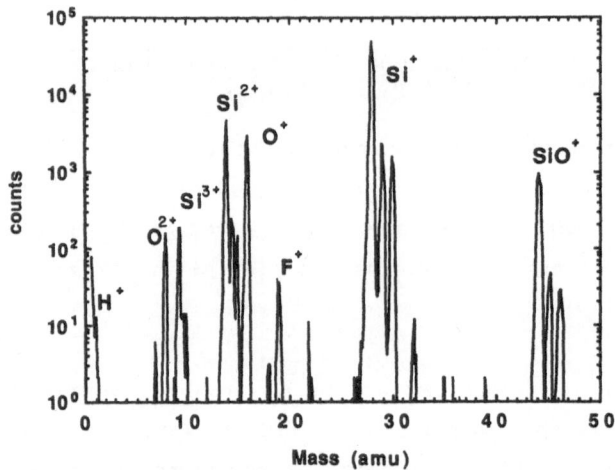

Figure 18. DIET mass spectra from a 5000 Å SiO_2 film bombarded by 2 keV
electrons. (From ref.[57]).

The problem of F⁺ emission has recently been attacked
theoretically [75], with the unexpected result that the equilibrium
distance of F⁺ from an Al surface is larger than for F⁻, due to partial
filling of the fluorine 3s and 3p levels. Although the theory explains how
a repulsive state can be formed, it is not clear how large F⁺ yields can
occur in the presence of strong resonant neutralization to n = 3 states,
which have a large overlap with metal states.

High ion yields are expected when an atom is ejected in an
autoionizing state (e.g., F $2p^3 3s^2$). Due to relatively long lifetimes,
autoionizing atoms will decay into ions far away from the surface, where
reneutralization is unlikely. This process has recently been identified to
explain anomalous high ion yields for Ne scattered from Mg surfaces [76],
but has not yet been proven for DIET.

6. DIET WITH BOMBARDING IONS

6.1 Simple Excitation

Ions carry large momentum, hence sputtering by momentum transfer is always possible above the surface binding energy (a few eV). At high velocities, elastic scattering cross sections become very small and inelastic processes dominate. Electronic desorption by ions can be identified clearly under conditions where sputtering is small. This is the case for light ions (H^+, He^+), or for heavy ions below the sputtering threshold or at very large velocities.

In the limit of very high velocities, ions induce desorption through dipole and direct excitations, in a similar way as electrons of the same velocity, but with cross sections which increase with the square of the charge of the projectile.

At low velocities, several differences exist between electrons and ions. At ion velocities approaching orbital velocities of target electrons, the collisions become nearly adiabatic, as the electrons adjust to the slowly varying atomic fields of the transient molecule formed during the collisions. Unlike electrons which easily scatter from atoms and other electrons, ions have straight trajectories. Also, threshold effects are different, since unlike electrons, ions can have a much smaller velocity than orbital electrons, and still produce excitations.

Heavy ions can increase ASD by several orders of magnitude through Pauli excitation [77]. However, this mechanism should be of secondary importance for desorption by heavy ions, since core excitations are accompanied by large momentum transfers, which are more effective to eject atoms by the normal sputtering mechanism. However, observations have been made of Ar induced H^+ and F^+ emission from impurities in silicon, correlated with the emission of Si2p Auger electrons. The interpretation of those observations is not clear and is still a matter of controversy [78].

6.2 Desorption by Fast Heavy Ions

For fast heavy ions, like fission fragments, the degree of ionization can be so large, keV/Å, that the projectile can produce ionization spikes, i.e., most or all of the atoms along the path are ionized. Since high-energy electrons (δ rays) are ejected outside the path of the projectile, this region can become unstable due to the excess positive charge, and explode by Coulomb-repulsion creating damage tracks in the bulk of the solid [79] and desorption [80]. Fission fragments also produce the remarkable observation that extremely large protein molecules can be desorbed intact from surface layers [81], as a result of expansion of the substrate around the track of the ion.

6.3 Desorption by Electron Capture.

An ion can capture electrons from a target atom, which ends up ionized. This is an additional ionization mechanism, not present in the case of electron and photon excitation, which can lead to desorption [82]. Cross sections for one-electron capture from the outer shells can normally be higher than ionization cross sections, especially when the energy defect in the collision is small [83]. The 2h state, which is more efficient in desorption, can be achieved by additional ionization during electron capture. An incident ion can capture not only bonding electrons, but also inner-shell electrons. This effect is more probable when the velocity of the ion is close to the orbital velocity of the electron to be captured [84]. Electron capture can also occur at low velocities, if a resonant process is possible, like process (C) in fig.3. Such is the case for He$^+$ bombardment of PbS. Oscillations in the damage as a function of

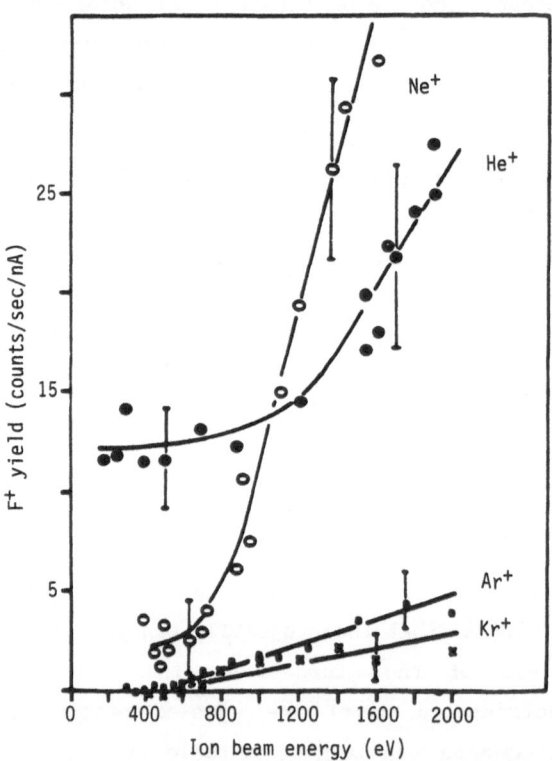

Figure 19. Relative yields of F$^+$ from LiF versus bombardment energy, for singly-charged noble-gas ions. (From ref.[86])

ion energy are correlated with oscillations of the total capture cross sections [85].

Another electron capture process leading to a 2h final state is Auger neutralization, discussed in sect.2.2, but in the context of neutralization of desorbing species. In this case, a slow ion approaching the surface can capture a bonding electron in an interatomic Auger process, where the excess energy is spent in exciting another surface electron, leading to DIET. This mechanism has been demonstrated to be responsible for desorption of F^+ from LiF by low energy He^+ ions [86], as shown in fig.19. Indications for a similar mechanism come also from recent measurements of enhanced ion emission in bombardment of CO/Rh(001) [87], elemental Si [88] and of CO/Ni [89] by multiply charged ions.

7. PROBLEMS AND OPPORTUNITIES

It has surely not escaped the reader that most arguments used to describe DIET are qualitative. In fact, we cannot yet make quantitative calculations of desorption yields, or of ion fractions. Although the intervening factors are identified, we cannot predict the quantitative aspects of ESDIAD patterns with precision, when they occur far from normal ejection. We have not been able to find the way to average out detail to obtain quantitative trends, as has been possible, in many cases, in other topics of particle-solid interactions.

All these failures point to the same culprit: our limited understanding of the details of electron-transfer processes at surfaces, which determine the main aspects of observations. On the other hand, this points out clearly the direction where one should focus the research, which happens to be the same as that followed by other efforts to understand secondary ion mass spectrometry, ion scattering from solids, chemisorption, catalysis, and a variety of photochemical processes at surfaces. Therefore, the opportunity exists for improving our knowledge in a wide variety of fields, by solving some of the questions central to DIET about how electrons are transferred between atoms, molecules, and solids.

REFERENCES

1. Proceedings Diet-I, N.H. Tolk, M.M. Traum, J.E. Tully, and T.E. Madey, eds. Springer Verlag, Berlin (1983).
2. Proceedings DIET-II, W. Brenig and D. Menzel, eds., Springer Verlag, Berlin (1985).
3. Proceedings DIET-III, R.H. Stulen and M.L. Knotek, eds., Springer, Berlin (1988).

4. Proceedings DIET-IV, Springer Verlag, in press.
5. T.E. Madey and J.T. Yates, J. Vac. Sci. Technol. *8* (1979) 525.
6. R.E. Johnson and W.L. Brown, Nucl. Instr. Meth. *198* (1982) 103.
7. M.L. Knotek, Rep. Prog. Phys. *47* (1984) 1499.
8. T.A. Tombrello, Nucl. Instr. Meth. *B2* (1984) 555.
9. T.E. Madey, D.E. Ramaker and R. Stockbauer, Ann. Rev. Phys. Chem. *35* (1984) 215.
10. D. Menzel, Nucl. Inst. Meth. *B13* (1986) 507.
11. T.E. Madey, Science *234* (1986) 316.
12. N. Itoh, Nucl. Instr. Meth. *B27* (1987) 155.
13. J. Schou, Nucl. Instr. Meth. *B27* (1987) 188.
14. W.L. Brown, Nucl. Instr. Meth. *B32* (1988) 1.
15. V.N. Ageev, O.P. Burmistrova, and Yu.A. Kuznetsov, Sov. Phys. Usp. *32* (1989) 588.
16. P. Avouris and R.E. Walkup, Ann. Rev. Phys. Chem. *40* (1989) 173.
17. H.S. Massey, E.H.S. Burhop, and H.B. Gilbody, *Electronic and Ionic Impact Phenomena*, Vol II, Oxford Univ. Press, London (1969).
18. Nuclei moving with relative velocities $< 10^5$ cm/s, proper of ground-state vibrations, displace < 0.001 Å in 0.1 fs; a negligible value, compared to interatomic distances.
19. T.E. Madey, Surface Sci. *36* (1973) 281.
20. W. Brenig, Z. Physik *B23* (1976) 361.
21. R.A. Baragiola, Radiat. Eff. *61* (1982) 47.
22. See for instance P.J. van den Hoek and A.W. Kleynss, Comments At. Mol. Phys. *23* (1989) 93.
23. P. Nordlander and J.C. Tully, Phys. Rev. *B* (in press).
24. H.D. Hagstrum, in *Inelastic Ion-Surface Collisions*, N.H. Tolk, J.C. Tully, W. Heiland, and C. W.White, eds. Academic, NY (1977), p.1.
25. T.E. Madey, J.T. Yates, D.A. King, and C.J. Uhlaner, J. Chem. Phys. *52* (1970) 5215.
26. M.Q. Ding, E.M. Williams, J.P. Adrados and J.L. de Segovia, Surf. Sci. *140* (1984) L264; M.Q. Ding and E.M. Williams, Surf. Sci. *160* (1985) 189.
27. T.R. Hayes and J.F. Evans, Surface Sci. *159* (1985) 466.
28. F.P. Netzer and T.E. Madey, J. Chem. Phys. *766* (1982) 710.
29. This may not hold for very small metal particles in conditions of large ionization density, cf. A. Howie, Nature *320* (1986) 684; I.V. Vorobeva, Ya.E. Geguzin and E.A. Ter-Ovanes'yan, Sov. Phys. Solid State *29* (1987) 1947.
30. R. Pedrys, D.J. Oostra, and A.E. deVries, in Ref.2, p.190.
31. D.J. O'Shaughnessy, J.W. Boring, S. Cui, and R.E. Johnson, Phys. Rev. Lett. *61* (1988) 1635.
32. H. Sambe, D.E. Ramaker, L. Parenteau, and L. Sanche, Phys. Rev. Lett. *59* (1987) 236.
33. P.A. Redhead, Can. J. Phys. *42* (1964) 886.
34. D. Menzel and R. Gomer, J. Chem. Phys. *41* (1964) 3311.
35. M.L. Knotek and P.J. Feibelman, Phys. Rev. Lett. *40* (1978) 964; Surface Sci. *90* (1979) 78.
36. J. Dresner and B. Goldstein, J. Appl. Phys. *47* (1976) 1038.
37. J.A. Kilner and L. Ilkov, Vacuum *34* (1984) 139.
38. D.E. Ramaker, J.S. Murday, N.H. Turner, G. Moore, M.G. Lagally and J. Houston, Phys. Rev. *B19* (1979) 5375; D.E. Ramaker, C.T. White, and J.S. Murday, Phys. Lett. *89A* (1982) 211; D.E. Ramaker, J. Vac. Sci. Technol. *A1* (1983) 1137.
39. L. Calliari, M. Dapor, L. Gonzo and F. Marchetti, in ref.[4].
40. P. Feulner, R. Treichler, and D. Menzel, Phys. Rev. *B24* (1981) 7427.
41. R. Jaeger, J. Stör, R. Treichler and K. Baberschke, Phys. Rev. Let. *47* (1981) 1300; R. Jaeger, R. Treichler, and J. Stör, Surf. Sci. *117* (1982) 133.
42. R. Baragiola, T. Madey and A-M. Lanzillotto, in ref.[4].

43. D.E. Ramaker, J.S. Murday, N.H. Turner, G. Moore, M.G. Lagally, and J. Houston, Phys. Rev. *B19* (1979) 5375. The authors state that the actual Si charge is closer to one.
44. D.E. Ramaker, T.E. Madey, R.L. Kurtz, and H. Sambe, Phys. Rev. *B33* (1988) 2099.
45. K.M. Gibbs, W.L. Brown, and R.E. Johnson, Phys. Rev. *B38* (1988) 11001.
46. C.U.S. Larsson, A.S. Flodström, R. Nyholm, L. Incoccia, and F. Senf, J. Vac. Sci. Technol. *A5* (1987) 3321.
47. G.G. de Souza, P. Morin and I. Nenner, Phys. Rev. *34* (1986) 4770.
48. P. Morin and I. Nenner, Phys. Rev. Lett. *56* (1986) 1913.
49. M. Salmerón, A.M. Baró, and J.M. Rojo, Surf. Sci. *53* (1975) 689.
50. J.A.D. Matthew and Y.Komninos, Surf. Sci. *53* (1975) 716.
51. T.A. Green and D.R. Jennison, in ref.3, p.185.
52. T.A. Carlson and M.O. Krause, Phys. Rev. Lett. *14* (1965) 390.
53. T.A. Carlson and M.O. Krause, J. Chem. Phys. *56* (1972) 3206.
54. T.A. Carlson, in ref.3, p.169.
55. P. Feulner, D. Menzel, H.J. Kreuzer, and Z.W. Gortel, Phys. Rev. Lett. *53* (1984) 671.
56. R.H. Stulen, in ref.2, p.130.
57. R.A. Baragiola, T.E. Madey, A-M. Lanzillotto, Phys. Rev. *B41* (1990) 9541.
58. J.I. Gersten and N. Tzoar, Phys. Rev. *B16* (1977) 945.
59. R.E. Walkup and P. Avouris, Phys. Rev. Lett. *56* (1986) 524; R.E. Walkup and R.L. Kurtz, in ref.3, p.160.
60. A. Friedenberg and Y. Shapira, Surf. Sci. *87* (1970) 581.
61. T.E. Madey, Science *234* (1986) 316.
62. M.D. Alvey and J.T. Yates, Jr., J. Am. Chem. Soc. *110* (1988) 1782.
63. M.D. Alvey, M.J. Dresser and J.T. Yates, Phys. Rev. Lett. *56* (1986) 367.
64. A.L. Johnson, S.A. Joyce, and T.E. Madey, Phys. Rev. Lett. *61* (1988) 2578.
65. F.P. Netzer and T.E. Madey, Surface Sci. *119* (1982) 422.
66. Z. Miscovic, J. Vukanic, and T.E. Madey, Surface Sci. *169* (1986) 405; ibid. *141* (1984) 285.
67. C.Z. Dong, P. Nordlander and T.E. Madey, in ref.[4].
68. T.E. Madey, M. Polak, A.L. Johnson and M.M. Walczak, in ref.3, p.120.
69. M.L. Yu, J. Cables, and D.J. Vitkavage, J. Vac. Sci. Technol. *A3* (1985) 1316.
70. T.E. Madey, S.A. Joyce and C. Benndorf, in *Physics and Chemistry of Alkali Metal Adsorption*, H.P. Bonzel, A.M. Bradshaw and G. Ertl, eds., Elsevier, NY (1989) p.185.
71. P. Varga, Comments At. Mol. Phys. *23* (1989) 111.
72. P.J. Feibelman, Surf. Sci. *102* (1981) L51.
73. A.M. Lanzillotto, R.A. Baragiola, and T.E. Madey, to be published.
74. M.L. Knotek and J.E. Houston, J. Vac. Sci. Technol. *20* (1982) 544.
75. P. Avouris, R. Kawai, N.D. Lang, and D.M. Newns, Phys. Rev. Lett. *59* (1987) 2215.
76. O. Grizzi, M. Shi, H. Bu, J. Rabalais, and R. Baragiola, Phys. Rev. *B41* (1990) 4789.
77. M. Barat and W. Lichten, Phys. Rev. *A6* (1972) 211.
78. P. Williams, in ref.1, p.184.
79. A. Mozumder, in *Advances in Radiation Chemistry*, Vol I, M. Burton and J.L. Magee, eds., Wiley, NY (1969) 1; R.L. Fleischer, P.B. Price and R.M. Walker, *Nuclear Tracks in Solids*, Univ. of California Press, Berkeley (1975).
80. P.K. Haff, Appl. Phys. Lett. *29* (1976) 473.
81. B.U.R. Sundqvist, Nucl. Instr. Meth. *B48* (1990) 517.
82. R. Baragiola, J. Nucl. Mater. *126* (1984) 313.
83. H.S. Massey, E.H.S. Burhop, and H.B. Gilbody, *Electronic and Ionic Impact Phenomena*, Vol IV, Oxford Univ. Press, London (1969).

84. N. Bohr, Kgl. Danske Vidensk. Selskab, Mat.-fys. Medd. *18* (1948) 8.
85. R.A. Baragiola, Proc. 7th. Int. Conf. Atom. Coll. Solids, Moscow State Univ. (1977) p.106.
86. J.A. Schulz, P.T. Murray, R. Kamur, H-K. Hu and J.W. Rabalais, in ref.1, p.191.
87. J. Möller, M. Neumann and W. Heiland, Physica Scripta *T6* (1983) 104.
88. S.T. deZwart, T. Fried, D.O. Boerma, R. Hoekstra, A.G. Drentje, and A.L. Boers, Surf. Sci. *177* (1986) L939.
89. U. Diebold and P. Varga, in ref.4.

344

INVITED LECTURES

IMPACT PARAMETER DEPENDENT ENERGY LOSS FOR HEAVY IONS:

AN OVERVIEW

W.N. Lennard[1] and H. Geissel[2]

1. Department of Physics
   The University of Western Ontario
   London, Ontario N6A 3K7 Canada

2. Gesellschaft für Schwerionenforschung, GSI
   D-6100 Darmstadt
   Federal Republic of Germany

ABSTRACT

The correlation between inelastic energy loss and scattering angle (or impact parameter) for heavy ions transmitted through thin foils is discussed. To date, experimental data indicate a large angular dependence of the energy loss exceeding that predicted by either pathlength enlargement or (nuclear) elastic loss for projectiles as heavy as ^{58}Ni. The relation between measured energy-angle distributions and single-scattering cross sections is elucidated, focusing on the effects of multiple scattering.

1. INTRODUCTION

In passing through matter, heavy ions lose energy and change their original direction of motion via collisions with atoms of the stopping medium. Generally, the stopping processes can be described by electronic interactions arising from excitation or ionization of the collision partners, and by elastic collisions with the atomic (screened) nuclei comprising the target. The latter collisions impart directional changes to the projectiles, while the former class of interactions are principally responsible for the slowing-down behavior. In has usually been assumed that these two processes are independent, i.e. uncorrelated [1]. In very close collisions, the electronic energy loss is essentially constant and the nuclear energy loss is strongly dependent on the scattering angle. Most data on heavy-ion stopping have been derived from so-called transmission experiments in which the energy loss (or velocity loss for time-of-flight measurements) of projectile ions is measured after traversing a target whose thickness is less than the particle range.

Interaction of Charged Particles with Solids and Surfaces
Edited by A. Gras-Martí *et al.*, Plenum Press, New York, 1991

347

Often, the acceptance angle of the detector is restricted and not all of the transmitted particles are detected. The stopping power, $-dE/dx$ or S, is given by the expression

$$-\frac{dE}{dx} = \lim_{\Delta x \to 0} \frac{\Delta \overline{E}}{\Delta x}, \tag{1}$$

where $\Delta \overline{E}$ is the mean energy loss after a pathlength Δx. The stopping power is therefore defined for conditions where *all* forward directed ions are collected. It is also tacitly assumed that the target thickness is *much smaller* than the particle range. For projectiles heavier than hydrogen, the degree of excitation and ionization plays an important role in the corresponding slowing-down rate.

For heavy ions, where the elastic interactions are *not* negligible compared to inelastic effects, it can often be experimentally quite difficult to retrieve the cross section for inelastic energy loss as a function of impact parameter, or scattering angle, from measurements. Multiple-scattering processes complicate the interpretation of thickness and angle dependent results in terms of single-collision phenomena. In this paper, we will discuss recent evidence for impact-parameter dependent electronic energy loss for ions traversing thin solid targets, and comment on the feasibility of extracting values for this dependence from experimental data.

2. HISTORICAL

For light ions, the impact-parameter dependence of the average energy loss by electronic excitation and ionization was first treated using Lindhard's local density approximation (LDA) of the free electron gas model [2,3]. This approach was reviewed recently by Ziegler et al.[4], and it has been shown to yield accurate values of stopping powers. Investigations of the stopping of energetic light ions in single crystals have suggested that the energy loss depends on the ion trajectory. In fact, measured ratios for the energy loss in channeling directions and in random directions vary from almost unity to a factor or four [5-8]. All of these data were obtained from large-angle scattering investigations. It is, in fact, these applications of ion-beam physics that provide the driving force to understand the details of inelastic electronic energy loss.

The energy loss for low-velocity heavy ions was described by Firsov [9] in an analytic fashion. He estimated the cross section for inelastic interaction as a function of both the relative ion velocity and the impact parameter in a two-body collision. Thomas–Fermi (i.e. statistical) atoms were used to obtain electron distributions. The Firsov description for the impact-parameter dependence of electronic stopping is given by

$$\Delta E_e(b) = \frac{9.5(Z_1 + Z_2)^{5/3} \, v/v_0}{\{1 + 0.31(Z_1 + Z_2)^{1/3} \, b\}^5},\tag{2}$$

where $0.25 \leq Z_1/Z_2 \leq 4$, $\Delta E_e(b)$, in eV, is the energy loss associated with the impact parameter b, and b is in Å. With the usual expression for the stopping power, $S_e = \int \Delta E_e(b) \, d\sigma$, and $d\sigma = 2\pi \, b \, db$, we obtain

$$S_e = 5.15 \times 10^{-15} (Z_1 + Z_2) \, v/v_0,\tag{3}$$

with S_e having the units of [eV cm^2/atom]. Obviously, Firsov's expression has appropriate boundary conditions, i.e., $\Delta E_e(b)$ approaches zero for large impact parameters, and is finite (and takes on its *maximum* value) at b = 0. In a solid medium, $\Delta E_e(b)$ must be finite for $b \rightarrow b_{max}$. The impact parameter cannot become infinitely large since the next atom is only a nearest-neighbor distance away. This raises the question of reasonable choices for this cutoff impact parameter, b_{max}. In integrating the Firsov expression (2) above, we have explicitly taken the upper limit to be infinite. The Firsov theory is successful in describing the general trends for low-velocity heavy-ion stopping. Obviously, the so-called Z_1 oscillations [10-12], or the Z_2 dependence of the Z_1 oscillations [13], are not reproduced by this formalism. Since Firsov did not intend for this theory to deal with fluctuations in energy loss, we will not pursue its application to the area of energy-loss straggling.

Oen and Robinson [14] suggested a different functional form for $\Delta E_e(b)$, since the Firsov form was believed to be unsatisfactory for light ions. They suggested

$$\Delta E_e(b) = \frac{0.045 \, k \, E^{1/2}}{\pi a^2} \exp(- 0.3 \, \frac{r_0}{a}),\tag{4}$$

where $k \, E^{1/2}$ is the electronic stopping cross section in the low-energy regime and r_0 is the distance of closest approach. The forefactor was chosen such that at higher energies, when $r_0(b)$ approaches b, the full Lindhard-Scharff (LS, see below) electronic stopping is retrieved. This particular treatment has no restriction on the ratio Z_1/Z_2, but is not so appropriate for small impact parameters. However, we note that small impact parameters are not weighted very heavily in calculating integrated values for ΔE_e.

The electronic stopping was also investigated in the LS approach [15,16]. Although more precise tabulations have emerged since that seminal treatment, the LS value remains an invaluable reference standard, and the predictions of that theory have always been shown to be correct with an accuracy better than a factor of two. The velocity-proportional electronic stopping may be expressed as

$$S_e(\varepsilon) = k_1 \, \varepsilon^{1/2}, \tag{5}$$

where S_e is the reduced (dimensionless) electronic stopping power,

$$k_1 = \frac{0.0793 \, Z_1^{2/3} \, Z_2^{1/2} \, (M_1 + M_2)^{3/2}}{(Z_1^{2/3} + Z_2^{2/3})^{3/4} \, M_1^{3/2} \, M_2^{1/2}}, \tag{6}$$

and M_1 and M_2 are in amu. In more familiar terms, (6) can also be written

$$S_e = \frac{8\pi e^2 a_0 \, Z_1^{7/6} \, Z_2}{(Z_1^{2/3} + Z_2^{2/3})^{3/2}} \frac{v}{v_0}. \tag{7}$$

Although experimental results have suggested that the exponent in equation (5) may differ slightly from 0.5, or that straight-line fits with

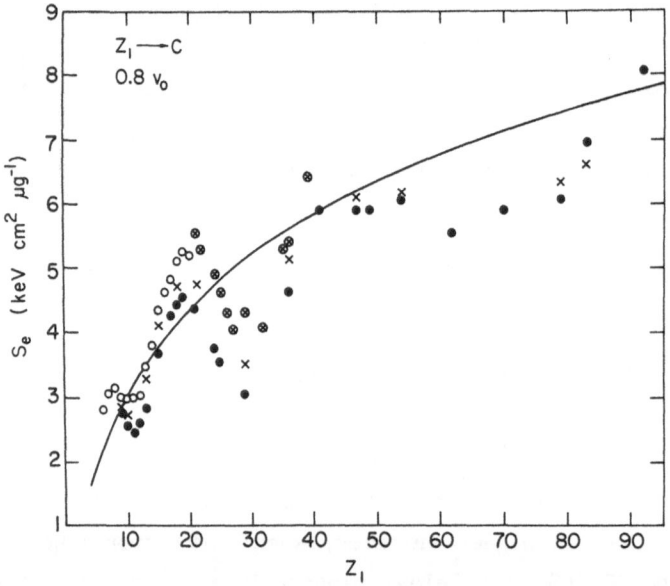

Figure 1. Derived electronic stopping power values for $v = 0.8 \, v_0$ projectiles incident on carbon targets: (•) target thickness ~ 5 μg cm^{-2}; (×) target thickness ~ 29 μg cm^{-2}; (o) data from Fastrup et al.[10]; ⊗ data from Hvelplund and Fastrup [11]. The solid curve shows the Lindhard-Scharff [15] result.

the velocity do not always extrapolate through the origin [17,18], the general features are well described by the LS treatment. Fig.1 shows derived S_e values for many low-velocity heavy ions measured in carbon targets; the LS treatment appears to describe the mean behavior. The observed oscillations are the result of atomic shell structure, see refs.[10-13].

For the medium-velocity region, the so-called Brandt-Kitagawa (BK) model has recently been developed [19-21]. This model implicitly contains the idea that the energy loss increases for small impact parameters due to reduced nuclear screening. The details of this description involve the Fermi velocity of the target electrons, the fractional ionization and the screening length of the projectile. Basically, the BK theory reduces to an effective-charge description.

The 1970's saw a resurgence in the study of inner-shell excitation phenomena (see ref.[22] for an excellent review of this topic). A plethora of evidence appeared which indicated that the inelastic energy loss was indeed correlated with projectile scattering angle, a detail that was, strictly speaking, ruled out when formally separating the nuclear and electronic energy loss. Kessel and co-workers published several works on gas-phase collisions between heavy collision partners where large probabilities for inner-shell excitation were observed, particularly for small impact parameters characteristic of the relevant shell radii. It is interesting to compare conceptually the idea of overlapping electron clouds, inherent in Firsov's description, with the idea of so-called "Pauli excitation", a term coined by Werner Brandt, due to overlap of e.g. L shells in inner shell excitation. The large Q values associated with these inner-shell events led to speculation by Thompson and Neilson [23] that such effects could have anomalous effects on particle ranges in amorphous media. A specific experiment by Mitchell and Lennard [24] for rare-earth projectiles incident on Al failed to confirm the initial suggestion.

3. IMPACT PARAMETER DEPENDENCE

Let us now focus on the impact-parameter dependence of electronic stopping for heavy charged particles in amorphous targets as studied in foil transmission experiments. We will distinguish two regions here: (i) more or less point charged particles, i.e. protons and alpha particles; and (ii) projectiles with $Z_1 > 2$. Experimentally, measurements of both the angle and thickness dependence of energy loss have been undertaken to investigate this problem. We note that for vanishingly small target thicknesses, the energy transfer in a single collision, $Q(\alpha)$, with corresponding scattering angle α, is given by

$$Q(\alpha) = \Delta E(\alpha) - \Delta E(0^\circ) - \Delta_n E(\alpha) - \frac{1}{4}\alpha^2\,\Delta E(0^\circ), \qquad (8)$$

where $\Delta_n E(\alpha)$ is the nuclear elastic loss corresponding to the emergent angle α and the last term in (8) represents the increase of the observed energy loss when going from $\alpha = 0^\circ$ to $\alpha > 0^\circ$ due to pathlength enlargement in the small-angle approximation. Thus, expression (8) represents the single-collision limit when multiple scattering can be ignored. Measurements are commonly made for the quantity $\Delta E(\alpha) - \Delta E(0^\circ)$. From there, one must try to determine $Q(\phi)$, where ϕ refers to a single-collision scattering angle which is not necessarily identical to the emergent angle, α.

We note here that it is, ironically, *not* necessary to have a detailed knowledge of angle-dependent electronic stopping in order to describe the angular, or multiple scattering, distribution of transmitted particles traversing thin amorphous targets at low velocity. We show in fig.2 results of measurements of some relevant distributions, together with those calculated using the Sigmund-Winterbon theory [25] or from Monte

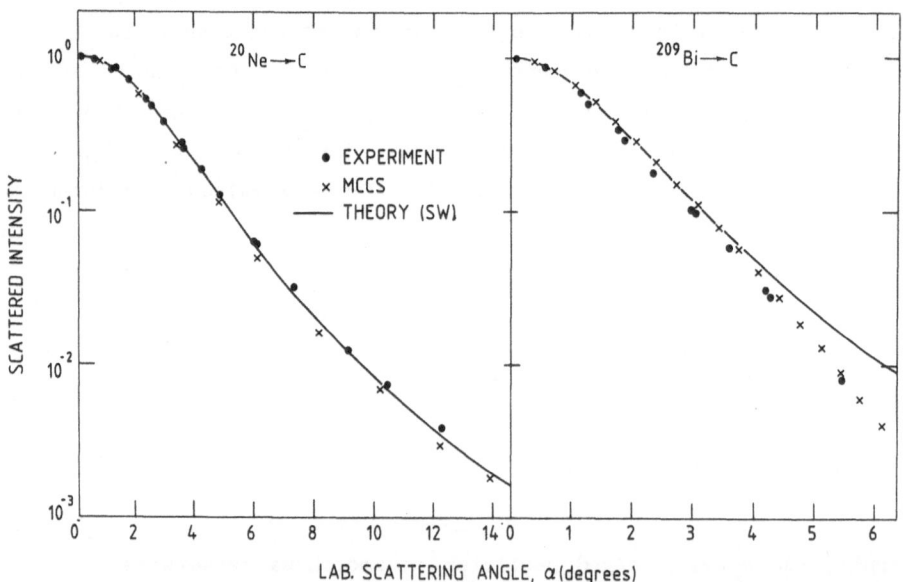

Figure 2. Measured scattering distributions ($\bullet$) for $v = 0.8\,v_0$ ^{20}Ne and ^{209}Bi ions transmitted through a carbon foil of thickness 13.7 μg cm^{-2}. The solid curves are the results of analytic calculations, ref.[25], and of Monte Carlo computer simulations, ref.[26].

Carlo computer simulations [26]. For all of the simulations (denoted by MCSS), a repulsive Molière potential has been used, yielding excellent agreement with the measured scattering intensities for a wide variety of collision systems. A very recent result by Winterbon [27] has been able to reproduce experimental data for scattering angles that *exceed* the maximum single scattering angle for very heavy projectiles (specifically ^{209}Bi) on a low-Z_2 target -in this case, carbon- see fig.3. An obvious reality is that, in a first approximation, those ions emerging from a thin foil in the forward direction (i.e., near $\alpha = 0°$) have *not* suffered collisions with scattering angles exceeding the multiple-scattering angular width.

The Ishiwari group [29-34] has reported several measurements of an anomalous angular dependence of proton energy loss in a host of target materials, thereby renewing interest in this topic. Gras-Marti [35] and Mikkelsen and Sigmund [36] have investigated those collision systems theoretically. Their findings, together with experimental evidence from

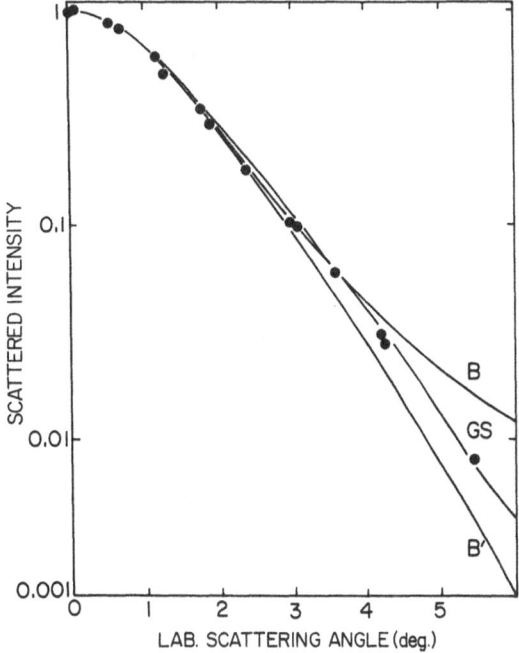

Figure 3. Calculation of the scattering distribution for $v = 0.8\ v_0$ ^{209}Bi ions transmitted through a carbon foil of thickness 13.7 μg cm^{-2}. The data and the calculated curves are normalized to a peak height of unity. The upper curve, (labelled B), is the ordinary Bothe theory calculation, the middle, (labelled GS), is from Goudsmit and Saunderson [28], and the lower, (labelled B'), is the Bothe theory with the single-scattering cross section cutoff at the maximum scattering angle.

Geissel et al.[37], suggest that the Japanese experimental data are not consistent with estimates for the impact-parameter-dependent electronic energy loss. Specifically, their angle-dependent data show a much larger increase with projectile scattering angle than is predicted theoretically. The source of this discrepancy is not clear at this time. Likewise, the data reported by Reid and Scanlon [38] concerning dE/dx values that *increased* with decreasing Au foil thickness were most probably the result of an experimental artifact.

Iferov and Zhukova [39] measured the angular dependence of energy loss for 100 - 400 keV protons transmitted through thin gold films. They observed a saturation for values of $\Delta E(\alpha) - \Delta E(0^\circ)$ at large angles; they erroneously equated α with ϕ, the single scattering angle. This interpretation was later shown to be incorrect [40]. The Russian group has since measured new data for both H and He^{+q} ions [41] and performed calculations [42-45] that reproduce their earlier proton data [39]. The general conclusion from their works is that the value for $\Delta E_e(b)$ varies very little at small impact parameters, thus leading to the saturation in measured values for $\Delta E(\alpha) - \Delta E(0^\circ)$. However, they also have presented evidence suggesting that charge-equilibration effects may be significant for He ions moving in solid targets.

Jakas and co-workers [46,47] have looked at H and He energy-loss data for thin solid targets as a function of both scattering angle and target thickness. They have also recently examined the variation of the peak energy loss with the angle of observation [48]. All of their results have been interpreted in terms of the varying participation of different target electrons, i.e., by an impact-parameter dependence of the mean energy loss. They have deconvoluted their experimental data to derive energy-angle information free from effects of multiple scattering. Here again, the angle-dependent loss appears to saturate for large scattering angles, similar to the Russian results, but they have not indicated any effects attributable to charge equilibration.

The situation for heavy ions is more ambiguous. In principle, we expect to observe an angle-dependent energy loss. As mentioned earlier, both Firsov [9] and Oen and Robinson [14] had attempted to incorporate this effect into theoretical treatments. An extensive set of measurements have been made for a variety of projectiles incident on thin carbon targets at low velocity, $v \sim v_0$ [13,49-51]. Fig.4 shows the experimental setup, and typical results for energy-angle measurements are shown in fig.5. If we examine only thickness-dependent data observed in the forward direction, i.e., at $\alpha = 0^\circ$, we see a marked effect: Fig.6 shows these

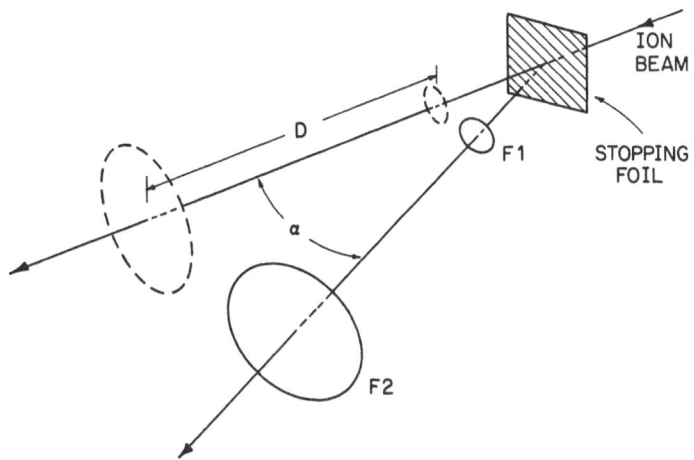

Figure 4. Experimental setup for energy-loss measurements by time-of-flight. The stopping foil is positioned normal to the beam direction. F1 and F2 are time-of-flight detector foils, located on an arm that rotates about the target ($\alpha = -10°$ to $+37°$). The solid angle subtended by F2 is slightly larger than that subtended by F1. The distance D is ~ 80 cm.

thickness-dependent results for the same collision system as shown in fig.5. We believe that charge-equilibration effects are *not* significant for this system, based on experimental measurements. The charge-equilibration length for the identical collision system was measured to be < 1 $\mu g \ cm^{-2}$ (~ 50 Å) [52]. Additionally, we have measured energy-loss values for 1 MeV $^{40}Ar^{q+}$ projectiles transmitted through thin

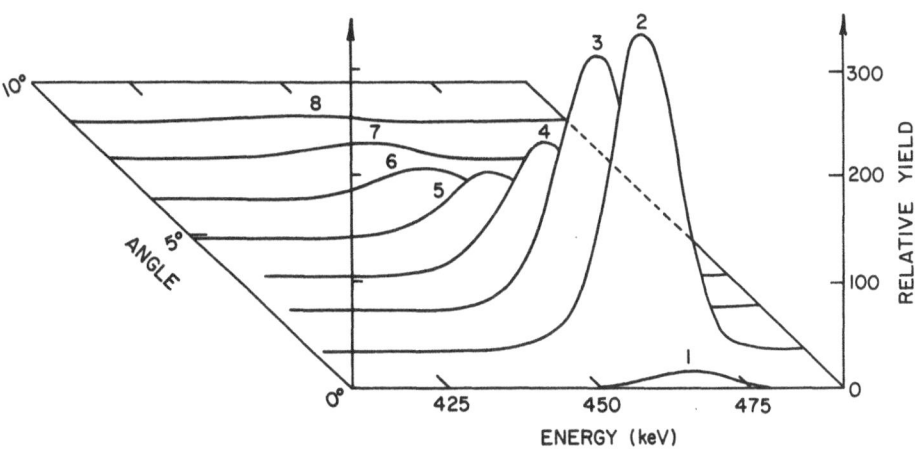

Figure 5. Energy-angle distribution measured for 0.8 v_0 ^{31}P projectiles after traversing a carbon foil of thickness 9.7 $\mu g \ cm^{-2}$. The angular range is from $0°$ to $8.6°$, corresponding to the curves from 1 to 8, respectively.

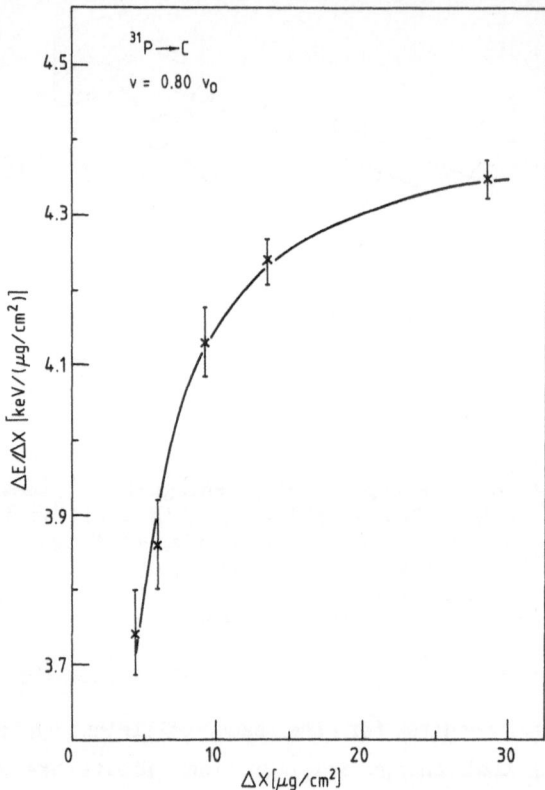

Figure 6. Most-probable specific energy loss measured at 0° for ^{31}P projectiles transmitted through a carbon foil, as a function of foil thickness. The beam velocity is 0.8 v_0.

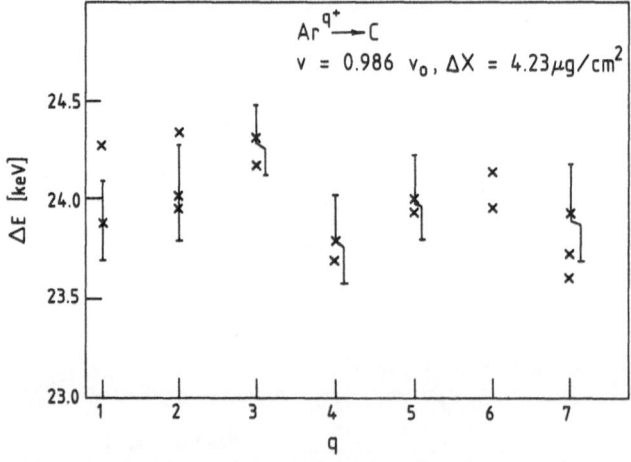

Figure 7. Measured energy loss values for ^{40}Ar^{q+} projectiles at $v = v_0$, transmitted through a thin carbon foil, as a function of the incident charge state. A time-of-flight system was used, wherein the detection system sums over all emergent charge states.

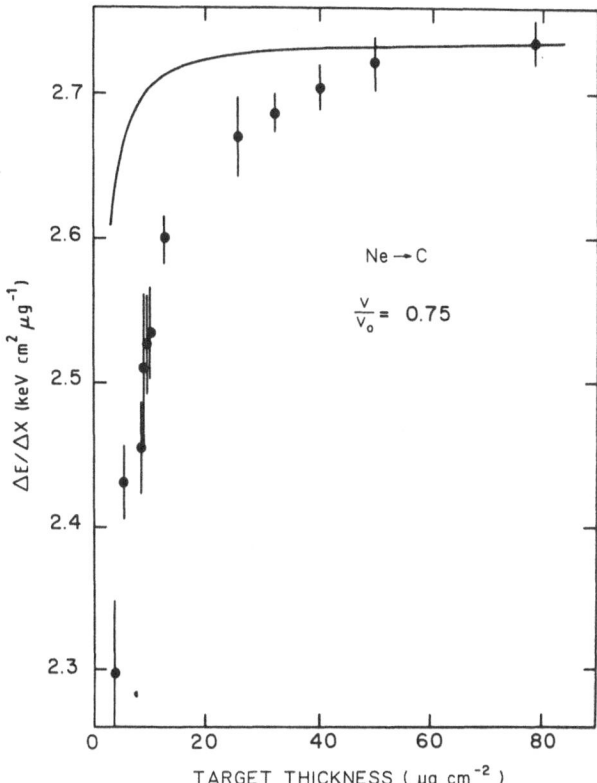

Figure 8. Measured data for the thickness dependence of the specific energy loss measured for ^{20}Ne ions transmitted through carbon foils at $v = 0.75\ v_0$. The smooth curve was produced by a Monte Carlo computer simulation using the Firsov expression for $\Delta E_e(b)$, and normalizing to yield the same value as measured for large foil thicknesses [54].

($\sim 4.2\ \mu g\ cm^{-2}$) carbon foils at 0°, where q was varied from 1 to 7. The results, shown in fig.7, exhibit no significant dependence on the incident charge state, which suggests again that charge-equilibration distances are small for these collision systems in this velocity region. Care must be exercised in the preparation and choice of target materials for experiments using low-velocity heavy ions: the integrity of the targets, particularly with respect to inhomogeneities, is of paramount importance, as emphasized by Mertens [53].

Ziegler et al.[4] have shown the thickness dependence for ^{20}Ne stopping in thin carbon foils at $v \cong v_0$. Early attempts to simulate the low velocity ^{20}Ne data using the Firsov expression were unsuccessful [54], see fig.8. If the observation angle is moved away from the 0° direction, we see a distinctive change in the thickness-dependent data, shown in

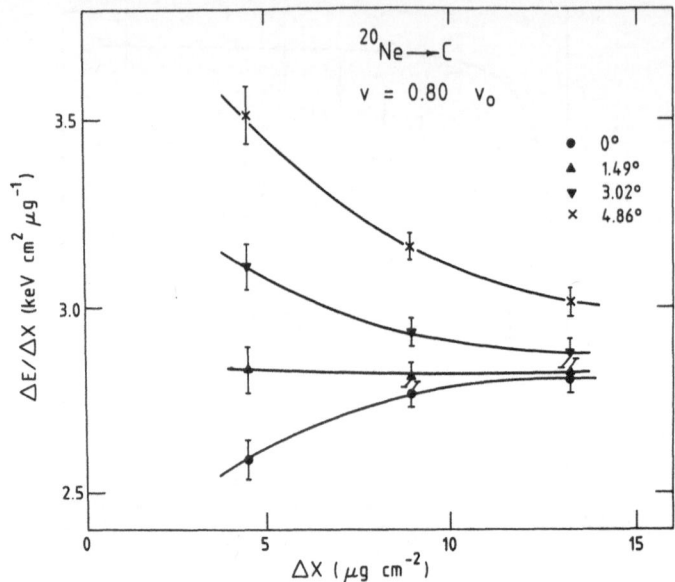

Figure 9. Most-probable specific energy loss as a function of carbon foil thickness measured at four different emergence angles for ^{20}Ne projectiles.

fig.9. For this relatively light system, there is good evidence for significant angle-dependent inelastic loss. Fig.10 shows results for $\Delta E(\alpha) - \Delta E(0^{\circ})$ as a function of emergent angle for target foils of different thickness. The lower panels show the discrepancies between the calculations (which have *no* angle-dependent loss included) and the measured values. As for the light ion (i.e., H or He) data, the inelastic loss appears to saturate with increasing emergence angle, which could be a consequence of the monotonic decrease of $\Delta E_e(b)$ with increasing b, from a finite value at b = 0. The increase of the energy loss with angle for all thicknesses is understood qualitatively by the contribution of small impact parameter collisions of projectiles which are subsequently multiply scattered into the detector.

For heavier incident projectiles, the thickness dependence becomes more enhanced, see figs.11 and 12, for ^{63}Cu and ^{209}Bi ions incident on carbon, respectively. For the very heavy projectiles at this velocity, however, there is no substantial evidence for energy-loss dependence on impact parameter within the experimental and computer simulation uncertainties. The *elastic* loss is very large for these systems, and will obscure smaller effects dependent on scattering angle, or equivalently, impact parameter.

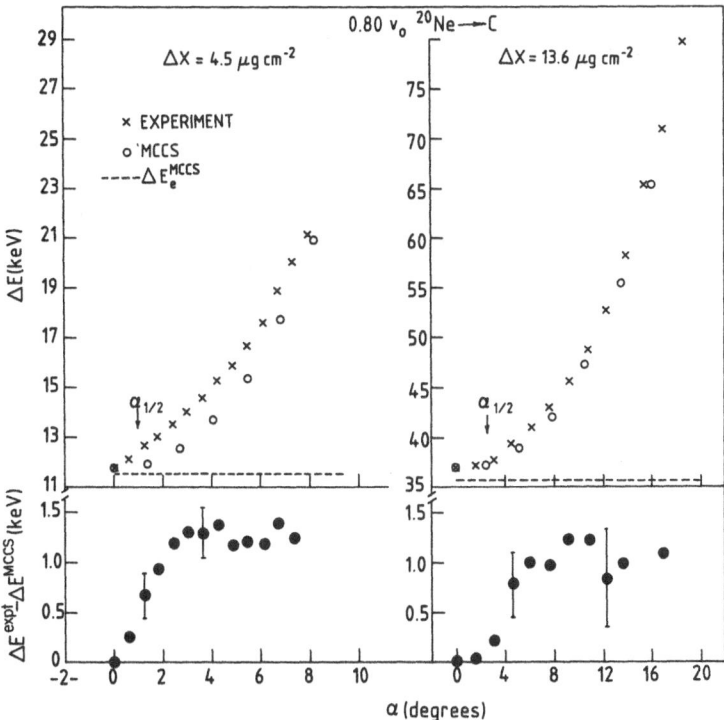

Figure 10. Experimental energy losses (x) for 0.8 v_0 ^{20}Ne projectiles transmitted through carbon targets; the corresponding Monte Carlo computer simulation (MCSS) results are shown by open circles (o), for *no impact-parameter dependence* of ΔE_e(b). The dashed curves show the path-dependent electronic energy loss. The lower panels show the differences between the experimental data and the MCSS calculations.

Finally, even for relatively fast heavy ions (80 MeV Ni^{18+}) traversing thin carbon foils, the thickness dependence of energy loss, shown in fig.13 [55], is qualitatively very similar to the lower velocity data shown in figs.9, 11 and 12 for $\alpha = 0°$. It is also similar to all of the light ion ($Z_1 = 1$, 2) data. The effect can be explained by an angle-dependent energy loss and the influence of multiple scattering for thicker targets, where projectiles with small impact parameter collisions are subsequently scattered into the spectrometer. Calculations suggest that there is approximately a 4 % contribution to the stopping power for large-angle scattering arising from impact parameters smaller that 0.25 a_0. However, equilibrium effects in the projectile excitation cannot be totally excluded in this explanation.

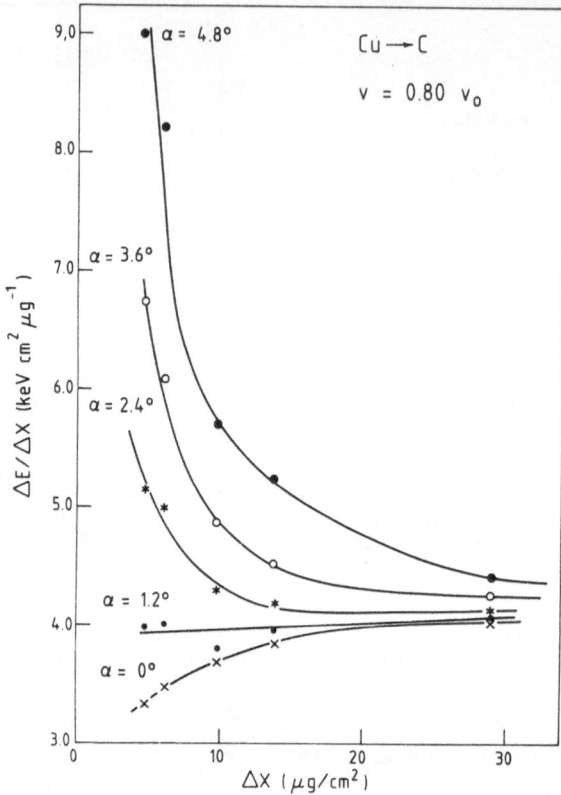

Figure 11. Same as fig.9, except for [63]Cu projectiles.

4. CONCLUSIONS

Based on the most recent experimental results of angle/thickness dependence of electronic energy loss, we conclude:

(1) a significant impact-parameter-dependent energy loss has been observed for a variety of heavy-ion collision systems in foil transmission experiments;

(2) we must learn more about charge-state equilibration effects, e.g., for He ions, in order to deduce the impact-parameter dependent energy loss from measurements of the thickness dependence;

(3) for slow heavy ions, where it appears that the charge-equilibration distances are $< 1 \mu g\ cm^{-2}$, the impact-parameter dependence can probably only be extracted via Monte Carlo computer simulations in conjunction with experimental measurements; and,

(4) the integrity of the targets is of vital importance for thin film experiments of the type discussed in this report. Experimenters must be

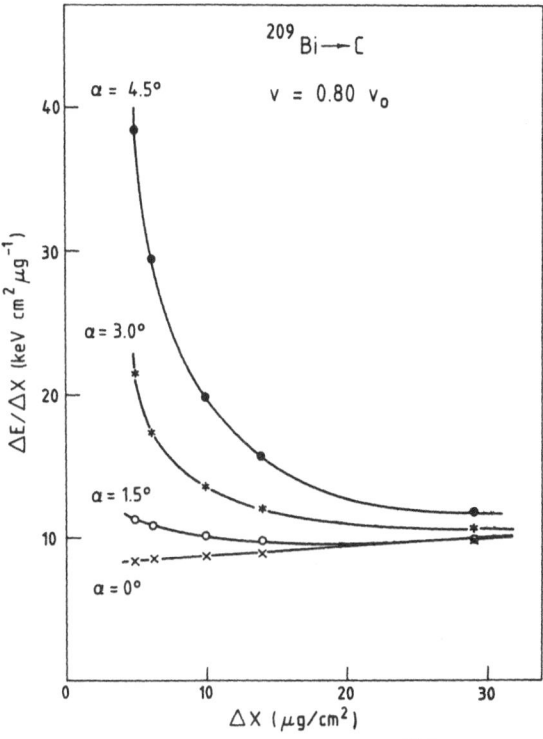

Figure 12. Same as fig.9, except for ^{209}Bi projectiles.

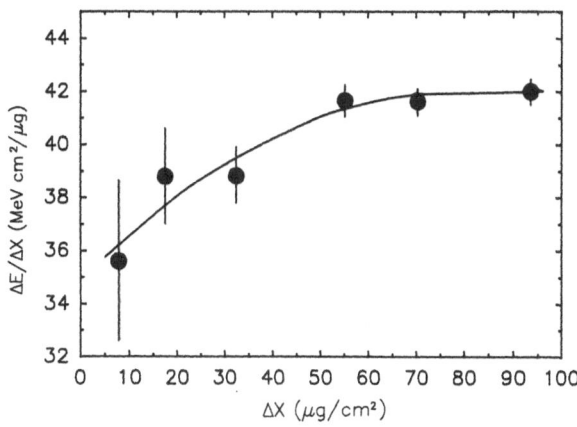

Figure 13. Measured thickness dependence of specific energy loss for 79.6 MeV Ni ions in carbon.

able to demonstrate that their data are free from experimental artifacts caused by, e.g., target thickness inhomogeneities.

Summarizing, investigators have *not* been successful to this point in describing experimental heavy-ion energy loss data using Monte Carlo computer simulations with stopping models that include only elastic and path-dependent electronic energy loss. The computer modeling necessary to derive reasonable impact-parameter dependent descriptions consistent with the data is formidable, since the distributions must be calculated out to rather large scattering angles in order to determine $\Delta E_e(b)$ for small b values. For low Z_1 (H or He) ions, it may be more instructive to study the impact-parameter dependence of the electronic loss in channeling-blocking experiments using high-resolution analyzers.

ACKNOWLEDGEMENT

One of us (WNL) is grateful to NSERC (Canada) for providing part of the financial support for this work.

REFERENCES

1. N. Bohr, Kgl. Dan. Mat. Fys. Medd. *18* (1948) No.8.
2. J. Lindhard and M.Scharff, Mat. Fys. Medd. Dan. Vid. Selsk. *27* (1953) No.15.
3. J. Lindhard, Mat. Fys. Medd. Dan. Vid. Selsk. *28* (1954) No.8.
4. J.F. Ziegler, J.P. Biersack and U.Littmark, *The Stopping and Range of Ions in Solids*, Volume 1, Pergamon Press, Inc. (1984) p.91.
5. E. Bøgh, Rad. Effects *12* (1972) 13.
6. W.F. van der Weg, H.E. Roosendaal and W.H. Kool, Rad. Effects *17* (1973) 91.
7. R.J. Culbertson, S.P. Withrow and J.H. Barrett, Nucl. Instr. and Meth. *B2* (1984) 19.
8. P.F.A. Alkemade, PhD. Thesis, University of Utrecht (1987)
9. O.B. Firsov, Zh. Eksp. Teor. Fiz., *36* (1959) 1517 (see Sov.Phys. JETP, *9* (1959) 1076).
10. B. Fastrup, P. Hvelplund and C.A. Sautter, Mat. Fys. Medd. Dan. Vid. Selsk. *35* (1966) No.10.
11. P. Hvelplund and B. Fastrup, Phys. Rev. *165* (1968) 408.
12. D. Ward, H.R. Andrews, I.V. Mitchell, W.N. Lennard, R.B. Walker and N. Rud, Can. J. Phys. *57* (1979) 645.
13. W.N. Lennard, H. Geissel, D.P. Jackson and D. Phillips, Nucl. Instr. and Meth. in Phys. Res. *B13* (1986) 127.
14. O.S. Oen and M.T. Robinson, Nucl.Instr. and Meth. *132* (1976) 647.
15. J. Lindhard and M. Scharff, Phys. Rev. *124* (1961) 128.
16. J. Lindhard, M. Scharff and H.E. Schiøtt, Mat. Fys. Medd. Dan. Vid. Selsk. *33* (1963) No. 14.
17. M.D. Brown and C.D. Moak, Phys. Rev. *B6* (1972) 90.
18. J.P. Biersack and P. Mertens, *Charge States and Dynamic Screening of Swift Ions in Solids*, p.131, Report CONF-820131, ORNL (1983).
19. W. Brandt and M. Kitagawa, Phys. Rev. *B25* (1982) 5631.
20. W. Brandt, Nucl. Instr. and Meth. *194* (1982) 13.
21. S. Kreussler, C. Varelas and W. Brandt, Phys. Rev. *B23* (1981) 82.
22. U. Wille and R. Hippler, Phys. Rep. *132* (1986) 131.
23. M.W. Thompson and G.C. Neilson, Phys. Lett. *49A* (1974) 151.

24. I.V. Mitchell, W.N. Lennard and O.M. Westcott, Phys. Lett. *60A* (1977) 337.
25. P. Sigmund and K.B. Winterbon, Nucl. Instr. and Meth. *119* (1974) 541.
26. D.P. Jackson, J. Nucl. Mat. *93/94* (1980) 507.
27. K.B. Winterbon, Nucl. Instr. and Meth. in Phys. Res. *B43* (1989) 146.
28. S. Goudsmit and J.L. Saunderson, Phys. Rev. *57* (1940) 24; Phys. Rev. *58* (1940) 36.
29. R. Ishiwari, N. Shiomi and N. Sakamoto, Phys. Rev. *A25* (1982) 2524.
30. N. Sakamoto, N. Shiomi and R. Ishiwari, Phys. Rev. *A27* (1983) 810.
31. R. Ishiwari, N. Shiomi and N. Sakamoto, Phys. Rev. *A30* (1984) 82.
32. N. Sakamoto, N. Shiomi and R. Ishiwari, Nucl. Instr. and Meth. in Phys. Res. *B2* (1984) 164.
33. R. Ishiwari, N. Shiomi, N. Sakamoto and H. Ogawa, Nucl. Instr. and Meth. in Phys. Res. *B13* (1986) 111.
34. R. Ishiwari, N. Shiomi-Tsuda, N. Sakamoto and H. Ogawa, Nucl. Instr. and Meth. in Phys. Res. *B48* (1990) 65.
35. A. Gras-Marti, Nucl. Instr. and Meth. *B9* (1985) 1.
36. H.H. Mikkelsen and P. Sigmund, Nucl. Instr. and Meth. in Phys. Res. *B27* (1987) 266.
37. H. Geissel, K.B. Winterbon and W.N. Lennard, Nucl. Instr. and Meth. in Phys. Res. *B27* (1987) 333.
38. I. Reid and P.J. Scanlon, Nucl. Instr. and Meth. *170* (1980) 211.
39. G.A. Iferov and Yu.N. Zhukova, Phys. Stat. Sol. *B110* (1982) 653.
40. M.M. Jakas, G.H. Lantschner, J.C. Eckardt and V.H. Ponce, Phys. Stat. Sol. *B117* (1983) K131.
41. A.A. Bednyakov, V.Ya. Chumanov, O.V. Chumanova, G.A. Iferov, V.A. Khodyrev, A.F. Tulinov and Yu.N. Zhukova, Nucl. Instr. and Meth. in Res. *B13* (1986) 146.
42. A.A. Bednyakov, G.A. Iferov, A.G. Kadmenskii, N.M. Naumova and Yu.N. Zhukova, Phys. Stat. Sol. *B140* (1986) 63.
43. N.M. Kabachnik, V.N. Kondratev and O.V. Chumanova, Phys. Stat. Sol. *B145* (1988) 103.
44. L.L. Balashova, V.Ya. Chumanov, O.V. Chumanova, G.A. Iferov, A.F. Tulinov, N.M. Kabachnik and V.N. Kondratyev, Nucl. Instr. and Meth. in Phys. Res. *B33* (1988) 168.
45. G.A. Iferov, Yu.N. Zhukova, V.Ya. Chumanov, O.V. Chumanova and N.M. Kabachnik, Nucl. Instr. and Meth. in Phys. Res. *B48* (1990) 43.
46. M.M. Jakas, G.H. Lantschner, J.C. Eckardt and V.H. Ponce, Phys. Rev. *A29* (1984) 1838.
47. J.C. Eckardt, G.H. Lantschner, M.M. Jakas and V.H. Ponce, Nucl. Instr. and Meth. in Phys. Res. *B2* (1984) 168.
48. G.H. Lantschner, J.C. Eckardt, M.M. Jakas, N.E. Capuj and H. Ascolani, Phys. Rev. *A36* (1987) 4667.
49. H. Geissel, W.N. Lennard, H.R. Andrews, D.P. Jackson, I.V. Mitchell, D. Phillips and D. Ward, Nucl. Instr. and Meth. in Phys. Res. *B12* (1985) 38.
50. W.N. Lennard and H. Geissel, Nucl. Instr. and Meth. in Phys. Res. *B27*(1987) 338.
51. H. Geissel, W.N. Lennard, H.R. Andrews, D. Ward and D. Phillips, Phys. Lett. *106A* (1984) 371.
52. W.N. Lennard, T.E. Jackman and D. Phillips, Phys. Lett. *79A* (1980) 309.
53. P. Mertens, Nucl. Instr. and Meth. in Phys. Res. *B27* (1987) 326.
54. J.P. Biersack, private communication (1986).
55. R. Mann, H. Geissel, R. Olson and Th. Schwab, GSI Annual Report (1988).

QUESTIONS CONCERNING THE STOPPING POWER

OF HEAVY IONS IN SOLIDS

H.D. Betz, R. Schramm and W. Oswald

Sektion Physik
Universität München
D 8046 Garching, Germany

ABSTRACT

The passage of heavy ions through solid targets is a topic which continues to puzzle investigators in various aspects. Historically, the problem was specifically treated first by Bohr [1] and Bohr and Lindhard [2]. Since that time, much progress in understanding has been achieved for both slow and fast collisions [3-5]. One of the remaining difficulties arises from the fact that the high spatial density of foil target atoms causes a high rate of electronic collisions which, due to the formation of copious excited states, cannot be treated independently. To circumvent this problem, successful theories of energy loss treat the target on the basis of an extended electron gas [3]. Despite the remarkable advantage for handling of atomic collision processes in this manner, we emphasize that a basic lack of understanding of the particular situation in solid targets still exists especially for heavy ions in certain velocity ranges. This situation is illustrated below.

1. EFFECTIVE CHARGE CONCEPT

In a quantitative calculation of the stopping power, dE/dx, the charge of the moving ion plays a dominant role. For point charges, Z, of not too slowly moving projectiles, the energy loss should be proportional to Z^2. Since heavy ions are not always fully ionized, it is not meaningful to use the nuclear charge; instead, a certain effective charge, Z_{eff}, has been introduced which takes into account all the relevant screening effects. In this way, dE/dx for a heavy ion is simply described by the corresponding stopping power of a proton moving with the same velocity through the same target, multiplied by Z_{eff}^2. Reversing this idea, Z_{eff} can then be determined experimentally from energy loss measured for both the proton and the heavy ion.

For a dilute gaseous target, Z_{eff} differs little from the average equilibrium charge state, Q_g, of the heavy ions. This fact can be easily

Interaction of Charged Particles with Solids and Surfaces
Edited by A. Gras-Martí *et al.*, Plenum Press, New York, 1991

365

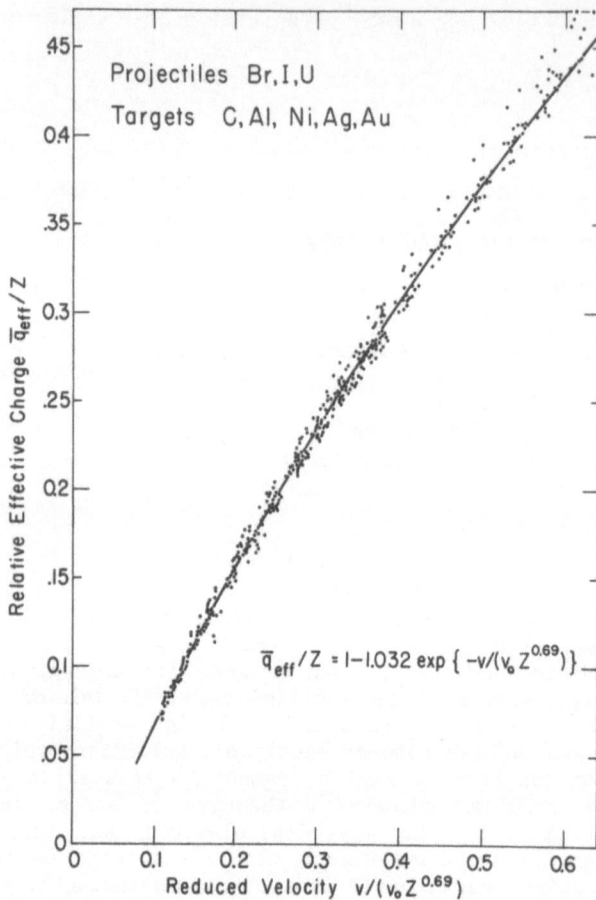

Figure 1. Relative average effective charge of various heavy ions, deduced
from measurements of the energy loss of these ions in solid
targets, plotted as a function of a reduced projectile velocity,
from refs.[6,8].

understood when one considers that (a) the projectile is usually in the
ground state prior to a collision, and (b) relevant impact parameters are
not smaller than the orbital dimensions of the ion.

It has been known for a long time that average equilibrium charge
states, Q_s, of heavy ions after penetration through thin solid targets can
exceed Q_g by a substantial amount. For this reason, Bohr and Lindhard
initially expected that dE/dx of such heavy ions is larger in solids than
in gas targets. However, the energy loss was observed to be not at all
greatly dependent on whether the heavy ion moved through a gas or a solid.
This result was concluded, for example, from a systematic experimental
study carried out by Brown and Moak [6] (fig.1); other similar

measurements were in agreement with those findings. One unexpected consequence is the observation that though Q_g and Q_s may greatly differ in gaseous and solid targets, Z_{eff} is nearly identical in the two cases.

For 120 MeV bromine ions, for example, Q_g and Z_{eff} are close to +17, while Q_s is approximately +23. The difference between Q_s and Q_g, therefore, is substantial and may be even larger in other cases, especially when the ions carry a large electron cloud under charge state equilibrium conditions. Due to the high collision rate inside solid targets, heavy ions are definitely in extremely high excitation states. This is experimentally evident from x-ray measurements. Unfortunately, theoretical treatments of heavy ion energy-loss have not yet taken into account the complex electronic structure of the moving ion. Instead, it is usually postulated from the outset, that Z_{eff} can be taken to be equal to Q_g, no matter whether a gas or a solid is considered as target [4]. Clearly, this procedure is less than satisfactory.

2. ELECTRONIC STATE OF HEAVY IONS INSIDE SOLIDS

There is no doubt that heavy ions reach highly excited states during their passage through a solid. It is less well known to what extent the charge state Q_s, which is measured after the ions exit from the foil target, is already developed inside the solid. Corresponding models have been forwarded by Bohr and Lindhard [2], who assumed that Q_s was indeed effective inside the solid, and by Betz and Grodzins [7], who proposed that the final value of Q_s is formed upon exiting from the solid. A final and unequivocal answer has not yet been given.

If it is assumed that heavy ions traverse the solid with charge Q_s, the approximate equality between Z_{eff} and Q_g is not easily understood. The charge difference $Q_s - Q_g$ must be compensated by screening. In principle, two possibilities can be envisaged. First, target electrons could be attracted by the ions to provide the extra shielding. Calculations of the disturbance of target electrons induced by moving ions do not readily explain the required screening effect. Since projectile velocities, v, may be of the order of 10 times the Bohr velocity, the response of target electrons will be too slow; the wave length of the induced wake, for example, amounts to $2\pi v/\omega_p$, where ω_p denotes the plasma frequency. There is no simple way to argue that, in the example above, the charge equivalent of some 6 extra target electrons moves close enough to the fast moving ion in order to produce the necessary screening to reduce the ion charge from +23 down to effectively +17.

Secondly, one could argue that extra electrons move along with the ion. These electrons could be loosely bound and become ejected from the ion upon emergence from the foil. Alternatively, these electrons could represent continuum states. In both cases, some 5 or 6 free electrons per bromine ion should be detectable behind the solid target. Corresponding measurements have been performed and are described below.

3. FORWARD ELECTRON SPECTROSCOPY

The Munich MP Tandem accelerator was used to produce 125 MeV bromine ions, directed onto thin carbon foils with thicknesses ranging from 10 to 180 $\mu g/cm^2$ (the latter value allows to provide charge state equilibration). The beam was collimated to a diameter smaller than 1 mm and a divergence of less than 0.1^o. A magnetic sector analyzer focusing in two planes was employed to deflect free electrons into a channeltron detector. By rotating the analyzer, electrons emitted into the entire forward cone could be measured. The semi-half-angle of the spectrometer was variable from 0.2^o to 2.5^o, and amounted to 1.2^o in this experiment; momentum resolution could be set from 0.1 to 0.8 % and was chosen to be 0.2 %. The resulting two-dimensional intensity spectrum is shown in fig.2. The large peak arises from convoy electrons, which exhibit velocities centered on the projectile velocity. The smaller and broader peak at approximately twice the beam velocity consists of target secondary electrons (binary encounters).

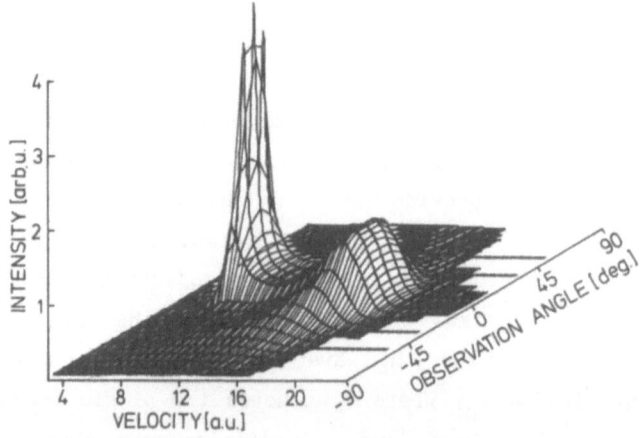

Figure 2. Velocity distribution of electrons emitted in the forward direction when 125 MeV bromine ions penetrate through a 20 $\mu g/cm^2$ carbon foil, plotted as a function of the observation angle relative to the beam direction. The sharp and broad peaks reflect convoy and binary-encounter electrons, respectively.

The sought-after electrons are expected at velocities not too different from the beam velocity, including Auger electrons, and must be contained in the collected data. Quantitative measurement of the detected electron yield reveals that we find no more than 1 electron per projectile ion. This yield does not greatly depend on the target thickness. Although these electrons represent a contribution to the solution of the question from above, their number is not sufficient to account for the screening effect equivalent to some 5 or 6 electrons.

These results suggest that the problem of heavy-ion energy loss should be re-examined, at least for those cases in which the observed average charge states Q_s differ significantly from Q_g and, thus, from the effective charges Z_{eff}. In particular, it might be interesting to give specific attention to the energy transfer associated with charge exchange processes. These events are generally neglected when projectiles are treated as structureless particles.

ACKNOWLEDGEMENT

This work is supported by BMFT (Bonn) under contract number 6ML177I.

REFERENCES

1. N. Bohr, Mat.-Fys. Medd., K. Dan. Vidensk. Selsk. *18* (1848) No. 8.
2. N. Bohr and J. Lindhard, Mat.-Fys. Medd., K. Dan. Vidensk. Selsk. *28* (1954) No. 7.
3. J. Lindhard, M. Scharff, and H.E. Schiott, Mat.-Fys. Medd., K. Dan. Vidensk. Selsk. *33* (1963) No 14; J. Lindhard and M. Scharff, Phys. Rev. *124* (1961) 128.
4. B.S. Yarlagadda, J.E. Robinson and W. Brandt, Phys. Rev. *B17* (1978) 3473; W. Brandt and M. Kitagawa, Phys. Rev. *B25* (1982) 5631; W. Brandt, Nucl. Instr. Meth. *194* (1982) 13; H. Geissel, W. N. Lennard and P. Armbruster, Radiat. Eff. and Defects in Solids *110* (1989) 7.
5. P.M. Echenique, F. Flores and R. H. Ritchie, Nucl. Instr. Meth. *132* (1988) 91.
6. M.D. Brown and C.D. Moak, Phys. Rev. *B6* (1972) 90.
7. H.-D. Betz and L. Grodzins, Phys. Rev. Lett. *25* (1970) 903.
8. H.-D. Betz, Rev. Mod. Phys. *44* (1972) 465.

SCATTERING OF FAST IONS AND ENERGY LOSS ON SURFACES

K. Kimura[1,2] and M. Mannami[2]

1. Dpt. of Physics, University of Tennessee
   Knoxville, Tennessee 37996-1200, U.S.A.
   and ORNL, Oak Ridge, Tennessee 37831-6377, U.S.A.

2. Department of Engineering Science
   Kyoto University
   Kyoto 606, Japan

ABSTRACT

The interaction of fast ions with a solid surface is studied by the specular reflection of MeV H^+ and He^+ ions from clean surfaces of SnTe and PbSe single crystals. Oscillatory structure is observed in the energy spectra of reflected He^+. This indicates that a part of the incident ions penetrate inside the crystal and travel for a few wavelengths of oscillatory motion in the (001) planar channel before reappearing at the surface. The position-dependent stopping power near the surface is derived from the observed energy loss of specularly-reflected ions. A large difference is observed in the charge-state distribution between the specular reflection and the transmission through a self-supporting foil. This is attributed to the fact that the specularly-reflected ions interact exclusively with the valence electrons. The technique of specular reflection of fast ions is applied to investigate the epitaxial growth of PbSe on a SnTe (001) surface. It is shown that the specular reflection is very sensitive to the surface distortion which is introduced in the first stage of the epitaxial growth.

1. INTRODUCTION

The interaction of ions with solid surfaces has been one of the main subjects in the research on ion-solid interactions [1-8]. Many experimental and theoretical studies have been performed on various phenomena such as the charge-exchange process at solid surfaces [1,2], the anisotropic distribution of the orbital angular momenta of excited states of ions produced by ion-surface interactions [3,4], and the dynamical response of target electrons to fast ions traveling near surfaces [5-8]. However, it is often difficult to eliminate bulk effects in experimental studies on ion-surface interactions.

When an ion is incident on a single crystal surface with a small

Interaction of Charged Particles with Solids and Surfaces
Edited by A. Gras-Martí *et al.*, Plenum Press, New York, 1991

371

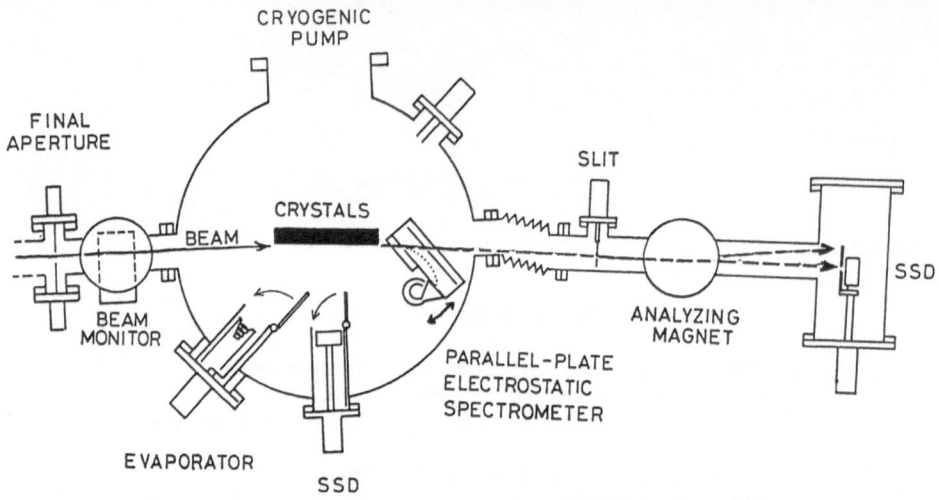

Figure 1. A schematic drawing of the experimental setup.

glancing angle, the ion experiences a correlated series of small-angle scattering with surface atoms, and the ion is reflected at a scattering angle of twice the glancing angle [9-12]. This phenomenon is called "specular reflection". The specularly-reflected ion does not penetrate inside the crystal. This situation is very favorable for the study of ion-surface interactions. In the present paper, we report our results of a series of experimental studies related to the specular reflection of MeV protons and He ions from clean (001) surfaces of SnTe and PbSe single crystals. Processes of energy and charge transfer between fast ions and solid surfaces are discussed.

2. EXPERIMENTAL PROCEDURE

The experimental setup is schematically shown in fig.1. A single crystal of KCl with a (001) cleavage surface was mounted on a high-precision goniometer in the scattering chamber whose base pressure was 3×10^{-10} Torr. A single crystal of SnTe (001), which has an NaCl-type crystal structure was prepared by epitaxial growth in situ, by vacuum evaporation on the cleavage KCl (001) surface at 500 K. A well defined (1×1) pattern was observed from the surface of SnTe (001) by reflection high-energy electron diffraction. A single crystal of PbSe (001), which also has an NaCl-type crystal structure, was prepared by epitaxial growth on a SnTe (001) surface at 500 K.

The incident beams of protons and He ions from 4 the MV Van de Graaff accelerator of Kyoto University were collimated by apertures to less than

0.1 mm × 0.1 mm, and to a divergence angle less than 0.5 mrad. The ions scattered at an angle θ_S, in the plane which contains the incident beam and the normal to the surface, were chosen by a movable aperture. The acceptance angle of this aperture for the scattered ions was 0.9 mrad. The ions passing through the aperture were resolved into their charge states by a magnetic analyzer and measured by a solid-state detector with an energy resolution of 14 keV.

3. SPECULAR REFLECTION

Fig.2(a) shows examples of the observed energy spectra of scattered He$^+$ ions for various scattering angles, for 0.7 MeV He$^+$ incidence on a SnTe (001) surface with a glancing angle of 4.9 mrad. The energy spectra consist of several well-defined peaks separated by equal energy spacings.

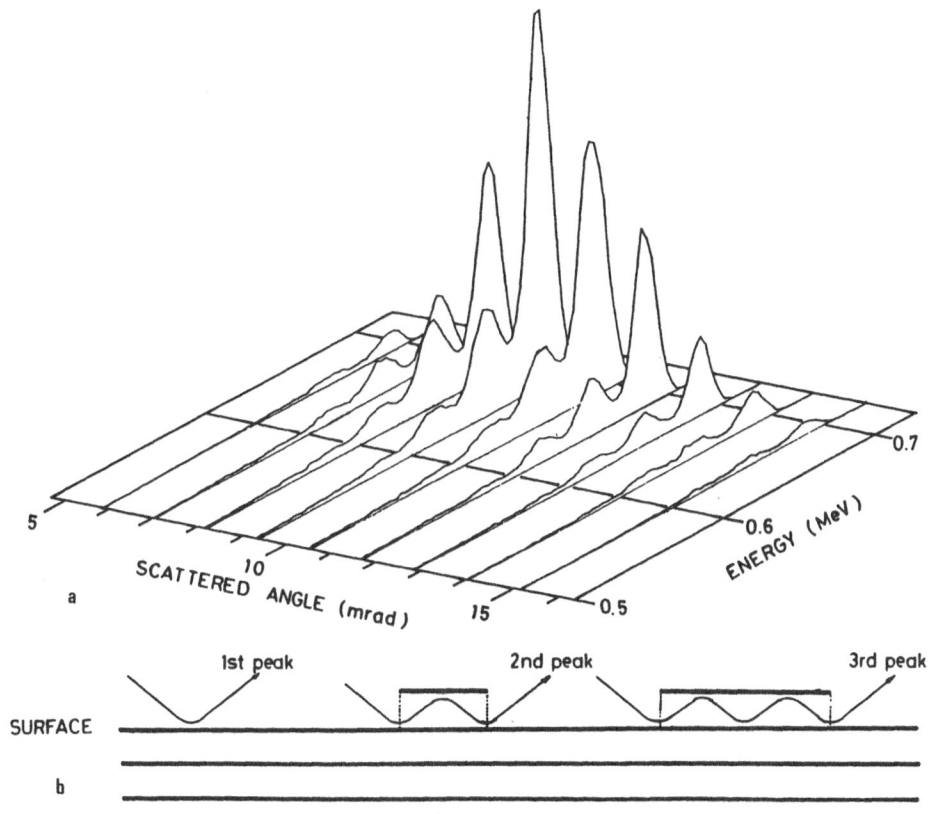

Figure 2. (a) Energy spectra of He$^+$ scattered at various angles, when 0.7 MeV He$^+$ ions are incident on the SnTe (001) surface with a glancing angle of 4.9 mrad. (b) Schematic illustration of the reflection of ions at a single-crystal surface with surface steps.

The ion yields have a sharp peak at a scattering angle of about twice the glancing angle. This indicates that the ions were specularly reflected in the continuum surface planar potential at the single-crystal surface. The observed energy spectra, having an oscillatory feature, resemble those of ions transmitted through a planar channel in the crystal [13]. The present oscillatory structure may be explained in terms of planar channeling. Fig.3 shows the energy spacing between adjacent peaks as a function of the ion energy. This energy spacing increases gradually with increasing energy and exhibits little dependence on either the glancing angle or the scattering angle. The energy loss of (100) planar channeled He ions in SnTe traveling for one wavelength of the channeling motion is also shown in fig.3, which was derived from the surface oscillation in the observed RBS spectra of (100) planar channeling. The energy losses for one wavelength agree with the energy spacing between adjacent peaks. This indicates that the ions of the first peak are reflected from the surface

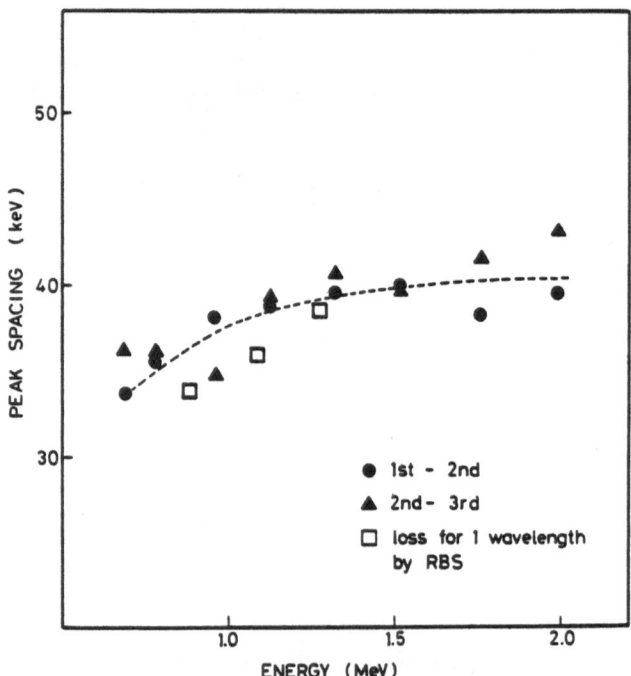

Figure 3. Dependence of the energy spacing between adjacent peaks on the incident energy for He$^+$ on SnTe (001), a glancing angle of 2.9 mrad and a scattering angle of 6.4 mrad. The spacing between the first peak and the second peak (●) and that between the second peak and the third peak (Δ,full), are shown. The energy loss of (100) planar channeled He ions in SnTe traveling for one period of the channeling oscillation is also shown (□).

374

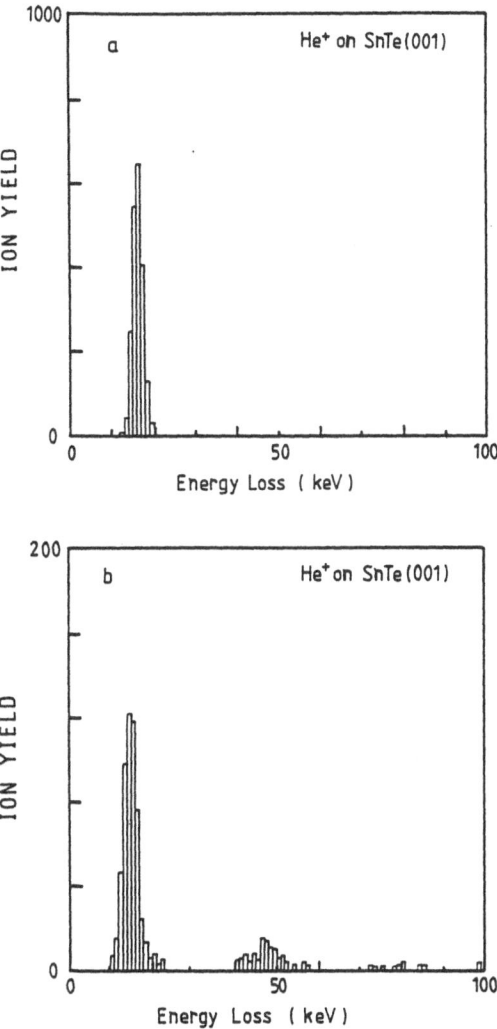

Figure 4. Computer simulation results of the energy spectra of reflected ions in a window 2×2 mrad^2 at the scattering angle 4 mrad, when 0.7 MeV He$^+$ ions are incident (a) on a flat (001) surface, and (b) on a stepped surface with mean step separation D_S = 300 a_0, with a glancing angle of 4 mrad.

atomic plane, and others penetrate inside the crystal and travel for a few wavelengths of oscillatory motion in the (001) channel before appearing at the surface.

If the continuum surface planar potential is used to calculate the ion trajectory, it is found that ions cannot penetrate inside the crystal for this experimental condition. A computer simulation of ion trajectories was performed using a binary collision approximation in order to reveal the mechanism of ion penetration inside the crystal. The simulation showed that the penetration of ions through the surface atomic plane cannot occur even if the thermal vibration of surface atoms, electronic scattering near the surface, and point defects in the surface are taken into account. Ion penetration was found to occur only when surface steps are introduced. Ions incident on the up-step side can penetrate inside the crystal. They have a chance to escape from the surface if they encounter a down-step as described in fig.2(b). Fig.4 shows calculated energy spectra of ions scattered from (a) a flat surface and (b) a stepped surface with mean step separation D_S = 300 a_0, where a_0 = 0.314 nm is the interatomic distance. This mean step separation is nearly equal to one wave length of the oscillatory motion of 1 MeV He ions in the (001) channel. The position-dependent stopping power derived in the following section is used in the simulation. The spectrum of fig.4(b) agrees well with the experimental one.

The ions lose their energy mainly at the apexes of the trajectories, i.e., the reflection points from atomic planes, because the position-dependent stopping power is large at a short distance from the atomic plane, as can be seen in the following section. Therefore, the energy loss of the ion is roughly proportional to the number of apexes and does not strongly depend on the length of the trajectory inside the

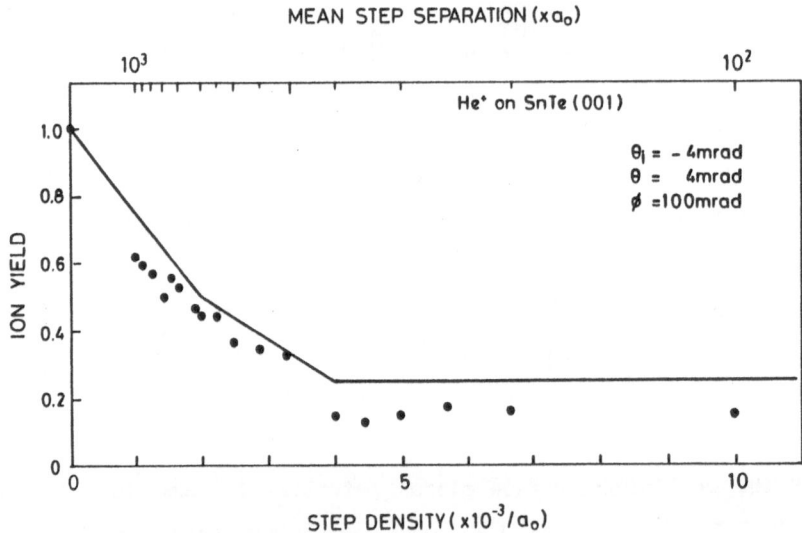

Figure 5. Calculated dependence of the yield of He ions at the specular reflection angle, on the step density. The detector window for accepting the scattered ions is 1x1 mrad². The glancing angle of the incident 0.7 MeV He⁺ ions is 4 mrad. The solid line shows the calculated yield using a simple optical model [15].

crystal. As a result, the energy spectrum of reflected ions shows well defined peaks. There may be some first-peak ions which penetrate inside the crystal through the side surface of an up-step and escape through the side surface of a down-step with only one apex, if the distance between the up-step and the down-step is smaller than one wave length of the oscillatory motion in the (001) channel. The fraction of these ions increases with increasing step density, so we must be careful in the analysis of the experimental data if the mean step separation is much smaller than one wave length.

The computer simulation was done for various step densities. The scattering-angle distribution of reflected ions depends on the step density. Fig.5 shows the dependence of the yield of He ions at the specular reflection angle on the step density. The yield becomes lower with increasing step density. The results show that the yield of ions at the specular reflection angle is sensitive to the step density. Therefore, it was proposed to use the ion reflectivity or yield of ions at specular reflection angle to detect the density of surface steps [15].

When an incident beam is directed towards a low-index crystallographic axis parallel to the surface, surface channeling occurs. Fig.6 shows the energy spectra of He ions scattered at 9.9 mrad for

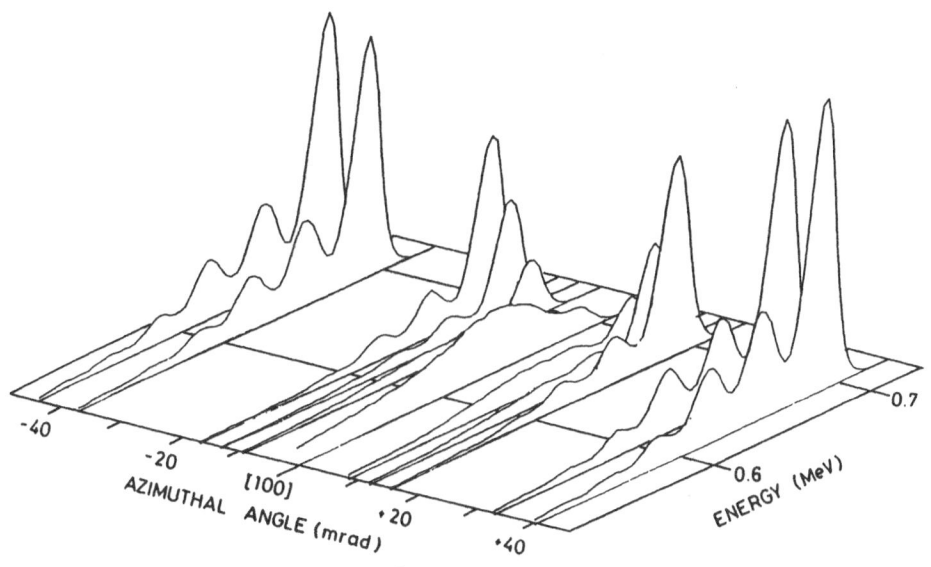

Figure 6. Energy spectra of He$^+$ ions scattered at 9.9 mrad, for various incident directions around the [100] axis, when 0.7 MeV He$^+$ ions are incident on a SnTe (001) surface with a glancing angle of 5.2 mrad.

various incident direction around [100] axis, when 0.7 MeV He[+] ions are incident on the SnTe (001) surface with a glancing angle of 5.2 mrad. Under the condition of surface channeling, the energy spectrum is very different from that of specularly-reflected ions. The surface-channeled ion feels the continuum string potential rather than the continuum surface-planar potential, so the oscillatory structure, which is caused by planar channeling, disappears. However, the first peak remains and so we can identify the ions reflected from the surface atomic rows, even if surface channeling occurs.

4. STOPPING POWER OF THE SURFACE

Fig.7 shows the most probable energy loss of the specularly reflected

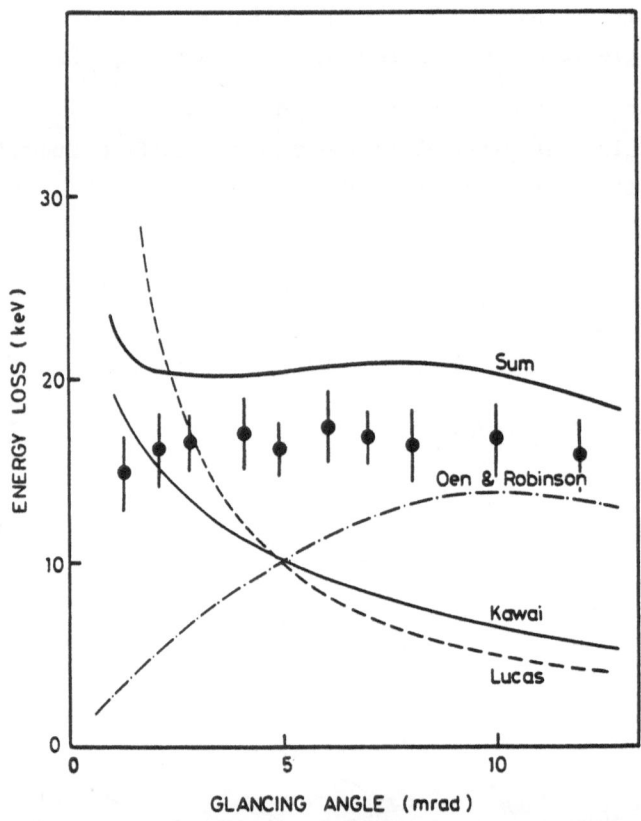

Figure 7. Dependence of the most probable energy loss of specularly reflected He ions on the glancing angle, when 0.7 MeV He[+] ions are incident on a SnTe (001) surface. The energy loss calculated with the model of Lucas (– – –), that of Kawai et al. (——), and that of Oen and Robinson scaled by the stopping power of Ziegler (– · –), are shown. The sum of the energy loss of Kawai et al. and that of Oen and Robinson is also shown by a heavy line.

He ions as a function of the glancing angle, when 0.7 MeV He$^+$ ions are incident on a SnTe (001) surface. The energy loss is independent of the glancing angle, within the experimental error. We also found that the energy loss of specularly reflected He ions is almost independent of the incident energy in the energy region 0.7 - 2 MeV [12]. The energy loss of specularly-reflected protons was also independent of either the glancing angle or the incident energy, although it was about one third of the energy loss of He ions [10].

The position-dependent stopping power near the surface can be derived from the present experimental results by the following procedures. Let $S(x)$ be the position dependent stopping power at a distance x from the surface; the energy loss of specularly reflected ions can be written by

$$\Delta E = \int S(x) \, dz, \tag{1}$$

where the integration is performed along the ion trajectory lying on the x - z plane. The trajectory of specularly reflected ions can be calculated with the use of a continuum surface potential. If a Moliére potential is employed, and the first and second terms of the Moliére potential are neglected (the error of this approximation is very small, e.g., less than 1 % for 0.7 MeV He ions incidence at a glancing angle of 7 mrad), the trajectory of an ion can be written as

$$x(z) = \frac{a_{TF}}{\beta_3} \ln\left[\frac{2\pi n_p Z_1 Z_2 e^2 a_{TF}\alpha_3}{E\theta_i^2 \beta_3}\right] \cosh^2\left[\frac{\beta_3 \theta_i z}{2a_{TF}}\right], \tag{2}$$

where n_p is the atomic density of the surface, θ_i is the glancing angle, and other symbols denote their usual meanings. Substituting eq.(2) into eq.(1), the energy loss can be written as

$$\Delta E = \frac{4a_{TF}}{\beta_3\theta_i} \int_{\xi_0}^{\infty} \Sigma\left(\frac{\xi}{\theta_i}\right) (\xi^2 - \xi_0^2)^{-1/2} \, d\xi, \tag{3}$$

where $\xi = \theta_i \exp(\beta_3 x/2a_{TF})$, $\xi_0 = (2\pi n_p Z_1 Z_2 e^2 a_{TF}\alpha_3/E\beta_3)^{1/2}$ and $\Sigma(\xi/\theta_i) = S(2a_{TF} \ln[\xi/\theta_i]/\beta_3)$. Because the observed energy losses were independent of θ_i, $\Sigma(\xi/\theta_i)$ must be proportional to θ_i/ξ, i.e., $\Sigma(\xi/\theta_i) = A(E)\theta_i/\xi$. Substituting this equation into eq.(3), the integration can be performed

$$\Delta E = A(E) \left[\frac{2\pi a_{TF}E}{\alpha_3\beta_3 n_p Z_1 Z_2 e^2}\right]^{1/2}. \tag{4}$$

Since the observed energy losses are independent of the ion energy, $A(E)$ must be proportional to $E^{-1/2}$. Thus the position-dependent stopping power near the surface is derived as

$$S(x) = CE^{-1/2} \exp\left[-\frac{\beta_3 x}{2a_{TF}}\right], \tag{5}$$

where C is a constant. From the experimental results, this constant C is determined to be 7700 $MeV^{3/2}$ cm^{-1} for He ion and 2000 $MeV^{3/2}$ cm^{-1} for protons. The obtained position-dependent stopping power for 1 MeV He ion is shown in fig.8.

Several theoretical studies have been performed on the energy loss for fast ions at solid surfaces. Lucas gave a formula for the energy loss

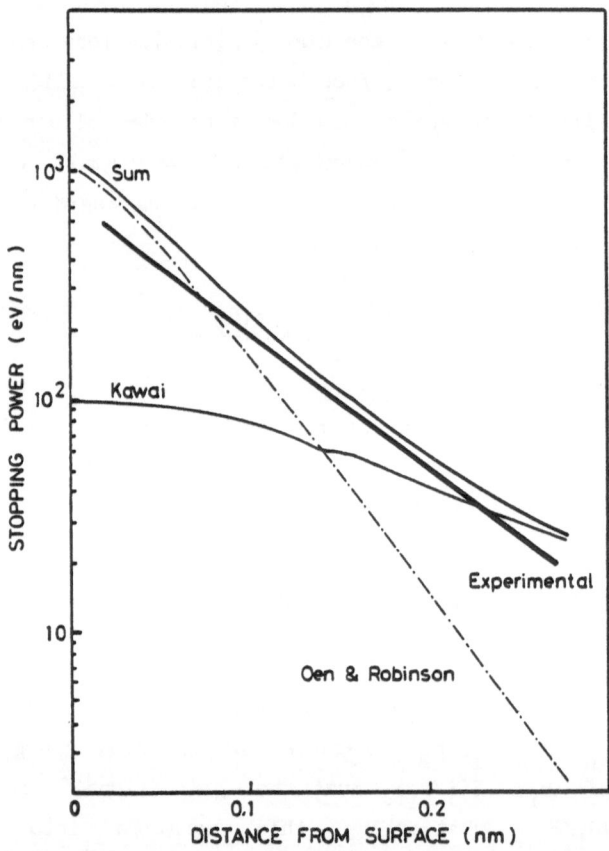

Figure 8. Position-dependent stopping power of 1 MeV He$^+$ near the surface of SnTe (001), derived from the experimental results. The calculated results with the model of Kawai et al. (——), that of Oen and Robinson (– · –) and the sum of those two models (heavy line) are shown.

of ions reflected at a solid surface due to excitations of surface plasmons, with an assumption of a straight path line of reflected ions [5]. Kawai et al. derived the position-dependent stopping power near the surface taking account of excitations of both surface and bulk plasmons [6]. The calculated position-dependent stopping power for 1 MeV He$^+$ near the SnTe (001) surface, using the formula of Kawai et al., is shown in fig.8. We used in the calculation the surface-plasmon energy 10.5 eV for SnTe, and the observed mean charge of reflected He ions. The calculated stopping is smaller than the experimental one, and the deviation is large for small x. This disagreement between the calculated and experimental stopping powers may be due to the single collision between the ion and electrons near the surface. We can evaluate this contribution using the formula given by Oen and Robinson for the inelastic energy loss of an ion scattered by an atom [16]. With the use of their formula, the position-dependent stopping power, averaged in the plane parallel to the surface, can be written as

$$S(x) = \frac{0.09 n_p s x}{\pi a_{TF}^2} K_1\left(\frac{0.3x}{a_{TF}}\right),$$ (6)

where s is the stopping cross section of the crystal atom, and $K_1(z)$ is a modified Bessel function. Fig.8 shows the calculated stopping power using eq.(6), for 1 MeV He ions. The sum of the stopping power derived from Kawai et al. and that of Oen and Robinson is also shown in fig.8. This sum of the calculated stopping powers agrees reasonably with the experimental one.

Recently, a similar measurement about the energy loss of specularly-reflected protons from a W(001) surface has been done, using an electrostatic analyzer with energy resolution $\delta E/E \geq 10^{-4}$ [17]. Although the observed ion energy dependence of the energy loss is the same as in the present result, the observed energy loss increases with increasing glancing angle. This glancing-angle dependence can be explained with only the Oen and Robinson formula, without the energy loss due to the excitation of plasmons. This difference on the glancing-angle dependence between their result and the present result may be explained by the difference of the surface structure.

5. CHARGE STATE DISTRIBUTION OF SPECULARLY REFLECTED IONS

We measured the fraction of He$^+$ ions, F_1 and that of He^{2+} ions, F_2, in the scattered ion beam, for various conditions [18]. Fig.9 shows the observed ratio F_1/F_2 for the ions of the first energy peak (these ions

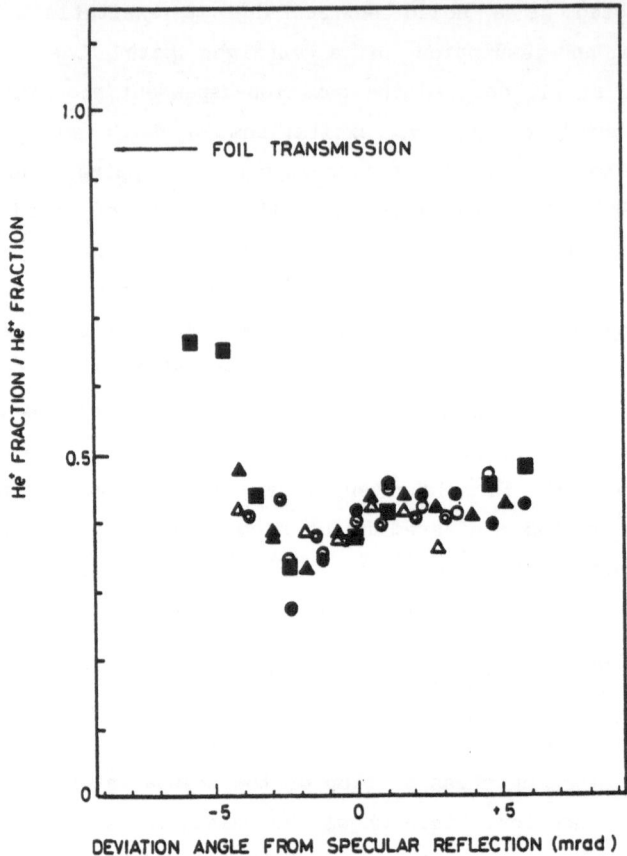

Figure 9. The ratio of the fraction of He$^+$ ions, F_1, to that of He^{2+} ions, F_2, for the ions in the first peak, as a function of the deviation angle from the specular-reflection direction. The ratios for 0.7 MeV He$^+$ incidence with glancing angles of 2.9 mrad (●), 4.9 mrad (Δ,full), 6.9 mrad (□), and 0.7 MeV He^{2+} incidence on SnTe (001) at a glancing angle 5.2 mrad (⊙), are shown. The ratio for 0.7 MeV He$^+$ incidence with glancing angles 2.9 mrad (○) and 4.9 mrad (Δ), in the condition of [210] surface channeling, are also shown. For comparison, the ratio for the He ions transmitted through a self-supporting SnTe foil, with exit energy 0.7 MeV, is also shown.

were reflected from the surface atomic plane). The abscissa of this figure is the deviation angle from the specular reflection, i.e., the scattering angle subtracted by twice the glancing angle. Shown are the results of the specular reflection (random incidence) and those of [210] surface channeling for the incidence of 0.7 MeV He$^+$ and He^{2+} ions. The observed ratios cluster into a near-universal curve, i.e., the charge-state distribution does not depend on the charge state of the incident ions, the glancing angle or the incident azimuthal angle with respect to the crystallographic axis. This indicates that the charge-state distribution becomes an equilibrium charge-state distribution after the interaction with the single-crystal surface.

We also measured the fractions F_1^T and F_2^T in the beam transmitted through a self-supporting SnTe foil. The foil was about 500 nm and thick enough for the attainment of equilibrium charge-state distribution. The observed ratio F_1^T/F_2^T for transmitted ions with an exit energy of 0.7 MeV is shown in fig.9. The ratio F_1^T/F_2^T is about twice the ratio F_1/F_2. This disagreement in the charge state distribution between specular reflection and transmission through a foil, is due to the ion-surface interaction. The velocity of an MeV He ion is approximately equal to the velocity of N shell electrons for the Sn and Te atoms. So the He ions passing through the SnTe foil capture electrons mainly from the N shells. However, this is not true for the specularly reflected He ions, which interact mainly with the tail of the valence electron distribution. These He ions have a small chance to capture electrons, as compared with the He ions passing through a foil, because the velocity of the valence electrons are much smaller than those of MeV He ions. Hence, the fraction of He^+ ions in the specularly reflected beam is smaller than that in the foil-transmitted beam.

6. SPECULAR REFLECTION FROM PbSe/SnTe(001)

It is known that a PbSe crystal grows epitaxially on SnTe (001), with a pseudomorphic structure, when the thickness is less than 3 ML [19]. At coverages larger than 3 ML, a cross grid network of misfit dislocations parallel to the <110> directions is formed in the (001) interface of PbSe/SnTe bicrystals. The initial stage of the epitaxial growth of PbSe on SnTe (001) was investigated using the specular reflection of He ions.

The scattering-angle distributions of reflected ions were measured for various PbSe thicknesses when 0.7 MeV He^+ ions were incident on PbSe/SnTe (001) surfaces with a glancing angle of 5 mrad. Fig.10 shows the FWHM of the scattering-angle distribution and the peak yield of scattered ions as a function of PbSe thickness. The FWHM begins to increase at about 3 ML and has a peak around 10 ML. The ion yield decreases with increasing thickness and has a dip around 10 ML. Both FWHM and ion yield become almost constant when the thickness is more than 50 ML. These thickness dependence may be explained by the effects of misfit dislocations.

Dislocations located near the surface induce surface distortion. The distortion can be calculated using elastic theory [20]. Calculations show that the surface of PbSe/SnTe (001) bicrystals has a cross grid network of

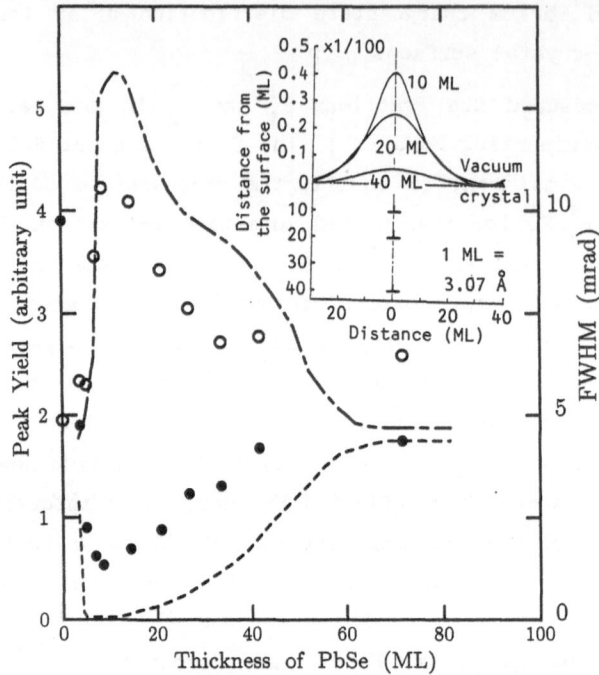

Figure 10. FWHM (o) and peak ion yield (•) of the scattering-angle
distribution of reflected He ions, as a function of PbSe
thickness. The glancing angle of the incident 0.7 MeV He$^+$ ions
is 5 mrad. Calculated results of FWHM (– · –) and peak ion
yield (– – –) are also shown. The inset shows the
cross-sectional views of the surface ridges induced by misfit
dislocations, for various thicknesses of the PbSe overlayers.

ridges along misfit dislocations, as described schematically by the inset
in fig. 10. The height and width of the ridges depends on the thickness of
the PbSe crystal, e.g., the height is 0.08 nm and the width is 8 nm at 20
ML, and the height becomes lower and the width becomes broader with
increasing PbSe thickness. This surface distortion changes the effective
glancing angle of the incident ions, and so the scattering-angle
distribution is changed. The scattering-angle distribution of ions
reflected from the distorted surface can be calculated by convoluting the
distribution of the effective glancing angle with the measured scattering
angle distributions of reflected ions from a flat PbSe (001) surface.

The FWHM of the calculated scattering angle-distribution and the
calculated peak yield are shown as a function of PbSe thickness in fig.10.
The calculated results agree qualitatively with the experiments. The PbSe
thickness dependence of the FWHM and ion yield can be explained as
follows: When the PbSe thickness is less than 3 ML, the PbSe crystal

overgrows the SnTe with a pseudomorphic structure. So the surface is flat and the specular reflection of ions is not affected by the overgrowth of PbSe until 3 ML. When the PbSe thickness becomes 3 ML, the cross grid network of misfit dislocations and that of surface ridges are formed. The scattering-angle distribution changes abruptly (the distribution becomes broader) at this thickness. After 3 ML the height of the ridges decreases and their width increases, i.e., the surface becomes flatter, with increasing PbSe thickness. Therefore, the FWHM is reduced and the ion yield is larger with increasing PbSe thickness. They are almost constant after 60 ML because the surface becomes almost flat at the thickness of 60 ML (the maximum surface gradient is less than 0.1 mrad).

Thus the experimental results can be explained by the effects of surface distortion induced by misfit dislocations. This indicates that the specular reflection is an excellent methods to detect small surface distortion (a surface gradient of the order of 1 mrad is detectable).

ACKNOWLEDGEMENTS

We are grateful to the members of the Department of Nuclear Engineering of Kyoto University for use of the 4 MV Van de Graaff accelerator. This work was supported by the Special Grant-in-Aid from the Ministry of Education, Science and Culture. One of the authors (K.K.) would like to acknowledge support by the National Science Foundation.

REFERENCES

1. E.W. Thomas, in *Applied Atomic Collision Physics* Vol. 4, Datz ed. Academic Press, Orland (1983) p. 299.
2. R. Haight, L.C. Feldmann, T.M. Buck and W.M. Gibson, Phys. Rev. *B30* (1984) 734.
3. H. Winter, Phys. Scripta T *6* (1983) 136.
4. H.J. Andrä, H. Winter, R. Fröhling, N. Kirchner, H.J. Plöhn, W. Wittmann, W. Graser and C. Varelas, Nucl. Instr. and Meth. *170* (1980) 527.
5. A.A. Lucas, Phys. Rev. *B20* (1979) 4990.
6. R. Kawai, N. Itoh and Y.H. Ohtsuki, Surf. Sci. *114* (1982) 137.
7. R.M. Nieminen and C.H. Hodges, Phys. Rev. *B15* (1978) 2568.
8. Y.H. Ohtsuki, K. Koyama and Y. Yamamura, Phys. Rev. *B20* (1979) 5044.
9. M. Mannami, K. Kimura, K. Nakanishi and A. Nishimura, Nucl. Instr. and Meth. *B13* (1986) 587.
10. K. Kimura, A. Nishimura and M. Mannami, Surf. Sci. *183* (1987) L313.
11. K. Kimura and M. Mannami, Nucl. Instr. and Meth. *B27* (1987) 442.
12. K. Kimura, M. Hasegawa and M. Mannami, Phys. Rev. *B36* (1987) 7.
13. S. Datz, C.D. Moak, T.S. Noggle, B.R. Appleton and H.O. Lutz, Phys. Rev. *179* (1969) 315.
14. Y. Fujii, K. Kimura, M. Hasegawa, M. Suzuki, Y. Susuki and M. Mannami, Nucl. Instr. and Meth. *B33* (1988) 405.
15. M. Mannami, Y. Fujii and K. Kimura, Surf. Sci. *204* (1988) 213.
16. O.S. Oen and M.T. Robinson, Nucl. Instr. and Meth. *132* (1976) 647.

17. H. Winter, J. Remillieux and J.C. Poizat, Nucl. Instr. and Meth. *B48* (1990) 382.
18. K. Kimura, Y. Fujii, M. Hasegawa, Y. Susuki and M. Mannami, Phys. Rev. *B38* (1988) 1052.
19. M. Suzuki, H. Kawauchi, K. Kimura and M. Mannami, Surf. Sci. *204* (1988) 223.
20. J.D. Eshelby, in *Dislocations in Solids*, Vol. 1, F.R.N. Nabaro, ed., North-Holland, Amsterdam (1979) p. 167.

NEUTRALIZATION OF FAST PROTONS
IN GRAZING COLLISIONS WITH AN Al(111) SURFACE

H. Winter

Institut für Kernphysik
Universitat Münster
Wilhelm-Klemm-Str. 9, D-4400 Münster, FRG

ABSTRACT

The neutralization of protons scattered from an Al(111) surface under grazing incidence, is investigated for projectile energies ranging from 50 keV to 1.25 MeV. The neutral fractions strongly depend on the state of preparation of the target, and show a monotonic decrease with increasing projectile energy. The data imply that electron loss plays an important role in charge exchange during grazing ion-surface scattering.

1. INTRODUCTION

Collisions of fast ions with surfaces under grazing incidence provide good conditions for well-defined studies of charge-exchange mechanisms between atoms and surfaces. Since experiments in grazing incidence geometry are run in terms of planar channeling, the interaction of projectiles are characterized by two vastly different time (velocity) regimes: the motion parallel to the surface plane, with projectile energy E (velocity v), whereas the motion along the surface normal is given by E $\sin^2\Phi$ (velocity v sin Φ). For, e.g., a grazing angle of incidence $\Phi = 0.2^o$, the energies of normal and parallel motion differ by about five orders of magnitude. As a consequence, even projectiles with large energies E cannot overcome the repulsive surface potential, and are predominantly reflected without penetration into the bulk of the solid, in terms of well defined trajectories. Furthermore very peculiar conditions with respect to charge-exchange processes at surfaces hold: a quasi-adiabatic regime with respect to the perpendicular motion, and fast ion-surface collisions with respect to parallel motion.

In a number of recent papers [1-6], it has been shown that experimental charge fractions, observed after grazing ion-surface scattering, can be well described by established concepts of

Interaction of Charged Particles with Solids and Surfaces
Edited by A. Gras-Martí *et al.*, Plenum Press, New York, 1991

387

charge-transfer. However, the modification of occupation of electronic target states, in the atomic rest-frame, due to the (fast) parallel motion, has to be incorporated by a Galilean transform [4,6]. Due to the adiabatic nature, charge-exchange processes of relatively long range are favored: (1) resonant one-electron tunneling and, (2) Auger neutralization/ionization. The contributions of both processes strongly depend on occupied and unfilled electronic metal states, available for charge transfer to/from the atom. Since this distribution of states can be modified to a large extent in the atomic rest-frame by the parallel motion, the neutralization in grazing incidence collisions shows a pronounced dependence on the parallel velocity of the projectiles.

First consequent experimental studies of neutralization of alkali ions, in grazing surface scattering, revealed two very different dependences of neutral fractions on projectile velocity [2]: (1) a monotonic decrease with increasing velocity and, (2) a kinematic resonance-type structure, with a threshold at the low velocity end. The data are consistently interpreted in terms of a model of electron capture and loss via resonant electron tunneling, where the relative size of the ionization energy of the populated atomic term, and the work function of the target, determine the velocity dependence of neutral fractions.

Neutralization of protons at an aluminum surface can be expected as a system which should allow detailed theoretical treatment; the one-electron system of a hydrogen atom has a simple (and well known) electronic structure, aluminum may be considered as the prototype of a "jellium" metal, with electronic states deduced from a free-electron model. However, the binding energy of hydrogen 1s, $E_a = -13.6$ eV, clearly exceeds the work function of Al(111), $W = 4.24$ eV [7]. Therefore, aside from resonant tunneling, neutralization at the surface will also proceed via the Auger process, which is more difficult to handle in the theoretical description. First steps to treat Auger neutralization in grazing scattering geometry have been performed by Miskovic and Janev [8], and Zimny and Miskovic [9]. By approximations of transition matrix elements, these papers discuss the effect of the parallel velocity on the density of unoccupied metal states, with respect to the relative contributions of resonant tunneling and Auger processes. In the range of velocities up to about the Fermi velocity v_F, these two processes dominate the neutralization of protons.

Above v_F, resonant neutralization, and above some v_F, Auger neutralization is strongly suppressed, because the large Doppler-shift of metal states ("Doppler Fermi-Dirac" distribution [3]) keeps contributions of occupied metal-states small. In analogy to ion-atom collisions, it is then possible that second order processes of charge transfer may dominate the neutralization. Very recently, Thumm and Briggs have presented an analysis of such higher-order processes in electron capture from a surface by grazing incidence impact, for protons with velocities greater than v_F [10]. Their calculations show that the second order Thomas-scattering process exceeds first-order electron capture by about two orders of magnitude. Since the Thomas-scattered electrons have typical energies used in LEED, structures in the monotonic decrease of the neutral fraction, with velocity, are expected. This paper describes experimental tests of the theoretical predictions in ref.[10].

2. EXPERIMENT AND RESULTS

The experiments are performed with protons of energies ranging from 50 keV to 1.25 MeV, at the 2.5 MV van de Graaff accelerator of IPN Lyon (France). The projectiles interact with a clean Al(111) surface, under a grazing angle of incidence $\Phi_{in} \simeq 0.2^{\circ}$, at UHV conditions. Differential pumping stages on both ends of the target chamber, allow to maintain base pressures below 10^{-9} mbar during the experiments. A set of diaphragms (0.2 mm diameter), at both ends of the pumping stage at the entrance, provide a collimation of the projectile beam to a divergence in the sub-mrad domain. The (111) face of an aluminum monocrystalline sample was polished with great care, by keeping the deviation between (111) plane and polishing plane as small as possible ($\leq 0.1^{\circ}$) to achieve long terrace widths. Contaminations at the surfaces, especially oxygen, are removed by grazing sputtering ($\Phi_{in} \simeq 1^{\circ}$), with 400 keV Ar^+ ions of about 2 $\mu A/mm^2$ current density. Preparation of the target over several days, by frequent cycles of sputtering and annealing by heating up to $560^{\circ}C$, finally yields a clean and flat surface of the target. After such a treatment, aside from a strong Al line at 68 eV, no traces of impurities can be found in the Auger spectra.

At the exit side of the target chamber, a slit of 0.2 mm width is positioned by a linear feed-through, about 150 mm behind the target. This slit is used to separate projectiles scattered at a selected angle for the analysis, with respect to charge states, by a pair of electric field plates and a KBL 210 (Dr. Sjuts, Katlenburg) channeltron. Since the neutral fractions are expected to be some orders of magnitude smaller than the positive fractions, great care has to be devoted to saturation effects which affect the ratio of high and low count-rates. The channeltron is mounted on a linear feedthrough, and can be moved via step-motor control along the polar direction of scattering. Thus the distribution of projectiles scattered in the polar plane can be recorded. With the help of the slit and the field plates, scattered projectiles with neutral and positive charge states are geometrically separated and analyzed by a scan of the detector.

During the sputtering process, the target is rotated in a range of 360° with respect to the azimuthal angle, under simultaneous recording of the effective (uncorrected) target current. Since this current is dominated by kinetic emission of electrons, which is observed to depend on the direction between projectile beam and low-indexed directions of the crystal surface, a simple on-line technique results for the positioning of the target relative to such directions. The experiments reported here are performed under "random" orientation of the crystal, i.e., the projectile beam is directed along a highly-indexed axis of the crystal surface.

In the first part of the experiments, the distributions of scattered projectiles is recorded in a polar plane. Fig.1 shows data for 525 keV protons, where the left peak stems from the residual beam that has passed above the target without interaction (direction of the projectile beam). The distribution of scattered protons is very well defined, a halfwidth of typically 0.2° is observed. In good approximation, the separation between both peaks in fig.1 defines the angle of scattering $\Phi_S \simeq 0.3^\circ$; assuming specular reflection, we then deduce $\Phi_{in} \simeq \Phi_{out} \simeq 0.15^\circ$.

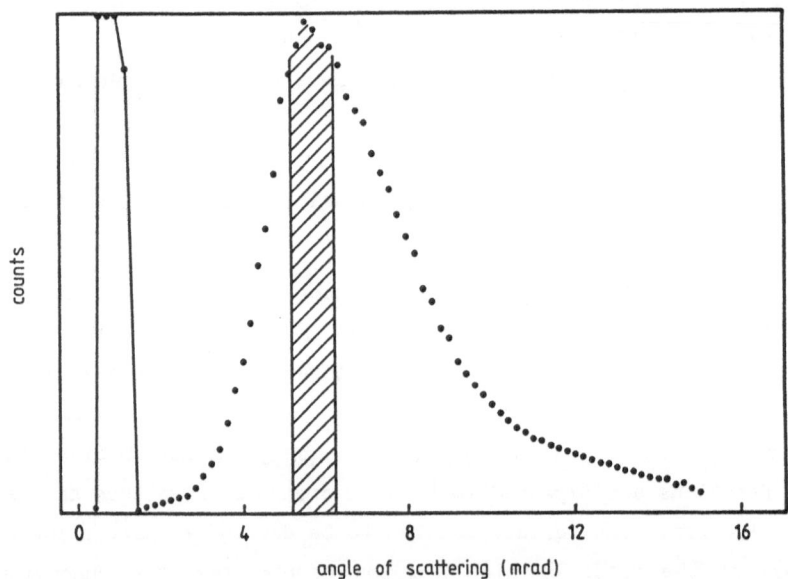

Figure 1. Polar distribution of 525 keV protons after scattering at Al(111).

A fraction of the distribution of scattered projectiles is selected by the slit behind the target, (see the marked part of the distribution in fig.1), and analyzed with respect to charge states. Only the neutral and positive charge states turned out to be relevant in these studies. Investigation of the neutral fraction $n^0/(n^0 + n^+)$, for different settings of the slit within the scattering distribution, shows no significant dependences on the scattering angle. The data discussed below are obtained for specularly-reflected projectiles, and $\Phi_{in} \simeq \Phi_{out}$ ranging from about 0.15° to 0.2°.

The neutral-fraction dependence on proton energy (velocity) is plotted in fig.2. The data show a pronounced decrease with increasing energy; however, no significant structures are observed. For comparison,

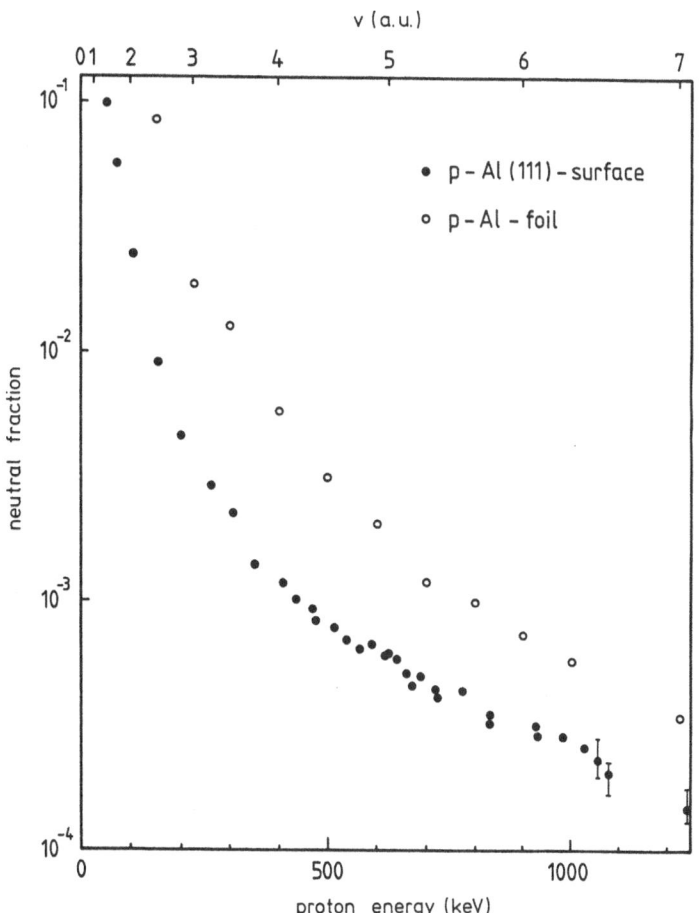

Figure 2. Experimental neutral charge fractions. The open circles represent data after transmission through thin Al foils, from ref.[11].

we have also given data obtained after proton transmission through thin Al foils (open circles) [9]. It is evident from the figure, that the neutral fraction observed after surface scattering is about factors 3 to 5 smaller than after beam-foil.

In the experiments a strong sensitivity of the neutral fractions on the state of target preparation is found. From initial to final states of preparation (sputtering and heating cycles) a reduction up to factors of ten are observed, indicating that impurities and defects at the target surface decisively affect proton neutralization in grazing surface scattering. A detailed discussion of this aspect will be given elsewhere.

3. CONCLUSIONS

Comparison of our data at low projectile energies shows smaller neutral fractions than observed in previous work [2]. Whereas in ref.[2] the neutral fraction is found to be comparable for surface and beam-foil interaction at velocities larger than 1 a.u., the data observed in the final state of preparation of the target are significantly smaller. Systematic studies of the dependence of neutral fractions on the "quality" of the target surface, imply that the former experiments are performed with targets of a higher density of defects. As a support of this assumption may serve the much broader polar distributions of scattered projectiles obtained in ref.[2].

In fig.3 we show a comparison of our data with recent calculations by Thumm and Briggs [10], which have partly motivated this study. The solid

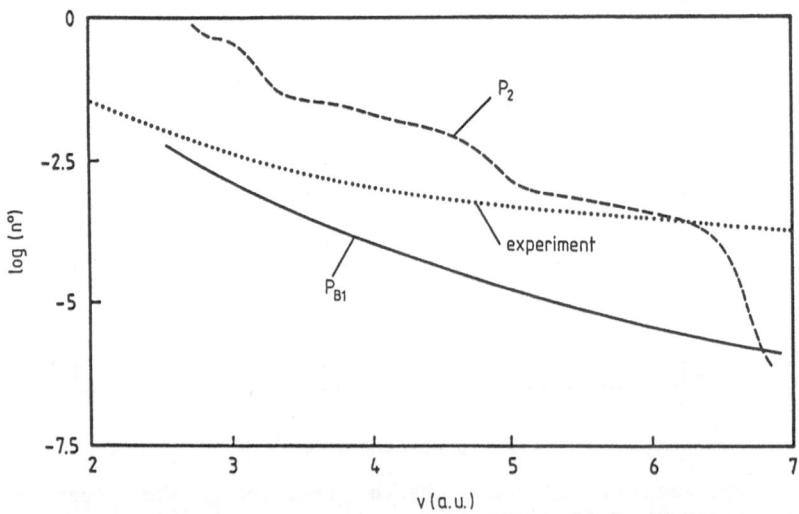

Figure 3. Comparison of our experimental data (dotted line) with recent calculations, refs.[10,12].

and dashed lines show results for contributions of first and second order, respectively [12], the experimental data are represented by the dotted line. The calculations show that second order processes dominate electron capture in the velocity interval from 3 to 7 a.u., and that LEED-type electron scattering results in step-like structures. Aside from the fact that the second-order contributions are found to be in poor agreement with the data, no step-like structures in the velocity dependence of the neutral fractions is observed. These findings may indicate that broadened resonances with localized inner-shell levels of the target, as well as electron-loss processes, which have relatively long ranges and modify the effects of capture processes close to the surface, are relevant at those velocities. Both processes are neglected in refs.[10,12].

ACKNOWLEDGEMENTS

The assistance of H.W. Ortjohann, R. Kirsch, and J. Kemmler, in preparation and running the experiments, is gratefully acknowledged. We thank U. Thumm and R. Zimny for many valuable discussions. This work was supported by Sonderforschungs bereich 216 (Bielefeld/Münster).

REFERENCES

1. H. Winter and R. Zimny, in *Coherence in Atomic Collision Physics*, H.J. Beyer et al., eds., Plenum, New York (1988) p.283.
2. R. Zimny, H. Nienhaus, and H. Winter, Rad. Effects and Defects in Solids *109* (1989) 9 and *109* (1989) 1.
3. D.M. Newns, Comments Cond. Mat. Phys. *14* (1989) 295.
4. J.N.M. van Wunnik and J. Los, Phys. Scripta *T6* (1983) 27.
5. J. Burgdörfer, E. Kupfer, and H. Gabriel, Phys. Rev. *A35* (1987) 4963.
6. H. Schröder, Nucl. Instr. Meth. *B2* (1984) 213.
7. J.K. Grepstad, P.O. Gartland, and B.J. Slagsvold, Surf. Sci. *57* (1976) 348.
8. Z. Miskovic and R. Janev, Surf. Science *221* (1989) 317.
9. R. Zimny and Z. Miskovic, Surf. Science, submitted for publication.
10. U. Thumm and J. Briggs, Nucl. Instr. Meth. *B43* (1989) 471.
11. A. Chateau-Thierry and A. Gladieux, in *Atomic Collisions in Solids*, S. Datz et al., eds. Plenum, New York (1975) p.307
12. U. Thumm, Thesis, Freiburg (1989).

ON THE STOPPING POWER OF AN ELECTRON GAS
FOR SLOW PROTON AND ANTIPROTON

I. Nagy[1], B. Apagyi[1] and K. Ladányi[2]

1. Quantum Theory Group, Institute of Physics
   Technical University of Budapest
   H-1521 Budapest, Hungary

2. Institute for Theoretical Physics
   Roland Eötvös University
   H-1088 Budapest, Hungary

ABSTRACT

The screening of a slow projectile of unit charge in an electron gas is discussed in analogy to the dielectric response formulation. A simple parametric form of the dielectric function is proposed which leads to analytic expressions for the effective potential and induced screening density. The parameter is fixed via the cusp condition for the total electron density at the position of the probe charge. The stopping power is calculated by making use of the transport cross section. The results are in fair agreement with those given by self-consistent calculations.

This work deals with the stopping of charged particles in a homogeneous electron gas. When the velocity of the projectile v is small compared with the Fermi velocity v_F of the electrons, the relevant excitations responsible for the energy dissipation are electron-hole pairs. The amount of phase space available for the creation of these pairs is determined by kinematical constraints. It is therefore independent of the details of the interaction between the probe charge and the system. These statements allow us to use the friction force formulation of the stopping power.

For low velocity of a massive projectile the energy loss per unit path length can be written as [1]

$$\frac{dE}{dR} = n_0 \, v \, v_F \, \sigma_{tr}(v_F),$$
(1)

where $n_0 = v_F^3/3\pi^2$ is the density of the host electron gas and σ_{tr} is the momentum-transfer cross section. (We use atomic units throughout this work).

Interaction of Charged Particles with Solids and Surfaces
Edited by A. Gras-Martí *et al.*, Plenum Press, New York, 1991

395

Eq. (1) is based on the adiabatic picture, i.e. the momentum transfer to the projectile per second is taken to act as a time-independent force. This picture is expected to be valid if the momentum transfer per collision is very small compared to the momentum of the probe charge. In terms of the phase shifts δ_l generated by a statically screened spherically-symmetric potential one can write for the cross section

$$\sigma_{tr}(v_F) = \frac{4\pi}{v_F^2} \sum_{l=0}^{\infty} (l+1) \sin^2(\delta_l(v_F) - \delta_{l+1}(v_F)). \qquad (2)$$

The problem of the low velocity stopping is then reduced to the determination of the effective scattering potential. Let us consider projectiles of negative and positive charge. The electron gas is repelled by an antiproton and a depletion hole is created. The electron density around a repulsive impurity may be reduced at any point only by an amount n_0 and the range of the induced hole is at least r_s, defined by $r_s = (3/4\pi n_0)^{1/3}$. On the other hand there is no restriction on the degree to which the density near an attractive impurity may be enhanced provided the induced screening density fulfils the charge neutrality condition. Furthermore the induced screening density must have finite value at the position of a probe charge independently of its sign.

The screening and stopping problem of a proton [2-4] and antiproton [5] have been discussed within the framework of density functional formalism. In this scheme the theory of potential scattering can be applied directly. The self-consistency condition of this formalism is the Friedel sum-rule [6] which relates the scattering phase shifts to the total impurity charge Z by the formula

$$Z = \frac{2}{\pi} \sum_{l=0}^{\infty} (2l+1) \delta_l(v_F), \qquad (3)$$

in which $Z = 1$ for a proton and $Z = -1$ for an antiproton. The density functional formalism provides an accurate method and its results may be used as references for model calculations.

Here we propose a simple model calculation for the induced density $\Delta n(r)$ and effective scattering potential $V(r)$. We use the following form for a parametric dielectric function [7]

$$\varepsilon(q) = 1 + \frac{3(\omega_p/v_F)^2}{q^2} \frac{1}{1+3\lambda(q/2v_F)^2} \qquad (4)$$

where $\omega_p = (4\pi n_0)^{1/2}$ is the classical plasma frequency and λ will be fixed below. By choosing this particular form we can reproduce, apart from the long range Friedel oscillation, the high density results in a very natural manner. The bare Coulomb potential $-Z/r$ of the projectile is screened [8] according to

$$V(q) = - Z \frac{4\pi}{q^2 \varepsilon(q)}. \tag{5}$$

The induced density is then calculated from the Poisson equation to be

$$\Delta n(r) = Z \int \frac{d^3 q}{(2\pi)^3} \, e^{i\vec{q}\vec{r}} \, \frac{\varepsilon(q)-1}{\varepsilon(q)}. \tag{6}$$

The free parameter λ is determined by making use of the cusp condition. This condition for the total electron density $n(r) = n_0 + \Delta n(r)$ at the position of an impurity with charge Z reads [9,10]

$$\left. \frac{n'(r)}{n(r)} \right|_{r=0} = - 2 Z \mu, \tag{7}$$

where μ denotes the reduced mass of the electron-impurity (two-body) system. (Note that the cusp condition is a consequence of the Coulomb interaction. It is always satisfied in self-consistent calculations [2-5]). It is easy to show that in our model calculation

$$\Delta n'(r = 0) = - \frac{2 Z n_0}{\lambda}. \tag{8}$$

Consequently, for $\mu = 1$ the cusp parameter λ tends to unity in the very high density limit ($n_0 \gg |\Delta n|$). This value of λ represents the lower bound for the antiproton case and the upper bound for the proton case. Furthermore, for very high density the Friedel sum-rule is satisfied exactly [11,12] by the present model, independently of λ. Note that $\lambda = 1$ leads to unphysical results for $r_s > 0.8$ yielding $|\Delta n(0)| > n_0$ at $Z = -1$. The numerical results we shall present will show that the correction to the linear response ($\lambda = 1$) is actually very large even for $Z = \pm 1$. Via the cusp condition we take into account the strong Coulomb interaction between the projectile and electrons. In other words the nonlinearity in the response of the electron density to the impurity charge Z enters into our treatment through the parameter $\lambda = \lambda(Z, r_s)$.

For completeness we note that from eqs.(5) and (6) together with eq.(4) one can obtain analytic expressions for the effective scattering potential and induced screening density at arbitrary values of λ. These are given by

$$V(r) = -\frac{Z}{r} e^{-\alpha r} \left\{ \frac{(\alpha+\beta)^2}{4\alpha\beta} e^{+\beta r} - \frac{(\alpha-\beta)^2}{4\alpha\beta} e^{-\beta r} \right\}, \tag{9}$$

and

$$\Delta n(r) = \frac{Z}{r} \frac{n_0}{\lambda\alpha\beta} e^{-\alpha r} (e^{+\beta r} - e^{-\beta r}). \tag{10}$$

In the above expressions the parameters α and β are defined as follows

$$\alpha = \alpha(Z, r_s) = \left\{ v_F^2/(3\lambda) + \omega_p/\sqrt{\lambda} \right\}^{1/2}, \tag{11}$$

$$\beta = \beta(Z, r_s) = \left\{ v_F^2/(3\lambda) - \omega_p/\sqrt{\lambda} \right\}^{1/2}. \tag{12}$$

Note the change in form of $V(r)$ and $\Delta n(r)$ for $\lambda \geq \pi v_F/12$.

We investigate the capability of our parametric formulation in the density range $0.2 \leq r_s \leq 4$. Table 1 contains the calculated values of λ and $Q = (dE/dR)v^{-1}$ as a function of the density parameter r_s at $Z = \pm 1$. The phase shifts have been calculated numerically by a high accuracy variational method [13,14]. For comparison we also include the values of the stopping power predicted by the self-consistent calculations [4,5]. Despite the simplicity of the model, and somewhat surprisingly, without

Table 1. Calculated values of the cusp parameter λ, and the stopping power Q for a proton (p) and an antiproton ($\bar{\text{p}}$). The values predicted by self-consistent calculations [4,5] and first Born approximations for linearly screened potentials are listed for comparison. See the text for details.

	Present results				Ref.4	Ref.5	First	Born
r_s	λ_p	$\lambda_{\bar{p}}$	$Q(\lambda_p)$	$Q(\lambda_{\bar{p}})$	Q_p	$Q_{\bar{p}}$	$Q(\lambda=1)$	$Q(\lambda=0)$
0.2	0.7488	1.315	0.590	0.550	0.604	0.547	0.574	0.524
0.5	0.5082	1.790	0.456	0.372	0.459	0.355	0.421	0.349
1.0	0.2786	2.610	0.363	0.242	0.377	0.220	0.317	0.232
1.5	0.1624	3.452	0.305	0.173	0.310	0.154	0.261	0.172
2.0	0.1013	4.304	0.257	0.131	0.269	0.114	0.224	0.136
3.0	0.0474	6.025	0.170	0.084	0.163	0.070	0.176	0.092
4.0	0.0265	7.758	0.102	0.059	0.098	0.048	0.145	0.067

forcing to satisfy the Friedel sum-rule, the results are in good agreement with those obtained by the more elaborated calculations. Consequently, it seems that the essential physics is treated properly by the cusp condition for the case of stopping power calculations.

The last two columns of table 1 refer to results obtained in first Born approximation for linearly screened scattering potentials. Because of this approximation these results are independent of the sign of the projectile. Within the framework of this scattering approach the following considerations may be done. The linearized Thomas-Fermi [8] description ($\lambda = 0$) yields acceptable results for the antiproton. The linearized Thomas-Fermi-von Weizsäcker [7] description ($\lambda = 1$) leads to similar conclusion for the proton. These practical statements might be useful for the stopping calculations to obtain exploratory results at different target conditions.

ACKNOWLEDGEMENTS

One of us (I.N.) is thankful for many useful discussions with Profs. P.M. Echenique and E. Zaremba.

REFERENCES

1. E.G. d'Agliano, P. Kumar, W. Schaich, and H. Suhl, Phys. Rev. *B11* (1975) 2122.
2. C.O. Almbladh, U. von Barth, Z.D. Popovic, and M.J. Stott, Phys. Rev. *B14* (1976) 2250.
3. P.M. Echenique, R.M. Nieminen, and R.H. Ritchie, Solid State Commun. *37* (1981) 779.
4. P.M. Echenique, R.M. Nieminen, J.C. Ashley, and R.H. Ritchie, Phys. Rev. *A33* (1986) 897.
5. I. Nagy, A. Arnau, P.M. Echenique, and E. Zaremba, Phys. Rev. *B40* (1989) 11983.
6. J. Friedel, Phil. Mag. *43* (1952) 153.
7. W. Jones and W.H. Young, J. Phys. C: Solid State Phys. *20* (1971) 1322.
8. G.D. Mahan, *Many-Particle Physics*, Plenum, N. Y. (1981).
9. E. Steiner, J. Chem. Phys. *39* (1963) 2365.
10. A. Kallio, P. Pietiläinen, and L. Lantto, Phys. Scr. *25* (1982) 943.
11. I. Nagy, A. Arnau, and P. M. Echenique, Phys. Rev. *A40* (1989) 987.
12. P.M. Echenique, I. Nagy, and A. Arnau, Int. J. Quantum Chem.: Quantum Chem. Symp. *23* (1989) 521.
13. K. Ladányi and T. Szondy, Nuovo Cimento *B5* (1971) 70.
14. B. Apagyi and K. Ladányi, Phys. Rev. *A33* (1986) 182.

ELECTRON CAPTURE

IN ATOMIC COLLISIONS

V.H. Ponce[1]

Dpto. Física de Materiales, Facultad de Química
Universidad del País Vasco
Aptdo. 1072, E-20080 San Sebastian, SPAIN

1. Permanent address: Centro Atómico Bariloche
8400 Bariloche, Argentina

ABSTRACT

The collision between two atomic systems is accompanied by electronic transitions, that are produced at finite distances R between the colliding partners. Nevertheless, the calculated cross sections will depend on the atomic interactions at $R \to \infty$, if the long-range nature of the Coulomb forces present is not taken into account by the scheme adopted to describe the collision.

A distorted-wave formalism, which explicitly considers the asymptotic R^{-1} part of the interatomic potentials, generates the correct boundary conditions for the process, and the resulting transition amplitudes are defined in terms of short-range interactions. First and second-order Born approximations within this scheme, give a reasonable description of total and differential cross sections in the medium and high collision-energy range.

1. INTRODUCTION: STATUS OF THE FIELD

Electron capture processes in high-energy atomic collisions has been a field of study since the early days of quantum mechanics [1], and even a description in terms of classical dynamics given by Thomas in 1927 [2] remains actual. Nevertheless, the subject is far from closed, and there are still questions to be answered and points to clarify.

This work will not present exhaustive references, since there are recent reviews [3,4] that give a detailed account of the many theoretical and experimental contributions in this area. Rather, it will concentrate in analyzing the consistency, merits and difficulties of the theoretical models used in actual calculations.

Atomic particles are subject to Coulomb forces whose long-range

Interaction of Charged Particles with Solids and Surfaces
Edited by A. Gras-Martí *et al.*, Plenum Press, New York, 1991

401

nature can not be neglected. This fact is the source of both interesting effects [4], and of complications and ambiguities in formal developments and calculations. The need to take into account the asymptotic nature of Coulomb forces was pointed out by Cheshire [5], who introduced a distorted-wave method in the frame of stationary scattering theory.

Channel Hamiltonians H_i describe the system well before and after the actual collision process $(R \rightarrow \infty)$; they include the electronic Hamiltonian and the interatomic kinetic energy. They should also incorporate the interatomic potential, if this is of Coulomb type. Distortion potentials are defined for all R values, and can be chosen with freedom once they reproduce the R^{-1} behavior of the interatomic potential when $R \rightarrow \infty$.

The need to incorporate Coulomb distortions, on one side, and the freedom to select the form of that distortion at finite R on the other, may be considered as the origins of two approaches to study electron capture processes. The distorted-wave approach [6] considers the correct boundary conditions for the channel potentials and wave functions, extending an asymptotically correct form of the interatomic potential to finite distances R. In this way the formalism satisfies the requirements of formal collision theory, and scattering amplitudes can be defined without ambiguities [6,7].

The form of the distortion at finite R gives a non-negligible contribution to the transition [8]. Since the distortion can be chosen with freedom in this region, as long as it approaches the correct asymptotic limit, the results obtained for the transition amplitudes will depend on the choice made, specially in first-order perturbative calculations. Independence of the results from this choice should be approached for higher orders of the perturbation procedure.

The formalism of distorted waves provides a frame where the boundary conditions are seen to be explicitly satisfied. The results obtained will depend on the choice made for the channel distortions. The physical soundness of this dependence may be analyzed within the formalism, by changing the form of the distortion and eventually by carrying out a variational optimization of the form of the distortion for finite R.

The other point of view to treat the capture process defines a first-order perturbation result that is independent of the choice of distorting potentials [9,10]. Even though it is recognized the necessity to satisfy the correct boundary conditions [11], these are assumed to be more a formal than a concrete requirement of the theory. The central elements of this approach are the off-shell channel wave functions

402

$|E'i,E\rangle$, that describe the relative motion between the two atomic systems with internal quantum numbers $i \equiv n_i, k'$, as originating in a free-particle state of energy E' for the interatomic motion, that evolves in a distorting potential V according to a Green's function of energy $E \neq E'$;

$$|E'i,E\rangle = |E'i\rangle + (E - E_{kin} + V + i\eta)^{-1} V|E'i\rangle \qquad (1)$$

These channel wave functions should describe the correct asymptotic states of the system in the on-shell limit $|Ei,E\rangle$. When V is a Coulomb distortion, the limit does not exist, and $|E'i,E\rangle$ is related to $|Ei,E\rangle$ by a factor $g_i(E',E)$ with a diverging phase. The formalism proposed assumes that transition amplitudes T_{if} can be defined in terms of the off-shell states $|E'i,E\rangle$, as long as they are multiplied by the factors $g_i^{-1}(E',E) \cdot g_f^{-1}(E'',E)$ before going to the limit $E',E'' \to E$,

$$g_f^{-1}(E'i,E) \langle E'f,E| \ T(E) \ |E''i,E\rangle \ g_i^{-1}(E'',E) \to T_{if},$$

$$E',E'' \to E. \qquad (2)$$

In this way, all the effect of the long-range potentials is reduced to the factors $g_{i,f}$, and the calculation is made with plane waves for the relative motion. This procedure was applied to electron capture in asymmetric systems, where the potential of the heavy nucleus (for example the target), was kept to all orders, and that of the light one (the projectile), was considered to first order [9,11]. The results obtained with this approximation, named strong-potential Born (SPB), compare well with measurements [10,12]. Nevertheless, there are some indications of anomalous behavior of the method: in the SPB approximation, all transitions to intermediate states on the target are considered, including the elastic channel, but the perturbation producing the transition is the pure Coulomb potential of the projectile, and its long-range nature produces a divergent contribution in this channel [14]. Furthermore, from the definition (2) of the transition matrix it is clear that the matrix elements of $T(E)$ should have divergent phases, to compensate those coming from the factors $g_{i,f}^{-1}$ [7,11]. To remedy these anomalies it is necessary to regularize the transition amplitudes, for example by explicitly considering the distortion potentials in the elastic channels, and by localizing the diverging phases of these amplitudes, seeing that they cancel with those provided by the factors $g_{i,f}$ [11]. These difficulties make the SPB method less transparent and practical than it was assumed to be.

2. THEORY

2.1 Defining the System

We consider the most elementary ion-atom collision system formed by a pair of nuclei P and T, massive enough to be described by the laws of classical dynamics, and an electron e initially bound to one of them. The time evolution is described by the Schrödinger equation for the electron coordinates

$$\left\{ H_e(\vec{r},t) - i\partial/\partial t \right\} \Psi(\vec{r},t) = 0, \tag{3}$$

with a time dependent Hamiltonian

$$H_e(\vec{r},t) = T_e + V_P(\vec{r}_P) + V_T(\vec{r}_T); \tag{4}$$

$\vec{r}_P$, $\vec{r}_T$ and $\vec{R}$ are respectively, the e-P, e-T and internuclear P-T relative coordinates; Z_P and Z_T , the nuclear charges. (Atomic units will be used).

Furthermore, let us restrict internuclear velocities v, so they are greater than the mean electron speed v_e in the bound orbitals that act as initial, intermediate or final electronic states during the collision. In this way, the parameters Z_P/v, Z_T/v will become natural perturbation terms for a time-dependent perturbation treatment of the evolution.

Because the momentum of the nuclear motion, proportional to the nuclear masses and velocities, is extremely large, $\vec{R}$ will be considered a known function of time: $\vec{R} = \vec{R}(t)$. This function, in principle, describes the classical motion of the charges Z_T, Z_P, but usually it is replaced by a straight line trajectory with constant velocity: $\vec{R} = \vec{b} + \vec{v}t$. Since in this case both T and P will move with constant speed, the inertial frame required to define the electron coordinate $\vec{r}$ can be any point along $\vec{R}$, including P and T. Internuclear potentials are not considered in the Hamiltonian (4), because they do not depend on the electron coordinates; they will produce a time-dependent phase on the wave function, which may be important only when analyzing the angular dispersion of the projectile.

2.2 Describing the Process

The electron is initially bound to the target charge Z_T in a state i described by

$$\left(H_T(\vec{r},t) - i\partial/\partial t \right) \phi_i(\vec{r},t) = 0, \tag{5}$$

where $H_T(\vec{r},t) = T_e + V_T(\vec{r}_T)$. The projectile interacts with this state through the time-dependent potential

$$V_P(\vec{r}) = - Z_P / |\vec{r}_T - \vec{R}(t)|. \tag{6}$$

For large negative times, well before the time $t = 0$ of closest approach, one may be tempted to assume that the action of V_P on ϕ_i is negligible. This expectation is not satisfied in this case of Coulomb potentials. We can perceive this anomalous behavior when describing the evolution of the electron in a frame centered on the projectile: there, the state ϕ_i is a wave packet centered at $-\vec{R}(t)$ and translating with velocity $-\vec{v}$. Now, it is well known that a particle moving in a Coulomb field, no matter how far away from the force center, never attains a free particle evolution. Therefore, the state $\phi_i(\vec{r}, t)$ not only acquires a plane wave phase $\exp(-i\vec{v}\vec{r}_P)$ when described from the projectile, but also a logarithmic phase $\exp\{-i(Z_P/v)\log(vR + \vec{v}\vec{R})\}$, characteristic of the long-range behavior of the potential.

The states of electron capture in a bound orbital centered at the projectile satisfy the Schrödinger equation

$$\left(H_P(\vec{r}, t) - i\partial/\partial t \right) \chi_f(\vec{r}, t) = 0, \tag{7}$$

where

$$H_P(\vec{r}, t) = T_e + V_P(\vec{r}_P); \tag{8}$$

now the channel interaction is

$$V_T(\vec{r}_T) = - Z_T / |\vec{r}_P + \vec{R}(t)|. \tag{9}$$

Here again, the asymptotic electron state is subject to the distortion set up by the target nuclear charge. Both initial and final orbitals turn out to be then distorted hydrogenic orbitals, that in the target frame may be given the form

$$\Phi_i(\vec{r}, t) = \exp\left\{-i \ (Z_P/v) \ \log(vR - \vec{v}\vec{R})\right\} \phi_i(\vec{r}, t), \tag{10}$$

$$X_f(\vec{r}, t) = \exp\left\{- i \ (Z_T/v) \ \log(vR + \vec{v}\vec{R})\right\} \chi_f(\vec{r}, t), \tag{11}$$

where

$$\phi_i(\vec{r}, t) = \phi_i(\vec{r}_T) \exp(-i\varepsilon_i t),$$

$$\chi_f(\vec{r}, t) = \chi_f(\vec{r}_P) \exp(i\vec{v}\vec{r}_T) \exp(- i \ \varepsilon_f t - i \ v^2 t/2). \tag{12}$$

405

2.3 Boundary Conditions

Given an initial state $\Phi_i(\vec{r}, t)$, there is a wave function $\Psi_i^{+}(\vec{r}, t)$ of the total system that evolves from it. The transition amplitude to a final capture state $X_f(\vec{r}, t)$ is given by

$$a(t) = \langle X_f(\vec{r}, t) | \Psi_i^{+}(\vec{r}, t) \rangle \to a, \quad t \to \infty, \tag{13}$$

and should approach a constant value for large t. For this to happen it is sufficient that the final channel interaction $V_C \to R^{-1-\varepsilon}$, ($\varepsilon > 0$), for large t. Using the time reversed form

$$a(t) = \langle \Psi_f^{-}(\vec{r}, t) | \Phi_i(\vec{r}, t) \rangle \to a, \quad t \to -\infty, \tag{14}$$

the same condition applies to the initial channel interaction V_D.

There are many ways to define the Hamiltonians H_D, H_C for the direct "D" and capture "C" channels. From

$$H(t) = H_D(t) + V_D(t) = H_C(t) + V_C(t), \tag{15}$$

the sufficient condition that produces well-defined transition amplitudes reads

$$V_D \approx V_C \approx R^{-1-\varepsilon}, \quad t \to \infty, \quad \varepsilon > 0. \tag{16}$$

In terms of distortion potentials $W(t)$, channel interactions and Hamiltonians are expressed as

$$V_D(t) = V_P(\vec{r}_P) - W_D(t), \quad H_D(t) = H_T(t) + W_D(t), \tag{17}$$

$$V_C(t) = V_T(\vec{r}_T) - W_C(t), \quad H_C(t) = H_P(t) + W_C(t). \tag{18}$$

Choosing W as the potential produced by a charge, on the electron localized on top of the other charge

$$W_D^{B}(t) = -Z_P/R, \quad W_C^{B}(t) = -Z_T/R, \tag{19}$$

(B: "basic distortion"), the asymptotic condition (16) is satisfied, and the wave functions of the channel Hamiltonians (17,18) are the hydrogenic orbitals of eqs. (10,11),

$$\Phi_i^{B}(\vec{r}, t) = \exp\left\{ -i (Z_P/v) \log(vR - \vec{v}\vec{R}) \right\} \phi_i(\vec{r}, t), \tag{20}$$

$$X_f^{B}(\vec{r}, t) = \exp\left\{ -i (Z_T/v) \log(vR + \vec{v}\vec{R}) \right\} \chi_f(\vec{r}, t). \tag{21}$$

A more precise description of the interaction of a bound electron with a bare charge is given by the mean electrostatic energy

$$W_D^A(t) = - Z_P <\phi_i(\vec{r}_T)|r_P^{-1}|\phi_i(\vec{r}_T)> = - Z_P \gamma_i(t). \qquad (22)$$

$$W_C^A(t) = - Z_T <\phi_f(\vec{r}_P)|r_T^{-1}|\phi_f(\vec{r}_P)> = - Z_T \gamma_f(t), \qquad (23)$$

(A: "average distortion").

For large R, these potentials reduce to the basic distortions (19); on the other hand, when $R \to 0$ they approach constant values, while those of (19) give an anomalous divergence of order R^{-1} [15]. The channel wave functions for the average distortion are

$$\Phi_i^A(\vec{r}, t) = \exp\left\{- i \ (Z_P/v) \int_{-\infty}^{\infty} ds \ \gamma(t)\right\} \phi_i(\vec{r}, t) \ , \quad z = v\vec{r}, \qquad (24)$$

$$\chi_f^A(\vec{r}, t) = \exp\left\{- i \ (Z_T/v) \int_{-\infty}^{\infty} dz \ \gamma_f(t)\right\} \chi_f(\vec{r}, t). \qquad (25)$$

Distortions of this type were first considered by Bates [16], and come naturally from a close-coupling calculation with a two-state atomic expansion of the scattering wave function.

Going on in order of increasing complexity, we may introduce local distortions, suffered by an electron moving with speed v in the field of a bare charge. If we keep only the asymptotic form of the distorted wave, the channel states are

$$\Phi_i^E(\vec{r}, t) = \exp\left\{- i \ (Z_P/v) \ \log(vr_P - \vec{v}\vec{r}_P)\right\} \phi_i(\vec{r}, t), \qquad (26)$$

$$\chi_f^E(\vec{r}, t) = \exp\left\{- i \ (Z_T/v) \ \log(vr_T + \vec{v}\vec{r}_T)\right\} \chi_f(\vec{r}, t); \qquad (27)$$

now the distortion potentials depend on the electron orbitals

$$W_i^E(t) = - \frac{Z_P}{r_P} - \frac{Z_P}{vr_P + \vec{v}\vec{r}_P} \left\{Z_P + i \ (v\vec{r}_P + r_P\vec{v}) \ \frac{\partial \log\phi_i(\vec{r}_T)}{\partial \vec{r}_T}\right\} \frac{1}{vr_P} \qquad (28)$$

$$W_f^E(t) = - \frac{Z_T}{r_T} - \frac{Z_T}{vr_T + \vec{v}\vec{r}_T} \left\{Z_T - i \ (v\vec{r}_T + r_T\vec{v}) \ \partial \ \frac{\log\chi_f(\vec{r}_P)}{\partial \vec{r}_P}\right\} \frac{1}{vr_P} \qquad (29)$$

(E: "eikonal phase").

The next step in sophistication results from considering the exact

Coulomb distortion produced by a charge, on an electron moving with velocity $\vec{v}$

$$\Phi_i^{CDW}(\vec{r},t) = N(\nu_P) \, L(\nu_P,\vec{r}_P) \, \phi_i(\vec{r}_T) \tag{30}$$

$$\chi_f^{CDW}(\vec{r},t) = N(\nu_T) \, L(\nu_T,\vec{r}_T)^* \, \chi_f(\vec{r}_P), \tag{31}$$

where $L(\nu,\vec{r}) = {}_1F_1(i\nu;1;ivr+i\vec{v}\vec{r})$ is the confluent hypergeometric function and $N(x) = \exp(i\pi x/2) \, \Gamma(1-ix)$. This continuum distorted wave (CDW) method was the original proposal of Cheshire [5]. The distorting potentials are defined by:

$$\left(V_P(\vec{r}_P) + W_D^{CDW}(t)\right) \Phi_i^{CDW}(\vec{r},t) =$$

$$\phi_i(\vec{r},t) \left\{ \frac{\partial \log\phi_i(\vec{r}_T)}{\partial\vec{r}_T} \frac{\partial L(\nu_P,\vec{r}_P)}{\partial\vec{r}_P} \right\}, \tag{32}$$

$$\left(V_T(\vec{r}_T) + W_C^{CDW}(t)\right) \chi_f^{CDW}(\vec{r},t) =$$

$$\chi_f(\vec{r},t) \left\{ \frac{\partial \log\chi_f(\vec{r}_P)}{\partial\vec{r}_P} \frac{\partial L(\nu_T,\vec{r}_T)}{\partial\vec{r}_T} \right\}^*. \tag{33}$$

The next level of sophistication would be to consider the momentum distribution of the electron orbitals $\langle\vec{k}|\phi\rangle$, and to apply to each component $\langle r|k\rangle$ the eikonal phase of (26,27), defined now as $\exp\{-i(Z/k)\log(kr - \vec{k}\vec{r})\}$. This same procedure could be carried out for the Coulomb distortion of the CDW method, eqs.(30,31).

3. RESULTS

3.1 Transition Amplitudes

The calculation of the transition amplitudes given by eqs.(13,14) is performed after writing the scattering wave functions Ψ_i^+, Ψ_f^- in terms of the channel interactions V_D, V_C. The resulting expressions will be the starting point for perturbation expansions in powers of that interaction, assumed to be weak. Green's and evolution operators are defined through

$$G_n^+(t,t') = \theta(t-t') \, U_n(t,t'), \quad G_n^-(t,t') = \theta(t'-t) \, U_n(t,t'), \tag{34}$$

$$U_n(t,t') = \exp\left\{-i\int_{t'}^{t} dt'' \, H_n(t'')\right\}. \tag{35}$$

From here, it is easy to relate Green operators for the full system with those of the scattering channels

$$G^+(t_2,t_1) = G_n^+(t_2,t_1) - i \int_{-\infty}^{\infty} dt \ G^+(t_2,t) \ V_n(t) \ G_n^+(t,t_1),$$ (36)

$$= G_n^+(t_2,t_1) - i \int_{-\infty}^{\infty} dt \ G_n^+(t_2,t) \ V_n(t) \ G^+(t,t_1),$$ (37)

with $n = D,C$. Integral equations for the scattering wave functions Ψ result from applying (36,37) to the channel wave functions Φ, X

$$|\Psi_i^+(t)> = |\Phi_i(t)> - i \int_{-\infty}^{\infty} dt' \ G^+(t,t') \ V_D(t') |\Phi_i(t')>,$$ (38)

$$<\Psi_f^-(t)| = <X_f(t)| - i \int_{-\infty}^{\infty} dt' <X_f(t')| V_C(t') \ G^+(t',t).$$ (39)

The transition amplitudes (13,14) can be expressed as functions of the evolution operators and interactions through eqs.(38,39). Using the orthogonality of Φ_i, X_f at large R, the result is

$$a(t) = -i \int_{-\infty}^{\infty} dt' <X_f(t)| G^+(t,t') \ V_D(t') |\Phi_i(t')>,$$ (40)

$$= -i \int_{-\infty}^{\infty} dt' <X_f(t')| V_C(t') \ G^+(t',t) |\Phi_i(t)>.$$ (41)

3.2 First Order Approximations

At high collision velocities, the time interval where there is an effective projectile-target interaction becomes small. In this regime, the simplest approximation is to neglect the channel interactions in the evolution of the system. G^+ is then replaced by G_C^+ in (40), and by G_D^+ in (41)

$$a_1 = -i \int_{-\infty}^{\infty} dt <X_f(t)| V_D(t) |\Phi_i(t)>,$$ (42)

409

$$= - i \int_{-\infty}^{\infty} dt \; \langle X_f(t) | V_C(t) | \Phi_i(t) \rangle, \qquad (43)$$

The expressions (42,43), called "prior" and "post" form respectively, give identical results in this approximation. This formulation satisfies the correct boundary conditions because, the approximation preserves the residual Coulomb distortion in the channel states Φ, X and interactions $V_{D,C}$. These two aspects of the approach have not always been satisfied in practice. The historical *Oppenheimer-Brinkman-Kramers* (OBK) approximation [1] to electron capture treated the interactions as if they were of short range, so Φ_i, X_f are reduced to bound orbitals centered on P and T, and $V_D(t) = - Z_P/r_P$ is a pure Coulomb potential. Obviously,

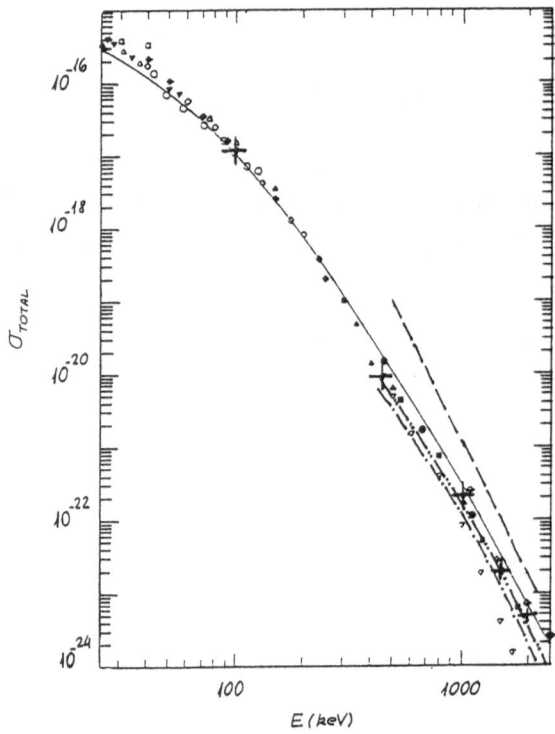

Figure 1. Total cross section for electron capture in H^+- H(1s) collisions.
Theory (——): B1B, Belkic et al.[17], corrected to describe capture to all bound states; (---): OBK; (-..-): CDW [6]; (+): points calculated for capture to the ground state, with the second order JS approximation, by Kramer [18], which for this system coincides with B2B0; (-.-): SE [19].
Experiments: Bayfield [20]; Fite et al.[21]; Gilbody and Ryding [22]; McClure [23]; Wittkower et al.[24]; Barnett and Reynolds [25]; Schryber [26]; Stier and Barnett [27]; Toburen et al.[28]; Welsh et al.[29]; Williams [30].

boundary conditions are not satisfied. Comparison of OBK predictions with measured total cross sections for H^+- H electron capture, show an overestimation in a factor ranging from 5 to 2 in the range of 0.5 to 2 MeV collision energies, (see fig.1). For H^+- Ar capture from the K-shell, the difference is several orders of magnitude in this energy range, but it should be remembered that those energies are too low for a first-order approximation to be reliable (Z_T/v is not small, as required).

The overestimate of the capture process by OBK was attributed to the neglect of the internuclear potential, which in the case of H^+- H exactly cancels the asymptotic Coulomb behavior of the projectile-electron interaction. The *Jackson-Schiff* (JS) approximation [31-32] conserved the internuclear potential in the interaction

$$V_D(t) = - Z_P/r_P + Z_p Z_T/R, \qquad (Z_P = Z_T = 1 \quad \text{for } H^+\text{- H}), \tag{44}$$

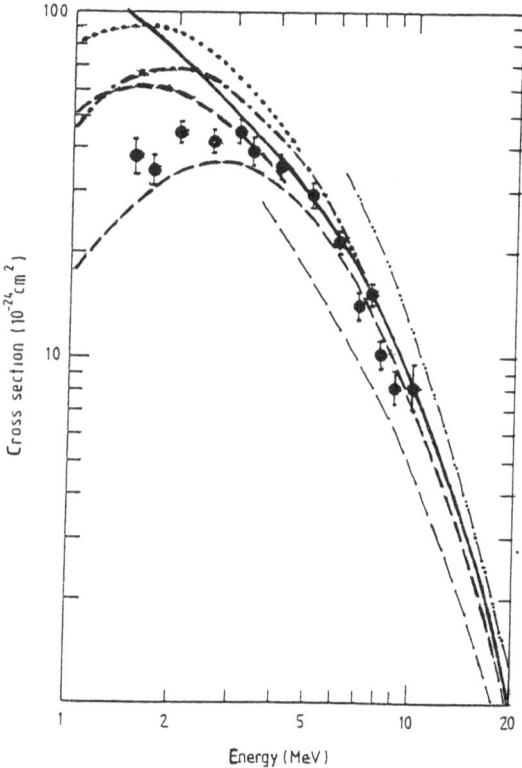

Figure 2. Total cross section for electron capture by protons from the K-shell of argon.
Theory (——): B1B, Dewangan and Eichler [33]; (---): SPB, lower curve from Alston [39]; upper curve including elastic channel distortion, from Taulbjerg et al.[13]; (...): IA, impulse approximation calculated by Alston [39]; (..-..): CDW [6]; (---): SE [19]; (-.-): CDW-EIS [40].
Experimental points of Horsdal-Pedersen et al.[41].

and used, as OBK, undistorted electron states. It is easy to see that a potential that only depends on time can not affect the value of the total cross section; this is not the case for JS, because it is a first-order approximation. The close agreement of JS with measurements for H^+-H collisions, seen in fig.1, seems surprising; nevertheless, for asymmetric systems the discrepancies are considerable, similar to those of OBK.

The first results of a *Boundary corrected first Born* (B1B) approximation [17,33-38] employed the basic time dependent distortion "B" of eq.(19). Channel interactions result

$$V_D(t) = - Z_P/r_P + Z_P/R, \tag{45}$$

$$V_C(t) = - Z_T/r_T + Z_T/R, \tag{46}$$

that behave as R^{-2} for large t. Results of B1B, presented in figs.1 and 2 for the cases of homonuclear and highly asymmetric collisions, show the excellent agreement reached for total cross sections.

The surprising good behavior of JS comes from the fact that for $Z_P = Z_T = 1$, the JS interaction (44) coincides with the boundary correct form, eqs.(17,19) and, furthermore, for $Z_P = Z_T$ the distorting phases of the electron orbitals shown in (21,22) cancel their overall time dependence. This makes JS equivalent to B1B for $Z_P = Z_T = 1$.

The average distortions "A", eqs.(22,23), go one step further and describe electron elastic scattering by the channel interactions. There are few applications of this *average potential first Born* (A1B) approximation [42-44], which has been generalized from its two-state close-coupling origin [16] to a multiple-state expansion in either target or projectile states, always conserving the correct asymptotic boundary conditions [45-47,4,7].

When the distortion of initial and capture channels is described by eikonal phases as defined in (26,27), the resulting model is called the *symmetric eikonal* (SE) approximation [48,19]. As a matter of fact, the *continuum distorted wave* (CDW) approximation, based on the introduction of exact continuum Coulomb waves as distortion factors, may be considered the primary model; SE results from reducing the Coulomb wave (30) to its logarithmic phase, and B1B from further reduction of this to a time-dependent-only phase. All these approaches produce results of similar accuracy for the total cross section, but it is worth noting that the simplest model, B1B, shows similar or better accord with measurements than the other, more complex approximations. A potential drawback of CDW resides in the loss of normalization of the channel states [49], which may become a serious defect at low collision energies.

Distortions can also be treated with different models in the incident and capture channels. Fig.2 shows the results of the *continuum distorted wave-eikonal initial state* (CDW-EIS) approximation [40], closely reproducing the measured values for the H^+- Ar capture cross section.

The high-energy behavior of the various first-order Born models just presented, show a large spread of values [15,50]. Using OBK as a reference, the H^+- H resonant-capture total cross sections become, in the limit $v \to \infty$,

$$\sigma_{B1B} \to 0.6614 \; \sigma_{OBK},$$

$$\sigma_{A1B} \to \sigma_{OBK},$$

$$\sigma_{CDW} \to (0.29 + 5\pi v/2^{12}) \; \sigma_{OBK}, \quad (47)$$

$$\sigma_{SE} \to 0.37 \; \sigma_{OBK},$$

with

$$\sigma_{OBK} \to 2^{18}\pi/(5v^{12}), \quad \text{for } H^+ + H(1s) \to H(1s) + H^+.$$

Even though these limits are relevant at velocities outside the range of experimental interest, they show the importance of considering second-order contributions.

3.3 Second Order Approximations

The differences just shown among first-order approximations indicate that second-order contributions are not negligible asymptotically. Furthermore, differential cross sections are not well described by first-order calculations, which may produce unphysical zeros at certain scattering angles due to cancellations between contributions of the potentials with those of the distortions. The v^{-11} behavior of the CDW cross section in (47) comes from the partial account of a process, Thomas double scattering [2,51], which naturally belongs to second-order processes, since it arises from a P-e collision where the electron acquires the speed v, followed by an e-T scattering where it emerges in the projectile direction, with the previous value v for its speed. The v^{-11} dependence indicates that the second order is really the dominating asymptotic term, but this happens at too high speeds to be described by a non-relativistic, non-radiative formalism. Nevertheless, the specific projectile angular dependence of Thomas scattering starts to be noticeable at energies of a few MeV in $H^+ + H$ collisions.

Starting from either of the forms (40,41) for the transition amplitude, high-order contributions can be explicited by using the integral eqs.(36,37) for the Green operators

$$a(t_f) = a_1(t_f) - \int_{-\infty}^{t_f} dt \int_{-\infty}^{\infty} dt' \ \langle X_f(t)|V_C(t) \ G^+(t,t') \ V_D(t')|\Phi_i(t')\rangle.$$

(48)

Second-order corrections are defined according to the potential that is assumed to be weak. If this corresponds to the projectile, $G^+(t,t')$ from (36,37) can be replaced by $G_D^+(t,t')$ at all times, since the correct asymptotic evolution is incorporated into G_D^+ through the distortion W_D. An explicit expression of G_D^+ can be obtained in terms of the basis of eigenstates $|\Phi_i(t)\rangle$, of H_D,

$$G_D^+(t,t') = G_D^+(t,t') \sum_n |\Phi_n(t')\rangle\langle\Phi_n(t')|$$

$$= \theta(t-t') \sum_n |\Phi_n(t)\rangle\langle\Phi_n(t')|;$$

(49)

and the *boundary-corrected second Born* (B2B) approximation to the transition amplitude becomes [13,52]

$$a_{B2B}(t_f) = a_1(t_f) - \sum_n \int_{-\infty}^{t_f} dt \ \langle X_f(t)|V_C(t)|\Phi_n(t)\rangle$$

$$\times \int_{-\infty}^{t} dt' \langle\Phi_n(t')|V_D(t')|\Phi_i(t')\rangle.$$

(50)

This approximation has several desirable features: it replaces the full evolution operator by a propagator that presents the correct boundary behavior, the states and interactions have the proper asymptotic forms, and effects of the strong potential are considered to all orders. The price to pay for this, is the infinite sum over intermediate states, that seems not reducible to a closed form, and has to be approximated by truncation [52].

Usually, the name of *second order Born* (B2B0) contribution is given to the approximation where G^+ is replaced by the free-particle propagator G_o^+ [53], or by the distorted free-particle propagator G_o^{D+} related to the Hamiltonian $H_D(t) = H_T(t) - Z_p/R$ [54], which presents the correct boundary

414

behavior at $t \rightarrow -\infty$. At the same time, these two forms of second-order approaches present the correct asymptotic behavior, assured by the distorted channel states and interactions [53,54].

The *strong potential Born approximation* (SPB) considers, as B2B, the propagation of intermediate states in the field of the strong potential V_C. In its original version, it used a formulation valid for short-range forces, so that states Φ, X and interactions $V_{D,C}$ are replaced by the undistorted forms ϕ, χ, and $V_{T,P}$, respectively. The SPB transition amplitude adopts a form similar to (48)

$$a_{SPB}(t_f) = \int_{-\infty}^{t_f} dt \ <\chi_f(t)|V_P(t)|\phi_i(t)>$$

$$- \int_{-\infty}^{t_f} dt \int_{-\infty}^{\infty} dt' <\chi_f(t)|V_T(t) \ G_T^+(t,t') \ V_P(t')|\phi_i(t')>. \tag{51}$$

Within this short-range scheme, states propagating in the potential V_T can be described according to an integral equation similar to (39)

$$<\Psi_{k,\varepsilon}^-(t)| = <\vec{k},\varepsilon;t| - i \int_{-\infty}^{\infty} dt' <\vec{k},\varepsilon;t'|V_T \ G_T^+(t',t), \tag{52}$$

where

$$<\vec{r}|\vec{k},\varepsilon;t> = (2\pi)^{-3/2} \ \exp(i\vec{k}\vec{r}_T - i\varepsilon t). \tag{53}$$

The final state (12), projected over the plane wave $<\vec{r}_T|\vec{k}>$, gives

$$<\chi_f(t)|\vec{k}> = \tilde{\chi}_f(\vec{k}-\vec{v})^* \ e^{i\vec{k}\vec{b}} \ e^{i\varepsilon t}, \qquad \varepsilon = \varepsilon_f - (\vec{k}-\vec{v})^2 + k^2/2, \tag{54}$$

where $\tilde{\chi}_f(\vec{k})$ is the Fourier transform of $\chi_f(t)$. From (51,52,54),

$$a_{SPB}(t_f) = \int_{-\infty}^{t_f} dt \int d^3k \ e^{i\vec{k}\vec{b}} \ \tilde{\chi}_f(\vec{k}-\vec{v})^* \ <\Psi_{\vec{k},\varepsilon}^- (t)|V_P(t)|\phi_i(t)>. \tag{55}$$

This expression indicates that the projectile excites the target orbital to a state $\Psi_{\vec{k},\varepsilon}^-$ and this, having a non-zero overlap with the projectile orbital $\tilde{\chi}_f$, feeds the capture amplitude according to the weight $\tilde{\chi}_f(\vec{k}-\vec{v})$. The intermediate states $\Psi_{\vec{k},\varepsilon}^-$ become asymptotic free waves at $t \rightarrow \infty$, but of a special type since the physical energy $k^2/2$ of a free particle is replaced by $\varepsilon = k^2/2 + \varepsilon_f-(\vec{k}-\vec{v})^2/2$. The first applications of (55), named then *impulse approximation* (IA) [55], replaced $\Psi_{\vec{k},\varepsilon}^-$ by the on-shell

$(\varepsilon = k^2/2)$ Coulomb wave function $\Psi_{\vec{k}}^{-}$, on the understanding that $\varepsilon - k^2/2$ is of order Z_P^2, the weak charge of the collision. But the difficulty comes from the on-shell limit of $\Psi_{\vec{k},\varepsilon}^{-}$, that presents a diverging phase and a change of amplitude [56,57]

$$\Psi_{\vec{k},\varepsilon}^{-}(\vec{r}) = g(k,p)\ \Psi_{\vec{k}}^{-}(r), \tag{56}$$

$$g_T(k,p) = (k - p/k + p)^{-i\eta(T)}\ \Gamma(1+i\eta)\ e^{-\pi\eta(T)/2}, \tag{57}$$

where $p = \sqrt{2\varepsilon}$ and $\eta(T) = Z_T/p$. Applications of the SPB approximations showed the influence of g on the results, even though differences with IA become small at high energies. The results of fig.2 show the close agreement of SPB with measurements, in the case of the asymmetric H^{+}- Ar system.

A problem of the SPB method, pointed out by Dewangan and Eichler [37], is the following. The off-shell states $\Psi_{\vec{k},\varepsilon}^{-}$ overlap with the bound target state ϕ_i, so elastic transitions are present in (55) and, due to the R^{-1} behavior of $V_P(t\to\infty)$, produce a logarithmic divergence of the amplitude. Explicit account of the residual interaction $- Z_P/R$, at least in the elastic channel, is necessary to eliminate this divergence [11,13].

The neglect of the distorting potentials in the incident and capture channels, also implies the neglect of the distorting phase $\exp\{i(Z_P/v)\log(vR+\vec{v}\vec{R})\}$ and $\exp\{i(Z_T/R)\log(vR-\vec{v}\vec{R})\}$ in the channel wave functions. From the definitions of eqs.(13,14) for the transitions amplitudes, we can infer the following relation for the B2B and SPB amplitudes

$$a_{B2B} = \lim_{t\to\infty} (vR + \vec{v}\vec{R})^{-iZ_P/v}\ ;\quad a_{SPB} = \lim_{t\to-\infty} (vR - \vec{v}\vec{R})^{-iZ_T/v}. \tag{58}$$

The diverging phases are exactly the inverse of those produced by the factors $g_{T,P}(k,p)$, connecting the off-shell plane wave with the on-shell Coulomb state for the interatomic motion, eq.(56). A corrected SPB amplitude is then proposed [11]: it results from multiplying the a_{SPB} amplitude by $g_T(k,p)^{-1}\ g_P(k',p)^{-1}$, $(k,k' \to p)$, as implied by eq.(58). For an application of the SPB approximation, once a_{SPB} is calculated, its diverging phases have to be isolated and they will be canceled by the phase factors appearing in eqs.(2) or (58).

Comparison of boundary-corrected B2B [52] and B2B0 [53,54] approximations, with measurements of differential capture cross-sections for H^{+}- H collisions at 125 keV, are presented in fig.3. The agreement

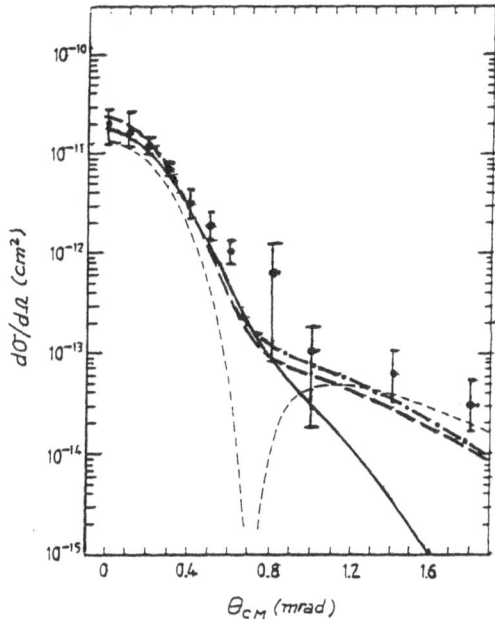

Figure 3. $H^+ + H(1s) \rightarrow H(1s) + H^+$ differential cross section at 125 keV.
Theory: (---): B1B [19]; (- - -): B2B0, Belkic [53] with $G^+ \approx G_o^+$;
(-.-): B2B0, Decker and Eichler [54], with $G^+ \approx G_o^+ - Z_p/R$; (——):
B2B [52], the drop in these results al large angles may be due
to numerical inaccuracies, see ref.[54].
Experimental results from Martin et al.[58].

reached suggests that the perturbative Born series may be well-behaved.
Values of the total cross sections for this collision reinforce these
expectations, being 0.109 and 0.14 a.u. for B1B and B2B respectively [54].

4. CONCLUDING REMARKS

The correct treatment of long-range Coulomb forces in atomic
collisions turns out to be something more than a formal requirement of the
formalism. The large discrepancies with measurements, of models like OBK
and JS, that neglect long-range effects, can be explained by this sole
omission.

Self-consistent perturbative approaches presented in the last years,
are able to reproduce measurements of total and differential capture cross
sections of homonuclear and heteronuclear systems. The first and
second-order Born approximations present a range of agreement with
experiments compatible with the condition $Z_i/v < 1$, with Z_i the charge
that is treated as a perturbation.

The freedom introduced by the use of distortion potentials generates a number of approximations, whose value and range of application needs to be clarified. Attempts to free the results from the selection of distorting potentials, may not be pursued through a neglect of the effects brought by long-range forces on the transition amplitudes.

ACKNOWLEDGEMENT

The present work has been performed while the author was in use of a Sabbatical stay sponsored by the Spanish Ministerio de Educación y Ciencia.

REFERENCES

1. J.R. Oppenheimer, Phys. Rev. *31* (1928) 349; H.C. Brinkman and H.A. Kramer, Proc. Acad. Sci. Amsterdam *33* (1930) 973.
2. L.H. Thomas, Proc. Roy. Soc. *A114* (1927) 561.
3. D.H. Jakubassa-Amundsen, Int. J. Mod. Phys. *A4* (1989) 769.
4. B.H. Bransden and D.P. Dewangan, Adv. At. Mol. Phys., (D.R. Bates & B. Bederson Eds.) Academic Press, Orlando, Florida. Vol. 25 (1988) 343.
5. I.M. Cheshire, Proc. Phys. Soc. *84* (1964) 89.
6. Dz. Belkic, R. Gayet and A. Salin, Phys. Rep. *56* (1979) 279.
7. D.P. Dewangan and J. Eichler, Comments At. Mol. Phys. *21* (1987) 1.
8. Dz. Belkic, Europh. Lett. 7 (1988) 323.
9. J.S. Briggs, J.Phys. *B10* (1977) 3057.
10. J.H. Macek and K. Taulbjerg, Phys. Rev. Lett. *46* (1981) 170.
11. J.H. Macek, Phys. Rev. *A37* (1988) 2365.
12. J.H. Macek and S.A. Alston, Phys. Rev. *A26* (1982) 250.
13. K. Taulbjerg, R.O. Barrachina and J.H. Macek, Phys. Rev. *A41* (1990) 207.
14. R.A. Mapleton, Proc. Phys. Soc. *91* (1967) 868.
15. R.D. Rivarola, J.M. Maidagan and J. Hanssen, Nucl. Instr. Meth. *B27* (1987) 565.
16. D.R. Bates, Proc. Phys. Soc. *A247* (1958) 294.
17. Dz. Belkic, S. Saini and H.S. Taylor, Phys. Rev. *A36* (1987) 1601.
18. P.J. Kramer, Phys. Rev. *A6* (1972) 2125.
19. J.M. Maidagan and R.D. Rivarola, J.Phys. *B17* (1984) 2477.
20. J.E. Bayfield, Phys. Rev. *185* (1969) 105.
21. L.W. Fite, R.F. Stebbings, D.G. Hummer and R.T. Brackman, Phys. Rev. *119* (1960) 663.
22. H.B. Gilbody and G.R. Ryding, Proc. Roy.Soc. (Lond.) *A291* (1966) 438.
23. G.W. McClure, Phys. Rev. *148* (1966) 47.
24. A.B. Wittkower, G. Ryding and H.B. Gilbody, Proc. Roy. Soc. (Lond.) *89* (1966) 541.
25. C.F. Barnett and H.K. Reynolds, Phys. Rev. *109* (1958) 355.
26. U. Schryber, Helv. Phys. Acta *40* (1967) 1023.
27. P.M. Stier and C.F. Barnett, Phys. Rev. *103* (1956) 896.
28. L.H. Toburen, M.Y. Nakai and R.A. Langley, Phys. Rev. *171* (1968) 114.
29. L.M. Welsh, K.H. Berkner, N.S. Kaplan and R.V. Pyle, Phys. Rev. *158* (1967) 85.
30. J.F. Williams, Phys. Rev. *157* (1967) 97.
31. D.R. Bates and A. Dalgarno, Proc. Phys. Soc. (Lond.) *A65* (1952) 919.
32. J.D. Jackson and H. Schiff, Phys. Rev. *89* (1953) 359.
33. D.P. Dewangan and J. Eichler, J. Phys. *B19* (1986) 2939.
34. Dz. Belkic, R. Gayet, J. Hanssen and A. Salin, J.Phys. *B19* (1986) 2945.
35. Dz. Belkic, S. Saini and H.S. Taylor, Z. Phys. *D3* (1986) 59.
36. Dz. Belkic and H.S. Taylor, Phys. Rev. *A35* (1987) 1991.
37. D.P. Dewangan and J. Eichler, Nucl. Instr. Meth. *B23* (1987) 160.
38. G.R. Deco, J. Hanssen and R.D. Rivarola, J. Phys. *B19* (1986) 3727.

39. S. Alston, Phys. Rev. *A27* (1983) 2342.
40. R.D. Rivarola and A.E. Martinez, private communication.
41. E. Horsdal-Pedersen, C.L. Cocke, J.L. Rasmussen, S.L. Varghesse and W. Waggoner, J. Phys. *B16* (1983) 1799.
42. R.H. Bassel and E. Gerjuoy, Phys. Rev. *117* (1960) 749.
43. B.H. Bransden and R.K. Janev, Adv. At. Mol. Phys., D.R. Bates and B. Bederson, eds., Academic Press, Orlando, Florida, vol. 19, (1983).
44. N. Toshima, T. Ishihara and J. Eichler, Phys. Rev. *A36* (1987) 2659.
45. A.L. Ford, R.L. Becker, G.L. Swafford and J.F. Reading, J. Phys. *B12* (1979) 491.
46. J.F. Reading, A.L. Ford, G.L. Swafford and A. Fitchard, Phys. Rev. *A20* (1979) 130.
47. R.L. Becker, A.L. Ford and J.F. Reading, J. Phys. *B13* (1980) 4059.
48. D.P. Dewangan, J. Phys. *B8* (1975) L119.
49. D.S.F. Crothers, J. Phys. *B15* (1982) 2061.
50. J.M. Maidagan, Ph D thesis, Universidad Nacional de Rosario (1986).
51. R. Shakeshaft, J. Phys. *B7* (1974) 1059.
52. D.P. Dewangan and B.H. Bransden, J. Phys. *B21* (1988) L353.
53. Dz. Belkic, Europh. Lett. 7 (1988) 323.
54. F. Decker and J. Eichler, J. Phys. *B22* (1989) L95.
55. M.R.C. McDowell, Proc. Roy. Soc. (Lond.) *A264* (1961) 277.
56. S.O. Okubo and D. Feldman, Phys. Rev. *117* (1960) 292.
57. R.A. Mapleton, J. Math. Phys. *2* (1961) 482.
58. P.J. Martin, D.M. Blankeship, J.J. Krale, E. Redd, J.L. Peacher and J.T. Park, Phys. Rev. *A23* (1981) 3367.

CHANNELING AND CHANNELING RADIATION THEORY

A.H. Sørensen

Institute of Physics
Aarhus University
DK-8000 Aarhus C, Denmark

In two lectures, a broad introduction was given to directional effects associated with the penetration of keV to TeV particles through single crystals near major crystallographic axial or planar directions.

In the first, the governing of the motion of charged particles by the crystal lattice was discussed. *Channeling* corresponds to such close alignments that correlated scattering on many target atoms prevents the projectiles from having uniform flux in the space transverse to the considered axial or planar direction. We discussed angles characteristic for channeling, angular dependence of close-encounter processes, negatively versus positively charged particle channeling, and touched briefly upon topics like dechanneling of initially channeled particles and doughnut scattering associated with correlated scattering at incident angles beyond the channeling regime. Further on, the question of a classical versus a quantal description of channeling was considered.

The second lecture was devoted to directional effects in bremsstrahlung emission by electrons and positrons. The coherence in scattering on target atoms under channeling conditions leads to a strong enhancement of the radiation as compared to that pertaining to directions of incidence far from axial and planar directions. We discussed the main features of *channeling radiation*, which in certain circumstances appear as intense X- and γ-ray peaks, and of *coherent bremsstrahlung*, which results due to coherent scattering outside the channeling region. Also the transition with increasing angle between the two types of radiation was considered. Further on, we singled out the special features associated with the extension of the coherent magnification of the bremsstrahlung spectrum all the way up to the primary energy which appears when the

Interaction of Charged Particles with Solids and Surfaces
Edited by A. Gras-Martí *et al.*, Plenum Press, New York, 1991

421

latter is beyond, typically, 10-100 GeV. In this region, also the electron-positron pair production by aligned photons is strongly enhanced.

REVIEWS

- D.S. Gemmel, *Channeling and Related Effects in the Motion of Charged Particles through Crystals*, Rev.Mod.Phys. *46* (1974) 129-227.
- L.C. Feldman, J.W. Mayer, and S.T. Picraux, *Materials Analysis by Ion Channeling*, Academic Press, New York (1982).
- J.U. Andersen, E. Bonderup and R.H. Pantell, *Channeling Radiation*, Ann. Rev. Nucl. Part. Sci. *33* (1983) 453-504.
- A.H. Sørensen and E. Uggerhøj, *Channeling and Channeling Radiation*, Nature *325* (1987) 311-318.
- A.H. Sørensen and E. Uggerhøj, *Channeling, Radiation and Applications*, Nucl. Sci. Appl. (1989) 147-205.

More specialized material (including a few reviews) can be found in:
- R.A. Carrigan, Jr., and J.A. Ellison, eds., *Relativistic Channeling*, NATO ASI Series B *165*, Plenum, New York (1987).

WAKES, DYNAMIC SCREENING

Y. Yamazaki

Institute of Physics
College of Arts and Sciences, University of Tokyo
3-8-1, Komaba, Meguro-ku, Tokyo, Japan 153

ABSTRACT

Two topics on "wake" phenomena are reported. The first one concerns the density fluctuation induced in solids by swift ions. For this purpose, the knock-on collision electrons ($v_e \sim 2v_p$) emitted at 0° from thin carbon foils under 4.5 MeV N_2^+ impact were measured as a function of the orientation θ of the molecular axis of the projectile with respect to the beam axis. A marked decrease in the yield has been observed near $\theta \sim 25^o$, which is a critical angle where the density fluctuation of target valence electrons changes its feature from collective to single-particle-like. The second topic concerns a search for wake-riding electrons employing slow antiprotons as well as equivelocity protons. Electrons emitted at 0° have been measured for projectile energies from 500 keV to 750 keV. Although a deep dip is predicted at $v_e \sim v_p$ for antiproton-gas collisions, the observed spectrum is rather smooth with indication of a bump at $\sim$ 50 eV below the energy where the dip is anticipated. The energy and the relative intensity of the bump are found to be consistent with those predicted for electrons released from a wake-riding state.

1. INTRODUCTION

A charged particle traversing a solid with velocity v_p induces an electronic polarization wake [1], which trails the particle with wavelength $2\pi v_p/\omega_{pl}$, where ω_{pl} is the plasmon frequency of the solid. Subsequently, various experiments have attempted to probe this phenomena through measurements of (i) energy and angular distribution of dissociated fragments of molecular ions [2], (ii) stopping power of molecular ions [3], (iii) energy shifts of bound electronic states of ions in solids [4], and (iv) the resonance pattern of the resonant coherent excitations [5]. It is noted that all these experiments are, in effect, related to measure the electric field strength of the wake at the position of the projectile. With the exception of heavy molecular ions [6], these observations have been successfully interpreted by linear-response theories.

Interaction of Charged Particles with Solids and Surfaces
Edited by A. Gras-Martí *et al.*, Plenum Press, New York, 1991

In the present report, we will discuss two different subjects in which new aspects of the wake are investigated. Sect.2 treats an experiment demonstrating the detection of the density fluctuation of target valence electrons, which is the origin of the dynamic screening and the oscillatory wake [7]. Sect.3 concerns a search for wake-riding electrons [8], which are trapped at the attractive part of the oscillating wake. This well-known but scarcely-understood topic has been studied experimentally, for the first time, employing slow antiprotons as projectiles.

Atomic units (e = m = ℏ = 1) are used in this report unless otherwise noted.

2. DETECTION OF DENSITY FLUCTUATION

A charged-particle traversing a solid induces the density fluctuation $\delta n(r,\theta)$ of target valence electrons, where r and θ are polar coordinates with respect to the beam with its origin at the particle. Such a fluctuation is the root of the wake potential and the dynamic screening. Fig.1 shows a theoretical prediction of the electron density fluctuation induced in carbon by a charged particle with $v_p = 2$, calculated by Echenique et al.[9]. It is seen that the most prominent enhancement of the electron density is localized in a conical region immediately behind the particle. This cone has a half angle $\theta_c \sim \sin^{-1}(\sqrt{3/5}\ v_F/v_p)$, where v_F is the Fermi velocity $[= (3\pi^2 n_0)^{1/3}]$ of the target valence electrons [9]. n_0 is the average density of valence electrons (~0.067 for carbon). For diatomic molecular projectiles, the electron density induced by a fragment ion in the vicinity of its companion ion may therefore be enhanced when it falls within this conical region. The yield of binary encounter ("knock-on") electrons produced by the companion ion should be sensitive to this enhancement.

We have measured knock-on electrons ($v_e \sim 2v_p$) emitted from carbon foils for 4.5 MeV N_2^+ ions ($v_p \sim 2.5$ a.u.), where v_e is the electron velocity [7]. The yield of these electrons was measured, at $0°$ with respect to the beam direction, as a function of the orientation of the nitrogen diclusters. Knock-on electrons from the valence band of carbon were adopted as a probe, because the effective impact parameter for producing such electrons (~ 0.2 a.u.), which gives a measure of the spatial resolution of the probe, is much smaller than $2r_N \sin\theta_c \sim 2$ a.u., the transverse extent of the conical region discussed above, where r_N is the internuclear separation.

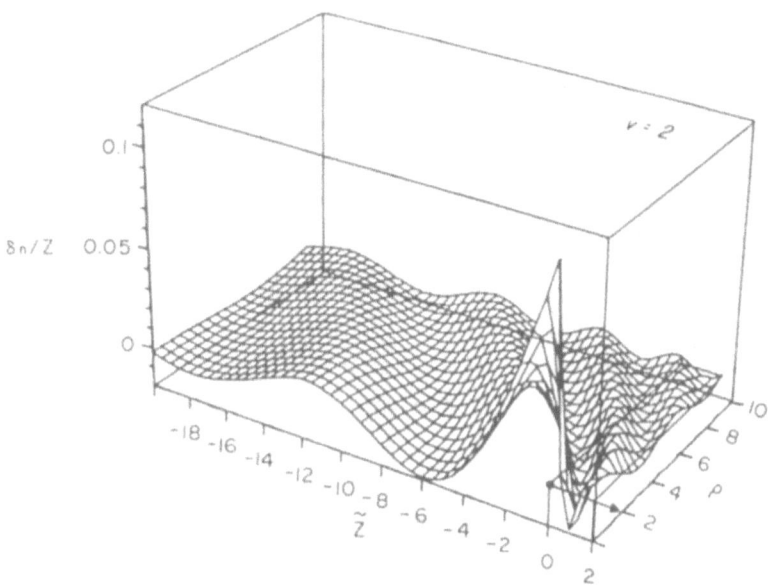

Figure 1. Electron-density fluctuations induced by a moving ion at v_p = 2, for carbon (n_o ~ 0.067, v_F ~ 1.26), ref.[9].

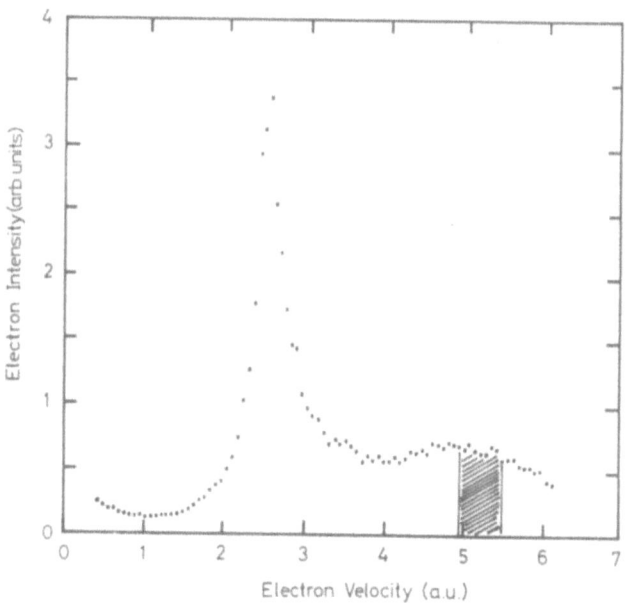

Figure 2. Electron spectrum at $0°$, for 4.5 MeV N_2^+ impacts on a 2 $\mu g/cm^2$ carbon foil.

2.1 Experimental

The experiment has been performed at Argonne National Laboratory using the MUPPATS (MUltiParticle Position And Time Sensitive detector), which provides three-dimensional imaging of molecular ions [10]. A well-collimated 4.5 MeV N_2^+ beam from the Argonne 4.5 MV Dynamitron accelerator was passed through thin (~ 2 $\mu g/cm^2$ or ~ 6 $\mu g/cm^2$) carbon foils, the center hole of the magnetic electron analyzer, and finally detected by the MUPPATS detector 6 m downstream from the target. In the foil, the binding electrons of the molecular ion are stripped and the resulting ions undergo a Coulomb explosion on a time scale of ~ 10^{-14} s. By measuring the positions and the relative arrival times of the fragment ions in the MUPPATS detector, the orientation of the molecular axis with respect to the beam axis is determined for every molecule. At the same time, the electrons emitted from the target foils at 0° with respect to the beam were detected by a magnetic electron analyzer ($\Delta v_e/v_e$ ~ 10 %, and $\Delta\theta_{FWHM}$ ~ 4.5^o).

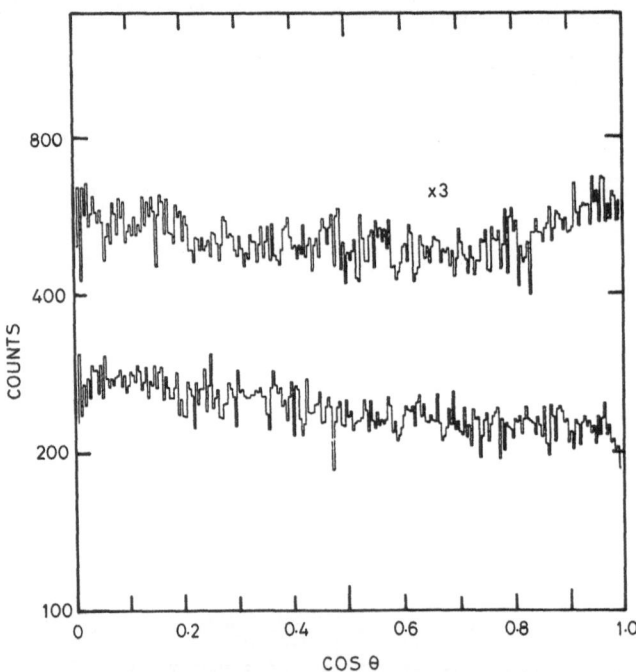

Figure 3. Angular distribution of the orientation of diclusters with (upper curve), and without (lower curve), coincidence with knock-on electrons emitted at 0° for 4.5 MeV N_2^+ impacts on a 2 $\mu g/cm^2$ carbon foil. Diclusters not accompanied by knock-on electrons are registered at every 100 counts.

2.2 Results and Discussions

Fig.2 shows an example of the electron spectra for 4.5 MeV N_2^+ impact on a 2 $\mu g/cm^2$ carbon foil. Crudely speaking, the electron spectra in the forward direction can be divided into three parts, i.e., a low energy part, a $v_e \sim v_p$ part, and a high energy part including a knock-on collision peak. At around the top of the knock-on peak ($v_e \sim 5$ a.u.), the electrons are kicked out mainly from the valence band, while the higher energy tail originates from inner shells. During the measurements of knock-on electrons, the magnetic analyzer was tuned to cover a velocity region of $1.95\ v_p < v_e < 2.15\ v_p$ (the shaded area of fig.2), i.e., to detect electrons kicked out from the valence band.

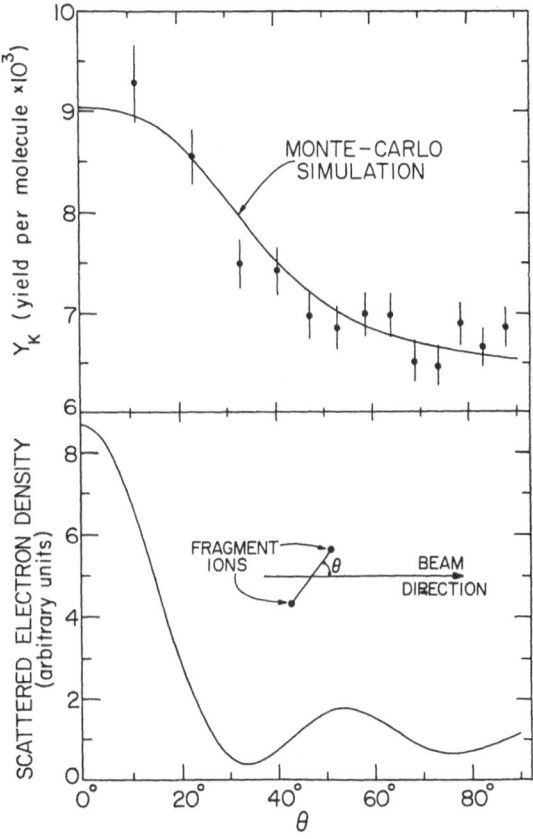

Figure 4. The knock-on electron yield $Y_K(\theta)$ as a function of the orientation of the dicluster for 4.5 MeV N_2^+ impacts on a 2 $\mu g/cm^2$ carbon foil. The solid line in the upper figure shows the results of the Monte Carlo simulation, which is arbitrarily normalized to the experimental results. The solid line in the lower figure shows the electron-density distribution at r_N, evaluated using the Coulomb wave function, ref.[7].

The lower curve of fig.3 shows the resultant angular distribution $I(\theta)$ of diclusters for the 2 $\mu g/cm^2$ carbon foil. The upper curve shows the angular distribution of ions accompanied by knock-on electrons $I_K(\theta)$. It is seen that $I_K(\theta)$ varies differently from $I(\theta)$ for $\cos\theta > 0.8$. The yield of knock-on electrons $Y_K(\theta)$ $[= \int_{\theta-\Delta\theta}^{\theta+\Delta\theta} I_K(\theta')d(\cos\theta')/\int_{\theta-\Delta\theta}^{\theta+\Delta\theta} I(\theta')d(\cos\theta')]$ is shown in the upper half of fig.4. It is seen that (i) $Y_K(\theta)$ is a monotonically decreasing function of θ with a decrease at $\theta \sim 25°$, and (ii) $Y_K(0°)$ is ~ 1.4 times larger than $Y_K(90°)$. Fig.5 shows $Y_K(\theta)$ for a 6 $\mu g/cm^2$ carbon foil, where $Y_K(0°)/Y_K(90°)$ is ~ 1.3, and the yield averaged over all orientations is ~ 95 % of that observed for the 2 $\mu g/cm^2$ foil.

The half-angle θ_c of opening of the conical region, where the strong electron-density enhancement is expected, is estimated to be $\sim 23°$ for a carbon target ($v_F \sim 1.26$ a.u.). The yields shown in figs.4 and 5 demonstrate an increase when the trailing ion is rotated across the boundary of that region, from outside to inside. Accordingly, the yield enhancement of the knock-on electrons for $\theta < \theta_c$ can be qualitatively explained as the enhanced density of target electrons by the leading ion, around the trailing ion position. As fig.1 indicates, $\delta n(r,\theta)$ decreases as the distance from the leading ion increases. The internuclear distance of the pair of ions near the exit surface for the 6 $\mu g/cm^2$ foil is larger

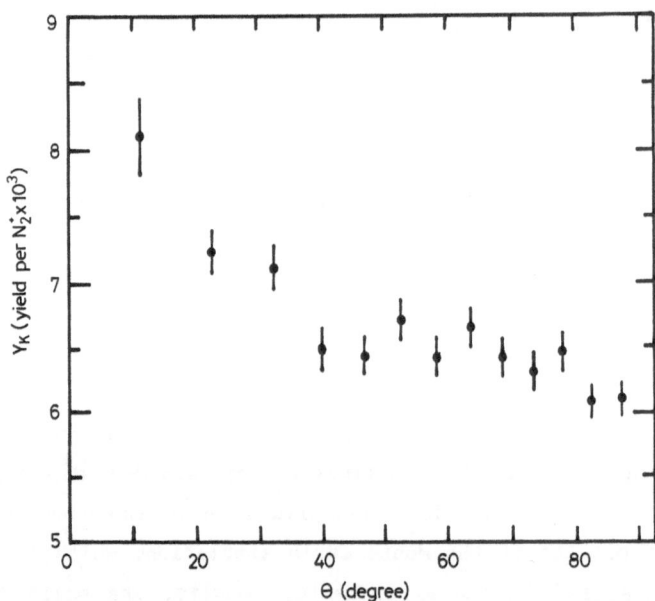

Figure 5. Same as the upper part of fig.4, except that the foil thickness is 6 $\mu g/cm^2$.

than that for the 2 $\mu g/cm^2$ foil. Accordingly, the decrease of the yield for the 6 $\mu g/cm^2$ foil (fig.5), compared to that for the 2 $\mu g/cm^2$ foil (fig.4), is reasonable. As the electron density enhancement at $\theta = 90^o$, $\delta n(r,90^o)$, is expected to be small, the yield ratio $Y_K(0^o)/Y_K(90^o) \sim 1.4$ for the 2 $\mu g/cm^2$ foil infers that $\delta n(r,0^o)$ is crudely $\sim 80 \%$, if we assume that $Y_K(0^o)$ is proportional to $\delta n(r,0^o)$. On the other hand, fig.1 shows that the theoretical value of $\delta n(r_N,0^o)$ is $\sim 200 \%$, when the effective charge of the leading ion is assumed to be equal to the mean charge state (~ 3.3) of the fragment nitrogen ions exiting the foil.

The discrepancy between the experimental result and the theoretical prediction is supposed to reflect the details of the momentum distribution of the density fluctuations. In order to incorporate such effects, the energy spectra of electrons scattered by a two-center Coulomb field, moving with v_p, were simulated by a classical trajectory Monte-Carlo calculation incorporating the initial velocity distribution of the target valence electrons [9]. The solid line in the upper half of fig.4 shows the results of the simulation, which reproduces the general trend of the experimental results fairly well. Since the collective aspects of the medium, which are essential to form the wake, are not taken into account in the simulation, the absence of a marked decrease near $\theta = 25^o$ is not surprising. The lower half of fig.4 shows $|\psi_c|^2$ as an alternative way of estimating $\delta n(r_N,\theta)$, where ψ_c is the Coulomb wave function of an electron with incident velocity $-v_p$ scattered by the leading ion. This description is expected to be effective particularly at $\theta > \theta_c$ and/or $r_N < v_p/\omega_{pl}$, where single-particle-like aspects dominate over collective aspects. Although the statistics of the data is not satisfactory, an indication of the fluctuating structure at $\theta > 40^o$ is more or less reproduced by $|\psi_c|^2$, inferring the importance of quantum interference effects (see also fig.1). However, the enhancement ($|\psi_c|^2 - 1$) at $\theta = 0^o$ is more than 800 %, four times larger than the prediction of the linear-response theory, which shows the importance of the screening of projectile charge and the initial velocity distribution. It is noted that $|\psi_c|^2$ is, in its nature, independent on the distance from the scattering center for $\theta = 0^o$, i.e., the rapid decrease of the density seen in fig.1 is a consequence of the screening of the projectile charge by valence electrons.

As a byproduct of the experiment, "two dimensional" charge-state distributions of diclusters have been obtained. Fig.6 shows R, the ratio of P_{ij} to P_iP_j, as a function of $q_i + q_j$, for 4.5 MeV N_2^+ bombarding the 2 $\mu g/cm^2$ carbon foil; P_{ij} is the probability that the charge states are q_i and q_j for diclusters, and P_i is the probability that the exiting charge

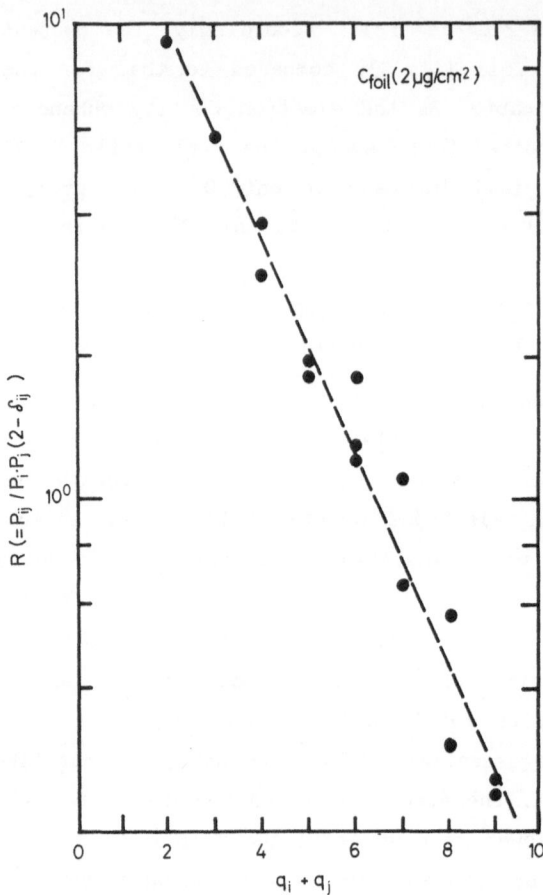

Figure 6. The ratio $R = P_{ij}/\{P_i P_j (2 - \delta_{ij})\}$, of the exit charge state fraction P_{ij} for 4.5 MeV N_2^+ ions to those for the equivelocity N^+ ions. The charge-state fractions for the atomic nitrogen were $P_1 = 0.01$, $P_2 = 0.13$, $P_3 = 0.45$, $P_4 = 0.35$, and $P_5 = 0.06$, independent of the foil thickness within the experimental uncertainty.

state is q_i for atomic ions. It is seen that (i) R decreases exponentially as a function of $q_i + q_j$, indicating that $q_i + q_j$ is an important parameter in describing the charge-state distribution of diclusters, and (ii) $R \sim 1$ occurs at $q_i + q_j \sim 6.5$, which is roughly twice the mean charge states of equivelocity atomic nitrogen ions, i.e., $P_{ij} = \exp[-\alpha(q_i + q_j - 2\bar{q})]\, P_i P_j$. For the 6 $\mu g/cm^2$ carbon foil, the exponential dependence has still been observed, but with gentler slope.

3. A SEARCH FOR WAKE-RIDING ELECTRONS

Neelavathi et al.[11] had first pointed out a possibility of trapping

430

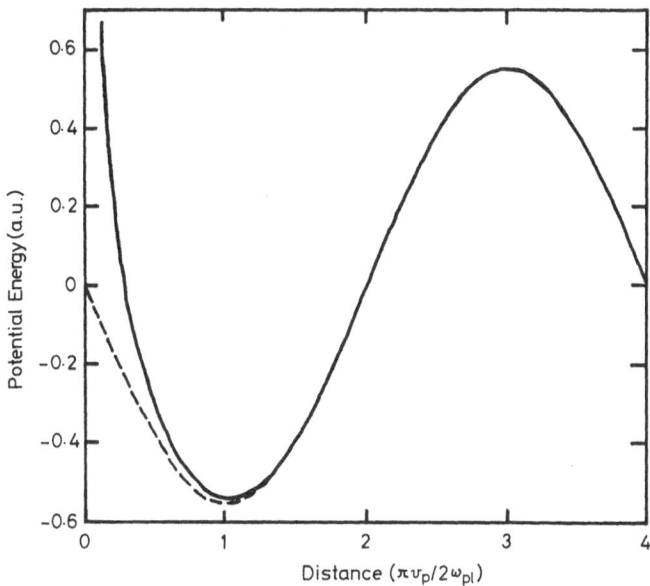

Figure 7. Schematic drawing of an effective potential energy between an antiproton and an electron in the solid (——). Also shown is the oscillatory wake part (---).

an electron at the attractive part of the wake potential, which proceeds in solids with the velocity of the projectile. The electron bound in the wake potential has been called a wake-riding electron [12]. This intuitive picture is extremely interesting because it predicts a new mechanism to induce dynamic electronic states in a solid. To search for a wake-riding electron, we have studied the energy spectra of electrons emitted from thin carbon foils at 0°, for several-hundred keV antiprotons as well as for equivelocity protons [8].

3.1 Needs for Negatively-Charged Particles

For a negatively-charged particle, the wake-riding electron is expected to be localized at $\sim \pi v_p/(2\omega_{pl})$ behind the projectile, see fig.7. The theoretical aspects of the wake-riding electron for a negatively-charged particle have first been studied by Rivacoba and Echenique [13]. When the projectile exits from a target, the wake-riding electron follows and finds itself in a state with a positive potential energy $\Delta E \sim 2\omega_{pl}/(\pi v_p)$. Consequently the electron is accelerated in the backward direction and finally gets this amount of kinetic energy in the projectile reference frame. The electron energy E_{wr} in the laboratory frame is, therefore, given by

$$E_{wr} \sim \frac{1}{2} (v_p - \sqrt{2\Delta E})^2 = \frac{1}{2} v_p^2 - 2(\sqrt{v_p \omega_{pl}/\pi} - \frac{\omega_{pl}}{\pi v_p}). \qquad (1)$$

Fig.8 shows ΔE_{wr} (= E_{wr} - $1/2$ v_p^2) as a function of antiproton energy. It is seen that, for a 0.5 MeV antiproton, ΔE_{wr} is about - 60 eV and - 50 eV for carbon and aluminum targets, respectively. As no cusp-shaped peak is expected to appear for antiproton impact, electrons released from a wake-riding state may be observed at E_{wr}, superimposed on a weak continuum.

For protons, on the other hand, the first minimum of the wake potential appears at ~ $3\pi v_p/(2\omega_{pl})$ behind the projectile, three times as far as for antiprotons. Upon exit, the attractive Coulomb potential of the projectile takes the place of the wake potential, i.e., the wake-riding electron finds itself in a state with a negative potential energy $\Delta E \sim - 2\omega_{pl}/(3\pi v_p)$. Due to the initial momentum spread of the wake-riding electrons, the final energies of the electrons may distribute above and below the vacuum level, in the projectile reference frame [14]. As a result, a trace of the wake-riding electron if any, will be dissolved into a huge background of ordinary convoy and/or Rydberg electrons.

As a target plasmon has a finite lifetime, the wake potential and hence the wake-riding state is more "well-defined" and stable as the distance from the projectile is closer, increasing the possibility of observing the wake-riding electrons for antiprotons rather than for protons.

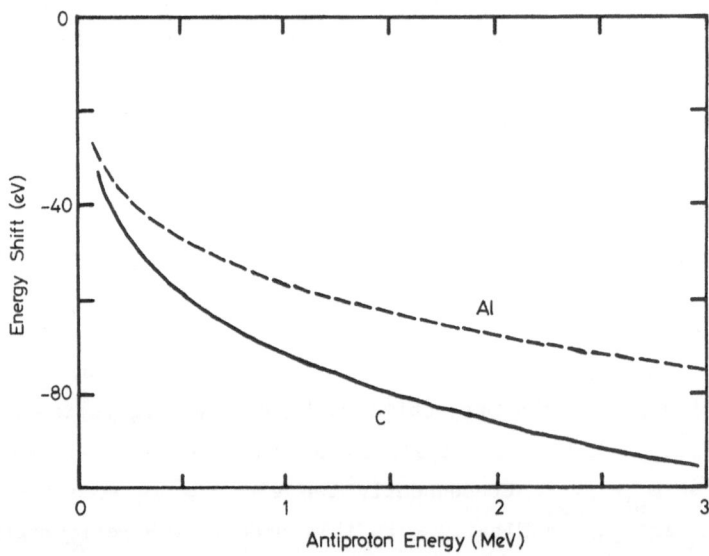

Figure 8. Energy shift of the wake-riding electrons, relative to the energy of electrons having the velocity of the projectile.

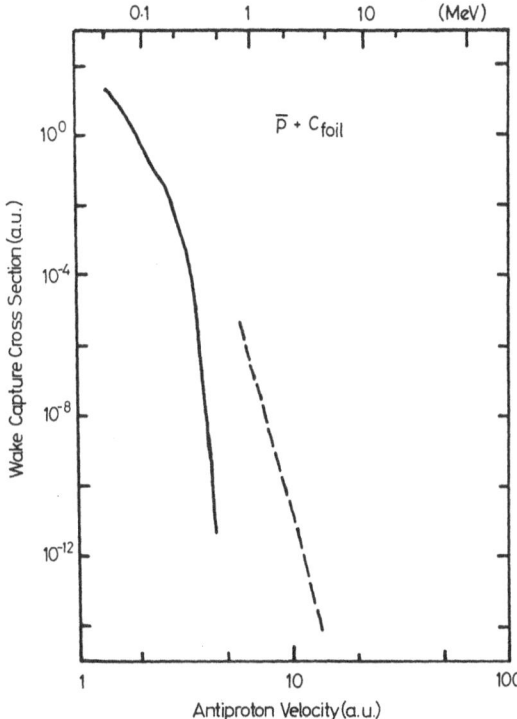

Figure 9. Capture cross section into the wake potential, as a function of an antiproton velocity evaluated by an OBK calculation (——), ref.[15], and by a second-Born calculation (---), ref.[18].

Another issue, which has not been discussed very often, is the way to trap an electron in the wake potential. The solid line of fig.9 shows cross sections of capturing target electrons into a wake potential, estimated by an OBK method (i.e., single step process) [15]. It is seen that the capture cross section is negligibly small at reasonable projectile velocities where the above simple-minded discussion is expected to be, at least qualitatively, effective. This negligible cross section occurs because the momentum spread of the wake-riding state is very narrow and the wake potential is a "soft" potential. Therefore, the requirement of momentum conservation before and after the electron capture is very hard to satisfy. Accordingly, the wake-riding electron is expected to be produced via a two-step process [16], i.e., the scattering of a target electron with the projectile followed by a second scattering with a target nucleus into the forward direction. In this case, the requirement on the momentum conservation is not so strict [17]. The dashed line in fig.9 shows the result of a second-Born calculation for antiproton impacts by Burgdoerfer et al.[18]. It is seen that, as expected, the capture cross section by a two step process is several orders of magnitude higher than by the single-step process. Further, Burgdoerfer et al. had predicted that the capture probability for proton impacts is still negligibly small, even when the two step process is incorporated, because the distance of the wake-riding site to the proton is large, suppressing the overlap between the intermediate (second-scattered) state and the final (wake-riding) state. In addition, they have simulated, by a Monte-Carlo method, the behavior of electrons accompanying antiprotons in a solid. A possibility to observe wake-riding electrons in antiproton-carbon-foil collisions has been demonstrated [18].

It is noted, however, that the theoretical understanding of wake-riding states is not necessarily satisfactory. Especially, the following arguments questioning the existence of the wake-riding electrons are worth considering; (i) perturbations such as the stopping force or stochastic collisions on the wake-riding electron may be comparable to, or larger than, its binding force, especially when the damping of the potential is large [19], (ii) the wake potential is an effective potential with its origin in the successive creation and annihilation of plasmons, i.e., the fluctuation of the potential may be too large to warrant a stationary wake-riding state [19], and (iii) the electron density of the wake-riding state is higher than the fluctuating density of valence electrons inducing the wake potential, i.e., a self-consistent treatment is essential to obtain a realistic prediction on wake-riding electrons.

3.2 Experimental

According to the discussions in sect.3.1, showing that negatively-charged heavy particles are promising tools in searching for

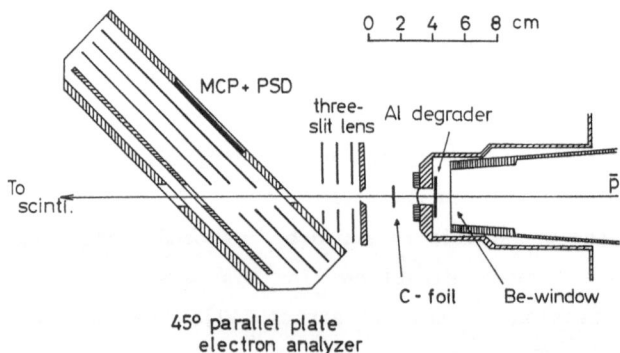

Figure 10. Schematic drawing of the experimental set-up to measure electron spectra in the forward direction with degraded proton and antiproton beams.

wake-riding electrons, an experiment has been conducted on electron emission from foils under antiproton impacts [8]. Fig.10 shows a schematic drawing of the apparatus used at LEAR of CERN for this purpose. High-quality beams of low-momentum antiprotons (105.5 MeV/c) passed through a 110 μm Be window at the end of the LEAR beamline, aluminum degrader foils, and a 22 μm mylar window at the entrance of the experimental vacuum chamber [20]. The thickness of the degrader was so determined, that the central energy of the beam at the target was ~ 600 keV. The beam intensity under the present experimental condition was, on the average, ~ 2×10^4 antiprotons/s. The degraded beam then passed through a target carbon foil, a three-slit lens, a center hole of an electron spectrometer, and finally was detected by a thin plastic scintillator 63 cm downstream from the target. Electrons emitted at $0°$ were energy analyzed by an electrostatic $45°$ parallel-plate spectrometer

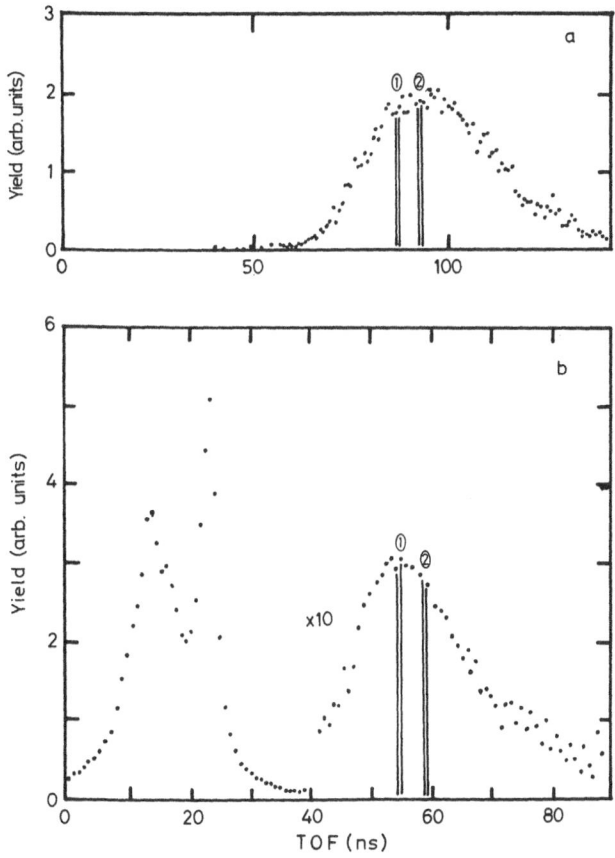

Figure 11. TOF spectra for (a) protons and (b) antiprotons degraded by aluminum foils. (1) and (2) show time gates to get spectra shown in figs.12 and 13, respectively.

with a position-sensitive MCP (micro-channel plate). The analyzer covers the electron energy from 0.6 V (eV) to V (eV), at the deflection potential of -V (V) with an energy resolution better than 4 % (FWHM). The acceptance angles parallel and perpendicular to the deflection plane of the analyzed electron were ~ 10° and ~ 2.5°, respectively. A plastic scintillator is set just behind the MCP to discriminate signals of annihilation products hitting the detector. The velocity of an antiproton which produced an electron was evaluated using the time difference of output pulses between the MCP (start) and the scintillator (stop). A reference experiment, using a degraded proton beam, has been performed at the EN Tandem Facility of the University of Aarhus. The same experimental set-up was used, except for the TOF length being 100 cm instead of 63 cm.

3.3 Results and Discussion

Fig.11 shows an example of TOF spectra for (a) proton- and (b) antiproton-impacts on carbon foils. The annihilation of an antiproton produces several pions, some of which may trigger both the MCP and the stop scintillator. Strong peaks observed at ~ 14 ns and ~ 23 ns, in fig.11(b), are supposed to be related with such annihilation products. The time resolution of the system was better than 2 ns. As the degraded antiproton beam suffers a big energy straggling (FWHM ~ 400 keV), an event-by-event recording technique based on a personal computer has been employed for the data acquisition. From the event data, electrons produced by antiprotons of 500 to 750 keV are divided into ten groups, each having an energy window of ~ 25 keV.

Fig.12 shows electron spectra in the forward direction, multiplied by the electron energy, for a 2 $\mu g/cm^2$ carbon foil, obtained with the time gate marked (1) in fig.11. The projectile energies are 670 keV for protons and 680 keV for antiprotons. Fig.13 shows electron spectra correspond to the time gate (2) in fig.11 (600 keV protons and 610 keV antiprotons). About 10^7 protons and antiprotons were used to get these spectra. Each spectrum is re-constructed from four narrow-energy-range spectra, independently measured applying different deflection voltages to the electron analyzer. The intensity of the binary-collision part is found to be roughly the same between protons and antiprotons. On the other hand, a big difference is observed in the $v_e \sim v_p$ part. For protons, the well-known convoy-electron peak is clearly visible. For antiprotons, no prominent dip is recognizable, in sharp contrast with theoretical predictions for gas targets [21]. This observation is in accord with the recent Monte Carlo simulation by Burgdoerfer et al.[18], and is understood

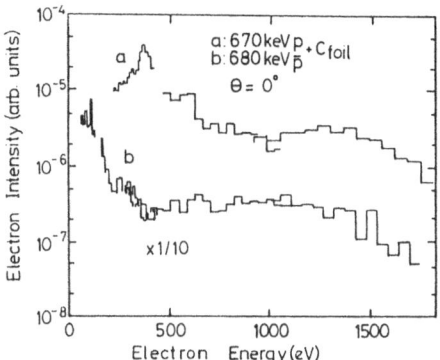

Figure 12. Wide-range electron spectra, multiplied by the electron energy, at 0° for (a) 670 keV protons, and (b) 680 keV antiprotons, bombarding a 2 μg/cm² carbon foil. Electron intensities are normalized to the same number of projectiles.

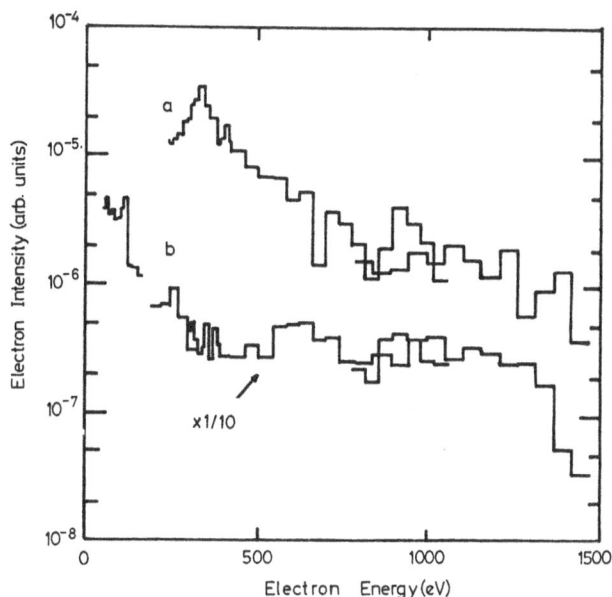

Figure 13. Same as fig.12, except that the proton and antiproton energies are 600 and 610 keV, respectively.

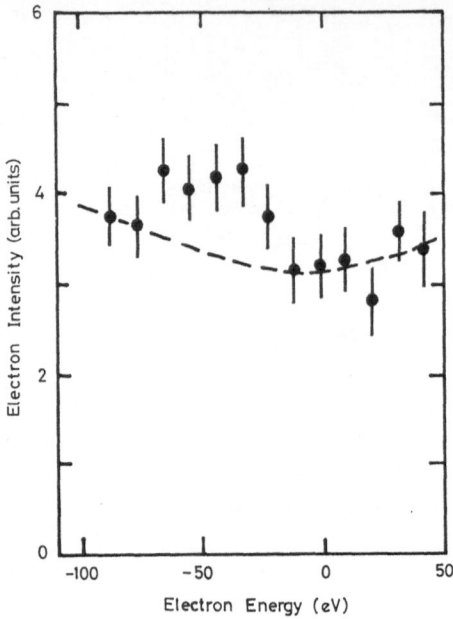

Figure 14. Electron spectra around $v_e \sim v_p$, observed at 0°, for 570 keV to 660 keV antiprotons bombarding a 2 μg/cm^2 carbon foil. The abscissa is given relative to $v_p^2/2$. The dashed line is drawn to guide the eyes only, to indicate the contribution of "normal" cascading electrons.

by taking into account a smearing of the sharp dip by cascading electrons, which maintain a very weak correlation with the projectile in the final state.

Fig.14 shows a forward-electron spectrum around the $v_e \sim v_p$ part, for antiprotons bombarding a 2 μg/cm^2 carbon foil. In order to study structures around $v_e \sim v_p$ with better statistics, several electron spectra with projectile energies from 570 to 660 keV are added up, after the abscissa is transformed into a relative energy with respect to $1/2 \, v_p^2$. In the present condition, as fig.8 shows, ΔE_{wr} ($= E_{wr} - 1/2 \, v_p^2$) varies only from -61 eV to -63 eV, i.e., a structure originating from the wake-riding electron is preserved after this transformation and addition. The error bars shown in the figure are statistical. There are indications of a bump at about -50 eV, i.e., ~ 10 eV higher than ΔE_{wr}. As the carbon KLL-Auger electrons have the energy of ~ 250 eV [22], those produced by ~ 570 keV antiprotons ($v_p^2/2 \sim 310$ eV) will overlap with the lower-energy tail of the bump, i.e., the real bump might be slightly narrower.

438

It is noted that the relative intensity of the bump to the continuum is in accordance with the theoretical prediction by Burgdorfer et al.[18] for ~ 900 keV antiprotons. The following arguments would be useful to further consider relations between the observed bump and wake-riding electrons: First, the distance between the wake-riding electron and the antiproton in solid is supposed to be different from that adopted in eq.(1), because of the repulsive Coulomb force, the damping of the wake, non-linear response of the media, etc. The solid line in fig.7 shows a crude feature of an effective potential energy estimated assuming a linear response of the media. The dashed line shows the oscillatory wake part only. It is seen that the screened-Coulomb part contributes to push out the wake-riding site by several %. As is seen from fig.1, the enhanced density δn evaluated by linear-response theory is, in a certain condition, as large as n_0, i.e., a realistic screening effect for an antiproton is expected to be weaker than for a proton, further pushing out the wake-riding site [23]. Secondly, ω_{pl} for a thin foil will be smaller than for bulk, because the density of a thin foil is normally smaller than the bulk density. In addition, the fraction of hydrogen contamination is usually higher for thinner carbon foils, resulting in the decrease of the effective ω_{pl}. The real value of ΔE_{wr} may further be modified by the momentum spread of the wake-riding state, the image force on the electron near the surface, etc. Although the quantitative evaluations of these effects are beyond the scope of the present report, some of them contribute to increase ΔE_{wr} toward the energy of the observe bump. Summarizing the above arguments, several aspects of the observed bump in fig.14 are found to be consistent with theoretical predictions on the wake-riding electrons.

4. CONCLUSION

It is demonstrated that the yield of knock-on electrons, produced by 4.5 MeV N_2^+, decreases monotonically when the orientation of the N_2^+ internuclear axis, θ, increases from 0° to 90° with respect to the beam direction. A steeper decrease is observed near $\theta \sim 25^\circ$. The overall behavior of the yield is fairly well-reproduced by a classical Monte-Carlo simulation of electrons scattered by a two-center Coulomb field, provided the initial velocity distribution of target electrons is taken into account. The decrease observed near $\theta \sim 25^\circ$ corresponds to the boundary where the density fluctuation of target valence electrons changes its feature from collective to single-particle-like. It is worth noting the possibility of investigating the relation of the charge state in solids, with that outside of it, by measuring the yield of the knock-on electrons as a function of exit charge state of ions for the longitudinal orientation.

Electrons emitted in the forward direction in antiproton-carbon foil collision have been measured for antiproton energies from 500 keV to 750 keV. No deep dip is seen at $v_e \sim v_p$, which is explained by a contribution of cascading electrons. Such electrons exit the foil at a relatively large distance from the projectile, resulting in a very weak correlation with the projectile in the final state, even when $v_e \sim v_p$. A bump is observed at ~ 50 eV below $1/2\ v_p^2$ for 570 - 660 keV antiprotons. The energy position of the bump is found to be consistent with the energy of electrons released from wake-riding states.

ACKNOWLEDGEMENTS

I am very much indebted to E.P. Kanter, W. Koenig, P.J. Cooney, A. Faibis, and B.J. Zabransky for their collaboration in work on the electron density fluctuation. I would like to express my deep thanks to K. Kuroki, K. Komaki, L.H. Andersen, E. Horsdal-Pedersen, P. Hvelplund, H. Knudsen, S.P. Møller, E. Uggerhøj, and K. Elsener for the collaboration in the antiproton work. Thanks are also due to P.M. Echenique, R.H. Ritchie, J. Burgdoerfer, and J. Lindhard for valuable and lively discussions on theoretical aspects of wake-riding electrons. The antiproton experiment is supported in part by the International Scientific Research Program from the Ministry of Education, Science and Culture in Japan and by the Danish Accelerator Committee.

REFERENCES

1. N. Bohr, K. Dan. Vidensk. Selsk. Mat.-Fys. Medd. *18* (1948) No.8; J. Neufeld and R.H. Ritchie, Phys. Rev. *98* (1955) 1632.
2. D.S. Gemmell, Nucl. Instrum. Meth. *194* (1982) 255.
3. W. Brandt, A. Ratkowski, and R.H. Ritchie, Phys. Rev. Lett. *33* (1974) 1325.
4. F. Bell, H.D. Betz, H. Panke, and W. Stehling, J. Phys. *B9* (1976) L433.
5. S. Datz et al., Phys. Rev. Lett. *40* (1978) 843.
6. M.F. Steuer, D.S. Gemmell, E.P. Kanter, E.A. Johnson, and B.J. Zabransky, Nucl. Instrum. Meth. *194* (1982) 277.
7. Y. Yamazaki, W. Koenig, P. Cooney, A. Faibis, E. Kanter, B.J. Zabransky, (to be published).
8. Y. Yamazaki, K. Kuroki, K. Komaki, L.H. Andersen, E. Horsdal-Pedersen, P. Hvelplund, H. Knudsen, S.P. Møller, E. Uggerhøj, and K. Elsener, J. Phys. Soc. Japan *59* (1990) (to be published).
9. P.M. Echenique, R.H. Ritchie, and W. Brandt, Phys. Rev. *B20* (1979) 2567.
10. W. Koenig, A. Faibis, E.P. Kanter, Z. Vager, and B.J. Zabransky, Nucl. Instrum. Meth. *B10/11* (1985) 259.
11. V.N. Neelavathi, R.H. Ritchie, and W. Brandt, Phys. Rev. Lett. *33* (1974) 302.
12. R.H. Ritchie, W. Brandt, and P.M. Echenique, Phys. Rev. *B14* (1976) 4808; and P.M. Echenique and R.H. Ritchie, Phys. Rev. *B21* (1980) 5854.
13. A. Rivacoba and P.M. Echenique, Phys. Rev. *B36* (1987) 2277.
14. M.H. Day, Phys. Rev. Lett. *44* (1980) 752; and M. Day and M. Ebel, Phys. Rev. *B19* (1979) 3434.

15. Y. Yamazaki and N. Oda, Nucl. Instrum. Meth. *194* (1982) 415.
16. Y. Yamazaki, K. Kurobi, F. Fujimoto, L.H. Andersen, P. Hvelplund, H. Knudsen, S.P. Møller, E. Uggerhøj, and K. Elsener, CERN-PSCC/87-39/ p.108.
17. R. Shakeshaft and L. Spruch, Rev. Mod. Phys. *51* (1979) 369; E. Horsdal-Pedersen, C.L. Cocke, and M. Stockli, Phys. Rev. Lett. *50* (1983) 1910.
18. J. Burgdoerfer, J. Wang, and J. Mueller, Phys. Rev. Lett. *62* (1989) 1599.
19. J. Lindhard, (private communication), and P.M. Echenique, R.H.Ritchie, and W. Brandt, Phys. Rev. *B33* (1986) 431; F.J. Garcia de Abajo, P.M. Echenique, and R.H. Ritchie, Nucl. Instrum. Meth. *48* (1990) 25.
20. E. Uggerhøj, Nucl. Instrum. Meth. *B33* (1988) 265; K. Elsener, Comments on Atomic and Molecular Physics *22* (1989) 263; L.H. Andersen, P. Hvelplund, H. Knudsen, S.P. Møller, K. Elsener, K.-G. Rensfelt, and E. Uggerhøj, Phys. Rev. Lett. *57* (1986) 2147; ibid. *62* (1989) 1731.
21. P.D. Fainstein, V.H. Ponce, and R.D. Rivarola, J. Phys. *B21* (1989) 2989; and C.R. Garibotti and J.E. Miraglia, Phys. Rev. *A21* (1980) 572; R.E. Olson and T.J. Gay, Phys. Rev. Lett. *61* (1988) 302; C.O. Reinhold and D.R. Schultz, Phys. Rev. *A40* (1989) 7373; G. Mehler, B. Mueller, and W. Greiner, Phys. Rev. *A36* (1987) 1454.
22. L.H. Toburen, W.E. Wilson, and H.G. Paretzke, Phys. Rev. *A25* (1982) 713.
23. I. Nagy and B. Apagyi, in this volume.

CURRENT TOPICS IN KINETIC ELECTRON
EMISSION FROM SOLIDS

R.A. Baragiola

Department of Physics and Astronomy
Rutgers University, Piscataway, NJ 08855, U.S.A.

Present address: Dept. of Nucl. Engn. and Engn. Physics
Univ. of Virginia, Charlottesville
VA 22901, U.S.A.

1. INTRODUCTION

This is a report of some of the current problems in heavy-particle-induced electron emission from solids; the choice of topics, one of many possible, represents the particular view of the author. For lack of space, I will not review here the literature on kinetic electron emission (KEE) from solids, but refer the reader to recent review articles [1-6], from which previous papers can be traced. In the following, I will start by describing the difference between potential electron emission (PEE) and KEE. I will then make a brief description of the basic physics in sect.2 and proceed with the discussion of some recent ideas and "hot" topics: estimates of yields and surface effects, sect.3, convoy electrons from foils, sect.4, emission at glancing incidence, sect.5, heavy-ion-induced Auger electron emission, sect.6, thresholds, sect.7, and the emission of high-energy electrons from slow collisions, sect.8. I will end in sect.9 by mentioning some of the other topics of current interest in the field, with references to the literature.

Heavy-particle-induced electron emission from solids results from two essentially different ways of delivering energy to the electrons in the solid. In PEE, a slow ion is neutralized to the ground state by an Auger electron capture from the solid; the potential energy released can liberate electrons from the solid. This process can occur if the ground-state neutralization energy of the ion, E_i, exceeds twice the work function of the solid ϕ. The magnitude of the PEE yield is determined by the overlap of electron states in the solid and the atom, and by the

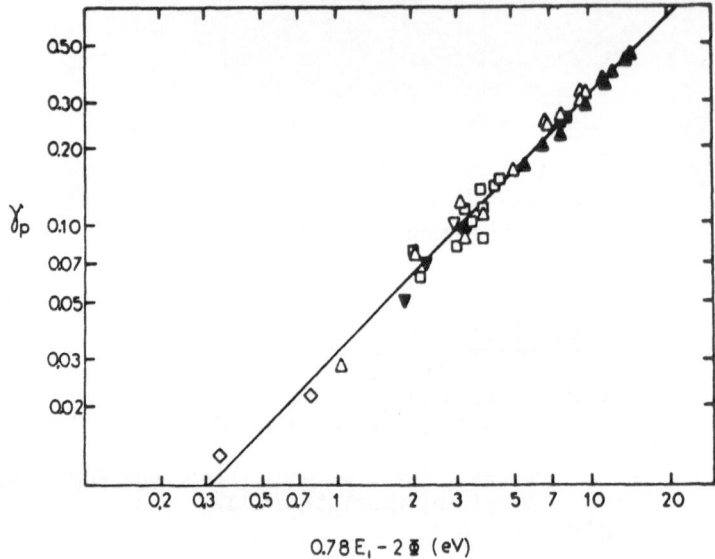

Figure 1. Potential electron emission yields versus $0.78\ E_i - 2\phi$, where E_i is the recombination energy of the ion and ϕ the work function of the solid. For references to the data, see Baragiola et al.[1].

availability of initial and final states. A compilation of PEE yields γ_p of clean surfaces under bombardment with single-charged rare-gas ions, shown in fig.1, is fitted within a narrow band by the relation [1]

$$\gamma_p = 0.032\ (0.78\ E_i - 2\ \phi),$$

where the yield is defined as the average number of ejected electrons per incident particle. More details on PEE by singly and multiply-charged ions can be found in review articles [7,8].

KEE is a different process, and is induced by the motion of a projectile outside or inside the solid. In this definition of KEE we exclude the variation of PEE with impact velocity. PEE, being exothermic, has no kinetic threshold. However, yields may depend on energy through differences in trajectories and distances of closest approach, and non-adiabatic excitations. These types of kinetic effects are usually small below an incident energy of a few hundred eV; they are mentioned to caution the reader that KEE and PEE are not always separable.

2. THEORETICAL CONCEPTS OF KEE

So far no theory of KEE exists which can calculate electron yields from first principles, and we have to rely on semi-empirical theories and

444

estimates based on theoretical concepts. Most of these concepts are nicely described in a compilation by Sigmund and Tougaard [9]. It is customary to describe KEE with a model which divides the process in three steps: electron excitation, transport, and escape. This heuristic model, which has been quite successful in explaining the majority of observations can be framed in the following general picture.

Heavy particles moving through matter transfer energy to electrons as a result of a time-varying Coulomb interaction; this can lead to excitation or ionization of the projectile and/or the target. The projectiles undergo multiple collisions, set recoiling target atoms into motion, and finally stop in the solid or are reflected into vacuum. On their paths, the projectiles and the recoils create many excited electrons (sources) in ionization events, which have initially a range of energies and directions.

Due to strong electron-electron interactions in solids, the electrons from primary excitations decay rapidly in energy exciting other electrons in the solid, within constraints imposed by existing band gaps. This continuous energy degradation and multiplication in the number of electrons persists until a large number of electrons end at the bottom of the conduction band, and eventually dissipates as heat.

The presence of the surface interrupts the energy-degradation process and causes a sample of the electronic cascade to be ejected outside the solid. At the time of ejection, the spatial extent of the low-energy part of the electron cascade is limited, both due to the strong electron-electron interactions, and to elastic scattering of low-energy electrons with ion cores, to distances L of typically 10 - 30 Å in metals and 30 - 100 Å in insulators. Electrons escaping the solid experience a potential-energy barrier which causes them to lose energy and be refracted.

If electron escape from the surface can be separated from the ionization events and the development of the electron cascade, we can write the electron yield as

$$\gamma = \int N(E) \, P(E) \, dE, \tag{1}$$

where E is the electron energy with respect to the vacuum level, $N(E)$ the energy distribution of electrons arriving to the surface, and $P(E)$ the escape probability through the surface barrier. $N(E)$ can in turn be approximated by

$$N(E) = f(E) \, L \, R, \tag{2}$$

where the function f accounts for the energy distribution of the electrons and includes angular factors, L is a mean electron escape depth, and R is the production rate of electrons per unit path length near the surface. R includes ionizations caused by the projectile, recoiling target atoms and secondary electrons.

Approximating the surface by a classical planar-potential barrier, and neglecting band-structure effects, the probability of escape is given by [10]

$$P(E) = \frac{E}{E + I}, \tag{3}$$

where I is the inner potential. For free-electron metals, $I = E_F + \phi$, where E_F is the Fermi energy and ϕ the work function.

For fast particles, R in eq.(2) is approximated by S_e/J, where S_e is the electronic stopping power dE/dx of the projectile, and J a mean energy to produce an excited electron. In this high-velocity approximation (electronic energy-deposition constant over L) we obtain

$$\gamma = \int f(E) \, L \, \frac{S_e}{J} \, \frac{E}{E + I} \, dE. \tag{4}$$

At lower velocities, recoil effects may be important, especially if the mass of the projectile is larger than the mass of a target atom. In this case, one has to add to S_e the contribution of recoils to ionization. However, at slow velocities it is no longer valid to assume that the number of excited electrons is proportional to a deposited energy divided by J. As we go down in energy, a larger and larger proportion of the inelastic energy deposited goes into sub-threshold events, with final electron energies below the vacuum level. This can be accounted for in several ways. One can use in eq.(4) a modified stopping power, one that takes into account only processes which end up with an electron above the vacuum level. We have found this approach to be successful in describing KEE by heavy projectiles and target recoils [11]. Another approach would be to use a modified J which changes with the velocity of the projectile. The simplest one possibly comes from recognizing that at low velocities, electron multiplication in the collision cascade is not very important. In this case R becomes σ_i N, where σ_i is the cross section for ionization with a final electron energy above the vacuum level, and N the target atom density.

446

3. YIELD ESTIMATES AND SURFACE EFFECTS

Electron yields are obtained by integrating eq.(4) over electron energies. One obtains

$$\gamma = K_m \langle P \rangle S_e, \tag{5}$$

where K_m is a materials' constant and $\langle P \rangle$ the average escape probability. Not too close to the threshold it is found [1] that electron yields for light ions can be estimated to within about 30 % by the expression

$$\gamma = 0.1 \, S_e \, (eV/\text{Å}), \tag{6}$$

a result also supported by measurements of the backward electron emission from foils, if bombardment conditions produce a clean surface [12]. The relatively small variations around the value of γ in eq.(6) come from target dependences [1]. Other deviations result when changing the atomic number of the projectiles [13] (Z_1 effects), but they are small, supporting the general proportionality between electron yields and electronic stopping powers.

Large variations in yields are produced by changes in surface properties [1]. Yields are usually higher for contaminated surfaces. Our study of the effect of oxidation showed that yields can increase or decrease accompanying changes in surface composition but in a manner which is not directly related to work-function changes [14]. KEE from thin oxide layers may be enhanced by a lowering of the inner potential from the high value for metals to the low electron affinity of oxides; this enhancement appears to decrease with bombarding energy [15].

The possibility of better estimates than eq.(6) is provided by the recent finding of Kirchhoff et al.[16], that data of KEE for 100 keV H^+ on metals [17] fall on a straight line when plotted as a function of $I^{-2.4}$. This shows that the material dependences of S_e and K_m cancel in eq.(5) and that KEE yields are governed by escape through the surface barrier. More work is needed to test the general validity of this finding, especially since it disagrees with extrapolated data for Li [48].

4. ORIGIN OF CONVOY ELECTRONS FROM SOLIDS

When fast ions pass through gases or thin foils, they cause an enhanced electron emission with velocity v_e close to that of the projectile, v_i, both in magnitude and direction [18-20]. In gases, this can be due to electron capture or loss to a continuum electron state centered around the positively charged projectile. In both cases, the effect is due to the long-range polarization of the continuum state by the Coulomb field of the projectile. The cross section diverges for v_e approaching v_i, but is expected to fall to zero at $v_e = v_i$ due to radiative recombination. Experimentally, a finite cusp-shaped peak is observed; its shape reflects the finite velocity resolution of the electron detector, the velocity spread of the beam resulting both from initial conditions in the ion accelerator, and from energy straggling and angular scattering of the projectile in the target.

Convoy electrons observed from the downstream side of foils have been explained to result from three different mechanisms: a) atomic collision with a surface atom, similar to gas-phase processes; b) the focusing action of the emerging ions on secondary electrons, and c) emission from the collapse of the polarization wake [21] trailing the ion in the solid. Process c) can only occur at velocities of the order or higher than the Fermi velocity, where plasma oscillations can be formed.

A recent result obtained by different authors [22,23] is that the position of the cusp does not depend on the work function of the sample (to within 0.3 eV). Thus the surface barrier does not retard convoy electrons, unlike the case of a bias voltage between the foil and the electron spectrometer. This suggest that the important region for determining the final electron energy distribution is that extending a small distance Δs from the formation of the state at the exit side of the foil. In this region, the average electric field between the electron and the proton is larger than the fields present in the experimental setup. The effect of external fields will be to mix Rydberg and continuum states; the value of Δs and the final outcome that can be observed will depend on the available velocity resolution and on the magnitude of external fields.

5. EMISSION AT GLANCING INCIDENCE

As the angle of incidence is increased away from the surface normal, the yield increases [24], as more inelastic energy is deposited in the electron escape zone. This is due to a longer path of the projectile near the surface, to a growing contribution of ionization by fast recoils, and to the anisotropy of the electron source. For angles close to 90°, the range of impact parameters available to the projectile decreases, due to shadowing by surface atoms. This effect will be particularly noticeable along certain directions of single-crystal targets. As more and more close collisions are excluded, there will be a drop in the probability of core excitations, and a drop in the inelastic energy loss, analogous to that found in channeling experiments. At some point, the softening of the average collision will more than compensate the increased path of the ion in the surface region and the KEE yield will fall [25].

At very small glancing angles, the projectile can be totally reflected from the planar surface potential. This occurs when the velocity component normal to the surface is lower than that required to penetrate the solid, in a way analogous to the case of planar channeling. In these conditions, the main excitations are the binary scattering of the projectile with the surface valence electrons. For projectile velocities $v_i > v_F$, the ejection will be peaked in the direction of the reflected projectile; this forward peaking will be further enhanced by the focusing action of the field of the projectile [26,27]. On the contrary, emission is suppressed for angles close to the surface normal, except for Auger emission from the decay of inner-shells.

Shallow core levels can still be excited by glancing protons of energies of 50 - 100 keV, as has just been reported for Si surfaces [28]. The detection of the resulting Auger electrons by bombarding at glancing incidence forms the basis of the technique we call GIAS, for Glancing Incidence Auger Spectroscopy, which can provide extreme surface sensitivity to the study of solids. Normal Auger electron spectroscopy (AES) studies of surfaces are limited by the escape depth of Auger electrons, which causes the Auger spectra to include contributions from both the surface and the bulk electronic structure. This can be avoided by GIAS; bulk transitions cannot be excited since glancing ions cannot penetrate the surface layer.

AES is also limited by the ability to separate the Auger features from a large background of secondary electrons. A recent discovery by Rau and co-workers [29] is that this background practically disappears when the observation is made normal to the surface, fig.2. The reason for this is that electrons from binary interactions between the proton and valence electrons, are emitted preferentially in the forward direction; in fact, from pure binary scattering, no energetic electrons can be emitted from the surface at 90° from the ion beam in this type of interactions.

The increased sensitivity of GIAS also results from the long path of an ion along the surface. On the other hand, GIAS requires very flat surfaces with large terrace lengths, to minimize the number of violent collisions with atoms at step sites. This requires unusual care in surface preparation which limits the use of the technique. In any event, GIAS can be used to measure the atomic composition, electronic structure, and defect density of the very surface of solids, a combination of abilities not present in any other technique.

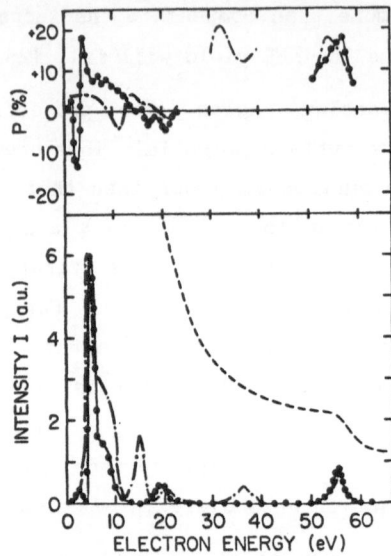

Figure 2. Electron energy distributions I and spin polarizations P for
H^+ (——) and He^+ (- · -) on Ni (110). Also shown is the energy
distribution obtained under electron impact (---). Adapted from
Rau et al.[29].

6. HEAVY ION INDUCED AUGER EMISSION

Holes in shallow atomic levels usually decay by the emission of Auger
electrons, easily identified in the high-energy tail of the energy
distribution of ejected electrons. Core excitations can occur in the
projectile and in the target atom, and the Auger decay can occur in the
bulk or in vacuum (for reflected projectiles or sputtered atoms). Auger
energies will be different in both cases, and reflect the different
distribution of valence electrons at the location of the core hole. The
Auger decay of excited projectiles outside the solid can give rise to high
ion/neutral ratios for projectiles traversing thin foils [30], and in
backscattering experiments [31,32].

Auger emission from target atoms can produce high ion yields in
secondary-ion mass spectrometry [33]. This subject is covered in several
reviews [17,34,35]. Auger emission consists of a broad structure, and
sharp (1 eV wide or narrower) peaks. While the broad structure is readily
associated to Auger transitions in the bulk, similar to those caused by
electron bombardment, there has been a controversy about the origin of the
sharp peaks. This has recently been resolved by the analysis of Doppler
shifts and broadenings [36,37], and by the observation of energy shifts
which accompany work-function changes induced by alkali adsorption [38].

450

It is now generally accepted that the sharp peaks correspond to transitions in core-excited sputtered atoms decaying in vacuum.

A topic still subject of controversy is the relative role of collisions between the projectile and a target atom (PT) and those between two target atoms (TT). Here the evidence is indirect, and comes from comparison of experiments to computer simulations, which have reached a high degree of development. Our Monte Carlo analysis of Be-K emission [39] showed an excellent agreement with the experimental energy dependence of the Auger yield. This provided firm grounds to determine the relative importance of PT and TT processes in X-ray emission yields, more accurately than possible using analytical theory. Recent simulations of L-shell excitations in bombardment of solid Al targets with Ar ions [40] conclude that PT collisions are largely responsible for Auger yields at very low energies. This is in striking contrast with other recent theoretical studies [41,42]. It is also contradicted by the observation that Auger yields in Al [43] and Si [44] compounds vary with the square of the concentration of the light element which is excited.

A recent discovery has been the strong enhancement of atomic Auger lines versus bulk lines for emission close to the surface [45]. This effect is particularly strong at low energies, as shown in fig.3. Very detailed studies of the dependence of ion-induced Auger spectra on angles of incidence and emission have just been published [46].

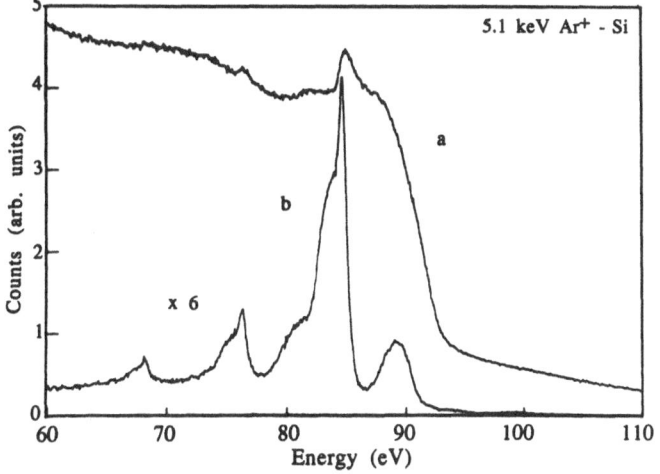

Figure 3. Electron energy spectra of ion-bombarded Si in the region of LVV and LMM Auger transitions, for a) normal and b) 2° glancing takeoff angles. The angle between the directions of the ion beam and the ejected electron is 146°.

A remarkable result which awaits independent experimental observation is the 1 - 4 eV shift of LVV Auger lines when the exciting fast ions impinge at glancing incidence [47].

7. KEE THRESHOLDS

For particles incident normally to the surface, KEE yields rise from a more or less well-defined threshold ion velocity, go through a maximum and then fall down with velocity. At low velocities, the stopping power is proportional to the velocity of the ion and electron yields also show the same behavior (fig.4), following eq.(4). However, extrapolation of the $\gamma(v)$ curve shows an apparent threshold, i.e., $\gamma \propto (v-v_{th})$. For light projectiles on clean metals, v_{th} is close to the value expected from binary excitations of electrons at the Fermi surface

$$v_{th} = \frac{v_F}{2} \left\{ \left(\frac{I}{E_F} \right)^{1/2} - 1 \right\}, \tag{7}$$

where v_F and E_F are the velocity and energy of an electron at the Fermi level [48]. For heavy ions on contaminated surfaces, such as the surfaces found in electron multipliers, the value of the extrapolated v_{th} has been

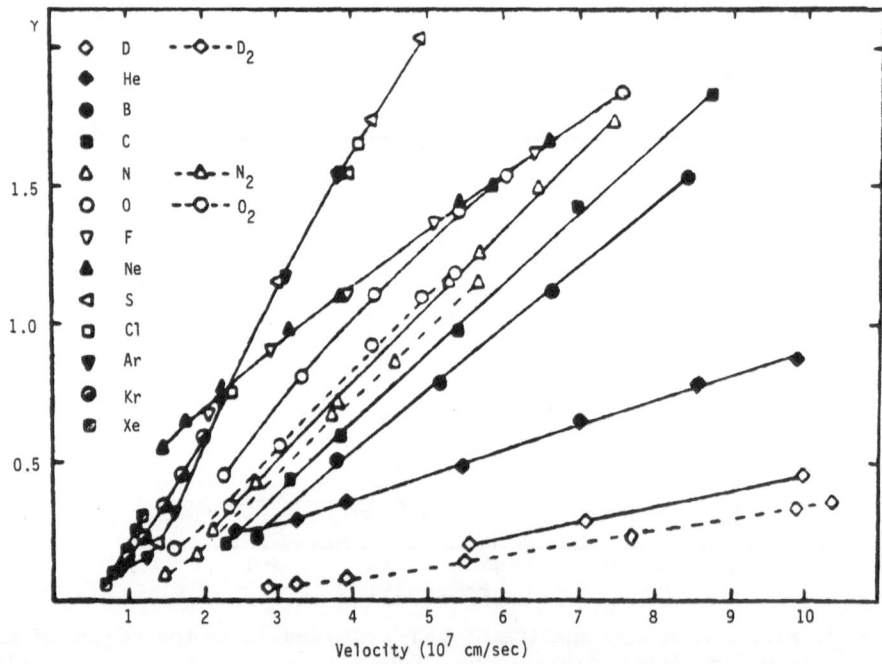

Figure 4. Electron emission yields for different ions on clean Al as a function of velocity. From Alonso et al.[13].

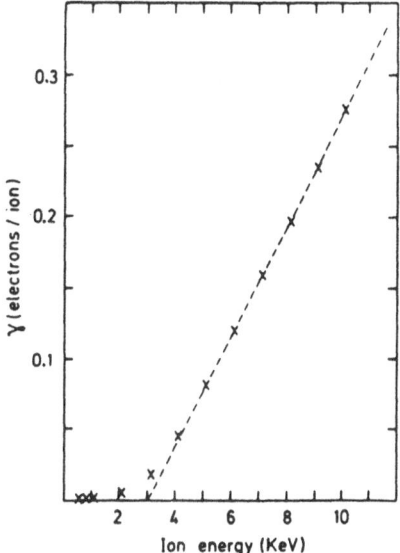

Figure 5. Electron yields versus ion kinetic energy in the threshold region for Xe[+] on Au. Notice the existence of a region where the yield is proportional to energy. From Alonso et al.[51].

found to lie at $4.3 - 5.5 \times 10^6$ cm/s [49] (10 - 16 eV/amu). Measurements closer to v_{th} show a departure from linearity with v, and measurable yields have been observed down to 1.8×10^6 cm/s (1.7 eV/amu) [50]. The absolute threshold for KEE should really occur when the incident-ion energy equals the work function of the target.

For clean solids, a linear dependence of γ with energy is seen for heavy ions in the keV range, of the type $\gamma \propto (E - E_{th})$, fig.5 [51]. Such a linear behavior is supported by theories which use *mean* energy losses, like the Firsov friction model. These theories give a threshold when the inelastic energy transfer in a collision equals the work function, and explains well the behavior of γ not too close to the threshold. Close examination of the "threshold" region shows, however, that yields decrease exponentially with decreasing ion energy, without a clear onset, fig.6. This "sub-threshold" behavior is most likely related to fluctuations in the inelastic energy loss in single collisions, not present in Firsov's model. In fact, electron energy distributions change rapidly in this energy range, contrary to their nearly invariance at high ion energies.

8. EMISSION OF FAST ELECTRONS IN SLOW COLLISIONS

If plotted in a log scale, long high-energy tails are apparent in the

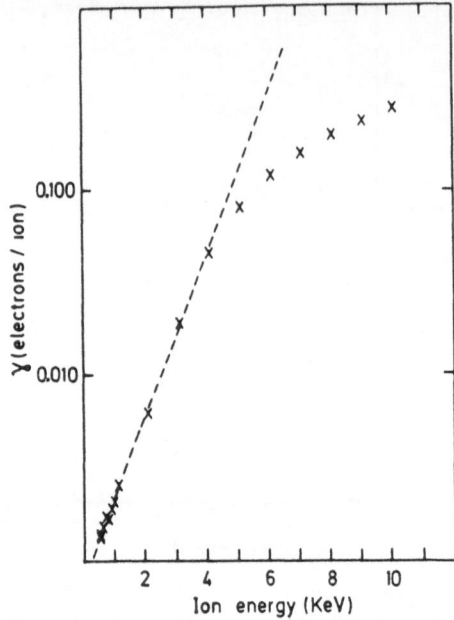

Figure 6. Electron yields versus ion energy in the threshold region for Xe[+] on Au (logarithmic scale). Notice the existence of a region where the logarithm of the yield is proportional to energy. From Alonso et al.[51].

electron energy distributions. At high impact velocities, these tails have been explained to arise from binary collisions between the projectile and a target electron. At low velocities, the explanation is still a matter of debate [52]. Our recent measurements for very low energy (0.5 - 6 keV)

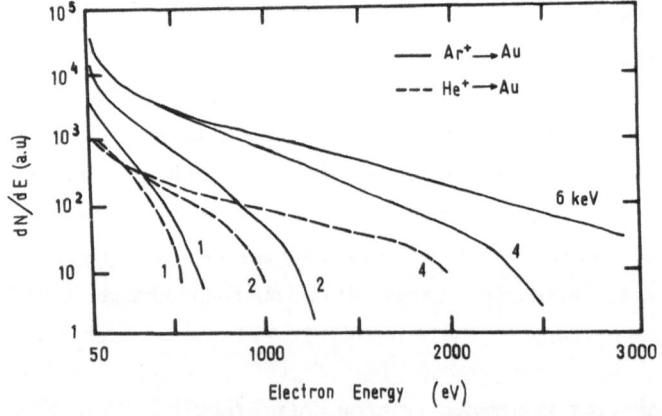

Figure 7. Electron energy distributions for He[+] and Ar[+] on Au. From Baragiola and Alonso [53].

ions in solids have shown that the continuum electron energy distributions extend up to a surprisingly large fraction of the center of mass energy [53]. This indicates that the high electron energies result from nearly head-on collisions between the projectile and a target atom.

Fig.7 shows that a continuum of very high-energy electrons is produced from solid Au by ion bombardment. This occurs not only for Ar^+ but also for a light ion like He^+. Continuum tails decay exponentially with electron energy, and a high-energy cutoff is apparent in most cases. This cutoff is higher for the heavier projectiles; both for Ar and He, the cutoff grows approximately linear with impact energy.

The results for Ar^+ on Mg, fig.8, show that high-energy tails are strongly dependent on the type of target. Measurements over these and other targets have shown that tails decay generally slower with electron energy the higher the atomic number of the target.

The origin of these super-delta electrons is not clear yet; it appears reasonable to associate higher measured electron energies with higher initial velocities of electrons, i.e., deeper shells. High frequency components of the dynamic perturbation occurring during the collision may produce the very low emission intensities of highly energetic electrons.

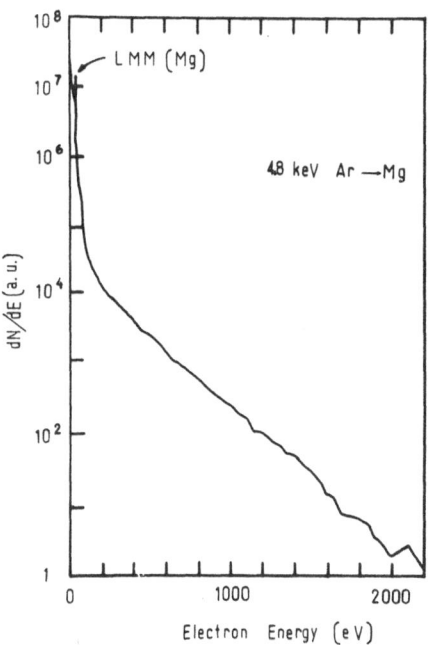

Figure 8. Electron energy distributions for 4.5 keV Ar^+ on Mg. From Baragiola and Alonso [53].

9. OTHER TOPICS

A topic that has received sustained attention is KEE by molecular ions. It is found that equal velocity H_2^+ ions have a slightly lower electron yield per atom than H^+. The origin of the effect is still a subject of controversy. Factors which contribute to the molecular effect in the total yields are differences in PEE yields, electron loss from the projectile, a lower yield for the neutral H constituent of the H_2^+ molecule [54], and the vicinage effect in the excitation of target electrons [55,56]. Although molecular effects have been observed in the energy loss of ions traversing foils [57], lower yields for atoms than ions on clean surfaces still await direct experimental verification.

Other topics of current interest include the influence of surface topography on electron yields [58], the statistics of KEE and PEE [59], differences in Auger spectra from crystals at normal and glancing incidence [60], and the emission of autoionization electrons [61-63].

REFERENCES

1. R.A. Baragiola, E.V. Alonso, J. Ferrón and A. Oliva-Florio, Surface Sci. *90* (1979) 240.
2. K.H. Krebs, Vacuum *33* (1983) 535.
3. D. Hasselkamp, Comments At. Mol. Phys. *21* (1985) 241.
4. N. Benazeth, J. Microsc. Spectrosc. Electron. *12* (1987) 235.
5. J. Schou, Scanning Microsc. *2* (1987) 607.
6. B.A. Brusilovsky, Appl. Phys. *A50* (1990) 111.
7. R.A. Baragiola, Radiat. Eff. *61* (1982) 47.
8. P. Varga, Comments Atom. Molec. Phys. *23* (1989) 111.
9. P. Sigmund and S. Tougaard, Springer Ser. Chem. Phys. 17 *Inelastic Particle-Surface Collisions*, E. Taglauer and W. Heiland, eds., Springer, Berlin (1981) 2.
10. M.S. Chung and T.E. Everhart, J. Appl. Phys. *45* (1984) 707.
11. J. Ferrón, E. Alonso, R.A. Baragiola and A. Oliva Florio, J. Phys. *D14* (1981) 1707.
12. H. Rothard, K. Kroneberger, A. Clouvas, E. Veje, P. Lorenzen, N. Keller, J. Kemmler, W. Meckbach, and K.-O. Groeneveld, Phys. Rev. *A41* (1990) 2521.
13. E.V. Alonso, R.A. Baragiola, J. Ferrón, M.M. Jakas, and A. Oliva-Florio, Phys. Rev. *B22* (1980) 80.
14. J. Ferrón, E. Alonso, R.A. Baragiola and A. Oliva-Florio, Surface Sci. *120* (1982) 427.
15. E.V. Alonso, R.A. Baragiola, J. Ferron and A. Oliva-Florio, Radiat. Eff. *45* (1979) 119.
16. J.F. Kirchhoff, J.J. Gay and E.B. Hale, Bull. Amer. Phys. Soc., *35* (1990) 538.
17. S. Hippler, D. Hasselkamp and A. Scharmann, Nucl. Instr. Meth. *B14* (1988) 518 and S. Hippler, Thesis (U. Giessen, 1988).
18. W. Meckbach and R. A. Baragiola, in *Inelastic Ion-Surface Collisions*, N.H. Tolk, J.C. Tully, W. Heiland and C.W. White, eds., Academic Press, N.Y. (1977), p. 283.
19. K.O. Groeneveld, W. Meckbach and I.A. Sellin, eds., *Forward Electron Ejection in Ion-Solid Collisions*, Springer, Berlin (1984).
20. W. Meckbach and P. Focke, Nucl. Instr. Meth. *B33* (1988) 255.

21. V.N. Neelavathi, R.H. Ritchie and W. Brandt, Phys. Rev. Lett. *33* (1974) 302.
22. S. Suárez, A.R. Goñi, W. Meckbach, and P. Focke, Z. Phys. *D6* (1987) 55.
23. L. de Ferrariis, F. Tutzauer, E.A. Sánchez, and R.A. Baragiola, Nucl. Instr. Meth. *A281* (1989) 43 and erratum in press.
24. J. Ferrón, E.V. Alonso, R.A. Baragiola and A. Oliva Florio; Phys. Rev. *B24* (1981) 4412.
25. M. Hasegawa, K. Kimura, Y. Fujii, Y. Susuki and M. Mannami, Nucl. Instr. Meth. *333* (1988) 334.
26. L. de Ferrariis and R. A. Baragiola, Phys. Rev. *A33* (1986) 4449.
27. H. Winter, P. Strohmeir and J. Burgdörfer, Phys. Rev. *A39* (1989) 3895.
28. J.W. Lee and R. Pfandzelter, Surface Sci. *225* (1990) 301.
29. C. Rau, K. Waters, and N. Chen, Phys. Rev. Letters (in press); see also these Proceedings.
30. R. Baragiola, P. Ziem and N. Stolterfoht, J. Phys., *B9* (1976) L447; P. Ziem, R. Baragiola and N. Stolterfoht, Proc, Int, Conf. Phys. X-Ray Spectra, US GPO, Washington, 1976; 278.
31. R.A. Baragiola, in Proc. 7th. Int. Conf. Atom. Coll. Solids, (Moscow St. Univ. 1977) p. 151.
32. O. Grizzi, M. Shi, H. Bu, J. Rabalais, and R.A. Baragiola, Phys. Rev. *B41* (1990) 4789.
33. I.F. Livschits, V.A. Pazdzersky and B.A. Tsipinyuk, this volume and references therein.
34. E.W. Thomas, Vacuum *34* (1984) 1031.
35. J. Mischler and N. Benazeth, Scanning Elect. Microsc. II 351 (1986).
36. R.A. Baragiola, Springer Ser. Chem. Phys. 17, *Inelastic Particle Surface Collisions*, E. Taglauer and W. Heiland, eds., Springer, Berlin (1981), p. 38.
37. R.A. Baragiola, E.V. Alonso and H. Raiti, Phys. Rev. *A25* (1982) 1969.
38. G.E. Zampieri and R.A. Baragiola, Phys. Rev. *B29* (1984) 1480.
39. O. Grizzi and R.A. Baragiola, Phys. Rev. *A30* (1984) 2297.
40. M.H. Shapiro and J. Fine, Nucl. Instr. Meth. *B44* (1989) 43.
41. O. Grizzi and R.A. Baragiola, Phys. Rev. *A35* (1987) 135.
42. M. Hou, C. Benazeth and N. Benazeth, Phys. Rev. *36* (1987) 591; C. Benazeth, N. Benazeth and M. Hou, Surface Sci. *151* (1985) L137.
43. P. Viaris de Lesegno and J.-F. Hennequin, Surface Sci. *103* (1981) 257.
44. S. Valeri, R. Tonini, and G. Ottaviani, Phys. Rev. *B38* (1988) 13282, and references therein.
45. L. de Ferrariis, O. Grizzi, G.E. Zampieri, E.V. Alonso and R.A. Baragiola, Surface Sci. *167* (1986) L175.
46. A. Bonanno, F. Xu, M. Comarca, R. Siciliano, and A. Oliva, Nucl. Instr. Meth. *B48* (1990) 371.
47. A. Koyama, H. Ishikawa and Y. Sasa, Nucl Instr. Meth. *B33* (1988) 308; A. Koyama, H. Ishikawa, K. Maeda, Y. Sasa, O. Benka, and M. Uda, Nucl. Instr. Meth. *B48* (1990) 608.
48. R.A. Baragiola, E.V. Alonso and A. Oliva-Florio, Phys. Rev. *B19* (1979) 121.
49. F. Thum and W. O. Hofer, Surface Sci. *90* (1979) 331, and references therein.
50. F.W. Geno and R.D. Macfarlane, Int. J. Mass Spectrom. Ion Phys. *92* (1989) 195.
51. E.V. Alonso, M.A. Alurralde and R.A. Baragiola, Surface Sci. *166* (1986) L155.
52. P. Clapis and Q. Kessel, Phys. Rev. *A41* (1990) 4766.
53. R.A. Baragiola and E.V. Alonso, Abstracts XIII Int. Conf. Phys. Electr. Atom. Coll., J. Eichler et al., eds., North Holland, Amsterdam (1983) 365.
54. G. Lakits, F. Aumayr and H. Winter, Europhys. Lett. *10* (1989) 679.
55. R. Baragiola, E. Alonso, O. Auciello, J. Ferrón, G. Lantschner and A. Oliva Florio, Phys. Lett. *67A* (1978) 211.

56. N.R. Arista, M.M. Jakas, G.H. Lantschner, and J.C. Eckardt, Phys. Rev. *A34* (1986) 5112.
57. J.C. Eckardt, G. Lantschner, N.R. Arista and R.A. Baragiola, J. Phys. *C21* (1978) L851.
58. J. Mischler, B. Maurel and N. Benazeth, Rad. Eff. Def. Sol. *108* (1989) 145.
59. G. Lakits, F. Aumayr and H. Winter, Phys. Lett. *A139* (1989) 395.
60. P.F.A. Alkemade, L. Wong, W.N. Lennard, and I.V. Mitchell, Nucl. Instr. Meth. *B48* (1990) 604, and this volume.
61. G.E. Zampieri, F. Meier and R.A. Baragiola, Phys. Rev. *A29* (1984) 116.
62. S.V. Pepper and P.R. Aron, Surface Sci. *169* (1986) 14.
63. P.A. Zeijlmans van Emmichoren and A. Niehaus, Comm. At. Mol. Phys. *24* (1990) 65.

CONVOY ELECTRONS

J. Burgdörfer

Department of Physics and Astronomy
University of Tennessee, Knoxville, TN 37996-1200, and
O.R.N.L., Oak Ridge, TN 37831-6377, U.S.A.

ABSTRACT

Recent developments in the theory of the production and of the transport of convoy electrons through solids are reviewed. Similarities and differences to cusp electron emission in binary ion-atom collisions, and to transport of "free" electrons through solids, are highlighted. We also discuss recent observations of convoy electron emission in ion-surface collisions at small glancing angles.

1. INTRODUCTION

In 1970 Crooks and Rudd [1] discovered a cusp-shaped peak in the doubly-differential cross section (DDCS) for electron emission, following ion-atom collisions, close to the direction of the outgoing projectile. This electron cusp appears when the electron velocity vector in the laboratory frame, $\vec{v}_e$, approximately equals the projectile velocity, $\vec{v}_p$, in both magnitude and direction. In the same year, Harrison and Lucas [2] observed a similar cusp in the electron distribution ejected by ions traversing solid foil targets. Cusp electrons produced in ion-solid collisions are usually referred to as "convoy" electrons. The underlying picture is that of a flow of electrons dragged along by the projectile ions through the solid and arriving at the exit surface in close correlation with it. Correlation refers here to close proximity in coordinate and velocity space, i.e., phase space. Convoy electrons represent one distinct component of the secondary-electron spectrum. The primary focus of this lecture is on their modes of production and transport. Cusp electrons represent excitations of final states in the low-lying continuum of the projectile with small velocities (in a.u.), $v \ll 1$, $(\vec{v} = \vec{v}_e - \vec{v}_p)$, as seen in the rest frame of the projectile. Since the asymptotic two-body final-state interaction between the electron and the projectile is independent of the target, the presence of a similar

Interaction of Charged Particles with Solids and Surfaces
Edited by A. Gras-Martí *et al.*, Plenum Press, New York, 1991

459

cusp in the emission spectrum of ion-solid collisions and of ion-atom
collisions is easily understood. However, numerous investigations of the
yield and shape have shown that the underlying collision dynamics
responsible for populating those states is vastly different. Several
reviews [3-6] are now available, which cover recent experimental
developments in this field.

Electronic excitation in ion-solid collisions are characterized by a
complex array of multiple scattering processes, fig.1. The presence of
dynamical screening, (dynamical screening length λ_D), in the solid
modifies the effective interactions relevant for the formation of
projectile-centered states. Furthermore, the passage through the exit

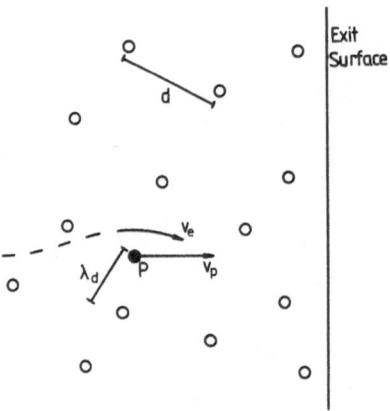

Figure 1. Propagation of an electron associated with the projectile P,
through the solid; d: nearest neighbor spacing; λ_D: dynamical
screening length.

surface causes a perturbation of the charge cloud moving in close
proximity of the projectile. The projection of the transient wavepacket
onto near-threshold states of the projectile, or classically speaking, the
phase-space distribution near zero energy upon exit, determines the
population of the cusp states in ion-solid collisions.

In the following we analyze the fundamental processes underlying the
various stages of convoy electron formation, their primary excitation,
their transport properties, and modifications due to exit-surface effects.
We concentrate in this lecture on simple physical pictures and refer to
references for mathematical details. Atomic units are used throughout,
unless otherwise stated.

2. PRODUCTION OF CONVOY ELECTRONS: CROSS SECTIONS FOR PROJECTILE-CENTERED STATES

We consider a projectile ion (or atom) of charge state q (usually q » 1) impinging on a target with large speed v_p » 1. The excitation of cusp states requires a large momentum transfer in a close collision, typically of the order of $\Delta p \approx v_p$ » 1. The production of convoy electrons inside the solid therefore closely resembles binary ion-atom collisions since condensed-matter effects are not expected to be relevant on this energy scale (ΔE » 1 a.u.), or length scale (impact parameter b « 1). Two processes are important for formation of final states near threshold: electron capture from inner shells of the target, or direct excitation of a low-lying projectile state of an electron either carried into the collisions or formed in an earlier stage of the transmission. The extension of these processes to continuum states are known as electron capture to continuum (ECC) and electron loss to continuum (ELC), respectively [3]. Neglecting multiple-scattering effects for the moment, the emission probability for a cusp electron is then proportional to the cross section for the excitation of projectile-centered continuum states near threshold. Near threshold, the velocity (v) or energy ($\varepsilon = v^2/2$) dependence of the DDCS is governed by threshold laws [7], characteristic for the two-body final state interaction, and is independent of the properties of the collision process. The threshold law can be expressed in terms of the Jost function for a given partial wave, f_l, as [8]

$$\frac{d^2\sigma}{d\varepsilon d\Omega} = v \sum_{l,l'} \frac{[(2l+1)(2l'+1)]^{1/2}}{4\pi} \frac{a_l}{f_l(v)} \frac{a_{l'}^*}{f_{l'}^*(v)} P_l(\cos\theta) P_{l'}(\cos\theta).$$

(1)

Roughly speaking, the Jost function is the reciprocal of the normalization constant of the continuum wave, in the limit of relative velocity $v \rightarrow 0$. From another, but equivalent, view $|f_l(v)|^{-2}$ describes the enhancement (or depletion) of probability density of the continuum wavefunction at small distances of the two-body system in the exit channel. All relevant velocity (or energy) dependence is contained in f_l near threshold. In turn, the reduced partial-wave amplitudes a_l are assumed to be velocity independent, over a range of velocities within which the threshold law (1) applies. The limitations for the validity of the threshold law are a priori not obvious. Detailed studies have been recently performed for short-range potentials [41]. In practice two criteria are very useful: The velocity v should be small compared to the initial orbital velocity v_0 of the electron prior to ionization and v should be small compared to the collisional momentum transfer, i.e., $v/\Delta p \ll 1$.

The DDCS as a function of v follows from the DDCS as a function of ε, eq.(1), as

$$\frac{d^2\sigma}{d\vec{v}} = \frac{1}{v}\frac{d^2\sigma}{d\varepsilon d\Omega}.$$ (2)

The solid angle in eq.(1) refers to the projectile frame. The integral over all Ω corresponds therefore, in the laboratory frame, to an integral over a small sphere in velocity space, with radius v centered about $\vec{v}_p$. The singly-differential cross section (SDCS), found by integrating (2) over the solid angle centered about $\vec{v}_p$, is given by

$$\frac{d\sigma}{d\varepsilon} = v \sum_1 \frac{|a_1|^2}{|f_1(v)|^2}.$$ (3)

We illustrate the implications of the threshold law with the help of a few important examples.

2.1 Attractive Coulomb Field

For an attractive Coulomb potential we have [8,9]

$$|f_1(v)|^2 = v,$$ (4)

leading to a constant SDCS near threshold, $d\sigma/d\varepsilon = \text{const.}$, and, in turn, to a $1/v$ singularity in the energy-differential SDCS in the lab frame. This singularity is the origin of the "cusp". Using the Galilei invariance of the velocity-differential DDCS, $d^2\sigma/d\vec{v} = d^2\sigma/d\vec{v}_e$, we find

$$\frac{d\sigma}{d\varepsilon_e} = v_e \frac{d\sigma}{v^2 dv} = \frac{v_e}{v}\left[\frac{d\sigma}{d\varepsilon}\right].$$ (5)

Therefore a $1/v$ singularity arises in $d\sigma/d\varepsilon_e$.

Three observations are of interest: The Jost function has a uniform velocity dependence for all 1, i.e., all partial waves are present immediately at threshold and contribute to the cusp. This leads to contributions from all Legendre polynomials, eq.(1), with $1 \neq 0$, and to a dependence of the DDCS on the emission angle in the projectile frame. In other words, even for small radii v, the sphere in velocity space centered about $\vec{v}_p$ will be anisotropically populated. Both electron capture to continuum (ECC) and electron loss to continuum (ELC) have been shown to display distinctly different anisotropic emission patterns. Multiple

462

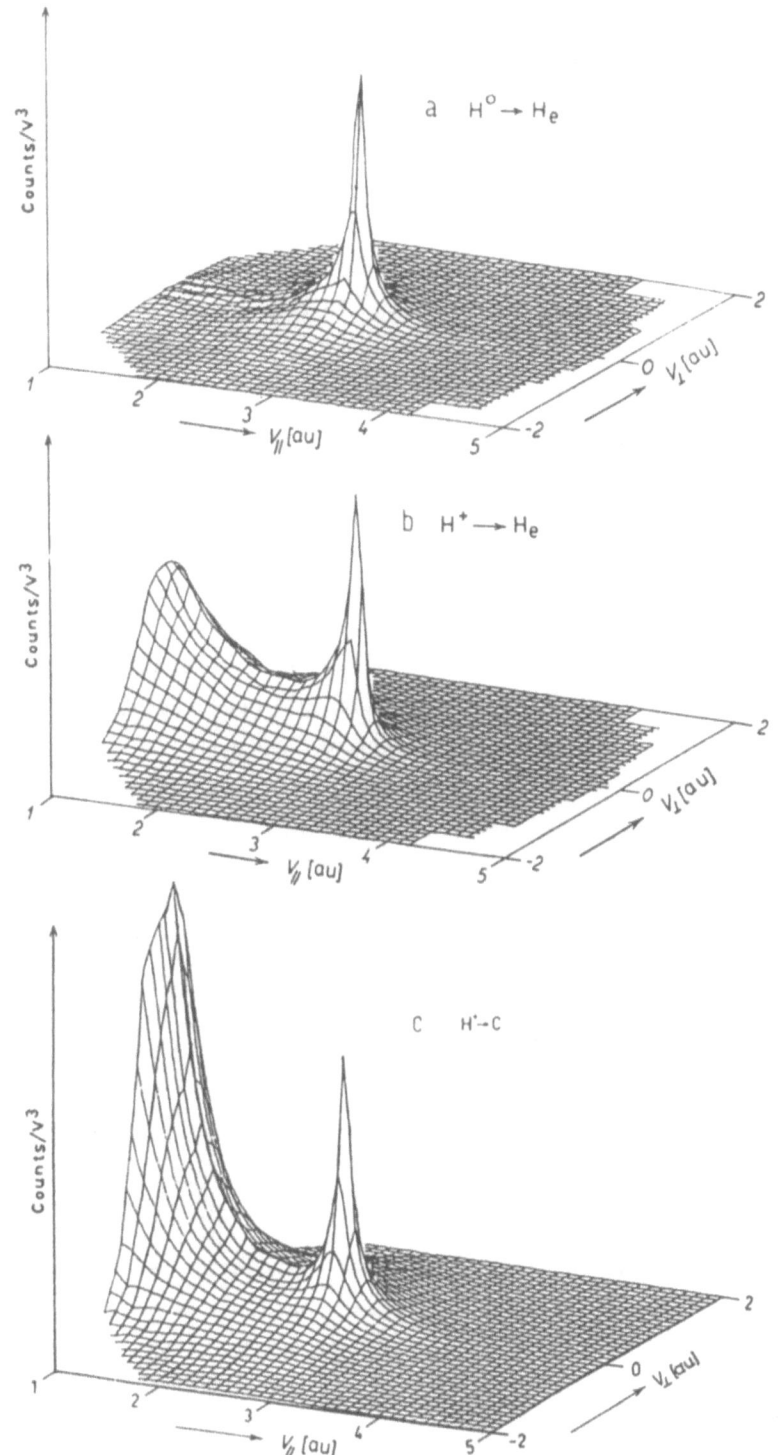

Figure 2. Doubly-differential electron distributions represented in v_e-space. Projectile energy: 170 keV, angular resolution $\theta_0 = 1^\circ$, from ref.[10].

collisions in the solids will modify this angular distribution. The characteristic shapes for the cusp peaks resulting from ECC, ELC, and foil transmission are illustrated by the data for the two-dimensional velocity distributions by Bernardi et al.[10], fig.2. A second observation refers to the radial distribution. The uniform singular behavior ($\propto v^{-1}$) for all l, does not provide a characteristic velocity scale for the cusp. The radial shape is therefore determined by the experimental resolution function, within the range of validity of the threshold law. Modifications of the radial distributions are possible only if the final-state interaction is not purely Coulombic at large distances, or if the scattering amplitudes show rapid variations over a small interval of v. The latter implies a break-down of the threshold law, for velocities smaller than estimated from the momentum transfer Δp in an energetic collision. This observation is of crucial importance for the modification of the convoy peak due to subsequent multiple soft collisions with small characteristic momentum transfers. Finally, the existence of the cusp is a simple consequence of the population of near-threshold states in the Coulomb field. The classical realization of near-threshold orbits in a Coulomb field are parabolas. The important conclusion is that a cusp is a classical phenomenon, and is present in classical treatments of ion-solid collisions [11] and of ion-atom collisions [12].

2.2 Repulsive Coulomb Field

For a repulsive Coulomb final-state interaction, the Jost function is given by

$$\left|f_1(v)\right|^2 = v \, \exp[2\pi q/v],\tag{6}$$

and the cross section near threshold becomes

$$\frac{d\sigma}{d\varepsilon} = \exp[-2\pi q/v] \sum_l \left|a_l\right|^2.\tag{7}$$

The cross section possesses an essential singularity at $v = 0$, i.e., the cusp-electron emission is exponentially suppressed at threshold, which is reflected in the forward ionization spectrum as a pronounced dip ("anti-cusp"). A comparison of theoretical forward spectrum for a cusp and for an anticusp, produced by protons and antiprotons on carbon under single-collision conditions [13], is shown in fig.3.

2.3 Short-Ranged Potential

For short-ranged final-state interaction we have, $\left|f_1(v)\right|^2 \propto v^{-21}$, and

464

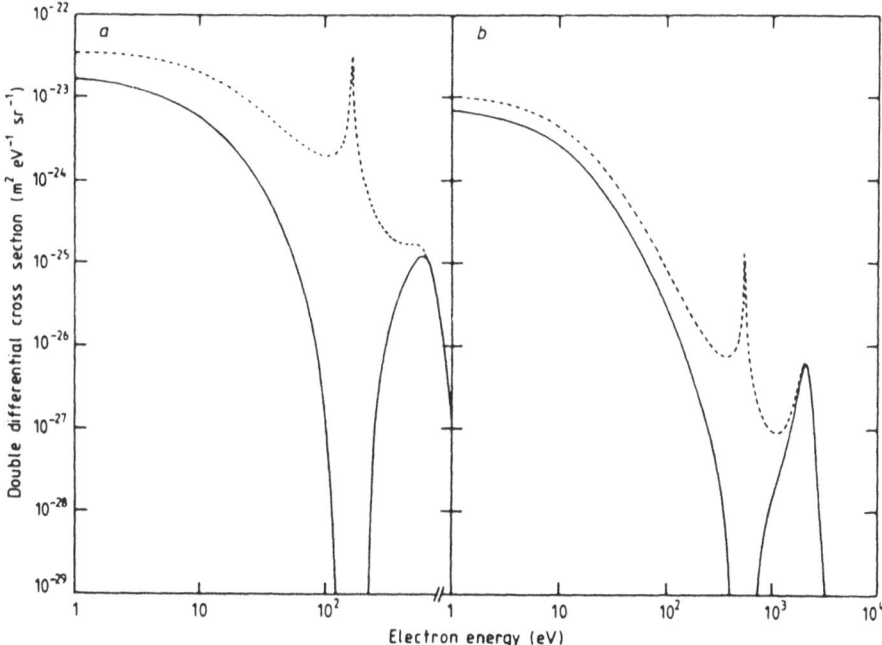

Figure 3. Theoretical doubly-differential cross sections for ionization of helium by impact of (a) 300 keV, and (b) 1 MeV, antiproton or proton, as a function of the electron-ejection energy for $0°$ ejection angle. (———) antiproton impact; (- - -) proton impact, from ref.[13].

$$\frac{d\sigma}{d\varepsilon} = v \left[\sum_{1} a_1 v^{21} \right].$$ (8)

The cross section approaches zero at threshold and higher partial waves ($1 \neq 0$) are strongly suppressed. The emission pattern is therefore isotropic. Eq.(8) is valid for electrons accompanying neutral projectiles in the absence of a permanent dipole moment of the projectile. There is, however, one notable exception to (8): If the strength of the potential is fine-tuned such that one of the finite number of bound states supported by a short-ranged potential crosses the threshold ("zero-energy resonance"), the velocity-differential cross section possesses a singularity $\propto v^{-2}$ which is even more strongly peaked than a Coulomb cusp.

Upon transforming into the laboratory frame, eq.(8) gives rise only to a weak enhancement of the ionization cross section near $\vec{v}_e \simeq \vec{v}_p$. A broad hump replacing the cusp has been indeed observed for forward electron emission in coincidence with neutral hydrogen after transmission of H^- through thin foils [14]. The recently observed sharp cusp for

465

electrons in coincidence with an outgoing neutral He projectile in He – Ar and He – He collisions [15] is, however, at variance with (8). Possible explanations include the presence of resonances sufficiently close to threshold which cannot be distinguished from a cusp within the experimental energy resolution, or an incomplete suppression of multiple scattering contributions in the experiment.

2.4 Dipole Field

For the projectiles possessing a dipole moment, or more generally, if the receding electron is subject to a long-ranged dipolar field ($\propto r^{-3}$), the behavior of the cross section depends sensitively on the strength of the dipole moment D and on the partial wave involved. For a given partial-wave, the cross section behaves as [9,16]

$$\frac{d\sigma_1}{d\varepsilon} \propto \begin{cases} v^{-1+2\beta(1)}, & (1 + 1/2) > D^{1/2}, \\ v^{-1}, & (1 + 1/2) < D^{1/2}, \end{cases} \tag{9}$$

where

$$\beta(1) = \left[(1 + 1/2)^2 - D \right]^{1/2}. \tag{10}$$

Oscillatory factors and the angular dependence have been neglected in (9) for simplicity. Eq.9 interpolates between the Coulomb ($1 \ll D^{1/2}$) and the short-ranged potential case ($1 \gg d^{1/2}$). Final-state interactions with a dipolar potential due to image charges play a role in ion-surface collisions at small glancing angles.

3. TRANSPORT OF CONVOY ELECTRONS

The binary ion-atom collisions discussed in sect.2 provide the initial conditions for the transport of the convoy electrons in solids. In the following we will briefly discuss the transport theory based on classical dynamics [11,17]. The justification lies in the fact that the projectile-centered states we are concerned with are either highly excited bound states or low-lying continuum states. Collisional broadening of excited states gives rise to a band of quasi-continuum states for which quantum effects are not likely to be important. An order-of-magnitude estimate may illustrate this point: Typical collision frequencies for highly excited quasi-free electrons are of the order of

$$\omega_c = 2\pi \, v_p / \lambda_f, \tag{11}$$

where λ_f denotes the mean free path of a free electron. In the velocity regime of interest, $\lambda_f \propto v_p^\gamma$, ($1 \le \gamma \le 2$). Typical numerical values of λ_f lie between 20 and 100 a.u. [18]. On the other hand, the orbital atomic frequencies are of the order of

$$\omega_0 = q^2/n^3, \tag{12}$$

where we have used a hydrogenic approximation. Collisional broadening exceeds the level spacing, i.e., a quasi-continuum exists for

$$\frac{\omega_0}{\omega_c} = \frac{q^2\lambda_f}{2\pi n^3 v_p} \le 1. \tag{13}$$

In other words, for $\omega_0/\omega_c < 1$ the evolution of a wavepacket is not dominated by the presence of a set of discrete Fourier components but is governed by a continuous ("noisy") Fourier spectrum. In the latter case classical dynamics is more likely to provide an approximate description of (transient) transport phenomena.

3.1 Initial Conditions

The binary atomic collision processes discussed in a previous section provide the initial conditions, a transient wavepacket, for the evolution under the influence of the dynamical screening potential in the solid and of multiple collisions. The cross sections for cusp electron production determines the initial distribution in energy and angular momentum. For the propagation of a classical phase-space distribution, the classical analog to the wavepacket, the complete set of phase-space coordinates is needed. Of importance is the spatial extent of the transient wavepacket formed in a violent collision. The radial extent of the wavepacket can be estimated as follows: In order to accelerate an electron, originally bound to the target, to a speed of $v_e \simeq v_p$, a close collision with a typical impact parameter $b \simeq 2q/v_p^2 \le 1$ is required. This implies that all trajectories are initially localized in close proximity to the projectile nucleus. The localization in coordinate space corresponds to a delocalization in energy, with an uncertainty $\Delta\varepsilon \simeq v_p/b \simeq v_p^3/(2q)$. This spread of the wavepacket in energy is closely linked to the applicability of the threshold laws to binary ion-atom collisions over a sizable interval of energies near threshold. We note that even for less stringent requirements on the energy transfer, the initial wavepacket (or classical phase-space contribution) is well-localized in coordinate space, since the dynamical screening length $\lambda_D = v_p/\omega_p$ (ω_p = plasma frequency of medium) confines the radius of initial conditions which are affected by the projectile field. The angular momentum distribution is peaked at low l

467

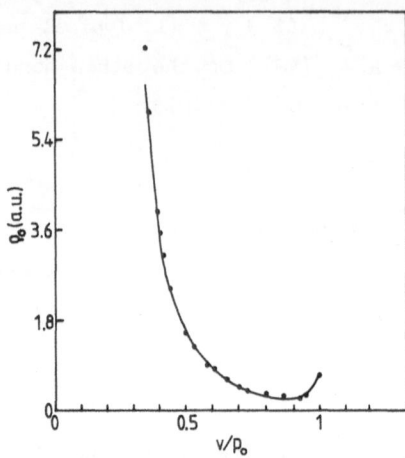

Figure 4. Momentum distribution $\rho_0(E = 0, l, v)$, for fixed l, of a classical cusp electron, in units of the reduced momentum variable $p_0 = 2q/l$. Scaling parameter for radial cut-off $\alpha = 10$. (——): eq.(8); (●) Monte Carlo simulation of initial conditions.

values, since only orbits with low l (or large eccentricities) come close to the nucleus. Initial values for phase-space distributions can also be taken from classical trajectory Monte Carlo (CTMC) calculations [19,12] for ion-atom collisions. The initial conditions corresponding to convoy electrons are trajectories with total energy $\varepsilon \simeq 0$, as seen in the projectile frame. Because of the spatial localization, $\varepsilon \simeq 0$ electrons have initially a large potential energy, and therefore a large kinetic energy $K = 1/2\, v(r)^2$. Fig.4 shows the momentum distribution for an ensemble of $\varepsilon = 0$ orbits of fixed angular momentum l. Because of classical scaling properties, the distribution function is a universal function of the reduced momentum variable $p_0 = 2q/l$. For a localized ensemble with distances less than a multiple α of the pericenter distance (distance of closest approach), the smallest available momentum in the ensemble is given by $v_{min} = p_0/\sqrt{\alpha}$. This results in a "hole" in the distribution function near $v \simeq 0$. Only as the ensemble expands radially outward, the kinetic energy is converted to potential energy, leading to true cusp electrons with $v(r) \simeq v \simeq 0$. The point to be noticed is that the dip in the velocity distribution near $v \simeq 0$ in the close collision region, as recently observed in the CTMC calculation [12], is a trivial effect of the finite radial extent of the charge cloud and is accounted for by two-body dynamics. The filling of the hole near $v \simeq 0$ at large projectile-target distances does not require any three-body scattering effects.

3.2 Evolution

The propagation of the initial wavepacket involves two major ingredients: the representation of the phase-space distribution function by an ensemble of representative test particles ("test particle discretization"), and the construction and solution of a stochastic differential equation. The stochastic differential equation is of the Langevin type,

$$\vec{v} = - \vec{\nabla} V_p + \vec{F}(t), \tag{14}$$

where V_p is the effective interaction potential between the electron and the projectile, and $\vec{F}(t)$ is the stochastic force. The picture underlying eq.(14) is that of a random walk in a Coulomb state space, perturbed by dynamical screening under the influence of multiple scattering. V_p is the dynamically-screened Coulomb potential of the projectile in the medium. The bare Coulomb potential, which is used in some of the calculations for illustrative purposes, will be denoted by V_p^0. The charge state q of V_p in eq.(14) shifts and fluctuates, as the projectile approaches charge-state equilibrium. Since the mean free path (mfp) for charge-changing collisions is large compared to λ_f, q will be assumed to be constant during convoy transport. The dynamically screened Coulomb potential is depicted in fig.5. At small distances it closely resembles the Coulomb potential, while screening becomes important at large distances. A prominent oscillatory structure ("wake") trailing the ion, may provide support for quasi-bound states of "wake-riding" electrons.

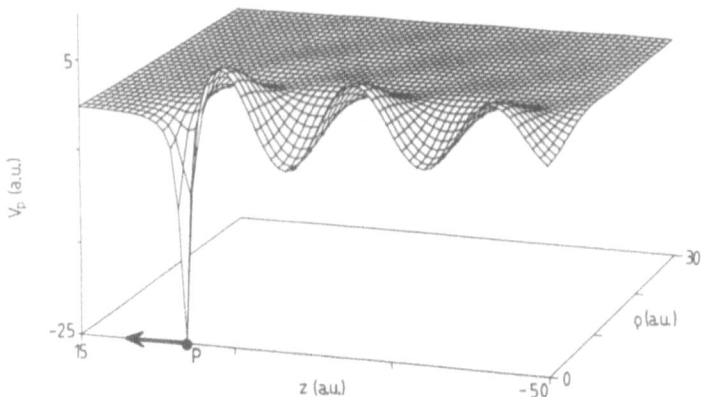

Figure 5. Dynamical screening potential of S^{16+} in Al, calculated in the plasmon-pole approximation to the dielectric function ($v_p = 1$ a.u.).

The stochastic force in (14) represents the multiple scattering of the fast electron in the solid. While its construction is not unique, an obvious strategy is to optimize the agreement with known multiple-scattering properties. We therefore describe the stochastic force in terms of a sequence of impulsive momentum transfers ("kicks"),

$$\vec{F}(t) = \sum_{\alpha=1,2} \sum_{i} \Delta\vec{P}_i^\alpha \, \delta(t - t_i^\alpha), \tag{15}$$

where $\Delta\vec{P}_i^\alpha$ is the stochastic momentum transfer per collision at the time t_i^α. The determination of $\vec{F}(t)$ is thereby reduced to that of a stochastic sequence of pairs $(\Delta\vec{P}_i^\alpha, t_i^\alpha)$. The approximation of the collisional interactions of fast electrons with target atoms, in terms of instantaneous momentum transfers, is based on the observation that the interaction is short-ranged and determined by the static screening length in the medium (typically of the order of 1 a.u.). The corresponding collision time $t_d \simeq 1/v_p$, is short compared to the orbital period $T_0 = 2\pi \, n^3/q^2$. In (15) we have decomposed the stochastic sequence of momentum transfers into two independent subsequences: One sequence ($\alpha = 1$), refers to elastic electron-target core scattering, while the other ($\alpha = 2$) refers to electron-electron scattering. The intervals between subsequent scattering events of the same kind, Δt^α, and the momentum transfer vectors, $\Delta\vec{P}^\alpha$, are chosen such that fundamental transport properties (mean free path, the stopping power and straggling) are reproduced for free electrons. For the propagation along a "random" direction in the crystal, or a homogeneous electron gas, the time intervals are Poissonian distributed,

$$P(\Delta t^\alpha) = \frac{1}{\langle\Delta t^\alpha\rangle} e^{-\Delta t^\alpha/\langle\Delta t^\alpha\rangle}, \tag{16}$$

where the mean time interval is given in terms of the corresponding mean free path λ^α by

$$\langle\Delta t^\alpha\rangle = \lambda^\alpha/v_e. \tag{17}$$

If the trajectory satisfies channeling conditions, the perturbation due to electron-target core interactions is periodic with fundamental period $\Delta t^1 = d/v_e$, (d: lattice spacing), rather than stochastic. The study of convoy electron emission under channeling conditions is presently underway [20,21]. A detailed description of the determination of the distribution of momentum-transfer vectors can be found in refs.[11,17].

In order to relate the dynamical evolution in the bulk to the post-foil experimental observation, two different modifications due to the penetration of the exit surface must be taken into account. The first one results from the sudden breakdown of the dynamical screening near the surface which leads to a redistribution of the final-state population. The sudden switch from a screened Coulomb potential, V_p, to a bare potential, V_p^0, closely resembles shakeup and shakedown processes in photoionization processes. The atomic orbitals of the old Hamiltonian (containing V_p) are redistributed among the eigenstates of the new Hamiltonian containing V_p^0. Shakeup and shakedown processes are easily incorporated in the classical formulation: The phase-space coordinates of the evolved orbit at the time t_s of the passage through the surface $(r(t_s), v(t_s))$ are used to construct orbits in the bare Coulomb field V_p^0, projecting thereby, within a classical framework, orbits in a screened potential onto orbits of the bare potential. We display all results for dynamical variables in terms of their asymptotic values in the Coulomb field using the projection of the phase-space coordinates onto dynamical variables in the bare potential. It should be noted that the potential step at the exit surface as required for the description of the surface in, for example, a Jellium model for a semi-infinite solid gives rise to additional "shake" contributions and can be treated along the same lines. A second modification results from the interaction with long-ranged surface image potentials. Long-range interaction cannot be treated in a sudden approximation. While the image potential perturbation is weak for normal-incidence foil transmission, because of short interaction times, it plays an important role in glancing-angle surface scattering.

3.3 Modification of Transport Properties

We highlight in the following the modifications of transport properties, by studying the evolution of rather specific initial ensembles. Comparisons with quantities of experimental interest will be presented in sect.4. For the latter, ensemble averages over path length and initial distributions are required. As a result, many subtleties of the evolution are no longer clearly visible.

The evolution of the distribution of convoy electrons ($\varepsilon = 0$ in the projectile frame), localized in the proximity of the projectile moving at a speed $v_p = 12.5$ a.u., and of a well-collimated beam of free electrons at the same speed, is compared in fig.6. The two-dimensional velocity distribution is shown at different times of evolution, t, which can be expressed in terms of the pathlength $Z = v_p t$ of the heavy projectile, which propagates along a (nearly) constant-velocity trajectory. The

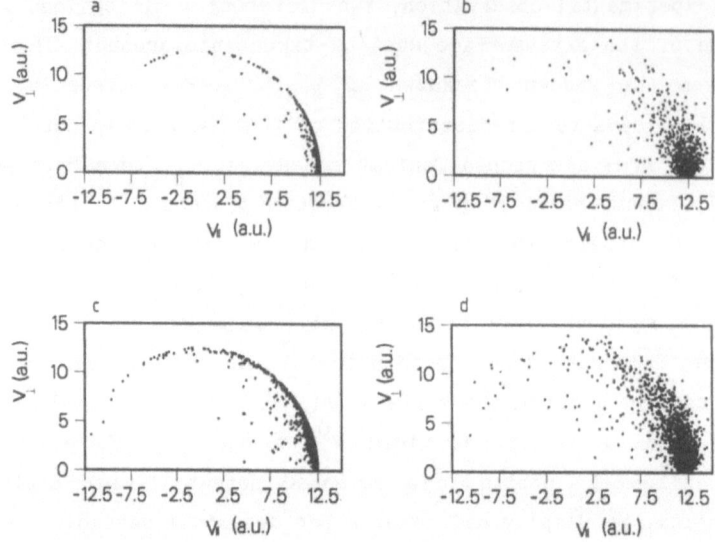

Figure 6. Two-dimensional velocity-space distributions after a path length of $Z = 42.5$, (a and b), and $Z = 127$ (c and d). Figs.(a) and (c) are for free electrons, (b) and (d) for convoy electrons with $p_0 = 12$. Note that all particles of the zero-scattering peak are represented by one dot.

electronic pathlength will be different, because of both deviations from a straight line due to both angular straggling and Coulomb trajectory effects. The present calculation is performed for a bare Coulomb potential V_p^0 and for a carbon target. The presence of dynamical screening does not, however, change the results significantly. The reason is that the most profound changes in the evolution due to the projectile field occur at small and intermediate distances, compared to the dynamical screening length λ_D, where screening effects are not yet dominant. Each diagram in fig.6 represents ensembles of initially 2000 test particles, each of which is marked by a dot. The scattered free-particle distribution displays the binary sphere of elastically scattered electrons. Particles in the interior of the sphere represent inelastic scattering events due to electron-electron collisions. In the presence of the Coulomb field, the binary sphere becomes "fuzzy". The most remarkable difference to free-electron transport is a rapid and almost "isotropic" diffusion in velocity space. This key observation leads to two consequences. The motion in the presence of a Coulomb field displays an increased *instability*, which we call Coulomb defocusing. The instability is related to the intrinsically chaotic dynamics of impulsively-perturbed atomic systems [22]. Secondly, the near isotropic expansion indicates that collisions can lead to both energy loss and energy gain. The underlying mechanism is the coupling of the collisional momentum and of the local orbit momentum. The change in energy per collision, as measured in the laboratory frame, is given by

$$\Delta\varepsilon_e = \frac{1}{2}\,(\vec{v}_p + \vec{v} + \Delta\vec{P})^2 - \frac{1}{2}\,(\vec{v}_p + \vec{v})^2 = \vec{v}_p\,\Delta\vec{P} + \frac{\Delta P^2}{2} + \vec{v}\,\Delta\vec{P}, \qquad (18)$$

or, equivalently, in the projectile frame, by

$$\Delta\varepsilon = \vec{v}\,\Delta\vec{P} + \Delta P^2/2. \qquad (19)$$

The first two terms in (18) are unchanged, compared to free electron scattering, while the third depends on the local momentum $\vec{v}(\vec{r})$ of the electron on the Coulomb orbit. The local momentum is of the order of the characteristic orbital momentum $p_0 = 2q/1$, which depends on the strength of the Coulomb field, and on the angular momentum for the orbit. Typical initial values for collisions involving fast highly-charged ions are $1 \approx 1$. In contrast to the first two terms which assure slowing down ($\Delta\varepsilon_e < 0$), the third term is not negative definite. For large p_0 the latter can change the overall sign of $\Delta\varepsilon_e$ resulting in an energy gain.

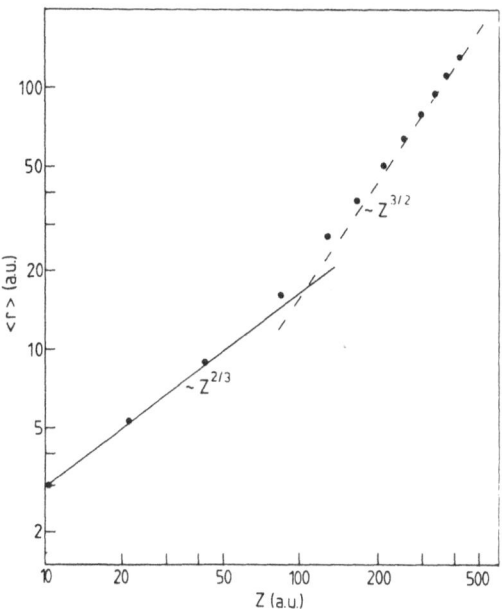

Figure 7. Evolution of the mean distance from the projectile, <r>, of convoy electrons with an initial distribution shown in fig.4, as a function of distance of propagation of projectile Z: (− − −) asymptotic $Z^{2/3}$ and $Z^{3/2}$ behavior, for Coulomb-like and free-particle-like expansion.

The comparison between Z = 42.5 and Z = 127 in fig.6 indicates that the strong influence of the Coulomb force is most pronounced at short distances of propagation. Due to the growth distance from the center of the Coulomb force, the phase-space evolution becomes increasingly free-particle like. Coulomb-defocusing effects are therefore transient. The "isotropic" redistribution, in particular the scattering to states with higher velocities in the laboratory frame (energy gain), is markedly different from treatments of convoy-electron transport neglecting the Coulomb field [23]. The rapid expansion of the cusp peak, due to Coulomb defocusing in both the parallel ($v_{\parallel}$) and traverse direction ($v_{\perp}$), is in qualitative agreement with the recent observation of increasing energy and angular width of the cusp peak as a function of the foil thickness [24].

The radial expansion of the classical ensemble in coordinate space is depicted in fig.7. The mean radial distance <r> from the projectile is measured as a function of evolution time, or equivalently, path length. The initial conditions correspond to those shown in fig.4. The trajectories are randomly distributed between one and ten times the pericenter distance ($\alpha = 10$). Two different regimes of the radial evolution can be distinguished. After a short transient period where the growth in <r> is comparable to the initial spread, $(<r^2> - <r>^2)^{1/2}$, (not shown in fig.7), the radial expansion is first Coulomb dominated. From the solution of Kepler's equation for a cusp electron, we expect

$$<r> \simeq const. \ Z^{2/3}. \tag{20}$$

At large distances, the radial expansion is dominated by multiple scattering and resembles free-particle diffusion in momentum space. In this regime we expect

$$<r> = const. \ Z^{3/2}. \tag{21}$$

The cross-over from the Coulomb-dominated to the free-particle-like regime, occurs at radial distances where the characteristic momentum transfer per collision $<\Delta P^{\alpha}>$ becomes larger than the momentum of the Coulomb orbit. Two observations are important. The cross-over occurs after a distance of propagation in the medium, Z, of the order of a few (typically 2 to 3) λ_f. This means that only a few collisions take place in a strong Coulomb field. Secondly, the cross-over occurs for distances <r> from the Coulomb center comparable to the dynamical screening length λ_D. This explains why some of the transport properties are not significantly altered by dynamical screening since the transport dynamics is already multiple scattering dominated when screening would become important. In

474

other words, at distances where differences between the bare and the screened Coulomb potential become significant, the momentum transfer due to collisions in the medium is large compared to the local momentum of both a bare and a screened Coulomb orbit and renders therefore these differences irrelevant.

3.4 Collisionally Induced Capture of Convoy Electrons

An important observation related to (18) and (19) is that negative energy transfers, $\Delta\varepsilon < 0$, as seen in the projectile frame, become possible. For convoy electrons with initial energies ($\varepsilon = 0$), this implies transitions to negative-energy states in the projectile, i.e., bound orbits in the moving projectile. Consequently, multiple scattering of convoy electrons does not necessarily lead to their defocusing and straggling, but can enhance their correlation with the projectile ion by a transient trapping in an excited state. This process possesses a well-known analog in astronomy: the injection of comets from the Oorth cloud into the inner solar system due to collisions with molecular clouds and field stars. The collisions reduce the comet's energy relative to the sun such that the comet falls into a bound orbit. In the present case, the projectile nucleus plays the role of the sun, and the encounters with molecular clouds are replaced by collisions with the electron gas and target cores. The capture of convoy electrons is of more than just academic interest: It changes the long-time behavior of convoy electrons. The electrons temporarily trapped into bound states can sustain complete correlation with the motion of the projectile, until the electron is finally reionized in a sequence of further collisions.

The rate of convoy attenuation depends on the experimental collection volume. In the following, a near-cubic collection volume with volume $\simeq (2\,\Delta v)^3$ centered about v_p,

$$v_p - \Delta v \le v_e \le v_p + \Delta v,$$

$$0 \le \theta_e \le \theta_0 = \Delta v / v_p,$$

(22)

is chosen. For free electrons, an exponential decay law is observed, fig.8. The slope gives an approximate attenuation length l_f. In the presence of a Coulomb field, a drastically different picture emerges. The convoy electron is initially depleted exponentially with a shorter attenuation length l_c than for free electrons which is a direct consequence of Coulomb defocusing (see eq.(18)). For large path lengths (or larger evolution times), the decay pattern changes dramatically. A

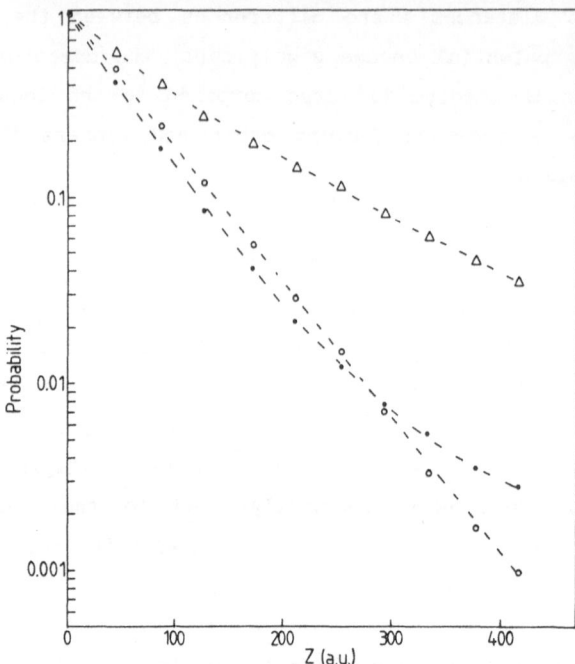

Figure 8. Attenuation of free electrons (o) and convoy electrons (●, p_0 = 12), for a collection volume with Δv = 0.25. Also shown is the attenuation of phase-space correlated electrons (Δ) (i.e, convoy electrons plus bound electrons).

long-time tail develops which is due to both the delayed release of the recaptured electrons in subsequent ionizing collisions and the redistribution and slowing down of high-energy continuum electrons with velocities $v_e > v_p$ produced in previous energy gaining collisions. Both effects are directly related to the presence of the strong Coulomb field and can be identified as a "focusing" effect. The result is the suppression of convoy electron attenuation at large path lengths. The effect of trapping can be made more obvious by studying the attenuation of all phase-space correlated electrons, i.e., the sum of all bound and convoy electrons, also shown in fig.8. The corresponding attenuation length l_{cc} is considerably larger than l_f. In the present case, we have $l_{cc}/l_f \simeq 2.6$. It should be noted, however, that the deviation from a monoexponential decay law for convoy electron occurs only at distances where the convoy intensity is reduced by a factor $\simeq 10^2$.

4. CASE STUDIES

In this section we focus on a few representative applications of the

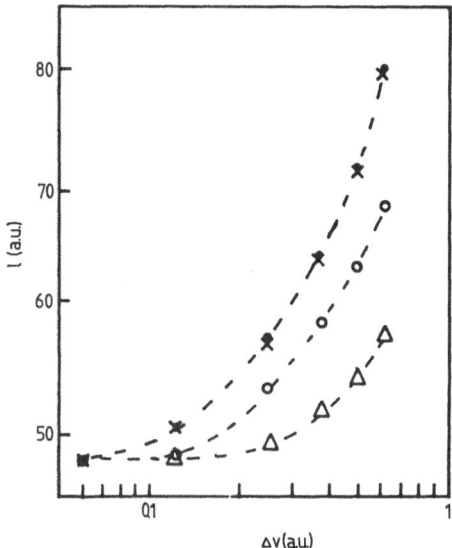

Figure 9. Attenuation length as a function of the linear dimension, Δv, of the collection volume, for different $p_0 = 2q/l$; ($\bullet$): $p_0 = 0$, ($\times$): $p_0 = 0.033$, (o): $p_0 = 0.67$; (Δ): $p_0 = 12$.

theoretical description of convoy electron production and electron transport to experimental observations.

4.1 Convoy Electron Attenuation

In recent experiments [24,25], the evolution of the convoy yield and of the convoy energy and angular distribution as a function of subequilibrium thickness of foils have been investigated. While the anisotropy of the angular distribution shows remarkably little dependence on the foil thickness, the convoy electron spreads out quickly in a radial direction, in qualitative agreement with fig.6. On a more quantitative level, this rapid diffusion can be expressed in terms of a reduction of the attenuation length. The dependence of l_c on the size of the collection volume is displayed in fig.9 for different characteristic momenta p_0 and for free electrons ($p_0 = 0$). The value $p_0 = 12$ corresponds to a hydrogenic carbon ion with $l \simeq 1$. The attenuation length for convoy electrons is significantly reduced compared to that for free electrons. Moreover, the functional dependence has changed. The attenuation length becomes independent of Δv for $\Delta v \leq 0.3$ a.u., and equals the mean free path for free electrons. In view of eq.(19), this can be easily understood in terms of Coulomb defocusing: The enhanced energy transfer per collision, due to the coupling to the local momentum in the Coulomb field, causes the removal of electron from the collection volume in a larger fraction of collision events than for free electrons. Accordingly, the loss rate entering the rate-equation model for the subequilibrium convoy yield, is given by a *convoy attenuation length* which is numerically close to the *mean free path of free electrons*. Preliminary data [24] seem to confirm this somewhat surprising prediction.

The experimentally observed slow approach to an equilibrium yield of convoy electrons as a function of the foil thickness [25] cannot be explained in terms of an enhanced mean free path or attenuation length. The characteristic distance for convoy yield equilibrium is rather determined by the approach of excitation equilibrium of the projectile core states, from which convoy electrons are generated by a ELC-like process. The latter can be a single-step or multiple-step ionization event [26].

4.2 Population of High l States

The initial phase-space distribution localized at small distances from the nucleus represents the excitation of low angular-momentum states. Typically, only states with $l \leq 2$ are strongly populated in binary ion-atom collisions in this energy regime. Multiple scattering in the solid, however, leads to a redistribution among many angular-momentum eigenstates. The measurement of the angular-momentum content of convoy electrons is complicated and only indirectly possible through the multipole content of the angular distribution [26,27]. Since the stochastic dynamics embodied in the Langevin equation, eq.(14), affects both highly-excited bound states and low-lying continuum states, electron or photon spectroscopy of highly-excited bound states can provide equivalent experimental information more directly.

Early measurements [28] based on a cascade analysis have revealed a dramatic shift of the l distribution, P(l), in highly charged projectiles excited in beam-foil collisions to high l states, compared to ion-atom collisions under single-collision conditions. A quantitative comparison between theory and experiment was, however, complicated by the fact that a cascade analysis allows only the determination of the global trends of the distribution P(l), but does not permit a state-selective detection of the population of individual l. Very recently, such a measurement has become available [29] for the 5l population in autoionizing $C^{2+}(2p5l)$ states, fig.10. Comparison with the simultaneously measured l distribution, for both excitation $C^{2+} \rightarrow$ He and electron capture $C^{3+} \rightarrow$ He in gas-phase collisions, clearly displays the shift to high-l states. In order to treat

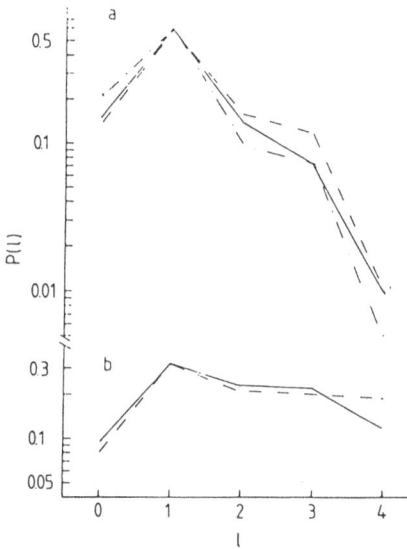

Figure 10. l-distribution, P(l), in n = 5 of C^{2+}(2p5l)(v_p = 2.25 a.u.).
(a) (---): experimental data, (ref.[29]), for C^{3+} + He →
C^{2+}(2p5l) + He^+; (-.-.-) experimental data (ref.[29]), for
C^{2+} + He → C^{2+}(2p5l) + He^+; (——): simulated initial
distribution for random walk. (b) (---): experimental data
(ref.[29]), for C^{2+}(2p5l) and a carbon foil; (——):
escape-depth averaged (steady state) solution the Langevin
equation, from ref.[30].

the l evolution within the framework of transport theory [30], we have
chosen an initial phase-space distribution uniform in energy corresponding
to an n^{-3} law, and an initial distribution in l, also shown in fig.10(a),
which is modeled after the experimental ion-atom collision data for He.
The mapping of the continuous classical l distribution onto discrete
quantum numbers, was made with the help of equally-spaced bins centered
around the semiclassical value l + 1/2. An equivalent method has been used
for n. The resulting stationary l distribution in n = 5, (averaged over
different escape depths), shows astonishingly good agreement with the
data. The only statistically significant deviation, by a factor ≈ 1.5,
from the experimental data, appears to occur in the g-state population. l
diffusion has been observed also in the limit of adiabatic collisions [31]
(i.e., Δn = 0). In the present case, the diffusion is strongly
non-adiabatic. The growth in l is strongly correlated with a growth in
energy. This is illustrated by the cross-correlation function <ΔεΔl>,
fig.11, for the C^{2+} data. The positive values indicate that growth in l,
Δl > 0, is positively correlated with a growth in energy. The decay of the

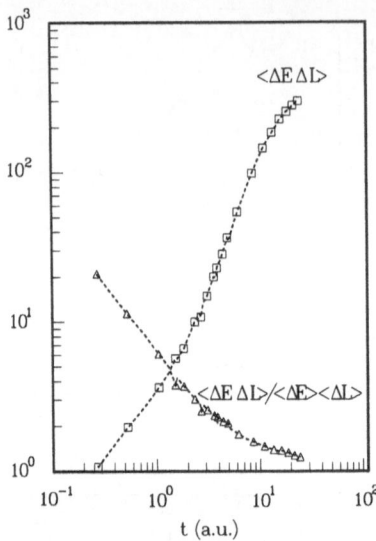

Figure 11. Normalized cross-correlation function $\langle\Delta\varepsilon\Delta L\rangle/\langle\Delta\varepsilon\rangle\langle\Delta L\rangle$ and cross-correlation function $\langle\Delta\varepsilon\Delta L\rangle$, as a function of evolution time; parameters as in fig.10.

normalized cross-correlation function, with an initial slope corresponding to an approximate power law t^{-1}, indicates that the stochastic character of the perturbation destroys the long-term correlation between growth in $\Delta\varepsilon$ and Δl.

4.3 Convoy Electrons Accompanying Antiprotons

Recently, the experimental study of antiproton transmission through carbon foils, using the Low Energy Antiproton Ring (LEAR) Facility at CERN, has been performed [32]. Projectile energies in the first experiments are of the order of 1 MeV (corresponding to a beam velocity of $v_p = 6$ a.u.). The primary goal of this study is the observation of the "anti-cusp" predicted by the threshold law for a repulsive two-body final state interaction, and the possible existence of a peak of "wake-riding" electrons which was predicted some time ago [33], but has not yet been unambiguously observed. Wake-riding electrons are electrons temporarily trapped in the oscillatory potential trailing the ions. Quantum mechanically, wake-riding states are not bound states but broad resonances in the continuum. The coupling to the continuum results from multiple scattering in the medium and the fluctuation of the wake potential around its average value [34].

The observation of wake-riding electrons is, for antiprotons, more likely than in the spectrum for positively charged ions, for two reasons:

(a) The presumably small peak of wake-riding electrons, which should appear in the $\vec{v}_e \simeq \vec{v}_p$ region of the emission spectrum, is not overshadowed by the "cusp", but instead accentuated by the anticusp predicted by the threshold law, eq.(7). (b) The wake-riding states are localized a factor of $\simeq 3$ closer to an antiproton than to a proton of the same speed, as can be seen from fig.5 when one rotates the figure by $180°$. Electron-capture probabilities into wake-riding states are therefore dramatically enhanced. The initial conditions are determined by electron capture. The transition amplitude for electron capture into wake-riding states is given in second-order Born (B2) approximation by [35]

$$t_{i,f} = <\Phi_{wake} \ |V_p + V_t \ G_0 V_p| \ \Phi_i>, \qquad (23)$$

where G_0 is the free-particle Green's function, and $V_{p,t}$ are the (effective) interaction potentials of the projectile and target. At high speeds, capture from the carbon K-shell dominates (i.e., $\phi = \phi_{1s}$). The evaluation of the capture cross section σ_c reveals the remarkable result that the first-order Born term (B1) is negligibly small compared to the second-order Born term. This is a simple consequence of the fact that, in the B1 approximation, capture is mediated by high-momentum components of the initial and final-state wavefunctions. However, the "soft" wake potential exponentially suppresses high-momentum components in Φ_{wake}, leading to an exceedingly small cross section [36]. The dominant contribution is therefore provided by the second-order Born term in eq.(23), which closely resembles the well-known Thomas double-scattering for ion-atom collisions. Here, an electron is first scattered off the projectile by $\simeq 60°$, followed by a second deflection at the target by about $60°$, such that the electron finally propagates in approximately the forward direction at zero speed relative to the projectile.

The crucial dependence on the sign of the projectile charge enters through the spatial displacement of the wake wavefunction relative to the Coulomb potential of the projectile. Since the size of the cross section depends, in addition to velocity matching, on the nonvanishing spatial overlap between the wavepacket of the scattered electron and the final-state Φ_{wake}, the factor 3 in distance of the wake-riding state translates into an order-of-magnitude larger cross section for antiprotons.

A realistic simulation of the forward spectrum of electrons ejected by antiprotons in the velocity regime $\vec{v}_e \simeq \vec{v}_p$, must take into account the "background" contribution due to multiple-scattered fast secondary

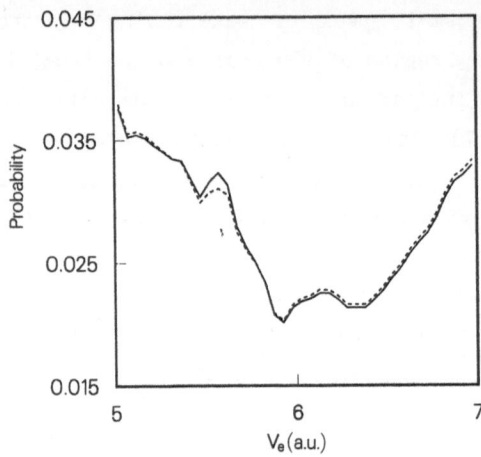

Figure 12. Normalized convoy-electron spectrum for antiprotons (v_p = 6 a.u.), emitted into a forward cone of half-angle of 5°: (- - -), and 2.5^o: (——), from ref.[35].

electrons ("binary encounter electrons"). As a result of slowing-down, energy straggling and angular straggling, they can contribute to the relevant spectral region. The transport and degradation of binary-encounter electrons has therefore been calculated using eq.(14), and the initial conditions derived from a first-order Born approximation for ionization. The resulting forward-electron spectrum, fig.12, shows a rather shallow dip near the projectile velocity, which is a remnant of the anticusp. Multiply-scattered binaries have almost completely filled the "hole" in the spectrum created by the anticusp in the primary ion-atom collision. In addition, a small peak at $v_e \simeq 5.6$ a.u., slightly below the speed of the antiproton, is visible, which is due to wake-riding electrons. The energy shift is determined by the Coulomb repulsion of the region where Φ_{wake} is localized. It should be noted that within our computer simulation, the origin of the "hump" can be unambiguously traced to the contribution from wake-riding electrons since the spatial localization of the electrons prior to foil exit can be determined on an event-by-event basis. The present prediction is currently being tested. Preliminary data [32] confirm that the spectrum is, indeed, flat near v_p, and that a small hump appears at lower velocities. It should be stressed that, because of the small cross section involved, the statistics of the experimental data, as well as the statistics of the theoretical Monte Carlo simulation, are almost inevitably poor. While both experiment and theory indicate the existence of a peak of wake-riding electrons, the evidence is not yet considered to be conclusive.

4.4 Convoy Electrons in Glancing Angle Surface Scattering

Recently, de Ferrariis and Baragiola [37] found first evidence for a cusp-like peak, after grazing-incidence scattering of protons at an Al surface. They observed a peak which was significantly broadened, compared to a corresponding cusp shape observed in ion-atom collisions as well as in foil transmission experiments. More recent data by Winter et al.[9] and by Hasegawa et al.[38] display a similar broadening also for semiconductor surfaces. The marked deviation of the shape of the convoy peak for surface scattering, compared to both foil-transmission and gas-phase collision data, suggests that the glancing-angle trajectory and the long interaction time with the target has a profound effect on the cusp electron formation. The final-state interaction (FSI) time for interaction with the surface increases with the glancing angle θ as

$$t_{FSI} = 2/v_p \sin \theta. \tag{24}$$

One model taking into account long-ranged final-state interaction treats the interaction of the receding electron in the Coulomb field of the proton and its image charge. Asymptotically, the electron moves in an effective dipole field. The threshold law for a dipole field, eq.(9), predicts a weakening of the Coulomb-cusp singularity. Approximating the threshold exponents for different partial waves by a single exponent,

$$\frac{d\sigma}{d\varepsilon} = \text{const. } v^{-1+2\beta}, \tag{25}$$

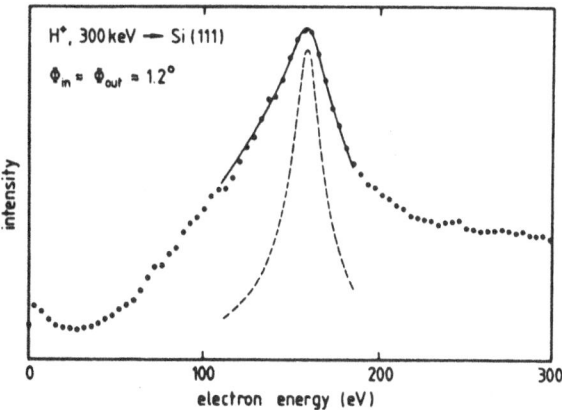

Figure 13. Electron spectrum after the interaction of 150 keV protons with a Si(111) surface, at $\Phi_{in} \simeq \Phi_{out} \simeq 1.2°$. (——) fit to the data, yielding $\beta = 0.42$; (- - -): spectrum with $\beta = 0$, at the same resolution with respect to angle and electron energy; from ref.[9].

allows the parameterization of the cusp shape in terms of a single parameter β. For several data sets taken at different energies [8], we have found consistently good agreement for a constant value $\beta \simeq 0.4$, fig.13. The extrapolation of the threshold behavior, eq.(25), with $\beta > 0$ across threshold, would predict vanishing cross sections for the formation of high-lying bound states, in agreement with experimental data [39]. Furthermore, the model of an image-charge interaction in the exit channel predicts the acceleration of convoy electrons accompanying projectiles with charge state $q > 1$, where the final-state interaction is governed by a net Coulomb force. Recent experiments by the Kyoto group show indeed an acceleration for He ions [38]. In spite of the good agreement with data, several conceptual difficulties should not be overlooked. The dielectric response of the surface generating the image charge should depend on the material. No conclusive evidence for such a target dependence has been found [40]. Furthermore, this model implicitly assumes a sudden electronic transition from the final-state interaction region, where the electron propagates under the combined influence of the proton and its image charge, to the asymptotic atomic regime where both the electron and the proton have left the surface region.

An alternative model would invoke the modification of a Coulombic cusp shape by multiple scattering on the "way out", in line with the transport theory outlined above. The glancing-angle geometry necessitates the following modifications to the transport model. The distribution of scattering events is anisotropic. The electron in an orbit around the projectile "sees" more target electrons in the "lower hemisphere" (towards the surface) than in the "upper hemisphere" (away from the surface). Another difference to the bulk transport is that primary production and transport take place on different time scales. The violent collision generating the isotachic component of the phase-space flow takes place at small distances from the surface layer, near the point of closest approach. Multiple soft collisions with the tail of the valence electron density will, however, continue for an extended period of time on the exit part of the glancing-angle trajectory. The fact that broadening occurs without a significant peak shift [9] does not necessarily argue against a multiple-scattering effect. For free electrons energy loss and straggling are intimately connected. However, multiple scattering in a Coulomb field leads to broadening of the convoy peak without a significant peak shift, as illustrated in fig.6. Further work is therefore necessary to delineate the relative importance of the dielectric response and of multiple scattering to the shape of convoy electron peak for grazing incidence collisions.

5. SUMMARY

The study of convoy-electron emission continues to challenge both experiment and theory. The shape and yield of convoy electrons has proven to be markedly different for ion-atom collisions, ion-solid collisions, and ion-surface collisions at grazing incidence. More detailed investigations are necessary to unravel the underlying collision dynamics, and to correlate in detail spectroscopic data with solid and solid-surface properties. The challenge to theory originates from the fact that (standard quantum coupled channel methods are extremely difficult to apply)for low-lying continuum states, as well as for high-lying bound states, even for binary ion-atom collisions. Classical dynamics seems to be better suited for this problem. However, its limitation due to quantum effects remains to be explored.

ACKNOWLEDGEMENTS

The author acknowledges important contributions made by many colleagues and collaborators. They include C. Bottcher, J. Gibbons, J. Kemmler, J. Müller, J. Wang, and H. Winter. This work was supported in part by the National Science Foundation and by the U.S. Department of Energy, Office of Basic Energy Division of Chemical Sciences, under Contract No. DE-AC05-84OR21400 with Martin Marietta Energy Systems, Inc.

REFERENCES

1. C.B. Crooks and M.E. Rudd, Phys. Rev. Lett. *25* (1970) 1599.
2. K.G. Harrison and M.W. Lucas, Phys. Lett. *33A* (1970) 142.
3. V.H. Ponce and W. Meckbach, Comments At. Mol. Phys. *10* (1981) 231.
4. M. Breinig, S. Elston, S. Huldt, L. Liljeby, C. Vane, S. Berry, G. Glass, M. Schauer, I. Sellin, G. Alton, S. Datz, S. Overbury, R. Laubert, and M. Suter, Phys. Rev. *A25* (1982) 3015.
5. K.O. Groeneveld, M. Meckbach, I.A. Sellin, J. Burgdörfer, Comments At. Mol. Phys. *14* (1984) 187.
6. W. Meckbach and P. Focke, Nucl. Instr. Meth. *B33*, (1988) 255.
7. E.P. Wigner, Phys. Rev. *73* (1948) 1002.
8. J. Taylor, *Scattering Theory*, Wiley, New York (1972).
9. H. Winter, P. Strohmeier, and J. Burgdörfer, Phys. Rev. *A39* (1989) 3895.
10. G. Bernardi, S. Suarez, P. Focke, and W. Meckbach, Nucl. Instr. & Meth. *B33* (1988) 321.
11. J. Burgdörfer in *Proceedings of the Third Workshop on High Energy Ion-Atom Collisions*, D. Berenyi and G. Hock, eds. Lecture Notes in Physics, *294*, Springer-Verlag, Berlin (1988) p.344.
12. C. Reinhold and R.E. Olson, Phys. Rev. *A39* (1989) 3861.
13. P. Fainstein, V. Ponce, and R. Rivarola, J. Phys. *B21* (1988) 2989.
14. Y. Yamazaki, L. Andersen, and R. Knudsen, to be published.
15. L. Sakardi, J. Palinkas, A. Köver, D. Berenyi, T. Vajnai, Phys. Rev. Lett. *62* (1989) 527.
16. W.R. Garrett, *Physics in Electronic and Atomic Collisions,* S. Datz, ed., North-Holland, Amsterdam (1982) p.65.
17. J. Burgdörfer and J. Gibbons, Phys. Rev. *A42* (1990) 1206.
18. M. Seah, W. Dench, Surface and Interface Analysis, *1* (1979) 1.

19. R. Abrines and I.C. Percival, Proc. Phys. Soc. (London) *88* (1966) 861 and 873.
20. K. Kimura et al., to be published.
21. J. Kemmler et al., to be published.
22. J. Burgdörfer, Nucl. Instr. Meth. *B42* (1989) 500; J. Müller and J. Burgdörfer, Phys. Rev. *A41* (1990) 4903.
23. J. Kemmler, S. Lencinas, P. Koschar, O. Heil, H. Rothard, K. Kroneberger, G. Szabo, and K.O. Groeneveld, Nucl. Instr. Meth *B33* (1988) 317; R. Barrachina, A. Goni, P. Focke, and W. Meckbach, Nucl. Instr. Meth. *B33*, (1988).
24. J.P. Gibbons, S.B. Elston, C.E. Gonzalez-Lepera, R. DeSerio, M. Breinig, O. Heil, H.P. Hülskötter, H. Rothard, I.A. Sellin, and C.R. Vane, Nucl. Instr. Meth. *B40/41* (1989) 53.
25. I.A. Sellin et al., *Forward Electron in Ion Collisions*, Lecture Notes in Physics *213*, Springer, Berlin (1984) p.109.
26. S.D. Berry, S.B. Elston, I.A. Sellin, M. Breinig, R. DeSerio, C.E. Gonzalez-Lepera, and L. Liljeby, J. Phys. *B19* (1986) L149.
27. J. Burgdörfer, Phys. Rev. *A33* (1986) 1578.
28. H.-D. Betz, D. Röschenthaler, and J. Rothermel, Phys. Rev. Lett *50* (1983) 34.
29. Y. Yamazaki et al., Phys. Rev. Lett *61* (1988) 2913.
30. J. Burgdörfer and C. Bottcher, Phys. Rev. Lett. *61* (1988) 2917.
31. I. Percival and D. Richards, J. Phys. *B12* (1979) 7051.
32. Y. Yamazaki et al., CERN Report No. CERN-PSCC *87-39* (unpublished), p.108; CERN Report No. CERN-PSCC *87-40* (unpublished), p.108.
33. V. Neelavathi, R. Ritchie, and W. Brandt, Phys. Rev. Lett. *33* (1974) 302.
34. J. Lindhard, private communication.
35. J. Burgdörfer, J. Wang and J. Müller, Phys. Rev. Lett. *62* (1989) 1599.
36. Y. Yamazaki and N. Oda, Nucl. Instr. and Meth. *194* (1982) 415.
37. L.F. de Ferrariis and R. Baragiola, Phys. Rev. *A33* (1986) 4449.
38. M. Hasegawa, K. Kimura, and M. Mannami, J. Phys. Soc., Japan *57* (1988) 1834.
39. J. Neumann, S. Schubert, U. Imke, P. Varga, and W. Heiland, Europhys. Lett. *3* (1987) 859.
40. As the referee of this article points out, such a material dependence of convoy electron acceleration has been very recently observed by A. Koyama et al. (unpublished).
41. J. Farley, Phys. Rev. *A40* (1989) 6286.

SURFACE ELECTRON SPECTROSCOPY; IMPORTANCE OF PARTICLE TRANSPORT

S. Tougaard

Fysisk Institut
Odense University
DK-5230 Odense M, Denmark

It was demonstrated that corrections for electron transport are highly important in quantitative analysis of surface electron spectra. This aspect has largely been ignored up to recently [1]. Central quantities are electron scattering cross sections and the path length distribution of emitted electrons.

Several examples of inelastic scattering cross sections evaluated by a dielectric response description [2] with model dielectric functions [3] were given. It was shown that for the large class of transition and noble metals, their alloys and many oxides, the cross sections are sufficiently similar that a "universal" cross section for energy loss < 100 eV can be defined [4,5]. In quantitative analysis of electron spectra it is usually sufficient to consider only an energy loss region up to ~ 100 eV from the peak energy. In this range it is a reasonable approximation to assume that the electrons follow a straight line on their way out of the solid [5,6]. In X-ray photoelectron spectroscopy (XPS), this leads to a direct correspondence between the in-depth concentration profile and the path length distribution of emitted electrons. In electron stimulated Auger electron spectroscopy (AES), the primary excitation may however vary slightly with depth within the surface region [5,6].

A simple formula is found which from a reflection electron energy loss spectrum (REELS) allows for a quantitative determination of the product of the inelastic mean free path and the inelastic scattering cross section [5,6].

A new simple formalism for quantitative analysis of AES and XPS is found [6,7]. It was shown that for inhomogeneous samples, this formalism can, in comparison with traditional methods, lead to improvements in

Interaction of Charged Particles with Solids and Surfaces
Edited by A. Gras-Martí *et al.*, Plenum Press, New York, 1991

487

accuracy by up to more than one order of magnitude. Simultaneously, the formalism gives information on the in-depth concentration profile.

REFERENCES

1. S. Tougaard and P. Sigmund, Phys. Rev. *B25* (1982) 4452.
2. J. Lindhard, Kgl. Dan. Vid. Selskab. Mat.-Fys. Medd. **28** no. 8 (1954).
3. R.H. Ritchie and A. Howie, Phil. Mag. *36* (1977) 463.
4. S. Tougaard, Solid State Commun., *61* (1987) 547.
5. S. Tougaard, Surf. Interf. Anal. *11* (1988) 453.
6. S. Tougaard, J. Electr. Spectr. *52* (1990) 243.
7. S. Tougaard, J. Vac. Sci. Technol. *A8* (1990) 2197.

STOPPING POWER OF HOT PLASMAS AND COLD SOLIDS:
DIELECTRIC AND TRANSPORT CROSS SECTION FORMULATIONS

N.R. Arista

Dpt. de Física Aplicada, Universitat d'Alacant
Apt. 99, E-03080 Alacant, Spain

Permanent address: División de Colisiones Atómicas
Centro Atómico Bariloche
C.C. 439, RA 8400 Bariloche, Argentina

1. INTRODUCTION

One of the relevant problems in studies of fusion plasmas is the energy loss of test particles to thermalized plasma electrons. Evaluations of the energy loss of charged particles of various energies are currently needed to describe the heating of dilute and dense fusion plasmas, as well as in studies of the energy transport in the medium.

To illustrate the wide range of plasma conditions of interest here, we show in fig.1 the range of temperatures and densities for the following systems: metals, inertially-confined plasmas (ICP), magnetically-confined plasmas (MCP), and the plasma conditions in the sun. The line $k_B T = E_F$ corresponds to the transition between degenerate and non-degenerate systems, while the line $v_e = v_0 = e^2/\hbar$ indicates the range where the mean electron velocity equals the Bohr velocity.

Conditions in the MCP range are currently obtained in tokamaks at various leading fusion laboratories. The more extreme conditions of high densities and temperatures in the ICP range are currently approached using either laser (*laser fusion*) or ion beams (*ion-beam fusion*).

The MCP and ICP ranges represent the most important ranges of application of stopping-power studies to dilute and dense plasmas of interest in fusion research. Thus, in the case of dilute plasmas, an important method of plasma heating consists in the use of neutral hydrogen beams, injected with energies of a few tens of keV [1]. In the ion-beam fusion approach [2], intense ion beams with energies of few MeV are used

Interaction of Charged Particles with Solids and Surfaces
Edited by A. Gras-Martí *et al.*, Plenum Press, New York, 1991

489

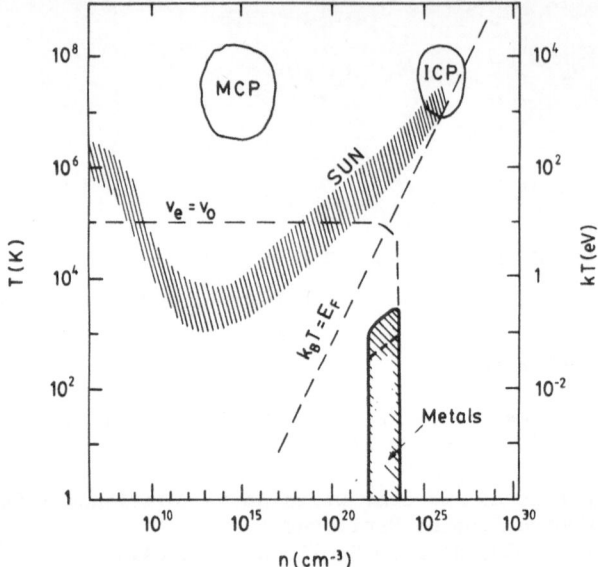

Figure 1. Range of plasma temperatures and densities of interest for stopping-power calculations, including metals, magnetically and inertially-confined plasmas (MCP, ICP), and the solar plasma.

as drivers for inertial-confinement fusion (ICF), through compression and heating of small solid pellets containing deuterium and tritium. Due to the property of efficient beam-target energy coupling, and the possibility of controlling and adjusting the energy-deposition profile, this approach is currently considered as an important alternative in ICF studies.

Evaluation of energy-loss rates for swift charged particles is also relevant for the study of energy transport in the plasma, particularly in relation with the emission of fast electrons, and with the alpha particles produced in the D-T fusion reaction.

We will discuss in this talk recent calculations of particle energy loss in dense plasmas, using two different formulations: (i) the dielectric-function formalism (DFF), and (ii) the transport cross-section formulation (TCSF).

The first approach permits a straightforward description of partial degeneracy effects in dense homogeneous plasmas. The model can also be extended to describe the inhomogeneous electron distributions of confined atomic systems in solids, using a generalization of the Thomas-Fermi model for finite temperatures. The results explain the enhancement of the stopping power of heated solids, an effect that has been observed in ICF experiments.

490

In the TCSF, the stopping power is expressed in terms of the transport cross section, and integrated over the velocity distribution of plasma electrons. The formulation applies both to dilute and dense plasmas. Classical, semiclassical, and quantum approximations are discussed, and compared with exact quantum-mechanical results based on phase-shift calculations.

2. DIELECTRIC FUNCTION FORMALISM

2.1 Homogeneous Electron Gas at Finite Temperatures

According to the dielectric-function formalism, the stopping power of a particle of charge $Z_1 e$ and velocity v is given, in terms of the dielectric function $\varepsilon(q,\omega)$, by the well-known expression (see, e.g., the lecture notes by F. Flores in this volume)

$$S = - \frac{dE}{dx} = \frac{2}{\pi} \left(\frac{Z_1 e}{v}\right)^2 \int_0^\infty \frac{dq}{q} \int_0^{qv} d\omega \ \omega \ \mathrm{Im}\left(\frac{-1}{\varepsilon(q,\omega)}\right). \tag{1}$$

The function $\mathrm{Im}[-1/\varepsilon(k,\omega)]$ is usually called the energy loss function. Extensive studies for slow and heavy particles in cold solids have been made on the basis of the DFF for a degenerate electron gas. In this discussion, we will concentrate on the extension of this formalism to describe partial-degeneracy effects, and related applications to real solids at high temperatures.

The extension of the DFF for the energy loss in plasmas of all degeneracies was considered by previous authors [3-6] based in analytical approximations or series expansions of the real and imaginary parts of the dielectric function, for low and high frequencies, and for long and short wavelengths [7-9].

With respect to the usual behavior of the energy loss function of a degenerate electron gas, the main feature observed at high temperatures is a thermal redistribution of the oscillator strength, which significantly modifies the energy-loss rate for slow ions [6,8]. Taking as a reference the mean thermal velocity of the plasma electrons, v_e, the plasma stopping power for all plasma degeneracies is given by

$$S(n,v,T) = \frac{4\pi n Z_1^2}{mv^2} \ L(n,v,T), \tag{2}$$

where n and T denote the density and temperature of the electrons. The *stopping number* $L(n,v,T)$ can be approximated, for low and high velocities, in the following way,

(a) for $v < v_e$,
$$L = \frac{1}{2}\left(\frac{v}{v_F}\right)^3 \frac{\ln \Lambda(n,T)}{[1 + \exp(-\mu/k_B T)]},$$
(3a)

(b) for $v > v_e$,
$$L = \ln\left(\frac{2mv^2}{\hbar\omega_P}\right) - \frac{\langle v_e^2\rangle}{v^2}.$$
(3b)

Here, ω_P is the plasma frequency, v_F is the Fermi velocity, and μ is the chemical potential. The low-velocity *collision logarithm*, $\ln \Lambda$, can be well approximated for all degrees of degeneracy by

$$\ln \Lambda(n,T) = \ln(1 + \zeta) - \zeta/(1 + \zeta),$$
(4)

with [6]

$$\zeta(n,T) = \frac{16}{3}\left(\frac{E_F}{\hbar\omega_P}\right)^2 + \frac{8}{\Gamma}\left(\frac{k_B T}{\hbar\omega_P}\right)^2.$$
(5)

The temperature dependence of L in eq. (3b) appears only in the $\langle v_e^2\rangle/v^2$ term. This is analogous to the way in which the so-called shell corrections appear in the high-velocity expansion for the degenerate electron gas [10]. We see also from eqs. (3a) and (3b) that the main temperature dependence arises in the low-velocity range. In particular, for high temperatures, $k_B T \gg E_F$, $\exp(-\mu/k_B T) \cong (3/4)\pi^{1/2}(k_B T/E_F)^{3/2}$, and the low-velocity stopping power becomes

$$S \cong \frac{16}{3}\pi^2 z_1^2\, n\, v \left(\frac{m}{2\pi k_B T}\right)^{3/2} \ln \Lambda(n,T),$$
(6)

which drops with T as $T^{-3/2}$. This result is the quantum-mechanical analog of the classical stopping-power result [11] (both results differ only in the form of the collision logarithm [12,13]).

As it can be seen from these expressions, and also from full calculations using the DFF [6], the stopping power of an homogeneous electron gas decreases with temperature. Moreover, since the low-velocity stopping decreases much more rapidly than the high-velocity stopping, the maximum of the stopping curve moves to higher ion velocities, following the thermal increase of the mean electron velocities (in general, the stopping-power maximum occurs when the ion velocity equals the mean electron velocity).

492

2.2 Inhomogeneous Electron Distributions: Energy Loss of Charged Particles in Heated Solids

Experiments at various laboratories [14-16] show the existence of a strong enhancement in the energy loss of swift proton and deuteron beams in heated solids, similar to those observed in hydrogen plasmas [17]. This effect can be explained in terms of changes in the contributions of bound and free electrons in dense ionized media [18,19]. The calculation of the energy loss by this approach requires evaluations of the degree of atomic ionization and average ionization potentials, or the use of approximate scaling laws for a wide range of temperatures.

It is of interest to show here how the dense-electron-gas approach can be generalized for confined atomic systems at finite temperatures, and how in this way the enhancement effect in the stopping power can be fully explained [20]. The model also provides a simple and accurate approach to calculate the energy loss of swift ions in heated media.

The description of electron density distributions in atomic systems is based on a generalization of the Thomas-Fermi model for confined atoms at finite temperatures [21]. Following this approach, for each atom in the solid we consider a Wigner-Seitz (WS) sphere of radius r_0. With the assumption of spherical symmetry, we express the electron density $n(r)$ by integrating over a Fermi-Dirac distribution (for a partially degenerate plasma) as follows

$$n(r)= \frac{1}{\pi^2 \hbar^2} \int_0^\infty \frac{p^2 \, dp}{1 + \exp\left[\beta\{\frac{p^2}{2m} - eV(r) - \mu\}\right]}, \tag{7}$$

where p denotes the electron momentum and $\beta = 1/k_B T$. The self-consistent potential $V(r)$ in the integrand is related to $n(r)$ through the Poisson equation. This leads to the following integro-differential equation for $V(r)$

$$\nabla^2 V(r)= \frac{2(2mk_B T)^{3/2}}{\pi \hbar^3} \, I_{1/2}[\beta\{eV(r)-\mu\}], \tag{8}$$

where $I_{1/2}(x)$ is a Fermi integral of order $1/2$ [8].

Eq.(8) is solved by numerical integration, starting from the cell border at $r = r_0$, and with the boundary conditions for a neutral system $(dV/dr = 0$ at $r = r_0)$.

Finally, we calculate the stopping power using the local approximation [22], by integrating over the atomic cell, viz.

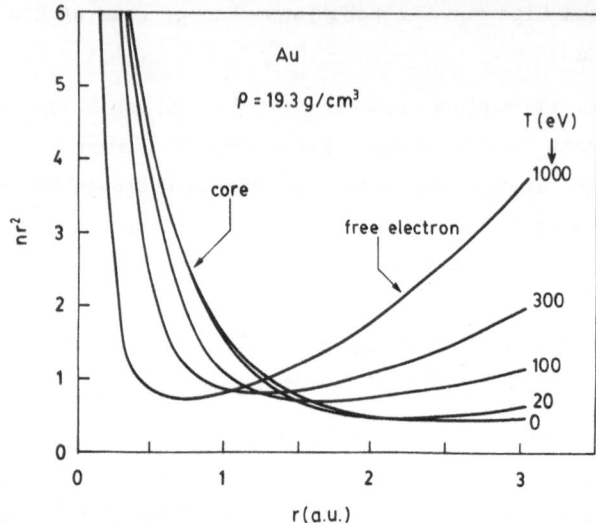

Figure 2. Calculated electron density profiles for confined Au atoms at various temperatures.

$$\frac{dE}{dx} = \frac{4\pi \ z_1^2 \ e^4}{mv^2} \int_0^{r_0} n(r) \ L(n(r),v,T) \ 4\pi r^2 \ dr \qquad (9)$$

with $L(n,v,T)$ given by eq.(3).

We show in fig.2 a set of results for the electron density profiles in Au at various temperatures. We observe clearly the separation of a core and a free electron component, and we see that for increasing temperatures more electrons are removed from the ion core, with a corresponding increase in the free-electron distributions.

The integrated stopping power is shown in fig.3, as a function of temperature, and for various target densities. A strong enhancement is observed, with a maximum at $T \simeq 400$ eV, followed by a decline that agrees with the asymptotic $T^{-3/2}$ dependence of eq.(6).

To understand the appearance of this enhancement (not present in the homogeneous electron-gas formulae) we refer again to fig.2. This shows that with increasing T, more electrons are removed from the high-density region of the atomic core, and distributed with a much lower density as nearly-free electrons, through the whole atomic cell. We are thus lead to consider the problem of energy absorption in an electron gas of varying degeneracy. Due to the exclusion principle, most of the transitions of electrons in the local Fermi sphere are suppressed, and this effect is much stronger at the inner high-degeneracy region of the ion core.

494

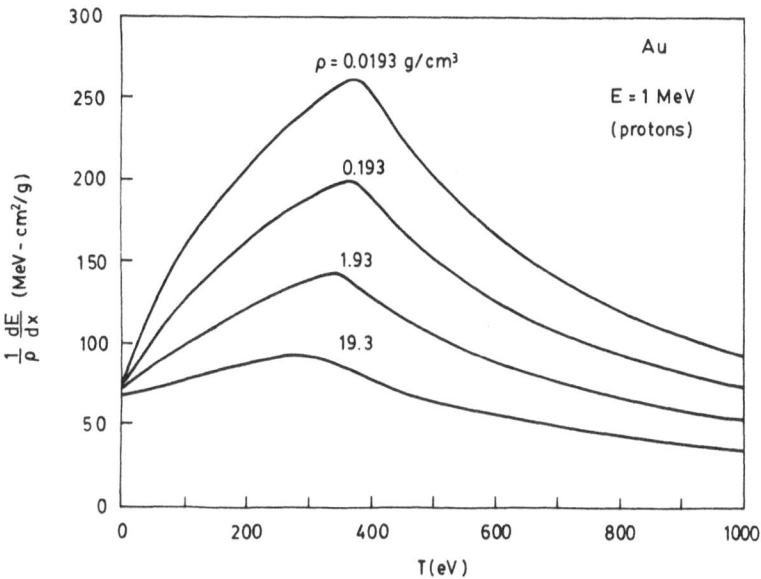

Figure 3. Temperature dependence of the stopping power of 1 MeV protons in Au, for various Au densities.

Therefore, the thermal redistribution of these electrons in regions of lower densities (and lower degeneracy), allows for an increase in the energy absorption of the whole system, and hence also of the stopping power. Obviously this effect is more pronounced the larger the volume of the WS cell, so that the largest enhancements are obtained for lower Au densities as shown in fig.3.

To conclude this section, we show that the confined-atom model permits also a straightforward calculation of the values of the mean excitation energies of all solids [23], without the use of atomic-structure parameters (in fact, the "atomic" parameter in this case is the Wigner-Seitz radius, determined from the density ρ and the atomic mass A, by $4\pi r_0^3/3 = A/\rho$).

Fig.4 shows the values of the mean excitation energy I, calculated according to the Lindhard-Scharff expression [22]

$$\ln I = \frac{1}{Z_2} \int d^3r \ n(r) \ \ln[\sqrt{2} \ \omega_P(r)]. \tag{10}$$

The calculated results for $I_0 = I/Z_2$ are compared in this figure with experimental results, and with the predictions of the Thomas-Fermi ($I_0 = 8.9$ eV) and Lens-Jensen ($I_0 = 10.7$ eV) models.

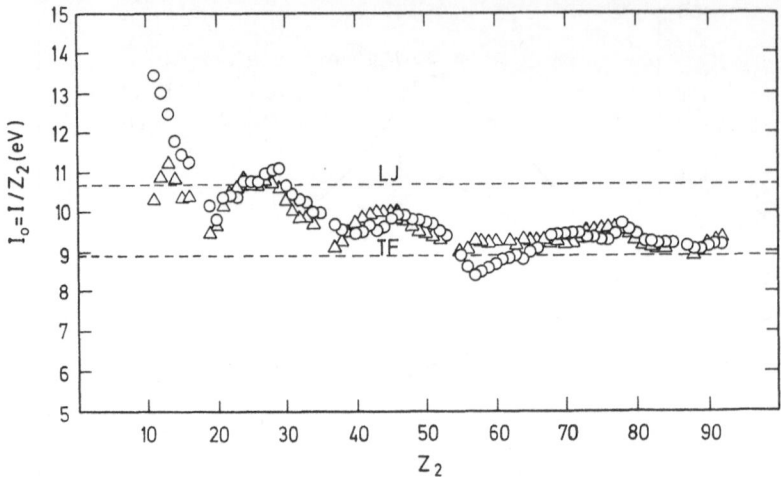

Figure 4. Z_2 dependence of mean excitation energy. Values calculated with the present model (triangles) and experimental results (circles).

It should be remarked that only the information on the atomic volume provided by the density ρ, is sufficient to explain with reasonable accuracy the oscillatory behavior of I/Z_2. The Z_2 structure of the mean excitation energy and stopping power is, in this model, produced by the fluctuations in the values of $n(r)$, $v_F(r)$, and $\omega_p(r)$, in the outer region of the atomic cells of different elements [23].

3. TRANSPORT CROSS SECTION FORMULATION

The previous description based on the standard DFF for the homogeneous electron gas at finite temperatures, and the extension to inhomogeneous electron distributions in confined atomic systems, are all based on the initial assumption that the coupling between the external particle and the system can be described in a perturbative way through the linear-response function of the medium.

For many cases of physical interest this approximation does not hold. This may be the case of static or slowly-moving ions in an electron gas at the metallic density range, as in the examples discussed by Echenique in this volume. Hence, it becomes of much interest to consider a different formulation of the energy-loss problem, such as the transport cross-section formulation (TCSF), where perturbative or linearized solutions can be compared with full quantum-mechanical calculations.

A general description of the energy loss of particles in plasmas, based on the TCSF was given before [12], based on previous binary-collision descriptions of particle energy loss in classical plasmas [24,25]. In general, the expression for the stopping power of a particle with mass m_1 and velocity v_1, is given by [12,26]

$$S(v_1) = \frac{\pi\,\mu}{v_1^2} \int_0^\infty dv_2\ v_2\ f(v_2) \int_{|v_1-v_2|}^{v_1+v_2} dv_r\ v_r^4\ \sigma_{tr}(v_r) \left[\frac{m_1-m_2}{m_1+m_2} + \frac{v_1^2-v_2^2}{v_r^2}\right],$$

(11)

where $\mu = m_1 m_2/(m_1+m_2)$ is the reduced mass, $f(v_2)$ is the velocity distribution of field particles with velocities v_2 and mass m_2, $v_r = |\vec{v}_1 - \vec{v}_2|$ is the relative velocity, and $\sigma_{tr}(v_r)$ is the so-called transport cross section, defined in the usual way by

$$\sigma_{tr}(v_r) = \int (1 - \cos\vartheta)\ d\sigma.$$

(12)

For the present case of ions moving in an electron gas, we approximate $m_2/m_1 \simeq 0$, and $\mu \simeq 1$ a.u..

The main ingredient in this formulation is the transport cross section, as a function of the relative velocity. The exact quantum-mechanical expression for σ_{tr} in terms of the phase shifts δ_l is

$$\sigma_{tr}(v_r) = \frac{4\pi}{k^2} \sum_{=0}^\infty (l+1)\ \sin^2(\delta_l - \delta_{l+1}),$$

(13)

where $k = mv_r/\hbar$. In the following, we will consider various model calculations of σ_{tr}, using as a reference a screened potential of the form

$$V(r) = \frac{Z_1 e^2}{r}\ \exp(-\alpha r).$$

(14)

In this case, the results for the transport cross section, calculated according to classical (CL), semiclassical (SCL), or plane-wave-Born approximations (PWBA), are the following (see ref.[12] for details)

$$\sigma_{tr}^{CL} = \frac{4\pi Z_1^2 e^4}{m^2 v_r^4} \left[\ln\left(\frac{2mv_r^2}{Z_1 e^2 \alpha}\right) - \gamma - \frac{1}{2}\right],$$

(15a)

$$\sigma_{tr}^{SCL} = \frac{4\pi Z_1^2 e^4}{m^2 v_r^4} \left[\ln\left(\frac{2mv_r}{\hbar\alpha}\right) - \operatorname{Re}\ \psi(i\eta) - \gamma - \frac{1}{2}\right],$$

(15b)

$$\sigma_{tr}^{PWBA} = \frac{4\pi Z_1^2 e^4}{m^2 v_r^4} \left[\ln\left(\frac{2mv_r}{\hbar\alpha}\right) - \frac{1}{2} \right], \tag{15c}$$

A point of particular interest here is the appearance of a "Bloch term" Re $\psi(i\eta)$ in eq.(15b), where $\psi(z)$ denotes the digamma function [27], and $\eta = Z_1 e^2/\hbar v_r$ is the well-known Bloch parameter. A very accurate approximation for the real part of the digamma function is given by [12]

$$\text{Re } \psi(i\eta) \cong \frac{1}{2} \ln(1 + \Gamma^2\eta^2) - \gamma, \qquad (\Gamma = e^\gamma = 1.781). \tag{16}$$

Thus, we see that the SCL approximation, eq.(15b), agrees with the classical CL result for $\eta \gg 1$, and with the quantum-mechanical PWBA result for $\eta \ll 1$, just in the same way as the Bloch expression connects the Bohr and Bethe formulae for the *high-velocity* stopping power in the usual treatment of atomic excitations. In fact, eqs.(15a-c) represent the analogous approximations to the transport cross section for the scattering of plasma electrons in a screened potential. The expressions apply here, in a more extended picture, as a function of the *relative* velocities between the test particle and the plasma electrons.

In the case of dilute fusion plasmas, where the semiclassical conditions apply, several analytical approximations can be used to integrate eq.(11), using eqs.(15a-c), in an almost exact way [12]. For particle velocities higher than the mean electron velocity, $v_e \simeq (2k_B T/m)^{1/2}$, the result agrees with the Bohr-Bethe-Bloch theory, and the transition between classical and quantum-mechanical regimes is reproduced as a function of particle velocity v_1 (since $v_r \simeq v_1$, and thus $\eta \simeq Z_1 e^2/\hbar v_1$). On the other hand, for low velocities, such that $v_1 < (2k_B T/m)^{1/2}$, the parameter that regulates the transition between classical and quantum-mechanical regimes is the temperature of the system ($v_r \simeq v_e$, and $\eta \simeq Z_1 e^2/\hbar (2k_B T/m)^{1/2}$ in this case). Therefore, in the low-velocity range one retrieves the classical plasma stopping power results for $T < T_c$, and the quantum mechanical PWBA results for $T > T_c$. The transition temperature is given by [13] $T_c \simeq Z_1^2 \times 10^6$ K, so that, for values of Z_1 between 1 and 10, T_c falls in the range of interest for fusion studies ($T_c \sim 10^6$ to 10^8 K). For higher Z_1 values the energy loss in dilute fusion plasmas can be fully described in classical terms. (For a detailed discussion of various classical and quantum approaches and results for dilute plasmas, see ref.[12] and other references therein).

Let us consider now the case of slow ions in dense plasmas, and approximate eq.[11] in the limit $v_1 \ll v_2$. An expansion of v_r and $\sigma_{tr}(v_r)$ in this formula yields the following expression for the low-velocity stopping power

$$S(v_1) = \frac{4\pi}{3} m \, v_1 \int_0^\infty dv_2 \left[-\frac{df}{dv_2} \right] v_2^4 \, \sigma_{tr}(v_2). \tag{17}$$

Eq. (17) applies for any velocity distribution. In particular, for a degenerate electron gas $(-df/dv) \propto \delta(v - v_F)$ (i.e., only electrons at the Fermi surface participate), and we retrieve the usual expression for the low-velocity stopping (see, e.g., the lectures notes by P.M. Echenique in this volume), namely

$$S_0(v_1) = n \, m \, v_1 \, v_F \, \sigma_{tr}(v_F). \tag{18}$$

This relation has been applied by various authors [28-34] to calculate the energy loss of slow ions in solids, as well as related quantities.

On the other hand, in the limit of dilute plasmas, by using any of the approximations in eqs. (15a-c), we obtain the semiclassical result [11,12,24,25] of eq. (6).

Comparison of eqs. (15-18) becomes useful to point out the differences between the CL, SCL or PWB approximations, and the exact quantum-mechanical calculation of the transport cross section. The PWBA result for σ_{tr}, eq. (15c), corresponds to the Bethe approximation (also a PWBA) for the energy loss in atomic systems, and to the linear-response approximations in the DFF for a degenerate electron gas, which in all cases lead to a simple Z_1^2 law. The CL and SCL results differ from the PWBA only in the form of the logarithmic terms, therefore, all these approximations show a smooth Z_1 dependence, contrary to the well-known structure of the low-velocity stopping powers in solids.

Eq. (17) shows in a general way the approximate proportionality of the low-velocity stopping power with particle velocity v_1. It is important to note here that this linear velocity dependence has no relation with the linearization assumed in the DFF or in the PWBA with respect to the ion charge Z_1. In particular, eq. (17) provides the basis for full non-linear quantum integrations of low-velocity stopping powers.

To proceed along this line, we calculate the phase shifts δ_l from the numerical integration of the radial Schrödinger equation [29] and then obtain the transport cross section $\sigma_{tr}(v_r)$ from the quantum-mechanical expression, eq. (13).

In figs. 5 and 6 we show the results for $\sigma_{tr}(v_r)$, represented by the logarithmic function $L(v_r)$, which is defined by

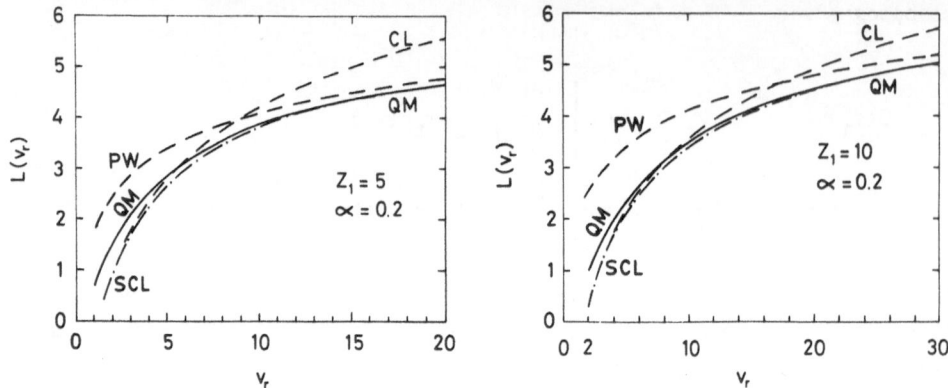

Figure 5. Transport cross section calculations for weak-screening conditions ($\alpha = 0.2$), versus relative velocity v_r.

$$\sigma_{tr}(v_r) = \frac{4 \pi Z_1^2 e^4}{m^2 v_r^4} L(v_r),\qquad (19)$$

The curves denoted QM are the full quantum-mechanical results from the phase-shift calculation. The curves denoted CL, SCL, and PW correspond to the previously discussed approximations of eqs.(15a–c).

Fig.5 shows a case of *weak screening* conditions (here $\alpha = 0.2$ a.u.), where the semiclassical (SCL) approximation agrees well with the QM result, except at low velocities. We also see that the SCL approximation connects with the CL and PW results in the appropriate limits of v_r (i.e., $\eta \gg 1$ and $\eta \ll 1$ respectively).

In the range of metallic densities, the value of α can be approximated by the RPA result, namely $\alpha_{RPA} = \sqrt{3}\ \omega_P/v_F$. However, a more

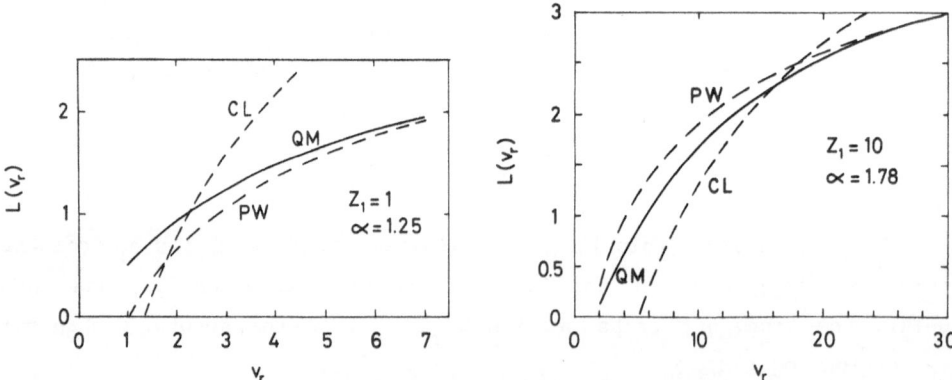

Figure 6. Transport cross section calculations for screening conditions in metals ($\alpha = \alpha_{FSR}$), versus relative velocity v_r.

accurate description is obtained if the value of α is adjusted so as to satisfy the Friedel sum rule for the phase shifts

$$\frac{2}{\pi} \sum_{1=0}^{\infty} (21+1) \; \delta_1(v_F) = Z_1. \tag{20}$$

It has been shown [32] that this procedure yields transport cross sections which are in good agreement with more elaborate calculations using the density-functional approach [31].

We show in fig.6 the results for $L(v_r)$ obtained in this way, for $Z_1 = 1$ and 10, using the values of $\alpha = \alpha_{FSR}$ obtained from the Friedel sum rule for an electron gas with $r_s = 2$ (i.e., in the range of solid-state electron densities). We observe here important differences among the results obtained by the previous approaches. The only point of agreement that remains is the asymptotic convergence of the exact QM result and the PWBA at high values of v_r. The SCL result (not shown here) also converges to these curves.

In conclusion, the CL, SCL or PWBA should not be expected to give accurate results for the transport cross section or stopping power for realistic screening conditions in metals.

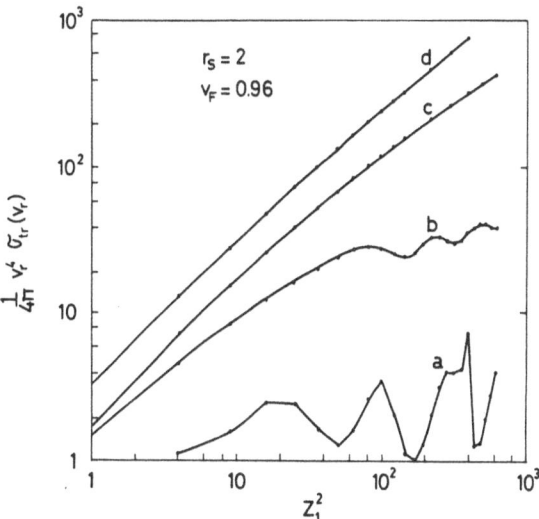

Figure 7. Z_1 dependence of the transport cross section, for various screening conditions and relative velocities. Curve a: $\alpha = \alpha_{RPA}$, $v_r = v_F$ (as expected for metals); curve b: $\alpha = 0.2$, $v_r = v_F$; curve c: $\alpha = \alpha_{RPA}$, $v_r = 5$; curve d: $\alpha = 0.2$, $v_r = 5$. (Values in a.u.).

Let us now consider the Z_1 dependence of the transport cross section. Fig.7 illustrates the dependence of the TCS with charge Z_1, for various velocities and screening conditions. Curve *a* corresponds to $v_r = v_F = 0.96$ a.u. and $\alpha = \alpha_{RPA} = 1.10$ a.u. (for $r_s = 2$), which approximately corresponds to the screening conditions in metals. Curve *b* corresponds to $v_r = v_F$ and $\alpha = 0.2$ (diminished screening), curve *c* to $v_r = 5$ a.u. and $\alpha = \alpha_{RPA}$ (increased velocity), and curve *d* is for $v_r = 5$ and $\alpha = 0.2$ (both diminished screening and increased velocity). The curves illustrate how the Z_1 oscillations disappear in the limits of low screening or high velocity; these are the limits where the SCL approximation applies. We also observe that curves *c* and *d* approach the Z_1^2 behavior of eq.(15); the slight bending below the straight line is in fact due to the Bloch correction term.

Thus, the figure also illustrates what was said before about the difference between the CL, SCL or PWBA, and the exact QM transport cross section.

4. SUMMARY

The dielectric-function formalism and the transport cross-section formulation provide two basic approaches for the study of the energy loss of particles in dense plasmas. The DFF has the advantage of its extended applicability through a wide range of test particle velocities, with the possibility of incorporating single particle and plasmon excitations at high velocities ($v > v_F$). The extension of the DFF to finite temperatures, and inhomogeneous electron distributions, permits to study partial-degeneracy effects, and energy deposition in heated solids. The model permits also to describe the Z_2 structure in the energy loss of particles in solids.

However, the DFF is a perturbative (or linearized) representation of the particle-plasma interaction, and thus fails to describe the cases where the perturbation induced by the external charge Z_1 is strong; in particular, in the low-velocity range.

On the other hand, the TCSF is based on binary-collision concepts, and therefore it can not describe dynamical-response effects, like collective excitations for velocities $v > v_F$. The TCSF can be best applied at low velocities ($v < v_F$), where the DFF generally fails.

In this energy range, the TCSF provides a powerful approach for quantum-mechanical calculations of low-velocity stopping powers, and as a test of various approximations to the transport cross section.

502

The SCL approximation (which includes the CL and the PWBA in the appropriate limits) applies for either *low screening* or *high relative velocities*. In particular, it applies always at very high temperatures, where the screening becomes weak and the electron velocities high.

In the important case of slow ions in solids, both the relative velocities ($\sim v_F$) and the screening constant ($\sim \alpha_{RPA}$) are close to atomic units, and in this case a full self-consistent quantum calculation of σ_{tr} is required. Here the TCSF provides the most convenient approach to represent the strong quantum effects that give rise to the Z_1 oscillations in the low-velocity stopping power.

In conclusion, the DFF and the TCSF can be used as complementary formulations of the energy loss problem, and its extensions to finite temperatures provide a convenient approach to study the energy loss of ions in dense quantum plasmas.

ACKNOWLEDGEMENTS

I would like to thank A. Gras-Martí and all the members of the Departament de Física Aplicada, for the warm hospitality received during my sabbatical stay at the Universitat d'Alacant, and to the Spanish Ministerio de Educación y Ciencia and Conselleria d'Educació i Ciència de la Generalitat Valenciana for economical support. Partial support from the DGICYT (projects PS88-0066 and 89-0065) is also recognized.

Discussions related to this talk with A. Gras-Martí, P.M. Echenique, A. Arnau and I. Nagy, are gratefully acknowledged.

REFERENCES

1. E. Speth, Rep. Prog. Phys. *52* (1989) 57.
2. T.A. Mehlhorn, J. Appl. Phys. *52* (1981) 6522.
3. N.R. Arista and W. Brandt, Phys. Rev. *A23* (1981) 1898.
4. G. Maynard and C. Deutsch, Phys. Rev. *A26* (1982) 665.
5. C. Deutsch, G. Maynard, and H. Minoo, J. Physique *44-C8* (1983) 67.
6. N.R. Arista, J. Phys. *C18* (1985) 5127.
7. C. Gouedard and C. Deutsch, J. Math. Phys. *19* (1978) 32.
8. N.R. Arista and W. Brandt, Phys. Rev. *A29* (1984) 1471.
9. R.G. Dandrea, N.W. Ashcroft and A.E. Carlsson, Phys. Rev. *B34* (1986) 2097.
10. J. Lindhard and A. Winther, K. Danske Vidensk. Selsk., Mat.-Fys. Medd. *34* (1964) No.4.
11. L. Spitzer, *Physics of Fully Ionized Gases*, Interscience, New York (1962).
12. L. de Ferraris and N.R. Arista, Phys. Rev. *A29* (1984) 2145.
13. N.R. Arista and W. Brandt, Phys. Rev. *A30* (1984) 630.
14. F.C. Young et al, Phys. Rev. Lett. *49* (1982) 549.
15. J.N. Olsen et al, J. Appl. Phys. *58* (1985) 2958.
16. B. Goel and H. Bluhm, J. Physique *12-C7* (1988) 169.
17. D.H.H. Hoffmann et al, Z. Phys. *A30* (1988) 339; D.H.H. Hoffmann, J. Physique *12-C7* (1988) 159; D. Gardes et al, Europhys. Lett. *8* (1988) 701; J. Physique *12-C7* (1988) 151.
18. E. Nardi, E. Peleg, and Z. Zinamon, Phys. Fluids *21* (1978) 574.
19. T.A. Mehlhorn, J. Appl. Phys. *52* (1981) 6522.

20. N.R. Arista and A.R. Piriz, Phys. Rev. *A35* (1987) 3450.
21. R.E. Marshak and H.A. Bethe, Astrophys. J. *91* (1940) 239; R.P.Feynman, N. Metropolis and E. Teller, Phys. Rev. *75* (1949) 1561; R. Latter, Phys. Rev. *99* (1955) 1854.
22. J. Lindhard and M. Scharff, K. Danske Vidensk. Selsk. Mat.-Fys. Medd. *27* (1953) No.15.
23. N.R. Arista, J. Phys. *C19* (1986) L841.
24. M. Gryzinski, Phys. Rev. *107* (1957) 1471.
25. S.T. Butler and M.J. Buckingham, Phys. Rev. *126* (1962) 1.
26. A somewhat similar expression was obtained by P. Sigmund, Phys. Rev. *A26* (1982) 2497, who expanded the solutions analytically, for low and high velocities.
27. M. Abramowitz and I.A. Stegun, *Handbook of Mathematical Functions*, Dover, New York (1970).
28. J.S. Briggs and A.P. Pathak, J. Phys. *C6* (1973) L153; ibid. *C7* (1974) 1929.
29. T.L. Ferrell and R.H. Ritchie, Phys. Rev. *B16* (1977) 115.
30. P.M. Echenique, R.M. Nieminen and R.H. Ritchie, Solid State Commun. *37* (1981) 779.
31. M.J. Puska and R.M. Nieminen, Phys. Rev. *B27* (1983) 612.
32. A. Cherubini and A. Ventura, Lett. Nuovo Cim. *44* (1985) 503.
33. J.C. Ashley, A. Gras-Marti and P.M. Echenique, Phys. Rev. *A34* (1986) 2495.
34. I. Nagy, A. Arnau, and P.M. Echenique, Phys. Rev. *B38* (1988) 9191; Phys. Rev. *A40* (1989) 987.

ION-INDUCED EMISSION FROM MAGNETIC MATERIALS
NEAR PHASE TRANSITION TEMPERATURE

V.E. Yurasova

Physics Faculty
Moscow State University
Moscow 117234 USSR

ABSTRACT

In this review, the results of investigations are presented dealing with atomic particle emission under ion bombardment of magnetic materials, in a temperature range around the Curie point. Besides, some other processes occurring near the Curie temperature are discussed, including sublimation, oxidation and structure phase transition.

The effects of magnetic and crystallographic phase transitions of metals on the ion-induced emission have received increasing attention in recent years. First of all, such interest is stimulated by intention to understand the mechanism of atomic particle interaction in a magnetic material during its phase transition. Besides, the further development of quantitative methods of surface analysis requires more detailed information about secondary-particle emission, in a wide temperature range and, in particular, near the phase-transition point.

In this report, some results are considered in brief, concerning the temperature dependence of atomic particle emission (in various charge states), under ion bombardment of magnetic materials.

1. SPUTTERING

The first work where ion-induced atomic particle emission was investigated near the Curie temperature T_c deals with sputtering [1]. A sharp increase in the sputtering yield S, upon passing from ferro to paramagnetic state, was reported in ref.[1], and then in refs.[2-4] for 3d and 4f metals (i.e., for Ni and Gd). The jump in sputtering yield was about 10 - 20 %. Moreover, a pronounced maximum of the sputtering yield was observed near the Curie temperature, fig.1.

An interesting type of temperature dependence S(T) was obtained for a single crystal of Ni_3Mn [5], fig.2. In general, for various Ni-Mn alloys the Curie temperatures are determined by the Ni percentage fig.2b; for Ni_3Mn, $T_c = 515\ ^{\circ}C$. However, in fig.2a a typical maximum of S(T) turns out

Interaction of Charged Particles with Solids and Surfaces
Edited by A. Gras-Martí *et al.*, Plenum Press, New York, 1991

505

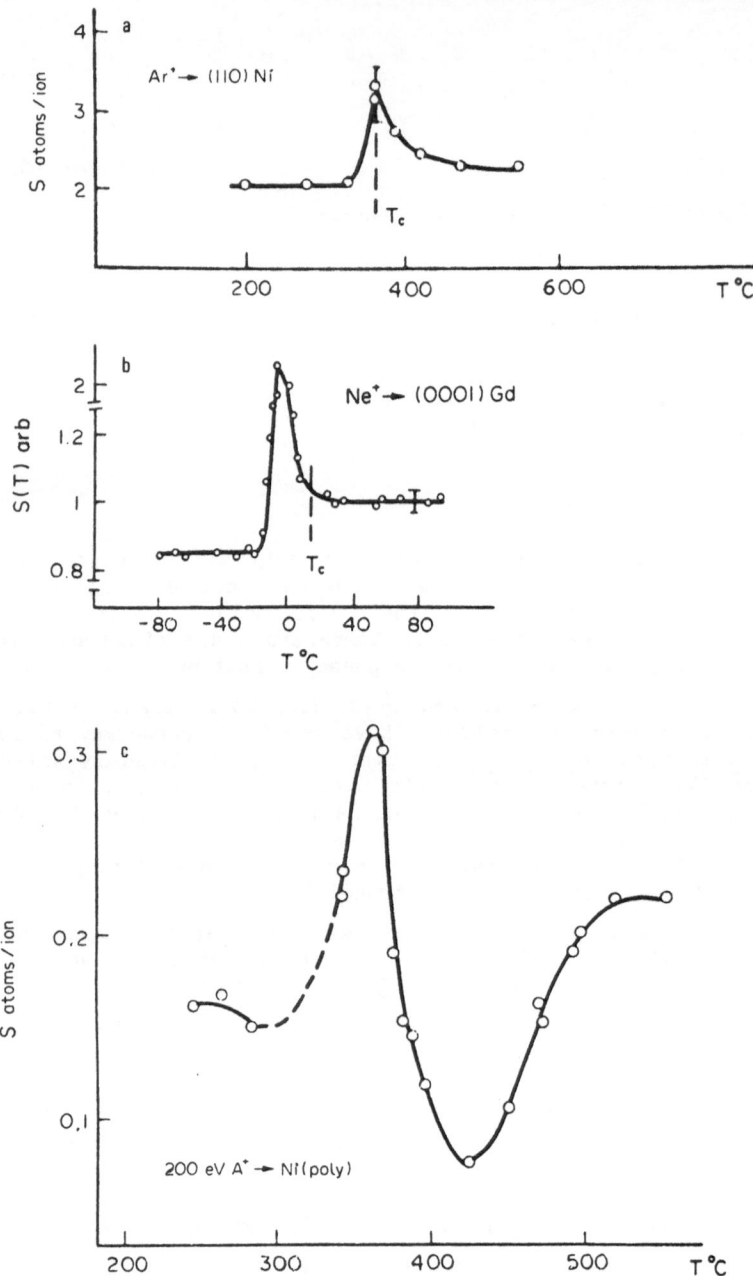

Figure 1. a) Variation of the sputtering yield for Ni targets near the Curie point. Bombardment with 15 keV Ar+ at an angle of 30° from the normal to the surface [1]. b) Temperature dependence of the sputtering yield from the (0001) face of a Gd single crystal for 20 keV Ne+, incident in the (11$\bar{2}$0) plane, at 30° measured from the normal to the surface [2]. c) Sputtering yield S, versus temperature, for polycrystalline Ni target under 200 eV Ar+ bombardment at normal incidence; ion current density j = 7 mA cm^{-2} [3].

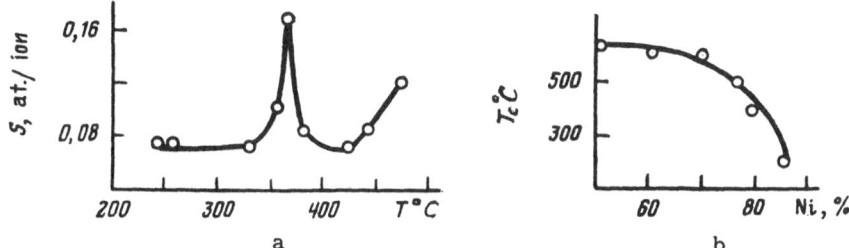

Figure 2. a) Temperature dependence of the sputtering yield for a Ni$_3$Mn (100) face, irradiated with 200 eV Ar$^+$ [5]. b) Curie temperature of Ni-Mn alloy versus Ni percentage.

to correspond to the value T = 365 $^\circ$C, which is merely the Curie temperature for pure nickel (or for a Ni-Mn alloy with a Ni portion exceeding 85 %). As Auger analysis has shown, this may be explained by Ni segregation on the specimen surface during sputtering.

Under special conditions of bombardment, (ion energies and masses, angle of incidence onto the single crystal, density of ion beam), a fine structure of the S(T) dependence arises, e.g., two minima are observed on both sides of the main peak near T$_c$, (see fig.1c and ref.[4]).

The decrease in the sputtering yield of Ni and Gd in the ferromagnetic state is connected with an increase of the potential energy of atomic interaction, fig.3. The theory developed in refs.[6,7], (see also ref.[8]), shows that a variation of the interaction potential for the f-state changes the collision cascades, and increases the binding energy of surface atoms, in such a way that the coefficient of sputtering S decreases by ~ 15 %. The fine structure of the S(T) dependence is shown to be caused by a peculiar relationship between the dimension of the collision cascade and the critical radii of correlation of magnetization fluctuations.

It is interesting to compare the temperature dependence of the sputtering yield with some other phenomena observed in the temperature range around T$_c$, for example, with sublimation and oxidation.

2. SUBLIMATION AND OXIDATION

Various processes such as desorption [9], sublimation [10], initial growth of oxide films [11], and chemical reaction [12], occurring near the T$_c$ of magnetic materials, prove to display anomalies in their rates K as a function of the temperature T. Namely, when T is approaching T$_c$ the process rate is not governed by the Arrhenius law any more, figs.4 and 5.

507

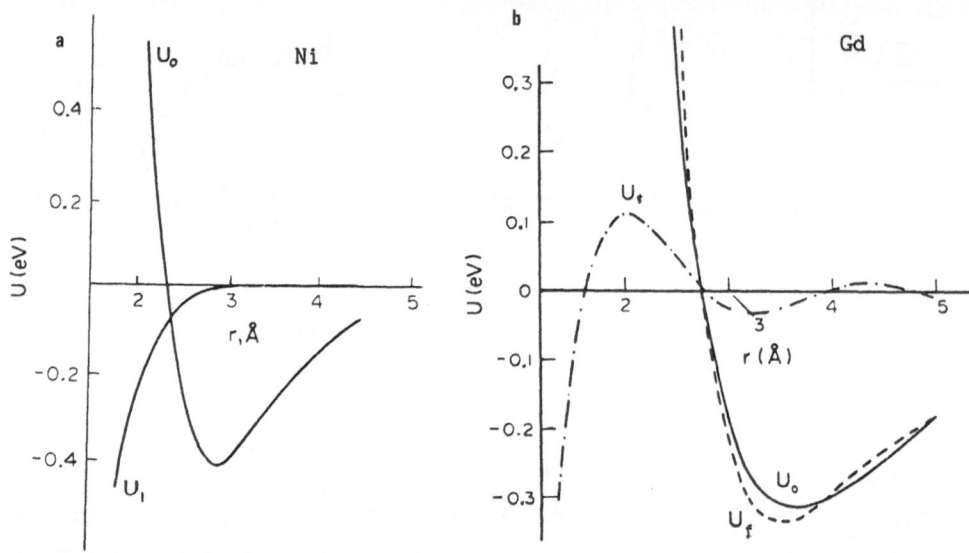

Figure 3. Interaction energy of Ni (a) and Gd (b) atoms, as a function of
the interatomic separation, for random (u_0) and parallel (u_f)
orientation of spins; u_1 is the energy of the direct (Ni) and
indirect (Gd) exchange interaction (magnetic addition).

In order to describe a non-equilibrium process like this, the
random-walk theory is usually applied. The process is treated, either as a
particle escape from the potential well through the barrier, or as a
particle transfer between two stable states, caused by random forces due
to fluctuations of the heat bath (inelastic photon scattering). The
temperature dependence of the particle transfer probability, and hence, of
the process rate, will be of the Arrhenius type. But this approach assumes
that the solid bulk, (the heat bath), is in equilibrium. It will be true
as long as the heat-bath relaxation time remains much less than the time
of particle transfer. If the temperature is close to T_c, however, this
condition may appear to be violated and the random walk theory may not be
valid. The reason is that, upon approaching T_c, the relaxation time of the
spin sub-system of the magnetic material is increasing.

The theoretical description of non-equilibrium processes on the
surface of a magnetic material, in the case of critical slowing down, was
developed in refs.[14,15]. It is based on the following arguments. In the
temperature range near T_c, an increase both in amplitude and life time of
fluctuations of the magnetic spin sub-system occurs. It is the result of
an anomalous enhancement of a random force applied to a surface atom. Due
to its exchange interaction with lattice atoms, the surface atom is

involved in collective motion, (vibration), caused by spatial and time fluctuations of the ordering magnetic field, i.e., of the spin density). Thus, collective vibrations of the spin subsystem of the solid have an effect on the probability of particle escape from the potential well. In order to characterize this dynamic collective process, a kinetic equation has been set up in refs.[14,15] for the distribution function of adparticles on the solid surface, in the case of slow relaxation of the heat bath. A frequency of adparticle relaxation, on the surface of the magnetic material, was determined by the dynamic correlator of acting random forces. And the random force, in its turn, was expressed using a magnetic order parameter. The time dependence of the order parameter is derived from models of critical dynamics. Hence, the effect of fluctuations of the magnetic spin sub-system was taken into account, when the probability of atom sublimation near T_c was derived from the kinetic equation obtained. The results are shown in figs.4 and 5, and compared with experimental data.

3. MORPHOLOGICAL TRANSITION IN THE VICINITY OF THE CURIE POINT

When ion-induced emission, and some other processes, are treated during a magnetic phase transition, a modification of surface structure near the Curie point was revealed in some experiments. In ref.[16], changes in the height of steps and terraces on the vicinal surface of Ni (111) 5° [1$\bar{1}$0], and Ni (111) 10^o [1$\bar{1}$0], upon passing through the Curie temperature, is reported. In the paramagnetic state each of the samples

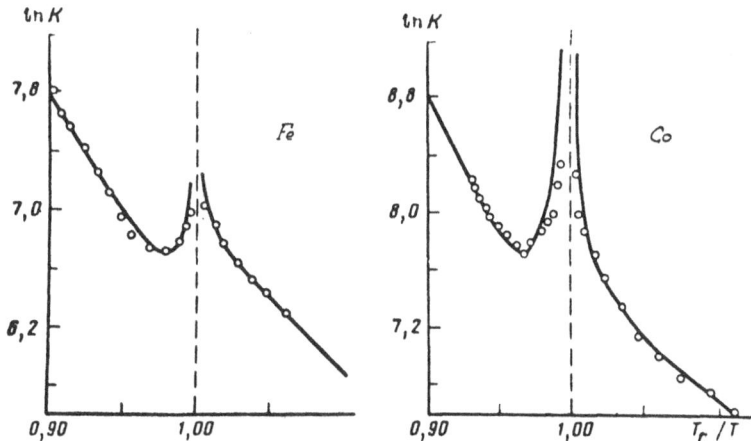

Figure 4. Rate of oxidation of Fe and Co, versus temperature. Points: experiment [11,13], curves theory [14].

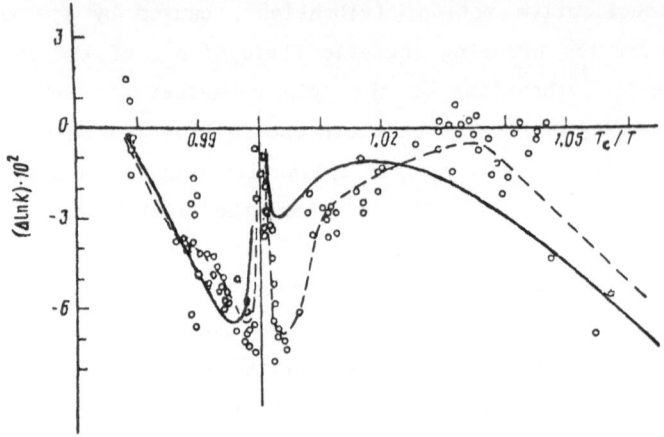

Figure 5. Change in the rate of Co sublimation, in the vicinity of Curie
temperature. Points and dashed curve, experiment [10]; solid
curve, theory [15].

exhibits single-atom height steps and narrow terraces, whereas in the
ferromagnetic phase two-atom height steps are observed for the Ni (111)
10° [1$\bar{1}$0] sample, and the Ni (111) 5° [1$\bar{1}$0] sample shows, on average, four
to five-atom height of steps and wider terraces.

A surface roughening transition of vicinal surface Ni (117) is
discussed in ref. [17]. Near the temperature of magnetic phase transition
of Ni, $T_c = 365\ ^{\circ}C$, a surface reconstruction was observed.

Recently, a computer simulation was used to describe the sputtering
of the (111) Ni face in both ferro and paramagnetic states, and different
types of upper-layer relaxation were found out [18]. Namely, a positive
relaxation, i.e., increasing of distances between upper layers proved to
occur for the paramagnetic phase (~ 4 % between the first and the second
layers), whereas for the ferromagnetic state the value of the relaxation
had an oscillating shape (~ 2.5 % between the first and the second layers,
and ~ - 1 % between the second and the third layers).

All the phenomena occurring near the Curie temperature, considered in
this section, may influence the character of the sputtering alteration
during the magnetic phase transition.

If emitted particles are charged, then, besides the processes
mentioned above, the change in electron structure must be significant as
the magnetic material passes through T_c. This question will be briefly
discussed in the following section.

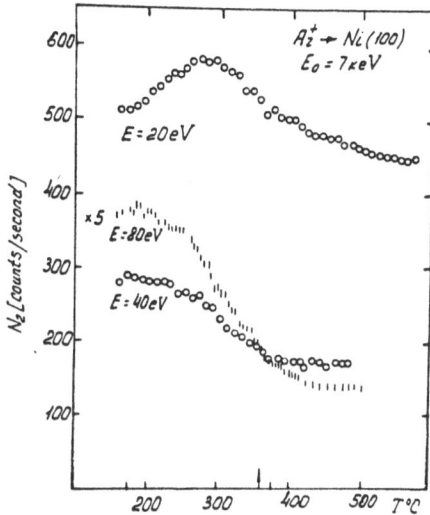

Figure 6. The variations of the 20 eV, 40 eV, and 80 eV Ni[+] secondary-ion emissions under the magnetic phase transition in Ni. For the (100) Ni face, under normal 7 keV Ar ion bombardment [19].

4. SECONDARY ION AND ION-PHOTON EMISSION

The data obtained for secondary Ni ion emission from the (100) Ni face, in the temperature range around the Curie point [19] are presented in fig.6. The shape of the curves is somewhat different for various energies of secondary Ni ions. A sharp decrease in the coefficient S^+ of secondary ion emission of Ni is observed, for high energies, in the paramagnetic state. For lower energies, a maximum emerges near the Curie temperature.

The variation of the secondary-ion emission during a phase transition, is determined by changes in the sputtering yield, binding energy, and ionization probability. As a rule, an increase in the sputtering yield is accompanied by decrease of the binding energy, and, as a result, by a decrease of the secondary-ion emission. However, if the sputtering yield variation is high, it will over-compensate for the opposite trend in the binding energy, and in the ionization probability. This is just the case for the S (T) behavior for Ni at T_c.

Ion scattering variation during the magnetic phase transition in Ni, in most cases, is of the same sign as the $S^+(T)$ variation, but is somewhat larger [8]. The decrease of ion scattering during the ferro-paramagnetic phase transition in Ni is due to the increased neutralization probability

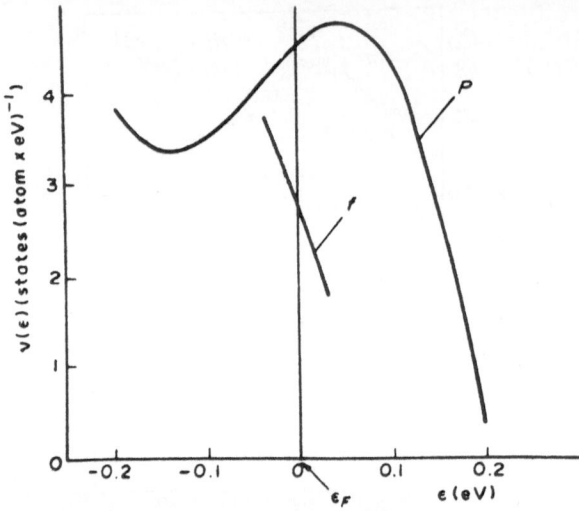

Figure 7. The variations of the electron density of states, $\nu(\varepsilon)$, near the Fermi level, ε_F, under magnetic phase transition in Ni. (From ref.[20]).

resulting from the increased density of electron states near the Fermi level [20], see fig.7.

A variation of ion-photon emission from Ni during the magnetic phase transition is observed. As a rule, the variation is of the same sign as for sputtering, but its value is lower [21,19]. It has been shown, in refs.[19,21], that the shape of the temperature dependence of ion-photon emission is very sensitive to the presence of residual oxygen in the sample chamber.

Ion-electron emission from Ni and Fe polycrystals has a minimum near the point of magnetic phase transition, and increases for the paramagnetic state [22,23]. As for the secondary-electron emission, a step increase was also observed in the Ni transition from ferro to paramagnetic state [24].

Thus, we have seen that magnetic phase transitions in metals is accompanied by various anomalous changes in surface processes, (oxidation, sublimation), and by a structure phase transition of the surface. So, ion-induced emission in the temperature range near the Curie point, proves to be a quite complicated phenomenon. In order to establish its features more clearly, and in more detail, further experimental research should be carried out, using a greater variety of investigated materials and of bombardment conditions.

REFERENCES

1. V.E. Yurasova, V.S. Chernysh, M.V. Kuvakin and L.B. Shelyakin, Pis'ma JETP (USSR) *21* (1975) 197.
2. V.I. Bachurin, E.S. Kharlamochkin, M.V. Kuvakin and V.E. Yurasova, Proc. of 15th ICPIG vol.*1* (1981) 465.
3. L.B. Shelyakin, T.P. Martynenko, A. Bischoff, V.E. Yurasova and G. Shaarschmidt, Poverkhnost' (USSR) N 6 (1983) 65.
4. V.E. Yurasova, Vacuum *33* (1983) 565.
5. N.E. Georgieva and T.P. Martynenko, Poverkhnost' (USSR) N 11, (1986) 112.
6. M.V. Kuvakin, Thesis, Moscow State University, 1975.
7. E.E. Karpova, Thesis, Moscow State University, 1984.
8. V.E. Yurasova, Vacuum *36* (1986) 609.
9. M.R. Shanabarger, Phys. Rev. Lett *43* (1979) 1964.
10. O.B. Khandros and N.A. Bogolyubov, Fiz. Tverd. Tela. (USSR) *14* (1972) 18; Teplofiz. Vysok. Temp. (USSR) *21* (1983) 600.
11. J.C. Measor and K.K. Afzulpurkar, Phil. Mag. *10* (1964) 817.
12. B. Delmon and A. Roman, J.C.S. Faraday I *69* (1973) 941.
13. B. Chatterdjee, Thin Sol. Films *41* (1977) 227.
14. V.D. Borman, A.N. Pivovarov and V.I. Troyan, JETF (USSR) *95* (1989) 1796.
15. V.D. Borman, A.N. Pivovarov and V.I. Troyan, Poverkhnost' (USSR) N *6* (1989) 13.
16. J.C. Hamilton and T. Jach, Phys. Rev. Lett. *46* (1981) 745.
17. D.L. Blanchard, D.F. Thomas, H. Xu and T. Engel, Surface Science *222* (1989) 477.
18. O.I. Ivanenko, A.S. Mosunov and M.V. Kuvakin, Izv.AN SSSR, Ser. fiz. *57* (1990) 1274.
19. V.A. Abramenko, A.A. Andreev and V.E. Yurasova, Nucl. Inst. Meth. *B13* (1986) 609.
20. B.K. Ponomarev and V.G. Tissen, JETP (USSR) *73* (1977) 332.
21. R. Pedrys and L. Gabla, J. Appl. Phys. *A32* (1983) 205.
22. I.N. Evdokimov and V.A. Molchanov, Izv. AN SSSR, Ser.fiz. *33* (1969) 762.
23. W. Soszka and T. Stepin, Nucl. Instr. Meth. *183* (1981) 163.
24. M.A. Vasiliev and S.D. Gorodetsky, Phys. Stat. Sol. *A102* (1987) k121.

CONTRIBUTED PAPERS: ION STOPPING

ACCURATE QUANTUM MECHANICAL CALCULATION OF STOPPING POWERS
FOR INTERMEDIATE ENERGY LIGHT IONS PENETRATING ATOMIC H AND He TARGETS

G. Schiwietz

Hahn-Meitner-Institut, Berlin GmbH, Dpt. P3
Glienicker Str. 100, D-1000 Berlin 39
Federal Republic of Germany

ABSTRACT

A single-center coupled-channel code, based on an expansion in terms of atomic wavefunctions, which includes dynamically-curved projectile trajectories, is applied to the calculation of stopping powers. Stopping powers and mean energy transfers are evaluated for protons, anti-protons, He^{2+} and Li^{3+} ions penetrating atomic H and He targets, at energies of 10 keV/u to 500 keV/u. The results are compared to experimental data, to predictions of the first-order plane-wave Born approximation, and to results of a calculation for excitation of a harmonic oscillator including Barkas corrections. The performance of the present model as compared to first-order or second-order perturbative treatments, as well as the effect of polarization on the projectile trajectories, is discussed.

1. INTRODUCTION

The interaction of ions with dense matter has been investigated for decades [1-3]. Classical [1,2] as well as quantum mechanical [3] theories have been applied to the calculation of electronic stopping powers. At low energy ion-atom collisions, the stopping power is dominated by electron capture and electron loss of projectile electrons [2]. At high incident energies, excitation and ionization of target electrons are the most important energy transfer mechanisms [3]. This was confirmed in numerous comparisons between experimental and theoretical stopping-power values [4]. However, it was shown recently [5] that exact classical three-body calculations are unable to predict cross sections for low-velocity ion-atom collisions where quasi-molecular effects come into play. For high incident energies [5] and low momentum transfers [6], classical collision theories tend to underestimate the dipole contribution of excitation and ionization cross sections. Quantal [3] and classical [6] models, based on first-order perturbation theory, predict all cross sections to be strictly proportional to Z_p^2, the squared projectile charge.

Interaction of Charged Particles with Solids and Surfaces
Edited by A. Gras-Martí *et al.*, Plenum Press, New York, 1991

517

However, Barkas and co-workers [7] found a difference between the ranges of π^+ and π^- particles, an indication for a contribution proportional to Z_p^3. Thus, there is a need for quantum-mechanical collision theories, which go beyond the perturbation limit. The aim of this work is to describe an improved quantal collision theory, applicable to the calculation of stopping powers for intermediate, (about 50 keV/u), and high incident energies. At high incident energies, ($\gg Z_p \times 300$ keV/u), the results are identical to those of the Bethe theory [8].

2. THEORY

The theoretical formulation of atomic excitation and ionization processes is conveniently discussed by introducing the quantum mechanical Hamilton operator. (Atomic units are used throughout this section). For one active electron, in the field of two nuclei and some spectator electrons, the Hamiltonian reads

$$H_e(t) = H_{te} + V_{pe}\left[\vec{R}(t), \vec{r}\right], \tag{1a}$$

$$V_{pe}\left[\vec{R}(t), \vec{r}\right] = -\frac{Z_p}{\left|\vec{R}(t) - \vec{r}\right|}, \tag{1b}$$

$$H_{te} \approx -V_t(\vec{r}) + T_e(\vec{r}), \tag{1c}$$

with the kinetic and potential energies denoted T and V. The subscripts p, t and e refer to the projectile ion, target core and electron. In the case of nonhydrogenic targets, the operator V_t is taken to be a single zeta Hartree-Fock potential, as defined by Clementi and Roetti [9]. In the case of hydrogen-like targets, V_t reduces to the pure Coulomb potential Z_t/r, for a target charge Z_t. The use of simple Hartree-Fock potentials excludes already initial and final-state electron correlation. Consequently, we adopt the independent-electron model, as described, e.g., by McGuire and Weaver [10], and exclude also dynamic correlation effects during the collision, as well as static Pauli correlation for intermediate excited states. It is noted that correlation effects and/or mean-field effects may become important for the excitation and ionization mechanisms, if multi-electron transitions come into play. However, the situation simplifies considerably in the case of H targets, where we have a well-defined target potential and only one active electron.

Before the solution based on the Hamiltonian in eq.(1) is explained in more detail, the classical path $\vec{R}(t)$ should be defined. Given the time-dependent electronic wave-function Φ_e, a classical Hamiltonian for the heavy particles may be defined

$$H_h = T_p(r_p) + T_t(r_t) + V_{pt}(R) + \langle \Phi_e | V_{pe}(R - r) | \Phi_e \rangle$$

$$+ \langle \Phi_e | V_{te}(r) | \Phi_e \rangle, \qquad (2)$$

with

$$V_{pt}(R) = Z_p V_t(R). \qquad (2a)$$

This equation is solved approximately by applying Newton's laws of motion. The last term in eq.(2) was neglected, because of its small influence on the motion of the target core, in case of a strongly target-centered wavefunction Φ_e. It is emphasized that the concept defined by eq.(2) introduces, for the first time, a dynamically curved projectile trajectory in the impact-parameter method. Thus, the projectile motion is coupled to the motion of the electrons.

Most of the previous calculations have been performed for straight-line paths, as given by the first two terms in eq.(2). Such calculations are equivalent to quantum-mechanical solutions of the three-body Schrödinger equation with plane projectile waves. Typical examples for such quantum-mechanical three-body theories are the plane-wave Born approximation [11], and its limiting form at high incident energies, the Bethe theory [8]. However, the main advantage of the present model, compared to previous stopping-power theories, is the exact description of the electronic motion at small internuclear distances. The corresponding calculation of excitation and ionization probabilities will be explained in the following paragraphs.

In this work we will use a coupled-channel theory, which allows for an infinite number of interactions between projectile, target and electron. The electron may be ionized in a first step, and may be accelerated or decelerated in a second step. It is also possible that an electron, after being ionized, is "thrown" back to the initial state. Furthermore, the probability for ionizing an electron is always less or equal unity. All this does not hold for perturbation theory.

The starting point of the present theory is an expansion of the time-dependent electronic wave-function Φ_e in terms of single-center eigenfunctions φ_i of the target Hamiltonian H_{te}

$$\Phi_e(\vec{r}, t) = \Phi_B(\vec{r}, t) + \Phi_C(\vec{r}, t), \qquad (3)$$

$$\Phi_B(\vec{r},t) = \sum_{n,1,m} a_{n,1,m}(t)\, e^{-iE_{n,1}t}\, \varphi_{n,1,m}(\vec{r}),\tag{3a}$$

$$\Phi_C(\vec{r},t) = \sum_{1,m} \int_0^\infty d\varepsilon\, b_{1,m}(\varepsilon,t)\, e^{-i\varepsilon t}\, \varphi_{\varepsilon,1,m}(\vec{r}).\tag{3b}$$

The eigenfunctions φ_i are solutions of the stationary Schrödinger equation, and $E_{n,1}$ are the corresponding eigenvalues. The radial wave-functions are calculated numerically using a third-order Runge-Kutta method, with variable step width, and the numerical uncertainty of the bound-state eigenvalues $E_{n,1}$ is about 10^{-6} eV. Partial waves up to orbital angular momenta of $1 = 8$, are considered for the continuum states. A problem arises, since the continuous energy variable of the free-wave functions is not easy to handle in a numerical calculation. Therefore, the continuum is represented by a sum over a few, (about 6, for each orbital angular momentum in the present work), pseudo-discrete radial wave-functions $\Psi_{1,m}$

$$\Phi_C(\vec{r},t) = \sum_{j,1,m} \frac{1}{r}\, \Psi_{1,m}\, (\varepsilon_j - \Delta\varepsilon/2, \varepsilon_j + \Delta\varepsilon/2, r, t)\, Y_{1,m}(\Theta,\phi),\tag{4}$$

with

$$\Psi_{1,m}(E_1,E_2,r,t) = \int_{E_1}^{E_2} d\varepsilon\, b_{1,m}(\varepsilon,t)\, e^{-i\varepsilon t}\, u_{\varepsilon,1}(r),\tag{4a}$$

$$\approx \frac{\bar{a}_{1,m}(\bar{\varepsilon},t)}{\sqrt{E_2 - E_1}}\, e^{-i\bar{\varepsilon}t}\, F(E_2 - E_1, t) \int_{E_1}^{E_2} d\varepsilon\, u_{\varepsilon,1}(r),\tag{4b}$$

and

$$F(\Delta E, t) = \begin{cases} \dfrac{2}{\Delta E \cdot t}\, \sin\left(\dfrac{\Delta E \cdot t}{2}\right); & \text{for continuum states and } t > 0, \\[2ex] 1 & ; \text{ otherwise.} \end{cases}\tag{4c}$$

This description is exact for small t, or small values of r, or if an infinite number of wave packets is considered. From the structure of Coulomb wave-functions, it is obvious that transition matrix-elements involving either a high Rydberg state or a low-energy continuum state, are identical when normalized per square root of energy. Thus, approximate completeness of the basis set may be achieved, although an infinite number of high-lying Rydberg states is not explicitly considered in Φ_e. This is done by renormalizing the energy width, $(E_1 - E_0)$, of the lowest energy continuum states, so that the corresponding wave packets include a contribution of high Rydberg states.

In the present work, the time-dependent Schrödinger equation is solved without any further approximations. Thus, probabilities and cross sections computed with this method should be highly accurate, as long as electron capture by the projectile is of minor importance. This corresponds to incident energies down to the maximum of the stopping-power curve (> 50 keV). However, reasonable convergence was achieved down to about 10 keV/u, by choosing appropriate basis sets to simulate the electron capture.

The solution of the only unknown variables, namely the coefficients $a_{n,l,m}$, is straightforward. The definition of Φ_e, eqs. (3,4), for small t, is inserted into the time-dependent Schrödinger equation. The resulting equation is multiplied with an arbitrary eigenfunction φ_f^* and integrated over the $\vec{r}$ coordinate. Furthermore, orthogonality and the eigenvalue equation are considered. Consequently, an infinite system of coupled first-order differential equations, (coupled-channel equations) defines the time evolution of $a_{n,l,m}$

$$i \frac{d}{dt} a_{j',l',m'}(t) =$$

$$= \sum_{j,l,m} a_{j,l,m}(t)\, e^{i(E_{j',l'} - E_{j,l})t}\, V_{pe}^{j,l,m \to j',l',m} \{\vec{R}(t)\}, \qquad (5)$$

with

$$V_{pe}^{i \to f} \{\vec{R}(t)\} = \langle \varphi_f | V_{pe}\{\vec{R}(t),\vec{r}\} | \varphi_i \rangle, \qquad (5a)$$

and

$$\lim_{t \to -\infty} a_{n,l,m}(t) = \delta_{1s_0, nl_m},$$

for H and He ground-state targets.

The interaction potential V_{pe} is replaced by its multipole expansion [12,13]. Thus, the Coulomb matrix elements $V_{pe}^{i \to f}$ are reduced to finite sums over weighted radial integrals, the so-called form factors. The form factors are computed numerically for all non-zero multipole terms. The weight factors are determined by Wigner 3j-symbols. For transitions from a continuum state to any state, the damping factor F, eq. (4c), was included. For those transition matrix elements where the final state is a continuum wave packet, the corresponding damping factor was set unity. This improves the coupled-channel results for small projectile charges. On the other hand the coefficients $a_{j,1,m}$ have to be unitarized after each integration step. It is noted that this unitarization affects the transition probabilities by about 5 % in the worst case considered, for incident protons (10 keV H^+ + H, He).

The Coulomb interaction does not lead to transitions between the, so-called, gerade and ungerade states. These states are linear combinations of two eigenstates with opposite angular-momentum projection numbers. In this paper, the initial state will be always 1s, which is gerade. Thus, the coupled-channel equations are solved in the present work for about 90 gerade states, (10 bound states and 80 continuum states), which replace about 150 eigenstates. These gerade states are chosen to yield optimum convergence for a certain regime of incident energies. The corresponding excitation and ionization probabilities are given by projection on asymptotic target-centered states,

$$P_{n,1,m}(b) = \lim_{t \to +\infty} \left| \langle \varphi_{n,1,m} | \Phi_e(b,t) \rangle \right|^2 = \lim_{t \to +\infty} \left| a_{n,1,m}(b,t) \right|^2. \tag{6}$$

If the sum in eq.(5) includes only the initial state φ_i, and the corresponding coefficient a_i is set equal to one, independent of time, the electronic motion is described in a perturbative way. If, furthermore, the last three terms in eq.(2) are neglected, the treatment would be reduced to the first-order Semi-Classical Approximation, SCA, with straight-line trajectories. This model yields the same cross sections as the first-order Plane-Wave Born Approximation, PWBA [11]. The corresponding range of validity is $v_p/Z_p \gg 3$ [14]. Since the Bethe theory [8], which is also a first-order theory, corresponds to the PWBA only in the limit of high projectile velocities, its range of validity is shifted to even higher incident energies. It is noted, that the present computer code may be restricted to perturbation theory (SCA-mode).

The nuclear energy transfer is determined from the solution of eq.(2), and the mean electronic energy transfer is obtained by summing over all terms $P_{n,1,m}(b) \Delta E_{n,1}$. The stopping power is obtained from the impact-parameter integration over the total energy transfer. It should be emphasized that the stopping powers given in this work are dominated by the electronic energy loss S_e ($S_e > 0.98$ S).

3. RESULTS AND DISCUSSION

Fig.1 displays SCA and coupled-channel (AO) results for stopping

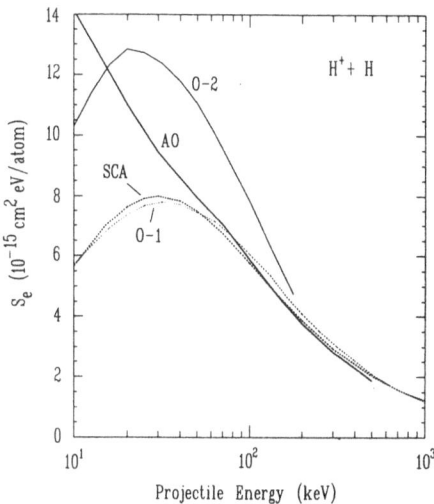

Figure 1. Theoretical electronic stopping power values for the H^+ charge-state fraction in collisions with hydrogen atoms. Present theory, full calculation (AO): thick solid line; restricted to first-order perturbation theory (SCA): thick dotted line. The results for an H target represented by an harmonic oscillator are taken from ref.[15]; first-order perturbation theory (O-1): dotted line; second-order perturbation theory (O-2): solid line.

powers of the H^+ charge-state fraction, in an atomic H target, calculated with the present code. These values are compared to first-order (O-1) and second-order (O-2) results for an H_2 model molecule described by a set of harmonic oscillators [15]. There is good agreement between both first-order results. It is noted that the molecular-binding correction, as included in ref.[15], would reduce the SCA stopping powers by about 3 %. However, there are significant discrepancies between the second-order (O-2) and the highest-order (AO) results. This is an indication of the importance of higher-order contributions to the stopping power. These higher-order contributions correspond to a z_p^n behavior, with n > 3, of the stopping power. It is thus interesting to investigate the projectile-charge dependence of the stopping power within the AO model, which includes all terms in Z_p.

Fig.2 displays the ratio of stopping powers calculated with the AO model, to the corresponding SCA values for protons, anti-protons, alpha particles and Li^{3+} ions. SCA, PWBA, as well as the Bethe approximation, do scale with the squared projectile charge. Hence, any deviation of the ratio from unity corresponds to a deviation from the z_p^2 law. First, it is noted that the anti-proton data are not equal to $1-R(H^+)$, where $R(H^+)$ is the ratio for protons. Also, the proton data below 200 keV correspond to

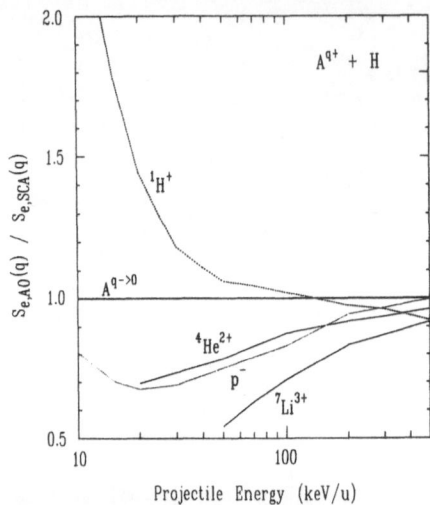

Figure 2. Ratio of stopping powers, as a function of the incident energy,
calculated with the present model, to the first-order results
computed with an identical set of wave-functions, for four
different bare ions penetrating an atomic hydrogen target.
Results for projectiles with hypothetical charge-states small
compared to one, fall onto the thick solid line for the ratio of
one.

ratios larger than unity, whereas R is smaller than one for the heavier
positively charged ions. Both facts clearly show that the Barkas term,
proportional to z_p^3, is not sufficient to describe the stopping powers
accurately. It is evident that, when there is a high probability for an
electron to undergo two-step transitions in a single collision, three-step
or multi-step processes might be also of importance.

The physical reason for the enhanced proton stopping power, compared
to the other projectiles, is the possibility of resonant electron-capture
processes, in addition to excitation and ionization of the target atom. In
the case of heavier positive ions there is a reduced capture probability,
since the collision systems are asymmetric with respect to the nuclear
charges. The increased binding due to the presence of a positively charged
particle, leads to a further reduction of the stopping power. In the case
of negative projectiles the target atom is polarized, so that there is a
reduced electron density on the projectile path and thus a reduced
stopping power. This is in accordance with the findings of Barkas and
co-workers, for π^+ and π^- particles [7]. In order to distinguish the
influence of polarization, electron capture or binding effect on the
stopping power, it is interesting to investigate the impact-parameter
dependence of the mean energy transfer in $H^+ + H$ collisions.

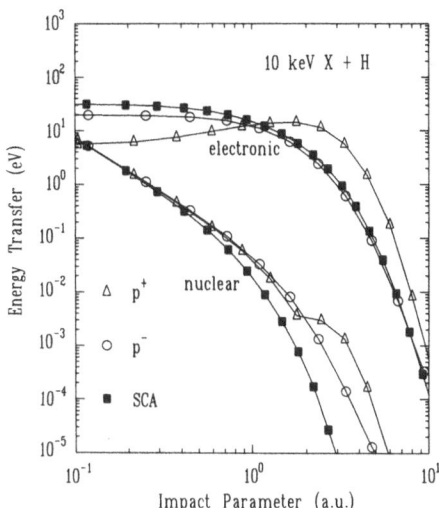

Figure 3. Theoretical results for the mean nuclear and electronic energy
transfer in 10 keV X^{q+} + H collisions, as a function of the
impact parameter. Present theory, for protons: triangles; for
anti-protons: circles; and restricted to first order
perturbation theory: solid squares.

In fig.3, theoretical values of the mean electronic and nuclear
energy transfer in 10 keV p^-, H^+ + H collisions are plotted versus the
impact parameter. As it may be inferred from two-center
calculations [5,16], at large impact parameters electron capture dominates
the mean electronic energy transfer for protons. This leads to a different
impact-parameter dependence of the electronic energy transfer for protons
and anti-protons. However, there are also significant deviations for the
nuclear energy loss. These deviations do result from the *dynamically*
curved projectile trajectories, as defined by eqs.(2) and (2a). The SCA
data correspond to scattering of a projectile in a *static* screened target
potential. The AO results for protons or anti-protons are strongly
influenced by polarization effects. In the case of anti-protons, the
electron density distribution is shifted away from the projectile path.
Thus, even at large impact parameters, the projectile interacts with a
partially unscreened target system. This leads to an enhanced nuclear
stopping, compared to the SCA results. In the case of protons at large
impact parameters (> 3 a.u.), the electron cloud is attracted by the
positively charged particle. Subsequently, there is a negative target
net-charge, and the projectile is deflected *towards* the target nucleus.
For small impact parameters the situation on the incoming and outgoing
path is similar, but for small internuclear distances both nuclei interact
with each other, independent of the electronic density distribution.

Because the target system is pulled by the projectile at large distances, and pushed away at small distances, both forces cancel partially, leading to a reduced nuclear stopping. The broad structure in the nuclear energy-transfer curve for protons at 3 a.u., corresponds to a zero crossing of the mean projectile scattering angle.

Finally, in fig.4, total model stopping-power values for an atomic hydrogen target are compared with experimental data of different groups [17-21], for H + H_2. It is noted that the experimental data were taken for thick gas targets, and correspond to an equilibrated charge-state

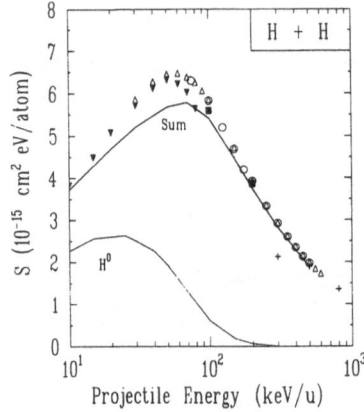

Figure 4. Equilibrium mean total stopping-power per atom, for H + H
         collisions. Exact PWBA results for single excitation and
         ionization in H^0 + H^0 collisions times the corresponding
         charge-state fraction: thin solid line. Sum of the present
         coupled-channel results, weighted with the H^+ charged-state
         fraction (taken from ref.[24]) and the H^0 contribution: thick
         solid line. Experimental data for H projectiles penetrating a
         thick H_2 target: ref.[17] (squares), ref.[18] (circles),
         ref.[19] (crosses), ref.[20] (open triangles) and ref.[21]
         (solid triangles).

distribution. Thus, the present AO results were weighted with the H^+ charge-state fraction, and summed up with the contribution from the reaction $H^0 + H^0 \rightarrow H^0 + H^+ + e$. The latter process involves a neutral collision system, and corresponds to a relatively small perturbation. Hence, excitation and ionization cross sections for this process were calculated within the PWBA, as described by Bates and Griffing [11].

Except for intermediate incident energies, there is agreement between experimental and theoretical data to within 5 % or better. At energies around 50 keV, the model predictions underestimate the experimental data by 10 to 15 %. Using PWBA calculations, it was checked that the molecular character of the H_2 targets will affect the stopping power values by less than 5 % at intermediate energies. Most likely, the deviation between experimental and theoretical results corresponds to underestimated ionization cross sections for $H^0 + H^0$ collisions. Electron correlation might lead to simultaneous ionization of both neutral collision partners, and may increase ionization cross sections by up to a factor of two [22]. This could enhance the stopping powers by 10 to 15 % at incident energies around 50 keV.

4. CONCLUSIONS

An improved stopping-power theory, with exact atomic target-centered wave functions for bound states and damped continuum wave-packets, is introduced in this work [23]. For the first time a coupled-channel theory is applied to the calculation of stopping powers. Dynamic curved projectile trajectories are introduced, which are sensitive to the time dependent electron density distribution [23].

Results from the present theory, agree reasonably well with experimental data for stopping powers for protons in thick H and He targets [23]. It is shown that protons suffer an exceptionally high energy loss compared to anti-protons, alpha particles, and Li^{3+} ions, at incident energies below 100 keV/u. This may be viewed as part of a Z_p oscillation of the stopping power, often discussed in the literature. Clearly, a Barkas correction is not able to describe the deviations from first-order perturbation theory found in this work. Furthermore, it is evident that polarization of the electron-density distribution leads to drastic changes of the nuclear energy loss, especially at large impact parameters.

However, at intermediate energies, (30 to 150 keV/u), there is a 10 % contribution to the stopping power which is consistent with ionization via electron-electron interactions in $H^0 + H^0$ collisions. Thus, it appears necessary to include electron correlation in the model, to further increase the accuracy at intermediate incident energies. In conclusion, the major problems connected with the slowing down of bare ions in gas targets, seem to be solved.

REFERENCES

1. N. Bohr, Philos. Mag. 25 (1913) 10.
2. O.B. Firsov, Zh. Eksp. Teor. Fiz. 9 (1959) 1076.
3. H. Bethe, Ann. Phys. 5 (1930) 325.

4. J.F. Ziegler, J.P. Biersack and U. Littmark, *The Stopping and Range of Ions in Solids*, (ISBN 0-08-021603), Pergamom Press (1985).
5. G. Schiwietz and W. Fritsch, J. Phys. *B20* (1987) 5463; V. Montemayor and G. Schiwietz, J. Phys. *B22* (1989) 2555; V. Montemayor and G.Schiwietz, Phys. Rev. *A40* (1989) 6223.
6. L.G.J. Boesten, T.F.M. Bonsen and D. Banks, J. Phys. *B8* (1975) 628.
7. W.H. Barkas, W. Birnbaum and F.M. Smith, Phys. Rev. *101* (1956) 778; W.H. Barkas, J.N. Dyer and H.H. Heckman, Phys. Rev. Lett. *11* (1963) 138.
8. M. Inokuti, Rev. Mod. Phys. *43* (1971) 297.
9. E. Clementi and C. Roetti, Atomic Data and Nucl. Data Tables *14* (1974) 177.
10. J.H. McGuire and L. Weaver, Phys. Rev. *A16* (1977) 41.
11. D.R. Bates and G. Griffing, Proc. Phys. Soc. *A66* (1953) 961.
12. F. Rösel, D. Trautmann and G. Baur, Nucl. Instr. Meth. *192* (1982) 43.
13. T.G. Winter, Phys. Rev. *A25* (1982) 697.
14. D. Schneider, D. DeWitt, A.S. Schlachter, R.E. Olson, W.G. Graham, J.R. Mowat, R.D. DuBois, D.H. Loyd, V. Montemayor, and G. Schiwietz, Phys. Rev. *A40* (1989) 2971.
15. H.H. Mikkelsen and P. Sigmund, Phys. Rev. *A40* (1989) 101.
16. W. Fritsch and C.D. Lin, J. Phys. *B15* (1982) 1255; Phys. Rev. *A27* (1983) 3361.
17. E. Bonderup and P. Hvelplund, Phys. Rev. *A4* (1971) 562.
18. S.D. Warshaw, Phys. Rev. *76* (1949) 1759.
19. R.A. Langley, Phys. Rev. *B12* (1975) 3575.
20. H.K. Reynolds, D.N.F. Dunbar, W.A. Wenzel, and W. Whaling, Phys. Rev. *92* (1953) 742.
21. J.A. Phillips, Phys. Rev. *90* (1953) 532; (d and t projectiles were used for incident energies below 40 keV).
22. R. Shingal, B.H. Bransden, and D.R. Flower, J. Phys. *B22* (1989) 855.
23. G. Schiwietz, Phys. Rev. A42 (1990) 296.
24. S.K. Allison, Rev. Mod. Phys. *30* (1958) 1137.

CHARGE EXCHANGE AND ENERGY LOSS

OF PARTICLES INTERACTING WITH SURFACES

A. Närmann[1,2], W. Heiland[1], R. Monreal[3],

F. Flores[3] and P.M. Echenique[4]

1. Universität Osnabrück, Fachbereich Physik
   Barbarastr. 7, D-4500 Osnabrück, FRG

2. Present Address: IBM, T.J. Watson Research Center
   P.O.B. 218-1261, Yorktown Hts., N.Y. 10598, U.S.A.

3. Universidad Autónoma de Madrid, Dept. Materia Condensada
   Cantoblanco, E-28049 Madrid, Spain

4. Euskal Herriko Unibertsitatea, Kimika Fakultatea
   Apdo.1072, E-20080 San Sebastián, Spain

ABSTRACT

Experiments on energy dissipation of He-particles interacting with metal surfaces have been performed. The energy range used was 1 - 5 keV. The particles were scattered off a Ni(110) surface at grazing incident angle (5°) and detected under specular reflection condition with a time-of-flight system.

The resulting spectra of backscattered neutral particles show an energy loss of about 5 % of the incident energy and an asymmetric shape. These features can be accounted for by two models: 1.- The friction coefficient method describes the interaction via friction coefficients depending on the charge state of the particle. 2.- The convolution method is based on the assumption that each impinging particle excites a number of elementary excitations in the target.

In both cases energy straggling plays an important role.

1. INTRODUCTION

We investigated experimentally and theoretically the energy loss of He ions interacting with a Ni(110) surface. The experiments were performed under grazing incidence condition (incident angle 5°, scattering angle 10°) in the energy regime of 1 - 5 keV incident energy.

The reflected neutral particles were recorded with a time-of-flight system. The energy spectra obtained in this way show an asymmetric shape and an energy loss of about 5 % of the incident energy.

Interaction of Charged Particles with Solids and Surfaces
Edited by A. Gras-Martí *et al.*, Plenum Press, New York, 1991

529

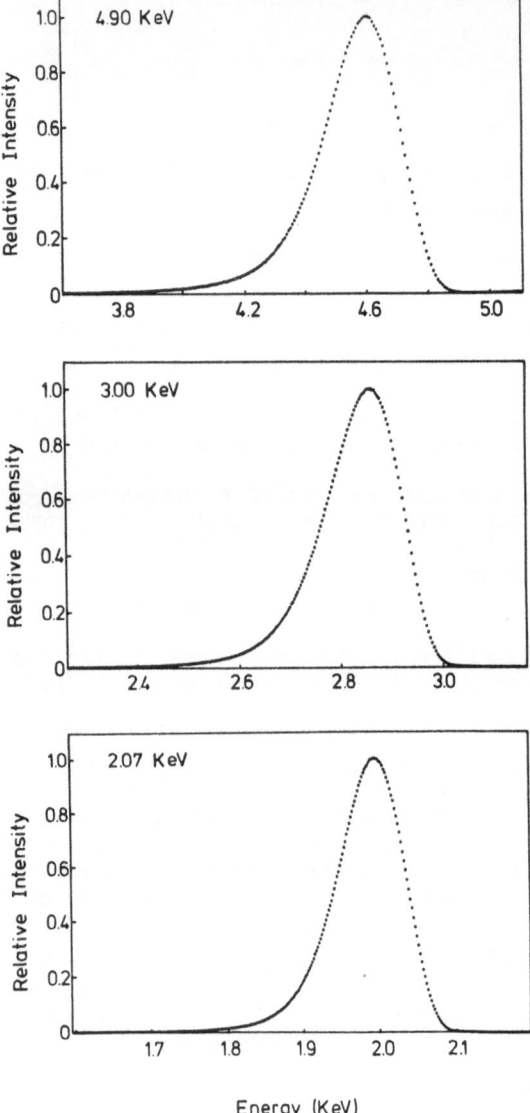

Figure 1. Energy spectra of reflected neutral particles for different incident energies in random direction. Incident angle of incoming He ions was 5°, scattering angle 10°.

We will explain these experimental results by two different methods: the friction coefficient method and the convolution method. For a detailed and very instructive introduction into the topic see the contributions of P.M. Echenique and F. Flores in these proceedings (and references therein).

2. FRICTION COEFFICIENT APPROACH

The energy spectra of reflected neutral He particles for three different incident energies are shown in fig.1. We will show that the shape of the curves, which is appreciably more asymmetric than expected in energy-loss experiments, can be accounted for by charge-exchange processes between the incoming ion and the target.

In the following we neglect the dynamic resonant-loss processes which have only a small contribution to the stopping power and account for Auger-neutralization only. Thus we write the rate equations for the ion and the neutral yield as

$$\frac{dn(He^+)}{dt} \simeq - \frac{1}{\tau^A} n(He^+), \tag{1a}$$

$$\frac{dn(He^0)}{dt} \simeq \frac{1}{\tau^A} n(He^+). \tag{1b}$$

Taking into account that the friction coefficient changes with the charge state of the He particle we write for the energy loss before and after neutralization, $Q_+(x) = \int_0^x \gamma_s^+ v \, ds$, and $Q_0(x) = \int_x^L \gamma_s^0 v \, ds$, so that the total energy loss Q is given by

$$Q = Q_+(x) + Q_0(x) = (\gamma_s^+ - \gamma_s^0) vx + \gamma_s^0 vL, \quad x \in [0,L], \tag{2}$$

where x is the free path for the particle being an ion and L the trajectory length (see fig.2 and table 1).

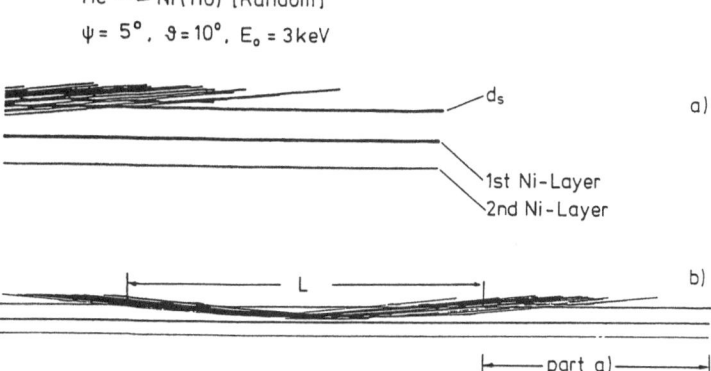

Figure 2. Sketch of how the trajectory length L was determined from the MARLOWE data [11,12]. The obtained values are listed in table 1.

Table 1. Trajectory lengths L (in Å) for different energies determined as indicated in fig. 2.

Energy	[110]	[001]	random
2.0 keV	– Å	– Å	26.5 Å
3.0 keV	54.5 Å	61.0 Å	32.4 Å
4.0 keV	59.7 Å	64.7 Å	43.0 Å

The second term in eq. (2) is obtained for $x = 0$, i.e., it corresponds to those particles which are neutral during the whole trajectory. Thus the first term gives the additional energy lost by the particle when being an ion with $(\gamma_s^+ - \gamma_s^o)$ as an effective friction coefficient. We define $Q_o = Q_o(0)$ as the energy loss of a neutral particle.

Solving eq. (1a) and inserting into eq. (1b) yields $n(He^o) = 1 - \exp(-x/\lambda_A)$. Transforming variables, see eq. (2), by means of $x = (Q-Q_o)/\{(\gamma_s^+ - \gamma_s^o)v\}$, and calculating the derivative with respect to Q, yields the energy-loss spectrum for the reflected neutral particles

$$\frac{dn(He^o)}{dQ} \propto \exp\left(-\frac{Q - Q_o}{(\gamma_s^+ - \gamma_s^o)v\lambda_A}\right) \Theta(Q - Q_o). \tag{3}$$

In fig. 3 we compare the experimental data to the calculated ones. It is important to note that we obtain an asymmetric shape of the distribution, by considering the change of the charge state only. But the position of the maximum is shifted with respect to the experiment.

In eq. (3) the straggling of the energy loss [1] has been neglected. We include the straggling by convoluting the right hand side of eq. (3) with the straggling distribution function $\exp\{-(Q' - Q)^2/(2\Omega^2)\}$ which yields

$$\frac{dn(He^o)}{dQ} \propto \exp\left(-\frac{Q - Q_o}{(\gamma_s^+ - \gamma_s^o)v\lambda_A}\right) \left\{1 + \frac{2}{\sqrt{\pi}}\int_o^y \exp\left(-t^2\right) dt\right\}, \tag{4}$$

where $y = \frac{1}{\sqrt{2}}\left(\frac{Q - Q_o}{\Omega} - \frac{\Omega}{(\gamma_s^+ - \gamma_s^o)v\lambda_A}\right)$.

The calculated spectrum depends on the following parameters: τ^A, γ_s^+, γ_s^o and the straggling parameter Ω^2. The Auger lifetime τ^A has been calculated in ref. [2]; the ratio γ_s^+/γ_s^o has been calculated in linear theory for an

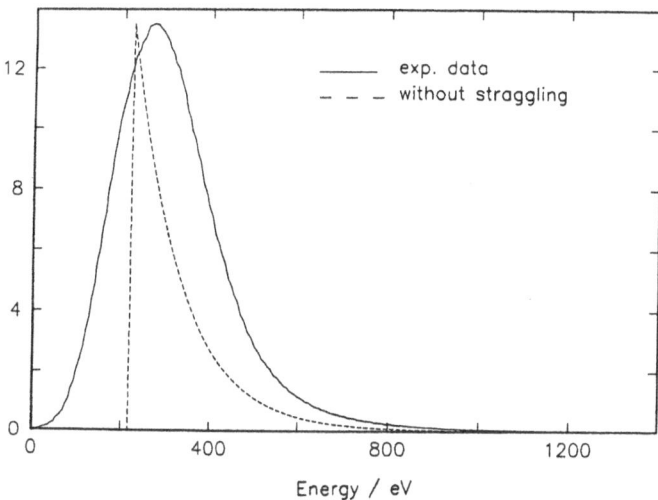

Figure 3. Comparison of the experimental energy loss with eq.(3) for the 5 keV case.

electron gas as a function of the electron density [3], and the results are given in table 2.

We find that a good fit to the experimental data is obtained by taking $\gamma_s^+/\gamma_s^o = 3.7$, corresponding to an r_s between 1.5 and 2, a density parameter appropriate for Ni. Then γ_s^o and Ω^2 have been chosen for each energy in such a way that eq.(4) gives the best fit to our experimental data. The results are shown in fig.4.

The inclusion of the straggling of the energy loss leads to good agreement between the theoretical and experimental curves. Only when Q_o and Ω are of similar magnitude - as is the case for 2 keV incident energy - do we find a slight disagreement for energies close to the primary energies.

Table 3 shows the values for Q_o, Ω and γ_s^o. We have also calculated Q_o and Ω using a local-density approach by means of

$$Q_o = \int_{-\infty}^{\infty} \gamma_s^o(1)\ v\ dl, \qquad \Omega^2 = \int_{-\infty}^{\infty} W(1)\ dl, \tag{5}$$

where γ_s^o (fig.3 in ref.[4]) and W (fig.3 in ref.[5]) were obtained locally assuming that they take in each point the values associated with the corresponding electronic local density. The results are included in table 3.

It is important to note that the energy loss experienced by the ions

Table 2. Ratio of the friction coefficients for He^{+} and He^{o}, as a function of the "one electron radius" r_s.

r_s/a.u.	1	2	3	4
$\gamma_s^{+}/\gamma_s^{o}$	1.64	4.45	11.95	26.4

before entering (and after emerging) the length L, i.e. the energy loss before (and after) penetrating the interaction region, is a small contribution to the total energy loss [6]. We estimate such a contribution to the loss to be 3 eV for an incident energy of 5 keV.

We mention that another source of straggling is due to the distribution of trajectory lengths around a mean value L. This effect can also be taken into account as a Gaussian broadening of the energy spectra [7] and will be more important for lower velocities, i.e. shorter trajectory lengths; this fact explains why in table 3 the fitted Ω values are larger than the theoretical ones.

It is also worth mentioning that $Q_o = \gamma_s^{o}vL$ changes with the particle energy due not only to v, but also to γ_s^{o} and L. In the energy range considered here, γ_s^{o} (see table 3) and L (see table 1) are roughly linear in v, suggesting a v^3 dependence of Q_o, a behavior that seems to be followed by the experimental data of table 4.

3. CONVOLUTION METHOD

Up to now, we have followed a quite simple approach by describing straggling effects by a Gaussian function. Since the particles may even gain energy from the target one might doubt the correctness of this mean statistical approach. We will demonstrate that the experimental data can also be explained using the so-called "convolution method" [8], thereby showing the validity of the previous method [9]. We introduce this method - adapted to our needs - in the following way. First, we assume that the particle loses a fixed amount of energy, say $\hbar\bar{\omega}$, by creating an excitation in the solid. In the following we will call such an event a "collision". Referring now to the notations introduced in fig.5, we are looking for an expression which gives us the probability P(n) of having n collisions. From this we could calculate the energy spectrum simply by identifying n with the corresponding energy loss $n\hbar\bar{\omega}$.

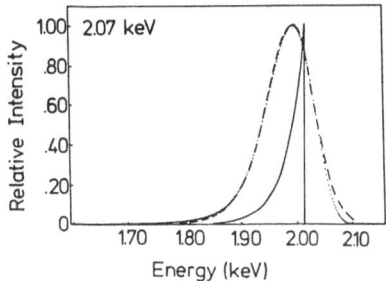

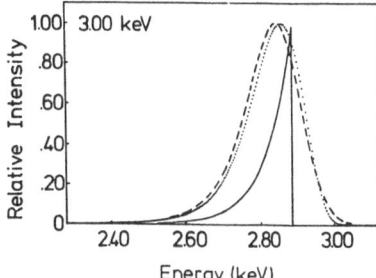

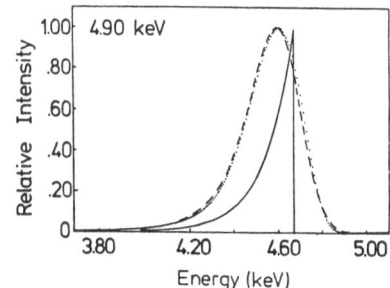

Figure 4. Energy-loss spectra for emerging He^o with incident He$^+$ scattered off Ni(110) for three different energies (angle of incidence 5°, scattering angle 10°). ($\cdot$): experimental results, (---): theoretical result including straggling and (——): theoretical result excluding straggling.

The probability of suffering n_+ collisions in the interval $[0,x]$, i.e., when being an ion, is given by $\dfrac{\exp(-x/\lambda_x)}{n_+!}\left(\dfrac{x}{\lambda_+}\right)^{n_+}$. This is the well-known Poisson law which holds in cases where large numbers of trials (i.e., collisions in our case) are involved with a small probability of occurrence each. With the same argument we get the probability that the then neutral particle suffers n_o collisions in the interval $[x,L]$

$$\frac{\exp\{-(L-x)/\lambda_o\}}{n_o!}\left(\frac{L-x}{\lambda_o}\right)^{n_o}.$$

Taking into account that the probability for the particle being neutralized in the interval $[x, x + dx]$ is given by $(dx/\lambda_A)\exp(-x/\lambda_A)$, where $\lambda_A = v\tau^A$ is the mean free path for the particle being an ion, we have now an expression which gives us the probability for the particle having n_+ collisions as an ion and n_o collisions as a neutral particle

$$P(n_+,n_o) = \int_o^L \frac{dx}{\lambda_A}\exp\left(-\frac{x}{\lambda_A}\right)\frac{\exp(-x/\lambda_+)}{n_+!}\left(\frac{x}{\lambda_+}\right)^{n_+}$$

$$\times \; \frac{\exp\{-(L - x)/\lambda_o\}}{n_o!} \left(\frac{(L - x)}{\lambda_o}\right)^{n_o}. \qquad (6)$$

From eq.(6) we then calculate the probability that the particle suffers n collisions during its interaction with the surface,

$$P(n) = \sum_{n_+=0}^{n} P(n_+, n - n_+).$$

We arrived at a discrete energy-loss spectrum $P(\omega)$ where ω only takes the values $\omega = n\bar{\omega}$. In one collision the particle actually does not lose a fixed amount of energy $\hbar\bar{\omega}$ but loses a value $\hbar\omega$ which is distributed around the mean energy loss $\hbar\bar{\omega}$ according to the distribution function $f(\omega)$. Similarly the energy loss for a particle which has suffered m collisions is distributed around the mean energy loss $m\hbar\bar{\omega}$ with the distribution function $f_m(\omega)$ given by the m-fold convolution of the distribution function for a single collision

$$f_m(\omega) = \int_o^\omega d\omega' \; f(\omega') \; f_{m-1}(\omega - \omega'),$$

with $f_o(\omega) = \delta(\omega)$ and $f_1(\omega) = f(\omega)$. In our case, for low-velocity particles, this single spectrum does not present strong asymmetries since tails in the spectrum going like $1/\omega^2$ do not appear (see chapter 5.2.4 of ref.[6]). Because of this, we describe $f(\omega)$ with good accuracy by a Gaussian, and accordingly f_m by the m-fold convolution of f

$$f_m(\omega) = \left(\frac{2}{\pi \; m\omega_o^2}\right)^{1/2} \exp\left(-\frac{(\omega - m\bar{\omega})^2}{2m\omega_o^2}\right),$$

where ω_o^2 defines the intrinsic straggling of the probability density

Table 3. Parameters used to fit the energy spectra, (marked with a superscript *fit*), and corresponding values from a Local Density Approximation.

Energy	Ω_o^{fit}	Ω_o^{LDA}	Ω^{fit}	Ω^{LDA}	γ_s^o
2.07 keV	49 eV	57.6 eV	38 eV	25.3 eV	0.26 a.u.
3.00 keV	102 eV	96.7 eV	54 eV	39.6 eV	0.37 a.u.
4.90 keV	217 eV	199.4 eV	84 eV	83.6 eV	0.44 a.u.

Table 4. Experimental and calculated values for the energy loss.

Energy	ΔE_{exp}	ΔE_{calc}	$\Delta E_{exp}/v^3$
2.07 keV	75 eV	80 eV	25.2
3.00 keV	136 eV	150 eV	26.1
4.90 keV	287 eV	280 eV	26.5

function $f(\omega)$. With this distribution we obtain the now continuous energy-loss spectrum

$$P(\omega) = \sum_{m=0}^{\infty} P(m) \, f_m(\omega).$$

In fig.6 we show the case of 5 keV, different curves corresponding to the experimental result and to the two different theoretical approaches discussed above.

The relations between stopping power S and mean energy loss per collision $\hbar\bar\omega$ on the one hand and between straggling per unit length W and intrinsic straggling $\hbar\bar\omega_o$ on the other are given by

$$S = \hbar\bar\omega/\lambda,$$

$$W = \left[(\hbar\bar\omega)^2 + (\hbar\omega_o)^2\right]/\lambda,$$

where λ_+ or λ_o is to be substituted for λ depending on the charge state of

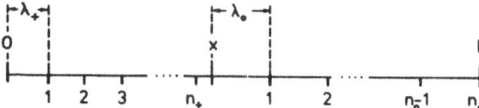

Figure 5. Sketch of an incoming ion's path near a surface. The interval [0,L] represents that part of the trajectory in which interaction with the surface takes place. At the point x the ion is neutralized. λ_+ and λ_o are the mean free paths between collisions produced by the ion and neutral particle, respectively.

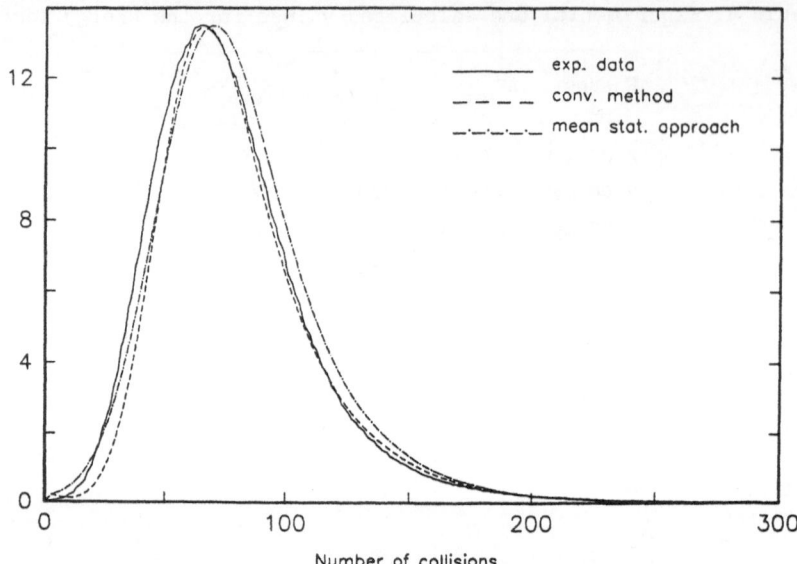

Figure 6. Energy-loss spectra for 5 keV incident He$^+$ on Ni(110) in random direction. (——): experimental values, (- - -): calculated via convolution method, and (-.-): calculated via mean statistical theory.

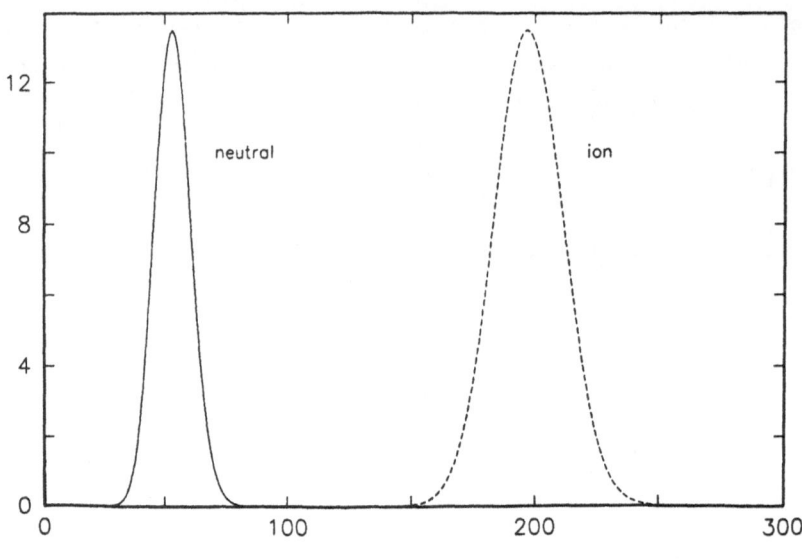

Figure 7. Energy-loss spectra calculated via the convolution method assuming that the particle is an ion or a neutral particle over the whole trajectory.

the He particle. The ratio $\lambda_o/\lambda_+ = 3.7$ is the same as for γ_+/γ_o given above.

In ref.[10], the differential probability for energy transfer ω in a single inelastic excitation process, (i.e., creation of an electron-hole pair), was integrated to give the width Γ of the quasi-particle states. From $\Gamma\tau = \Gamma\lambda_o/v \approx 1$, we determined (with $r_s = 1.5$ a.u.) $\lambda_o = 1.613$ a.u.

In fig.7 we verified that taking only a single charge state yields Gaussian energy-loss distributions, showing again that the asymmetry in the energy-loss spectra has as origin the change of the charge state during the interaction with the surface.

REFERENCES

1. M.A. Kumakhov and F.F. Komarov, *Energy Loss and Ion Ranges in Solids*, Gordon and Breach, Science Publ. Inc., New York, (1981).
2. R. Monreal, E.C. Goldberg, F.Flores, A. Närmann, H. Derks, and W. Heiland, Surf. Sci. *211/212* (1989) 271.
3. T.L. Ferrell and R.H. Ritchie, Phys. Rev. B *16* (1977) 115.
4. P.M. Echenique, R.M. Nieminen, J.C. Ashley, and R.H. Ritchie, Phys. Rev. A *33* (1986) 897.
5. J.C. Ashley, A. Gras-Martí, and P.M. Echenique, Phys. Rev. A *34* (1986) 2495.
6. F. García-Moliner and F. Flores, *Introduction to the Theory of Solid Surfaces*, Cambridge University Press, Cambridge, 1979.
7. M. Kato. J. Phys. Soc. Jpn. *55* (1986) 1011.
8. H. Bichsel, Rev. Mod. Phys. *60* (1988) 663.
9. A. Närmann, R. Monreal, P.M. Echenique, F. Flores, W. Heiland, and S. Schubert, Phys. Rev. Lett. *64* (1990) 1601.
10. I. Nagy, A. Arnau, and P.M. Echenique, Phys. Rev. B *38* (1988) 9191.
11. M.T. Robinson and I.M. Torrens, Phys. Rev. B *9* (1974) 5008.
12. H. Derks, A. Närmann, and W. Heiland, Nucl. Instrum. Methods *B44* (1989) 125.

EFFECTIVE CHARGE OF HELIUM IONS IN SOLIDS

M. Peñalba, A. Arnau and P.M. Echenique

Departamento de Física de Materiales
Universidad del País Vasco, Facultad de Química
Apartado 1072, San Sebastián 20080, Spain

ABSTRACT

The effective charge of Helium ions in different materials has been calculated. A charge-state approach to stopping power has been used. In a first step linear theory and interaction with the valence electrons have been taken into account. In order to improve on these results we introduce the interaction with the lattice ion cores and a self-consistent non-linear calculation of the stopping power for the charge state relevant at low velocity. Comparison with experimental data is presented.

INTRODUCTION

The mechanisms that contribute to the energy loss of energetic ions require a different description of the interaction depending on the energy and charge (Z_1) of the moving ion. At low velocities the ion is surrounded by a cloud of electrons and these are strongly perturbed by the presence of the intruder ion. In this case a non-linear treatment of the screening process is adequate to describe the interaction, as has been done in the stopping power calculations by Echenique and coworkers [1]. When the ion is moving much faster than the mean orbital velocity of its electrons $(Z_1^{2/3} v_0$, v_0 being the Bohr velocity), it is stripped [2] of its electronic charge and the energy loss process is well described by the theories of Bethe and Bloch. In this limit the perturbation introduced by the incoming ion is small and so a first Born description of the interaction may be used. However at intermediate velocities $(v \cong Z_1^{2/3} v_0)$ there is a succession of electron capture and loss events by the ion that complicates the treatment of the interaction process. In order to describe the energy loss in this velocity range semi-empirical effective-charge theories have been proposed [3] to explain experimental data. In these approaches the stopping power for an ion charge Z_1 is given by

$$S(Z_1) = (Z_1^*)^2 \, S(Z_1 = 1),$$

Interaction of Charged Particles with Solids and Surfaces
Edited by A. Gras-Martí *et al.*, Plenum Press, New York, 1991

541

where $S(Z_1)$ is the stopping power for the ion at a given velocity and $S(Z_1 = 1)$ is the stopping power for an equal velocity proton. In this work we calculate Z_1^* for Helium ions moving through solids using a charge-state approach to calculate the stopping power.

THEORETICAL BACKGROUND

We have considered the He ion to be in three possible charge states. The bound states are described by a hydrogen-like wave function that is calculated minimizing the total energy of the ion using a variational procedure. A screened Coulomb interaction is used. The screening parameter depends on the ion velocity and reproduces the static (Thomas-Fermi) and high velocity (ω_p/v) limit. It reads $\lambda = \left[\dfrac{1}{\lambda_{TF}^2} + \dfrac{v^2}{\omega_p^2} \right]^{1/2}$. The obtained values are very close to those obtained in a more sophisticated self-energy approach by Guinea et al. [4]. Several mechanisms can contribute to the charge-exchange processes [5,6]:

(i) A transition may occur between the valence band states and the ion bound states with an electron-hole or plasmon creation. These are called Auger processes.

(ii) In the frame of reference of the moving ion the lattice potential is seen as a periodic time-dependent potential of frequency $\omega \cong v/a$ (a is the lattice constant). This can originate transitions between the ion bound states and the valence band states. We call these, resonant processes.

(iii) Shell processes. When the ion velocity is high enough, Coulomb interaction with lattice ions can produce a direct capture of an inner-shell electron by the moving ion.

In a first approach we have included only Auger processes. However in a better description the interaction with the target lattice ions is taken into account.

The equilibrium charge state fractions can be obtained from the electron capture and loss cross sections [6,7]. The stopping power is then calculated by summing the stopping powers for each charge state weighted with the respective charge-state fractions and adding the energy loss per unit path length in capture and loss processes [5,8,9]

$$S_p(He) = \Phi^0 \, S_p^0 + \Phi^+ \, S_p^+ + \Phi^{++} \, S_p^{++} + S_p^{C\&L},$$

where Φ^0, Φ^+ and Φ^{++} are the equilibrium fractions for the charge states

He^0, He$^+$ and He^{++}, respectively; S_p^0, S_p^+, S_p^{++} are the stopping powers of each charge state, and $S_p^{C\&L}$ is the energy loss per unit path length in the charge exchange processes.

The proton stopping power $S_p(H)$ can be calculated in a similar way by considering the relevant charge states at each velocity [9]. Thus we define the Helium effective charge as

$$Z_1^*(He) = \left(\frac{S_p(He)}{S_p(H)}\right)^{1/2}.$$

RESULTS

As a first approach we have considered only the interaction between the ion and the target valence electrons. These are represented by a free electron gas of homogeneous density n_0. The stopping power for each charge state is calculated in the dielectric formalism [10], taking into account the charge density around the nucleus of the ion by means of the substitution $Z_1 \rightarrow |Z_1 - \rho_e(q)|$, where Z_1 is the bare ion charge and $\rho_e(q)$ is the Fourier transform of the electronic charge density surrounding the ion.

We show in fig.1 the Helium effective charge as a function of the ion

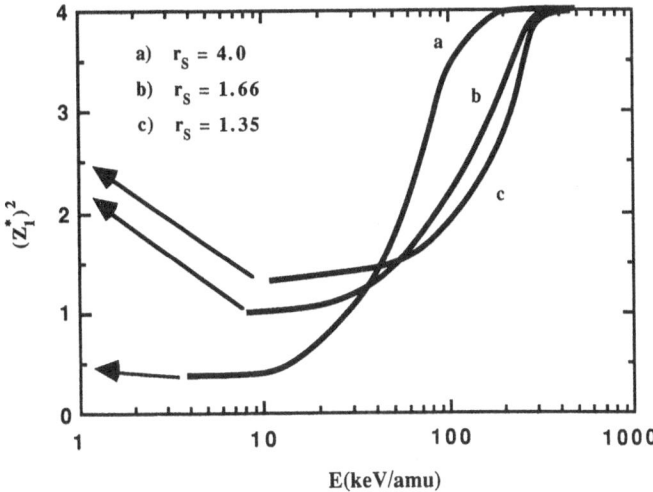

Figure 1. Square of the effective charge of He ions as a function of E, for three values of r_s. The stopping power of each charge state is calculated, in the dielectric formalism, taking only into account the interaction with the electron gas. The arrows represent the low-velocity limit obtained in a self-consistent non-linear model [1] for the corresponding r_s.

energy for several values of r_s ($n_0 = 3/\{4\pi r_s^3\}$). Typical r_s values in the metallic density range are $1.5 \leq r_s \leq 4$. It is also shown by an arrow in the figure the low-velocity limit obtained in a self-consistent non-linear screening model [1]. Our calculations reproduce the trends observed in the experiments [11] when r_s varies. At low velocities Z_1^* decreases when the density decreases. The stopping power can be greater for protons than for Helium ions for high r_s. This be explained because at low velocities the relevant charge state for Helium is He^0, and its stopping power is smaller (at this velocities and with high r_s) than the corresponding to H. At intermediate velocities ($v \cong 2v_0$) the opposite behavior is seen: Z_1^* grows when increasing r_s. This can be understood in terms of the Pauli exclusion principle which restricts the number of available final states for the electron loss by the ion; the restriction is weaker at lower densities where the Fermi level is small (narrow band) and so the charge state of the moving ion is higher in this case.

Although the general behavior of Z_1^* is well described by this model, the experimental values [11] of Z_1^* are systematically higher than the ones calculated with this simple model. So we have done a more accurate treatment of this question in aluminum.

For aluminum we have taken into account the interaction of the

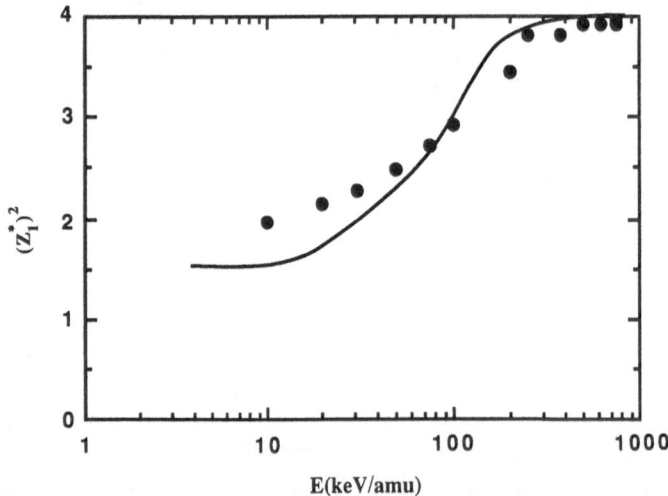

Figure 2. Square of the effective charge of He ions in aluminum as a function of E. The low-velocity stopping power is calculated in a self-consistent non-linear model. The interaction of the ion with the lattice ion cores is taken into account. The dots represent the effective charge of Helium calculated from the experimental data of reference [13,14].

incoming ion with the lattice ion cores. The aluminum target is characterized by an f.c.c. lattice of ion cores with ten electrons per ion. The remaining three electrons per atom form a free-electron-like band with density parameter $r_s = 2$. To describe the interaction of the ion with the lattice in the resonant processes we use a crystal pseudopotential [5,6]. The capture cross sections for the shell processes are calculated in the Brinkman-Kramers approximation.

The stopping power for the charge states relevant at high energy are calculated in linear theory, as we did before. However, for the charge state with highest equilibrium fraction at low velocities, we have calculated the stopping power from the phase-shifts at the Fermi level for the scattering of electrons by the self-consistent potential, created by the ion, calculated in density functional theory [1,12]. The results are shown in fig.2. We also show the Z_1^* of He deduced from experimental data of stopping power in aluminum [13,14]. It can be said that the agreement with experimental values is quite good. The improvement with respect to the results obtained with the first approach, fig.1, is due both to the low-velocity non-linear calculations of the stopping power, and to the interaction with the lattice ion cores that produces higher loss cross-sections and so higher effective charge.

ACKNOWLEDGEMENTS

The authors gratefully acknowledge help and support by Eusko Jaularitza, Gipuzkoako Foru Aldundia, the Spanish Comisión Asesora Científica y Técnica (CAICYT) and Iberduero S.A..

REFERENCES

1. P.M. Echenique, R.M. Nieminen, J.C. Ashley and R.H. Ritchie, Phys. Rev. A *33* (1986) 897.
2. N. Bohr, Dan. Mat. Fys. Medd. *18* (1948) num.8.
3. H. Betz, Rev. Mod. Phys. *44* (1972) 465; W.Brandt and M.Kitagawa, Phys. Rev. B *25* (1982) 5631.
4. F. Guinea, F. Flores and P.M. Echenique, Phys. Rev. B *35* (1982) 6106.
5. P.M. Echenique, F. Flores and R.H. Ritchie, Nucl. Sci. Appl. *3* (1989) 293.
6. F. Sols and F. Flores, Phys. Rev. B *30* (1984) 4878; F. Sols, Ph. D. Thesis, Universidad Autónoma de Madrid, 1985 (unpublished).
7. M.C. Cross, Phys. Rev. B *15* (1977) 602.
8. S.K. Allison, Rev. Mod. Phys. *30* (1958) 1137.
9. P.M. Echenique, I. Nagy and A. Arnau, Int. J. Quant. Chem. *23* (1989) 521.
10. J. Lindhard, K. Dan.Vidensk, Selsk. Mat. Fys. Medd. *28* (1954) num.8
11. S. Kreussler, C. Varelas and R. Sizmann, Phys. Rev. B *26* (1982) 6099; P. Bauer, Nucl. Instr. Methods B *45* (1990) 673.
12. H.H. Andersen and J.F. Ziegler, *Stopping Powers and Ranges in all Elements: Hydrogen*, Pergamon, New York (1977); A. Seilenger, D. Semrad and H. Paul, to be published.

13. D.E. Watt, *Stopping Cross-Sections, Mass Stopping Powers and Ranges in 30 Elements for Alpha Particles (1 keV to 100MeV)*. Internal report, University of St.Andrews, Fife KY169SS, U.K. (1988).
14. A. Arnau, M. Peñalba, P.M. Echenique, F. Flores and R.H. Ritchie, Phys. Rev. Letters *65* (1990) 1024; F. Sols, Ph. D. Thesis, Universidad Autónoma de Madrid (1985, unpublished).

WAKE POTENTIAL IN THE VICINITIES OF A SURFACE

F.J. Garcia de Abajo[1] and P.M. Echenique[2]

1. Departamento de CCIA
   Facultad de Informática, Universidad del País Vasco
   Apartado 649. 20080 San Sebastián, Spain

2. Departamento de Física de Materiales
   Facultad de Química, Universidad del País Vasco
   Apartado 1072. 20080 San Sebastián, Spain

ABSTRACT

An expression for the potential set up by a swift ion when it crosses a solid-vacuum interface is presented. The evaluation of this expression is carried out by using the plasmon-pole approximation for the dielectric function.

The wake phenomena associated to the motion of a swift charged point particle in an electron gas has been extensively studied in the past [1]. Many authors have studied the dynamical corrections to the image potential [2-4]. The surface potential when the particle moves parallel to the surface and with constant velocity has also been considered [2,6]. It has also been shown that the image potential is mainly due to the interaction with surface plasmons [7,8]. The self-energy of a moving charge near a surface has also been studied in detail [3].

The problem we want to deal with consists in deriving expressions for the potential induced by the moving charge, i.e., finding the response of the semi-infinite medium. We shall adopt the specular reflection model (SRM), first introduced by Ritchie and Marusak [9] to study the dispersion relation for surface plasma modes. In this model each semi-infinite medium is treated as if it was infinite under a symmetrisation of the charges being inside it, and adding a charge distribution on the surface which is eliminated after imposing the matching conditions for the potential. Some authors have applied this model to similar physical problems [4,5]. In order to have a good knowledge of the induced potential near the surface it is necessary to study the effect of the surface on electron production induced by swift ions, as is the case in convoy electrons and in particular the wake-riding electrons, which has been the subject of intense experimental [10] and theoretical work [11].

Let us consider a swift charged point particle which is travelling in the vicinities of a flat solid-vacuum interface with velocity $\vec{v}$, and let Z_1 be the charge of the particle. We shall use the notation $\vec{r} = (\vec{R}, z)$, $\vec{k} = (\vec{Q}, k_z)$, $\vec{v} = (\vec{V}_\parallel, v_z)$, where $\vec{R}$, $\vec{Q}$ and $\vec{V}_\parallel$ are the components parallel to the surface, and the z-axis is chosen in the perpendicular direction. The trajectory of the particle, whose recoil is negligible, is described by the equation $\vec{r} = \vec{v}t$, so that it reaches the surface, located at $z = 0$, when $t = 0$. The bulk solid ($z < 0$) is taken to be described by its dielectric function $\varepsilon(k, \omega)$.

Let us concentrate on the outgoing particle case ($v_z > 0$). The charge stays in the metal side ($z < 0$) while t is negative. Thus, the symmetrised charge density [4] for the extended metal should read

$$\rho^{metal}(\vec{r}, t) = Z_1 \left[\left\{ \delta(\vec{r} - \vec{v}t) + \delta(\vec{r} - \vec{v}'t) \right\} \theta(-t) + \rho^s(\vec{R}, t)\delta(z) \right], \tag{1}$$

with $\vec{v}' = (\vec{V}_\parallel, -v_z)$ the specular reflection of $\vec{v}$, and $\theta(x)$ the Heaviside step function. The ingredients of this charge density are the real charge and its specular image, (which are only present for negative times, i.e., while the real charge is in the metal side), and a surface-charge distribution, that is used to match the potential at the surface. After inserting this charge density into Poisson's equation and expressing the potential in terms of its Fourier transform

$$\phi(\vec{r}, t) = \frac{1}{(2\pi)^4} \int d\vec{k} \int d\omega \ e^{i(\vec{k}\vec{v} - \omega t)} \ \tilde{\phi}(\vec{k}, \omega), \tag{2}$$

we obtain

$$\tilde{\phi}^{metal}(k, \omega) = \frac{4\pi Z_1}{k^2 \varepsilon(k, \omega)} \left\{ \int_{-\infty}^{0} \left[e^{i(\omega - \vec{k}\vec{v})t} + e^{i(\omega - \vec{k}\vec{v}')t} \right] dt \right.$$

$$\left. + \int d\vec{R} \int_{-\infty}^{\infty} dt \ e^{i(\omega t - \vec{Q}\vec{R})} \ \rho(\vec{R}, t) \right\}.$$

After undoing the Fourier transform in the variable z, it becomes

$$\tilde{\phi}^{metal}(\vec{Q}, \omega, z) = 4\pi Z_1 \left[U_0(z) + \rho^s(\vec{Q}, \omega) \ I_0(z) \right], \quad (z < 0) \tag{3}$$

where $\varepsilon(k,\omega)$ is assumed to depend on k_z only through $k^2 = Q^2 + k_z^2$,

$$U_0(z) = \frac{1}{\pi} \int_{-\infty}^{\infty} dk_z \int_{-\infty}^{0} dt \; e^{i(\omega - \vec{Q}\vec{V}_\| - k_z v_z)t} \left(\frac{\cos(k_z z)}{k^2 \varepsilon(k,\omega)} \right), \qquad (4)$$

and

$$I_0(z) = \int_{-\infty}^{\infty} dk_z \; \frac{\cos(k_z z)}{k^2 \varepsilon(k,\omega)}. \qquad (5)$$

We proceed in the same way with the vacuum side $(z > 0)$, where now

$$\rho^{vacuum}(\vec{r},t) = Z_1 \left[\left\{ \delta(\vec{r} - \vec{v}t) + \delta(\vec{r} - \vec{v}'t) \right\} \theta(t) - \rho^S(\vec{R},t) \, \delta(z) \right]. \qquad (6)$$

The complete charge scheme is depicted in fig.1. The term proportional to $-\rho^S(\vec{R},t)$ provides the continuity of the perpendicular component to the surface of the electric displacement. Solving again Poisson's equation it is easily found that

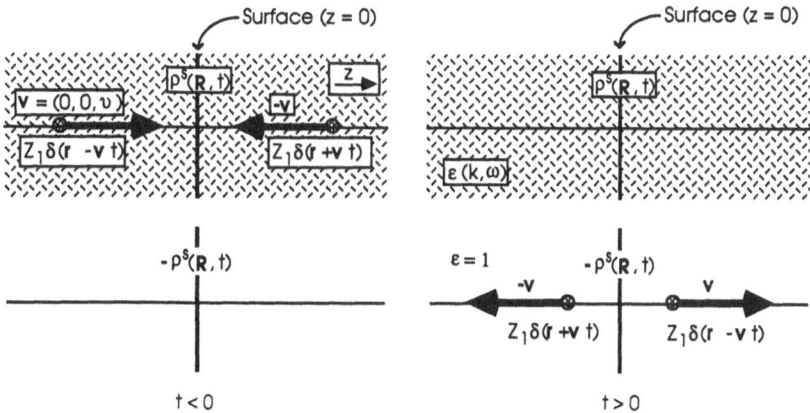

Figure 1. Schematic representation of the charges used in the specular reflection model to calculate the induced potential created by a charged point particle which crosses a solid-vacuum interface from the solid side. Z_1 is the charge of the particle. For the sake of simplicity we plot the case of perpendicular motion, i.e., the velocity of the particle is given by $\vec{v} = (0,0,v)$. The particle leaves the solid at $t = 0$. The two drawings at the left hand side show the charge situation when $t < 0$ and the ones at the right are used when $t > 0$. The upper drawings stand for $z < 0$, that is, inside the solid, $\varepsilon(k,\omega)$ being its bulk dielectric function, while the lower ones represent the vacuum, $z > 0$, where $\varepsilon(k,\omega) = 1$. The surface is placed at $z = 0$.

$$\tilde{\phi}^{\text{vacuum}}(\vec{Q},\omega,z) = 4\pi Z_1 \left[U_1(z) - \rho^s(\vec{Q},\omega)\, \frac{\pi}{Q}\, e^{-Qz} \right], \quad (z > 0) \tag{7}$$

where $\displaystyle U_1(z) = \frac{1}{\pi} \int_{-\infty}^{\infty} dk_z \int_{0}^{\infty} dt\; e^{i(\omega - \vec{Q}\vec{V}_\parallel - k_z v_z)t} \left(\frac{\cos(k_z z)}{k^2} \right) =$

$$= \frac{e^{i\tilde{\omega}|z|/v_z}}{v_z\, \tilde{K}^2} + \frac{i\tilde{\omega}\, e^{-Q|z|}}{v_z^2\, \tilde{K}^2\, Q}, \quad \text{with} \quad \tilde{\omega} = \omega - \vec{Q}\vec{V}_\parallel, \quad \text{and} \quad \tilde{K}^2 = Q^2 + \frac{\tilde{\omega}^2}{v_z^2}.$$

The continuity of the potential at the surface $(z = 0)$ fixes the value of $\rho^s(\vec{Q},\omega) = Q(U_1^+ - U_0^-)/(\pi + QI_0^-)$, where the upper signs $+$ and $-$ stand for the limits when z approaches 0 while remaining positive or negative respectively (both limits are equal because U_0, U_1 and I_0 depend only on $|z|$).

The analysis of the incoming particle case follows the same calculations but taking

$$\rho^{\text{metal}}(\vec{r},t) = Z_1 \left[\left\{ \delta(\vec{r}-\vec{v}t) + \delta(\vec{r}-\vec{v}'t) \right\} \theta(t) + \rho^s(\vec{R},t)\, \delta(z) \right]$$

$$\rho^{\text{vacuum}}(\vec{r},t) = Z_1 \left[\left\{ \delta(\vec{r}-\vec{v}t) + \delta(\vec{r}-\vec{v}'t) \right\} \theta(-t) - \rho^s(\vec{R},t)\, \delta(z) \right],$$

instead of eqs.(1) and (6). This assumes the particle is traveling from positive to negative z's. The important point is to realize the difference in the signs of the step functions.

A first evaluation of the wake potential may be carried out by using the classical frequency-dependent dielectric function $\varepsilon(\omega)$. We find the same results for the surface wake than those in ref.[12], while for the image potential in the vacuum our results reproduce those of ref.[3] for a non-dispersive medium.

Let us now consider the solid to be described by the plasmon-pole dielectric function, which has been previously used to calculate the bulk wake [1],

$$\varepsilon(k,\omega) = 1 + \frac{\omega_p^2}{\dfrac{k^4}{4} + \beta^2 k^2 - \omega(\omega + i\eta)},$$

with $\beta = \sqrt{3/5}\, v_F$, (v_F is the Fermi velocity) and $\omega_p = \sqrt{3/r_s^3}$ the plasmon frequency. η is a positive constant that represents the damping of the plasma modes. For the sake of simplicity we consider an outgoing particle

550

in a perpendicular trajectory, $\vec{v} = (0,0,v)$. Similar expressions are found when $\vec{V}_\parallel$ is not set to 0. Then $I_0(z)$ is obtained from eq.(5) and the result is [3]

$$I_0(z) = \frac{\pi}{Q} \frac{\omega(\omega + i\eta)}{\Omega} e^{-Q|z|}$$

$$+ \frac{\pi\omega_p^2}{\Lambda_+ - \Lambda_-} \left\{ \frac{\exp\left[-|z|\sqrt{Q^2 + 2\Lambda_-}\right]}{\Lambda_- \sqrt{Q^2 + 2\Lambda_-}} - \frac{\exp\left[-|z|\sqrt{Q^2 + 2\Lambda_+}\right]}{\Lambda_+ \sqrt{Q^2 + 2\Lambda_+}} \right\},$$

where $\Omega = \omega(\omega + i\eta) - \omega_p^2$ and $\Lambda_\pm = \beta^2 \pm [\beta^4 + \Omega]^{1/2}$. From this we may calculate $U_0(z)$, defined in eq.(4), after realizing that (for $\vec{V}_\parallel = 0$)

$$U_0(z) = \frac{1}{2\pi} \int_{-\infty}^{0} dt\, e^{i\omega t} \left\{ I_0(z - vt) + I_0(z + vt) \right\},$$

and we find

$$U_0(z) = \frac{\exp(-i\omega|z|/v)}{vk^2 \varepsilon(k,\omega)} - \frac{\omega(\omega + i\eta)}{\Omega} \frac{i\,\omega}{v^2 k^2 Q} e^{-Q|z|}$$

$$+ \frac{i\,\omega\,\omega_p^2}{\Lambda_+ - \Lambda_-} \left\{ \frac{\exp\left[-|z|\sqrt{Q^2 + 2\Lambda_+}\right]}{\Lambda_+ \sqrt{Q^2 + 2\Lambda_+}\left[v^2(Q^2 + 2\Lambda_+) + \omega^2\right]} \right.$$

$$\left. - \frac{\exp\left[-|z|\sqrt{Q^2 + 2\Lambda_-}\right]}{\Lambda_- \sqrt{Q^2 + 2\Lambda_-}\left[v^2(Q^2 + 2\Lambda_-) + \omega^2\right]} \right\}.$$

The final expression for the potential comes from eqs.(2), (3) and (7)

$$\phi(\vec{r},t) = \frac{1}{2\pi^2} \int_0^\infty dQ\, Q\, J_0(RQ) \int_0^\infty d\omega\, \mathrm{Re}\left\{ e^{-i\omega t}\, \tilde{\phi}(\vec{Q},\omega,z) \right\},$$

where ($\vec{V}_\parallel = 0$ and $v_z = v$),

$$\tilde{\phi}(Q,\omega,z) = \tilde{\phi}^{metal}(Q,\omega,z)\, \theta(-z) + \tilde{\phi}^{vacuum}(Q,\omega,z)\, \theta(z).$$

We have evaluated ϕ^{ind}, the wake potential minus the bare Coulomb potential, for $\vec{r} = \vec{v}t$, $v = 4$ a.u. and $Z_1 = 1$. We have done this for an incoming ion as well. The results are plotted in fig.2. In both cases (incoming and outgoing ion) the induced potential at the origin presents

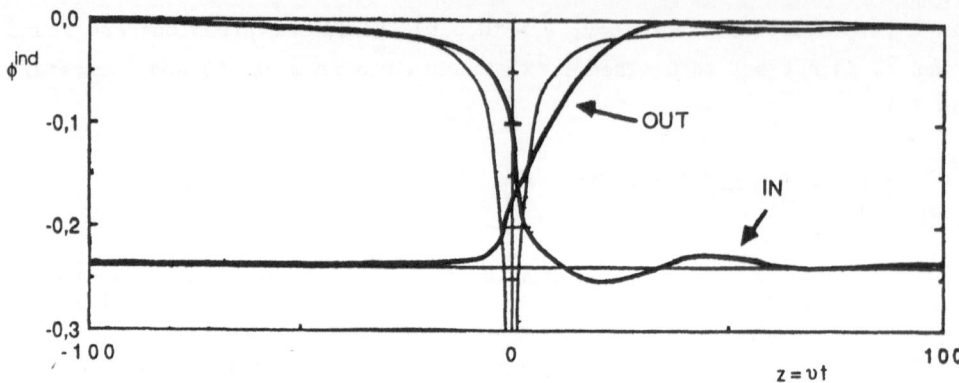

Figure 2. Induced potential at the position of the charge set up by a charged particle which is crossing a solid-vacuum interface in a direction perpendicular to the surface. IN is the situation where the particle is coming from the vacuum and penetrates into the solid, and OUT represents the particle coming from the solid. In both cases the particle is moving from left to right with a velocity $v = 4$ a.u. The equation of motion is $z = vt$, and this is the quantity that appears in the horizontal axis. The bulk medium is described by the plasmon-pole dielectric function with $r_s = 2$ and a damping $\eta = 0.015$ a.u. (aluminum). The potential presents oscillations after the particle has crossed the surface in both cases. The period of these oscillations is approximately given by $2\pi v/\omega_s$. The classical induced potential and the bulk induced potential at the position of the particle are also plotted.

oscillations once the particle has crossed the surface. The period of these oscillations is approximately given by $2\pi v/\omega_s$, where $\omega_s = \omega_p/\sqrt{2}$ is the surface plasma frequency [3].

Fig.3 shows the wake formation and destruction process for an incoming and outgoing proton respectively along the points of the trajectory. The potential remains almost unchanged until the particle crosses the surface (t = 0). In fig.4 we plot the wake destruction process for $v = 6$ a.u. and for points not only in the trajectory.

To summarize, we have found expressions for the induced potential created by an ion when it crosses a solid-vacuum surface. A numerical evaluation has been carried out for a perpendicular trajectory and with the plasmon-pole dielectric function to describe the bulk metal. We find that in the case of an outgoing trajectory the potential remains nearly unchanged until the projectile crosses the surface. The induced potential at the position of the ion presents oscillations, due to the surface plasmon excitations, once the projectile has crossed the surface in both the outgoing and the incoming trajectories.

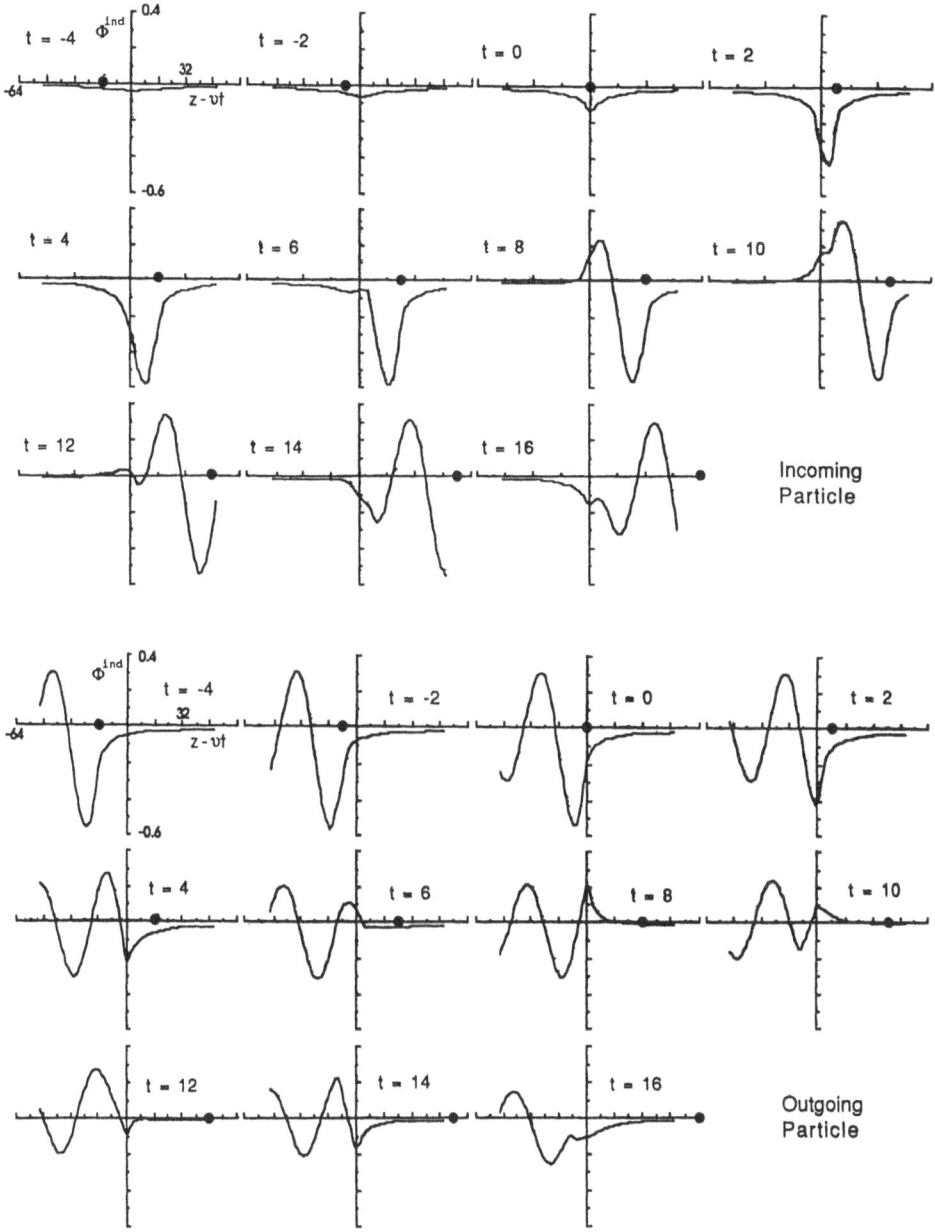

Figure 3. Wake destruction and formation process for a proton crossing an Al-vacuum interface under the same conditions as in fig.2. Here we plot the induced potential for points along the trajectory and t = -4, -2, 0, ..., 16 a.u. when the particle is incoming and outgoing. The proton is shown as a black circle.

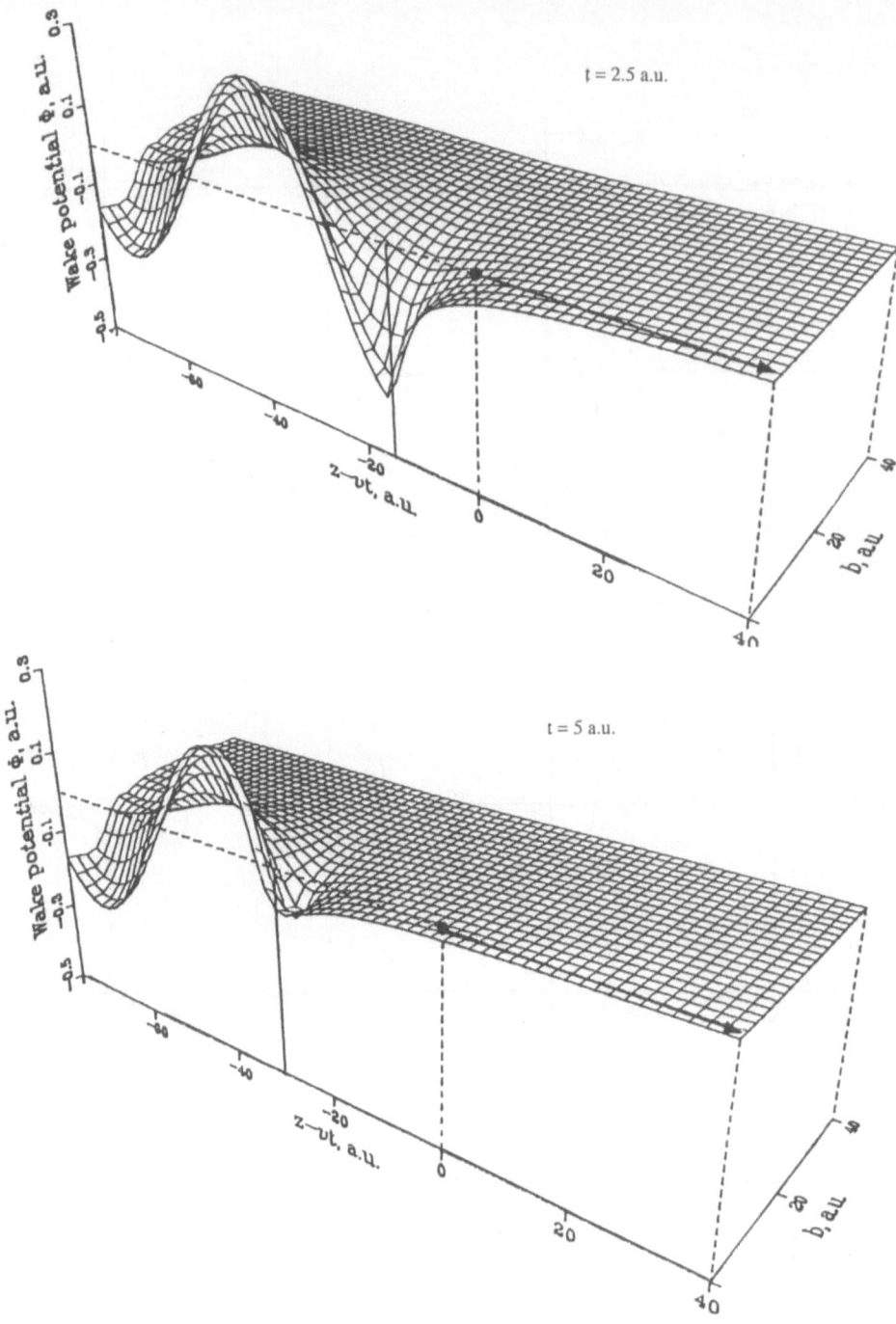

Figure 4. Induced wake potential created by an outgoing proton under the same conditions as in fig.2 except that the velocity is now $\nu = 6$ a.u. The wake is plotted on a grid in $z - \nu t$ and b (the direction perpendicular to the trajectory) for t = 2.5, 5, 7.5 and 10 a.u. The surface is represented by a vertical solid line. It comes out clearly how the wake begins to vanish only once the proton has crossed the surface.

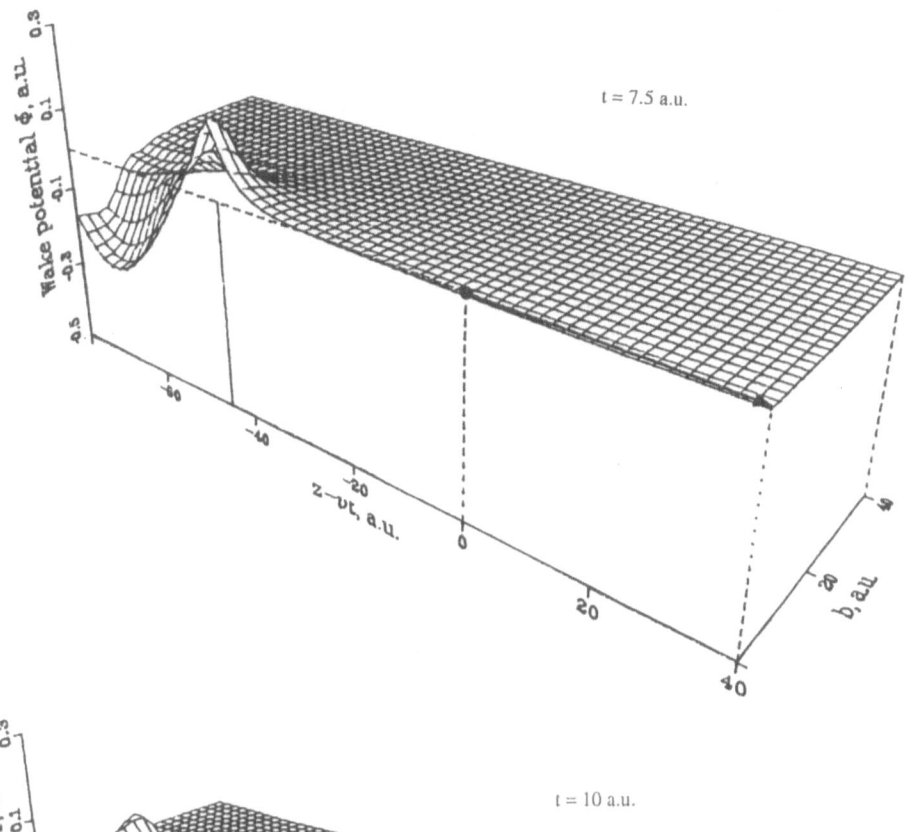

$t = 7.5$ a.u.

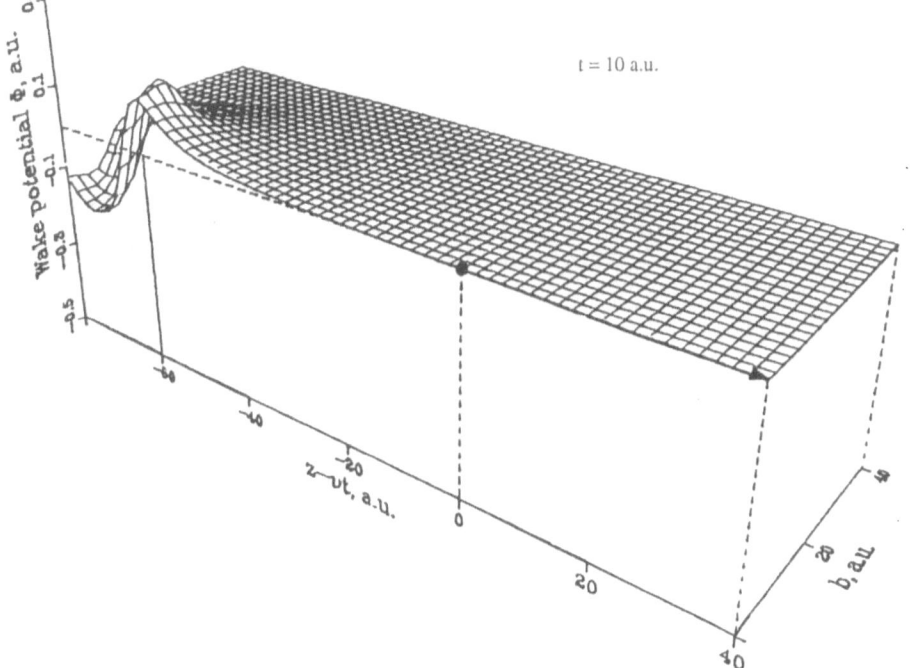

$t = 10$ a.u.

Figure 4. part 2

ACKNOWLEDGEMENTS

The authors gratefully acknowledge help and support by Eusko Jaularitza, Gipuzkoako Foru Aldundia, the Spanish Comisión Asesora de Investigación Científica y Técnica (CAICYT) and Iberduero S.A.

REFERENCES

1. J Neufeld and R.H. Ritchie, Phys. Rev. *98* (1955) 1632; *99* (1955) 1125; V.N. Neelavathi, R.H. Ritchie and W. Brandt, Phys, Rev. Lett *33* (1974) 302, *34* (1975) 650(E); P.M. Echenique, F. Flores and R.H. Ritchie, *Dynamic Screening of Ions in Condensed Matter,* Solid State Physics Series. Vol. *43*, p.229, Erhenreich and Turnbull, eds, Academic (1990); P.M. Echenique, R.H. Ritchie and W. Brandt, Phys. Rev. *B20* (1979) 2567.
2. D. Chan and P. Richmond, J. Phys. *C9* (1976) 163; N. Takimoto, Phys. Rev. *146* (1965) 366; J. Heinrichs, Phys. Rev. *B8* (1973) 1346; Yu.A. Romanov and V.Ya Aleshkin, Sov. Phys. Solid State *23* (1981) 147; B.N. Libenson and V.V. Rumyantsev, Sov. Phys. JEPT *59* (1984) 999; M.L.Glasser and G. Gumbs, Phys. Rev. *B35* (1987) 7490; N.J.M. Horing, H.C. Tso and G. Gumbs, Phys. Rev. *B36* (1987) 1588; K. Suzuki, M. Kitagawa and Y.H. Otsuki, Phys. Status Solidi *B82* (1977) 643; J.P. Muscat and D.M. Newns, Surface Sci. *64* (1977) 641.
3. F. Flores and F. Garcia-Moliner, J. Phys. *C12* (1979) 907; P.M. Echenique, R.H. Ritchie, N. Barberán and J.C. Inkson, Phys. Rev. *B23* (1981) 6486.
4. F. García-Moliner and F. Flores, *Introduction to the Theory of Solid Surfaces,* Cambridge University Press, (1979).
5. R. Núñez, P.M. Echenique and R.H. Ritchie, J. Phys. *C13* (1980) 4299.
6. Y.H. Ohtsuki, *Charged Beam Interaction with Solids*, London, Taylor & Francis Ltd., (1983).
7. R.H. Ritchie, Phys. Lett. *A38* (1972) 189.
8. A.A. Lucas, Phys. Rev. *B4* (1971) 2939.
9. R.H. Ritchie and A.L. Marusak, Surf. Sci. *4* (1966) 234.
10. Y. Yamazaki, Proc. 10th W. Brandt Workshop, Alicante, Spain (1987).
11. J. Burgdörfer, Nucle. Inst. and Meth. *B24/25* (1987) 139.
12. P.M. Echenique, Ph.D. Thesis, Universidad Autónoma de Barcelona (1977) (unpublished).

STOPPING POWER OF LARGE MOLECULAR CLUSTERS
IN COLD AND HEATED SOLIDS

N.R. Arista[1] and A. Gras-Martí

Dpt. de Física Aplicada,
Universitat d'Alacant
Apt. 99, E-03080 Alacant, Spain

1. Permanent address:
   Centro Atómico Bariloche
   RA-8400 S.C. Bariloche, Argentina

1. INTRODUCTION

The energy lost per particle and per travelled path length for a cluster of ions moving in a solid target, shows important differences -usually called *vicinage effects*- with respect to the energy loss of the separated ions [1-5]. The origin of this effect is the interference in the electronic excitations of the target due to the correlated motion of the penetrating ions. The effect has been theoretically explained [1,2] as interference effects in plasmon and single-particle excitations, but current research on inertial-confinement fusion using ion beams [6,7], and recent beam-target experiments with large molecular clusters [8-10], rise new interest in the evaluation of correlation effects in the stopping of ion clusters in cold and heated solids.

Here we present a description of the vicinage effects in the stopping power of large clusters containing up to several hundred molecules.

2. CLUSTER STOPPING POWER

Consider a cluster consisting of N ions of atomic charge Z_i and relative positions $\vec{r}_{ij}$. A general expression for the energy loss per unit pathlength of such a cluster penetrating a medium with velocity $\vec{v}$, is given by [2]

$$S_{cl} = \left\{ \sum_{i=1}^{N} Z_i^2 + \sum_{i \neq j}^{N} Z_i Z_j I(\vec{r}_{ij}) \right\} S_p,$$

(1)

Interaction of Charged Particles with Solids and Surfaces
Edited by A. Gras-Martí *et al.*, Plenum Press, New York, 1991

where S_p is the proton stopping power, and $I(\vec{r})$ is an interference function. Both S_p and $I(\vec{r})$ can be written in terms of the dielectric function of the medium $\varepsilon(k,\omega)$ as follows [2],

$$S_p = \frac{2}{\pi v^2} \int_0^\infty \frac{dk}{k} \int_0^{kv} d\omega \; \omega \; \mathrm{Im}(\frac{-1}{\varepsilon(k,\omega)}),$$ (2)

$$<I(r)> = I(r) = \frac{2}{\pi v^2 S_p} \int_0^\infty \frac{dk}{k} \frac{\sin(kr)}{kr} \int_0^{kv} d\omega \; \omega \; \mathrm{Im}(\frac{-1}{\varepsilon(k,\omega)}),$$ (3)

where we have approximated $I(\vec{r})$, in the case of large clusters, by its angular average (assuming random orientation of the internuclear axis).

Eq.(1) is to be compared with the stopping corresponding to independent particles (this is also the limit of large internuclear distances, where $I(r) \to 0$), viz.

$$S_{ind} = \left(\sum_i^N z_i^2\right) S_p.$$ (4)

Hence, we will describe the vicinage effect on the stopping power in terms of the ratio $\gamma = S_{cl}/S_{ind}$.

The vicinage effect in the energy loss has been studied extensively for small ion clusters, mainly for diatomic and triatomic clusters. Here we consider only the case of large clusters.

Since we are interested in the case of slow ion clusters ($v < v_F$), the energy loss function of an electron gas at finite temperature T can be approximated, for all degrees of plasma degeneracy, by [11]

$$\mathrm{Im}(\frac{-1}{\varepsilon(k,\omega)}) \simeq \frac{2\omega}{k^3} f_s(k),$$ (5)

where $f_s(k)$ a temperature-dependent screening function.

2.1 Homonuclear Clusters

The general formulation sketched before can be cast in a form appropriate to discuss the energy loss of large clusters. The stopping power of each ion in the cluster will be influenced by interference with all the other ions, so that, using eq.(1), we can write the modified stopping power for each ion in the cluster as follows

$$S_1^* = \left\{ z_1^{*2} + z_1^{*2} n_1 \int d^3r \; g_{11}(r) \; I(r) \right\} S_p,$$ (6)

where we have introduced the effective ion charge Z_1^*, the average ion density n_1 in the cluster, and the pair-correlation function $g_{11}(r)$, that describes the structure of the cluster [12,13]. The cluster stopping power is simply given by $S_{cl} = N_1 S_1^*$, where N_1 is the total number of ions in the cluster.

Next we introduce the cluster vicinage function

$$G_{11} = n_1 \int d^3r \, g_{11}(r) \, I(r), \tag{7}$$

so that the cluster stopping power takes the simple form

$$S_{cl} = N_1 \, Z_1^{*2} \, S_p \, (1 + G_{11}). \tag{8}$$

The factor before the parenthesis in eq.(8) is the stopping of N_1 uncorrelated atomic ions. Thus, the cluster stopping ratio becomes

$$\gamma = \frac{S_{cl}}{S_{ind}} = 1 + G_{11}, \tag{9}$$

which is independent of the value of the effective charge.

2.2 Heteronuclear Clusters

Let us consider now a large cluster containing two different types of ions, with effective charges Z_1^* and Z_2^*, and let $g_{11}(r)$, $g_{22}(r)$ and $g_{12}(r) = g_{21}(r)$ denote the corresponding pair-correlation functions [13]. The average density for each kind of particle in the cluster will be denoted by n_1 and n_2.

By a slight generalization of the previous formulation, we write the modified stopping power for each ion, 1 or 2, as follows

$$S_1^* = (Z_1^{*2} + Z_1^{*2} \, G_{11} + Z_1^* \, Z_2^* \, G_{12}) \, S_p, \tag{10a}$$

$$S_2^* = (Z_2^{*2} + Z_2^{*2} \, G_{22} + Z_2^* \, Z_1^* \, G_{21}) \, S_p, \tag{10b}$$

where

$$G_{ij} = \frac{4v}{3\pi S_p} \int_0^\infty \frac{dk}{k} \, f_s(k) \, g_{ij}(k), \tag{11}$$

and

$$g_{ij}(k) = n_j \int d^3r \, g_{ij}(r) \, \frac{\sin(kr)}{kr}. \tag{12}$$

Then, if N_1 and N_2 are the total numbers of ions of type 1 and 2 in the cluster, the cluster stopping power will be given by

$$S_{cl} = N_1 S_1^* + N_2 S_2^*, \tag{13}$$

while

$$S_{ind} = (N_1 Z_1^{*2} + N_2 Z_2^{*2}) S_p. \tag{14}$$

Finally, we define the ratio of the effective charges

$$\zeta = \frac{Z_2^*}{Z_1^*}, \tag{15}$$

and then write the stopping ratio γ, in terms of ζ, as follows,

$$\gamma = \frac{S_{cl}}{S_{ind}} = \frac{N_1 (1 + G_{11} + \zeta G_{12}) + N_2 \zeta(\zeta + \zeta G_{22} + G_{21})}{N_1 + N_2 \zeta^2}. \tag{16}$$

3. CALCULATIONS

To illustrate the magnitude of the vicinage effect for large clusters, we have calculated the values of γ for hydrogen and water

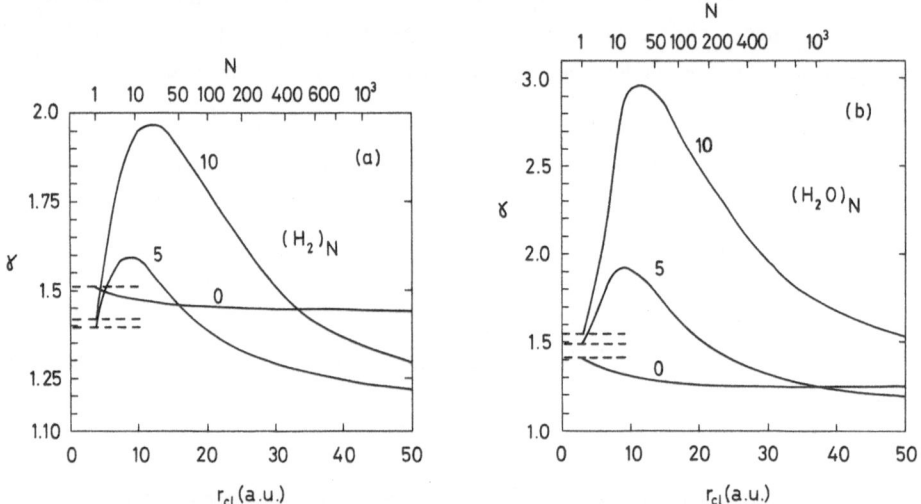

Figure 1. Stopping-power ratio γ as a function of cluster size, for (a) hydrogen molecules, and (b) water molecules, and for three temperatures $\tau = k_B T/E_F = 0$, 5 and 10. The figure shows the contributions to the vicinage effect due to the intramolecular terms (dashed lines), as well as the total value of γ. The upper scale shows the number of particles in the cluster.

clusters. The functions g_{ij} for these cases have been obtained from refs. [12,13], and the effective charge-ratio ζ (for slow oxygen and hydrogen) was obtained from the calculations of ref. [14].

We show in fig. 1 the stopping-power ratio γ for hydrogen and water clusters, as a function of cluster radius r_{cl}, for temperatures $k_B T/E_F$ = 0, 5 and 10. The solid lines give the total vicinage effect resulting from *intramolecular* terms (i.e., self interactions of the ions within each molecule) and *intermolecular* terms (interactions with ions in all other molecules in the cluster). The contributions of the intramolecular terms are indicated here with dashed lines.

We observe in figs. 1a and 1b that for cold media the intramolecular terms are dominant (of the order of 40 to 50 %), whereas the intermolecular terms are negative, so that the values of γ decrease with cluster size.

At high temperatures of the medium we observe an enhancement in γ, for intermediate values of the cluster size, $r_{cl} \sim 10$ a.u. It can be shown that this is related to the increase of the plasma screening length with temperature, so that the value of r_{cl} for which γ is maximum corresponds to the number of ions that interfere constructively in the cluster vicinage function, eq. (1).

4. CONCLUSIONS

The calculations presented here provide the first estimate of the vicinage effects on stopping power for large molecular clusters.

In the analysis of these effects we distinguish between the intramolecular and the intermolecular interference terms. In the range of ordinary internuclear distances in molecules, the intramolecular interference terms give rise to an *enhancement* of up to 50 % in the cluster stopping power.

The intermolecular vicinage effects can be formulated in terms of the pair-correlation functions $g_{ij}(r)$ of the cluster components. In the range of normal intermolecular distances and target temperatures, this vicinage effect is found to be *negative*, due to negative values of the interference function $I(r)$. Further calculations of vicinage effects on the stopping power of homonuclear and heteronuclear clusters are given in ref. [15].

These results should be of interest for further studies on the penetration of large molecular clusters in solids, as well as for possible applications to inertial-confinement fusion research.

ACKNOWLEDGEMENTS

The present collaboration has been possible through a Sabbatical stay of N.R. Arista sponsored by the Ministerio de Educación y Ciencia and the Conselleria d'Educació i Ciència de la Generalitat Valenciana. Partial support from the Dirección General de Investigación Científica y Técnica (DGICYT, projects number PS88-0066 and PS89-0065) is also recognized.

REFERENCES

1. W. Brandt, A. Ratkowsky, and R.H. Ritchie, Phys. Rev. Lett. *33* (1974) 1329; N.R. Arista and V.H. Ponce, J. Phys. *C8* (1975) L188.
2. N.R. Arista, Phys. Rev. *B18* (1978) 1.
3. J.C. Eckardt, G. Lantschner, N.R. Arista, and R.A. Baragiola, J. Phys. *C11* (1978) L851.
4. Proc. of the 9th Int. Conf. on Atomic Collisions in Solids, J. Remilleux et al, eds., Nucl. Instrum. Meth. *194* (1982).
5. G. Basbas and R.H. Ritchie, Phys. Rev. *A25* (1982) 1943; A. Arnau, P.M. Echenique, and R.H. Ritchie, Nucl. Instrum. Meth. *B40/41* (1982) 329.
6. T.A. Mehlhorn, J. Appl. Phys. *52* (1981) 6522; F.C. Young, D. Mosher, S.A. Goldstein, and T.A. Mehlhorn, Phys. Rev. Lett. *49* (1982) 549; J.N. Olsen, T. A. Mehlhorn, J. Maenchen, and D.J. Johnson, J. Appl. Phys. *58* (1985) 2958.
7. E. Nardi, E. Peleg, and Z. Zinamon, Phys. Fluids *21* (1978) 574; Phys. Rev. Lett. *49* (1982) 1251; C. Deutsch, G. Maynard and H. Minoo, J. Physique (Paris) *C8* (1983) 67; N.R. Arista and A.R. Piriz, Phys. Rev. *A35* (1987) 3450.
8. R.J. Beuhler and L. Friedman, Chem. Rev. *86* (1986) 521.
9. R.J. Beuhler, G. Friedlander, and L. Friedman, Phys. Rev. Lett. *63* (1982) 1292.
10. P.M. Echenique, J.R. Manson, and R.H. Ritchie, Phys. Rev. Lett. *64* (1990) 1413.
11. N.R. Arista and W. Brandt, Phys. Rev. *A23* (1981) 1898; J. Phys. *C18* (1985) 5127.
12. I.F. Silvera, Rev. Mod. Phys. *52* (1980) 393.
13. A.K. Soper and M.G. Phillips, Chem. Phys. *107* (1986) 47.
14. P.M. Echenique, R.M. Nieminen, J.C. Ashley, and R.H. Ritchie, Phys. Rev. *A33* (1986) 897.
15. N.R. Arista and A. Gras-Marti, to be published.

A STOPPING POWER FORMULA FOR HELIUM IONS

G. Lapicki

Department of Physics
East Carolina University
Greenville, NC 27858, USA

ABSTRACT

The empirical ratio of He to H stopping powers from Ziegler, Biersack, and Littmark [1] is compared with the Brandt-Kitagawa theory [2], trying various choices for the screening length Λ and the fractional number of electrons on the projectile ion q. The best agreement is obtained when $\Lambda = \langle r \rangle$ and $q = 1 - \exp[-v_r/(Z_1 v_o)]$ are chosen for hydrogenic helium ion. The stopping power ratios are furthermore corrected because the stopping power is not simply proportional to Z_1^2 as assumed in Ziegler's scaling. A formula, which interpolates between a Lindhard-Scharff-like formula at low velocities and the Bethe-Bloch expression of high velocities, is used to make this correction.

Assuming that the effective charge, $Z_1^* = \gamma Z_1$, for an ion of atomic number Z_1 is independent of the target atomic number Z_2, i.e., $\gamma = \gamma(v_1)$ being only a function of the projectile ion velocity v_1, Ziegler et al.[1] fitted the scaled ratio of He to H stopping powers,

$$(S_{He}/S_H)/4 = \gamma_{He}^2, \tag{1}$$

as given by eq.(3-10a) of ref.[1]. The solid curve in fig.1 represents this fit, while the dashed curves bracket the calculations of γ_{He}^2 for all targets, according to the formulas of Brandt and Kitagawa (BK) [2]. Eq.(31) of ref.[2] is used when v_1 is less than the target's electron Fermi velocity v_F, and eq.(23) of ref.[2] is employed when $v_1 > v_F$. In both equations, the screening length Λ, derived variationally in a statistical model by Brandt and Kitagawa, is taken as given by eq.(15) of ref.[2].

The BK stopping-power theory underestimates Ziegler's semiempirical

Interaction of Charged Particles with Solids and Surfaces
Edited by A. Gras-Martí *et al.*, Plenum Press, New York, 1991

563

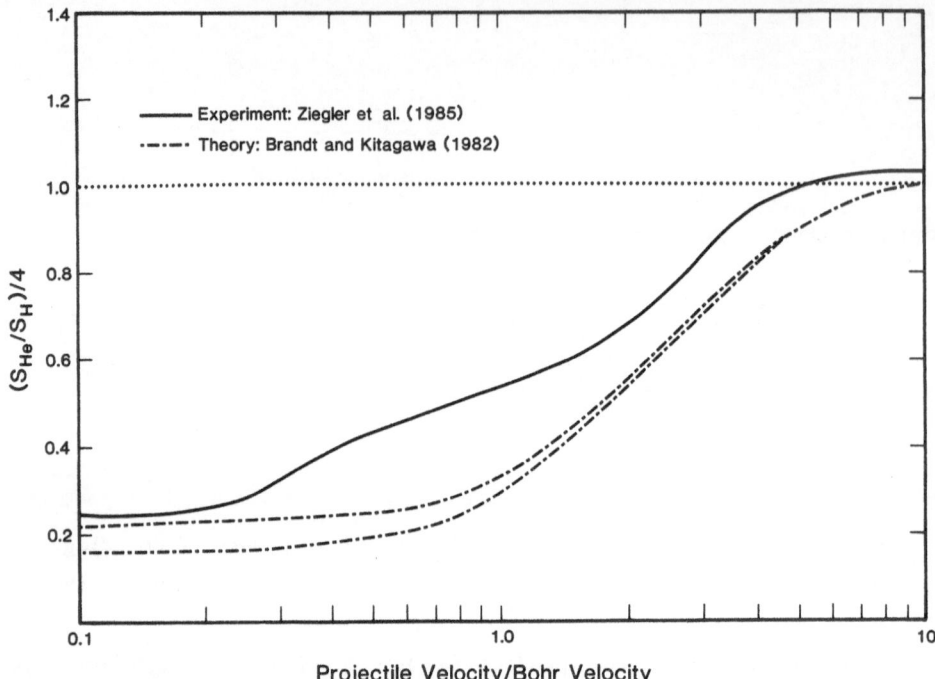

Figure 1. Ratio of helium to proton stopping powers divided by 4, according to the empirical fit of Ziegler et al.[1], is shown by the solid curve. The dashed curves represent the theory of Brandt and Kitagawa [2] for the lightest, upper curve, and middle-of-the-periodic table, lower curve, target elements; the calculated scaled ratios of He/H stopping powers in heavy targets are typically between these curves.

values at all velocities; around $v_1 = v_F$, the BK model exceeds the data by as much as a factor of 2. This is clearly visible in fig.2, where $\gamma^2_{He}(BK)/\gamma^2_{He}$(Ziegler et al.) is plotted for the whole range of targets, $Z_2 = 1, \ldots, 92$; the upper dashed curve corresponds to the lightest targets, the lower dashed curve is typical of the elements from the middle of the periodic table, and the theory-to-experiment ratios for the heavy targets fall between these curves.

Brandt subsequently suggested [3] that Λ should be identified with the average radius of the K-shell electron on the projectile, $\langle r \rangle_K$. In terms of the Bohr radius a_o, this Λ for helium is $1.5\,a_o/(Z_1 - 0.3125)$, in an He atom, and $1.5\,a_o/Z_1$ in an He^+ ion. In conjunction with the formula for the fractional number of electrons on the ion q,

$$q = 1 - \exp\left(-\frac{0.92\,v_r}{Z_1^{2/3}\,v_o}\right), \qquad (2)$$

where $v_r \equiv \langle |\vec{v}_1 - \vec{v}_2| \rangle$ is the average speed of the ion relative to the

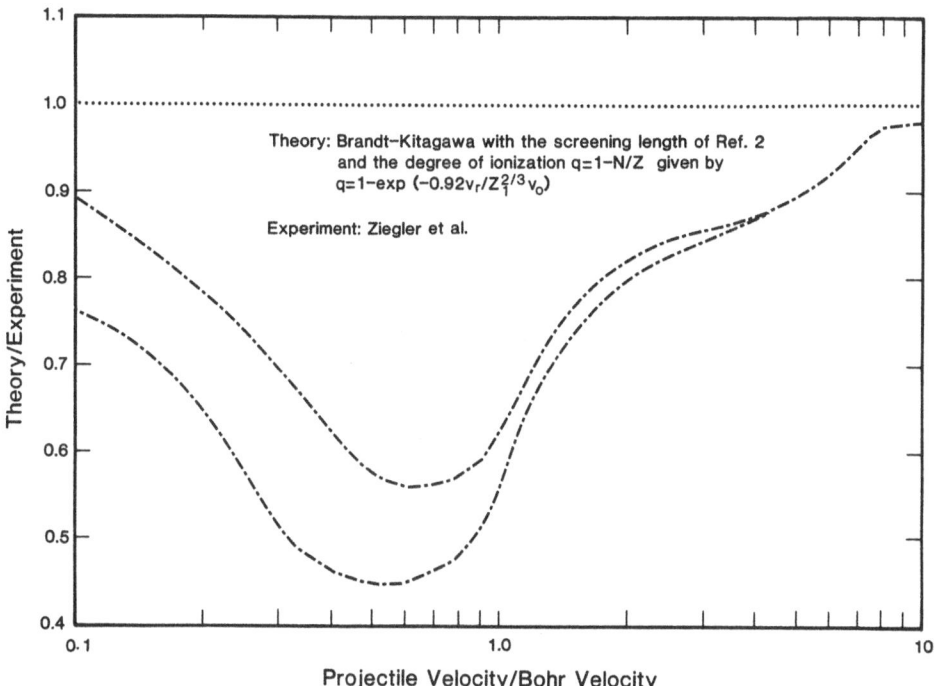

Figure 2. The theory of Brandt and Kitagawa [2], normalized to the empirical fit of Ziegler et al.[1]; the curves are described in the caption of fig.1.

velocity of the target's electrons [4], such Λ leads, however, to unphysical values of γ_{He}. In fact, at certain velocities and for some Z_2 targets, the effective charge would be significantly larger than Z_1. The BK formulas can be used, nevertheless, once $1/\Lambda$ is equated with $<1/r>_K$, i.e., $(Z_1 - 0.3125)/a_o$ in a He atom. These ratios are shown by the dashed curves in fig.3. The solid curves are obtained when, in eq.(2), the statistical $Z_1^{2/3}$ is also replaced with $Z_1 - 0.3125$.

The dashed curves in fig.4 result when Brandt's original choice for $\Lambda = 1.5\ a_o/Z_1$ is retained, but with

$$q = 1 - \exp(-\frac{v_r}{Z_1 v_o}). \tag{3}$$

This equation is more consistent with Bohr's stripping criterion applied to a *hydrogenic* helium ion, where electrons of the speed less than the orbital velocity $Z_1 v_o$ are stripped. By contrast, the stripping criterion applied to a Thomas-Fermi atom - the scheme on which eq.(2) was based - is clearly unrealistic for a single-electron ion like helium. It should be also noted that around $v_1 \simeq v_o$ (or where, according to eq.(6) of ref.[4], $v_r = 1.2\ v_o$) the ions of Bohr's velocity have sufficiently low velocities

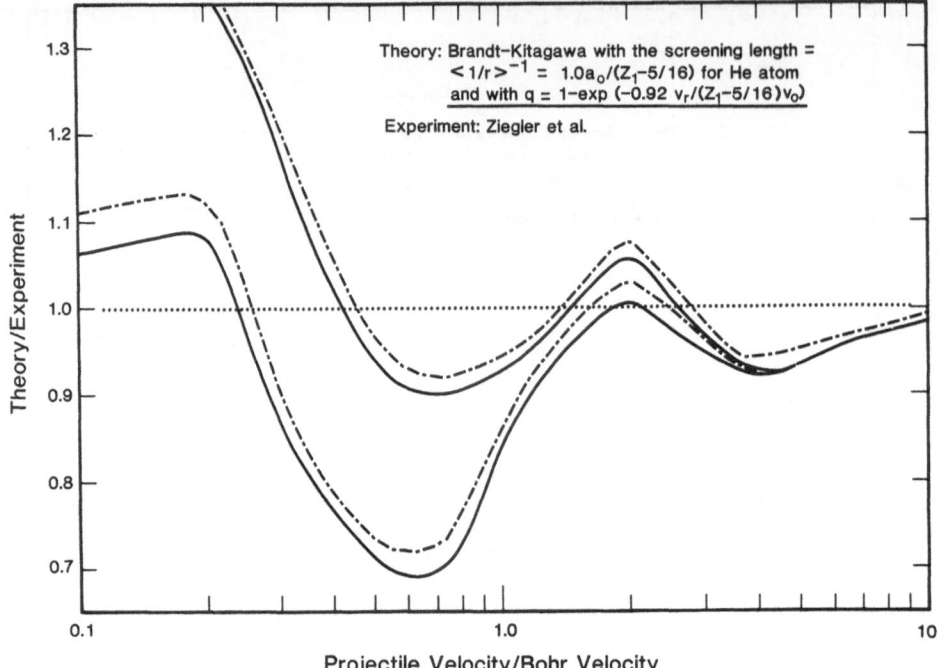

Figure 3. As in fig.2, but with a different screening length:
1.0 $a_o/(Z_1-5/16)$, (---); and also with
$q = 1 - \exp[- 0.92 \, v_r/(Z_1-5/16)v_o]$, (——).

for γ^2_{He} to be meaningfully less than one, and yet they are in the velocity range where data for He/H stopping power ratios are still available for a reliable fit. Eq.(3) gives $q \simeq 0.5$ when $v_1 \simeq v_o$, so that indeed He ions are singly ionized in this velocity regime.

Finally, an additional correction is proposed here. Ziegler's scaling of eq.(1) presumes that the stopping power is simply proportional to Z_1^2. Many stopping-power theories go beyond this linear-response treatment. In the Lindhard-Scharff approach to stopping powers at low velocities [5], the reduced stopping power $L \propto S/Z_1^2$ *does* depend on the projectile. This formula was recently [6] modified to

$$L_{low} = 5.55 \left[(v_1/v_o)/(Z_1^{1/3} + Z_2^{1/3}) \right]^3, \qquad (4)$$

which indeed reproduces Andersen and Ziegler's [7] semiempirical tables of stopping powers for protons excellently at low velocities. These tables are also well reproduced at all proton velocities once L_{low} of eq.(4), with $Z_1 = 1$, is joined at high velocities with the Bethe-Bloch formula [8]: $L_{high} = \ln[2(v_1/v_o)^2/I] - \beta^2 - \ln(1-\beta^2)$, where I is the mean

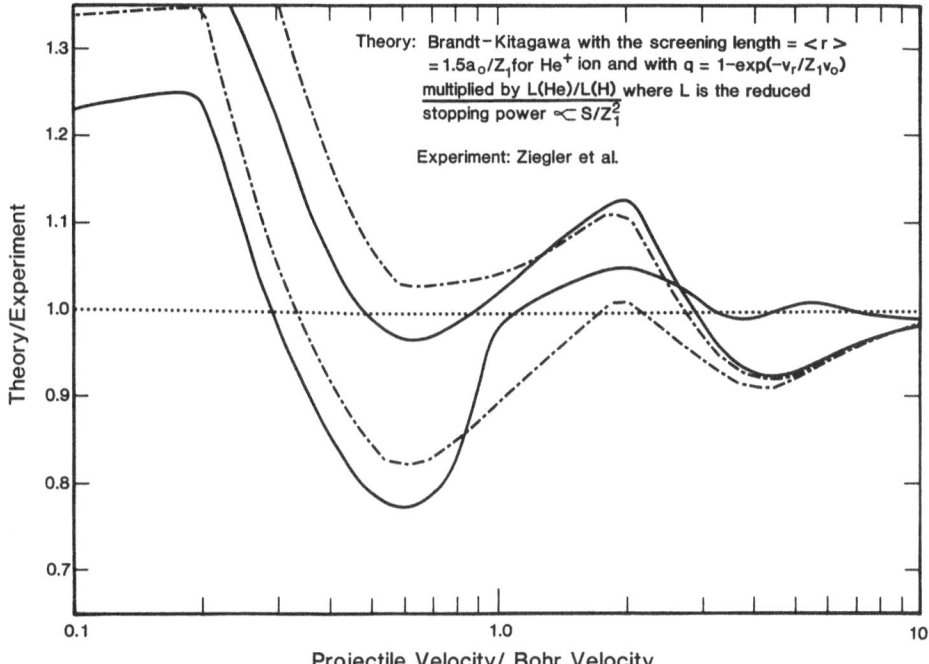

Figure 4. As in fig.2, but with the different screening length, 1.5 a_o/Z_1, and the different $q = 1 - \exp[- v_r/(Z_1 v_o)]$, (dashed curves). The solid curves (——) result after multiplication of the dashed curves by $L(He)/L(H)$, given in this work.

ionization energy and $\beta \equiv v_1/c$. By fitting $I = 0.2756 \; Z_2 \; (1 + Z_2^{-1/3})$, for the mean ionization energy in atomic units, the argument of the Bethe-Bloch logarithm was obtained [6] as

$$ARG = 7.26 \; (v_1/v_0)^2/[Z_2(1 + Z_2^{-1/3})]. \tag{5}$$

The reduced stopping powers for protons and helium ions, $L(H)$ and $L(He)$, can be calculated at all velocities using an interpolation formula,

$$L = \frac{1}{3} \left\{ \ln\left[\{1 + 3 \; L_{low} + 0.01 \; (ARG)^5\}/\{1 + 0.01 \; (ARG)^2\}\right] \right\}$$

$$- \beta^2 - \ln(1 - \beta^2), \tag{6}$$

that converges to L_{low} and $L_{high} = \ln(ARG) - \beta^2 - \ln(1 - \beta^2)$ at low and high velocities, respectively, and that is more compact than eq.(8) of ref.[6]. For these light projectiles, the L_{low} of eq.(6) is calculated according to eq.(4), with Z_1 set always as 1 for $L(H)$, and equal to $2(1-q)$ for $L(He)$. We find

$$(S_{He}/S_H)/4 = \gamma_{He}^2 \, L(He)/L(H), \qquad\qquad (7)$$

which appears to yield the best overall fit to Ziegler's data base, for the scaled ratio of He to H stopping powers (see the solid curves in fig.4). Note that our L(He)/L(H) correction does not change the dashed curves for these ratios in an appreciable manner. It decreases $(S_{He}/S_H)/4$, at the lowest velocities of the low velocity range $(0.1 < v_1/v_o < 1.0)$, by 10 %; it increases these ratios, at moderate velocities $(1.0 < v_1/v_o < 10)$, by a few percent, and subsides to 1 at high velocities $(v_1/v_o > 10)$. The enhancement of stopping powers is in accord with corrections that would follow as an z_1^3 effect. The reduction of stopping power at low velocities is reminiscent of the binding effect in inner-shell ionization; this effect decreases ionization cross sections when the projectile moves slower than the inner-shell electron. It would be very interesting to include an $L(Z_1)/L(Z_1=1)$ correction for heavy ion $(Z_1 > 2)$ stopping powers, when $[S(Z_1)/S(Z_1=1)]/Z_1^2$ are fitted to the data. In this manner, an empirical Z_1^*, unspoiled by Z_1^3 effects, might emerge in some universal form; perhaps this would give a better insight into the effective charge of heavy ions.

REFERENCES

1. J.F. Ziegler, J.P. Biersack, and U. Littmark, *The Stopping and Range of Ions in Solids*, Vol.I, Pergamon Press, New York (1985).
2. W. Brandt and M. Kitagawa, Phys. Rev. B *25* (1982) 5631.
3. W. Brandt, Nucl. Instr. Meth. *194* (1982) 13.
4. A. Mann and W. Brandt, Phys. Rev. B *24* (1981) 4999.
5. J. Lindhard and M. Scharff, Phys. Rev. *124* (1961) 128.
6. G. Lapicki, *Proc. of the 12th Werner Brandt Int. Conf. on the Penetration of Charged Particles in Matter*, 1989, San Sebastian, Spain. (National Tech. Infor. Service, Springfield, VA, April (1990); printed copy: A-99, microfiche: A01, p.85).
7. H.H. Andersen and J.F. Ziegler, *Hydrogen Stopping Powers and Ranges in All Elements*, Vol.III, Pergamon Press, New York (1977).
8. H.A. Bethe, Ann. Phys. *5* (1930) 325.

SURVEY OF HEAVY-IONS RESEARCH AT CIRIL/GANIL

A. Cassimi, E. Balanzat, S. Bouffard, E. Doorhyee,
J. Dural, J.P. Grandin, J.C. Jousset, E. Paumier and
M. Toulemonde

C.I.R.I.L.
rue Claude Bloch, BP 5133 CAEN Cedex
FRANCE

ABSTRACT

Some of the experimental programs realized at CIRIL in collaboration with many experimental groups are reviewed. They concern two fields of physics:

- Solid-state physics: how is the large amount of energy deposited during the slowing down of a swift heavy ion in matter converted into the form of lattice defects or structural modifications?

- Atomic physics: what are the elementary phenomena of the interaction between swift heavy ions and target electrons?

1. SOLID STATE PHYSICS (Radiation damage)

The specific interest of GeV heavy ions, in the field of radiation damage, comes from the fact that during 90 % of their large range (about 100 μm), the electronic stopping power dominates the nuclear stopping power by a factor of 2×10^3, whatever the ion or the target. Moreover, the electronic stopping power is very large: a few keV/Å. This energy spreads radially around the ion path, and it can be assumed that a few tens of eV/atom are transferred to the target electrons during 10^{-15} s in a cylindrical volume containing some tens of atoms. This unusual situation constitutes the very interest of swift heavy ion irradiation experiments.

The main question to be answered is: how can the high-density electronic excitation created in the wake of the projectile ion be converted into atomic motion, leading to induced stable lattice defects? None of the models which have been so far proposed are really able to account for the whole of the observed effects.

We shall present here numerous results obtained with various types of targets and projectile ions.

Interaction of Charged Particles with Solids and Surfaces
Edited by A. Gras-Martí *et al.*, Plenum Press, New York, 1991

569

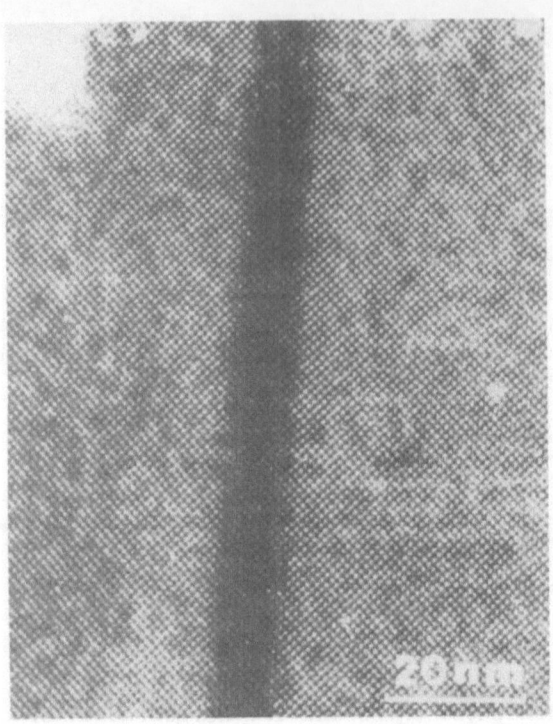

Figure 1. High Resolution Electronic Microscopy observation of induced
latent tracks in $Y_3Fe_5O_{12}$ for an electronic stopping power of
32 MeV cm^2 mg^{-1} .

1.1 Insulators

It was clearly demonstrated that in almost all the insulators the
stopping power has to be higher than a certain threshold to increase the
damage efficiency. This efficiency varies dramatically with the electronic
stopping power S_e: for SiO_2 quartz, it is increased by four orders of
magnitude when S_e varies from 1.5 to 104 MeV cm^2 mg^{-1} [1]. Moreover, it
has been shown that in magnetic oxide yttrium iron garnet $Y_3Fe_5O_{12}$ [2] -
above threshold - the latent ion track could be described as a continuous
cylindrical damaged zone, fig.1. Its radius increases with S_e. At low S_e ,
the latent track consists of isolated extended defects with spherical
shapes, fig.2. For a mean value of S_e, the spherical defects form a
discontinuous cylindrical track.

Comparisons of quartz crystals irradiated in channeling and
non-channeling conditions have indicated that K-shell target excitation
plays no significant role in the latent-track creation [1].

Recently, experiments were devoted to study, by time-resolved

Figure 2. High Resolution Electronic Microscopy observation of induced latent tracks in $Y_3Fe_5O_{12}$ for an electronic stopping power of 13 MeV cm^2 mg^{-1}.

spectroscopy, the transient electronic defects such as self-trapped excitons or excited color centers. These electronic excited states are created in about 10^{-11} s by the incident ion. Relaxation to the ground state takes place in the nanosecond to millisecond range by photon emission or atomic displacement.

1.2 Electrical conductors

Is it possible to create some damage in a metal by electronic energy deposition? All the models which were proposed to explain track formation predict that no damage process could exist in a material presenting a metallic character, since free electrons of high mobility damp any electronic perturbation in a time of 10^{-15} to 10^{-16} s. But recently it has been shown that metallic alloys [5,6] and some pure metals [3,4] can be damaged by electronic excitation.

1.2.1. Pure metals. Some pure metals, such as aluminum and copper, are insensitive to electronic excitations, while nickel and iron exhibit a decrease of the defect production rate when irradiated with ions with S_e values up to 4 keV/Å (fig.3). When S_e increases, the energy released in the material induces an annealing of point defects. Thus, the number of displaced atoms per incident ion is reduced [4]. Moreover, in pure gallium an anomalous enhancement of the defect creation rate is observed, fig.3, after xenon irradiation [3]. This enhancement of the defect production

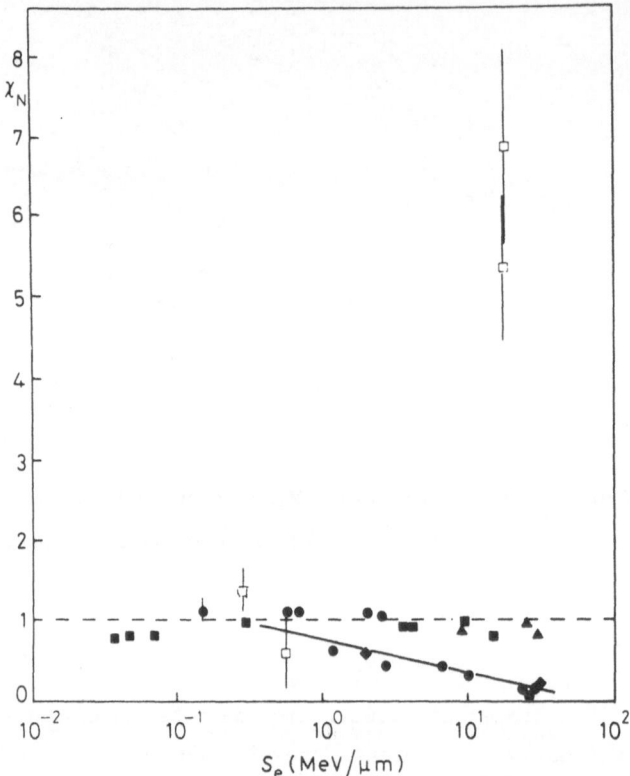

Figure 3. Variation of the damage efficiency normalized to the number of displacements per atom (χ_N) as a function of the electronic stopping power S_e. It is clearly seen that Al (full squares) and Cu (full triangles) are insensitive to S_e. Ga (open squares) has the opposite behavior with respect to Ni (full circles) and Fe (full diamonds). The solid line is traced to guide the eye.

rate is assumed to be the result of damage induced by the high electronic excitation density which follows the slowing down of the projectile ion. This effect has been also observed in iron for the highest values of S_e (uranium irradiation) [4].

1.2.2 Metallic alloys. Low-temperature irradiation of amorphous metallic alloys ($Fe_{85}B_{15}$ [5], Ni_3B [6]) shows that the electronic energy loss effects are:

- a defect production observed when S_e reaches the threshold value of 1.4 keV/Å. For $S_e \simeq 4$ keV/Å the defect-creation cross section is more than two orders of magnitude higher than the displacement cross section due to elastic collisions, fig.4;

- the existence of a large anisotropy in the macroscopic deformation

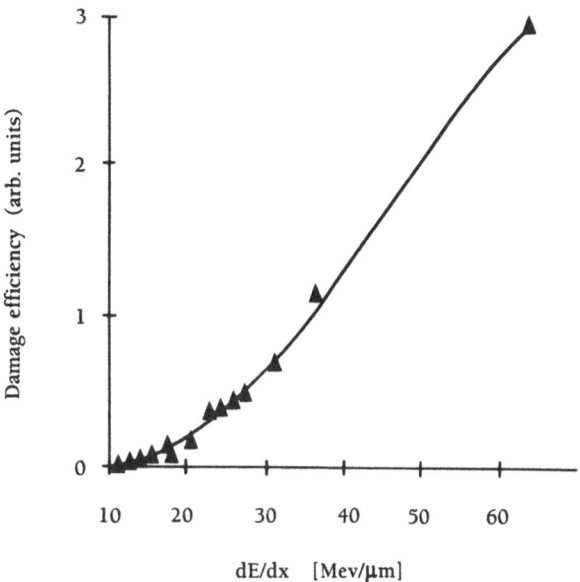

Figure 4. Damage efficiency in $Fe_{85}B_{15}$ as a function of the electronic stopping power.

of the sample above a critical "incubation" fluence (the sample dimensions perpendicular to the ion beam increase, while the dimension parallel to it shrinks).

2. ATOMIC PHYSICS

Heavy-ion interactions with matter permit to study elementary phenomena such as electron excitation, ionization and capture which determine finally such general quantities as the energy loss. The aim of the atomic physics experiments performed at G.A.N.I.L. is to study these phenomena and, starting from gas targets, to follow their evolution up to solid state targets.

2.1. Highly-charged recoil ions

Collisions of swift heavy ions with free atoms produce efficiently low-energy multicharged recoil ions. These ions are very interesting to perform low Doppler-spread spectroscopy and to produce high-quality secondary beams for slow ion-atom collision experiments.

The experimental determination of their characteristics, such as the recoil energies, in the eV range, and the production cross section of ions in a given charge state, is of fundamental interest for the understanding

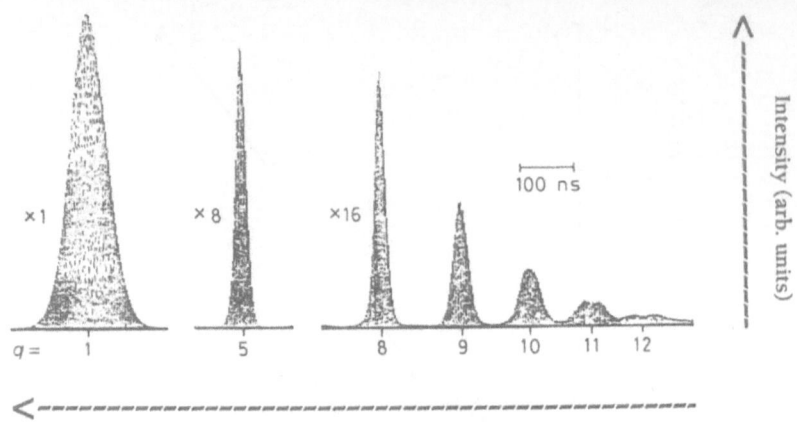

Time of flight

Figure 5. Part of a time-of-flight spectrum of Ar recoil ions produced by 27 MeV/u Xe^{52+} incident ions.

of the multielectronic processes involved in the production of these recoil ions.

The experimental results, obtained by time-of-flight spectrometry [7] are shown in fig.5 and table 1. These results are of interest for stopping power and angular straggling calculations since we measure simultaneously, under single-collision conditions, the energy transferred to the target atom ("nuclear stopping") and the degree of ionization, which is, besides the energy of the ejected electrons, an important part of the "electronic stopping power".

2.2 Comparison between single collisions in solids and gas targets

It is possible to study, with a beam of bare projectiles, the

Table 1. Mean kinetic recoil energies in meV for Ar and Ne ions of charge q produced by 27 MeV/u Xe^{52+}.

Argon		Neon	
q	$m\bar{v}_0^2/2$	q	$m\bar{v}_0^2/2$
9	79	7	102
10	216	8	257
11	515	9	791
12	993		
13	2208		
14	3970		
15	6823		
16	7800		

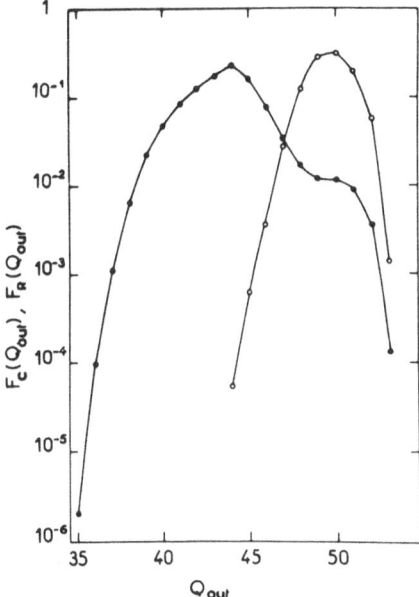

Figure 6. Charge-state distributions measured for 27 MeV/u Xe^{35+} incident ions transmitted through a 21 μm Si crystal under random condition (open circles) and along the <110> direction of a 21 μm Si crystal.

one-electron capture process in measuring the radiative decay of hydrogen-like ions produced in single ion-atom collisions. At very high energy, the single-collision condition can be fulfilled not only in gas targets but also in thin foils. It was found [8] that the total population of all the excited states of Kr^{35+} is the same for gas and foil targets, but a clear difference appears in the relative populations of the 2p, 3p and 4p substates.

The proposed interpretation for this solid-state effect is founded on the Stark mixing of the substates in the strong electric field induced by the polarization wake of the target electrons.

2.3 Channeling

Another kind of experiments, namely the channeling of swift multicharged heavy ions in thin crystals, is also in between atomic and solid-state physics. In these experiments, the charge distribution of the emerging ions, the energy-loss distribution of the ions emerging in a given charge state, the target and projectile X-ray emission, and the convoy electrons are simultaneously measured.

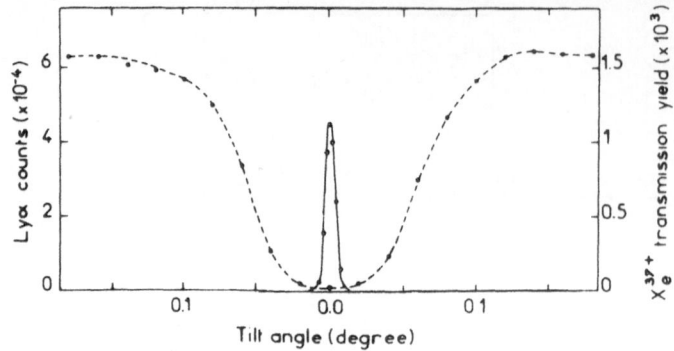

Figure 7. Angular scans of Xe Lyman-α yield (full circles) and 37+
fraction in the transmitted beam (open circles) for
27 MeV/u Xe^{35+} ions incident in the vicinity of the <110>
direction of a 21 μm Si crystal.

In particular, incident 27 MeV/u Xe^{35+} ions, far from the equilibrium
charge state in matter, have been transmitted through a 20.7 μm-thick
crystal, along the <110> axis [9]. Out-of-equilibrium emerging
charge-state distributions, fig.6, arise mainly from electron-impact
ionization. The particles with low emerging charge state Q_{out} are
"hyperchanneled" and have experienced a low electron density inside the
crystal. This is evidenced in fig.7 by the very narrow peak of the
fraction of emerging Xe^{37+} ions measured as a function of tilt angle. In
the same figure is also given for comparison the normal channeling dip
corresponding to the Lyman-α emission yield.

REFERENCES

1. M. Toulemonde, E. Balanzat, S. Bouffard, J.J. Grob, M. Hage-Ali and
   J.P. Stoquert, Nucl. Instrum. Methods Phys. Res. *B46* (1990) 64.
2. M. Toulemonde and F. Studer, Phil. Mag. *A58* (1988) 799.
3. E. Paumier, M. Toulemonde, J. Dural, F. Rullier-Albenque, J.P. Girard
   and P. Bogdanski, Europhys. Lett. *10* (1989) 555.
4. A. Dunlop, D. Lesueur, J. Morillo, J. Dural, R. Spohr and J. Vetter,
   C.R. Acad. Sci. Paris *309* (1989) 1277.
5. A. Audouard, E. Balanzat, J.C. Jousset, G. Fuchs, D. Lesueur and L.
   Thome, Nucl. Instrum. Methods Phys. Res. *B39* (1989) 18.
6. A. Audouard, E. Balanzat, J.C. Jousset, A. Chamberod, G. Fuchs, D.
   Lesueur and L. Thome, to be published in Phil. Mag.
7. J.P. Grandin, D. Hennecart, X. Husson, D. Lecler, I. Lesteven-Vaisse
   and D. Lisfi, Europhys. Lett. *6* (1988) 683.
8. J.P. Rozet, A. Chetioui, P. Bouisset, D. Vernhet, K. Wohrer, A.
   Touati, C. Stephan and J.P. Grandin, Phys. Rev. Lett. *58* (1987) 337.
9. A. L'Hoir, S. Andriamonje, R. Anne, N.V. Castro-Faria, M. Chevalier,
   C. Cohen, J. Dural, M.J. Gaillard, R. Genre, M. Hage-Ali, R. Kirsch,
   B. Farizon-Mazuy, J. Mory, J. Moulin, J.C. Poizat, Y. Quere, J.
   Remillieux, D. Schmaus and M. Toulemonde, Nucl. Instrum. Methods Phys.
   Res. *B48* (1990) 145.

INFLUENCE OF THICKNESS FLUCTUATIONS ON THE ENERGY-LOSS SPECTRA
OF PROTONS TRANSMITTED THROUGH THIN FILMS

N.E. Capuj[1] and M.M. Jakas

Departament de Física Aplicada
Universitat d'Alacant
Apt. 99, E-03080 Alacant, Spain

1. Permanent address: Centro Atómico Bariloche
C.C. 439
RA-8400 Bariloche, Argentina

ABSTRACT

Structures appearing in the energy-loss spectra of energetic protons transmitted through thin Al films, such as peaks and shoulders, are studied as a function of both the bombarding energy and the tilt angle of the foil with respect to the beam direction. The results not only confirm previous evidences indicating that the foils contain thickness inhomogeneities, but also suggest that such inhomogeneities can be attributed to a discrete set of thicknesses.

Evidences accumulated during the past years suggest that films employed in beam-foil transmission experiments are frequently not so homogeneous as one would desire. Results indicate that foils do not have a well-defined thickness but a whole distribution of them [1,2]. Thus, the final state of the transmitted ions becomes *a priori* uncertain owing to both the statistical nature of the ion-target interactions *and* the fluctuations in the amount of stopping material that each ion may have "seen" during its passage through the inhomogeneous foil. Hitherto we shall refer to thickness as the amount of the stopping material along the ion trajectory.

In this paper we show energy-loss spectra of protons transmitted through thin Al films containing features such as peaks and shoulders. The spectra are analyzed by means of data-processing techniques borrowed from digital imagery which allow us to enhance and identify unambiguously those structures in the energy-loss spectra. We study the behavior of such structures with both the tilt-angle of the foil and the bombarding energy.

Interaction of Charged Particles with Solids and Surfaces
Edited by A. Gras-Martí et al., Plenum Press, New York, 1991

577

In this experiment the original equipment [3] was modified in order to make it possible to tilt the foil with respect to the beam direction. The measured energy resolution of the whole system, including the energy spread of the incoming beam, is 0.3 %.

The target foils are obtained by an evaporation technique described in detail elsewhere [4]. TEM studies of the foils indicate the existence of micro-crystals of 50 - 350 Å of lateral dimension, but the diffraction patterns show no evidence of preferential orientations.

In order to isolate the mentioned structures in the spectra we use three different procedures: a correlation technique [5], and two contrast-enhancement algorithms developed for digital-image processing [6,7]. Since they all lead to identical results, in the following we shall describe only that of ref.[6] which showed the best performance. In this technique a new spectrum (contrast-enhanced spectrum or, simply enhanced data) is obtained from the old one according to the equation,

$$ c_j^{(new)} = A\,m_j + B + K\left(c_j^{(old)} - m_j\right), \tag{1} $$

where m_j is the local mean value, i.e., the average count between channels j-n to j+n, and n is the integer part of (WIDTH/2). A, B and K are parameters whose values depend on the type of information one is looking for. In our case A = 0.02, B = 0 and K = 1 provide the best enhancement of the structures in the spectrum. The WIDTH value is chosen equal to the full width at half maximum (FWHM) of the ion-beam energy, typically WIDTH = 5. Furthermore, the enhancement procedure appears to be relatively insensitive to previous smoothing of the data. In fact, unless a severe smoothing was applied to the raw data, one ends up with the same enhanced data in most cases.

Figs.1a and 1b show the energy-loss spectra for 100 keV H^+ on 315 Å and 250 Å Al foils [8], respectively. In both cases the foils are tilted 45° with respect to the beam direction. The corresponding enhanced data, obtained from the previous spectra after applying eq.(1), are also plotted in the same figure.

In the first place, one can observe that the largest FWHM belongs to the thinnest foil. This is a clear indication that the thinner foil contains more inhomogeneities than the thick one. In fact, if these foils were uniform the FWHM's would be proportional to their thicknesses. The measured/theoretical FWHM [9], in eV, in these foils are 1990/1430 and 2680/1260, for the 315 Å and 250 Å foil, respectively.

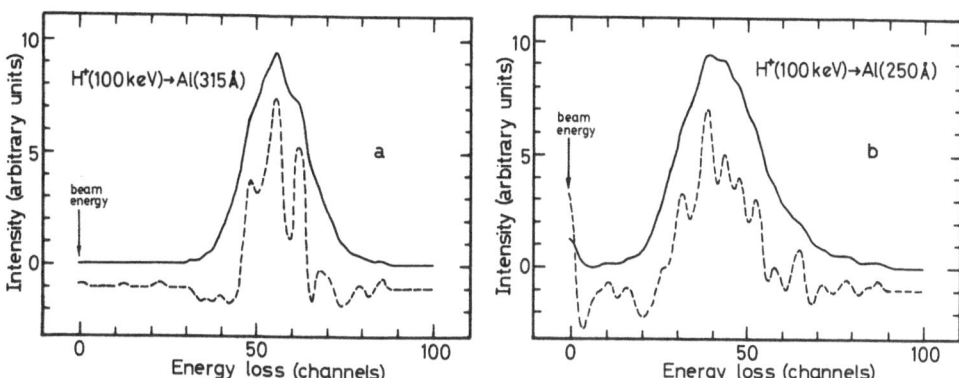

Figure 1. Energy-loss spectra (——) and enhanced spectra (---), for 100 keV H$^+$ on Al (a) 315 Å, ρ = 0.03 and (b) 250 Å, ρ = 0.3. Both targets with a tilt-angle ϕ = 45°. ρ is the inhomogeneity coefficient of the foils (see ref.[2]). The enhanced spectra have been shifted along the vertical scale to avoid superpositions with the energy-loss spectra. The mean energy-loss and FWHM corresponding to these foils are: (a) 5370 eV and 2000 eV and, (b) 4180 eV and 2680 eV, respectively.

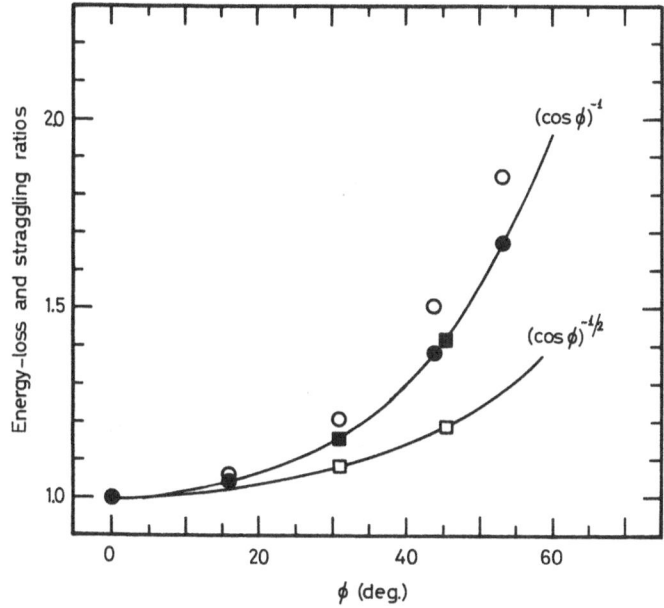

Figure 2. Mean energy-loss $\langle\Delta E\rangle(\phi)/\langle\Delta E\rangle(0)$ (full symbols), and FWHM $W(\phi)/W(0)$ (open symbols), for different tilt-angles of the foil ϕ, relative to those at ϕ = 0. Results are for 100 keV H$^+$ on 250 Å (circles), and 315 Å (squares) aluminum foils, respectively.

Secondly, it can be clearly seen in fig.1 that by applying the enhancement algorithm, the little "bumps" in the energy-loss spectra become more pronounced and isolated as peaks in the enhanced data. Further, the enhanced data corresponding to the thinner foil show a larger number of such peaks than that of the thick one. As we shall see later, this indicates that the thinner foil contains at least as many different thicknesses as peaks appear in the enhanced spectrum.

In fig.2 we plot the mean energy-loss, $\langle\Delta E\rangle(\phi)$, and FWHM, $W(\phi)$, as functions of the tilt-angle of the foil, ϕ, and normalized to the normal-incidence values ($\phi = 0°$). There we can see that the normalized FWHM's of the thinner foil increase with ϕ as $\cos^{-1}\phi$, whereas those of the thicker foil behave like $\cos^{-(1/2)}\phi$. The latter is the result that one expects for a uniform foil. In fact, for an inhomogeneous foil we have FWHM $\propto [x\ \Omega^2 + (\rho\ \Delta E)^2]^{1/2}$, where Ω is the intrinsic energy-loss straggling, x the thickness of the foil, ΔE the mean energy-loss and ρ the roughness coefficient [2]. Hence, given that $x \propto \cos^{-1}\phi$ and assuming that all other parameters are independent of ϕ, it thus follows that FWHM(ϕ)/FWHM$(0) \approx \cos^{-(1/2)}\phi$ implies that $x\ \Omega^2 \gg (\rho\ \Delta E)^2$.

The position of the structures along the energy-loss spectra, ΔE_i, for $i = 1, \ldots, n$, can be easily obtained from the enhanced data. The

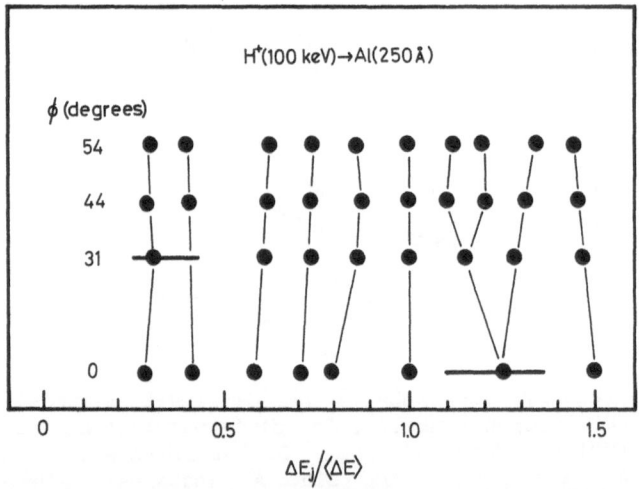

Figure 3. Position of peaks and shoulders ΔE_j, in the energy-loss spectrum corresponding to the foil in fig. 1b, and for four tilt-angles: $0°$, $31°$, $44°$ and $54°$. The positions are obtained from the enhanced spectrum (see text) and given relative to the mean energy loss $\langle\Delta E\rangle(\phi)$.

resulting set of ΔE_i constitutes a pattern characteristic of the foil which is studied as a function of tilt angle and bombarding energy.

In fig.3 we plot $\Delta E_j/<\Delta E>$ for $\phi = 0$, 31, 44, and 54 degree tilt-angle of the foil, where $<\Delta E>$ is the mean energy-loss. One can establish a correlation between the patterns corresponding to different ϕ, as shown by vertical lines. Also, as ϕ increases, new structures appear in the pattern which can be explained by the higher thickness resolution achieved at larger tilt angle [10]. It is worth noticing that the smallest separation between two adjacent structures, $\Delta E_{j+1} - \Delta E_j$, is comparable to the resolution of the detection system, including the beam spread. Also, the separation between peaks can be attributed to differences in thicknesses between 20 and 30 Å.

A study of the patterns obtained for protons transmitted through 280 Å Al foil tilted 44° and bombarding energies ranging from 60 to 100 keV, shows correlations similar to those discussed above. They are not shown here due to space limitations.

For a foil with a well-defined thickness we have an energy loss $\Delta E(E,\phi) = S(E) \; N \; x(\phi)$, where $x(\phi) = x(0)/\cos\phi$, $S(E)$ is the stopping cross-section, and N is the atomic density of the target. One may conceivably assume that a non-homogeneous foil has several thicknesses x_i entering with weight w_i and hence, the mean energy loss can be obtained as

$$<\Delta E>(E,\phi) = \sum w_i \; \Delta E_i (E,\phi),$$

where $\Delta E_i (E,\phi) = S(E) \; N \; x_i (0)/\cos\phi$. Accordingly, $\Delta E_i (E,\phi)/<\Delta E>(E,\phi)$ should be independent of both the tilt-angle and the bombarding energy.

The results in this experiment, for $E \approx 60 - 100$ keV and $\phi \le 60^{\circ}$, do agree with the conclusion above. We may thus conclude that the structures observed in the energy-loss spectra are produced by the interaction of the beam with regions in the foil containing different thicknesses, i.e., as a collection of flat tiles, each one having a well-defined thickness but different to one another. The very origin of such a distribution of thicknesses is not clear at all. One may wonder why a discrete set is being favored over a continuous distribution of thicknesses. To asses whether or not it is linked to the growth mechanism of the foil, or to deviations from a perfectly flat surface in the substrate, or else, further investigation is needed.

ACKNOWLEDGEMENTS

This work has been partially supported by CONICET (Argentina). One of us (MMJ) is grateful to the Spanish Ministerio de Educación y Ciencia MEC for a sabbatical grant, and to the Spanish DGICYT for support (project numbers PS88-0066 and PS89-0065). The authors would like to thank Prof. A. Gras-Marti for a critical reading of this manuscript.

REFERENCES

1. P. Mertens, Thin Sol. Films *60* (1979) 259.
2. N.E. Capuj, J.C. Eckardt, G.H. Lantschner, N.R. Arista and M.M. Jakas, Phys. Rev. *A39* (1989) 1049 (and references therein).
3. G.H. Lantschner, J.C. Eckardt, M.M. Jakas, N.E. Capuj and H. Ascolani, Phys. Rev. *A36* (1987) 4667.
4. A. Valenzuela and J.C. Eckardt, Rev.Sci. Instrum. *42* (1971) 127.
5. W.W. Black, Nucl. Instr. and Meth. *71* (1969) 317.
6. Jong-Sen Lee, IEEE Transactions on Pattern Analysis and Machine Intelligence, Vol. PAMI-2 (1980), 165.
7. R. Wallis, *An Approach to the Space Variant Restoration and Enhancement of Images*, in Proc. Symp. on Current Mathematical Problems in Image Science, Naval Postgraduate School, Monterey, CA, Nov. 1976.
8. The mean target thicknesses were evaluated using standard stopping-power tables: H.H. Andersen and J.F. Ziegler, *Hydrogen Stopping Power and Ranges in All Elements*, Pergamon, New York (1977).
9. W.K. Chu, Phys. Rev. *A13* (1976) 2057.
10. The thickness resolution increases with angle because the energy difference between two adjacent structures, $\delta E(\phi) = \Delta E_{j+1}(\phi) - \Delta E_j(\phi)$, is proportional to $1/\cos\phi$, while the energy width of each structure turns out to be $\propto [1/\cos\phi]^{1/2}$.

CONTRIBUTED PAPERS: STOPPING OF ELECTRONS

INELASTIC INTERACTIONS OF ELECTRONS WITH SOLIDS:
LPA AND PENN'S STATISTICAL MODEL

R. Mayol, J.M. Fernandez-Varea, F. Salvat and
D. Liljequist[1]

Facultat de Física (ECM), Universitat de Barcelona
Societat Catalana de Física (IEC)
Diagonal 647. 08028 Barcelona Spain

1. University of Stockholm, Department of Physics
Vanadisv. 9. S-113 46 Stockholm Sweden

ABSTRACT

Inelastic mean free paths and stopping powers of low-energy electrons in aluminum and copper have been calculated by using the statistical model proposed by Penn [Phys. Rev. B 35(1987)842] and the Local Plasma Approximation (LPA). LPA optical oscillator strengths (OOS), computed from Dirac-Hartree-Slater self-consistent densities under Wigner-Seitz boundary conditions, are compared with experimental OOS. The origin of systematic differences between the LPA and Penn's mean free paths and stopping powers, already found by other authors, is discussed in terms of the discrepancies between the LPA and the experimental OOS.

1. THEORY

Inelastic interactions of electrons in matter can be described by considering the medium as an admixture of free electron gases of different densities with weights determined by the optical oscillator strength (OOS). This idea, formulated in a slightly different way, was used by Tung et al.[1] to compute mean free paths and stopping powers of electrons in solids employing the OOS provided by the Local Plasma Approximation (LPA) of Lindhard and Scharff [2]. The LPA is based on the assumption that the electrons in each volume element of the target respond as if they were in a free electron gas of the same density. Since the OOS of a free electron gas reduces to a delta function at the plasma energy, the LPA allows the calculation of the OOS from only the knowledge of the local electron density. A refinement of this method was developed by Penn [3] (see also

Interaction of Charged Particles with Solids and Surfaces
Edited by A. Gras-Martí *et al.*, Plenum Press, New York, 1991

585

refs. [4–6]), who used experimental OOS to avoid the uncertainties introduced by the LPA.

For the present purposes, it is convenient to characterize a free electron gas by the plasmon energy E_p defined by (we use atomic units, $e = \hbar = m = 1$, unless otherwise specified)

$$E_p^2 = 4\pi\rho, \tag{1}$$

where ρ is the electron density (i.e., number of electrons per unit volume). The Fermi energy is $E_F = 0.8853\ E_p^{4/3}$. The generalized oscillator strength (GOS) per electron in the gas is given by

$$F_L(E_p;Q,W) = \frac{2W}{\pi E_p^2}\ \mathcal{I}m\left[\frac{-1}{\varepsilon_L(E_p;Q,W)}\right], \tag{2}$$

where ε_L is the Lindhard dielectric function [7]. The variables W and Q are the energy loss and recoil energy ($Q = q^2/2$) respectively.

In order to facilitate the calculations, it is convenient to use a simple approximation to the GOS (2). In the present work we use the following 'two-modes' model

$$F_L(E_p;Q,W) \simeq [1-h(Q)]\ \delta\{W-[E_p+BQ]\} + h(Q)\ \delta(W-Q), \tag{3}$$

where

$$B = \frac{6}{5}\frac{E_F}{E_p}\ , \qquad h(Q) = \min\left\{\frac{A}{E_p^2}\frac{Q^3}{E_p+Q},1\right\}, \tag{4}$$

and A is a parameter determined in such a way that the energy-loss differential cross section obtained from (3) agrees exactly with the one derived from ε_L in the low-W limit [7]. Furthermore, the plasmon dispersion relation contained in (3), i.e. $W = E_p + BQ$, also agrees with the low-Q limit of ε_L. Stopping powers and inverse mean free paths from (3) can be computed analytically and are in excellent agreement with the values obtained from the Lindhard dielectric function. Simpler approximate GOS have been used before [3,5,6], however they fail to reproduce the differential cross section for energy losses below the plasma energy.

Tung, Ashley and Ritchie [1] adopted the OOS obtained from the LPA by using Dirac-Hartree-Slater (DHS) atomic electron densities $\rho(r)$ computed

586

using Wigner–Seitz (WS) boundary conditions. The LPA sets the OOS spectrum as

$$\left[\frac{df}{dW}\right]_{LPA} = \int \rho(r)\ \delta\left[W - \chi\ E_p(r)\right]\ dr, \tag{5}$$

where $E_p(r)$ is the local plasmon energy, i.e., $E_p^2(r) = 4\pi\rho(r)$, and χ is a parameter that Tung et al.[1] set equal to unity. Their approach is equivalent to using the following GOS

$$\left[\frac{df(Q,W)}{dW}\right]_{LPA} = \int_0^\infty \left[\frac{df}{dW'}\right]_{LPA} F_L(W';Q,W)\ dW', \tag{6}$$

where $F_L(W';Q,W)$ is the GOS per electron in an electron gas whose plasmon energy is W'.

The procedure proposed by Penn [3] goes beyond the LPA by incorporating experimental OOS. The GOS used in Penn's model is given by

$$\left[\frac{df(Q,W)}{dW}\right]_{Penn} = \int_0^\infty \left[\frac{df}{dW'}\right]_{exp} F_L(W';Q,W)\ dW', \tag{7}$$

where $[df/dW]_{exp}$ stands for the OOS derived from available experimental optical data [8]. It is assumed to be consistent with currently accepted values of the mean excitation energy I, defined by

$$Z\ \ell n(I) = \int_0^\infty \ell n(W)\ \left[\frac{df}{dW}\right]_{exp}\ dW. \tag{8}$$

Therefore, the stopping power derived from the GOS (7) will agree with the Bethe formula at high enough kinetic energies. This is not the case for the LPA with $\chi = 1$. The accepted I value, and hence the high–energy stopping power, is reproduced within the LPA by using the value of χ given by

$$\ell n(\chi) = \ell n(I) - \frac{1}{Z}\int \rho(r)\ \ell n\left[E_p(r)\right]\ dr. \tag{9}$$

The mean free path λ and the stopping power S for electrons of kinetic energy E in a medium containing N scattering centers per unit volume are

$$\lambda^{-1} = N \int dW \int dQ \, \frac{\pi}{E} \, \frac{1}{WQ} \, \frac{df(Q,W)}{dW},$$

$$\text{(10)}$$

$$S = N \int dW \int dQ \, \frac{\pi}{E} \, \frac{1}{Q} \, \frac{df(Q,W)}{dW},$$

where the integrations extend over the kinematically allowed values of Q and W,

$$\{\sqrt{E}-\sqrt{(E-W)}\}^2 \leq Q \leq \{\sqrt{E}+\sqrt{(E-W)}\}^2, \quad 0 \leq W \leq E. \tag{11}$$

2. RESULTS AND DISCUSSION

Experimental OOS for Al and Cu [8] are compared with the OOS obtained from the LPA in fig.1. The values of the mean excitation energy, obtained by integrating the experimental OOS, are I = 165 eV and 321 eV for Al and Cu, respectively. The value of χ, as given by (9), is 1.31 for Al and 1.26 for Cu. LPA-OOS, with this value of χ, have been calculated from DHS atomic densities with WS boundary conditions by using numerical differentiation of the integrated radial density. The LPA-OOS always drop

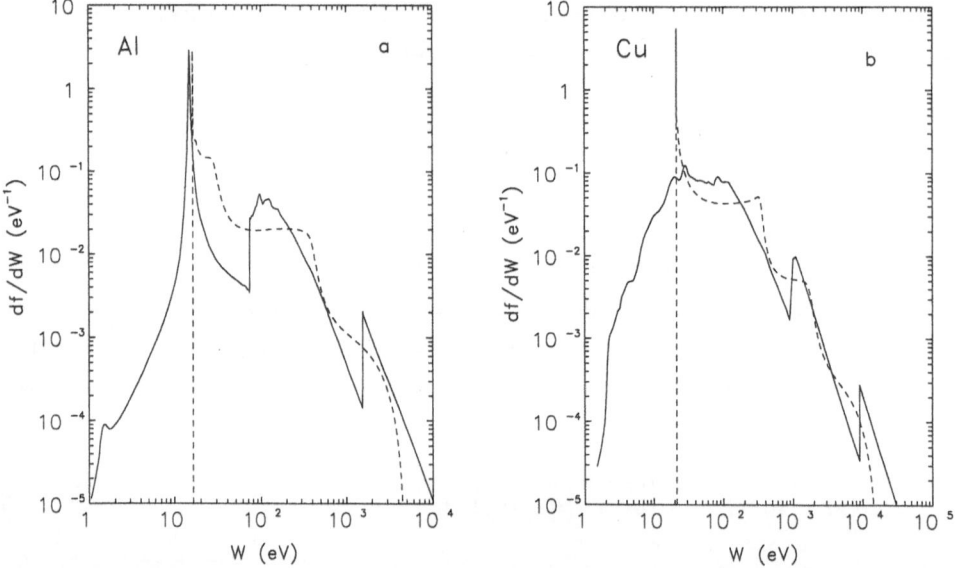

Figure 1. OOS derived from experimental optical data (continuous curves) and obtained from the LPA formula (5) with DHS densities. (Dashed curves: χ = 1.31 for Al and χ = 1.26 for Cu).

to zero at some maximum energy loss which is determined by the value of the density at the nucleus. For free-electron-like materials, such as Al, the LPA roughly reproduces the plasmon line (which is located at the lower end of the excitation spectrum) since the density at the WS radius nearly coincides with the conduction electron density (bound electrons do not move so far from the nuclei). For other materials, as exemplified by Cu, the LPA leads to an unrealistic plasmon-like peak due to the non-vanishing value of ρ at the WS radius. This behavior is different from what we have in gases [9], where the LPA gives an OOS spectrum starting from W = 0

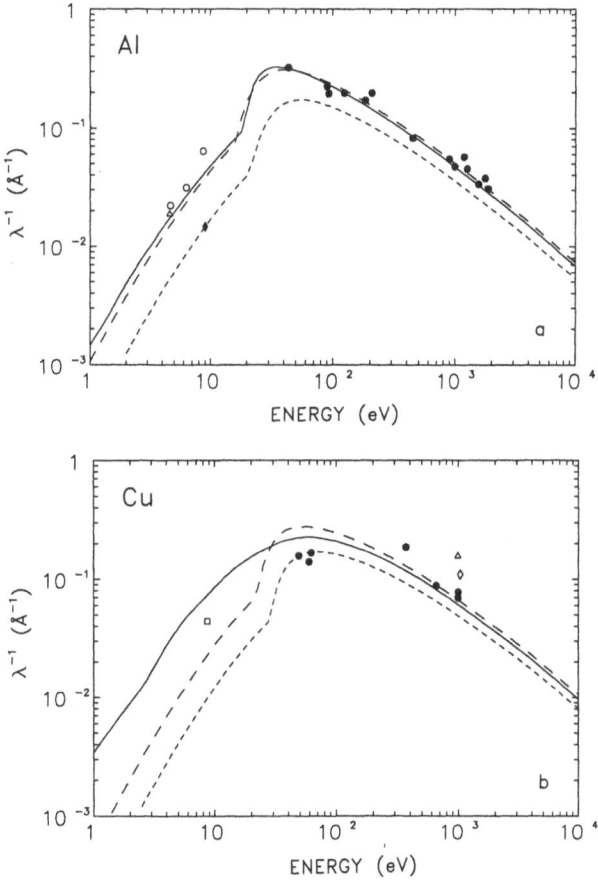

Figure 2. Inverse mean free paths for electrons in Al and Cu as a function of the electron energy above Fermi level. The continuous curves are the results from Penn's model using the experimental OOS given in fig.1. The short-dashed curves are LPA results obtained from the LPA-OOS with $\chi = 1.31$ for Al and 1.26 for Cu. LPA mean free paths calculated with $\chi = 1$ are also given (long-dashed curves). Experimental data from ref.[1].

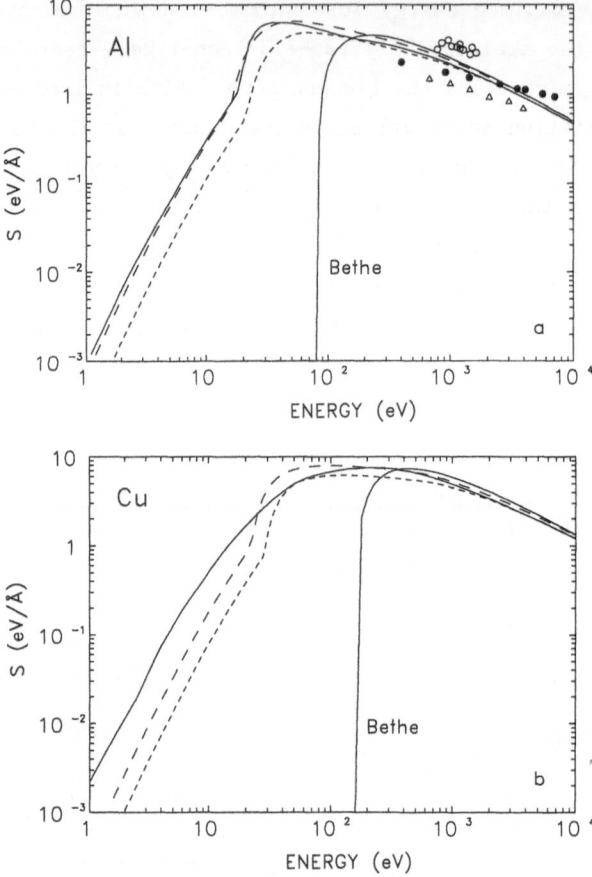

Figure 3. Stopping powers for electrons in Al and Cu. The curve labelled
'Bethe' is the stopping power computed from the non-relativistic
Bethe formula. Other details are the same as in fig.2.

(since then the density decreases smoothly when r goes to infinity).
Although the LPA OOS tries to roughly reproduce even the absorption edges,
it differs in shape from the experimental OOS and hence, it may give
sizable errors when used to calculate mean free paths and stopping powers.

Inverse mean free paths and stopping powers for Al and Cu computed
from Penn's model (7) and from the LPA (6) with the adjusted value of χ
(see eq.(9)) and with $\chi = 1$ are given in figs.2 and 3 as a function of the
electron energy above Fermi level. Expressions (6) and (7) have been
evaluated using the two-modes model (3); notice that exchange and
correlation effects have not been introduced (see, e.g., ref.[10]). For
low energies, the LPA results lie below the results of Penn's model. This

fact indicates that the LPA tends to concentrate the OOS at too high energy losses. In the case of Al, there is a clear excess of the LPA OOS between the plasmon line and the L absorption edge. Indeed, the result of using the adjusted value of χ (>1) is to shift the OOS towards higher energy losses and therefore, the resulting LPA inverse mean free paths and stopping powers are even smaller than those for $\chi = 1$.

As a matter of fact, the LPA OOS (5) with the value of χ determined from (9) fulfills the Bethe sum rule and, when used in (8), it yields the accepted value of I. These two properties altogether determine the high-energy stopping power (Bethe formula). Therefore, the good agreement between the stopping powers computed from Penn's model and from the LPA, with the adjusted χ value, found for energies above 1 keV was to be anticipated. This is not the case for other quantities, e.g. the mean free path, which follow high energy formulas determined by other moments of the OOS which are more or less distorted by the LPA [9]. From the present results it is quite evident that the LPA with $\chi = 1$ gives more accurate mean free paths than those obtained with the adjusted χ value. Hence, the use of the LPA with the adjusted χ value should be restricted to stopping power calculations (and to energies above 100 eV). For the evaluation of inelastic mean free paths within the LPA it is preferable to use $\chi = 1$ (as it was done by Tung et al.[1]).

ACKNOWLEDGEMENT

This work has been partially supported by the Spanish Comisión Asesora para la Investigación Científica y Técnica, project no. PB86-0589.

REFERENCES

1. C.J. Tung, J.C. Ashley and R.H. Ritchie, Surf. Sci. *81* (1979) 427. (The references in this paper include earlier applications of the LPA).
2. J. Lindhard and M. Scharff, Kgl. Danske Videnskab. Selskab., Mat.-Fys. Medd. *27* (1953) num. 15.
3. D.R. Penn, Phys. Rev. *B35* (1987) 842.
4. S. Tanuma, C.J. Powell and D.R. Penn, Surf. Interf. Anal. *11* (1988) 577.
5. J.C. Ashley, J. Electron Spectrosc. Related Phenomena *46* (1988) 199.
6. Z-J. Ding and R. Shimizu, Surf. Sci. *222* (1989) 313.
7. J. Lindhard, Kgl. Danske Videnskab. Selskab., Mat.-Fys. Medd. *28* (1954) num. 8; J. Lindhard and A. Winter, Kgl. Danske Videnskab. Selskab., Mat.-Fys. Medd. *34* (1964) num. 4.
8. E.D. Palik (Ed.), *Handbook of Optical Constants of Solids*, Academic Press, New York (1985); H.-J. Hagemann, E. Gudat and C. Kunz, Deutsches Electronen-Synchrotron Report SR-74/7, Hamburg (1974); J.H. Hubbell, Atomic Data 3 (1971) 242.
9. P.E. Jonhson and M. Inokuti, Comm. Atom. and Mol. Phys. *14* (1983) 19.
10. D.R. Penn, Phys. Rev. *B13* (1976) 5248.

ENERGY DISTRIBUTION OF ELECTRONS
EXCITED BY SLOW PARTICLES IN SOLIDS

Z. Sroubek[1] and G. Falcone[2]

1. Czechoslovak Academy of Sciences URE
   Lumumbova 1, 18251 Prague 8, Czechoslovakia

2. Dipartimento di Fisica
   Universita della Calabria
   I-87036 Rende (CS), Italy

ABSTRACT

An expression for the energy distribution $W(\varepsilon)$ of electrons excited by slow particles in solids is derived using the local representation for the electron wave functions. The expression for $W(\varepsilon)$ is then extended to include inner shells by orthogonalization. Possible modification of the shape of $W(\varepsilon)$ by the electron-electron interaction is shortly discussed.

1. INTRODUCTION

The interaction of very slow particles in condensed matter results in production of excited electrons by induced transitions in the valence band of the solid. Detailed knowledge of the excitation process and of the temporal and spatial evolution of excited electrons is needed for better understanding of secondary electron energy spectra [1], ionization of secondary atomic particles [2] and electron-hole production in semiconductors [3]. In this contribution a method is described for evaluating the energy distribution $W(\varepsilon)$ of excited electrons in solids before the distribution is altered by transport and by electron-electron interaction. In the method the electrons of the solid are represented by localized pseudofunctions and the moving particle by the corresponding pseudopotential. The approach, based on local representation, allows to identify contributions to $W(\varepsilon)$ from individual atomic collisions. The use of a pseudopotential is justified for slow collisions when the inner shells are not excited by the level-crossing mechanism. However, even for such slow processes, where only valence electrons are involved, the calculation of high energy parts of $W(\varepsilon)$ may be strongly affected by the admixture of core states in the valence band wave functions. A formula is

Interaction of Charged Particles with Solids and Surfaces
Edited by A. Gras-Martí *et al.*, Plenum Press, New York, 1991

593

presented which allows to evaluate this contribution of core states quantitatively. Finally, the application of the method is illustrated using as an example an Al atom moving in aluminum metal. The theoretical $W(\varepsilon)$ is compared with recent experiments on electron emission, and possible reasons for the differences are shortly discussed.

2. THE CALCULATION OF THE ELECTRON ENERGY DISTRIBUTION $W(\varepsilon)$

The distribution $W(\varepsilon)$ is defined as the probability that an electron is excited by the moving particle in the energy levels between ε and $\varepsilon+d\varepsilon$ above the Fermi level $\varepsilon_f = 0$. The distribution is calculated in first-order perturbation theory using as the perturbation the pseudopotential $V(r-R(t))$ of the moving particle at $R(t)$. Then the formula for $W(\varepsilon)$ is [4]

$$ W(\varepsilon_{mk}) = \frac{1}{2} \sum_n \int_{-b}^{0} d\varepsilon_{nk'} \ \rho \ \frac{d\Omega_k}{4\pi} \frac{d\Omega_{k'}}{4\pi} $$

$$ \times \left| \int_{-\infty}^{\infty} dt \ \langle \psi_{mk} | V(r-R(t)) | \psi_{nk'} \rangle \ \exp\left[i(\varepsilon_{mk}-\varepsilon_{nk'})t \right] \right|^2 , \qquad (1) $$

where ρ is the electronic density of states and the bottom of the valence band is denoted by $-b$. The surfaces of constant ε are assumed to be spherical.

The electron wave function ψ_{nk} is usually represented in the free electron approximation. However, for low energy processes it may be more appropriate to describe the electrons in the local representation. The electronic excitation caused by individual binary atomic collisions can then be better identified. Furthermore the formulation in the local representation facilitates possible application of higher-order perturbation theories.

Generally, the wave function of a solid is given in the local representation by

$$ \psi_{nk}(r) = A_{nk} \sum_l \varphi_{nkl}(r-R_l) \ \exp(ik.R_l), \qquad (2) $$

where the summation is over all lattice points R_l and A_{nk} is the normalization factor. At each lattice point the wave function φ_{nkl} is a linear combination of s, p, and possibly also of d-like local pseudofunctions. This representation is well known from the field of semiconductor band structure calculation [5] but it is less popular in the description of metals. The correctness of the local pseudopotential approach can only be estimated by comparing the calculated and real band structures for any particular case. If (2) is substituted in (1) the following relation is obtained

$$W(\varepsilon_k) = \frac{\rho}{2} \int_{-b}^{0} d\varepsilon_{k'} \, S_{kk'}$$

$$\left| A_k A_{k'} \int_{-\infty}^{\infty} dt \sum_{ll'} \langle \varphi_{kl} | V(r-R(t)) | \varphi_{k'l'} \rangle \, \exp\left[i(\varepsilon_k t - \varepsilon_{k'} t - k.R_l + k'.R_{l'}) \right] \right|^2 ,$$

$$(3)$$

where $S_{k,k'}$ denotes the averaging over all k and k' vectors which correspond to the same ε_k and $\varepsilon_{k'}$ respectively. To simplify the notation the band indices n and m in (3) have been dropped and ρ is set in good approximation equal to ρ at the Fermi energy. The actual evaluation of (3) is simplified by the observation that only atoms (and thus φ_{kl} in (3)) nearest to the moving particle contribute significantly to $W(\varepsilon)$.

To be able to analyze in more detail the shape of $W(\varepsilon)$ at high ε, i.e., in the region which is important in electron emission, the following extension of (3) (still within first-order perturbation) is necessary: The pseudofunction describes correctly the electron distribution only outside the core region. To make it correct in the whole region the pseudofunction must be orthogonalized to the core functions φ_{cl} and possibly also renormalized. The true wave function χ_{nkl} (except for the renormalization factor) is then

$$\chi_{nkl} = \varphi_{nkl} - \sum_c \langle \varphi_{cl} | \varphi_{nkl} \rangle \, \varphi_{cl}. \tag{4}$$

Also the pseudopotential $V(r)$ must be replaced in the core region by the coulomb-like potential. The new potential will be denoted by $V_t(r)$. Usually only one of the most violent collisions gives significant contribution to $W(\varepsilon)$ and thus only one diagonal matrix element of V needs to be considered in (3). If we denote by χ_{nkp} the true wave function localized on the recoiling atom at the position $R_p = 0$ the expression (3) yields

$$W(\varepsilon_k) = \frac{\rho}{2} \int_{-b}^{0} d\varepsilon_{k'} \, S_{k,k'} \times$$

$$\times \left| A_k A_{k'} \int_{-\infty}^{\infty} dt \ <\chi_{kp}| V_t(r-R(t))|\chi_{k'p}> \ \exp\left(i(\varepsilon_k - \varepsilon_{k'})t\right) \right|^2. \tag{5}$$

The band indices have been again dropped. It is important to note that (5) contains the Fourier transform of matrix elements like $<\varphi_c|V_t|\varphi_c>$. Due to the strong localization of the deep core states φ_c these matrix elements change very rapidly with time, certainly faster than $<\varphi_{kl}|V|\varphi_{kl}>$ in (3). Thus eq.(5) may give a significantly larger amplitude of $W(\varepsilon)$ than eq.(3) at large energies ε.

3. APPLICATION FOR Al

To illustrate the method based on eq.(3) we have calculated the energy distribution $W(\varepsilon)$ of electrons excited by an Al atom moving in the fcc lattice of Al from its original position at a00 among four nearest neighbors to the next nearest atom at 000. By the collision with the atom at 000 the moving particle is stopped. The trajectory has been deduced from molecular-dynamics calculations [6]. The pseudopotential has been fitted by two gaussians in such way that the band structure calculated with help of this pseudopotential reproduces the real band structure within a few percent. The pseudopotential has the form

$$V(r)= 92.1 \ \exp(-2.7r^2) - 79 \ \exp(-1.35r^2), \tag{6}$$

with r in Å and V in eV.

The wave functions are described by linear combinations of s and p gaussians $\exp(-Ar^2)$ and $x_i\exp(-Ar^2)$, respectively, where A is equal to 0.62 Å^{-2}. The details of the calculation will be published elsewhere, here only some results regarding $W(\varepsilon)$ will be quoted.

The calculated energy distributions for average kinetic energies of the moving atom equal to 15, 30 and 50 eV are shown in fig.1.

The energy distributions are very narrow, with negligible amplitude above $\varepsilon = 2k_f v$, in general agreement with the calculation based on the free-electron approximation, also as far as the shape is concerned. This finding is in accord with the estimation made previously [4]. The local approach is however less k-dependent, giving nonzero contributions to $W(\varepsilon)$, even for k = 0, and gives slightly larger energy losses.

The extension of the results to higher kinetic energies of the moving particle (but energies still below the level-crossing excitation threshold) shows that $W(\varepsilon)$ remains very narrow. For example, for a kinetic energy of 500 eV the value of $W(\varepsilon)$ would be negligible above $\varepsilon = 1.3$ eV.

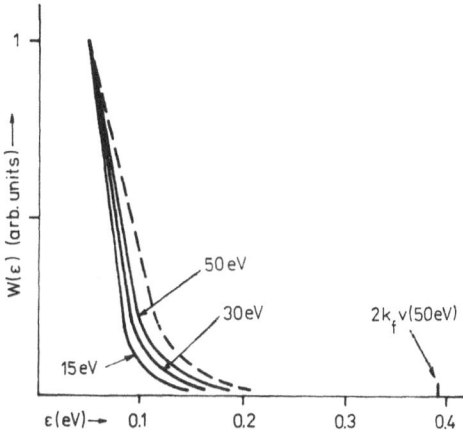

Figure 1. The energy distributions of electrons excited by an Al atom moving with average kinetic energies 15, 30 and 50 eV calculated with the use of eq.(3). $W(\varepsilon)$ calculated in the free-electron approximation for the energy of 50 eV is shown by a dashed line. The distributions are normalized to one at $\varepsilon = 0.05$ eV.

This result is in disagreement with recent experiments on electron emission from Au [1] and W [7], which indicate nonzero $W(\varepsilon)$ still at $\varepsilon > 15$ eV, with a rather slow dependence on ε described approximately by the relation

$$W(\varepsilon) \propto \exp(-\varepsilon/Q), \tag{7}$$

where Q has a value around 2 eV. Such experimental observation can hardly be explained using the relation (3). It is conceivable that core states need to be included as it is done in eq.(5) to obtain a more pronounced tailing of $W(\varepsilon)$.

Another possibility which needs to be studied in more detail is the role of the electron-electron interaction in the broadening of the distribution $W(\varepsilon)$. At this stage the estimation of this effect can be done only qualitatively and is rather speculative. As an example we may consider a local electronic excitation due to a typical electronic energy loss process of $\Delta E = 20$ eV. If this excitation remains sufficiently localized during the electronic thermalization process ($< 10^{-14}$ s), so that only about 50 electrons are involved, the effective electronic temperature T_e can be obtained from the electronic specific heat relation

$$k_B T_e = (2\ \Delta E\ \varepsilon_f / N\ \pi^2)^{1/2}. \tag{8}$$

Taking N = 50 and $\varepsilon_f = 13.5$ eV for the Fermi energy of Al, the relation (8) yields $kT_e = 1$ eV, which is of the same order of magnitude as Q in eq.(7). This estimate illustrates that under conditions of a strong localization the e-e interaction may indeed considerably influence the shape of $W(\varepsilon)$.

4. CONCLUSION

An expression for the electron energy distribution $W(\varepsilon)$ is derived using the local representation for the electron wave functions. The use of the local representation enables to identify contributions of binary collisions and facilitates the use of higher-order perturbation. The expression for $W(\varepsilon)$, within the first-order perturbation, has then been extended to include inner shells by orthogonalization. Finally the possible effect of the electron-electron interaction on the shape of $W(\varepsilon)$ is shortly discussed. Precise analysis of higher-order perturbation processes and of the e-e interaction is needed to clarify the experimentally observed high-energy tail of $W(\varepsilon)$.

REFERENCES

1. E.V. Alonso, M.A. Alurralde and R.A. Baragiola, Surf. Sci. Lett. *166* (1986) L155.
2. G. Falcone and Z. Sroubek, Rad. Effects and Defects in Solids *109* (1989) 253.
3. A. Amirav and M.J. Cardillo, Phys. Rev. Lett. *57* (1986) 2299.
4. G. Falcone and Z. Sroubek, Phys. Rev. *B39* (1989) 1999.
5. E.O. Kane, Phys. Rev. *13* (1976) 3478.
6. B.J. Garrison, private communication.
7. A. Oliva, private communication.

ION-INDUCED AUGER ELECTRON EMISSION FROM SINGLE CRYSTALS:

EXPERIMENT AND SIMULATION

P.F.A. Alkemade[1], W.N. Lennard and I.V. Mitchell

Interface Science Western, Dpt. of Physics
The University of Western Ontario
London, Ontario, N6A 3K7, Canada

1. Present address: Delft Inst. for Micro Electronics &
   Submicron Technology (DIMES)
   Delft University of Technology, Lorentzweg 1
   2628 CJ Delft, The Netherlands

ABSTRACT

The energy spectra of the Si KLL and KLM Auger electrons produced by bombardment of a Si(100) single crystal with a 1 MeV He$^+$ beam, are measured. The incident beam is aligned with a major crystallographic direction or with a random direction. The shapes of the measured Auger spectra are successfully explained using an efficient Monte Carlo simulation code for the electron transport. Different models for the energy loss distribution are compared with each other. In the comparison with the experiment, the local density approximation of the dielectric response model for the inelastic scattering cross section and the quantum mechanical phase-shift model for the elastic scattering cross section are used. The experiments and the Monte Carlo simulations show an enhancement of the spectral intensity within 150 eV of the initial Auger electron energy. The enhancement is caused by a reduction of the effective stopping power for the majority of the electrons created in the surface region.

1. INTRODUCTION

In recent years, it has been realized that elastic scattering of electrons affects the intensity of the signal measured in techniques such as photoemission spectroscopy (XPS) and Auger electron spectroscopy (AES) [1]. In order to study this effect theoretically, several authors have used analytical models for the transport of electrons in solids [2-4]. Others, however, preferred to use Monte Carlo simulations [5-8]. The main disadvantage of the analytical approach is the need for approximations in order to avoid calculations that are too complicated. The main disadvantage of the Monte Carlo approach is the need

Interaction of Charged Particles with Solids and Surfaces
Edited by A. Gras-Martí *et al.*, Plenum Press, New York, 1991

599

for long computation times and, sometimes, a lack of clarity in the interpretation of the results.

It has been demonstrated that ion-induced Auger spectroscopy, in combination with ion channeling, can be used to study the transport of electrons in solids [9,10]. Energy spectra of KLL electrons emitted from Al and Si single crystals, have been analyzed using analytical models described by Tougaard and Sigmund [2]. However, the results indicated that the (transport) mean free path for elastic scattering of the electrons was about an order of magnitude smaller than that expected theoretically. It has been suggested that the analytical approach was oversimplified, and that Monte Carlo simulations may prove to be more useful [10].

In the present study, the energy spectra of ion-induced KLL Auger electrons, emitted by 1 MeV He$^+$ bombardment of a Si(100) single crystal, are reexamined using Monte Carlo simulations, instead of analytical models. After a brief description of the experimental results, we describe the Monte Carlo code. Subsequently, we discuss the effects of various models, used in our simulations, on the shape of the Auger electron spectra. Finally, we compare simulated with measured spectra.

2. EXPERIMENTAL RESULTS

The experiments were performed in a UHV chamber, connected to the

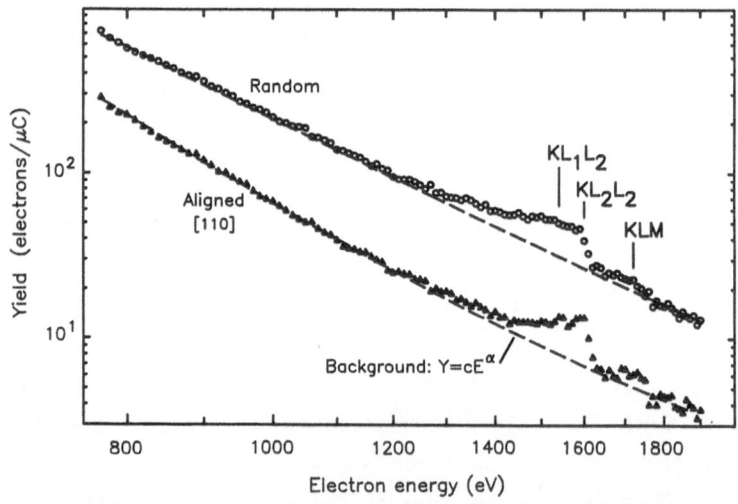

Figure 1. Secondary electron spectra from Si(100), measured under random and channeling conditions of the incident 1 MeV He$^+$ beam. (Note the double logarithmic scale).

2.5 MV Van de Graaff accelerator of the University of Western Ontario. The chamber was equipped with an argon sputter gun for surface cleaning, a surface-barrier particle detector, an electron gun, and a hemispherical electron analyzer. The target, a Si(100) single crystal, was mounted on a five-axis goniometer and was heated by a filament at the back side of the crystal.

The incident beam, 1 MeV He$^+$, was aligned with the [110] crystallographic direction of the crystal, or was incident along a random direction. The angle between the incident beam and the surface normal was in all cases 45°. Both backscattered ions and secondary electrons, in the 800 to 2000 eV energy range, were collected. The angle between the electron analyzer and the surface normal was 45°. The energy resolution of the analyzer was 12 eV (fwhm). The measured electron spectra were corrected for the energy dependence of the (relative) sensitivity of the analyzer. Details about the equipment and the measurements can be found in previous publications [9-11].

Fig.1 shows secondary electron spectra. At about 1600 eV and below, the KLL-Auger peak is visible, superposed on background of electrons emitted by direct Coulomb ionization. The shape of the background spectrum is fitted with a function of the form $Y_{bgr}(E) = c\ E^{\alpha_0 + \alpha_1 E}$ (where c, α_0 and α_1 are fitting parameters). After subtraction, the spectra of the Auger electrons are obtained as shown in figs.2 and 3. For more details about the method of background subtraction, see ref.[10]. The area of the

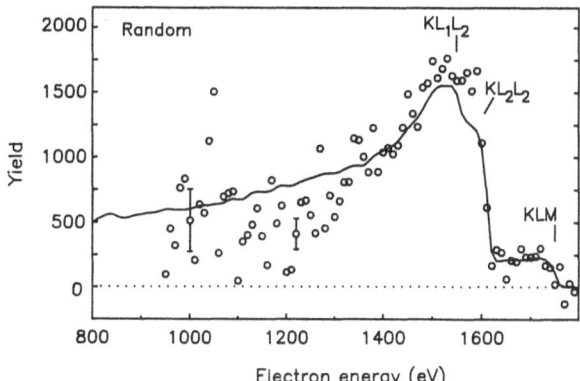

Figure 2. The Si KLL and KLM Auger-electron spectrum produced by a 1 MeV He$^+$ beam, incident in a random direction. The beam dose is 84 µC. The error bars at 1000, 1200, and 1500 eV indicate the uncertainty in the data, due to the method of background subtraction. The full curve is the simulated spectrum.

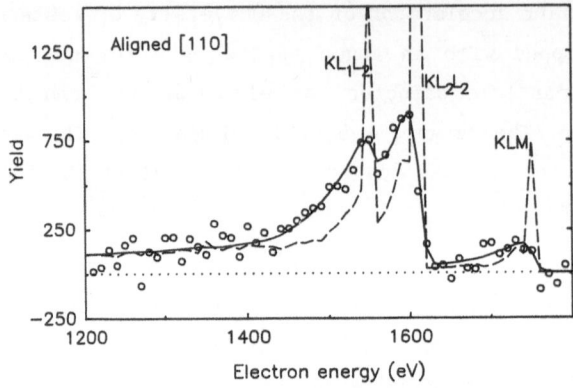

Figure 3. The Auger-electron spectrum measured with the 1 MeV He$^+$ beam incident along the [110] direction of the Si(100) crystal. The beam dose is 140 μC. The full curves show simulated spectra.

surface peak in the ion backscattering spectrum corresponds to an effective layer thickness visible to the beam of 22 ± 4 Å; the minimum yield, χ_{min}, is 6 % [10]. Full curves in figs.2 and 3 are calculated spectra. The method of calculation is discussed in detail in the next section. The Auger electron spectrum measured under channeling incidence conditions differs markedly from that measured under random conditions: the maximum intensity per unit of beam dose of the former is 3 to 4 times lower than that of the latter; furthermore, the intensity drops faster with decreasing electron energy. Obviously, these differences reflect the difference in the regions where the detected Auger electrons are produced: mainly the surface region (the outermost few tens of Å) for the channeling case, and throughout the "whole" solid (the outermost hundreds of Å) for the random case.

3. THE MONTE CARLO SIMULATION METHOD

Many investigators have used Monte Carlo simulations to study the transport of electrons in solids [5-8,12-17]. For our type of experiments, ion-induced Auger-electron emission from a homogeneous solid, we have developed a simulation code which is based partially on codes described in the literature [5-8,12-17]. In addition, it contains features that enhance the speed of computation considerably.

3.1 Construction of the Trajectories

In a Monte Carlo simulation, the trajectories of a large number of

602

electrons, often many tens or several hundreds of thousands, are followed. A trajectory starts at a source, e.g., an external electron beam or an internal electron emitter, and ends in an analyzer. In our simulation code, each trajectory starts at the origin of a coordinate system. The source is isotropic, and located inside the medium. For the moment, the medium is considered to be infinitely large in all directions. The initial directions of a trajectory are chosen isotropically. The mean free path for scattering, $\lambda(E)$, (*elastic* plus *inelastic*), a function of the electron energy E, is calculated using models for the inelastic and elastic scattering cross sections. The mean free path, and a random number, determine the actual distance to the first scattering point. Another random number determines whether the scattering process is elastic or inelastic. In the case of elastic scattering, the scattering angle, ϑ, is chosen according to the differential elastic scattering cross section, $d\sigma_e(\vartheta)/d\Omega$. In the other case, the energy loss, T, is chosen according to the differential inelastic scattering cross section, $d\sigma_i(T)/dT$. The mean free paths, λ, and the differential cross sections, σ, are related by

$$\lambda_i^{-1} = N \int \sigma_i(T)dT, \qquad \text{or} \qquad \lambda_e^{-1} = N \int \sigma_e(\vartheta)d\Omega, \tag{1}$$

where N is the atomic density of the solid.

After the scattering event, the position and the energy of the electron are stored in an array, the mean free paths are updated, and the process is repeated. This process continues until the energy of the electron reaches a preselected minimum value. The whole trajectory is now characterized by the above mentioned array with positions and energies.

Subsequently, we assume that the surface of the solid coincides with the X-Y plane of the coordinate system, and that the solid coincides with the region where the z-coordinate is negative. The starting point of the trajectory is now varied from the surface to some large depth where there is no chance for escape. For each depth element dz, the energy E and the polar and azimuthal angles ϑ and ϕ of possible escape are determined. This result, together with the production rate, r(z)dz, i.e., the number of electrons produced per unit of beam dose in the depth element dz at depth z, determines the contribution, $dY(E,\vartheta,\phi)$, of the depth element, to the spectrum at energy E measured at escape angles ϑ and ϕ. Hence, one trajectory does not provide one single number to one particular spectrum, but many contributions to many spectra. By rotating the trajectory, it can be used many times more to obtain even better statistics. It is noted that the translations are performed continuously (dz is infinitesimally small) while the rotations are performed in discrete steps. Since the construction of the trajectories is the most time-consuming part of the simulation, the calculation time is greatly reduced. However, one cannot reduce the number of trajectories and increase the number of transformations indefinitely: correlations in the results will decrease the extra benefit. It is noted again, that the method of translating and rotating the trajectories to increase efficiency works only for homogeneous solids, or solids with only a small, depth dependent concentration of foreign atoms.

3.2 Models for Inelastic Scattering

Fast electrons lose energy in solids by interactions with the atomic electrons of the solid. Several models for the energy loss can be used in our simulation code. In the so-called *continuous slowing-down approximation* (CSDA), the energy loss, T, is simply proportional to the path length, p, travelled by the electron:

$$T(p) = S(E) \, p, \tag{2}$$

where $S(E)$ is the stopping power of the medium at electron energy E. In reality, electrons lose their energy in discrete amounts, by separate inelastic scattering events. If $\lambda_i(E)$ is the mean free path between two successive inelastic scattering events, the mean energy loss, <T>, per event is

$$<T> = S(E) \, \lambda_i(E). \tag{3}$$

This model is called the *mean energy-loss approximation* (MELA). A better model includes a distribution for the possible energy loss in an inelastic scattering event. Tougaard *et al.*[2] have suggested a distribution function f(T) for the energy loss T,

$$f(T) = \frac{4T}{S(E)^2 \, \lambda_i^2} \, \exp\left(-\frac{2T}{S(E)\lambda_i}\right). \tag{4}$$

This model is called the *exponential energy-loss distribution* (EELD). It is shown in fig.4.

The model that describes the energy loss of a charged particle in a medium much better is based on the dielectric-response theory of Lindhard [18]. Tung *et al.* used this model, and the local electronic density, to calculate the inelastic mean free path and the stopping power

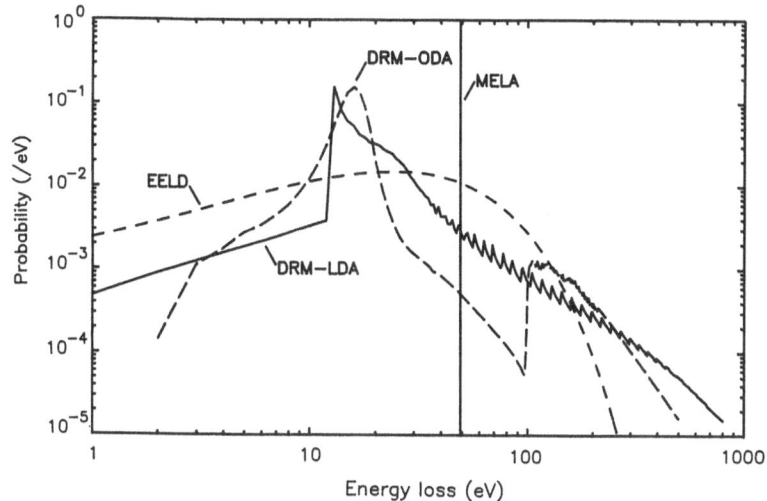

Figure 4. Probability distribution for the energy loss of a 1600 eV
electron in Si, in a single inelastic scattering event. Full
curve: DRM-LDA; broken curve: DRM-ODA; and EELD: T exp(-T);
delta function at 49 eV: MELA. The distributions shown are
normalized to an integrated probability of unity. The irregular
region of the DRM-LDA curve, between 40 and 400 eV, is an
artifact of the calculation.

for various materials [19]. This approach is called the *local-density
approximation* of the *dielectric-response model* (DRM-LDA). The
dielectric-response model can also be used in combination with the
dielectric function of the medium derived from optical data, as is shown
by Powell [20] and by Ding and Shimizu [16]. This approach is called the
optical-data approach of the dielectric-response model (DRM-ODA). Fig.4
shows also the DRM distribution, calculated according to the LDA method of
Tung [19], full curve, or using optical dielectric data, broken curve,
labelled DRM-ODA. The two distributions are peaked near 16 eV,
corresponding to the plasmon energy of silicon. It is noted that the LDA
scheme does not include any shell structure, while the ODA does: the ODA
peak above 100 eV loss represents the energy loss by ionization of the
silicon L shell.

When electrons lose energy, they are deflected as well. The
relationship between deflection angle, ϑ, and energy loss T is [14]

$$\sin\vartheta = \sqrt{T/E}. \qquad (5)$$

This deflection can be included or neglected in our simulations.

3.3 Elastic Scattering

Electrons scatter elastically from the atomic nuclei in the medium. If the interaction is treated as a classical scattering process, the differential scattering cross section, $d\sigma_e/d\Omega$, is the well-known Rutherford formula

$$\frac{d\sigma_e^R(\vartheta)}{d\Omega} = \frac{Z^2 e^4}{16\ E^2} \sin^{-4}(\vartheta/2), \tag{6}$$

where Z is the atomic number of the atom, and ϑ is the scattering angle. Eq.(6) is improved by including the screening of the nucleus by the atomic electrons [14]. However, for energies below several tens of keV, the quantum-mechanical cross section must be used [12]. The calculation of the latter is based on the phase shifts of the different angular-momentum states of the incident plane wave. Tables for the quantum mechanical, or phase-shift, differential cross sections, $d\sigma_e^{PS}/d\Omega$, are available in the literature [21-23]. Our Monte Carlo code uses interpolations of these tables.

3.4 Simulations

We have performed Monte Carlo simulations, using the above mentioned models for inelastic and elastic scattering cross sections. In a simulation, the number of trajectories constructed was 1600. Each trajectory was translated and rotated to improve statistics. The initial energy of the electrons was either 1600 eV, or was chosen to be one of the KLL and KLM Auger energies of silicon [24]: 1555, 1615 or 1755 eV, with relative probabilities of 0.20, 0.67 and 0.13. (The $KL_1L_{2,3}$ - $KL_2L_{2,3}$ ratio stems from [24], the $KL_{2,3}M$ - $KL_{2,3}L_{2,3}$ ratio from the measured intensities, see figs.2 and 3). The energy at which the calculation of a trajectory was terminated was 800 eV. The calculated energy spectra of the

Table 1. Stopping power and mean free paths for electrons in silicon.

Electron energy (eV)	Stopping Power (eV/Å)	DRM λ_i (Å)	Screened Rutherford λ_e (Å)	Phase-shift λ_e (Å)
800	2.20	17.8	7.8	13.1
100	1.84	24.4	11.6	17.5
100	1.59	30.7	15.5	21.6

escaping electrons were combined into groups depending on the polar angle of escape, ϑ. Each group was 10° wide. The spectra were convoluted with a Gaussian function representing the energy resolution of the electron analyzer. The energy of the incident 1 MeV He^+ beam changes by less than 20 keV in the outermost 500 Å of Si, a thickness in which the energy loss of escaping KLL electrons is at least 800 eV. Therefore, the KLL Auger production rate, a function of beam energy, does not change appreciably (not more than 7 % [11]) and we can approximate the random incidence production rate, $r(z)$, by a constant, r_o. In the channeling case, the production rate, $r(z)$, was obtained from analysis of the spectra of the backscattered ions, and by a Monte Carlo simulation of the shadowing effect of the incident ion beam [10]. Note that at $z = 0$, $r(z)$ is equal to r_o.

The stopping power and the mean free paths used in our simulations are summarized in table 1. The values for inelastic scattering were calculated using the local-density approximation of the dielectric-response model, as described by Tung *et al.*[19].

The spectral intensities per energy and per solid-angle interval, $d^2Y/dE/d\Omega$, were calculated for all exit angles. Spectra for different azimuthal angles were combined. The solid was divided into layers with 50 Å thickness each. Spectra were calculated for each layer separately. Table 2 summarizes the conditions for the various simulations.

3.5 Results of the Simulations

Fig.5 shows the spectrum for the continuous slowing-down approximation, without elastic scattering, and for a monoenergetic

Table 2. Overview of all Monte Carlo simulations. Ruth. = Rutherford-scattering cross section; P.S. = phase shift model for elastic scattering; rnd = random; alg = aligned; mult.= the 3 KLL and KLM Auger peaks are considered.

Fig.	Energy loss model	Elast. scatt. model	Incid. beam direc.	Init. energy	Mult. ioniz.	Defl. by elect.	fwhm anal. (eV)
4	CSDA	none	rnd	1600	no	no	1
5	CSDA	Ruth.	rnd	1600	no	no	1
5	CSDA	P.S.	rnd	1600	no	no	1
6	EELD	P.S.	rnd	1600	no	yes	1
7	DRM-LDA	P.S.	rnd	1600	no	yes	1
2	DRM-LDA	P.S.	alg	mult.	n/y	yes	12
1	DRM-LDA	P.S.	rnd	mult.	yes	yes	12

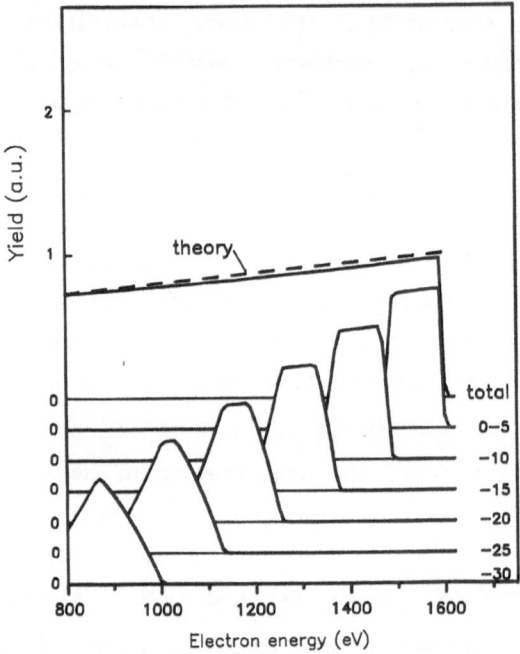

Figure 5. Calculated spectrum for a homogeneous and monoenergetic electron source, according to the CSDA. Elastic scattering is neglected. The separate contributions of various depth intervals are indicated in units of nm.

electron source. Thus, all electron trajectories are straight lines. For this case, the spectral intensity, $d^2Y/dE/d\Omega$, should theoretically be

$$\frac{d^2Y(E)}{dE\ d\Omega} = \frac{1}{4\pi}\ \frac{r_o}{S(E)}\ <\cos\vartheta>, \tag{7}$$

where $<\cos\vartheta>$ is the average cosine of the escape angle. The theoretical intensity is shown as a broken curve. We see that the agreement between the Monte Carlo and the theoretical result is good. The small deviation at higher energies is due to the statistical character of the Monte Carlo simulation. Because of the finite acceptance angle, the separate contributions of the layers are truncated triangles. Obviously, the energy straggling in this model, i.e., the spread in energy of electrons originating from the same depth, is small, especially for small energy losses.

Fig.6 shows spectra calculated when elastic scattering is included. The upper full curve, PS, is calculated using the phase-shift cross

608

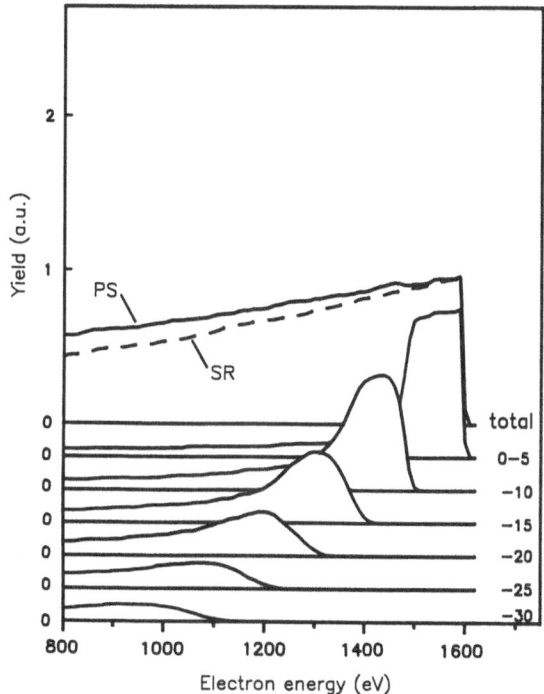

Figure 6. Calculated spectrum for a homogeneous and monoenergetic electron source, according to the CSDA, plus elastic scattering according to the (screened) Rutherford model (SR curve), or the phase-shift model (PS curve). The contributions for the various depth intervals are calculated using the phase-shift model.

section; the broken curve, SR, using the screened Rutherford cross section. Since the path length for most electrons is now larger than the shortest path length possible, the spectral intensity is lower, at all energies, than that calculated for the CSDA. Furthermore, the separate contributions have tails extending over hundreds of eV to lower energies. These tails represent the so-called *backscattering* contributions by those electrons initially emitted inwards, but subsequently scattered back towards the surface. At low energies, electrons are detected that originate from a wide range of depths: the energy straggling is large, especially for large energy losses. Since λ_e^{SR} is smaller than λ_e^{PS}, see table 1, the 'Rutherford' spectrum drops faster with decreasing energy than the 'phase-shift' spectrum.

For small energy losses, the CSDA cannot be used: the discrete character of the energy loss must be taken into account. If we use the exponential distribution function for the energy loss, eq.(4), we get the

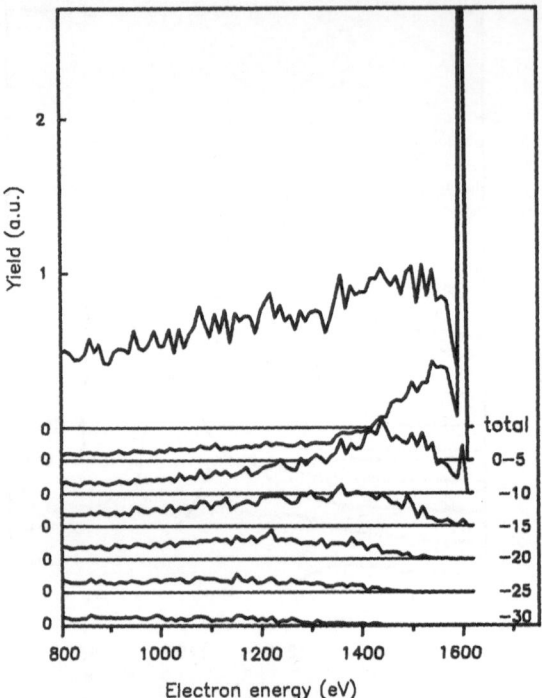

Figure 7. Calculated spectrum for a homogeneous and monoenergetic electron
source according to the EELD, plus elastic scattering according
to the phase-shift model.

result shown in fig.7. The overall shape of this EELD spectrum is similar
to that of fig.6, except near the initial energy at 1600 eV. A peak, the
zero-energy loss peak, is visible, followed by a dip. Since inelastic
scattering is a statistical process, and since the energy lost per
scattering event varies, the energy straggling is large over the whole
range of the spectrum. (See the broad contributions of the separate
layers). The similarity between the total spectra in figs.5 and 6 at
intermediate and low energies, < 1400 eV, show that the CSDA is valid for
an energy loss several times the mean energy loss per scattering event.

Spectra calculated for the DRM-LDA are shown in figs.8, 2, and 3. The
spectrum of fig.8 shows a large zero-energy loss peak and a smaller peak,
the plasmon loss peak, about 20 eV lower in energy. At energies below 1450
eV, the spectrum falls off in a manner similar to that observed in the
previously calculated spectra. However, the spectral intensity is
appreciably larger between 1450 and 1600 eV. It is noted that this
spectrum looks very similar to a typical XPS spectrum. At first sight, the
enhancement near the initial energy is puzzling, but it will be shown here

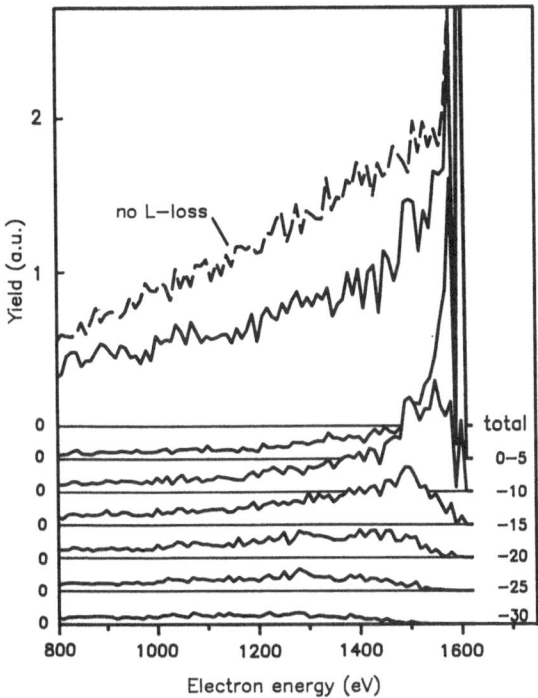

Figure 8. Calculated spectrum for a homogeneous and monoenergetic electron source according to the (DRM-LDA) model, plus elastic scattering according to the phase-shift model. The 'no L-loss' spectrum is calculated by suppressing all energy losses of more than 100 eV.

that it is a consequence of the character of the energy loss distribution. According to the DRM-LDA model, the relative contribution to the stopping power of all energy losses larger than 100 eV is about 50 %. However, the mean free path for such losses is 330 Å, about ten times larger than that for all losses, see table 1. For small path lengths, p < 330 Å, most of the electrons lose energy only in small amounts, and the average energy loss per Å is less than the stopping power. Since the spectral intensity is inversely proportional to the mean energy loss per Å, it is larger for small path lengths (near the initial energy of 1600 eV) than for large path lengths (at 1450 eV and below). To illustrate this effect, we have performed a simulation in which energy losses of more than 100 eV were suppressed. The broken curve in fig.8, labelled 'no-L loss', shows the result. As expected, this spectrum falls off much slower with decreasing energy than the normal DRM-LDA spectrum. However, at small energy losses the two spectra coincide.

In fig.3, a measured channeling Auger spectrum is shown, together with two simulated spectra. The broken curve is a spectrum calculated for the case where the initial Auger electron spectrum consists of three monoenergetic peaks. Since most Auger electrons are produced near the surface, the calculated spectrum consists of three peaks, plus their backscattering tails. Obviously, this calculated spectrum does not resemble the measured one very well. The probability for multiple ionization of the target atoms by MeV energy ions, is relatively large, > 50 % [25]. Multiple ionization causes the Auger energies to shift to lower energies, by at most several tens of eV [26]. This explains the above mentioned dissimilarity between the measured and calculated spectra. Since no data exist for our case, 1 MeV He$^+$ on Si, we have performed simulations and optimized the value for the width of the initial energy distribution. The full curve in fig.3 is the spectrum calculated assuming an initial energy distribution given by [10]:

$$g(E') = \frac{\beta}{(\delta + \gamma\beta)} \, e^{(E' + \delta)\gamma}, \qquad \text{for } E' < -\delta,$$

$$= \frac{1}{(\delta + \gamma\beta)}, \qquad\qquad \text{for } -\delta < E' < 0, \qquad\qquad (8)$$

$$= 0, \qquad\qquad\qquad \text{for } 0 < E';$$

δ, γ, and β are fitting parameters and $E' \equiv E - E_A$. The choice for the shape of the function $g(E')$ is based on data for 6 keV electrons on Ar [26], neglecting any fine structure. Since the solid angle of the aperture of the analyzer is not known precisely, the vertical scaling of the spectrum is another fitting parameter. We found good agreement between the simulation result and the measured spectrum for $\delta = 22$ eV, $\gamma = 40$ eV, and $\beta = 0.6$; see the full curve in fig.3. The vertical scaling corresponds to 3.8 ± 0.6 detected Auger electrons, per μC and per Å thickness of Si.

Fig.2 shows the random Auger-electron spectrum, plus the spectrum calculated using the initial Auger linewidth and the vertical scaling factor, as determined above. The calculated and the experimental spectra are in good agreement. The uncertainty in the calculated intensity stems from (a) an uncertainty of about 20 % in the vertical scaling factor; (b) an estimated uncertainty of about 20 % in the DRM-LDA model; and (c) at energies below 1200 eV, the method of background subtraction, see the vertical error bars.

5. DISCUSSION AND CONCLUSION

In two previous studies [9,10], we have tried to analyze ion-induced

Auger-electron spectra using analytical models for electron transport in matter. These models, based on the work of Tougaard and Sigmund [2], predict a spectral intensity that is constant for small energy losses, and decreases proportional to $(E_A - E)^{-1/2}$ for larger losses. The energy loss function applied was the exponential energy loss distribution, eq.(4). Although the model-spectra agreed qualitatively with the measured spectra, the width of the spectrum with constant intensity was about an order of magnitude too narrow, i.e., about 50 eV instead of an expected value of 700 eV. The present Monte Carlo simulations, with the energy-loss distribution given by the dielectric-response model in the local-density approximation, show that there is an enhancement of the spectral intensity for energy losses less than 150 eV. This enhancement is caused by a lower effective stopping power for most of the electrons created in the outermost $\approx$ 100 Å, where the average path length to the vacuum is less than the mean free path for large (> 100 eV) energy loss by ionization of the silicon L shell. The probability that an electron loses energy in such an event increases with depth of creation. Therefore, the spectral intensity drops exponentially with increasing energy loss (i.e., depth of creation) to a lower, approximately constant level. We conclude that the rapid fall-off in spectral intensity is not a consequence of the increase in path length by elastic scattering, but rather of the L-shell contribution to the energy loss distribution. Of course, if the DRM-LDA or DRM-ODA loss distribution is used in the analytical calculation, instead of the exponential distribution (EELD), a similar enhancement at small energy losses should be obtained. The use of these distribution function is, however, mathematically rather awkward.

Below 1400 eV, the measured and calculated decrease in spectral intensity with decreasing energy, is indeed due to the increase in path length by elastic scattering, and to the increase in stopping power. The uncertainty in the spectral intensity between 800 and 1200 eV is too large to deduce an accurate value, i.e., within a factor of 2, for the (transport) mean free path, as was done, albeit unsuccessfully, in previous publications [9,10]. Nevertheless, the relatively good agreement between experiment and simulation shows that the model used for the elastic scattering, i.e., the phase-shift cross section, is appropriate.

Finally, the method of transforming the electron trajectories to increase computation efficiency, can also be used for simulating electron-induced AES and XPS spectra; though in the former case, the energy loss and angular spread of the incident electron beam must be taken into account as well. The method, however, cannot be applied if the solid is not homogeneous or if it contains a non-negligible and depth-dependent concentration of foreign atoms.

REFERENCES

1. C.J. Powell, J. Vacuum Sci. and Technol. *A4* (1985) 263.
2. S. Tougaard and P. Sigmund, Phys. Rev. *B25* (1982) 4452.
3. A.L. Tofterup, Surf. Sci. *167* (1986) 70.
4. V.M. Dwyer and J.A.D. Matthew, Surf. Sci. *193* (1988) 549.
5. S. Ichimura and R. Shimizu, Surf. Sci. *112* (1982) 386.
6. J. Ferrón, E.C. Goldberg, L.S. de Bernardez and R.H. Buitrago, Surf. Sci. *123* (1982) 239.
7. M. Cailler and J.P. Ganachaud, Surf. Sci. *154* (1985) 524.
8. A. Jablonski, Surf. Sci. *188* (1987) 164.
9. L. Wong, P.F.A. Alkemade, W.N. Lennard and I.V. Mitchell, Nucl. Instr. and Meth. *B45* (1990) 637.
10. P.F.A. Alkemade, W.N. Lennard and I.V. Mitchell, submitted to Surf. Sci.
11. P.F.A. Alkemade, L. Wong, W.N. Lennard and I.V. Mitchell, Nucl. Instr. and Meth. *B48* (1990) 604.
12. S. Ichimura, M. Aratama and R. Shimizu, J. Appl. Phys. *51* (1979) 2853.
13. J.P. Ganachaud and M. Cailler, Surf. Sci. *83* (1979) 498.
14. S. Valkealahti and R.M. Nieminen, Appl. Phys. *A32* (1983) 95.
15. J. Ferrón, L.S. De Bernárdez, E.C. Goldberg and R.H. Buitrago, Surf. Sci. *163* (1985) 266.
16. Z.-J. Ding and R. Shimizu, Surf. Sci. *197* (1988) 539.
17. A. Jablonski, J. Gryko, J. Kraaer and S. Tougaard, Phys. Rev. *B39* (1989) 61.
18. J. Lindhard, Kgl. Danske Videnskab. Selskab, Mat.-Fys. Medd. *28* (1954) 3.
19. C.J. Tung, J.C. Ashley and R.H. Ritchie, Surf. Sci. *81* (1979) 427.
20. C.J. Powell, Surf. and Interf. Anal. 7 (1985) 263.
21. M. Fink and J. Ingram, Atom. Data 4 (1972) 129.
22. D. Gregory and M. Fink, Atom. Data and Nucl. Data Tabl. *14* (1974) 39.
23. M.E. Riley, C.J. MacCallum and F. Biggs, Atom. Data and Nucl. Data Tabl. *15* (1975) 443.
24. L.E. Davis, N.C. MacDonald, P.W. Palmberg, G.E. Riach and R.E. Weber, *Handbook of Auger Electron Spectroscopy*, 2nd ed., Physical Electronics Industies Inc., (1976).
25. P. Richard, R.L. Kauffman, J.H. McGuire, C.F. Moore and D.K. Olsen, Phys. Rev. *A8* (1973) 1369.
26. M.O. Krause in *Atomic Inner-shell Processes*, vol.2, B. Crasemann, ed. Academic Press, New York (1975).

ELECTRON SPECTRA FROM ION BOMBARDMENT OF CLEANED SURFACES
COVERED WITH FROZEN GASES AT LOW SAMPLE TEMPERATURES

H. Rothard, M. Schosnig, K. Kroneberger and K-O. Groeneveld

Institut für Kernphysik
Johann-Wolfgang-Goethe Universität
August-Euler-Strasse 6, D-6000 Frankfurt am Main 90
Germany

1. INTRODUCTION AND EXPERIMENT

Secondary electrons (SE) from ion bombardment of solids are messengers of the physical processes occurring during the interaction of swift ions ($v_p > v_B$) with condensed matter [1]. In particular, experimental studies of doubly differential SE spectra $d^2n/dEd\Omega$ in forward direction (observation angle $\theta = 0^o$) induced by ion penetration of thin foils yield information on such important phenomena as low energy "true" SE (E < 10 eV), electrons from the decay of plasmons (E < 30 eV), convoy electrons (CE, $v_e = v_p$) and high-energy binary encounter electrons ($v_e = 2 v_p$) [2]. Moreover, characteristic target Auger electrons can be observed in SE spectra.

Since the shape of SE spectra strongly depends on the contamination of the surface with adsorbates such as hydrocarbons or water [3], meaningful measurements must be performed in ultrahigh vacuum (UHV, $p < 10^{-7}$ Pa) with controlled surface conditions. During the last years, first studies of $\theta = 0^o$ electron spectra in UHV became possible [3-6]. Two different types of experiments have been performed until now:

(1) The first evidence for a strong dependence of CE emission on surface properties was obtained in the following way [3,5]: Thin foils produced by standard evaporation techniques and thus contaminated with $\approx 1 - 2$ monolayers (ML) of adsorbed hydrocarbons [7] were sputter-cleaned with noble gas ions. A strongly enhanced CE yield from cleaned surfaces as well as a reduction of the low energy "true" SE compared to untreated foils was reported [3-5] for MeV H^+, H_2^+ and Kr^+ ion impact. The target surfaces were controlled simultaneously by a variety of independent

Interaction of Charged Particles with Solids and Surfaces
Edited by A. Gras-Martí *et al.*, Plenum Press, New York, 1991

615

methods such as Auger electron spectroscopy (AES), secondary electron spectroscopy (SES), Rutherford backward/forward scattering spectroscopy (RBS/RFS) and elastic recoil detection (ERD) [7,8]. The residual coverage, β, with carbon and oxygen atoms, was estimated to be lower than $\beta(C) < 0.2$ ML and $\beta(O) < 0.1$ ML [7].

(2) In the second kind of experiments, an ion source was used to deposit Na on the surface of an Al foil. In good agreement with the results of type (1) experiments, a reduction of the CE yield and an increase of the "true" SE yield induced by 60 keV H^+ was found for Na-contaminated surfaces, $\beta(Na) \approx 0.15$ ML [6]. Unfortunately, the Al foils were not sputter-cleaned before the Na deposition, and no independent means of surface control (such as AES) could be applied.

In this paper, we introduce a third kind of experiments involving the surface contamination dependence of SE emission from foils. A new UHV setup including a sample cooling device (T < 30 K) and a gas inlet nozzle has been installed to study SE spectra from untreated and sputter-cleaned surfaces of thin foils at different observation angles θ. Gases such as CO_2 and Xe can now be condensed on the cold surfaces of sputter-cleaned targets. The combination of AES and RBS (which allows us to control the thickness of the frozen gas layer) enables us to study the dependence of SE spectra on defined surface contamination β in a controlled, quantitative manner. Also, SE spectra taken at different sample temperatures can be studied to elucidate the origin of a possible dependence of the total SE yield γ on the target temperature [9,10].

The SE spectra presented here were obtained with an UHV chamber at our 2.5 MV Van-de-Graaff accelerator. A hydrocarbon-free UHV with a residual pressure of $p < 9 \times 10^{-8}$ Pa is obtained with two 1500 l/s cryopumps. The composition of the residual gas can be controlled with a quadrupole mass spectrometer. A detailed drawing of the experimental setup is shown in fig.1.

Details of the ion beam collimation with a set of three apertures (B1- 3), the measurement of the beam current with a Faraday cup (FC) and an SE repeller (RP) as well as the connection of the UHV system to the standard vacuum ($p = 10^{-4}$ Pa) of the beam line with a differential pumping stage (PS) have been described in a previous paper [11]. Magnetic fields inside the chamber are compensated to less than 10 mGauss with a set of three pairs of Helmholtz coils [11].

A 45° electrostatic parallel-plate spectrometer [4] (S1, energy resolution $\Delta E/E \approx 3$ %) is used to record SE energy spectra

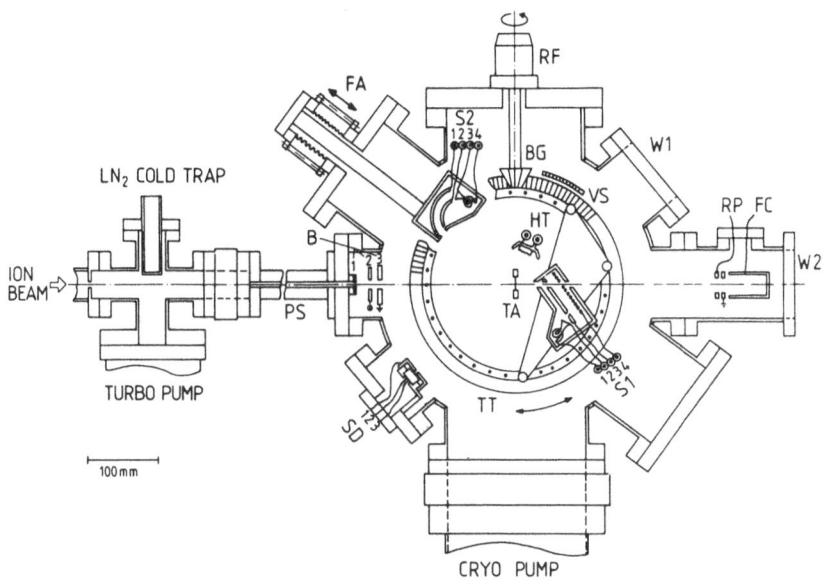

Figure 1. Detailed drawing of the experimental setup. The components of the UHV system are explained in the text.

(0 eV < E < 2500 eV) at different observation angles ($\theta_1 = 0°$, $35° \leq \theta_1 \leq 145°$ and $\theta_1 = 180°$). The spectrometer (S1) is mounted on a turntable (TT) and can be turned around the target (TA) with a bevel gear (BG) mounted on a rotational feedthrough (RF). A vernier scale (VS) can be read through a window (W1) and allows us to adjust the observation angle θ_1 with an accuracy of $\pm 0.05°$.

An electrostatic spherical-sector spectrometer [7] (S2, $\Delta E/E < 0.8$ %, $\theta_2 = 143°$) was used for the spectroscopy of Auger electrons (AES) under ion impact. A bakeable silicon detector mounted either in forward direction ($\theta = 40°$) or backward direction ($\theta = 140°$) allowed the measurement of scattered projectiles or recoiled target ions (RFS/RBS, ERD).

The target foils are mounted on a copper sample holder at the end of a copper rod connected to the cold head of a closed-loop helium refrigerator. This design has the big advantage of being more economic and more convenient to use compared to a liquid helium cryostat [12]. The sample temperature can be measured with a standard platinum resistor (Pt 100). Temperatures as low as T < 28 K have been obtained. The cooling device is integrated in an XYZ manipulator with additional LN_2 shielding. Also, the target can be tilted with respect to the beam axis (tilt angle δ, $0° \leq \delta \leq 360°$).

Furthermore, the experimental setup is equipped with a novel gas-inlet system including a UHV fine-leak valve [13] to solidify uniform layers of frozen Xe or CO_2 on the cold foil surfaces. The selection of gases for these experiments is determined by their vapor pressure at the lowest available target temperature (T ≤ 30 K). The target surfaces were sputter-cleaned with noble gas ions (Ar^+ or Kr^+ at 20 keV/u) and the composition of the surfaces could be controlled by AES, SES, RFS, RBS and ERD [7-8].

2. ELECTRON SPECTRA FROM UNTREATED AND SPUTTER-CLEANED SURFACES AT LOW SAMPLE TEMPERATURES

The strong dependence of the shape of SE spectra on the surface coverage is demonstrated in fig.2 which presents $\theta = 0^o$ doubly differential SE spectra (not corrected for the transmission of the spectrometer), $n = E\ d^2n/dEd\Omega$, as a function of the electron velocity v_e from 1.2 MeV H^+ penetration through a Cu foil (d = 1500 Å). The dash-dotted lines belong to spectra from untreated foils, and the dashed lines to spectra from sputter-cleaned foil surfaces (type (1) experiment,

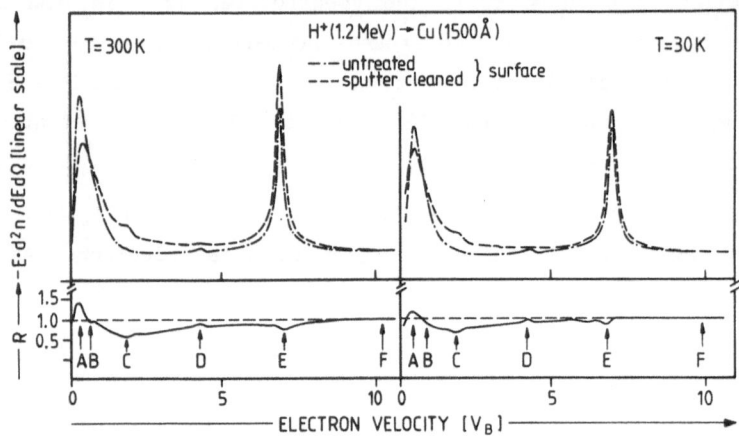

Figure 2. Doubly differential SE spectra ($\theta = 0^o$) $n = E\ d^2n/dEd\Omega$ as a function of the electron velocity v_e, from 1.2 MeV H^+ penetration through a Cu foil (d = 1500 Å). The dash-dotted lines are spectra from untreated foils, and the dashed lines are spectra from sputter-cleaned foil surfaces. The left-hand side shows spectra taken at a sample temperature of T = 300 K (room temperature), and the right-hand side shows spectra taken at a low sample temperature, T = 30 K. The bottom presents the ratio R = {n(untreated surface) / n(sputter-cleaned surface)} of the spectra. Intensity enhancements after cleaning can be observed as dips, intensity reductions as peaks in the lower part of the figure. The distinct structures labeled A ... F which can be observed in the spectra are discussed in the text.

as mentioned in the introduction). Fig.2 presents spectra taken at a sample temperature of T = 300 K (room temperature) and at low sample temperatures of T = 30 K and also the ratio R = {n(untreated surface) / n(sputter-cleaned surface)} of the spectra. Intensity enhancements after cleaning can be observed as dips, intensity reductions as peaks in the lower part of fig.2.

The distinct structures which can be observed in the spectra are labeled A ... F. Except for the high-energy binary-encounter electrons (F), originating from close collisions between the projectile and target electrons, all other structures change drastically after sputter-cleaning of the surfaces. The (fast) binary-encounter electrons mainly originate from a depth of several hundred Å and thus are not affected strongly by changes of the near-surface layers.

First of all, since the surface is cleaned from carbon adsorbates, no more carbon Auger electrons at E = 265 eV (D) can be observed, whereas the copper MVV Auger electrons (C) at E = 63 eV appear from the clean copper surface. The maximum residual coverage can be estimated to be $\beta(C) < 0.2$ ML and $\beta(O) < 0.1$ ML [7]. The thickness of the carbon layer before cleaning is $\beta(C) \approx 1 - 2$ ML.

Furthermore, the height of the low-energy "true" SE distribution (A) decreases, its FWHM increases, and the energy E_{max} of the "true" SE peak maximum (A) is shifted to higher energies, E_{max}("dirty" surface) < E_{max}("clean" surface). This behavior can be explained by the following arguments:

(i) A layer of adsorbates on a clean metal surface can lead to a reduction of the electron work function Φ. This leads to an enhanced surface-transmission probability, in particular for low-energy SE [3,14].

(ii) Layers of adsorbates or oxides lead to a larger electron escape depths λ and thus to a higher SE yield [14].

(iii) The change of the composition (i.e., the target material Z_T) of the near-surface layers affects the stopping power and thus the production of SE.

(iv) It can be shown that a rough, uncleaned surface is associated with an enhanced electron escape probability compared to a smooth, planar surface [15]. The sputtering process applied here both cleans and smoothes the surfaces by preferential sputtering [16].

One of the most striking results of this type (1) experiments is that both the yield of CE and the shape (full width at half maximum, FWHM) of the CE peak (E) are significantly enhanced [3,5]. This shows that a large fraction of CE are formed at the last layers of the solid or even when the ion leaves the solid. Their yield is influenced by changes of the target material Z_T [3] leading to changes of the electron capture and loss cross sections σ_C and σ_L as well as the CE transport lengths λ. Also, the surface transmission probability related to the work function Φ depends on the surface coverage β, in particular for low energy CE (E < 50 eV) [3,5-6].

The dip labeled (B) in the ratio spectrum of fig.2 may result from collective excitations (plasmons) of the metal. Electrons from the decay of plasmons can only be observed from clean surfaces [14,17]. Surprisingly, this dip can be observed at T = 300 K, whereas it is hardly or not visible at T = 30 K.

A closer inspection of fig.2 reveals that the shape of the spectra also depends on the sample temperature T. The overall intensity of the low-temperature spectra is reduced. This is seen more clearly in fig.3, where $\theta = 0^o$ spectra from 1.6 MeV H_2^+ bombardment of Au (d = 1000 Å) at

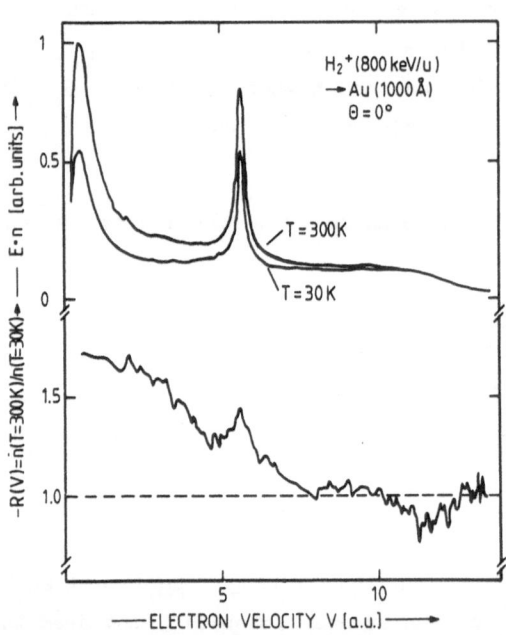

Figure 3. Doubly differential spectra ($\theta = 0^o$) n = E $d^2n/dEd\Omega$, from 0.8 MeV/u H_2^+ bombardment of Au (d = 1000 Å) at T = 300 K (upper curve) and T = 30 K (lower curve) as a function of v_P. The bottom part shows the ratio of the spectra, R = {n(T=300K) / n(T=30K)}.

T = 300 K and at T = 30 K are compared. The bottom part of fig.3 shows the ratio of the spectra, R = {n(T=300K) / n(T=30K)}. One observes a ca. 70 % reduction of the low energy "true" SE distribution, and, surprisingly, a ca. 40 % reduction of the CE yield. A small (around 10 %) enhancement of the binary encounter electrons around $v_e \approx 2 v_p$ may be a hint for a small density effect.

These results may indicate that the interaction of the ions with nearly free electrons connected to collective excitations such as plasmons or the ion-induced "wake" [16] depends on the temperature. Possibly, the stopping power, which is partly determined by the interaction of the ions with nearly-free target electrons (distant collisions, plasmon excitation, wake) is reduced at low temperatures [10]. The T-dependence of the possible plasmon-decay dip B in fig.2 may indicate that the interaction of the ions with the nearly free electron gas is reduced at low temperatures. Further studies of e.g. the T-dependence of wake-induced shock-electron emission [16], in particular from high-temperature superconductors [10], may help to elucidate these findings.

Possibly, transport properties such as T-dependent changes of electron attenuation lengths λ or surface properties such as T-dependent changes of the work-function Φ have to be considered. Although no surface contamination was detected, we have to keep in mind that the surfaces may still be contaminated with water or hydrocarbons. A possible influence of a coverage of $\beta < 0.1$ ML on the shape of the spectra cannot be excluded. However, the dependence of "true" SE and CE emission on T (fig.3) clearly differs from the dependence on the surface coverage β (fig.2). In the later case, the CE yield from "dirty" surfaces is reduced and the intensity of "true" SE is enhanced, whereas in the first case, at low T both CE and "true" SE yields are reduced. Also, non-equilibrium charge and heat transfer between the highly ionized track of the ions and the cold environment have to be taken into account. Both further experimental and theoretical investigations are necessary.

3. ELECTRON SPECTRA FROM SURFACES COVERED WITH FROZEN GASES

The preparation of the cooled target is divided in two stages. First, the surface is cleaned by sputtering with 0.8 MeV Ar^+ ions using a typical ion fluence of ~ 15 nA min/mm^2. The cleaning procedure is supervised by AES and SES, as demonstrated in fig.3, as well as by RBS or RFS. Then, by condensing CO_2 on the cold, cleaned targets we contaminate the surface in a controlled manner.

Fig.4 demonstrates the dependence of the SE spectra on the

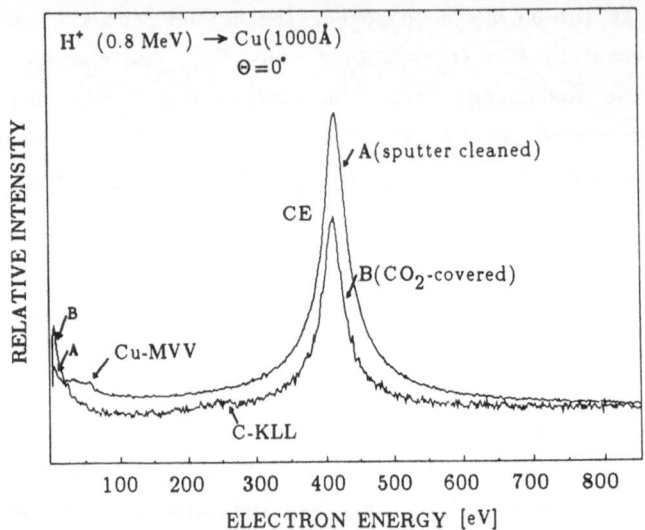

Figure 4. (0.8 MeV) H$^+$-induced SE spectra ($\theta = 0^o$) from a sputter-cleaned (A) and subsequently CO_2-contaminated (B) copper target (d = 1000 Å) at T = 30 K.

contamination with a frozen gas layer. The spectra of electrons ejected under $\theta = 0^o$ from 0.8 MeV H$^+$ bombardment of Cu (d = 1000 Å) are not corrected for the spectrometer transmission and can directly be compared to those shown in fig.2. The spectra from the sputter-cleaned (A) and subsequently contaminated (B) copper target differ significantly:

(a) The yield of convoy electrons (CE) in spectrum B is reduced as a result of CO_2 contamination, i.e. the CE yield is higher for a clean Cu surface than for a CO_2-covered surface. This result is in good agreement with the results of the type (1) [3-5] and type (2) [6] experiments.

(b) The Cu MVV Auger electrons of the target material at an energy of 63 eV disappeared, but a weak structure at 265 eV indicates Carbon KLL Auger electrons. The Auger electron intensity is surprisingly low compared to the intensity from hydrocarbon adsorbates (fig.2). A similar result has recently been reported [18], but remained unexplained: Up to a threshold fluence of 9×10^{14} ions/cm^2, no C-KLL Auger electron emission from 1 MeV H$^+$ irradiated frozen CO_2 was observed.

(c) The low-energy "true" SE are enhanced. The possible explanations (i - iv) as, e.g., a reduction of the work function Φ of the solid have already been mentioned above in the discussion of fig.2. On the other hand, in both spectra, no significant dependence of the intensity of the higher-energy electrons (E > 750 eV) on the surface coverage is observed, because the influence of the surface conditions on the emission of these electrons is negligible.

In conclusion, we have presented a new technique to study the dependence of SE emission on controlled surface contamination by condensing gases such as CO_2 and Xe on the cold surfaces of sputter-cleaned targets. This will enable us to directly correlate SE yields with the thickness of the layer of physisorbed gases, which can be determined by RFS. Thus, information about escape depths of electrons as well as the influence of the coverage-dependent change $\Delta\Phi$ of the work function can be obtained. The measurement of doubly differential SE spectra as a function of the observation angle θ should allow us to test refined theoretical models which take into account the quantum-mechanical surface transmission probability for low-energy electrons [2,13,19]. In particular, it will be interesting to study the dependence of the shape (E_{max}, FWHM) of the "true" SE distribution on $\Delta\Phi$ and θ. In this connection, it is important to know the erosion yield of frozen gases under ion irradiation [20]. This nonlinear effect changes the experimental conditions during the measurement of SE spectra. Also, the study of SE spectra from various gas/target combinations like Xe or CO_2 on C, Cu or other metals can yield further information on the dependence of SE emission on Z_T and Φ [13].

ACKNOWLEDGEMENTS

This work has been funded by the German "Bundesminister für Forschung und Technologie" (BMFT, Bonn), under contract number 06 OF 173/II Ti 476. Two of us (H.R. and K.K.) acknowledge a grant from "Deutscher Akademischer Austauschdienst" (DAAD, Bonn).

REFERENCES

1. K.O. Groeneveld, Rad. Eff. and Defects in Solids 110 (1989) 121; Nucl. Tracks Radiat. Meas. 15 (1988) 51.
2. K.O. Groeneveld, R. Maier, H. Rothard, Il Nuovo Cimento 12D (1990) 843; H. Rothard, K. Kroneberger, E. Veje, M. Schosnig, P. Lorenzen, N. Keller, J. Kemmler, C. Biedermann, A. Albert, O. Heil, K.O. Groeneveld, Nucl. Instrum. Meth. B48 (1990) 616.
3 H. Rothard, M. Burkhard, C. Biedermann, J. Kemmler, K. Kroneberger, P. Koschar, O. Heil, D. Hofmann, K.O. Groeneveld, J. Physique (Paris) 48 (1987) C9-215; H. Rothard, M. Burkhard, C. Biedermann, J. Kemmler, P. Koschar, K. Kroneberger, O. Heil, D. Hofmann, K.O. Groeneveld, J. Phys. C21 (1988) 5033.
4. W. Lotz, M. Burkhard, P. Koschar, J. Kemmler, H. Rothard, C. Biedermann, D. Hofmann, K.O. Groeneveld, Nucl. Instrum. Meth. A245 (1986) 560.
5. M. Burkhard, H.Rothard, C. Biedermann, J. Kemmler, P. Koschar, K.O. Groeneveld, Nucl. Instrum. Meth. B24/25 (1987) 143.

6. E. Sanchez, L.F. de Ferrariis, S. Suarez, Nucl. Instrum. Meth. *B43* (1989) 29; P. Focke, W. Meckbach, Nucl. Instrum. Meth. *B33* (1988) 255.
7. P. Lorenzen, H. Rothard, K. Kroneberger, O. Heil, J. Kemmler, N. Keller, C. Biedermann, M. Schosnig, A. Albert, R. Maier, K.O. Groeneveld, *Annual Report IKF-D49* (1989) p. 33, Institut für Kernphysik, Frankfurt, FRG; P. Lorenzen, H. Rothard, K. Kroneberger, J. Kemmler, M. Burkhard, K.O. Groeneveld, Nucl. Instrum. Meth. *A282* (1989) 213.
8. M. Burkhard, H. Rothard, J. Kemmler, K. Kroneberger, K.O. Groeneveld, J. Phys. *D21* (1988) 472.
9 T.J. Gay, H.G. Berry, Phys. Rev. *A19* (1979) 952; J. Physique (Paris) *40* (1979) C1-298.
10. H. Rothard, P. Lorenzen, N. Keller, O. Heil, D. Hofmann, J. Kemmler, K. Kroneberger, S. Lencinas, K.O. Groeneveld, Phys. Rev. *B38* (1988) 9224.
11. M. Burkhard, W. Lotz, K.O. Groeneveld, J. Phys. *E21* (1988) 759.
12. K. Wandelt, S. Daiser, R. Miranda, H.-J. Forth, J. Phys. E17 (1984) 22.
13. M. Schosnig, H. Rothard, K. Kroneberger, P. Lorenzen, A. Albert, R. Maier, O. Heil, K. O. Groeneveld, *Annual Report IKF-D49* (1989) p. 35, Institut für Kernphysik, Frankfurt, FRG.
14. D. Hasselkamp, Comments At. Mol. Phys. *21* (1988) 241, and references therein.
15. J.E. Borovsky, D.J. McComas, B.L. Barraclough, Nucl. Instrum. Meth. *B30* (1988) 191.
16. H. Rothard, K. Kroneberger, M. Burkhard, C. Biedermann, J. Kemmler, O. Heil, K.O. Groeneveld, J. Physique (Paris) *50* (1989) C2-105.
17. M. Burkhard, H. Rothard, K.O. Groeneveld, Phys. Stat. Sol. (b) *147* (1988) 589.
18. W.M. Ariyasinghe, R.D. McElroy Jr., D. Powers, Nucl. Instrum. Meth. *B44* (1989) 61.
19. J. Mischler, N. Benazeth, M. Nègre, C. Benazeth, Surf. Sci. *136* (1984) 532.
20. J. Schou, Nucl. Instrum. Meth. *B27* (1987) 188.

EMISSION OF SPIN-POLARIZED ELECTRONS TO STUDY
SURFACE MAGNETISM AND PARTICLE-SURFACE INTERACTION

C. Rau and K. Waters

Dept. of Physics and Rice Quantum Institute
Rice University
POB 1892, Houston, TX 77251, USA

ABSTRACT

Ion-induced, angle and energy resolved, spin-polarized electron emission spectroscopy (SPEES) has been used to probe the electron spin polarization (ESP) at the surface of Ni(110) picture-frame single crystals. The energy distribution of the emitted electrons, which includes spin-polarized Auger electrons, is significantly different from that of electron-induced secondary electrons and contains fundamental information on the local surface magnetic structure of Ni(110) and on particle-surface interaction mechanisms.

Quite recently, we reported on first experimental data using spin-polarized, angle-resolved electron emission spectroscopy (SPEES) during reflection of hydrogen ions at various magnetic surfaces [1,2]. In SPEES, we employ grazing-angle ion-surface interaction [3] as a means to study surface magnetic and electronic properties of magnetic materials. At small angles of incidence, ions cannot penetrate an atomically flat surface, they are specularly reflected and, therefore, probe the physical properties of the topmost surface layer. Grazing angle particle-surface interaction experiments allow us to study electronic transition processes occurring near surfaces. Many of these processes should depend on the local surface electronic structure and on the ion species used in the experiments.

In this paper, we present detailed information about SPEES experiments performed at Ni(110) surfaces where, besides 15 keV to 30 keV H^+ ions, 15 keV to 30 keV He^+ are used for the grazing-angle surface reflection. Using different ion species for SPEES, allows us to separate "true" Auger electron peaks from other Auger-type transitions occurring at Ni(110) surfaces. We take advantage of the well-known

Interaction of Charged Particles with Solids and Surfaces
Edited by A. Gras-Martí *et al.*, Plenum Press, New York, 1991

625

energetic shift in atomic or ionic final electron states when incident hydrogen ions are replaced by helium ions.

Investigating the energy and angle-resolved electron spin polarization (ESP) of electrons emitted from Ni(110), we find that the energy distribution of the emitted electrons is completely different from that of electron-induced secondary electron spectra. It consists of a series of characteristic peaks which can be linked, either to various electronic processes where spin-slit, local surface electronic states of Ni(110) and electronic states of the reflecting particles are involved, or to spin-polarized, element-specific Auger electrons emitted from Ni or, in case of a C or O surface contamination, from C or O atoms. In the latter case, the ionic species are solely used to produce (ion induced) holes in electron shells of Ni, C or O atoms.

Using an einzel-lens system, we detect electrons emitted along the surface normal (emission cone angle: 11°) of a remanently magnetized Ni(110) picture-frame crystal. For energy analysis, we use an electrostatic energy analyzer and for spin analysis, we employ a calibrated detector for ESP [1,2,4].

A magnetizing field is applied in the Ni(110) surface plane to

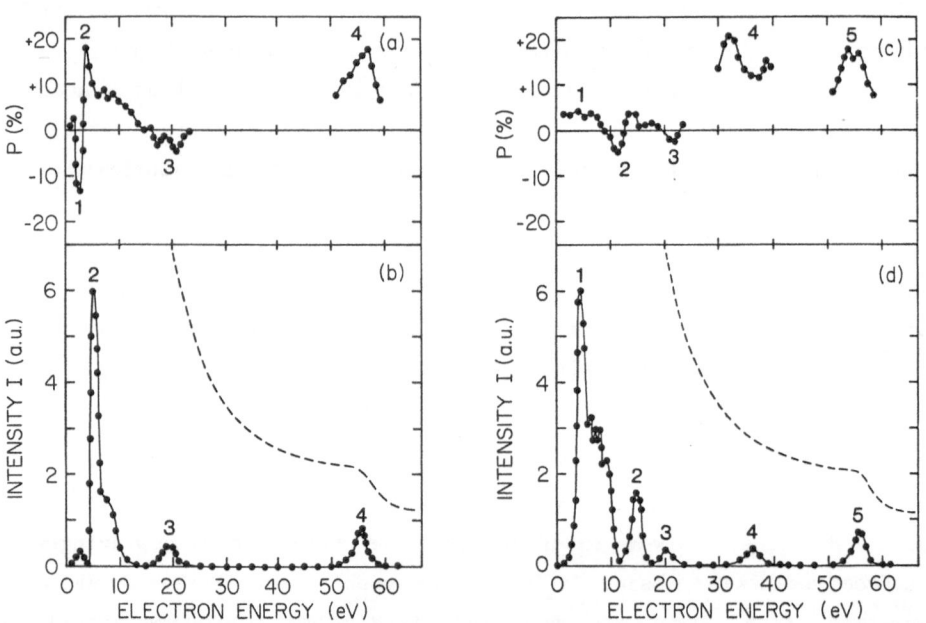

Figure 1. Intensity distribution I(E) and electron spin polarization P(E) of spin-polarized electrons emitted during grazing-angle surface reflection of 25 keV H+ (a and b) and of 25 keV He+ ions (c and d) at magnetized Ni(110) at 300 K as function of electron energy E.

produce a macroscopic remanent magnetization along which sign and magnitude of the ESP are measured. With P the ESP along the magnetizing axis, we obtain $P = (n^+ - n^-)/(n^+ + n^-)$, where n^+ and n^- are the fractional numbers of electrons emitted with moment parallel (majority spin electrons) and antiparallel (minority spin electrons), respectively, to this axis.

Ni(110) surfaces are prepared at 1×10^{-10} mbar as described in ref.[3]. The samples are characterized using AES, LEED, RHEED and STM. In order to be able to show that SPEES is sensitive to residual surface contaminants, the final cleaning of the Ni(110) surfaces, using Argon ion sputtering, is terminated when a residual C and O surface coverage of a few % is reached.

Figs.1a and 1b show the spin P(E) and the intensity I(E) distribution of electrons emitted during grazing angle (1°) surface reflection of 25 keV H^+ ions at magnetized Ni(110) surfaces as function of the electron energy E.

Replacing, for grazing-angle surface reflection at Ni(110), the 25 keV H^+ ions by 25 keV He^+ ions, results in the I(E) and P(E) data given in figs.1c and 1d [5]. We observe four low-intensity peaks (located at 14 eV, 20 eV, 36 eV and 56 eV; see labels 2, 3, 4 and 5 in fig.1d) and one high-intensity peak (see label 1 in fig.1d) located with a maximum at around 4.5 eV.

From figs.1b and 1d, it is obvious that I(E) is completely different from that of electron-induced secondary electron spectra (see dashed lines [6] in figs.1b and 1d) which shows that, using grazing angle ion-surface reflection, cascading effects, caused by energetic secondary ions or energetic secondary electrons, are nearly totally suppressed.

From fig.1b, we observe three low-intensity peaks (located at 2.5-3 eV, 20 eV, and 56 eV; see labels 1, 3, and 4 in fig.1b) and one high-intensity peak (see label 2 in fig.1b) located with a maximum at around 4.5 eV. By varying the primary-beam energy of the H^+ ions between 15 keV and 30 keV, we observe no measurable shift in the energetic location of these peaks.

The corresponding spin distributions P(E) of the emitted electrons (see figs.1a and 1c) show several pronounced, characteristic peaks (see labels 1 ... 4 in fig.1a, and labels 2 ... 5 in fig.1c). We use the sign and magnitude of the ESP as additional "label" to identify the underlying charge-exchange processes.

From Ni(110) ground state, one-electron band structure

calculations [7] and from spin-sensitive experiments at Ni(110) surfaces [3,7-10], it is known that valence band (V) electrons, originating from energy levels near (< 0.5 eV) the Fermi energy, possess a predominant minority spin orientation (negative ESP) of nearly -100 %, whereas V electrons from energy levels below this energy range overwhelmingly possess a majority spin orientation (positive ESP). These informations can be used to identify various physical processes involved in SPEES. V electrons originating in k-space from energy levels located 0.5 eV or less near the Fermi level, being emitted without any spin-flip process involved, would contribute to the negative part of the P(E) curve. On the other side, V electrons originating from levels located at least 0.5 eV below the Fermi level would contribute to the positive part of the P(E) curve.

We discuss the electron emission in terms of two-electron charge exchange XVV Auger electron emission denoting by X an electronic level either of the incoming ion, process (a), or of atoms located at the surface, process (b).

Process (a): Neutralization of the incoming ion and ejection of an Auger electron from the Ni(110) surface. For the work function of Ni(110), we use 5.04 eV and for the 1s ground state of hydrogen -13.58 eV, which gives a maximum kinetic energy of the ejected Auger electrons of 13.58 - 2×5.04 eV = 3.5 eV. Taking into account the surface band structure of Ni(110), P(E) should exhibit a negative value at around 3 eV followed by an immediate change to positive P values with decreasing electron energy (see fig.1a). This is indeed observed in the experiment.

In order to be able to possibly identify XVV Auger processes where process (b) is involved, we used He^+ ions instead of H^+ ions for SPEES (see figs.1c and 1d).

We assume that XVV Auger peaks, where X is a level of the incoming ion (process (a)), undergo a shift in energy when H^+ ions are replaced by He^+ ions. XVV Auger transitions, where X is an empty state of a surface atom (process (b)), should not undergo a shift. Therefore, replacing H^+ ions by He^+ ions should cause an energetic shift of peak 1, fig.1b, by 11 eV, which is the difference between the He^+(1s) (E_X = -24.6 eV) and H^+(1s) ground state energies. This is indeed observed in the experiment (see peak 2 in fig.1d) and corroborated by the occurrence of a change of P from negative to positive values (peak 2 in fig.1c) as found for the H^+ data.

The electron peak located at 36 eV (see peak 4 in fig.1c) was already observed by Hagstrum and Becker [11] and by Niehaus and coworkers [12] and interpreted by the latter group as a two-electron capture process from He^{++} - Cu(110) collisions leading to $He^{**}(2p^2)$ decaying to $He^+(1s)$ followed by the emission of an electron of 36 eV kinetic energy. We remark that, in the present experiments, a residual part of the ionic He beam consists of He^{++} ions. We observe nonzero ESP values (see peak 4 in fig.1c) which is consistent with the findings of Hagstrum [11] that this peak contains information on the local surface density of states, which for Ni should exhibit nonzero ESP values.

Process (b): Comparing figs.1b and 1d, we observe that the three peaks located at around 4.5 eV, 20 eV and 56 eV (see peaks 2, 3 and 4 in fig.1b and peaks 1, 3, 5 in fig.1d) are independent of the ion species used in the experiments and are not shifted to higher energies by the use of He^+ ions instead of H^+ ions. Therefore, we attribute these peaks to XVV Auger transitions where the ions are solely used to produce ion-induced holes in the electron shells of the surface atoms. Comparing our data with presently existing Auger electron data for Ni and with data obtained by Landolt and coworkers [6] using spin-polarized Auger electron spectroscopy at Ni(100) surfaces, we identify the peak located at 56 eV (see peak 4 in fig.1b and peak 5 in fig.1d) with a $M_{23}VV$ Ni Auger decay. We note that, after background subtraction, these authors find similar P values for this peak.

Regarding the peak (see label 3 in figs.1b and 1d) located at around 20 eV, we remark that the intensity of this peak depends strongly on surface cleanness, and therefore, can be attributed to Auger decays where X is a hole level in the surface contaminants, induced by the incoming ions.

Due to similarities of the electron spectra (peak 2 in fig.1b and peak 1 in fig.1d) at around 4.5 eV, induced by H^+ and He^+, respectively, we correlate this feature with the substrate. A possible mechanism could be an ion induced, kinematically enhanced XVV Ni Auger decay with X the level in the lower part of the valence band.

In conclusion, we note that, provided the future availability of a quantitative theory of SPEES and more refined experimental data, SPEES will advance towards an experimental technique of great potential for studies on the local surface electronic structure of magnetic materials and the physics of ion-surface interactions.

ACKNOWLEDGEMENTS

This work was supported by the National Science Foundation, by The Welch Foundation and by the Texas Higher Education Coordinating Board.

REFERENCES

1. C. Rau and K. Waters, Nucl. Instr. Meth. *B33* (1988) 378.
2. C. Rau and K. Waters, Nucl. Instr. Meth. *B40/41* (1989) 127.
3. C. Rau, J. Magn. Magn. Mat. *30* (1982) 141.
4. J. Unguris, D.T. Pierce and R.J. Celotta, Rev. Sci. Instr. *59* (1986) 1314.
5. C. Rau, K. Waters and N. Chen, Phys. Rev. Lett. *64* (1990) 1441.
6. M. Landolt, R. Allenspach, and D. Mauri, J. Appl. Phys. *57* (1985) 3626.
7. H. Krakauer, A.J. Freeman and E. Wimmer, Phys. Rev. *B28* (1983) 610; C.L. Fu., A.J. Freeman and T. Oguchi, Phys. Rev. Lett. *54* (1985) 2700.
8. *Magnetic Properties of Low-Dimensional Systems*, L.M. Falicov and J.L. Moran-Lopez, eds., Springer Proceedings in Physics, Vol. 14, Springer-Verlag, Berlin (1986).
9. W. Gudat, E. Kisker, E. Kuhlmann, and M. Campagna, Phys. Rev. *B22* (1980) 3282.
10. E. Kisker, W. Gudat, E. Kuhlmann, R. Clauberg, and M. Campagna, Phys. Rev. Lett. *45* (1980) 2053.
11. H.D. Hagstrum and G.E. Becker, Phys. Rev. *B8* (1973) 107.
12. P.A. Zeijlmans Van Emmichoven, P.A.A.F. Wouters and A. Niehaus, Surf. Sci. *195* (1988) 115.

POSITRON THERMALISATION IN ALUMINUM

K.O. Jensen and A.B. Walker

School of Physics
University of East Anglia
Norwich NR4 7TJ, U.K.

ABSTRACT

A brief description is given of our studies of the energy loss and transport of positrons in solids. Results are presented from an investigation of the slowing down and thermalisation of positrons implanted into metals. In this study we have solved the Boltzmann equation for the positron momentum distribution in a homogeneous medium allowing the positrons to scatter off electrons and phonons. We obtain both the time-dependent and steady-state solutions. Our results give a more detailed and accurate description of positron thermalisation than earlier work.

Positron annihilation is a well established nondestructive technique for studying defects in bulk materials [1,2]. The advent of slow positron beams has, furthermore, in the last decade enabled defect profiling near surfaces and studies of surface phenomena [3]. In a typical experiment, positrons are injected into the sample either directly from a radioactive source (implantation energies peaking at $\approx$ 200 keV), or as a monoenergetic beam (energies from a few eV to $\approx$ 50 keV). Once injected, the positrons scatter elastically and inelastically off ion cores and conduction electrons, and within a picosecond reach energies of about 1 eV followed by a slower approach towards thermal equilibrium. On thermalisation, the positrons diffuse until they (in bulk studies) either annihilate or are trapped at open volume defects where they subsequently annihilate. Positron defect studies utilize the fact that the annihilation characteristics such as the lifetime and γ ray energy spectrum are different for positrons annihilating from the bulk compared to the strongly localized states formed in defects. In surface studies the surface acts as an additional sink for the positrons and the positron-surface interaction leads to a number of novel surface spectroscopies based on re-emitted positrons and positronium atoms [3].

Interaction of Charged Particles with Solids and Surfaces
Edited by A. Gras-Martí *et al.*, Plenum Press, New York, 1991

631

A detailed understanding of the slowing down and transport of positrons is important for both bulk and surface studies. For example, in bulk studies it has been an open question whether trapping prior to thermalisation is important for the analysis of positron defect measurements. Furthermore, the implantation profile of positrons is of vital importance for defect profiling near surfaces since the depth resolution is achieved by varying the implantation profile by changing the energy of the incident positron beam and in the interpretation of the positron surface spectroscopies it is important to know the energy distribution of positrons as they return to the surface after implantation. These questions have prompted us to embark on an extensive theoretical investigation of positron transport in metals, semiconductors, and insulators, employing techniques such as Monte Carlo simulations and direct solutions of the Boltzmann transport equation. In the present paper we give a brief description of one of the first stages of this project, namely the solution of the Boltzmann equation for positrons in homogeneous bulk metals in which we aim to obtain an improved description of the evolution of the positron momentum profile at low energies and positron thermalisation. We only give a brief description of our work here and refer the reader to ref.[4] for further details.

The Boltzmann equation describing the evolution with time t of the positron momentum distribution $f(p,t)$ in a homogeneous medium is [4,5]

$$\frac{d}{dt} f(p,t) = \int dq\, K(q,p)\, f(q,t) - H(p)\, f(p,t) + f_1(p,t). \qquad (1)$$

The term $f_1(p,t)$ represents the external source of positrons. The kernel $K(q,p)$ accounts for scattering of positrons from state $\hbar q$ into states in the interval dp around $\hbar p$. The function H is the rate of loss of positrons from state p due to scattering, annihilation, and trapping, $H(p) = H_{sc}(p) + \lambda + k(p)$, where H_{sc} is the total scattering rate, λ is the annihilation rate (which is independent of p at low momenta) and $k(p)$ is the momentum-dependent trapping rate into defects [6,7]. We assume that no detrapping occurs, i.e. trapped positrons are removed from the system and annihilate from a trapped state.

Earlier studies of positron energy loss [8,9] have shown that the energy loss down to 5 - 10 eV is so rapid (less than 0.1 ps) that the fraction of positrons being trapped or annihilated at higher energies is negligible. Thus, to investigate the effects of non-thermal trapping we need only consider low energies of the order of 1 eV and below. The dominant scattering mechanisms for positrons below 10 eV are acoustic phonon and conduction electron scattering [6]. Phonon scattering is included in a deformation potential model assuming a Debye model for the phonon dispersion relation. Conduction electron scattering is included in the Random Phase Approximation (RPA) assuming a free-electron gas at temperature T. The low positron energies allows us to use the low-energy and low-momentum limit of the RPA [5].

We have solved the Boltzmann equation, eq. (1), for the cases where the positron source term, $f_i(p,t)$, is a) a delta function in time and b) independent of time. In both cases the equation can be solved numerically without any additional approximations.

If f_i is a delta-function in time, $f_i(p,t) = f_1(p) \delta(t)$, eq. (1) describes the time development of the positron momentum distribution from the initial distribution $f_1(p)$. For zero trapping, $k(p) = 0$, the solution can be written in the form [5]

$$f(p,t) = \exp(-\lambda t) \ g(p,t), \tag{2}$$

where the normalized momentum distribution $g(p,t)$ satisfies

$$\frac{d}{dt} \ g(p,t) = \int dq \ K(q,p) \ g(q,t) - H_{sc}(p) \ g(p,t), \tag{3}$$

with the initial condition $g(p,0) = f_1(p)$.

Eq. (3) was integrated numerically, and fig.1 shows a series of

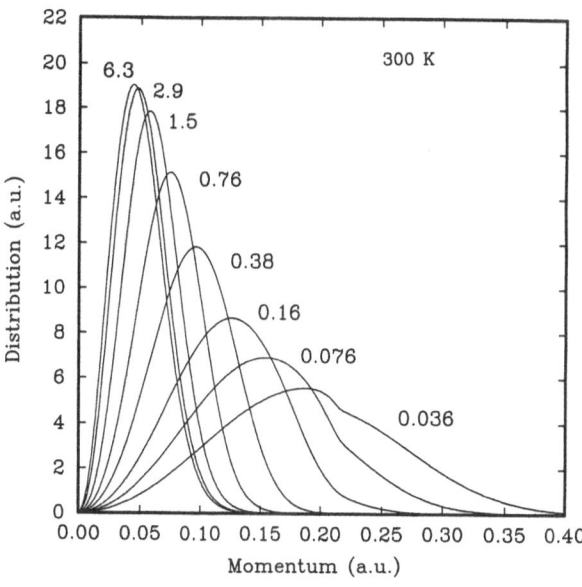

Figure 1. Positron momentum distributions in Al at 300 K. The time after implantation, in picoseconds, is indicated for each curve.

positron momentum distributions g(p,t) at different times. The initial distribution $f_1(p)$ was chosen as a narrow Gaussian centered at $p = 0.857$ a.u. corresponding to an average energy of 10 eV. However, all memory of the starting distribution is rapidly lost as the scattering takes effect and the distributions when the positrons approach thermalisation are independent of the choice of $f_1(p)$ [5]. The distributions shown in fig.1 all correspond to the regime where the effects of the initial distribution are no longer discernible. The distributions converge towards a thermal Maxwell-Boltzmann (MB) distribution as time increases and the curve for t = 6.3 ps is virtually indistinguishable from the MB curve.

Fig.2 shows the average energy of the positrons as a function of time, $\bar{E}(t)$, evaluated from the calculated momentum distributions. It is seen that the slowing down initially when the energy is above ≈ 1 eV is independent of temperature followed by an asymptotic approach of the mean energy towards the thermal energy $E_{th} = (3/2) k_B T$ (k_B is the Boltzmann constant) as $t \to \infty$. Note that the thermalisation times are short compared to λ^{-1} which is 165 ps for Al.

In most earlier studies, e.g. [6,8], the slowing-down processes have been described in terms of averaged scattering parameters such as the mean free path and the average energy loss rate. This ignores the fact that the positrons have an energy distribution of non-negligible width which leads to considerably higher energy loss rates compared to when all positrons

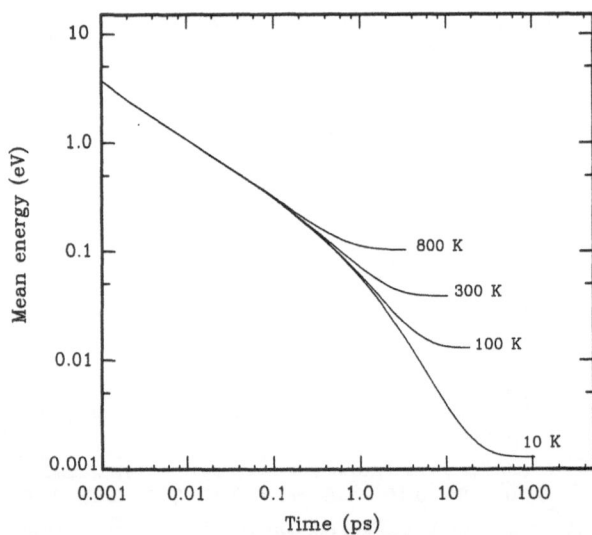

Figure 2. Mean energies $\bar{E}$ as a function of time after implantation for positrons slowing down in Al at different temperatures.

are assumed to have a single energy at a given time [5]. Only in the case where phonon scattering is neglected, i.e. electron scattering only, has the slowing down been followed in detail by calculations of the time dependent momentum distribution from the Boltzmann equation [5]. Our work thus goes beyond earlier treatments by including both electron and phonon scattering in a formulation based on the Boltzmann equation and the energy loss rates that can be obtained by differentiating $\overline{E}(t)$ with respect to t (results presented in ref.[4]) therefore are the most accurate calculated to date.

When f_i is independent of t, $f_i(p,t) = f_i(p)$, we get a steady state with $df/dt = 0$. Eq.(1) thus reduces to an integral equation for $f(p) \equiv f(p,t)$,

$$\int dq\ K(q,p)\ f(q) - H(p)\ f(p) + f_i(p) = 0. \qquad (4)$$

We have solved this equation iteratively. Fig.3 shows the steady-state momentum distributions at 10 and 300 K when no traps are present. In both cases the source distribution $f_i(p)$ is a narrow Gaussian corresponding to an energy of 3 eV but the distributions in the momentum range shown in the figure are not influenced by the choice of f_i as long as the initial mean energy is higher than $\approx$ 2 eV. The thermal Maxwell-Boltzmann (MB) distributions are also shown. The figure shows that $f(p)$ consists of a MB-like distribution with a tail of low intensity extending to high momenta. The tail represents positrons annihilating before thermalisation.

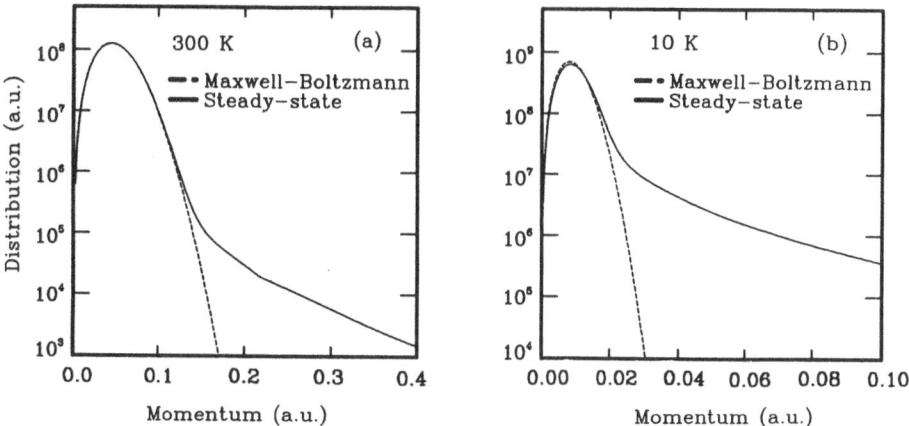

Figure 3. Steady-state positron momentum distributions in Al calculated from the Boltzmann equation at T = 300 K (panel a) and T = 10 K (panel b), in both cases without any defect trapping. Also shown are the thermal Maxwell-Boltzmann distributions. The curves in each panel are normalized to have same total area.

The tail means that the average positron energy $\bar{E}$ is higher than the thermal energy E_{th}. However, the low intensity of the tail means that at high temperatures $\bar{E}$ is extremely close to the thermal energy E_{th}. The relative difference between the two increases with decreasing T but only at very low temperatures, below 50 K, is $\bar{E}$ substantially higher than E_{th}. As T approaches zero $\bar{E}$ converges towards a minimum of about 2×10^{-3} eV corresponding to an effective temperature of 15 K.

By including calculated momentum-dependent trapping rates into defects [7] in the steady-state Boltzmann equation we can evaluate the influence of trapping of positrons before thermalisation on measurable parameters in positron defect experiments. Our results (see ref.[4] for details) clearly show that the trapping probabilities into vacancies from the full calculation including non-thermal effects are so close to those obtained for a thermal distribution of positrons as to have negligible impact on experimental measurements.

In summary, we have briefly described our solutions of the Boltzmann equation for a positron in a homogeneous metal. Future extension of this work will include the incorporation of spatial transport and extensions to semiconductors and insulators. We are also pursuing Monte Carlo simulations to examine positron (and electron) slowing down from energies of $\approx$ 5 keV down to $\approx$ 10 eV as well as low energy transport in a range of systems.

ACKNOWLEDGEMENT

KOJ would like to the Commission of the European Communities for a Research Grant under the Stimulation Action scheme contract no. SCI-0163.

REFERENCES

1. W. Brandt and A. Dupasquier, eds., *Positron Solid-State Physics*, North-Holland, Amsterdam (1983).
2. P. Hautojärvi, ed., *Positrons in Solids*, Springer, Berlin (1979).
3. P.J. Schultz and K.G. Lynn, Rev. Mod. Phys. *60* (1988) 701.
4. K.O. Jensen and A.B. Walker, submitted to J. Phys. Condensed Matter (1990).
5. E.J. Woll and J.P. Carbotte, Phys. Rev. *164* (1967) 985.
6. T. McMullen and M.J. Stott, Phys. Rev. *B34* (1986) 8935.
7. M.J. Puska and M. Manninen, J. Phys. F. Met. Phys. *17* (1987) 2235.
8. A. Perkins and J.P. Carbotte, Phys. Rev. *B1* (1970) 101.
9. R.M. Nieminen and J. Oliva, Phys. Rev. *B22* (1980) 2226.

CONTRIBUTED PAPERS: LOW ENERGIES

INTERACTION OF A CHARGED PARTICLE

WITH A SEMI INFINITE NON POLAR DIELECTRIC LIQUID

M.L. Forcada[1] A. Gras-Martí[1], N.R. Arista[1,2],
H.M. Urbassek[3] and R. Garcia-Molina[4]

1. Departament de Física Aplicada, Universitat d'Alacant
   Apartat 99, E-03080 Alacant, Spain

2. Permanent address: Centro Atómico Bariloche
   RA-8400 Bariloche, Argentina

3. Institut für Theoretische Physik, Technische Universität
   Mendelssohnstrasse 3, D-3300 Braunschweig, Germany

4. Departamento de Física Aplicada, Universidad de Murcia
   E-30071 Murcia, Spain

1. SUMMARY AND INTRODUCTION

We investigate the classical interaction between a charged particle and a semi-infinite non-polar dielectric liquid. The classical results, in the long-distance regime, agree with a previous quantum-mechanical approach [1], which was used to describe the energy loss of an ion travelling parallel to the liquid surface (see also ref.[2]). The results in the (non-linear) short-distance regime provide information that was not available in the linearized quantum-mechanical calculations of the energy loss.

The understanding of the basic static and dynamic interactions between charged particles and liquid surfaces may be of particular interest in fields such as aerosol physics [3] and atomic force microscopy [4]. The main motivation of this work was to analyze, with the least number of mathematical approximations, the classical aspects of these interactions.

2. DESCRIPTION OF THE PROBLEM

Our first aim is to compute the equilibrium shape of the surface of a non-polar dielectric liquid, bounded by an otherwise planar surface, when

Interaction of Charged Particles with Solids and Surfaces
Edited by A. Gras-Martí *et al.*, Plenum Press, New York, 1991

639

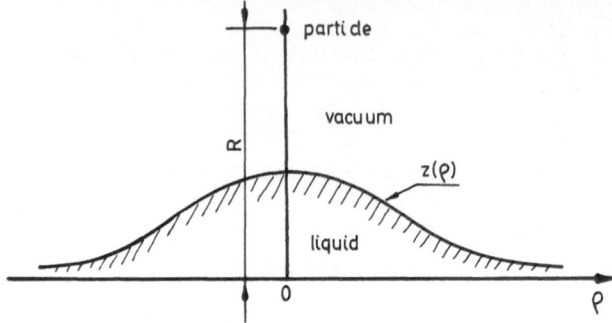

Figure 1. Schematics of the interaction between the charged particle and the liquid, and the resulting surface profile.

a point-charged particle is held in the vacuum at a certain distance. The interaction energy may then be computed when the shape is known. The situation is sketched in fig.1. An attractive interaction between the charge and the liquid arises due to the dipoles induced by the charge on the constituents (atoms or molecules) of the liquid. This attractive interaction, which tends to elevate the surface of the liquid near the charge, is balanced by the opposing forces arising from the surface tension of the liquid and gravity. In this work, the liquid is considered to be incompressible and is treated as a continuous medium. Screening inside the liquid is neglected. Axial symmetry holds for a static charge.

Let $z(\rho)$ be the shape of the liquid surface, where z is the vertical displacement from the original planar surface and ρ is the horizontal distance from the symmetry axis (see fig.1). The total energy of the system containing the gravitational, surface-tension and electrostatic energies, respectively, is

$$\Delta E = \int_0^\infty 2\,\pi\,\rho\,d\rho\left\{\frac{1}{2}\,M\,n\,g\,z^2(\rho)\right.$$

$$\left. + \sigma\left[\left(1 + \left(\frac{dz}{d\rho}\right)^2\right)^{1/2} - 1\right] - \frac{n\,\alpha\,Q^2}{2}\int_{-\infty}^{z(\rho)}\frac{dz'}{[\rho^2 + (z'-R)^2]^2}\right\}. \qquad (1)$$

Here, M is the mass of a constituent of the liquid (atom or molecule), g the gravity constant, σ the surface tension coefficient, n the number density, α the atomic or molecular static polarizability of the liquid, R the height of the particle and Q its charge. The electrostatic energy term has been constructed as a pairwise summation of interactions between the charge and the constituents of the liquid. The energy in eq.(1), which is a functional of $z(\rho)$, may be minimized to obtain the equilibrium shape function, as will be shown in the following section.

3. BASIC DIFFERENTIAL EQUATION

Now, we look for the stationary points of ΔE, i.e., those satisfying

$$\delta(\Delta E) = 0, \tag{2}$$

for a small displacement of the surface $\delta z(\rho)$ (see, e.g., refs.[5,6]). This leads to a nonlinear differential equation in $z(\rho)$,

$$0 = M \, n \, g \, z(\rho) - \sigma \, \frac{1}{\rho} \frac{d}{d\rho} \, \frac{\rho \, \frac{dz}{d\rho}}{\left[1 + \left(\frac{dz}{d\rho}\right)^2\right]^{1/2}} - \frac{n \, \alpha \, Q^2}{2} \, \frac{1}{[\rho^2 + (R - z(\rho))^2]^2}. \tag{3}$$

The solutions of eq.(3) may be minima (stable equilibrium), but also maxima or saddle points. We solve eq.(3) with the boundary conditions

$$\left.\frac{dz}{d\rho}\right|_{\rho=0} = 0 \tag{4}$$

and

$$\lim_{\rho \to \infty} z(\rho) = 0. \tag{5}$$

The first condition, (4), accounts for the smoothness of the surface, due to surface tension, and the second, (5), is just a consequence of our choice of origins (the unperturbed surface).

We have not been able to find an analytical solution for eq.(3), so we resort either to approximations or to a numerical treatment in the following.

4. DISTANT PARTICLE-SURFACE INTERACTION

For large values of the charge-surface distance R,

$$z(\rho) \ll R, \tag{6}$$

one may assume that the resulting surface is very flat, i.e.,

$$\left(\frac{dz}{d\rho}\right)^2 \ll 1, \tag{7}$$

for all values of ρ. This linearizes the basic equation to give a linear second-order differential equation

$$0 = M n g z - \sigma \left[\frac{1}{\rho} \frac{dz}{d\rho} + \frac{d^2 z}{d\rho^2} \right] - \frac{n \alpha Q^2}{2 (\rho^2 + R^2)^2} \qquad (8)$$

The solutions of eq. (8) that satisfy the boundary conditions (4) and (5) have the form

$$z^{lin}(\rho) = \frac{n \alpha Q^2}{4 \sigma R^2} \int_0^\infty du \; K_1(u) \; J_0\left(u \frac{\rho}{R}\right) \frac{u^2}{u^2 + \frac{R^2 Mng}{\sigma}}, \qquad (9)$$

where K_1 and J_0 are Bessel functions. The width of the profiles predicted by these solutions is of the order of the capillary length $(\sigma/Mng)^{1/2}$ (9.55×10^6 a.u. for liquid He), as shown in the solid line of fig.2. The maximum height of the profile, $z(0)$, induced by an electron on liquid helium is plotted as a function of the distance R in fig.3, curve 1. This curve may thus be used to check the range of validity of the distant-interaction approximation, eq. (6), which is exactly equivalent to the corresponding approximation made by Gras-Martí and Ritchie [1]; in fact, if one computes following their procedure the displaced equilibrium position of the capillary waves on a liquid surface (*ripplons*) when a point charge is present, one gets eq. (9). Thus, the range of R for which the linear treatment of ref. [1] remains valid may be assessed with the present calculations.

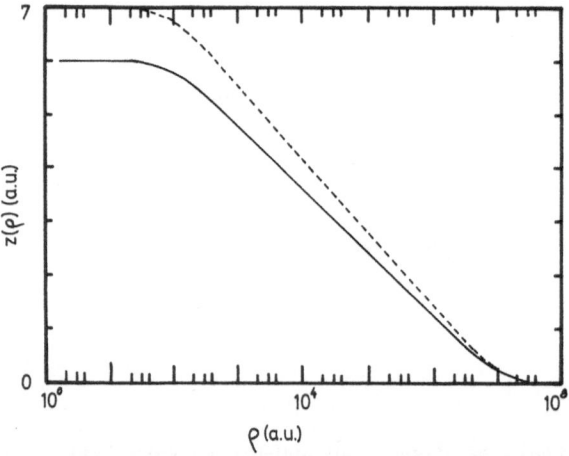

Figure 2. Comparison between a linear (——) and a non-linear (---) profile of a liquid in front of a charge. For liquid He at 3 K, and (in a.u.) R = 96, |Q| = 1; $\sigma = 2.25 \times 10^{-7}$, M = 7310, n = 0.00312, $\alpha = 1.37$, $g = 1.08 \times 10^{-22}$.

5. NUMERICAL RESULTS

The basic eq.(3) may be solved numerically to yield exact results for the induced surface shapes. Numerically computed maximum heights z(0) are shown in fig.3, curve 3. Some interesting features may be inferred from these results. As expected, as R gets smaller, z(0) departs from the linear result represented by curve 1. However, a minimum value of R is reached, below which no solutions are found. This may be interpreted as the point where the liquid *pops up* to incorporate the charge (the parts of curves 2 and 3 above the arrow correspond to stationary points of the energy functional that are not minima, i.e., they do not correspond to stable equilibrium).

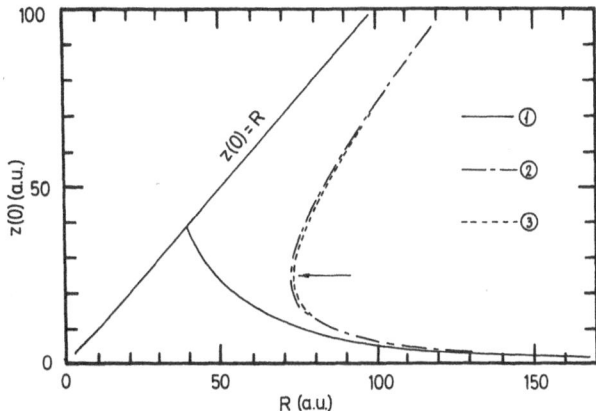

Figure 3. Heights z(0) of profiles induced in liquid He at 3 K by a particle of |Q| = 1 a.u. at a distance R; curve 1: linear approximation; curve 2: numerical results; curve 3: constant-shape approximation. The arrow indicates the point where the minimum value of R is reached and above which the results do not correspond to minima in the energy functional.

However, the profile given by the exact solutions is not very different from that of the linear ones, as it is shown in fig.3. In fact, the stable branch of the non-linear solution may be approximated with high accuracy by just scaling the linear one, and this holds for all the allowed values of R. Moreover, all exact solutions show more or less the same profile, except for their height.

Numerically computed charge-surface interaction energies are well approximated by the image potential of the unperturbed planar surface, but vertically displaced to a height around 0.95 z(0), i.e.,

$V = - \pi n\alpha Q^2/\{2[R - 0.95\ z(0)]\}$. This is consistent with the extended, flat shape adopted by the liquid.

The results of further calculations also suggest an interesting liquid-independent approximate relation for the highest stable profile (corresponding to the minimum distance),

$$R \simeq 3\ z(0). \qquad (10)$$

An *exact* relation $R = 3\ z(0)$ may be obtained by using a simple model where the following assumptions are made: (a) the surface is so flat that a *displaced* planar image potential $V = - \pi n\alpha Q^2/\{2[R-z(0)]\}$ describes the electrostatic interaction properly; (b) horizontal distances are equivalent for every liquid when scaled with the capillary length $(\sigma/Mng)^{1/2}$; and (c) all profiles for the same liquid have the same shape and differ only in height $z(0)$. This model allows us also to obtain an approximate equation for R as a function of $z(0)$,

$$R = z(0) + \left[\frac{n\ \alpha\ Q^2}{4\ \sigma\ K\ z(0)}\right]^{1/2}, \qquad (11)$$

where K is a constant related to the common shape of all profiles. An excellent fitting of (11) to the numerical results (including the non-minimum results) is shown in fig.3, curve 2.

6. CONCLUSIONS

We have computed classically the static interaction between a charged particle and the surface of an incompressible, non-polar, dielectric liquid. The results in the long-distance regime agree with previous treatments [1], that were applied also to the dynamic interaction. The results we present in this paper suggest that, at short distances, where the perturbation of the liquid is larger, a quantum treatment necessarily has to include approximations higher than first-order.

ACKNOWLEDGEMENTS

The collaboration was facilitated by the German DAAD and the Spanish MEC program of Acciones Integradas. One of us (NRA) thanks MEC and the Generalitat Valenciana for support through their sabbatical and visitor programs. Partial support from the Spanish DGICyT (projects PS88-0066 and PS89-0065) is acknowledged.

The authors thank Prof. F. Flores for having suggested the calculation of the classical deformation of liquids, and Nuria Barberan for encouragement. We also thank M. Vičánek for fruitful discussions, and the support of N.E. Capuj and V. Vermeta.

REFERENCES

1. A. Gras-Martí and R.H. Ritchie, Phys. Rev. *B31* (1985) 2649.
2. N. Barberan, R. Garcia-Molina and A. Gras-Marti, Phys. Rev. *B40* (1989) 10
3. W.H. Marlow, ed., *Aerosol Microphysics I: Particle Interactions*, Springer, New York (1980).
4. G. Binnig, C.F. Quate and C. Gerber, Phys. Rev. Lett. *56* (1986) 930.
5. R.P. Feynman, R.B. Leighton and M. Sands, *The Feynman Lectures on Physics*, Addison and Wesley, Reading, Mass., U.S.A (1964) chapter II-9.
6. L.D. Landau and E.M. Lifshitz, *Fluid Mechanics*, Pergamon, Oxford (1959).

ON THE EXPERIMENTAL VERIFICATION OF VARIOUS MECHANISMS

OF SECONDARY ATOMIC AND MOLECULAR ION PRODUCTION

UPON ION BOMBARDMENT OF PURE METALS

I.F. Livshits, V.A. Pazdzersky and B.A. Tsipinyuk

Applied Physics Institute
Tashkent State University
Tashkent 700095, USSR

ABSTRACT

Various mechanisms for the production of secondary positive ions upon ion bombardment of pure metals will be considered. The attention will be focused on the feasibility of experimental verification of the mechanisms, rather than on a detailed theoretical description of one or another model of the phenomenon.

1. MECHANISMS OF SPUTTERED-ION PRODUCTION

Well known mechanisms of secondary-ion emission (SIE) of pure metals can be divided into three types according to the role of the parameters, such as the primary ion energy E_o, electron-energy structure of a metal surface and sputtered atoms, as well as the energy E (or velocity v) of these atoms.

First of all one should distinguish the so-called kinetic mechanism [1,2] according to which atom collisions inside a metal result in the production of electron vacancies in the inner shells of the colliding atoms. Atoms with such a vacancy and having acquired momentum favorable for their escape from the metal into vacuum can, with a certain probability, retain this vacancy and leave the metal as a neutral atom (ion) in the overexcited state. Outside the metal such an atom (or ion) experiences Auger relaxation to produce a singly-charged (or doubly charged, respectively) ion in the ground or excited state and an electron with a characteristic energy [2-6]. Double-Auger processes leading to the production of singly-charged ions in highly excited states as well as doubly-charged ions and electrons with a continuous energy distribution are also feasible [6]. With a certain probability, de-excitation of

Interaction of Charged Particles with Solids and Surfaces
Edited by A. Gras-Martí *et al.*, Plenum Press, New York, 1991

647

sputtered overexcited atoms and ions can proceed without a change in the charge of the particles to be accompanied by soft X-ray radiation [7]. The probability of production of inner-shell vacancies highly depends on the collision energy, hence on the energy of primary ions E_0.

Another type of SIE mechanisms comprises various models according to which a charged state of sputtered atoms is defined by the electron exchange between the metal and sputtered atom, moving away from the metal surface with the velocity v (see reviews [8-11]). In these models, that can be referred to as electron exchange or tunneling models, the probability of sputtered-atom ionization depends only on the electron and energy features of the metal and the sputtered atoms, as well as the value and direction of the atom velocity. The energy and kind of primary ions, as well as the process itself of atomic collisions in the metal, giving rise to the ejection of atoms into vacuum, play no role in these models.

For the probability P^+ of sputtered-atom ionization such models yield either a power dependence on the normal component of the velocity v_n [8,12] or an exponential dependence on the inverse velocity [9]. For small v_n one has to take atom slowing down into account due to the action of the surface potential-barrier forces [11].

To this type of SIE mechanisms one can also attribute a mechanism for the production of transition and noble metal ions [13]. By this mechanism, sputtered atoms, while crossing the "metal-vacuum" interface, are excited into the autoionizing state with a low overexcitation energy (~ 0.1 eV). Outside the metal such atoms are autoionized to produce an ion and a low energy electron.

Recently other mechanisms [14-17] that can be also related to those of electron exchange have been proposed. The distinction between the approach used in [14] and the traditional approaches [8-10] lies in taking account of the influence of the last collision on the electron exchange in the "metal-sputtered atom" system. Such a collision in the case, for instance, of sputtered Cu atoms leads to the overlapping of partially localized d-orbitals of the colliding atoms. This, in its turn, leads to the promotion of the 4s energy level of a sputtered atom, the population probability of which determines the value of P^+. Even though such situations are rather rare, and contribute little to the yield of sputtered neutral atoms, they can make an essential contribution to the yield of sputtered ions owing to the high values of P^+.

In refs.[15-17], various channels of sputtered-atom ionization and molecular-ion production are considered, in the framework of the so-called collisional mechanism. According to this mechanism, atomic and molecular ions can be produced during the collision of two or more atoms ejected by a primary ion near and far from the surface. A quasimolecule, formed as a result of this collision, can follow the repulsive term $E_r^o(R)$ which with the decrease of R approaches the coupling term $E_c^+(R)$ of the molecular ion, and in some cases crosses it. If in this case the energy of quasimolecule ionization from the repulsive term, that is equal to $I_{mr}(R) = E_c^+(R) - E_r^o(R)$, appears lower than the metal work function, the electron can tunnel from the quasimolecule into the metal. As a result, once the collision is over, either an ionized molecule or a neutral atom and ion (in case of the collision between two atoms), depending on the relation between the collision energy and dissociation energy of the molecular ion, is formed. Provided the terms $E_r^o(R)$ and $E_c^+(R)$ are crossed, the quasimolecule can be ionized far away from the metal surface as well. Ultimately, an atomic or molecular ion and a free electron are produced (associative ionization) [17]. A similar process was previously considered [18,19] to interpret the production of excited N_2^{+*} ions upon ion bombardment of nitrogen-implanted Si.

Finally, there is a group of SIE mechanisms for which both the energy and kind of primary ions and the parameters determining the electron exchange between a sputtered atom and metal can play an important role. In some of these mechanisms the features of the metal in the area of the collision cascade may be distinct from those of the unperturbed metal.

One of the mechanisms of this type is that associated with the temporal ($10^{-12} - 10^{-13}$ s) lattice-structure distortion at the place of ion impact [8]. In this region the energy band structure of the metal can be distinguished from that of the ordered crystal. In particular, the increase of the degree of localization of the electron states is not ruled out. This can hinder the electron exchange with the ion sputtered from this region, thereby decrease the probability of its neutralization.

On the contrary, in refs.[12,20,21] the change of the metal band structure in the region of the collisional cascade was ignored and the increase of P^+ was related to the excitation of the electron gas in that region due to the inelastic energy loss while the particles were moving in the metal. That excitation was assumed to lead to the rise of the effective "temperature" T_e in the cascade region. Additionally, assuming that at the metal surface the energy level of the sputtered atom coincides with the Fermi level, the power dependence $P^+ \propto v^n$ with the exponent $n \propto 1/T_e$ has been obtained [11]. The calculated dependence $P^+(v)$ turned out to give a good fit to an experimental one for Cu (n = 1.9 and

$P^+ = 1.6 \times 10^{-4}$ for $v = 7 \times 10^5$ cm/s) for $T_e = 2200$ K. The concept of increasing the effective temperature in the region of the primary ion impact was also used previously in the framework of local thermodynamical equilibrium or nonequilibrium plasma formed in this region as the source of secondary ions [22-24]. However, in these papers much too high values of T_e were used (~ 5000 K).

A specific mechanism for the production of atomic ions is a monomolecular decay of secondary cluster ions of the type

$$M_{n+1}^+ \rightarrow M_n + M^+. \tag{1}$$

Up to now, as far as we know this type of decays was only observed for sputtered cluster Al_n ions with $n \leq 6$ [25]. For this n the decay via the channel

$$M_{n+1}^+ \rightarrow M_n^+ + M, \tag{2}$$

took place with much lower probability. Conversely, for $n > 6$ a much higher probability was observed via the channel (2). This is due to the fact that for $n \leq 6$ the ionization energy $I(Al)$ of Al neutral atoms is lower than $I(Al_n)$ for the neutral cluster Al_n [26,27]. And, vice versa, for $n > 6$, i.e., $I(Al) > I(Al_n)$. Since in the process of the cluster ion decay the resulting fragments are separated with a small relative velocity, the electron has a higher probability to reside in the fragment with high ionization energy. It is to be noted that decays of cluster ions of many transition metals fail to occur via the channel (1) with an appreciable probability, since for such metals $I(M) > I(M_n)$ for all $n > 2$ [28]. Therefore, this mechanism cannot contribute noticeably to the yield of atomic ions of transition metals.

2. ON THE EXPERIMENTAL VERIFICATION OF THE SIE MECHANISMS

As seen from the previous section, the yield of secondary ions may be due to the action of a number of mechanisms. The determination of the contribution by each mechanism is a very involved problem that has not been solved so far.

To evaluate the role of the kinetic mechanism in the production of sputtered ions we can make use of the fact that the ejection of particles having vacancies in the inner electron shells, from the metals, is a common reason for a few phenomena such as: the secondary ion emission (including the emission of multicharged ions); emission of excited atoms and ions and accompanying photon emission (including soft X-ray radiation); emission of electrons with characteristic values of energies (Auger electrons). Unfortunately, complex investigations of all these phenomena were not carried out under identical experimental conditions. In refs.[2,6] a comparative quantitative analysis of the results of investigations of the above phenomena for ion bombardment of Al and Mg was attempted. When required, corrections allowing for the differences between experimental parameters, such as the energy and kind of the primary ions and the angle of incidence on the metal surface, were introduced. In the range of E_o = 6 - 10 keV a major part of Al and Mg excited ions was shown to be produced by the kinetic mechanism. As regards the total ion yield, both in the ground and excited states, the kinetic mechanism provides ~ 70 - 80 % of the total Mg^+ ion yield [29] at the normal primary-ion incidence, and ~ 95 % at the incidence angle of $\theta = 60^o$ to the surface normal [2]. For the case of Al^+ ions [29] this contribution is much lower: ~ 5 % for the normal incidence and ~ 20 % for $\theta = 60^o$. This allows us to understand such an experimental fact observed in ref.[29] as a strong distinction between the θ-dependences of the Mg^+ and Al^+ ion yields. For Mg^+ ions the progression from the normal incidence of primary ions to the oblique one at $\theta = 60^o$ leads to the increase of Mg^+ ion yield approximately by an order of magnitude, whereas for Al^+ ions the yield increases only by a factor of ~ 2. These results will be clear if we take into consideration that for $\theta = 60^o$, the atoms with L_{23} vacancies have a more favorable direction of momentum for escaping into vacuum than for $\theta = 0^o$. Hence, they will keep this vacancy with greater probability when escaped into vacuum. Therefore, the yield of Mg^+ ions that are largely produced by the kinetic mechanism greatly increases at $\theta = 60^o$, while that of Al^+ ions, only a small part of which is produced by the kinetic mechanism, changes as a sputtering coefficient $K_s (\sim 1/\cos\theta)$.

Clearly, in order to rule out the contribution of the kinetic mechanism to the total yield of secondary ions, hence, more unambiguously to analyze the contribution of other mechanisms, E_o ought to decrease below the energy threshold of the vacancy creation in the inner electron shells, that in most cases is ~ 1 keV. It is also unlikely that in the range of $E_o < 1$ keV the SIE mechanism, related to the radiation damage, can be effective [8]. As far as the mechanism related to the concept of the effective electron "temperature" T_e in the collisional cascade [11] is concerned, the question about the E_o dependence of T_e is of importance. A rough estimate of the value of T_e , averaged over the collision cascade, indicates that T_e is practically E_o independent in the linear cascade region [30] (the nonadiabatic electron-exchange models [8-10,12,13] are strictly E_o independent). In this case the mechanism under consideration results in the dependence $P^+(v) \propto v^n$, with n independent of E_o, i.e., it can be attributed to the mechanisms of electron-exchange type whose role will be discussed below. The role of T_e in the collisional cascade can be expected to increase, in progressing from the atomic primary ions to cluster ions of the same kind and the same energy per nucleus, due to nonlinear effects in the evolution of the collisional cascade. This increase of T_e must affect both the yield of secondary ions and that of secondary electrons. Unfortunately, no similar experiments for secondary ions were carried out. At the same time it was stated [31] that nonlinear effects were missing for secondary electrons, for instance, when primary cluster V_n^+ and Nb_n^+ ions with n = 1 - 9 and energies of ~ 1.4 - 25 keV per nucleus were used.

Now let us consider the problem of the experimental verification of the electron-exchange mechanism. A common assumption, for different models of this mechanism, is that each sputtered atom with energy E may leave a metal as an ion with probability $P^+(E)$. Then, denoting the energy distributions of ions and atoms, ejected within a unit solid angle by primary ions with the energy E_o , via $N^+(E,E_o)$ and $N^o(E,E_o)$, one can write down the expression ($N^+ \ll N^o$)

$$N^+(E,E_o) = P^+(E) \, N^o(E,E_o), \qquad (3)$$

i.e., for a fixed E, the E_o dependence of the distributions $N^+(E,E_o)$ and $N^o(E,E_o)$ should be identical. The experimental verification of the validity of this conclusion is of crucial importance for the determination of the contribution of the electron-exchange mechanism to the secondary ion yield in the range of $E_o < 1$ keV. At the same time up-to-date direct experiments using SIMS and SNMS techniques to determine $P^+(E)$ [32,33] were carried out for some fixed values of E_o.

In refs.[15,16] it was pointed out that in the range of $E_o < 1$ keV the yield N^+ of secondary V^+, Nb^+, Ta^+ ions with the fixed value of E = 5 eV (i.e., near the maximum of $N^+(E)$) increases with the growth of E_o much faster than the integral sputtering coefficient $K_s(E_o)$. This is also the case with the sputtered Nb^+ ions with energies in narrow ranges of ΔE = 5 - 2.5 eV, 7.5 - 5 eV and 10 - 7.5 eV [32]. The analysis performed in ref.[15,16] revealed that the relevant E_o dependences of N^o, for the same values of E, should be still more gentle sloping than the $K_s(E_o)$. These facts are inconsistent with the expression (3), hence with the nonadiabatic electron-exchange models [8-10,12,13] or with the assumption that T_e is E_o-independent for $E_o < 1$ keV. Nevertheless, SIMS and SNMS experiments seem to be quite significant to verify the above conclusion more reliably.

Expression (3) cannot be used, either, to determine $P^+(E)$ in order to verify a model of the electron-exchange mechanism put forward in ref.[14], since in (3) the distribution $N^o(E)$ is characteristic for all sputtered atoms. In this connection the question arises: how can $P^+(E)$ be determined experimentally in the cases when it is specially large in certain, though rare, situations in the collisional cascades. These situations contribute little to the yield of neutral atoms, but do make an essential contribution to the secondary ion yield. In such cases the way of determining $P^+(E)$ should not be based on the yield of all neutral atoms. In ref.[35] an unusual procedure for determining $P^o(E)$ using only measurements of normalized families of the energy spectra $N^+(E,E_o)$ was put forward. This procedure relies on the experimentally verified assumption [35] that in the range of $E_o \sim 0.1 - 1$ keV, atomic collisions inside a solid are rather accurately described in the framework of the hard-sphere approximation. In this approximation the energy distribution $N^+(E,E_o)$ of ions sputtered in the normal direction to the surface may be represented in the form

$$N^+(E,E_o) \, dE = P^+(E) \, g(E,E_B) \, F[(E+E_B)/E_o] \, (dE/E_o), \tag{4}$$

where E_B is the potential barrier for sputtered ions at the metal-vacuum interface; $F[(E + E_B)/E_o](dE/E_o) = F(x)dx$ is the energy distribution, (before the potential barrier has been overcome), of those sputtered ions that, with the probability $P^+(E)$, are ionized while leaving the surface; $g(E,E_B)$ is a barrier function equal to $E/(E + E_B)$ for the case of a flat barrier. If in eq.(4) $E = E_k = $ const. and E_o changes, from the experimental family of $N^+(E,E_o)$ one can define the function $F(x) = F[(E_k+E_B)/E_o]$ with an accuracy to the constant factor. By varying E_k, we obtain $F(x)$ that in the overlapped intervals of the argument x should coincide with an accuracy to constant factors. This will be the case only if the function $P^+(E)$ is independent of E_o, i.e., one of the electron-exchange mechanisms works. Thus, once $F(x)$ has been determined and the $g(E,E_B)$ has been given, one can derive $P^+(E)$ from eq.(4). This procedure can be applied to the determination of $P^+(E)$ in the cases when the relation (3) is inapplicable. One of the difficulties for applying this procedure is a proper choice of the value of E_B and the function $g(E,E_B)$.

As far as the collisional SIE mechanism is concerned, it agrees unambiguously with known experimental results on the emission of atomic and molecular ions. So, in its framework one can qualitatively interpret the above mentioned more rapid increase of the sputtered-ion yield, with the increase of E_o, as compared with that of neutral atoms (for E = const). The yield of ions produced by this mechanism ought to grow with the increase of E_o not slower than the probability of the emission of more than one atom sputtered by one primary ion. In the range of $E_o < 1$ keV, this probability ought to depend on E_o more strongly than $K_s(E_o)$. In particular, for small K_s this probability is proportional to $(K_s)^2$.

The occurrence of crossing of quasimolecular terms $E_r^o(R)$ and $E_c^+(R)$ in the case of transition metals, and its absence in the case of noble metals, as well as high values of the dissociation energy of transition metal dimers as compared with that of noble metal dimers, allows us to understand an essentially higher yield of atomic ions and high values of the ratio between the yields of dimer and atomic ions for transition metals as compared with noble ones [17].

Estimates of the efficiency of the collisional mechanism, performed in refs.[15,16], showed that this mechanism can essentially contribute to the secondary ion yield. However, a quantitative solution of this problem needs further investigations. Important information on the role of the collisional mechanism can be obtained from complex experimental investigations of the emission of atomic and molecular ions and secondary electrons under the bombardment of metals by atomic and multiatomic ions with energies of ~ 0.1 - 1 keV per nucleus.

REFERENCES

1. P. Joyes, J. Phys. *30* (1968) 243 and 365.
2. V.A. Pazdzersky and B.A. Tsipinyuk, Poverkhnost *1* (1987) 19.
3. R. Whaley and E.W. Thomas, J. Appl. Phys. *56* (1984) 1505.
4. G. Zampieri and R. Baragiola, Phys. Rev. *B29* (1984) 1480.
5. C. Benazeth, M. Hou, C. Mayoral and N. Benazeth, Nucl. Instr. Meth. *B18* (1987) 555.
6. V.A. Pazdzersky and B.A. Tsipinyuk, Vacuum *35* (1985) 255.
7. W. Degel and W. Hink, Nucl. Instr. Meth. *B13* (1986) 619.
8. V.I. Veksler, Rad. Eff. *51* (1980) 129.
9. G. Blaise and A. Nourtier, Surf. Sci. *90* (1979) 495.
10. M.L. Yu and N.D. Lang, Nucl. Instr. Meth. *B34* (1986) 403.
11. Z. Sroubek, Spectrochimia Acta *44B* (1989) 317.
12. V.A. Pazdzersky and B.A. Tsipinyuk, Fiz. Tverdogo Tela *20* (1978) 3283.
13. G. Blaise and G. Slodzian, J. Phys. *31* (1970) 93.
14. A. Nourtier, J. Quazza and J-P. Jardin, *Second. Ion Mass Spectrometry SIMS VI*, A. Benninghoven, A.M. Huber and H.W. Werner, eds., p.147, Wiley, Chichester (1988).
15. A.S. Avakov, V.A. Pazdzersky and B.A. Tsipinyuk, Poverkhnost *12* (1987) 13.
16. A.S. Avakov, V.A. Pazdzersky and B.A. Tsipinyuk, *Second. Ion Mass Spectrom. SIMS VI*, A. Benninghoven, A.M. Huber and H.W. Werner, eds., p.125, Wiley, Chichester (1988).

17. A.S. Avakov, V.A. Pazdzersky and B.A. Tsipinyuk, Nucl. Instr. Meth. *B48* (1990) 562.
18. K.J. Snowdon, W. Heiland and E. Taglauer, Phys. Rev. Lett. *46* (1981) 284.
19. K. Snowdon, R. Hentschke, W. Heiland and P. Hertel, Z. Physik *A318* (1984) 261.
20. Z. Sroubek, Nucl. Instr. Meth. *194* (1982) 533.
21. J. Zavadil, Surf. Sci. *143* (1984) L383.
22. C.A. Andersen and J.R. Hintorn, Science *175* (1972) 853.
23. A.G. Lototsky and F.A. Gimelfarb, Zh. Analit. Khim. *31* (1976) 433.
24. K.J. Snowdon, Rad. Eff. *40* (1979) 9.
25. A.D. Bekerman, N.Kh. Dzhemilev, V.M. Rotshtein, Pis. Zh.Techn. Fiz. to be published.
26. L. Hanley, S.A. Ruatta and S.L. Anderson, J. Chem. Phys. *87* (1987) 260.
27. M.F. Jarrold, J.E. Bower and J.S. Kraus, J. Chem. Phys. *86* (1987) 3876.
28. M.D. Morse, Chem. Rev. *86* (1986) 1049.
29. J.-F. Hennequin, J. Phys. *29* (1968) 957.
30. Z. Sroubek, Appl. Phys. Lett. *45* (1984) 849.
31. W.O. Hofer, Nucl. Instr. Meth. *170* (1980) 275.
32. T.R. Lundquist, J. Vac. Sci. Technol. *15* (1978) 684.
33. A. Wucher and H. Oechsner, Surf. Sci. *199* (1988) 567.
34. A.S. Avakov, I.F. Livshits, V.A. Pazdzersky and B.A. Tsipinyuk, to be published.
35. B.A. Tsipinyuk and V.I. Veksler, Vacuum *29* (1979) 155.

CASCADE EVOLUTION AND SPUTTERING IN CONDENSED RARE GASES

K.T. Waldeer and H.M. Urbassek

Institut für Theoretische Physik
Technische Universität
PO Box 3329, D-3300 Braunschweig, FRG

1. INTRODUCTION

The sputtering of strongly-bonded materials, like metals and semiconductors, by energetic ions can in many cases be understood by the so-called linear cascade theory [1,2]: The projectile ion shares its energy in collisions with target atoms; these recoil from their original positions, collide with other target atoms; etc. Atoms which cross the target surface and are able to overcome the surface barrier, are sputtered. In this approach, the cascade is assumed to be *dilute*, and collisions between moving atoms can be neglected. It is clear that after some time, practically all particles in the cascade volume have been energized; these then have so small energies, however, that they do not contribute to sputtering.

In condensed rare-gas targets with their extremely low surface-binding energy U, however, sputtering can indeed take place at late times also. Thus here the evolution of *dense* cascades, and the sputtering therefrom, have to be studied. This shall be done in the present note. In order to identify the differences between dilute and dense cascades most clearly, cascade evolution in a specific system will be treated using alternatively the linear Boltzmann equation, which forms the basis of the linear cascade theory, and the nonlinear Boltzmann equation [3]. This latter equation shows the following three features.

(i) It contains the linear Boltzmann equation in the limit of dilute cascades. Hence both approaches should give identical results for the high-energy spectrum of cascade particles, which are necessarily dilute, and possibly at early times.

(ii) It is able to exhibit gas-dynamical structures like collective

Interaction of Charged Particles with Solids and Surfaces
Edited by A. Gras-Martí *et al.*, Plenum Press, New York, 1991

657

particle flows, the build-up of shock waves, and also thermal effects like thermalization, heat conduction, and surface evaporation.

(iii) In contrast to the linear equation, it conserves the particle number.

Thus a description of sputtering using the nonlinear Boltzmann equation incorporates many of the concepts which have been employed to understand the sputtering of condensed-gas targets like a contribution from the linear collision cascade, evaporation from spikes and gas flow [4]. In contrast to a full molecular-dynamics calculation, however, it does not incorporate the multiple interaction of one particle with several others simultaneously, nor energy transport without mass transport like in focused collision sequences. Furthermore attractive forces, which are responsible for, e.g., cluster formation in the sputtered flux, phase transitions in the target and the formation of the surface barrier have no natural place in this approach and need additional ad hoc assumptions. In summary, particle and energy flows are modeled in our approach as if occurring in the gas phase; effects which are typical of transport phenomena in dense (solid or liquid) systems are neglected.

2. THE MODEL

As a specific system we model the sputtering of a condensed Ar target (number density $n = 0.024$ $Å^{-3}$, temperature $T = 0$ K) by a 1 keV Ar atom. All particles interact via the Kr-C potential [5], which is cut off at the interparticle distance 1.9 Å. The target is represented by a cubic box with depth 121 Å and lateral width 170 Å, containing 84569 simulated particles. Particles can leave the target through the surface, if they can overcome a planar surface barrier of height $U = 0.08$ eV; otherwise they are reflected specularly. With an initial stopping power of the Ar projectile of 14.5 eV/Å, its mean projected range is 75 Å. The effects of the boundaries of our box are therefore negligible; this was also checked by simulations using larger box volumes.

This system is modeled in two ways: On the one hand, we solve the linear Boltzmann equation using a Monte Carlo simulation code [6]. The nonlinear Boltzmann equation on the other hand is also solved by a simulational procedure, which roughly proceeds as follows: Space is divided into cubic cells of volume $(5.5 \ Å)^3$. In each cell, 4 particles are placed randomly. Within each cell, particles may collide; collision partners are chosen in a Monte Carlo procedure based on the statistics of the collision frequencies in the cell and the collision probability of the

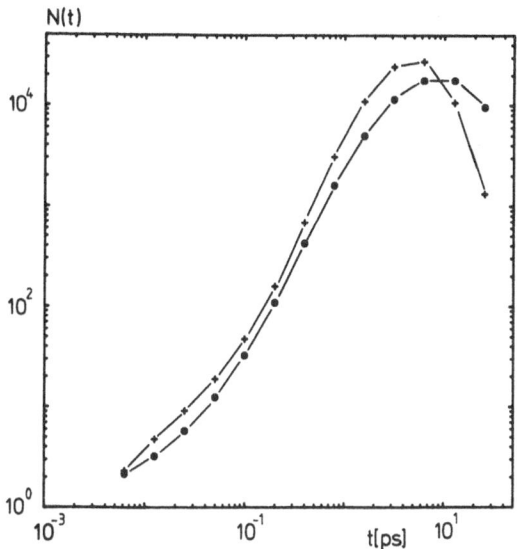

Figure 1. Time evolution of the number N(t) of particles in the cascade.
(+): linear simulation; (●): nonlinear simulation.

particle pairs available. After a time step Δt every particle is moved a
distance $\vec{v}\Delta t$ according to its individual velocity $\vec{v}$, and may leave the
cell. Then the collision procedure restarts, etc. In this simulation, no
collisions above the surface were considered. Particles with energy
$E < E_c$, where E_c is a cutoff energy, were set to zero velocity. We use
$E_c = 0.01$ eV, which corresponds to the energy $3k_B T_m/2$ of Argon at its
melting temperature $T_m = 88$ K [4]. The code is adapted from conventional
Monte Carlo algorithms available for the solution of the Boltzmann
equation [7,8]. Further details of the simulation will be reported
elsewhere [9]. We note that in both simulation procedures, roughly 10^3
runs were performed to obtain sufficient statistics.

3. RESULTS

The results of the two simulation procedures are contained in
fig.1-3. We wish to state the following three points.

(i) The number N(t) of particles in the cascade at time t, with
energy above E_c, evolves in three stages, fig.1. Initially, until 0.01 ps,
linear and nonlinear evolutions coincide, since the cascade is dilute at
this stage. In the second stage, until the maximum of the linear curve at
7 ps, the linear cascade produces more particles than the nonlinear one.
The reason hereto clearly is the fact that in a linear cascade, in every
collision a new cascade particle is born, whereas in the nonlinear case,
two cascade particles may collide, and no new particle is born. Note that

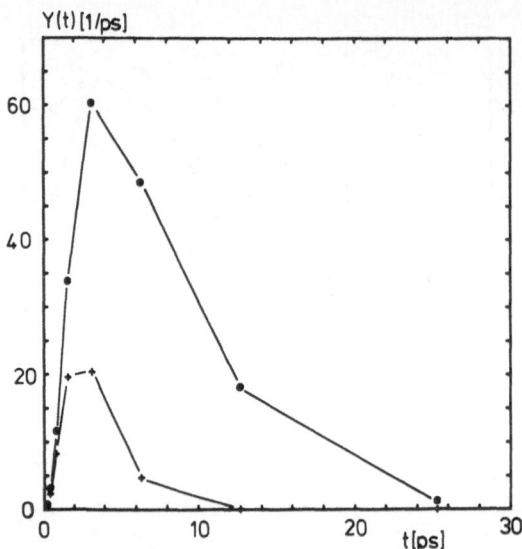

Figure 2. Time evolution of sputter yield Y(t). (+): linear simulation;
(●): nonlinear simulation.

this second stage begins rather early; this is due to the forward-scattering nature of the Kr-C potential, which produces predominantly low-energy recoils which soon start colliding which each other. In the third stage, after roughly 10 ps, the linear cascade quickly dies out, since all particles lose their energy in collisions, whereas the nonlinear cascade keeps alive to considerable longer times, until heat conduction carries all energy away from the cascade volume. We note that if a higher cutoff energy E_c is used, the cascade lives correspondingly shorter times.

(ii) Let Y(t)dt be the number of particles sputtered between time t and t + dt. Fig.2 shows that the time distribution Y(t), as calculated by the nonlinear Boltzmann equation, exceeds strongly the distribution calculated in the linear model. Furthermore sputtering lasts a factor of 2 longer in the nonlinear calculation.

At very short times after the ion impact, nonlinear and linear simulations coincide. Both distributions show a maximum at approximately 4 ps but in the nonlinear case it is much higher than in the linear one. In the linear case the maxima of Y(t) and N(t) occur at approximately the same time, whereas in the nonlinear simulation the sputtering maximum is earlier than the maximum of N(t). The reason hereto is that in the nonlinear case collisions between moving particles can contribute to the sputter process. Particles with energies lower than the surface binding

660

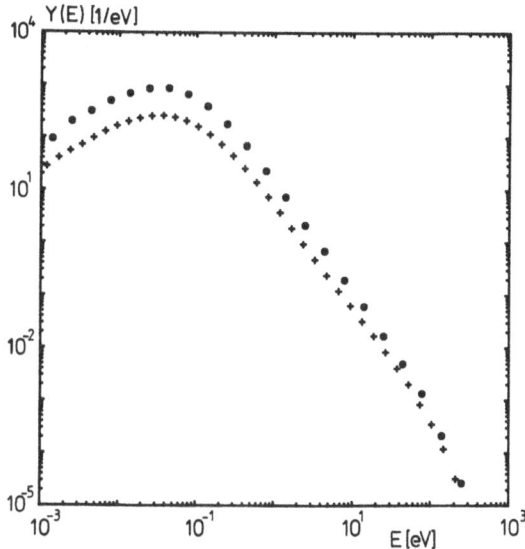

Figure 3. Energy distribution Y(E) of sputtered particles. (+): linear simulation; (●): nonlinear simulation.

energy U, can obtain energy by collisions with other moving particles, so that they may overcome the surface barrier [10]. So the total yield Y, i.e., the time integrated Y(t), in the nonlinear simulation is larger than the linear yield by a factor of 3.5.

(iii) The energy distributions Y(E) of sputtered particles are displayed in fig.3. At the high-energy side, above around 30 eV, the linear and nonlinear calculations yield identical results, as it may be expected. At low energies, considerably more particles are sputtered in the nonlinear calculation. This is in agreement with the observations made above, (i) and (ii), and stresses the importance of collisions between moving particles in the late *hot* phase of the cascade. It is striking, however, that the *shape* of the energy distribution does not deviate much from the linear result. It cannot be decided from the present calculations whether the inclusion of proper binding forces between the particles will change the energy distribution at the very low-energy side (E ≤ 0.1 eV).

To understand how typical these mean total sputter yields Y are, we study the probability distribution of the sputter yield F(Y) for two different times in the nonlinear model, fig.4. Here F(Y) is the probability to obtain the sputter yield Y in a single ion impact. The mean sputter yield is denoted by <Y>, its standard deviation by ΔY. Whereas for t = 0.2 ps, F(Y) shows the characteristic behavior of the linear regime [11], with relative standard deviation ΔY/<Y> = 1.36, at

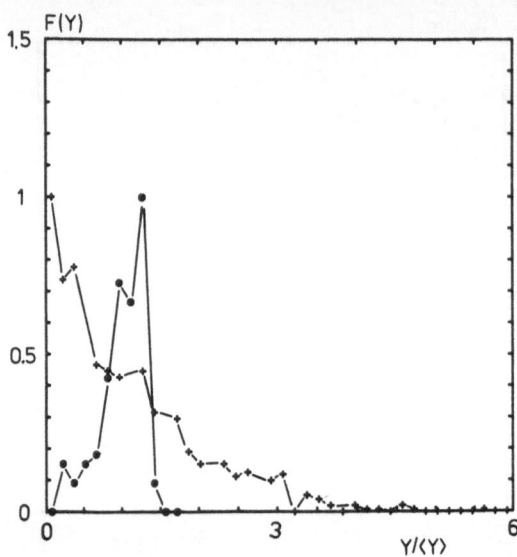

Figure 4. Probability distribution F(Y) of the sputter yield, normalized
to its maximum, versus the sputter yield Y, normalized to its
mean value <Y>, for the times t = 0.2 ps (+) and
t = 25.6 ps (●).

t = 25.6 ps it is well concentrated around the mean value <Y>, with
$\Delta Y/<Y> = 0.28$. This result shows that the long-time behavior of the
sputter yield is well described by its mean value.

4. CONCLUSIONS

A definite difference between the linear and nonlinear cascade
evolution and sputtering of our model system has been observed. The main
feature seems to be the long duration of the nonlinear cascade, in which
the energy remains concentrated in the cascade region for quite a long
time. This leads in particular to a definitely higher value of the
sputtering yield in the nonlinear case and an excess of particles
sputtered at low energies, below 10 eV. The form of the energy
distribution of sputtered particles is less affected. The long-time
behavior of the total sputter yield of each individual cascade is well
described by the mean value over many cascades.

REFERENCES

1. P. Sigmund, Phys. Rev. *184* (1969) 383; *187* (1969) 768.
2. P. Sigmund, in *Sputtering by Particle Bombardment I*, R. Behrisch, ed.,
   Springer, Berlin (1981) p. 9.
3. L. Boltzmann, Ber. Wien. Akad. *66* (1872) 275.
4. H.M. Urbassek and J. Michl, Nucl. Instr. Meth. *B22* (1987) 480.

5. W.D. Wilson, L.C. Haggmark and J.P. Biersack, Phys. Rev. *B15* (1977) 2458.
6. M. Vicanek and H.M. Urbassek, Nucl. Instr. Meth. *B30* (1988) 507.
7. G.A. Bird, *Molecular Gas Dynamics*, Clarendon Press, Oxford (1976).
8. O.M. Belotserkovskiy and V.Y. Yanitskiy, Fluid Mech. Soviet Res. 7 (1978) 42.
9. K.T. Waldeer and H.M. Urbassek (in preparation).
10. R.A. Weller and M.R. Weller, Nucl. Instr. Meth. *194* (1982) 573.
11. U. Conrad and H.M. Urbassek, Nucl. Instr. Meth. *B48* (1990) 399.

CONTRIBUTED PAPERS: APPLICATIONS

MONTE CARLO SIMULATIONS OF DEPOSITION PROCESSES

J.C. Moreno-Marín

Dept. Física Aplicada
Universitat d'Alacant
Apt. 99, E-03080 Alacant, SPAIN

1. INTRODUCTION

Under ordinary thin-film deposition conditions, where atoms arrive at the substrate with thermal energies, and at a substrate temperature where the atom mobility is restricted, the film may develop an inhomogeneous and porous structure. If additional energy is provided to the growing film, either directly by the depositing atoms or through simultaneous bombardment with energetic ions, the (crystal) structure is substantially modified, and becomes more dense.

In this communication, the process whereby atoms are directly deposited on single-crystal substrates has been studied. By using a growth model to describe *epitaxial* thin-film formation, a Monte Carlo code has been developed that establishes an adequate tool to obtain a qualitative understanding of the physical parameters involved in the deposition process. The structural modifications in ion-assisted deposition and the effect of a post-deposition annealing treatment, due to thermally activated atomic hopping, are modeled.

Sect.2 deals with the deposition process, and sect.3 introduces thermal redistribution of deposited atoms. Taking into account the coordination number, the deposition model includes the possibility of atom migration before finding a certain site. In a second stage in the simulation, further migration is modeled, using a thermal activation model.

2. MODEL OF VAPOR DEPOSITION

Direct epitaxial growth by vapor deposition has been simulated by a simple kinetic model [1] with the following assumptions: (1) Atoms are permitted only at the crystal lattice sites; (2) only one atom is allowed

Interaction of Charged Particles with Solids and Surfaces
Edited by A. Gras-Martí *et al.*, Plenum Press, New York, 1991

667

to occupy a site; (3) attractive interactions extend to nearest-neighbor atoms, and the total potential energy is the sum of all such pair-interaction energies.

Beginning with an atomic substrate, as a seed crystal, the vapor atoms collide with the growing film, and some of them stick. When an atom arrives to a site, one of three things can happen: (a) the atom can stick at that site; (b) the atom can hop off the substrate altogether; or (c) the atom can move sideways to a neighboring position in the lattice. The choice of one of these results is through a Monte Carlo rejection procedure, and is determined by the absolute temperature T and the potential energy U of the atom at the surface by means of the Boltzmann's factor $\exp(-U/kT)$, where k is Boltzmann's constant. Therefore, the probability of the atom to remain in its present site or in one of the adjoining sites, is given by the canonical distribution [2]

$$P(i) = \exp(-U_i/kT) / \sum_j \exp(-U_j/kT).$$

This model, called the Single Atom Approximation, includes the following important features. First, there is a nonzero probability that the atom hops to a site without neighboring atoms and thus returns to the gaseous phase. Second, the variables of physical interest are the temperature and the relative potential energies of different geometric configurations of nearest neighbors (atomic sites and active chemical bonds). As this hopping procedure can be iterated until the atom comes to rest at a site or hops off the surface, it is necessary to set a maximum number of iterations to simulate the deposition process in the presence of thermal motion.

3. THERMAL MODEL OF ATOMIC MOBILITY

If, during crystal growth, the film is bombarded with energetic atoms, its structure is modified and with it the physical properties of the film. When a deposited film is subject to an annealing treatment, its structure also changes with the diffusion of vacancies and their coalescence into larger voids.

Both situations are partially due to thermally-activated atomic mobility, whose effects can be modeled [3]. The atoms find a site using the model in sect.2, and while the local rearrangement produced by energetic ions is applied in the model only to the atoms in a small volume, in the annealing process, developed over all the grown film, the possibility of thermal migration is tested for all film atoms.

From the probability distribution of the Boltzmann statistics one assigns a random thermal energy ε to an atom which is calculated as [3]

$$\varepsilon = - kT \ln(1-R^{\gamma}),$$

where $\gamma = e^{-Q/kT}$, R is a computer-generated random number in the interval $(0,1)$, and Q is the minimum activation energy for atomic diffusion. In a slightly different model, the exponent γ is taken as one [4]. The influence of the power γ, for the parameters used in the calculations reported in this paper, is negligible.

The energetic barrier ΔE_{ij} which an atom has to overcome to hop from site i to an empty neighboring site j, depends on the surrounding atomic configurations of both sites, and can be described [3] as

$$\Delta E_{ij} = \infty, \quad \text{if } j \text{ is occupied or has no neighboring atoms,}$$

$$\Delta E_{ij} = \begin{cases} Q, & \text{if } N_i \leq N_j, \\ (N_i - N_j)\phi + Q, & \text{if } N_i > N_j, \end{cases}$$

where ϕ is the bonding energy of a single bond, and N_i and N_j are the number of neighbors at sites i and j. If the thermal energy of the atom is larger than the activation barrier, the atom hops to the site with largest coordination number. The choice between several equivalent sites is made randomly. It is possible to define in a different manner the sites where the atom can hop, and the contribution of the geometrical surrounding to the potential energy of the atom in a site, taking into account nearest neighbors and smaller contributions of atoms in more distant positions. One may also consider the possibility of the atom to diffuse on a step, where through an intermediate unstable single-bonded site, the atom relaxes into an accessible site.

4. EXAMPLES OF SIMULATIONS

In the calculations, the vapor atoms arrive at normal incidence, sequentially and randomly on the growing film surface. To reproduce the deposition process first one sets up the crystal lattice by means of a logical matrix, whose elements are considered as the atom locations, in correspondence with the atomic spatial coordinates, and showing their occupancy state. The bottom row is full to represent the substrate, and the incidence of one atom is simulated by choosing a random column, putting the atom at the top of the column, and letting it fall until it

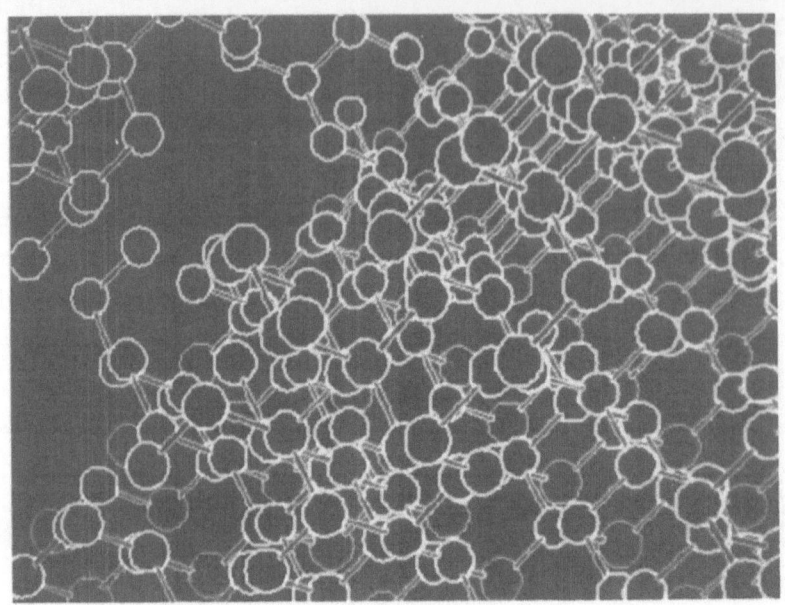

Figure 1. Simulation of epitaxial diamond—film growth where vacancies in
the film are illustrated. The three-dimensional graphic image
has been obtained with the MOLGRATH code [5].

touches another atom, i.e., until there is at least an occupied neighbor
site.

With the previous deposition model, only the first site selection is
allowed without considering any activation barrier, thereby delaying
thermal motion in the solid.

The sticking coefficient of the vapor atoms is assumed to be unity,
and in all cases periodic boundary conditions are taken to eliminate edge
effects.

Fig.1 shows the geometrical structure obtained in a three—dimensional
simulation of epitaxial film growth of a diamond lattice for a high
temperature substrate with $kT/\phi = 2$. To obtain a diamond-like film it is
necessary to account properly for the atomic positions in the cubic
lattice. The same lattice developed, with unit distance $L = D/\sqrt{3}$, where D
is the inter-atomic bond distance, may be used to reproduce Si and Ge
deposition.

When the deposition is assisted by energetic ions, these and the
vapor atoms arrive with a certain ratio. To include the effect of
impinging ions, a spike region centered where the ion hits on the surface
is defined, and all atoms in this volume are examined for atomic hops in a

random sequence several times. The number of possible hops and the geometric shape of the spike are essential input of the simulation and depend on the energy and the atomic species. After each hop the new configuration of occupied sites is revised. The incorporation of the energetic ions and the thermal atoms in the growing film is treated equally.

This spike model tries to reproduce the structural modifications due to the collision cascade generate by the energetic bombardment. In hexagonal and tetrahedral lattices all the nearest positions of a given atom are equidistant, and this spike mechanism is considered non-directional except for the definition of the spike volume, which is surely dependent on the incident atom direction.

With this procedure one simulates the ion-assisted deposition process and the occurrence of low-temperature epitaxy. Fig.2 shows the variation of the occupation degree of the crystal lattice (in this case a hexagonal bidimensional structure) with respect to the fraction of energetic incident ions. The dose gives information about the relocation frequency,

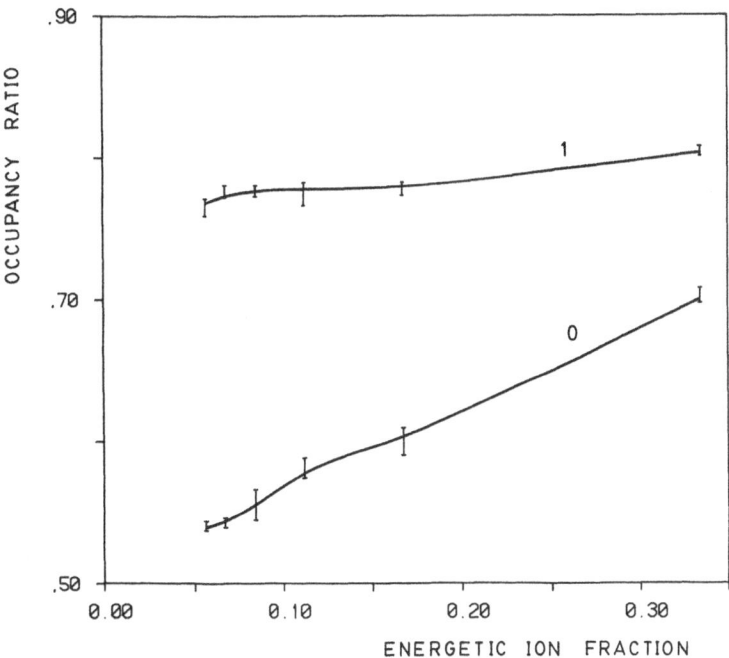

Figure 2. Occupancy ratio of the thin film as a function of the fraction of energetic incident ions versus thermally deposited atoms. The parameter on the curve represents the maximum number of possible hops allowed during the deposition process.

which is the parameter on the curves. In this example, a deposition temperature of 200 K and a hopping temperature in the spike of 1000 K are taken. The value of the minimum activation barrier is Q = 0.86 eV [3,4], and the bonding energy ϕ = 0.6 eV. The spike volume includes first and second neighbors of the incidence site with a maximum of five hops allowed. The value Q = 0.86 eV may be large in comparison with bulk vacancy migration energies in some systems [6].

To obtain reasonable statistics, three different film samples of 400×100 sites were considered. For diffusionless deposition (curve 0 in fig.2) a densification of the films is observed and the number of vacancies decreases when the energetic-to-thermal ratio increases. The effect of the energetic ions is weaker when one thermal hop is allowed after deposition (curve labelled 1 in fig.2). In ref.[7] the occupancy ratio of Ge films grown under simultaneous irradiation was compared with a model. The ion-to-atom ratios used in the present simulation are larger than the critical arrival ratio for stress annealing.

Fig.3 shows structural variations after different times of isothermal annealing on a vapor-deposited bidimensional film. The effect of this treatment is obtained by checking for atomic hops in a random sequence. All atoms in the film are allowed to relocate with the model explained above. As seen in the figure, this process tends to eliminate loosely bound atoms.

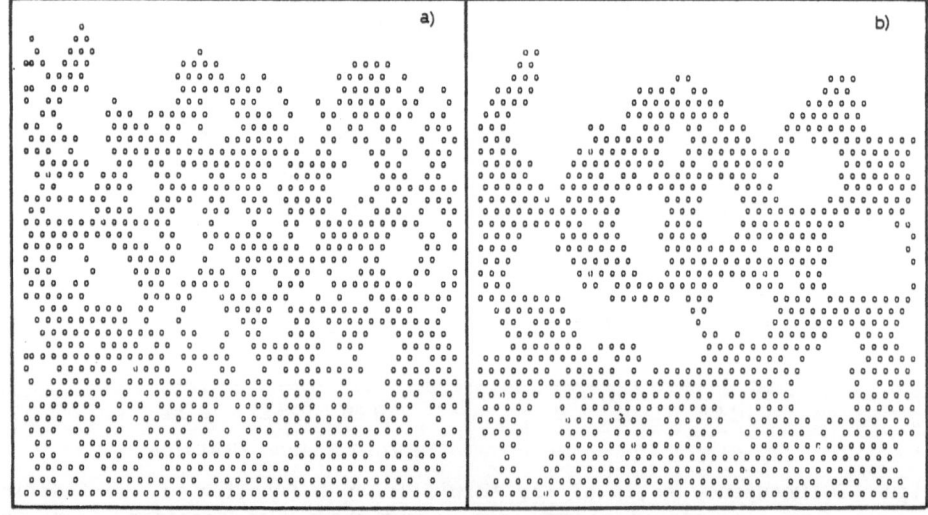

Figure 3. Effect of thermal annealing, after vapor deposition, on a bidimensional structure, with kT/ϕ = 2.5; a) before, and b) after 20 relocation processes at T = 1000 K.

5. SUMMARY

Simulations of epitaxial growth, including high-temperature vapor deposition process and the modifications in ion-assisted deposition and post-deposition annealing are developed.

The present model describes qualitatively the thin-film densification process observed, for example, in ref.[8]. It is possible to use a standard code, like TRIM [9], to generate more realistic collision cascade volumes. This shall provide a better description of the effect of energetic ion bombardment. Coupled with the code TRAP [10] describing the transport and thermalization of sputtered atoms in glow discharges, the simulations described in this communication can be used as a modeling tool in deposition processes. In particular, the deposition of amorphous material (like Ge, Si) may be modeled with an extension of the present program, including interstitial positions in the lattice.

ACKNOWLEDGEMENTS

The author is grateful to Prof. Alberto Gras-Martí and Lcdo. Mikel L. Forcada for help and stimulating discussions during the course of this work. Partial support from the Dirección General de Investigación Científica y Técnica (DGICYT, projects number PS88-0066 and PS89-0065) is recognized. The assistance of Prof. R.P. Webb in the codes was facilitated by a Link sponsored by the British Council.

REFERENCES

1. G.H. Gilmer, SCIENCE 208 (1980) 355.
2. S.J. Berkowitz and D.G. Haase, Phys. Educ. 22 (1987) 380.
3. K.H. Müller, J. Vac. Sci. Technol. A3(6) (1985) 2089.
4. K.H. Müller, J. Vac. Sci. Technol. A4(2) (1986) 184.
5. J.M. Perez-Jordá and E. San-Fabián, Quantum Chem. Progr. Exch. 5/5 vol. 8 (1989) 64.
6. Rad. Eff. 46 (1980) 157.
7. D.R. Brighton and G.K. Hubler, Nucl. Instr. Meth. B28 (1987) 527.
8. J.E. Greene, S.A. Barnett, J.E. Sundgren and A. Rockett, Proceedings of the NATO-ASI on Plasma-Surface Interactions and Processing of Materials, Spain (1989), O. Auciello et al., eds., series E, 176 Kluwer Acad. Publ. (1990) 281.
9. J.F. Ziegler, J.P. Biersack and U. Littmark, Stopping Powers and Ranges of Ions in Matter, Pergamon Press, New York (1985), vol. 1.
10. A. Gras-Martí, J.A. Vallés-Abarca and J.C. Moreno-Marín, Proceedings of the NATO-ASI on Plasma-Surface Interactions and Processing of Materials, Spain (1989), O. Auciello et al, eds., series E, 176 Kluwer Acad. Publ. (1990) 135.

GROWTH OF THIN FILMS BY ION BEAM SPUTTERING:

STUDY OF RARE GAS INCORPORATION

C. Desgranges, C. Pellet, C. Schwebel, A. Chabrier and
G. Gautherin

Institut d'Electronique Fondamentale
Unité associée au CNRS (U.R.A. D022)
Université PARIS XI, Bât. 220, F 91505 ORSAY CEDEX, FRANCE

ABSTRACT

This paper reports the results on a study of rare-gas incorporation in films deposited by ion beam sputter deposition (IBSD). It has been observed that electrical properties of metallic and semiconductor films depend on the nature and concentration of rare gas trapped in the layers. The origin of this rare gas is discussed. Angular distributions of rare gas emitted from a sputtered target are shown.

1. INTRODUCTION

To date the ion beam sputtering deposition (IBSD) technique has been used to deposit metallic, dielectric, superconductor and semi-conductor films. Many promising results have been obtained, for example good mechanical characteristics of the films. In the demanding domain of microelectronics, a thorough knowledge, theoretical as well as experimental, of the sputtering and deposition processes seems to be necessary for determining the potential for using the IBSD technique to deposit thin films.

In the first part of this paper we report the main results of our studies in the field of epitaxial and metallic films. In the second part, a discussion of these results makes evident the rare-gas effect on the properties of ion-beam deposited layers.

2. GROWTH OF THIN FILMS BY ULTRA-HIGH VACUUM ION-BEAM SPUTTERING (UHV IBS)

The principle of the method consists in the sputtering of a target with an energetic ion beam. The sputtered particles condense on the substrate giving a deposit (fig.1). When rare-gas ions are used, the

Interaction of Charged Particles with Solids and Surfaces
Edited by A. Gras-Martí et al., Plenum Press, New York, 1991

675

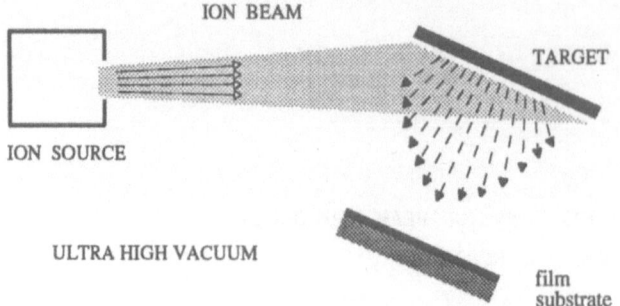

ION BEAM

ION SOURCE

TARGET

ULTRA HIGH VACUUM

film
substrate

Figure 1. Principle of the ion beam sputter-deposition technique.

deposit has the same nature as the target; contrarily, the use of reactive ions (O_2^+, N_2^+, ...) leads to the growth of oxide and nitride films.

We have deposited homoepitaxial silicon and metallic films using an ultra high vacuum apparatus previously described [1]. In the experimental set-up neon, argon, krypton and xenon ion beams of 10 - 20 keV energy and 50 - 500 $\mu A/cm^2$ maximum current density at the target are used. For these beams the pressure in the deposition chamber is of the order of 10^{-5} Pa and consists mainly (99 %) of rare gas.

The silicon homoepitaxy has been studied because it is a good test vehicle in the microelectronics field [1]. The beginning of single-crystal growth occurs at deposition temperature as low as 250° C. Perfect crystalline structure is obtained above 750° C. The doping of the layers was achieved by sputtering of a boron-doped target. For films grown at 600 - 900°C, room-temperature mobilities fall within 20 % of the bulk value at the same doping level.

On the other hand tungsten films were also deposited [2,3]. The IBS technique is well adapted to the deposition of low vapor-pressure material. All the deposits grown at room temperature exhibit the bcc structure (αW).

3. RARE GAS INCORPORATION IN THE FILMS

The major impurity in our films is rare gas. The concentrations in films deposited at room temperature (measured by Rutherford Backscattering) are reported table 1. The table shows that even when backscattering via a single collision process is reduced, (R < 1), the rare-gas concentration in the deposits is still high. Since thermal rare-gas atoms are only weakly physisorbed on the silicon surface at room

Table 1. Rare-gas concentration C (%) in films deposited at room temperature.

target	rare gas	concentration (%)	$R = \dfrac{\text{target atomic mass}}{\text{ion atomic mass}}$
silicon	argon	0.7	0.7
	xenon	0.24	0.2
tungsten	argon	2.5	4.6
	xenon	< 0.3	1.4

temperature, the trapped rare-gas atoms in our films could only result from sputtering of atoms implanted in the target or from backscattering of the primary ions on the target surface due to multiple collisions [4]. As a result of these mechanisms, the presence of rare gas in the films and the interaction of fast rare-gas and/or energetic sputtered particles with the growing films, could explain the unusual low temperature mobility of the carriers in the silicon films [1]. Rare gas effects on the electrical properties of tungsten films has been also observed. The resistivity increases when: a) the rare-gas concentration in the films increases, b) R decreases [3].

With the goal of determining the relative responsibility of each emission process, a study of rare-gas incorporation mechanisms during growth was carried out.

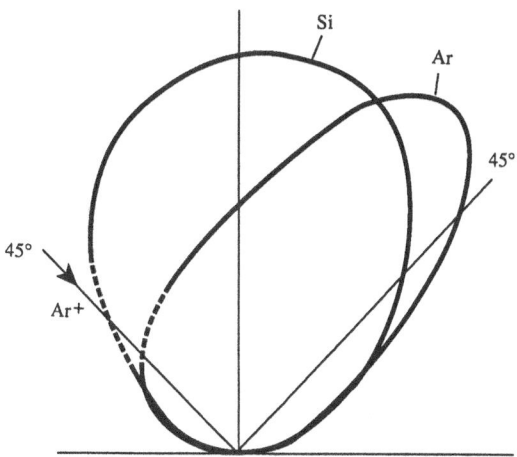

Figure 2. Angular distributions of sputtered silicon and trapped argon emitted from a silicon target bombarded at oblique incidence (45°) with a 20 keV argon ion beam.

4. RARE-GAS EMISSION FROM A SPUTTERED TARGET

The first stage of this study was the determination of the angular distribution of the rare gas emitted from a sputtered target [5]. This was obtained by composition analysis of films deposited on nine substrates set on a cylindrical collector centered at the point where the beam axis intercepts the target. The rare-gas distribution was calculated from the target-material distribution by using the following relation

$$I_{rg}(\vartheta) = I_{ta}(\vartheta)\, C_{rg/ta}(\vartheta),$$

where $I_{rg}(\vartheta)$ and $I_{ta}(\vartheta)$ are emission intensities in direction ϑ (measured with respect to the target normal) for rare-gas and target material respectively, and $C_{rg/ta}(\vartheta)$ the rare gas concentration in a film grown at direction ϑ. Typical angular distributions are shown in fig.2 for 20 keV argon ions bombarding a silicon target at $45°$ beam incidence. Contrary to the silicon angular distribution, the rare gas distribution is not symmetric with respect to the target normal. The sharp shape distribution points in a forward direction near the specular reflection direction of the beam. Similar results were obtained for Xe^+/Si and Kr^+/Si. This seems to prove that the major part of the incorporated rare-gas results from backscattering by a multiple collisions process [5]. Fig.3 shows angular distributions for 20 keV argon ions bombarding a tungsten target at $45°$ beam incidence. In this case the tungsten and the argon angular distributions are similar in shape and in direction of maximum intensity. Let us note that this sputtering event is close to the spike regime [6].

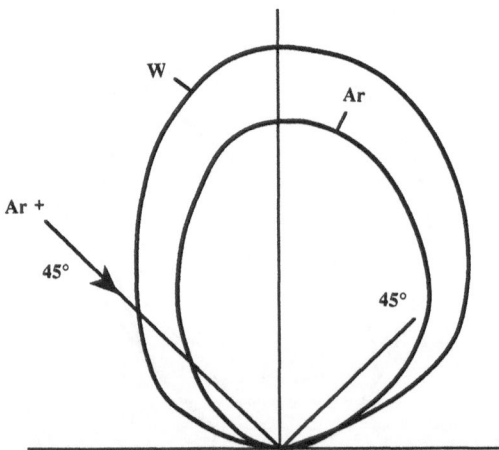

Figure 3. Angular distributions of sputtered tungsten and trapped argon emitted from a silicon target bombarded at oblique incidence ($45°$) with 20 keV argon ion beam.

This work is an important stage towards an understanding of mechanisms which govern rare gas incorporation in films. Nevertheless, it does not enable us to conclude categorically about the role played by the rare gas in film properties. Knowledge of the energy distribution of backscattered rare gas may be a decisive element in this research. With the purpose of analyzing the angular-resolved energy distribution of the rare-gas emitted by a sputtered target, we have built a specific set-up. First experiments are underway.

REFERENCES

1. C. Shwebel, F. Meyer, G. Gautherin and C. Pellet, J. Vac. Sci. Technol. *B4* (1986) 1153.
2. F. Meyer, C. Schwebel, C. Pellet and G. Gautherin, Appl. Surface Sci. *36* (1989) 231.
3. F. Meyer, D. Bouchier, V. Stambouli, C. Schwebel and G. Gautherin, Appl. Surface Sci. *38* (1989) 286.
4. A.L. Boers, Surface Sci. *63* (1977) 475.
5. C. Schwebel, C. Pellet and G. Gautherin, Nucl. Instr. and Meth. *B18* (1987) 525.
6. H.H. Andersen and H.L. Bay, Topics in Applied Physics 47 (R. Behrisch, ed., *Sputtering by Particle Bombardment*, Springer Berlin (1981).

EXPERIMENTAL EVIDENCE OF THE BOND BREAKING MECHANISM
IN OXYGEN SECONDARY ION EMISSION ENHANCEMENT

J. Ferrón

INTEC, CONICET
Universidad Nacional del Litoral
Guemes 3450, RA-3000 Santa Fe, Argentina

ABSTRACT

We have measured the Ti^+ and O^- secondary ion emission yields of a Ti sample exposed to oxygen. The oxygen amount at the surface was determined by measuring the O_{KLL} Auger signal, and the chemical effect of oxygen at the surface by measuring the Ti_{LMV} and Ti_{LMM} Auger line shapes. We found that the oxidation process of Ti presents two clearly differentiated regimes characterized by a different electronic transfer mechanism from titanium to oxygen. These two regimes are also observed in the behavior of the Ti^+ and O^- yields showing a correlation between the ionization mechanism and the chemical surrounding of the ejected atom.

The impact of energetic ions on a solid surface causes the emission of electrons, photons, atoms and ions. The mass analysis of these emitted ions constitutes the basis of one of the most sensitive surface analysis techniques, Secondary Ion Mass Spectrometry (SIMS). It is well known that secondary ion emission is particularly large in the case of ionic solids, and that positive ion emission of metals and semiconductors can be strongly enhanced by the presence of electronegative elements at the surface. In fact, oxygen bombardment is currently used in commercial SIMS equipment to improve the sensitivity. On the other hand, the basic mechanisms of the ejection of sputtered ions, and particularly the chemical enhancement of positive secondary ion yield, are still a matter of discussion [1-4].

The different models for SIMS have recently been reviewed by Yu and Lang [4], stressing the differences between alkali and oxygen secondary-yield enhancement. While the alkali enhancement appears to be well described by models using extended surface properties, like in the tunneling models [5,6], the chemical effect is better understood in terms

Interaction of Charged Particles with Solids and Surfaces
Edited by A. Gras-Martí *et al.*, Plenum Press, New York, 1991

681

of a more localized picture, like in the bond breaking model, hereinafter BB model [7-13].

Although the bond breaking is the most successful model for explaining the chemical ion emission, quantitative calculations are available only since recent years [9,10,12,13]. On the other hand, direct experimental evidence for the bond-breaking mechanism is rather scarce, and sometimes contradictory. The BB model is compatible with at least two different microscopic mechanisms, i) the ionization of the ejected atom by the breaking of the atom-surface bond [1,9,10,12,13], hereinafter called the surface process, and ii) the ionization of a neutral excited sputtered oxide molecule (MeO) outside the surface region [8,12], (the molecular emission model). This last model predicts a ratio Me^+/O^- equal to unity, and in fact such a constant ratio has been measured for Si^+ [8] and Sn^+ [11]. In the present work we present measurements of the Ti^+ and O^- ion yields which give direct evidence for the bond breaking process understood as a surface process.

The measurements were performed in a UHV surface analysis equipment with AES and SIMS facilities. The base pressure was in the low 10^{-10} Torr range. The secondary ion emission was induced by a 2 keV, 5 nA Ar^+ ion beam, rastered over a 2×2 mm^2 surface area in order to achieve the conditions of static SIMS. The secondary ions were analyzed by an energy-filtered quadrupole mass spectrometer. The SI yield was gated to the central half of the bombarded area to prevent edge effects. The exposures were performed at 10^{-8} Torr. The measurements were taken during oxidation and, with the exception of Ti^+, during the cleaning of the surface using higher ion-current density.

The surface state was monitored through AES in two different forms: i) as usual, through the O_{KLL} Auger yield, and ii) through the shape variation of the Ti_{LMM} and Ti_{LMV} Auger lines. Since the M level of Ti may be treated as an inner level [14], the electron transfer between titanium and oxygen affects preferentially the LMV Auger transition, and therefore the ratio between both Ti Auger yields ($R = Ti_{LMV}/Ti_{LMM}$) is a measure of this transfer [15].

In fig.1 we show experimental data for Ti^+ and O^- yields and Ti_{LMM} and Ti_{LMV} Auger peak to peak heights as a function of the oxygen coverage given by the O_{KLL} Auger yield. The behavior of the two ion yields is different in a qualitative way. While the Ti^+ yield exhibits a monotonic increase with oxygen coverage, the O^- yield presents a relative maximum. This result is in clear contradiction to a constant Me^+/O^- ratio as expected from the molecular emission model.

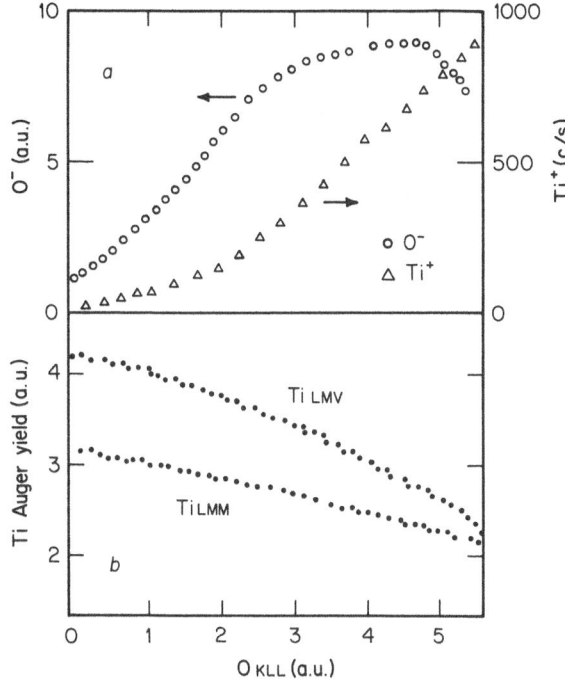

Figure 1. (a) O^- and Ti^+ secondary ion yields (a), and (b) Ti_{LMV} and Ti_{LMM} Auger yields versus O_{KLL} yield.

Our results may be discussed in a more quantitative way by using instead of the ion yields, the relative ionization probabilities RTi^+ and RO^- defined as the ratio between the ion yields and the oxygen coverage given by the O_{KLL} yield. In fig.2, we show these relative probabilities together with the ratio between Ti_{LMV} and Ti_{LMM} Auger yields as a function of the oxygen coverage. All curves show two clearly differentiated regimes, since the increase of RO^- and RTi^+ observed as the oxygen coverage tends to zero is due to the different sensitivities of SIMS and AES, i.e., secondary ions are detected although oxygen coverage is below the AES detection capability.

The curve Ti_{LMV}/Ti_{LMM} versus O_{KLL} shown in fig.2a reveals the existence of two regimes for different oxygen coverages and characterized by different electron-transfer mechanisms between Ti and O. Following Bignolas et al.[16], these regimes may be associated with changes in the oxygen stoichiometry, i.e., the appearance of TiO_2 in a matrix with TiO. With this explanation for the Ti_{LMV}/Ti_{LMM} behavior, the O^- and Ti^+ dependence can be consistently understood. Within this picture, the ionization probability is clearly related to the titanium oxidation state.

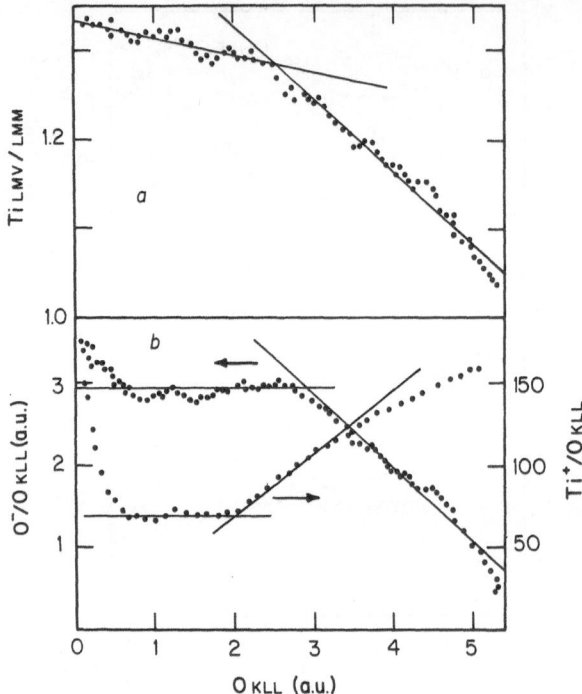

Figure 2. (a) Ti_{LMV}/Ti_{LMM} Auger yield ratio (a), and (b) the relative ionization probabilities RTi^{+} and RO^{-} versus O_{KLL} yield.

During the first stage, the positive ionization probability shows a linear dependence with oxygen coverage (RTi^{+} nearly constant), in agreement with previous measurements on Si and Ge [9]. In this stage the O^{-} emission also shows a linear dependence with oxygen coverage, in good agreement with results of Benninghoven et al.[17]. The dependence of both kinds of ion emission, i.e., positive and negative ions, in this stage gives support to the surface model as well as to the molecular interpretation of the BB model [9]. The second stage appears as more effective in the production of positive ions. It coincides with a larger Ti electron depletion and also with the decrease of the O^{-} emission, giving in this case evidence against the molecular emission interpretation of the BB mechanism. In this regime RTi^{+} shows a linear dependence on the O_{KLL} yield, i.e., a quadratic dependence of the ionization probability on oxygen coverage. This behavior, restricted to the beginning of the second stage, is consistent with a random distribution of the oxygen atoms and a strong dependence of the ionization probability on the number of oxygen atoms located as near neighbors of Ti [12,13,18]. Finally, the decrease of the O^{-} yield with oxygen coverage may be also rationalized within the BB model. The larger electron depletion around each Ti atom, in going from TiO to TiO_2, is consistent with a smaller density around each oxygen atom, because in the case of TiO_2 there are two oxygen atoms to share the Ti electrons.

In summary, we have measured the behavior of Ti^+ and O^- secondary ion yields of a titanium sample exposed to oxygen. We monitored the oxidation state of Ti through the shape variations of the Ti_{LMV} and Ti_{LMM} Auger lines. We found a strong correlation between secondary ion yields and the Ti oxidation state as revealed by the Auger analysis. We found that the behavior of both ion yields may be explained within the BB model, understood as a surface mechanism, throughout the entire exposure range.

ACKNOWLEDGMENTS

I am deeply indebted to Drs. E.C. Goldberg and M.C.G. Passeggi from INTEC, Universidad del Litoral, Argentina, and Dr. F. Flores from Universidad Autónoma de Madrid, for helpful discussions. This work has been partially supported by CONICET under Grant PID 75300 and Fundación Antorchas through Grant no. 11790.

REFERENCES

1. P. Williams, Appl. Surface Sci. *13* (1982) 241.
2. P. Williams, Surface Sci. *90* (1979) 588.
3. G. Blaise and A. Nourtier, Surface Sci. *90* (1979) 495.
4. M.L. Yu and N.D. Lang, Nucl. Inst. Meth. *B14* (1986) 03.
5. J.K. Norskov and B.I. Lundqvist, Phys. Rev. *B19* (1979) 5661.
6. M.L. Yu and N.D. Lang, Phys. Rev. Lett. *50* (1983) 127.
7. G. Slodzian and J.F. Hennequin, C. R. Acad. S. C. Paris *263B* (1966) 1246.
8. K. Wittmaack, Surface Sci. *112* (1981) 168.
9. M.L. Yu and K. Mann, Phys. Rev. Lett. *57* (1986) 1476.
10. M.L. Yu, Nucl. Inst. Meth. *B18* (1987) 542.
11. W. Gerhard and C. Plog, Surface Sci. *152/153* (1985) 127.
12. M.C.G. Passeggi, E.C. Goldberg and J. Ferron, *Lectures Notes in Surface Sci.*, M. Cardona and G. Castro, eds., *319*, Springer-Verlag, (1987).
13. M.C.G. Passeggi, E.C. Goldberg and J. Ferron, Phys. Rev. *B35* (1987) 8330.
14. F.J. Szalkouski and G. Somorjai, J. Chem. Phys. *56*, (1972) 6097.
15. C.N. Rav, D.D. Sarma and M.S. Hedge, Proc. R. Soc. Lond *A370* (1980) 269.
16. J.B. Bignolas, M. Bujor and J.Bardolle, Surface Sci. *108* (1981) L453.
17. A. Benninghoven, H. Bispinck, O. Ganschow and L. Wiedmann, Appl. Phys. Lett. *31* (1977) 341.
18. H. Oechsner and Z. Sroubek, Surface Sci. *127* (1983) 10.

ION BEAM INDUCED Ni-Ag MIXING

A. Miotello[1,2], D.C. Kothari[1] and L. Guzman[3]

1. Department of Physics, University of Trento
   I-38100 Trento, Italy

2. Unità I.N.F.M. di Trento, Italy

3. I.R.S.T., I-38050 Povo (Trento), Italy

ABSTRACT

Ion-induced mixing is a complex phenomenon where many processes are involved: collisional effects, vacancy-atom transport correlation effects, surface segregation, preferential sputtering and chemical effects. Here we describe some experiments on Ni-Ag ion beam mixing, taken from the literature, which give information concerning the relevant mechanisms involved. To this purpose we discuss also a model for describing the mixing process where the vacancy-atom transport correlation effects play a major role in the dynamics of the mixing process. Finally we present some new experimental results which confirm at least qualitatively the model.

Ion-induced mixing is a complex phenomenon where many processes dependent on either chemical or collisional effects are involved and which take place within the time period ranging from a picosecond to a few hours. Recent experimental [1,2] results indicate that the radiation effects in Ag/Ni multilayer films are persistent for a few tens of minutes after the ion beam is switched off. This suggests that the major factor in the transport of silver in silver/nickel multilayer configurations could be the delayed effects, taking place after about a nanosecond [3], involving the interaction of silver atoms with the vacancies produced by ion bombardment. The prompt effects taking place before a nanosecond, like replacement collisions and cascade mixing, could play a minor role in the transport of silver atoms. Therefore it would be proper to consider the vacancy-silver atom correlation effects while describing the transport of silver atoms in bombarded Ni-Ag multilayered films. These structures consist of five thin (3 - 4 nm) Ag layers alternately deposited among six thick (50 nm) Ni layers [1,2].

In the present paper the microscopic mechanisms of the mixing

Interaction of Charged Particles with Solids and Surfaces
Edited by A. Gras-Martí *et al.*, Plenum Press, New York, 1991

687

phenomena occurring in Ag/Ni multilayers (under 1 - 4 keV Ar$^+$ bombardment) are described on the basis of transport equations [4], for the silver depth-concentration profiles evolution into Ni, as detected by Auger technique, and reported by Marton et al. in ref.[1].

Briefly, the transport equations [4] are numerically integrated by treating the diffusion coefficients (various D_{ii}), as parameters in the following coupled continuity equations

$$\partial N_{Ag}/\partial t = D_{11} \; \partial^2 N_{Ag}/\partial x^2 + D_{12} \; \partial^2 N_V/\partial x^2,$$

$$\partial N_V/\partial t = D_{22} \; \partial^2 N_V/\partial x^2 + D_{21} \; \partial^2 N_{Ag}/\partial x^2 + f(x), \tag{1}$$

where N_{Ag} is the Ag atom concentration at depth x (x = 0 is the actual surface position) and time t, N_V is the vacancy concentration, D_{11} is the Ag diffusion coefficient, D_{22} is the Ni vacancy diffusion coefficient, D_{12} is the correlation coefficient of Ag atoms with vacancies, D_{21} is the correlation coefficient of vacancies with Ag atoms, and f(x) is the vacancy generation rate.

A silver-atom preferential sputtering process is considered using the following boundary condition [4]

$$- D_{11}(\partial N_{Ag}/\partial x) - D_{12}(\partial N_V/\partial x) = \sigma \; \Delta x_0 \; I \; N_{Ag}, \text{ at x = 0,} \tag{2}$$

where σ is the sputtering cross-section, I is the incident particle flux and Δx_0 is the depth interval from which the sputtering event may occur ($\Delta x_0 \approx 1$ nm).

The Auger profile is computed using the following relation

$$C_A(t) = \int_0^\infty N_{Ag}(x,t) \; \exp(-x/L_e) \; dx, \tag{3}$$

where L_e = 0.5 nm is the electron inelastic mean-free-path (for an Auger electron of Ag at 375 eV). The vacancy source term f(x) is obtained using a computer code available from Implant Science Corp., U.S.A. Eqs.(1) - (3) are solved numerically for a chosen set of D_{ii} and ϕ to fit the experimental data (where $\phi = \sigma \; \Delta x_0 \; I$) in a moving frame whose velocity is the erosion speed. Although the number of parameters looks large, the most relevant parameter is only D_{12} in the present case, as can be seen later. Figs.1a and 1b show the theoretical fits to the experimental results of Marton et al.[1] for 4 keV Ar ion implantation energy.

It is found that the silver diffusion coefficient is moderate

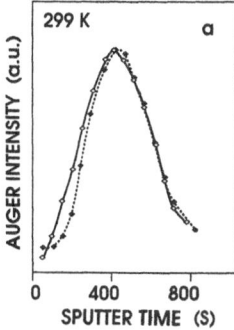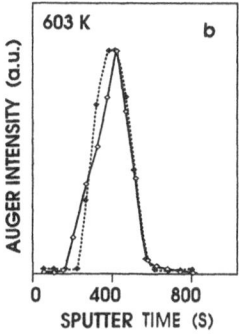

Figure 1. Theoretical (- - -) and experimental (——) normalized Ag Auger
profiles recorded at (a) 299° K and (b) 603° K. Experimental
results from Marton et al.[1].

$(D_{11} = 1.0 \times 10^{-18}$ cm^2 s$^{-1})$ at room temperature and does not increase
considerably at higher temperature (603° K) indicating that the transport
of silver atoms is governed by a rather low activation energy. The vacancy
diffusion coefficient is observed to be high $(D_{22} = 4.1 \times 10^{-15}$ cm^2 s$^{-1})$ in
the nickel system. Such a high mobility of vacancies and a strong silver
atom-vacancy coupling allows movements of silver atoms in the same or in
the opposite direction of the vacancy flux depending on the sign of the
coupling coefficient D_{12}. It was found that the coupling coefficient is
positive $(D_{12} = 1 \times 10^{-15}$ cm^2 s$^{-1})$, when the ion bombardment is carried out
at room temperature, which induces the silver-atom flux in the direction
of the vacancy flux. At higher temperatures (603° K) the coupling
coefficient D_{12} is negative $(D_{12} = -8.5 \times 10^{-15}$ cm^2 s$^{-1})$, which induces the
silver atom flux in the direction opposite to that of the vacancy flux.
The change of sign of D_{12} between 300° K and 600° K is obviously due to a
different interaction of Ag atoms with vacancies. This fact is probably
connected to the increased density of vacancies with increasing
temperature. However, at this stage, it is difficult to understand, at a
microscopic level, such a behavior. The vacancy coupling coefficient D_{21}
is always found to be very low $(D_{21} \leq 1.0 \times 10^{-20}$ cm^2 s$^{-1})$ and is not
expected to have any effect on the silver-atom flux. The above set of
diffusion and coupling coefficients accounts for the sharpening of silver
profile observed at higher temperatures (603° K). The vacancy profile is
gaussian in nature and the mobility of vacancies causes the profile to
spread in the width of the gaussian. Since at higher temperatures the flux
of silver atoms is opposite to the vacancy flux, the silver profile
sharpens. Thus the so-called vacancy flux induced de-mixing of silver
atoms from silver-nickel multilayer ion bombarded films is observed.

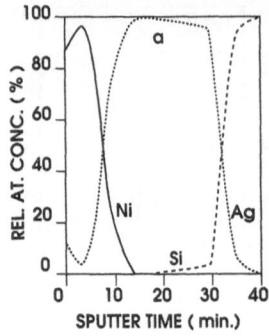 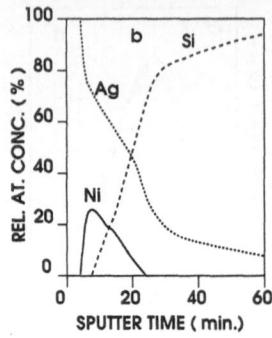

Figure 2. Auger concentration profiles of Ni and Ag for (a) Ni/Ag as
evaporated on copper, and (b) for the same sample, implanted
with 100 keV Ar+ at a dose of 7.0×10^{16} atoms/cm².

Since the vacancies are highly mobile and the surface acts as a sink
for vacancies, a vacancy flux occurs in the direction towards the surface.
At room temperature, where the D_{12} coefficient is positive, the silver
atomic flux is in the same direction than the vacancies, and the silver
atoms are observed to segregate at the surface (see fig.2 of ref.[2]).

New experiments which support the above model proposed for the silver
atom transport into nickel during ion-induced mixing have been performed.
Auger concentration depth profiles were measured before and after ion
bombardment on a nickel-silver bilayer deposited on a Cu substrate
(figs.2a and 2b). The Ni layer thickness is 20 nm while the Ag layer is
200 nm. The expected surface accumulation of silver is observed in the
profile measured after ion bombardment with 100 keV Ar+ at a dose of
7.0×10^{16} atoms/cm² and at a current density of 70 μA/cm². X-ray

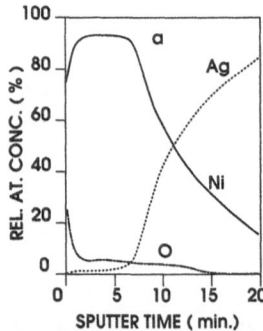 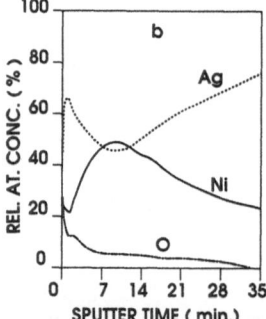

Figure 3. Auger concentration profiles of Ni and Ag for (a) Ni/Ag bilayer
evaporated on silicon, and (b) for the same sample, implanted
with 100 keV N+ at a dose of 7.0×10^{16} atoms/cm².

690

diffraction measurements showed the peaks to be always due to pure Ag and Ni metals indicating that neither a compound nor a solid solution are formed. The Ag atoms could be in the form of dispersed precipitates. This also shows that cascade-mixing effects have not taken place.

In figs.3a and 3b the results for Ni/Ag bilayers on a silicon wafer substrate are shown. In this case the ion bombardment was done by using N^+ ions with the same energy, current and dose as in the previous experiments, but due to the experimental set-up chosen, a higher temperature (exceeding 600° K, according to our estimate [5]) was reached here. It is observed that the segregation of silver atoms is so high that the nickel layer which was at the surface before bombardment now appears to have been hidden in the bulk covered by the layer of segregated silver atoms.

Thus, in conclusion, the silver-atom transport in silver/nickel multilayered films under ion bombardment is mainly controlled by the silver atom-vacancy transport correlation effects. Cascade-mixing effects probably play a minor role in this system. The processes observed in a Ni/Ag bilayer bombarded at high temperature ($\geq 600^{\circ}$K) are, however, too complex to be described completely by our set of equations.

REFERENCES

1. D. Marton, J. Fine and G.P. Chambers, Phys. Rev. Lett. *61* (1988) 2697.
2. D. Marton, J. Fine, and G.P. Chambers, Mat. Sci. Engineer. *A115* (1989) 223.
3. S. Matteson and M.A. Nicolet, Annual Rev. Mater. Sci. *13* (1983) 339.
4. D.C. Kothari and A. Miotello, J. Phys. Cond. Matter. *1* (1989) 10619.
5. L.M. Gratton, L. Guzman, A. Miotello, C. Tosello and G. Wolf, Nucl. Instr. Meth. *209/210* (1983) 1117.

SIMS SURFACE ANALYSIS OF PET:

DEGRADATION DEVELOPMENT

D. Léonard, Y. De Puydt and P. Bertrand

PCPM Laboratory
Université Catholique de Louvain (U.C.L.)
1, Place Croix du Sud, B-1348 Louvain-la-Neuve, Belgium

ABSTRACT

Semi-crystalline industrial polyethylene terephthalate (PET) has been analyzed, directly or after an ionic prebombardment, by Static SIMS spectrometry using 4 keV Xe ions. The prebombardment doses cover approximately one order of magnitude from 1.8×10^{13} ions/cm^2 (dose for a direct SSIMS spectrum) to 1.8×10^{14} ions/cm^2. By comparing the spectra using differences it is shown that, for the mentioned doses, the major features of the appearing degradation are, on the one hand, crosslinking and/or graphitization and, on the other hand, unsaturation. These features are studied as a function of the ionic dose and the polymer characteristics.

1. INTRODUCTION

Since about ten years, Secondary Ion Mass Spectrometry (SIMS) has become a very powerful method used for the analysis of the polymers surface. The molecular information obtained from the SIMS spectra is very useful, and permits to resolve problems related to their very surface (< 10 Å), as for example the influence of surface treatments on their surface chemistry [1,2] and the semi-quantification of the precise composition of the surface of copolymers [3,4].

It has been rather soon emphasized, that the main problem of the SIMS technique applied to polymers is the degradation of the samples following ionic bombardment [5-7]. Indeed, polymers are known to be very sensitive to ion bombardment [1,8,9]; this implies that so-called "static" conditions have to be used. That means that the ion dose used for the SSIMS (or Static SIMS) analysis must be lower than the dose leading to damage [10]. The "static" conditions are opposed to the so called "dynamic" ones, where the intensity of a few number of masses are followed as a function of the ion dose (a clear and intentional degradation is then observed). In the case of polymers, it is really difficult to obtain strict "static" conditions, because it is clearly impossible that two successive SSIMS analyses give the same spectra. However the term "static" is used in polymer surface analysis, to characterize conditions supposing to use the lowest possible dose, i.e., the compromise between (unavoidable) degradation and good statistics.

The recent investigations on the use of SSIMS to study the polymers surface emphasize the importance of a more precise analysis of the spectra, in order to obtain more information about the fragmentation patterns [11], the precise identification of ions [12], and the degradation mechanisms [13].

2. EXPERIMENTAL

Ion bombardment and analysis conditions have already been presented elsewhere [14]. 4 keV Xe ions are used, with the ion beam being rastered on ~ 33 mm^2 areas; the sample current is 10 nA, and the time required for the acquisition of a 160 - 10 amu spectrum is 102 s; charge neutralization with low-energy electrons is used. A Riber Q-156 spectrometer is used. The ion doses for which the degradation has been studied are 1.8×10^{13} ions/cm^2 (normal dose used for the acquisition of a SSIMS spectrum), and 3.6, 5.5, 9.1, and 18×10^{13} ions/cm^2.

3. RESULTS AND DISCUSSION

Figs.1 and 2 show the spectra of semi-crystalline industrial PET for

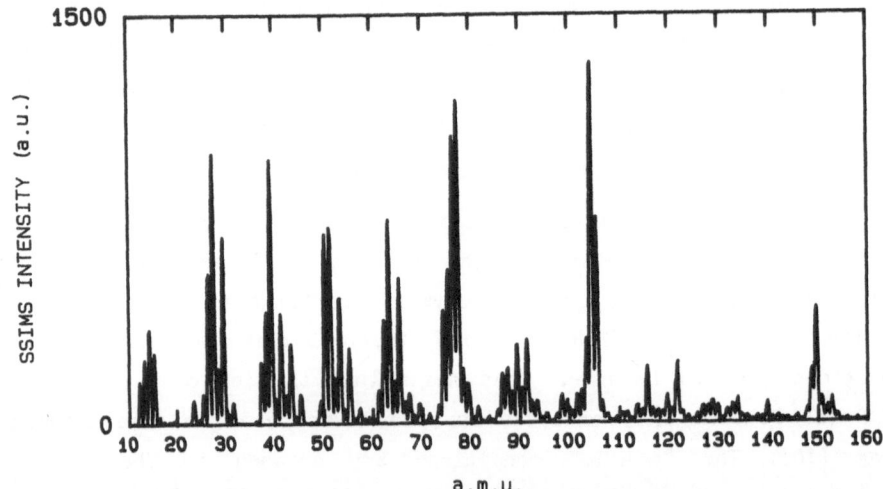

Figure 1. Static SIMS spectrum of semi-crystalline Polyethylene terephthalate, acquired with an ion dose of 1.8×10^{13} Xe^+ ions/cm^2.

total ionic doses of 1.8×10^{13} and 1.8×10^{14} ions/cm^2. It is clearly possible to identify the PET in fig.1: all the clusters (i.e., sets of peaks corresponding to ions with the same number of C atoms) are present with their relative intensities, and also those of their components, in good agreement with spectra from the literature [15]. The peaks at masses 104 and 149 are amongst the most intense, and are directly related to very well-known parts of the PET backbone [15]. The study of the evolution of the relative intensities for the different clusters, and for the peaks characteristic of the PET backbone, as a function of the dose, indicates a rapid degradation. From fig.2, it is no more possible to identify the PET structure: the overall spectrum shows only the set of hydrocarbon clusters typical of a completely degraded polymer [16]. This confirms our previous results [14], and is in agreement with the ion-induced degradation of PET essentially marked by crosslinking [17]. Indeed, the absolute secondary ion yield in fig.2 has drastically decreased compared to that of fig.1; this is known to be related to the development of crosslinking and/or graphitization [1].

The evolution of the SSIMS spectra, for increasing preirradiation doses, has been analyzed more precisely in terms of the polymer degradation, using normalized spectra differences performed with respect to a characteristic peak not undergoing too large intensity variations with increasing preirradiation doses. The analysis of these difference spectra has appeared to be well suited for the purpose of the present

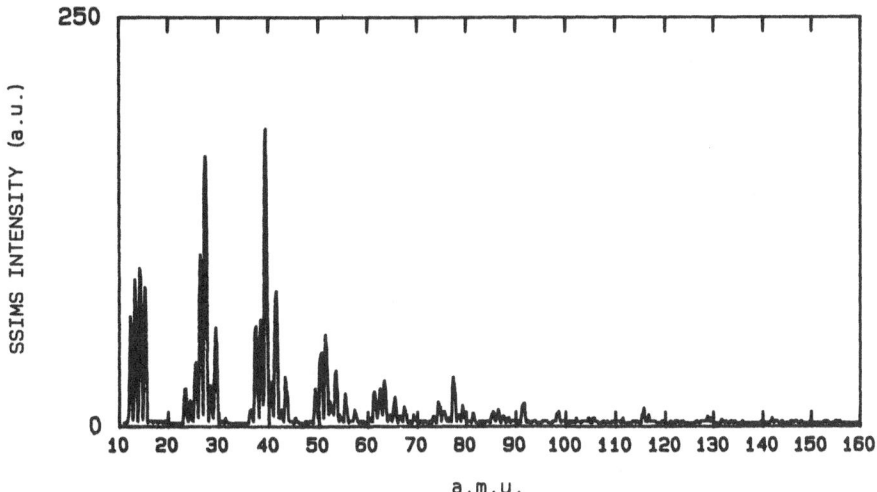

Figure 2. Static SIMS spectrum of preirradiated semicrystalline Polyethylene terephthalate, acquired under the same conditions than in fig.1; the overall ion dose is 1.8×10^{14} Xe$^+$ ions/cm^2.

study. For example, it has to be noted a change in the relative intensity of the different peaks into the first hydrocarbon clusters: peaks at low masses (i.e., with low hydrogen content) become relatively more intense for the degraded polymer spectrum than in the initial one. This is a sign of the development of unsaturation, which is clearly seen in fig.2. Another principal sign of the unsaturation development, for increased pre-irradiation doses, is the relative increase of the peaks at masses 77, 91, 105, 115, 128 and 141, which could easily be assigned to aromatic structures [13]. Since the PET molecule contains some unsaturated parts (benzenic rings), these unsaturated peaks are of course present in the original SSIMS spectrum but, after ion induced degradation (fig.2), they are the only ones, amongst the high masses, to overlap the background noise.

From these results, it appears that the peaks characteristic of the PET structure result from chain scissions which avoid the direct identification of the degraded polymer, even after low preirradiation doses; this confirms the degradation scheme proposed by Hearn and Briggs [18], where higher molecular weight fragments are suggested as a result of the ion-induced degradation of polymers (i.e., still with an important and significant part of the polymer structure, and not reduced to the only monomer). The importance of chain scissions has also been emphasized by Venkatesan et al.[19].

4. CONCLUSION

The degradation of PET, induced by ion bombardment, is characterized at low ion doses by scissions in the backbone (not far from a depolymerization, but not exactly so). Hence, crosslinking and/or graphitization is also observed as the degradation of the polymer proceeds. For higher preirradiation doses, all the polymers [20] should give Static SIMS spectra typical of a standard hydrocarbon structure with a high level of unsaturation, as PET shows it.

In the case of PET, the observation of the characteristic peaks in the Static SIMS spectra is related to groups present in the polymer backbone; chain scissions break the links between these groups, and destruct very rapidly this characteristic structure. Only the original SSIMS spectrum, (without any preirradiation dose), permits to identify the PET structure without any doubt.

ACKNOWLEDGEMENTS

The authors are sincerely indebted to Claude Poleunis for its technical assistance. This work is partially supported by the PAI program of the Belgian Science Policy Programming Office and the BRITE program of the European Economic Community.

REFERENCES

1. D. Briggs and M.J. Hearn, Vacuum *36* (1986) 1005.
2. Y. De Puydt, P. Bertrand, Y. Novis, M. Chtaib, P. Lutgen and G. Feyder, Proc. of NATO ASI *Plasma-Surface Interactions and Processing of Materials*, O. Auciello et al., eds., Kluwer Academic Publishers, Dordrecht (1990) 179.
3. D. Briggs, Surface and Interface Analysis *8* (1986) 133.
4. D. Briggs and B.D. Ratner, Polymer Communications *29* (1988) 6.
5. D. Briggs, Surface and Interface Analysis *4* (1982) 151.
6. J.A. Gardella Jr. and D.M. Hercules, Anal. Chem. *52* (1980) 226.
7. J.E. Campana, J.J. DeCorpo and R.J. Colton, Applications of surface science *8* (1981) 337.
8. S. Storp and R. Holm, J. Electron. Spectrosc. Relat. Phenom. *16* (1979) 183.
9. D.M. Ullevig and J.F. Evans, Anal. Chem. *52* (1980) 1467.
10. A. Benninghoven, F.G. Rüdenhauer and H.W. Werner, in *Secondary Ion Mass Spectrometry*, P.J. Elving and J.D. Winefordner, eds., John Wiley, New-York (1987) 671.
11. M.J. Hearn and D. Briggs, Surface and Interface Analysis *11* (1988) 198.
12. R.H.G. Brinkhuis and W.J. Van Ooij, Surface and Interface Analysis *11* (1988) 214.
13. W.J. Van Ooij and R.H.G. Brinkhuis, Surface and Interface Analysis *11* (1988) 430.
14. Y. De Puydt, D. Leonard and P. Bertrand, Proc. of the *International Symposium on the Metallization of Polymers*, ACS books, to be published.
15. D. Briggs, A. Brown and J.C. Vickerman, *Handbook of Static Secondary Ion Mass Spectrometry*, John Wiley and Sons, Chichester (1989).
16. M.J. Hearn and D. Briggs, Surface and Interface Analysis *9* (1986) 411.
17. L. Reich and S. Stivala, in *Elements of Polymers Degradation*, Mc Graw Hill Book Company, New York (1971) p. 38.
18. M.J. Hearn and D. Briggs, Surface and Interface Analysis *11* (1988) 198.
19. T. Venkatesan, L. Calcagno, B.S. Elman and G. Foti, in *Ion Beam Modification of Insulators*, P. Mazzoldi and G.W. Arnold, eds., Elsevier, Amsterdam (1987) p. 322.
20. D. Leonard, Y. De Puydt and P. Bertrand, submitted to Nuclear Instruments and Methods in Physics Research (sect.B).

CLOSING REMARKS

CLOSING REMARKS:

PERSPECTIVES IN PARTICLE-SOLID INTERACTIONS

R.A. Baragiola[1]

Department of Physics and Astronomy
Rutgers University, Piscataway, NJ, U.S.A.

1. Present address: Dept. of Nucl. Engn. and Engn. Physics
University of Virginia, Charlottesville
VA 22901, U.S.A.

The Advanced Study Institute (ASI) has included a wide variety of aspects of particle-solid interactions. At its end, the students have been exposed to the basic interactions and the processes which determine the fate of the projectile and ejected particles.

Due to the current limitations in the understanding of this field, the picture that the students could assemble from the classes, hot topics and posters was incomplete. The topics which are easier to explain in a good class are normally those better understood, a familiar situation in university classrooms. However, those are not the grounds on which most scientific explorations are made. The more interesting topics are those about which we are more uncertain, and usually unable to explain in a clean presentation. They should nevertheless be spelled out since it is with the appreciation of what is unknown that the newcomers to the field can more fruitfully apply their imagination. My comments have this aim, but should be taken *cum grano salis*, since they are just some of many possible, and cannot be complete. I have chosen here to mention the gaps we have in our understanding of the following topics: electronic stopping and charge states at low velocities, energy loss variations with impact parameter, the surface barrier in sputtering, ways to efficiently process data from computer simulations, bombardment of insulators, electron transfer at surfaces, models of solids appropriate for strong excitations, and particle-molecule collisions.

In the field of stopping powers, most developments have been made for high-velocity light ions, a case quite well understood for many years.

Interaction of Charged Particles with Solids and Surfaces
Edited by A. Gras-Martí *et al.*, Plenum Press, New York, 1991

701

Efforts have been made to apply relatively small corrections to expectations of first-order perturbation theory and the agreement between theory and experiment has reached a few percent. This has been possible due to the averaging which smears the unknown details about individual interactions, which have been swept under the rug. Most of the challenges are not in stopping of fast light ions, but in the case of strong perturbations, where the parameter Z/v (in atomic units) is large (Z is an effective charge and v the relative collision velocity). It is not at all clear, for instance, that the proportionality of stopping power with v for slow particles, implicit in Ohm's law, should also hold for atoms which move through the lattice, undergoing collisions not only with delocalized electrons, but also with lattice atoms. New experimental and theoretical approaches are needed to solve this question, and the related one on the charge state of ions in solids.

Some of the other areas, where the need for more detail has uncovered deficiencies or holes in our understanding, are ion scattering at surfaces, channeling, electron emission and computer simulations of particle ranges and of sputtering. The standing questions here are the dependence of inelastic energy losses on impact parameter and the distributions of final electronic states.

It was made clear at the ASI that a central issue in the understanding of particle ejection from solids, (sputtering and electron emission), is the nature of the surface barrier, or the forces which retard the emitted particles. Little progress has been made in the last two decades in the case of sputtering. Here, large efforts have been applied to the description of the elastic collision processes, but most treatments use simplistic planar or spherical surface barriers. We need to understand how the probability of ejection depends on the coordination of the surface atom and on the atomic motion surrounding it.

Many advances in particle-solid interactions have been made using computer simulations, which have flourished due to the exponential growth of the speed/price ratio of computers. There are two important challenges here. One is to obtain realistic input quantities and descriptions, like many-body potentials and inelastic effects. The other is to device ways to process the enormous flow of computed data, in order to get trends and regularities which can be used to make estimates and to identify the main physical mechanisms.

So little has been said about insulators at the ASI, and at ongoing conferences on atomic collisions in solids, that we have to remind ourselves that life, and most of our planet, are made of solids and liquids which are neither metals nor semiconductors. Insulators are difficult to work with in the laboratory, due to problems with charging and with sample preparation. Their theoretical description is also more difficult than for solids with delocalized electrons, and one also needs to deal with long-range Coulomb tails in the potentials, and with strongly directional bonding forces. Large, and not well understood differences with metals exist in the behavior of insulators in sputtering and electron emission. It is easy to predict that many physical phenomena remain to be uncovered in this almost virgin area.

Of all obstacles to our understanding of particle-solid interactions, I feel that the principal one is our limited knowledge of electron transfer processes at surfaces. They are important in determining the charge state of backscattered and transmitted projectiles, and of sputtered or desorbed ions. They are also the dominant factor in determining the probability of desorption of neutral species. At present, there is no other topic visible where a quantitative understanding could have so much of a beneficial influence. We would be able to make surface analysis techniques quantitative, like secondary-ion mass spectrometry, low-energy ion scattering spectroscopy, and electronically stimulated desorption. The possible impact of a better understanding is huge in technology, since electron transfer processes at surfaces are fundamental in determining the behavior of surfaces to oxidation, corrosion and catalysis. So this is an area not only interesting for exploration, but one in which there should be no significant restrictions in funding in the foreseeable future.

A question that often arises in the study of inelastic processes is: what is a good description of a solid, when excitations are intense and localized? The traditional approach to solid-state physics is to use models which describe well the average properties of electrons away from the lattice sites. This approach has been very fruitful in explaining transport properties of solids, but I feel it is not satisfactory in our case, and that it is necessary to introduce an accurate description of the ion cores in electron gas models.

Finally, it has become more and more evident over the years that our picture of the physics of ion-solid interactions has to rest on conceptual frames both of atomic and of solid-state physics. This means, for instance, that much can be gained by looking closely at the ongoing research on free-particle collisions, and that a fruitful approach would be to study in detail elastic and inelastic collisions with molecules of increasing degree of complexity. The combination of ideas in atomic and

703

solid-state physics has not only been a useful exercise over the years, it also makes particle-solid interactions a very intellectually stimulating field of research.

PARTICIPANTS

I. Abril
Dept. Física Aplicada
Univ. Alacant
Apartat 99
E-03080 ALACANT
SPAIN

D. Aiello
Instituto de Fisica
Universita della Calabria
COSENZA
ITALY

P. Alkemade
University of Western Ontario
London
N6A 3K7 ONTARIO
CANADA

J. Alvarellos
Dept. Física Fundamental
U.N.E.D.
Apartado 60141
E-28080 MADRID
SPAIN

H. Andersen
Pysisk Lab.
Univ. Kobenhavn
Universitetsparken 5
DK-2100 KOVENHAVN O
DENMARK

T. Andreadis
Code 4651
Naval Research Laboratory
DC-20375 WASHINGTON
U.S.A.

P. Apell
Institute of Theoretical Phys
Chalmers Univ. of Technology
S-41296 GöTEBROG
SWEDEN

N. Arista
Dept. Física Aplicada
Univ. d'Alacant
Apartat 99
03080 ALACANT
SPAIN

R. Averback
Dept. of Mats. Sci. & Engn.
Univ. of Illinois
1304 W. Green St.
IL-61801 URBANA
U.S.A.

R. Baragiola
Dept. of Physics
Univ. Rutgers
NJ-08855 PISCATAWAY
U.S.A.

H. Betz
Sektion Physik
Universitat München
AM Coulombwall 1
D-8046 GARCHING
GERMANY

J. Biersack
HMI
Hamh Meitner Institut
Glienicker Str. 100
D-1000 BERLIN 37
GERMANY

H. Brongersma
Dept.of Physics
Eindhoven University
P.O. Box 513
5600 MB EINDHOVEN
NETHERLANDS

H. Burgdöfer
Department of Physics
University of Tennessee
POB 1200
TN-37996 KNOXVILLE
U.S.A

J. Calera
Dept. Física Aplicada
Universidad de Alicante
Apartado 99
E-03080 ALACANT
SPAIN

N. Capuj
Centro Atómico Bariloche
8400 BARILOCHE
ARGENTINA

A. Cassimi
C.I.R.I.L.
Rue Claude Bloch, B.P. 5133
F-14040 CAEN CEDEX
FRANCE

J. Castellá
Dept. Física Aplicada
Univ. d'Alacant
Apartat 99
E-03080 ALACANT
SPAIN

R. Cavell
Department of Chemistry, Fac.
University of Alberta Edmonton
Chemistry Building
T6G 2G2 EDMONTON
CANADA

U. Conrad
Inst. Theoret. Physik
Tech. Univ. Braunschweig
D-3300 BRAUNSCHWEIG
GERMANY

S. Cruz-Jimenez
Inst. de Investigaciones en Ma
Univ. Autónoma Mexico
Apartado Postal 70-360
04510 MEXICO DF
MEXICO

R. Daniele
Institute of Physics
Via Archirafi 36
90123 PALERMO
ITALY

C. Desgranges
Institut d'Electronique Fondam
Universite Paris-Sud
BAT 220
F-91405 ORSAY CEDEX
FRANCE

P.M. Echenique
Dept. Física, Fac. Química
Universidad Pais Vasco
Apartado 1072
E-20080 DONOSTIA
SPAIN

J. Diaz Medina
IFIC-Facultat de Física
Univ. Valencia
E-46100 BURJASSOT (VALENCIA)
SPAIN

L. Feldman
AT&T Bell Labs
RM. 1D-145
NJ-07974 MURRAY HILL
U.S.A.

D. Fenyo
Dept. Radiation Sciences
Uppsala University
Box 535
S-751 21 UPPSALA
SWEDEN

J. Ferron
INTEC
CC91
3000 SANTA FE
ARGENTINA

F. Flores
Dept. Materia Condensada
Univ. Autónoma Madrid
Cantoblanco
E-28049 MADRID
SPAIN

M. Forcada
Dept. Física Aplicada
Univ. d'Alacant
Apartat 99
E-03080 ALACANT
SPAIN

F. Garcia de Abajo
Dept. Física de Materiales
Univ. País Vasco
Apartado 1072
E-20080 DONOSTIA
SPAIN

R. Garcia-Molina
Dept. Física Aplicada
Universidad de Murcia
E-30071 MURCIA
SPAIN

F. Garcia-Vidal
Dept. Materia Condensada, C-XI
Univ. Autónoma de Madrid
Cantoblanco
E-28049 MADRID
SPAIN

H. Gecim
Electrical & Electronics Eng.
Hacettepe Univ.
Beytepe
06532 ANKARA
TURKEY

A. Gras-Martí
Dept. Física Aplicada
Univ. D'Alacant
Apartat 99
E-03080 ALACANT
SPAIN

O. Grizzi
Centro Atómico Bariloche
8400 BARILOCHE
ARGENTINA

L. Guzman
I.R.S.T.
I-38050 POVO (TRENTO)
ITALY

E. Hale
Graduate Ctr. for Mater. Res.
Univ. of Missouri-Rolla
Martin E. Straumanis Hall
MO-65401 ROLLA
U.S.A.

M. Hautala
Dept. of Physics
Univ. Helsinki
SF-00170 HELSINKI 17
FINLAND

W. Heiland
Fach. Physik
Univ. Osnabrueck
D-4500 OSNABRUECK
GERMANY

P. Hendrickx
Inst. Kern-Stralingspysika
Katholieke Univ. Leuven
Celestijnenlaan 200 D
B-3030 LEUVEN
BELGIUM

W. Hetterich
FB Physik
Univ. Osnabrueck
Barbarastrasse 7, POB 4469
D-4500 OSNABRUECK
GERMANY

H. Jacobsson
Dept. of Physics
Chalmers, Univ. Technol.
S-412 96 GOTEBORG
SWEDEN

M. Jakas
Dept. Física Aplicada
Univ. d'Alacant
Apartat 99
E-03080 ALACANT
SPAIN

K. Jensen
School of Physics
Univ. of East Anglia
NR4 7TJ NORWICH
ENGLAND

J. Jimenez-Rodriguez
Dept. Electricidad y Electrónica
Universidad Complutense
E-28040 MADRID
SPAIN

J. Kemmler
Inst. Physique Nucleaire
Univ. Claude Bernard
43 BD DU 11 Novembre 1918
69622 VILLEURBANNE
FRANCE

K. Kimura
Oak Ridge Nat. Lab.
POB. 2008
TN-37831 OAK RIDGE
U.S.A.

K. Kroneberger
Inst. F. Kernphysik
August Euler Strasse 6
D-6000 FRANKFURT
GERMANY

G. Lapicki
Dept. of Physics
East Carolina University
NC-27858 GREENVILLE
U.S.A.

W. Lennard
Dept. of Physics
Univ. Western Ontario
N6A 3K7 LONDON, ONTARIO
CANADA

J. Lindhard
Institute of Physics
Aarhus University
DK-8000 AARHUS C
DENMARK

A. Lucas
Department de Physique
Fac. Univ. Notre Dame de la Paix
Rue de Bruxelles 61
B-5000 NAMUR
BELGIUM

R. Mayol
Solid State Grp. the Blackett
Imperial College
Prince Consort Road
SW7 2BZ LONDON
ENGLAND

A. Miotello
Department of Physics
University of Trento
I-38100 TRENTO
ITALY

I. Nagy
Institute of Physics
Thecnical University
1521 BUDAPEST
HUNGARY

R. Nieminen
Laboratory of Physics
Helsinki University of Technol.
02150 ESPOO
FINLAND

T. Lamy
C.E.N.G.
L.A.G.R.I.P.P.A.
F-38041 GRENOBLE CEDEX, 85X
FRANCE

J. Laszlo
Max Planck Inst. Plasmaphysics
Max Planck Str. 7
D-8046 GARCHING
GERMANY

D. Leonard
Unite de Physico-Chimie & Phys.
Univ. Catholique Louvain
1, Place Croix du Sud
B-1348 LOUVAIN-LA-NEUVE
BELGIUM

U. Littmark
IGV-KFA
Postfach 1913
D-5170 JUELICH
GERMANY

A. Martinez Torregrosa
Departamento de Física
Univ. de Murcia
E-30071 MURCIA
SPAIN

H. Mikkelsen
Physik Inst.
Univ. Odense
Campusvej 55
DK-5230 ODENSE M
DENMARK

J. Moreno-Marin
Dept. Física Aplicada
Univ. d'Alacant
Apartat 99
E-03080 ALACANT
SPAIN

A. Närman
IBM Thomas J. Watson Res. Ctr.
PO Box 218-011261
NY 10598 YORKTOWN HEIGHTS
U.S.A.

J. Ortega-Mateo
Dept. Materia Condensada, C-XI
Univ. Autónoma de Madrid
Cantoblanco
E-28049 MADRID
SPAIN

M. Peñalba Otaduy
Dept. Física de Materiales
Facultad de Químicas
Apartado 1072
E-20080 SAN SEBASTIAN
SPAIN

V. Ponce
Dept. de Física, Fak. Kimika
Univ. País Vasco
Apartado 1072
E-20080 DONOSTIA
SPAIN

J. Remillieux
Institut de Physique Nucleaire
Univ. Claude Bernard, Lyon 1
43 BD DU 11 Novembre 1918
69622 VILLEURBANNE CEDEX
FRANCE

F. da Silva
L.N.E.T.I.
E.N.10
2685 SACAVEM
PORTUGAL

H. Rothard
Inst. F. Kernphysik
August Euler Str. 6
D-6000 FRANKFURT 90
GERMANY

L. Salmi
Inst. Theoret. Phys.
Univ. Thechnol. Chalmers
S-41296 GÖTEBORG
SWEDEN

G. Schiwietz
Physics Dept.
Hahn Meitner Inst.
Glienicker Str. 100
D-1000 BERLIN 39
GERMANY

D. Sibold
Inst. Theoret. Physiks
T.U. Braunschweig
D-3300 BRAUNSCHWEIG
GERMANY

A. Sorensen
Institute of Physics
Univ. Aarhus
DK-8000 AARHUS
DENMARK

J. Pitarke
Fisika Teorikoa
Zientzi Fakultatea, Euskal Her
E-48080 BILBO
SPAIN

C. Rau
Physics Dept.
Rice University
POB 1892
TX-77251 HOUSTON
U.S.A.

R.H. Ritchie
H.A.S.R.D.
O.R.N.L. and Univ. Tennesse
MS-6123, POB 2008
TN-37381 OAK RIDGE
U.S.A.

M. Rosen
Code 4651
Naval Research Laboratory
DC-20375 WASHINGTON
U.S.A.

A. Rubio
Dept. Física Teórica
Univ. de Valladolid
E-47011 VALLADOLID
SPAIN

F. Saris
FOM
Kruislaan 407
NL-1098 SJ AMSTERDAM
NETHERLANDS

L. Serra
Dept. de Física
Univ. Islas Baleares
E-07071 PALMA DE MALLORCA
SPAIN

P. Sigmund
Pysisk Inst.
Univ. Odense
Campusvej 55
DK-5230 ODENSE M
DENMARK

Z. Sroubek
Czech. Academu of Sciences
Blanicka 17
12000 PRAGUE 2
CZECHOSLOVAKIA

C. Stein
Air Force Weapons Lab/Cal
Kirtland Air Force Base
NM-87117 ALBURQUEQUE
U.S.A.

C. Stone
School of Engineering
U.C.L.A.
6291 Boelter Hall, UCLA
CA-90024 LOS ANGELES
U.S.A.

S. Tong
Dept. Physics, Interface Sci.
Univ. of Western Ontario
London
N6A 3K7 ONTARIO
CANADA

S. Tougaard
Pysisk Institut
Univ. Odense
DK 5230 ODENSE
DENMARK

D. Tsatis
Physics Dept.
Univ. of Patras
PATRAS
GREECE

B. Tsipinyuk
Institute of Applied Physics
Tashkent State University
700095 TASKENT
U.S.S.R.

H. Urbassek
Inst. Theoret. Physik
Univ. Braunschweig
Mendelssohnstrasse 3
Post 3329
D-3300 BRAUNSCHWEIG
GERMANY

A. Vantomme
Inst. Kern-Stralingspysika
Katholieke Univ. Leuven
Celestijnenlaan 200 D
B-3030 LEUVEN
BELGIUM

M. Vicanek
Univ. California Los Angeles
6275 Boelter Hall
CA-90024 LOS ANGELES
U.S.A.

T. Waldeer
Inst. Theoret. Physik
T.U. Braunschweig
D-3300 BRAUNSCHWEIG
GERMANY

R. Webb
Dept. Electronic & Electrical
Engineering
University of Surrey
Guildford Surrey GU2 5XH
ENGLAND

H. Winter
Inst. Kernphysik
W. Klemmstr 9
D-4400 MUNSTER
GERMANY

D. Wutte
Technische Universitat Wien
Wiedner Hauptstr 8-10
1040 WIEN
AUSTRIA

Y. Yamazaki
Inst. of Physics
Univ. Tokio
Komaba, Meguro-Ku
153 TOKYO
JAPAN

F. Yubero Valencia
Dept. Física Aplicada-CXII
Univ. Autónoma Madrid
Cantoblanco
E-28049 MADRID
SPAIN

V. Yurasova
Faculty of Physics
Moscow state University
117234 MOSCOW
U.S.S.R.

R. Zangara
Dip. Di Energetica ed Applicaz
Viale Delle Scienze
90128 PALERMO
ITALY

E. Zironi
Instituto de Física
Univ. Autónoma de México
Apartado Postal 20-364
01000 MEXICO DF
MEXICO

There are three types of manuscripts in these Proceedings, according to the presentations that took place in the Institute: Mainstream, Invited, and Contributed papers. We have indicated them by the initials M, I and C.

N
Neutralization 317, 387
   Auger 280, 318
   resonant 283, 317
   quasi-resonant 280
NRA (Nuclear Reaction Analysis)
   309

O
Oscillator strength,
   optical 585
   generalized 586

P
Penn's statistical model 585
Phase shifts 46, 609
PIXE (Particle Induced X-ray
   Emission) 165, 309
Plasmons 7, 423
Poisson statistics 75
Polariton 194
Positrons 631
Potentials, interatomic 228
   screened Coulomb 261
Pseudopotentials 593

Q
Quasimolecule 648

R
Radiation damage, by swift ions 569
Ranges 245
Rare gas incorporation 675
RBS (Rutherford Backscattering
   Spectrometry) 309
REELS (Reflection Electron Energy
   Loss Spectrometry) 487
Repulsive state, in desorption 321
Resonant electron tunneling 388
Ripplon 642

S
Scattering, of ions (ISS) 271
   rainbow 267
Screening parameter 261
Secondary electron spectra, 459, 593,
   615
Secondary ion emission 505, 647, 681
   of molecules 647
Shadow cone 267
Shell processes 23
SIMS (Secondary Ion Mass
   Spectrometry) 681
   of polymers 693
   Static 693
Specular reflection,
   model 547
   of ions from surfaces 371,
      391, 529

SPEES (Spin-Polarized Electron
   Emission Spectroscopy) 625
Spin-dependent interactions 204
Spin polarized electrons 625
Sputter deposition 675
Sputtering 248
   of magnetic materials 505
   of rare gases 657
   preferential 688
   direct recoil 264
STM (scanning tunneling microscope)
   194
Stopping cross section 105
Stopping, electronic 122

Stopping power,
   at surfaces 371, 529
   dependence on impact parameter
      347, 701
   electronic 570
   in H, He 517
   measurements 176
   of antiprotons 55, 395, 517
   of clusters 557
Stopping power (continued)
   of electron gas 517
   of electrons 585
   of H 34, 53, 385, 517
   of heavy ions 365
   of He 28, 53, 529, 563
   of light ions 28, 517
   of molecular ions 423
   of slow ions 45, 395, 701
   position dependent 371
   Z oscillations 62, 349
Straggling 65, 79, 529
   measurements 176
   Bohr 86
Strained-layer structures 309
Sum rules 198
Surface barrier,
   for electrons 445
   in sputtering 701
Surface magnetism 625

T
Thickness fluctuations 178, 577
Thin-film formation 309, 667, 675
Thomas-Fermi model 41, 493
Thomas scattering 388
Threshold laws, for electron
   emission 461
Time of flight (TOF) measurement
   181, 271

Track formation 570
Transport cross section 395, 490
Transport equations 92, 232